ESTUARINE AND COASTAL MODELING

PROCEEDINGS OF THE FIFTH INTERNATIONAL CONFERENCE

October 22-24, 1997
Alexandria, Virginia

EDITED BY
Malcolm L. Spaulding and Alan F. Blumberg

ASCE American Society of Civil Engineers

1801 ALEXANDER BELL DRIVE
RESTON, VIRGINIA 20191–4400

Abstract: This volume contains the proceedings of the 5th International Conference on Estuarine and Coastal Modeling held in Old Town Alexandria, Virginia, October 22-24, 1997. The conference included oral and poster sessions describing recent model developments, applications, and interpretations and visualization of results. Special sessions were held on the state of the art in nowcasting and forecasting of the currents for the sailing events associated with the 1996 Olympic Games. The poster sessions featured the presentation of a wide variety of model results for hydrodynamic and water quality simulations. Each paper in the publication was presented orally or in a poster session at the conference and was peer-reviewed prior to acceptance. The authors represent a well-balanced mix of professionals from private industry, government, and academic institutions from the US and abroad.

Library of Congress Cataloging-in-Publication Data

Estuarine and coastal modeling: proceedings of the 5th international conference/editors, Malcolm L. Spaulding and Alan F. Blumberg.
    p.   cm.
   "Proceedings of the 5th International Conference on Estuarine and Coastal Modeling, held in Old Town Alexandria, Virginia, October 22-24, 1997"–Abstract.
   ISBN 0-7844-0350-3
   1. Estuaries–Mathematical models–Congresses. 2. Coast changes–Mathematical models–Congresses. 3. Hydrodynamics–Mathematical models–Congresses. 4. Water quality–Mathematical models–Congresses. 5. Sediment transport–Mathematical models–Congresses. I. Spaulding, Malcolm L. II. Blumberg, Alan F. III. International Conference on Estuarine and Coastal Modeling (5th: 1997: Old Town Alexandria, Va.)
GC96.5.E74   1998
551.46'09–dc21   98-22682
                                                   CIP

FOREWORD

This conference represents the fifth in a biennial series to explore the development, application, calibration, validation, and visualization of predictions from estuarine and coastal models. Application of the models to problems in hydrodynamics, water quality, and sediment transport were included. Special sessions were held on nowcast modeling and forecasting for the sailing events of the 1996 Olympic Games held in Wassaw Sound, Georgia. Attendance at the conference was about 135 and included representatives from both the US and many foreign countries. Attendance was predominantly from government and academic engineers and scientists with industry representatives comprising the remaining 15 %.

The goal of the conference, as for past conferences in the series, was to bring together a diverse group of model researchers, users, and evaluators to exchange information on the current state of the art and practice in marine environmental modeling. The primary focus was on the application of models to bays, sounds, estuaries, embayments, lagoons, and coastal seas to solve engineering and environmental impact and assessment problems. Examples of the application of models to meet regulatory requirements were also given.

The conference included a total of 12 oral and 2 poster sessions held over the two and one half day meeting. Papers from both the poster and oral sessions are included in the conference proceedings. Each paper in the proceedings was presented at the meeting, subjected to two peer reviews and accepted, after appropriate revision, by the proceedings editors.

The help of the organizing and advisory committees, whose names are listed below, and Joseph Pittle and his staff at the University of Rhode Island Conference and Special Programs Development Office contributed greatly to the success of the meeting. Our sincere thanks is extended to them.

ORGANIZING COMMITTEE

Dr. Alan F. Blumberg, HydroQual, Mahwah, NJ
Mr. H. Lee Butler, Veritech, Vicksburg, MS
Prof. Keith W. Bedford, Ohio State University, Columbus, OH
Dr. Ralph T. Cheng, U.S. Geological Survey, Menlo Park, CA
Prof. Malcolm L. Spaulding, University of Rhode Island, RI

ADVISORY COMMITTEE

Prof. Eric Adams, Massachusetts Institute of Technology, Cambridge, MA
Dr. Frank Aikman, NOAA National Ocean Service, Silver Spring, MD
Dr. Mark Dortch, U.S. Army Corp of Engineers, Vicksburg, MS
Prof. John Hamrick, Tetra Tech, Gloucester Pt., VA
Prof. Nikolaos Katopodes, University of Michigan, Ann Harbor, MI
Prof. Ian King, University of California at Davis, Davis, CA
Prof. W. Lick, University of California at Santa Barbara, Santa Barbara, CA

Prof. Daniel R. Lynch, Dartmouth College, Hanover, NH
Prof. Y. Peter Sheng, University of Florida, Gainesville, FL
Dr. Richard P. Signell, U.S. Geological Survey, Woods Hole, MA
Mr. Peter E. Smith, U.S. Geological Survey, Menlo Park, CA
Prof. Guus Stelling, Delft Hydraulics, The Netherlands
Dr. David J. Schwab, NOAA/GLERL, Ann Arbor, MI
Dr. Roy Walters, U. S. Geological Survey, Tacoma, WA

Special thanks are extended to Dolores Provost of the University of Rhode Island Ocean Engineering department who generously provided her time to perform the many administrative tasks necessary to plan the meeting and produce the conference proceedings. Thanks are also due to Larry Simoneau who developed and maintained the conference web site.

Given the enthusiastic response of the participants, planning is in progress for the 6th International Estuarine and Coastal Modeling conference to be held in New Orleans, Louisiana in 1999.

Malcolm L. Spaulding
Narragansett, Rhode Island

Alan F. Blumberg
Mahwah, New Jersey

Conference Co-chairmen

Contents

Pollutant Load Reduction Models for Estuaries

Y. Peter Sheng[1], Member, ASCE

Abstract

To develop scientifically-defensible management strategy to regulate pollutant loading into estuaries, pollutant load reduction models are being developed and applied. These pollutant load reduction models relate the loading of pollutants from various sources to quantifiable estuarine health indices (e.g., water quality, habitat biomass and productivity, or fishery production), and allow the determination of pollutant load reduction goals which are required to reach or retain target estuarine health indices. This paper reviews the scientific aspects of pollutant load reduction models. Three different types of pollutant load reduction models: regression-based models, box models, and multi-dimensional process-based models are presented and compared. Development and application of a process-based pollutant load reduction model are then presented. Since the model has been developed primarily for shallow sub-tropical and tropical estuaries (e.g., Roberts Bay, Indian River Lagoon, Tampa Bay, and Florida Bay), the process-based model includes the following processes: hydrodynamics (circulation and wave), sediment transport, nutrient cycling, micro-algae, light attenuation, and seagrass dynamics.

Introduction

Human activities related to population growth and development of industry and municipality have led to increased loadings of various pollutants (e.g., nutrients, suspended solids, waste heat, and even freshwater) into estuaries during the past few decades. These increased pollutant loadings have caused declined estuarine health which can be measured by a variety of indices (e.g., dissolved oxygen concentration, phytoplankton biomass, seagrass, and fish population). Pollutant load reduction models are models that can relate estuarine health indices to pollutant loading, and hence can be used to determine the required pollutant load reduction in order to achieve target estuarine health indices.

Pollutant load reduction models can be classified into three groups. The first group makes use of a simple regression technique to relate pollutant loading to various estuarine quality indices (e.g., chlorophyll-a concentration, dissolved oxygen concentration, light attenuation, and seagrass productivity). The second group utilizes a box-type water quality model to develop the relationships between pollutant loading and estuarine health indices. The third group uses a process-based multi-dimensional hydrodynamics model in conjunction with a water quality model and an ecological model to determine the response of estuarine health indices to various load reduction strategies.

[1]Professor, Coastal & Oceanographic Engineering Department, University of Florida, Gainesville, Florida 32611-6590, pete@coastal.ufl.edu

1

Regression-based models often resort to relating estuarine parameters which contain very different temporal and spatial scales. Box models usually use both very large spatial grids and time steps, and are often unable to represent important processes which have smaller temporal and spatial scales. Neither the regression-based models nor the box models contain the detailed cause-effect relationships included in the process-based pollutant load reduction models. Process-based pollutant load reduction models generally include component models of the dominant hydrodynamic, chemical, and biological processes. Details of a process-based pollutant load reduction model depend on the dominated processes in a particular estuary. In the Chesapeake Bay, which is relatively deep with temperate climate, the dominant processes are circulation driven by tide, wind, and density gradient, nutrient cycling, and phytoplankton dynamics. In most Florida estuaries, which are generally very shallow with subtropical or tropical climate, the dominant processes generally include wave, sediment transport, and seagrass dynamics. Hence a process-based pollutant load reduction model may include several component models of the following processes: hydrodynamics (circulation and wave), sediment transport, nutrient cycling, micro-algae and macro-algae dynamics, light attenuation, and seagrass dynamics.

Various water management agencies have adopted different pollutant load reduction models to manage their particular estuaries. The Chesapeake Bay National Estuary Program has developed and is using multi-dimensional hydrodynamics and water quality models to manage the loadings into Chesapeake Bay and its tributaries. The Tampa Bay National Estuary Program has developed regression-based models to manage the nitrogen loadings into Tampa Bay.

In the following, a brief review of the three types of models is first presented, followed by a description of a process-based pollutant load reduction model and its application to several estuaries: Roberts Bay, Indian River Lagoon, and Tampa Bay. Conclusions are then given.

Regression-Based Models

Vollenweider (1968) developed the first loading plot for north temperate lakes by compiling data of total areal phosphorus loading (L_p) and mean depth (H). He found that the mean depth of the lake is positively correlated to the total phosphorus loading for a particular trophic state. As the mean depth of the lake increases, it requires more phosphorus loading for the lake to reach eutrophic condition. Straight lines on the log-log plot of L_p vs. H actually correspond to the various trophic states (e.g., oligotrophic, mesotrophic, eutrophic) of the lake. Vollenweider (1975) later modified his loading plot by adding the inverse of the residence time into the abscissa, hence $\log(L_p)$ is correlated to $\log(H/T)$ where T is the residence time. Vollenweider's loading plot can be used to predict the trophic state based on known H/T and L_p. It can also used to estimate the loading required to attain a desired trophic state for a particular lake. Hence, Vollenweider's loading plot can be regarded as a simple regression-based pollutant load reduction model.

Boynton and Kemp (1993), in their study of the Chesapeake Bay, found significant regressions between river flow and algal biomass, particulate organic matter deposition, and algal primary production. These regression relationships improved after temporal lags on the order of weeks to months are built into the model. Based on the significant regressions, Boynton and Kemp (1993) suggested a series of conceptual models with potential specific cause-effect relationships. The regression-based models were not used as pollutant load reduction model for Chesapeake Bay. Instead, the presently used Chesapeake Bay pollutant load reduction model is based on a three-dimensional hydrodynamics model coupled to a three-dimensional water quality model (Cerco and Cole 1994).

Tampa Bay is a large coastal plain estuary located on the southwest coast of Florida. There has been a considerable increase in population, development, and pollutant loading over the past four decades. As a result, seagrass biomass and productivity have declined significantly in Tampa Bay (Figure 1). Recently the Tampa Bay National Estuary Program decided to

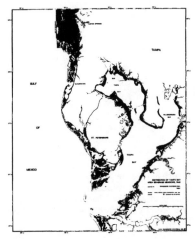

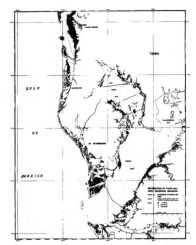

Figure 1(a). Extent of seagrass meadow in Tampa Bay in 1943.

Figure 1(b). Extent of seagrass meadows in Tampa Bay in 1983 (Lewis *et al.*, 1985).

establish seagrass restoration targets for various segments of Tampa Bay in order to restore seagrass in areas where seagrass was lost during 1950-1990. The target depth for each segment in Tampa Bay was established, e.g., 1.5m for the Hillsborough Bay and 3m for the Old Tampa Bay. A light limitation study found that the average annual percentage of light present in the water column was 22.5% of subsurface irradiance at the maximum depth limits of *Thallassia*. Subtracting 2% bottom reflectance, a new target of 20.5% was established for the segments. In order to achieve the required light condition at the target depth, water quality in Tampa Bay must be improved by pollutant load reduction.

Recent studies on Tampa Bay found significant relationship between nitrogen loading and algal biomass and production. Johansson (1991) examined the relationship between estimated nitrogen loadings and average chlorophyll *a* concentrations in Hillsborough Bay during 1968-1990 when a very large decrease in nitrogen loading was reported. He found a significant correlation between the annual total nitrogen loadings and the annual average chlorophyll *a* concentrations three years later. Coastal (1996) used more accurate loading estimates for each of the segments of Tampa Bay (Hillsborough Bay, Old Tampa Bay, Middle Tampa Bay, and Lower Tampa Bay), and found a significant correlation between the monthly average chlorophyll *a* concentrations and total nitrogen loadings during the present plus two previous months. This relationship was produced with the assumption that phytoplankton populations in Tampa Bay are nitrogen limited and that external nitrogen loadings to the bay can be managed to affect desirable average water quality conditions.

A functional nitrogen loading vs. chlorophyll *a* model was developed based on 256 sets of data which were provided by the Hillsborough County Environmental Protection Commission (HPEPC). These data were collected every month at the stations shown in Figure 2. The model-predicted chlorophyll *a* concentrations compare well against measured values with an R^2 value of 0.73, as shown in Figure 3. The lag in the response of the bay to nitrogen loading was also found in the Chesapeake Bay by Malone *et al.* (1988), and was hypothesized to be due to the fact that much of the nitrogen loading is in the particulate form, hence requiring remineralization via microbial activity before becoming available for uptake by water-column phytoplankton. To further develop the load reduction model, a model that relates the light attenuation coefficient to water quality was developed for Tampa Bay following the approach of McPherson and Miller

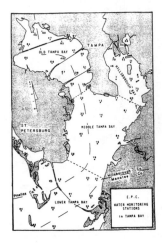

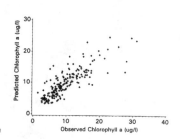

Figure 3. Predicted vs. measured chlorophyll *a* concentrations based on a regression model relating measured total annual nitrogen loading to chlorophyll *a* concentration. (From Coastal 1996).

Figure 2. Monthly water quality monitoring stations in Tampa Bay maintained by Hillsborough County.

(1993). The model is based on the good correlation between light attenuation coefficient versus chlorophyll *a* and turbidity as shown in Figure 4.

Based on the target depths for each segment of Tampa Bay and the regression models for nutrient loading vs. chlorophyll *a* concentration and chlorophyll *a* concentration vs. light attenuation, it was found that if the total annual nitrogen loading is held at the present level, the total seagrass acres in Tampa Bay excluding Lower Tampa Bay will be at 94% of the target 22,350 acres. With a 2% nitrogen load reduction, the seagrass acreage can reach 97% of the target acres (Coastal 1996). The estimations obtained from the regression models for Tampa Bay are quantitative and appear to be reasonable. However, many simplifying assumptions were invoked in arriving at the estimations. The partitionining of Tampa Bay into a few segments is rather arbitrary.

The estimations are based on a few simple correlations between a few parameters. However, "correlation" can only *suggest* but not *guarantee* "cause-effect relationship". Hence the regression model cannot be regarded as a *predictive* model. The fundamental processes involved in the complex ecosystem have been either ignored or grossly oversimplified. The effects of circulation and transport on water quality dynamics have been completely ignored. Spatial and temporal dynamics cannot be represented. The light attenuation model, which assumes that the effects of various water quality parameters are cumulative and linearly independent, is inherently incorrect. The regressions could be improved if more than two parameters are used. For example, in the Tampa Bay case, the regression between nitrogen loading and chlorophyll *a* concentration could be improved by including additional parameters such as salinity, flow, and rainfall. However, inclusion of additional parameters which cannot be controlled will make the simple regression model much less useful as a management tool. Thus, although regression models can be useful in suggesting potential cause-effect relationship, using such regression models to set a pollutant load reduction goal is questionable.

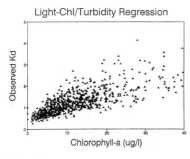

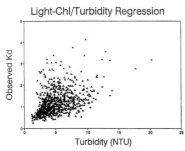

Figure 4. Light attenuation coefficient versus chlorophyall *a* concentration (upper panel) and turbidity (lower panel) in Tampa Bay. (From Coastal 1996).

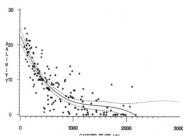

Figure 5. Relationship between measured salinity at Shell Point in tidal Peace River vs. 45-day preceding total gauged flow rates at upstream. (From EQL 1996).

Another example where regression models have been used is the determination of "minimum flow" in rivers where freshwater is being diverted for multiple uses before entering into an estuary. Recently, however, this type of regression model has been replaced by a coupled flow-salinity model (Chen and Flannery 1997) in setting the minimum flow for Hillsborough River which flows into Tampa Bay. In the case of Peace River which flows into Charlotte Harbor, regression models have been developed to relate the gauged flow rates upstream versus the salinity at various locations in the tidal river and estuary down stream. For example, regression between the salinity at Shell Point (Station 12) and the 45-day preceding total gauged flow rates is shown in Figure 5. Again, although such regression can be used to suggest possible cause-effect relationship in the model development stage, it is questionable to use the simple regression as a management tool to set minimum flow. More robust models which resolve the fundamental processes are needed.

Box Models

Box models consist of relatively simple mechanistic models for water quality processes with rather coarse spatial and temporal resolutions. For example, DiToro and Connoly (1980) simulated the eutrophication dynamics in Lake Erie with a box model which includes simple mechanistic processes for nutrients, dissolved oxygen, and phytoplankton, and three boxes representing the Western Basin, Central Basin, and Eastern Basin of Lake Erie. The basic model has now become the WASP model (e.g., Ambrose *et al.* 1991) which has been applied to many lakes and estuaries. For example, the WASP model was applied to Tampa Bay by Martin *et al.* (1996).

The basic processes included in the WASP model are shown in Figure 6. The model includes a mass conservation equation (which includes transport and kinetics terms) for each of the following eight constituents (with kinetic processes in parentheses): Ammonia (mineralization, phytoplankton death, algal uptake, nitrification, benthic release), Nitrate Nitrogen (nitrification, algal uptake, denitrification), Organic Nitrogen (Phytoplankton respiration and death, mineralization), Orthophosphate (mineralization, phytoplankton death, algal uptake, benthic flux), Organic Phosphorus (Phytoplankton respiration and death, mineralization), Dissolved Oxygen (reaeration, phytoplankton growth and respiration, nitrification, CBOD oxidation, sediment oxygen demand), CBOD (Phytoplankton death, oxidation, and CBOD denitrification), and Phytoplankton (growth and respiration). A sediment

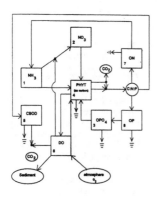

Figure 6. State variables included in the WASP Eutrophication model. (From Martin *et al.* 1996)

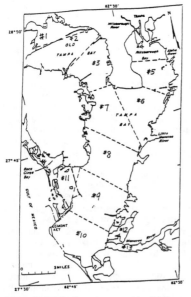

Figure 7. Revised segments used by the Tampa Bay WASP Model.

layer is included in the WASP model. The kinetic processes in the WASP model are somewhat simplified. For example, equilibrium partitioning is assumed instead of considering the desorption and absorption/adsorption as kinetic processes. The benthic (diffusive and resuspension) fluxes of nutrients have not been rigorously modeled. A sediment transport model is not part of the overall model. The model has not been linked to any ecological model, e.g., seagrass model.

The transport terms in the mass conservation equations include advection terms and diffusion terms which represent the effects of circulation and mixing on the transport of water quality constituents. Since the model generally uses only a few boxes (Figure 7) with one vertical layer, the mass conservation equations are practically one-dimensional, i.e., each box is connected to the other two boxes along two sides. However, advective fluxes and diffusive fluxes along the box boundaries must be determined in order to solve the mass conservation equations in each box. For the boxes adjacent to tributaries, flow rates and loading rates of constituents can be measured. For the boxes in the interior of the estuary, the advective and diffusive fluxes should be determined from the momentum equations subjected to the proper forcing functions (wind, tide, and river inflow). However, within the framework of WASP, the advective fluxes are generally obtained by simply solving the mass conservation equation for water, instead of the momentum equations. The diffusive fluxes are obtained by solving the mass conservation equation for salinity to achieve a best fit between simulated and measured salinity while adjusting the diffusion coefficient (or the so-called dispersion coefficient) for every box.

The basic grid for the Tampa Bay application, as shown in Figure 7, has a total of 13 segments. Although this grid does not overlay exactly with the curvilinear grid (Figure 8) developed by Sheng and Yassuda (1995) for the three-dimensional model of Tampa Bay, it does conform to the physical structures (causeways and shorelines) and represents an improvement over the original grid. Preliminary model calibration runs were conducted in the 13-segment grid with a daily time step during the period of January 1985 to December 1994. Data collected at the EPC

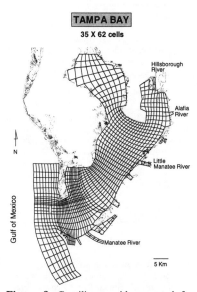

Figure 8. Curvilinear grid generated for three-dimensional circulation model CH3D. (From Sheng and Yassuda 1995).

stations (Figure 2) from January 1985 to December 1986 were used for model calibration to determine site-specific values of transport and kinetic coefficients for Tampa Bay. Sediment oxygen demand and nutrient release rates for the various segments were calibrated in the model calibration process, due to the lack of data and understanding of these processes in Tampa Bay. The model was then used to simulate the period of January 1987 to December 1991. The simulated chlorophyll *a* concentration in high concentration segments (e.g., Hillsborough Bay) was consistently lower than field measurements, while those predictions in low concentration segments (e.g., Lower Tampa Bay) were higher. To achieve a better fit between model results and data, it was necessary to calibrate the net algal settling velocity within the various segments. The simulated and measured monthly dissolved oxygen concentrations in Hillsborough Bay (Figure 9), exhibit seasonal variations. The predicted and observed chlorophyll *a* concentrations in Hillsborough Bay (Figure 10) are also reasonable.

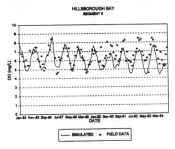

Figure 9. Simulated and measured monthly DO concentrations in Hillsborough Bay (From Martin *et al.* 1996).

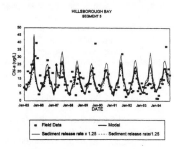

Figure 10. Simulated and measured monthly chlorophyll *a* concentrations in Hillsborough Bay. (From Martin *et al.* 1996).

The Tampa Bay WASP model revealed that, over a one-year time period, internal cycling and sediment processes are more important than the external loadings in affecting the water quality. Maximum algal growth rate, algal respiration rate, and sediment nutrient release rate were the most sensitive model parameters. The model was used to predict the response of the bay water quality to changes in external nutrient loadings. With sustained reductions in nutrient loads, the model predicted an increase in DO concentration and decrease in chlorophyll *a* concentration. However, since the WASP model does not explicitly resolve the complex processes governing the sediment oxygen demand and sediment nutrient release, the model predicted response of the bay (particularly the magnitude and time scale of the response) may contain large uncertainty and must be interpreted with caution. There are additional limitations of the Tampa Bay Box Model. For example, suspended sediment concentration, which can affect light attenuation as well as nutrient dynamics, was not included as a state variable. Epiphytic algae, macro algae, seagrass, and grazers were also not considered in the model.

Multi-Dimensional Process-Based Models

In order to predict the response of an estuary to changes in pollutant loadings, the author believes it is essential to go beyond the regression models and box models, which contain many simplifying assumptions and hence limitations, to multi-dimensional process-based models. Anticipating applications in the shallow sub-tropical and tropical estuaries, the relevant processes in the present pollutant load reduction model include the following:

- hydrodynamic processes (circulation and wave),
- sediment transport processes,
- water quality processes (salinity and temperature dynamics, nutrient cycling processes, dissolved oxygen dynamics, phytoplankton dynamics, zooplankton dynamics),
- light processes, and
- seagrass processes.

For example, Figure 11 shows the hydrodynamic processes, sediment transport processes, and water quality processes (except that dissolved oxygen dynamics are not shown) contained in a coupled hydrodynamics-sediment-water quality model (Sheng 1994, Chen and Sheng 1995). The hydrodynamics models include a 3-D rectangular-grid model (Sheng and Chen 1993, Chen and Sheng 1995) and a curvilinear-boundary-fitted-grid model (Sheng 1989, Sheng and Yassuda 1995). In the water quality model, the state variables and processes are more detailed than the WASP eutrophication model. The desorption and absorption/adsorption processes are treated as kinetic processes instead of assuming equilibrium partitioning. A sediment transport model (Sheng 1993), which includes resuspension, deposition, settling/flocculation, advection, and mixing, is explicitly included in the overall model. Diffusive as well as resuspension fluxes of nutrients are both incorporated into the nutrient model. Figure 12 shows the additional processes involving light attenuation in the water column and seagrass processes. The additional effects of vegetation on hydrodynamics (e.g., the additional vegetation-induced drag on flow in the air and water columns, the additional vegetation-induced turbulent kinetic energy, etc.) and on sediment transport (e.g., increased sediment deposition and reduced sediment resuspension) must be included in the overall model. Other processes that are important in Florida include evaporation and precipitation. In Florida Bay (Sheng and Davis 1996) where evaporation is often more important than precipitation, salinity inside the bay can reach 40-50 psu which is much higher than that (35 psu) in the Gulf of Mexico and Florida Strait. Even in Indian River Lagoon, during months with high evaporation, the northern part of the lagoon can reach 30 psu in spite of the poor tidal flushing (Sheng 1997).

Light attenuation processes and seagrass processes (see e.g., Virnstein 1993) are much more important in the shallow sub-tropical and tropical Florida estuaries than the deeper temperate estuaries in the northern U.S., e.g., Chesapeake Bay. Seagrass meadow can grow over a much larger portion of the estuary bottom in a Florida estuary than in Chesapeake Bay. Light models in practice include the simple regression model developed by McPherson and Miller (1993) and the spectral light model developed by Galegos (1993). A conceptual seagrass model was developed by Fong and Harwell (1994) with data from Florida Bay. The model includes the seagrass growth limitation by light, salinity, temperature, sediment nutrient, as well as seagrass density. These light models and seagrass models have been incorporated into the present pollutant load reduction model and are in the process of further development (Sheng 1997).

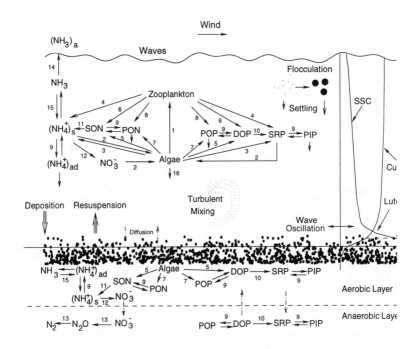

Pathways:

1: Uptake of Algae by Zooplankton
2: Uptake of SRP, ammonium, and nitrate by Algae
3: Release of SRP and ammonium due to algal
   excretion and respiration
4: Release of SRP and ammonium due to zooplankton
   excretion and respiration
5: Release of DOP and SON during mortality of algae
6: Release of DOP and SON during mortality of zooplankton
7: Mortality of algae

8: Mortality of zooplankton
9: Adsorption-Desorption
10: Mineralization of DOP
11: Ammonification
12: Nitrification
13: Denitrification
14: Volatilization of ammonia
15: Instability of ammonium
16: Settling of algae

Figure 11. Nutrient dynamics and the relevant hydrodynamics and sediment transport processes in lakes and estuaries. (From Chen and Sheng 1995).

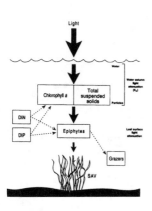

Figure 12.. Light attenuation by chlorophyll-a and total suspended solids (from Dennison *et al.* 1993).

The multi-dimensional process-based model as described above uses the same spatial grid and time step for all the process models. This eliminates the need for ad-hoc parameter tuning which would be required if different time steps and spatial grids were used for the various models. After calibration and validation of the various process models, the overall pollutant load reduction model can then be used to conduct predictions for management decisions.

Example Applications of a Process-Based Pollutant Load Reduction Model

The processes shown in Figures 11 and 12 are site-specific and vary from one estuary to another. Hence, models developed for one estuary may not be readily applicable to another estuary with new data collection and model calibration and validation. More often than not, the models have to be further enhanced to take into account the additional processes based on observation. For example, although the CH3D model developed by Sheng (1987) has been used as the foundation of the Chesapeake Bay circulation model, the more recent versions of CH3D have been substantially enhanced for applications in shallow sub-tropical and tropical waters (e.g., Sheng and Lee 1991, Sheng and Peene 1992, Sheng et al. 1995, Sheng and Yassuda 1995, Sheng and Davis 1996, Sheng 1997). The enhancements incorporated the following processes into the model: moving boundary, vegetation, evaporation, rainfall, watershed, and shelf-estuary interaction.

The water quality and light and seagrass models are even more site-specific. As mentioned earlier, in shallow estuaries waves can often reach the bottom to cause resuspension of sediments and nutrients. The hydrodynamic processes can thus more significantly affect the water quality processes in shallow versus deeper estuaries. The limiting nutrients for phytoplankton growth may also vary. In addition to chlorophyll *a*, various additional algal species (e.g., epiphytic algae, macro algae) and seagrass may compete for nutrients in the water and sediment columns. The dominant seagrass species, the dominant limiting factor for seagrass growth, and the dominant light-controlling water quality parameter all vary significantly from one estuary to another. In the following, some example studies using multi-dimensional process-based models are described.

Roberts Bay, Florida

Roberts Bay is a small embayment in the southern portion of the Sarasota Bay estuarine system, and receives a large amount of nitrogen loading from the septic tanks through the Phillippe Creek. Resource managers, including Sarasota Bay National Estuary Program and Southwest Florida Water Management District, were interested in estimating the effect of nitrogen load reduction due to the conversion of septic tanks into sewage treatment plant. Sheng *et al.* (1995) developed a coupled hydrodynamics-water quality-light-seagrass model to predict the response of the water quality and seagrass biomass to the anticipated nitrogen load reduction.

The overall model includes the CH3D model as the hydrodynamics model, a water quality model based on the WASP kinetics but include the advection and diffusion terms computed using the simulated flow field and diffusion coefficients. Moreover, a sediment transport model is included. The seagrass model and the light model are based on the Fong and Harwell model (1994) and the McPherson and Miller model (1993), respectively, with the proper coefficients for Roberts Bay. A curvilinear boundary-fitted grid with a maximum spacing of 1km and a minimum grid of 30m was developed. Model simulations of water quality constituents compare well with limited bay-wide water quality data. Model simulations of reduced load scenarios showed that a 31% reduction in nitrogen loading from the Phillippe Creek will lead to a 5-10% recovery in seagrass biomass. This prediction has recently been validated by field observation which showed a 5-10% recovery in seagrass biomass with approximately 30% nitrogen load reduction (Tomasko *et al.* 1997).

Indian River Lagoon, Florida

Indian River Lagoon is an elongated coastal lagoon along the Atlantic coast of Florida. It is approximately 250km in length from Ponce de Leon Inlet in the north to St. Lucie Inlet in the south, but varies between 2-10km in width. The average depth is around 2m. Tides enter the estuary through Ponce de Leon Inlet, Sebastian Inlet, Ft. Pierce Inlet, and St. Lucie Inlet. Circulation in the estuary is driven by tide, wind, and density gradient, and is influenced by evaporation and rainfall and the complex geometry and bathymetry. The pollutant load reduction model for the Indian River Lagoon (Sheng 1996) consists of a hydrodynamics model (Sheng 1989), a wave model, a sediment transport model (Sheng 1993), a water quality model, a light model (McPherson and Miller 1993), and a seagrass model (Fong and Harwell 1994). For water quality simulation, as indicated in Sheng (1996), the IRL model can use either the enhanced WASP model used in the Roberts Bay application or the more robust model shown in Figure 11.

The Indian River Lagoon Pollutant Load Reduction (IRL-PLR) Model study is in its second phase (Sheng 1997). Presently, various process models (including a circulation model, a wave model, a sediment transport model, a water quality model, a light model, and a seagrass model) are being calibrated, refined, and validated with data which have been and are still being collected. A horizontal grid with 25-50m spacing has been developed. The circulation model is being validated with extensive data of water level, currents, and salinity inside the estuary and in the offshore region. A wave model based on the SWAN model is being applied to investigate the wind and swell induced wave conditions inside the estuary. An extensive sediment mapping program which includes analyses of sediment type, sediment size distribution, sediment erosion rate, and sediment chemistry is being implemented. A lagoon-wide water quality monitoring program is being conducted every 2-4 weeks. Preliminary water quality simulations showed a south-to-north increase in nitrogen concentrations whereas a north-to-south increase in phosphorus concentrations, in agreement with the field data. The model results are shown in Figure 13 in terms of the Organic Nitrogen and Organic Phosphorus concentrations. These results suggest the possibility of nitrogen limitation for primary and/or secondary production in a certain part of the estuary whereas phosphorus limitation occurs in another part of the estuary. Field and laboratory experiments are being conducted to check out the nutrient limitation. Extensive concurrent light and water quality data have been collected. Regression and spectral light models are being developed for various segments of the estuary. A more robust seagrass model is also being developed.

Tampa Bay, Florida

The same basic multi-dimensional process-based pollutant load reduction model has been applied to Tampa Bay as well. The circulation model CH3D was validated with extensive water level and current data collected by NOS (Yassuda and Sheng 1996). Sheng and Yassuda (1995) simulated the residual flows in Tampa Bay during a variety of hydrologic and climatologic

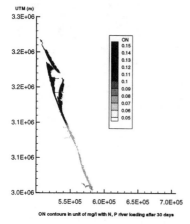

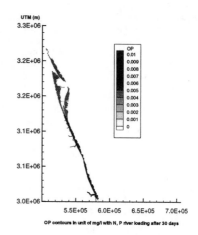

Figure 13 (a). Simulated vertically-averaged Organic Nitrogen concentrations in the Indian River Lagoon.

Figure 13 (b). Simulated vertically-averaged Organic Phosphorus concentrations in the Indian River Lagoon.

conditions. For example, the 30-day residual flows (after a 5-day barotropic spin-up and a 30-day baroclinic spin-up) during a rainy month are shown in Figure 14. Since the results showed a distinct two-layer structure of the flow field, Sheng and Yassuda (1995) suggested that the two-layer flow and salinity structure be incorporated into the Tampa Bay Box Model. Advective fluxes for the two-layer boxes in the grid shown in Figure 7 were computed. Additional diffusive fluxes could also be computed from the results of the 3-D model simulations. The coupled circulation-water quality model successfully simulated the observed development and recover of hypoxia in Hillsborough Bay during the summer of 1991 (Yassuda and Sheng 1997), an event which could not be simulated with either the simple regression model or the box model. The model has been used to develop a conceptual nutrient budget for Tampa Bay. The coupled circulation-water quality-light-seagrass model was used to simulate the response of Tampa Bay to reduced nitrogen loading.

Conclusions and Recommendations

This paper presents a brief review on three different types of pollutant load reduction models: regression-based models, box models, and multi-dimensional process-based models. Major conclusions and recommendations are:

- Regression models are simple and useful statistical tools for relating measured variables and suggesting potential cause-effect relationships in the process of developing more robust models. However, correlation between two parameters does not guarantee cause-effect relationship. Regression models which do not contain any mechanistic process cannot be used as stand-alone predictive tools for predicting the response of an estuary to pollutant load reduction. Use of regression models for management prediction is highly risky.
- Box models, which contain simplified mechanistic processes and use coarse spatial and temporal grids, are much more useful than the regression models in providing insight in understanding the relative importance of various processes in an estuary. However, due to the lack of understanding and resolution of the benthic processes, box models may contain large uncertainty and should be used with caution for management decision.

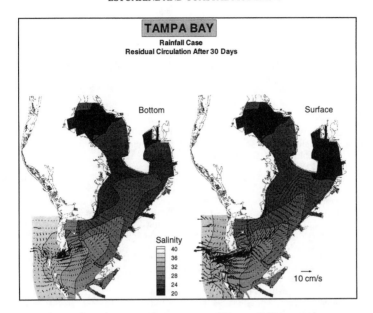

Figure 14. Simulated 30-day residual circulation in Tampa Bay during a rainy month.

- Both regression and box models should be used to aid the development of multi-dimensional process-based model, instead of being used as management tools.
- Multi-dimensional process-based models for estuaries have been developed and applied to derive quantitative pollutant load reduction goals. For shallow sub-tropical and tropical Florida estuaries, a pollutant load reduction model may include models for the following processes: hydrodynamics (circulation and wave), sediment transport processes, water quality dynamics (nutrient cycling, dissolved oxygen dynamics, phytoplankton and zooplankton dynamics), light attenuation dynamics, and seagrass dynamics. For a deeper temperate estuary such as Chesapeake Bay, such processes as wave, resuspension flux, and seagrass dynamics are not as important.
- The multi-dimensional process-based pollutant load reduction model has been applied to Roberts Bay, Indian River Lagoon, and Tampa Bay. Various process models have been validated with field data. Because of the comprehensive nature of the various process models, this type of model can be used to give much more scientifically defensible predictions.
- In order to develop pollutant load reduction goals, indices (or indicators) of estuarine health/quality need to be defined for individual estuaries. Depending on the estuary, it may be necessary to consider part or all of the following indices: dissolved oxygen concentration, chlorophyll a concentration, seagrass biomass and productivity, and fish population and diversity.
- To allow long-term simulations on the order of decades, additional work (e.g., high performance computing) is needed to further improve the efficiency of the multi-dimensional process-based pollutant load reduction model. Research is needed to address the uncertainties of the various process models as well as the overall pollutant load reduction model.

Acknowledgment

    The development of pollutant load reduction models has been supported by the St. Johns River Water Management District, U.S. Environmental Protection Agency National Center for Environmental Research, Sarasota Bay National Estuary Program, Southwest Florida Water Management District, Tampa Bay National Estuary Program, and the National Park Service.

References

Ambrose, R.B., T.A. Wool, J.P. Connoly and R.W. Schanz, 1991: "WASP4, A Hydrodynamic and Water Quality Model-Model Theory, User's Manual, and Programmers' Guide," USEPA Environmental Research Laboratory, Athens, GA.
Boynton, W.R. and W.M. Kemp, 1993: "Relationships Between River Flow and Ecosystem Processes/Properties in Chesapeake Bay," Abstract, Estuarine Research Foundation Biannual Meeting, Hilton Head, SC.
Cerco, C.F. and T. Cole, 1994: User's Guide to the CE-QUAL-ICM Three-Dimensional Eutrophication Model," *Tech. Rept. EL-95-15,* Waterways Experiment Station.
Chen, X. and Y.P. Sheng, 1995:"Application of A Coupled 3-D Hydrodynamics-Sediment-Water Quality Model," in *Estuarine and Coastal Modeling, IV,* American Society of Engineers, 325-339.
Chen, X. and S. Flannery, 1997: "Setting Minimum Fresh Water Flow Rate to a Stratified Narrow Estuary with Help of a Hydrodynamics Model," in *Estuarine and Coastal Modeling, V,* ASCE.
Coastal, Inc., 1996: "Estimating Critical External Nitrogen Loads for the Tampa Bay Estuary: An Empirically Based Approach to Setting Management Targets," *Technical Publication #06-96,* Tampa Bay National Estuary Program, St. Petersburg, FL.
Dennison, W.C., R.B. Orth, K.A. Moore, J.C. Stevenson, V. Carter, S. Kollar, P.W. Bergstrom, and R.A. Batiuk, 1993: "Assessing Water Quality with Submerged Aquatic Vegetation". *BioScience* 43(2):86-94.
DiToro, D.M., and J.P. Connoly, 1980: "Mathematical Models of Water Quality in Large Lakes, Part 2: Lake Erie," EPA-600/3-80-065.
Environmental Protection Commission (EPC), 1996: Personal Communication.
EQL, 1966: Personal Communication.
Fong, P. And M.A. Harwell, 1994: "Modeling Seagrass Communities in Tropical and Sub-Tropical Bays and Estuaries," *Bull. Mar. Sci.* 54(3):757-781.
Galegos, C.L., 1993: "Development of Optical Models for Protection of Seagrass Habitat," in *Proceedings and Conclusions of Workshops on Submerged Vegetation Initiative and Photosynthetically Active Radiation* (L.J. Morris and D. Tomasko, eds.), St. Johns River Water Management District, Palatka, FL, pp.77-90.
Johansson, J.O.R., 1991: "Long-term Trends of Nitrogen Loading, Water Quality and Biological Indicators in Hillsborough Bay, FL," in *Tampa BASIS 2* (S.F. Treat and P.A. Clark, eds.), pp.157-176.
Lewis, R.R. III, M.J. Durako, M.D. Moffler, and R.C. Phillips, 1985: "Seagrass Meadows of Tampa Bay - A Review," in *BASIS,* Proceedings Tampa Bay Area Scientific Information Symposium (S.F. Treat, J.L. Simon, R.R. Lewis, and R.L. Whitman, Jr., eds.) Tampa, FL, pp. 210-246.
Malone, T.C., L.H. Crocker, S.E. Pike, and B.W. Wendler, 1988: "Influences of River Flow on the Dynamics of Phytoplankton Production in a Partially Stratified Estuary," *Marine Ecology Progress Series,* 48:235-249.
McPherson, B.F. and R.L. Miller, 1993: "Causes of Light Attenuation in Estuarine Waters of Southwest Florida," in *Proceedings and Conclusions of Workshops on Submerged Aquatic Vegetation Initiative and Photosynthetically Active Radiation (Ed: L.J. Morris and D.A. Tomasko),* Melbourne, FL, pp.227-236.
Martin, J., P.F. Wang, T. Wool, and G. Morrison, 1996: "A Mechanistic Management-Oriented Water Quality Model for Tampa Bay," AScI Corporation and Southwest Florida Water Management District.

Sheng, Y.P., 1987: "On Modeling Three-Dimensional Estuarine and Marine Hydrodynamics", In *Three-Dimensional Models of Marine and Estuarine Dynamics*, Elsevier Oceanographic Series, Elsevier, pp. 35-54.

Sheng, Y.P., 1989: "Evolution of a 3-D Curvilinear-Grid Hydrodynamic Model: CH3D", in *Estuarine and Coastal Modeling, I*, ASCE, pp. 40-49.

Sheng, Y.P., 1993: "Hydrodynamics, Sediment Transport, and Their Effects on Phosphorous Dynamics in Lake Okeechobee," in *Nearshore, Estuarine and Coastal Sediment Transport* (A.J. Mehta, ed.), Coastal and Estuarine Studies, 42, American Geophysical Union, pp. 558-571.

Sheng, Y.P., 1994: "Modeling Hydrodynamics and Water Quality Dynamics in Shallow Waters," *Keynote Paper, Proceedings of First International Symposium on Ecology and Engineering (ISEE),"* University of Western Australia and Technical University of Malaysia.

Sheng, Y.P., 1996: "A Preliminary Hydrodynamics and Water Quality Model of Indian River Lagoon," *Technical Report*, Coastal & Oceanographic Engineering Department, University of Florida, Gainesville, FL.

Sheng, Y.P., 1997: Ongoing work on Indian River Lagoon Pollutant Load Reduction Model.

Sheng, Y.P., and H.-K. Lee, 1991: "The Effect of Aquatic Vegetation on Wind-Driven Circulation in Lake Okeechobee," *Report No. UFL/COEL-91-018*, Coastal and Oceanographic Engineering Department, University of Florida.

Sheng, Y.P., and S.J. Peene, 1992: "Circulation and Its Effects on Water Quality in Sarasota Bay," in *Framework for Action, Sarasota Bay*, Sarasota Bay National Estuary Program.

Sheng, Y.P., and X. Chen, 1993: "Lake Okeechobee Phosphorus Dynamics Study: A Three-Dimensional Numerical Model of Hydrodynamics, Sediment Transport, and Phosphorus Dynamics: Theory, Model Development, and Documentation," Final Report to South Florida Water Management District, University of Florida. 207 pp.

Sheng, Y.P., and E.A. Yassuda, 1995: "Application of 3-D Circulation Model to WASP Modeling of Tampa Bay Water Quality," *Final Report*, Tampa Bay National Estuary Program, University of Florida.

Sheng, Y.P., E.A. Yassuda, and C. Yang, 1995: "Modeling the Effect of Nutrient Load Reduction on Water Quality," in *Estuarine and Coastal Modeling, IV*, American Society of Civil Engineers, 644-658.

Sheng, Y.P., and J. Davis, 1996: "Circulation and Transport in Hypersaline Florida Bay," in *Proceedings of the 25th International Conference on Coastal Engineering*, ASCE, pp. 4247-4252.

Tomasko, D, M.O. Hall-Ruark, and J.R. Hall, 1997: "Abundance and Productivity of the Seagrass (Thalassia testudinum) Along Gradients of Freshwater Influence in Charlotte Harbor, FL" in *Proceedings of Charlotte Harbor Symposium*.

Virnstein, R., 1993: "Seagrass Response to Water Quality," In *Proceedings of An IRL Lagoon-Wide Modeling Workshop* (Sheng et al., eds.), University of Florida, pp. 137-147.

Vollenweider, R.A., 1968: "The Scientific Basis of Lake and Stream Eutrophication with Particular Reference to Phosphorus and Nitrogen as Eutrophication Factors," *Technical Report DAS/DSI/68.27*, Organization for Economic Cooperation and Development, Paris.

Vollenweider, R.A., 1975: "Input-Output Models with Special Reference to the Phosphorus Loading Concept in Limnology," *Schweiz. Z. Hydrologie*, 37:53-84.

Yassuda, E.A. and Y.P. Sheng, 1996: "Integrated Modeling of Tampa Bay Estuarine System," *Technical Report*, Coastal & Oceanographic Engineering Department, University of Florida.

Yassuda, E.A. and Y.P. Sheng, 1997: "Modeling Dissolved Oxygen Dynamics in Tampa Bay during the Summer of 1991," in *Estuarine and Coastal Modeling, V*, ASCE.

THE CHESAPEAKE BAY EXPERIENCE

Harry V. Wang,[1] M.ASCE, Billy H. Johnson,[2] M.ASCE, and
Carl F. Cerco,[3] M.ASCE

ABSTRACT

Due to the decline in the water quality of the Chesapeake Bay over the past 30 years or so as a result of excess loading of nutrients, the Chesapeake Bay Program (CBP) has been involved in the development of numerical modeling tools to aid in the assessment of various nutrient controls. The CBP is a voluntary partnership that includes the states of Maryland, Virginia, and Pennsylvania; the District of Columbia; the Chesapeake Bay commission (a tri-state legislative body); and the US Environmental Protection Agency (EPA). The CBP's initial strategy was to develop a quasi-three dimensional (3D) steady state hydrodynamic and water quality model along with the watershed model. Following this effort, the US Army Corps of Engineers jointly funded with the CBP the development of fully 3D time varying hydrodynamic and water quality models with output loads obtained from the watershed model. That effort focused on the main Bay with the tributaries rather coarsely resolved in the numerical grid. In addition to the main bay, current effort focuses on the tributaries. This effort has seen the continued development of the 3D hydrodynamics and water quality models and the watershed model along with the development of another modeling component, i.e., an airshed model, to provide nutrient loading from the atmosphere. The integrated suite of models will be used to simulate the impact of nutrient control scenarios. The various modeling

[1] Assistant Professor, Virginia Institute of Marine Science, The College of William and Mary, P. O. Box 1346, Gloucester Point, VA 23062

[2] Research Hydraulic Engineer, Army Engineer Waterways Experiment Station, 3909 Halls Ferry Road, Vicksburg, Mississippi 39180

[3] Hydrologist, Army Engineer Waterways Experiment Station

components are discussed and results from the hydrodynamic model are presented. Calibration of the current water quality model, which includes submerged aquatic vegetation and higher trophic levels, is ongoing.

INTRODUCTION

As discussed by Linker (1996), Chesapeake Bay (Figure 1) is the largest estuary in the United States and until recent years has possessed an abundance of fish, shellfish, and waterfowl. However over the past 30 years or so the Bay's resources have declined due to excess amounts of nutrients such as nitrogen and phosphorus. This excess of nutrients resulted in an overgrowth of algae blooms which ultimately sink to the bottom and decompose, and deprive the bottom water of dissolved oxygen. Areas of low dissolved oxygen are no longer able to support fish and other aquatic life.

With the Bay being fed by runoff from a 64,000 square mile watershed encompassing several surrounding states, no one state could adequately develop solutions for reversing the decline in the Bay's resources. Thus, the Chesapeake Bay Program (CBP), which includes all surrounding states and the EPA, was created in 1983 to develop meaningful controls for reducing the input of nutrients into the Bay.

To make these controls quantifiable and defendable, the CBP decided early on that a numerical water quality model of the Bay was needed to predict the response of the Bay to various nutrient control scenarios. Over the past 15 years, there have been three separate modeling efforts. The first model, developed in the early to mid-1980's, was a steady state model which simulated summer average conditions using simplified loading estimates (HydroQual Inc., 1987).

After this effort, the CBP realized that to adequately address the problem a fully three-dimensional time-varying water quality model was needed (Cerco and Cole, 1993). To provide the time-varying flow fields required by the water quality model, a 3D hydrodynamic model was also developed as part of the research effort (Johnson, et al,1991). Along with the development of the 3D time varying models, a watershed model was developed to provide magnitudes and sources of nutrient loads into the Bay resulting from runoff (Donigan, et al. 1991). Atmospheric nutrient loads were based on long-term averages from observations. The watershed and hydrodynamic models were run for the years of 1984, 85, and 86. These years represented wet, dry, and average hydrology years, respectively. The water quality model then cycled the three years of hydrodynamics and water shed and atmospheric loadings to make 30 year simulations to assess the impact of various nutrient control scenarios. A major component of the water quality model was the development of a sediment sub-model to allow the exchange of nutrients between the sediment and the water column. The long time scale for the sediment processes necessitated the 30 year simulations of the water quality model.

The second effort outlined above focused primarily on processes in the main Bay. Thus, the tributaries were coarsely resolved in the model. The planform grid is shown in Figure 2(a). There are approximately 4,000 total cells with about 750 planform cells. The maximum number of cells in the vertical direction is 15, with each layer being 1.52m thick except for the top layer which varies with the tide. Since there are a minimum of 2 vertical layers, some areas of the Bays are represented with too much depth.

In 1994 the CBP saw the need to focus on nutrient controls in the tributaries. Thus, third modeling effort was initiated by the CBP, which requires additional grid resolution and allowance of one vertical layer in the tributary. In addition, to move uncertainty in the water quality boundary conditions at the Bay mouth out into the ocean, and to perhaps better compute the exchange between the Bay and the Atlantic ocean, the Bay grid was extended onto the shelf. The resulting planform grid is shown in Figure 2(b). In addition to the grid changes, living resources were added to the suite of water quality parameters. These include submerged aquatic vegetation, zooplankton and benthos. Furthermore, since a significant portion of the total loading of nutrients into the Bay comes from the atmosphere, modeling of the Chesapeake Bay airshed was included in the third and latest modeling effort.

In summary, the latest modeling strategy of the CBP involves the integration of a suite of models. These include a watershed model, an airshed model, a 3D water quality model containing living resources, and a 3D hydrodynamic model of the bay, its tributaries, and the connecting shelf. These models are briefly discussed below. This discussion is followed by a presentation of typical hydrodynamic model results. Calibration of the water quality model is ongoing.

WATERSHED MODEL

As illustrated in Figure 3, the latest version of the watershed model divides the drainage basin into about 90 segments. The model is referred to as HSPF (Hydrological Simulation Program-Fortran) (Donigan et al, 1991), and is a refinement of the Stanford Watershed Model. The model contains a hydraulic submodel, a non-point source submodel, and a river submodel. Rainfall, evaporation, and meteorological data are used in the hydraulic submodel to calculate runoff and subsurface flow. Basin land uses modeled include forest, agricultural lands, and urban lands. The surface and subsurface flows are input to the non-point source submodel for the simulation of soil erosion and movement of pollutant loads from the land to the rivers. These are then routed through the rivers, lakes and reservoirs in the system by the river submodel until they are finally input to the 3D water quality model. The watershed and 3D hydrodynamic models are linked through the runoff flows computed by the hydraulic submodel.

WATER QUALITY MODEL

The latest version of the 3D and time-varying water quality model is based on CE-QUAL-ICM (Cerco and Cole, 1993) and incorporates 22 state variables plus living resources such as submerged aquatic vegetation, zooplankton and benthos. Kinetic interactions among the state variables are described in 80 partial differential equations that employ more than 140 parameters. The state variables can be categorized into six groups; namely, the physical group consisting of salinity, temperature, and inorganic solids; the carbon cycle; the nitrogen cycle, the phosphorus cycle; the silica cycle; and the dissolved oxygen balance.

The water quality model is directly coupled to a predictive benthic-sediment model which is an expansion of diagenetic principles established for freshwater sediments into the estuarine environment. The water quality and sediment models interact on a time scale equal to the integration step of the water quality model. Sediment-water fluxes of dissolved nutrients and oxygen based on predicted diagenesis and concentrations in the sediments and water are computed in the sediment model. These are then passed to the water quality model and incorporated into the appropriate mass balances and kinetic reactions. In the latest version of the water quality model, a benthos model that computes the impact on dissolved oxygen of bottom dwelling critters such as clams is coupled to the water quality model.

AIRSHED MODEL

The airshed model known as RADM (Regional Acid Deposition Model) is primarily developed by Air Resources Laboratory, NOAA for the EPA (Dennis, 1994). The Chesapeake Bay airshed covers the Eastern United States from Texas and North Dakota eastward to Maine and Florida. The 3D grid of this airshed measures 20 km square and contains 15 vertical layers spread over a vertical distance of about 15 km. The total number of cells is about 22,000. The nutrient loads in the atmosphere are transported from their sources and linked to the watershed model through deposition on land. In addition, there is a direct linkage with the water quality model through deposition onto the surface waters of the bay.

HYDRODYNAMIC MODEL

The 3D hydrodynamic model is a variable density model since both salinity and temperature are modeled and linked to the water density through an equation of state. The model is referred to as CH3D-WES (Curvilinear Hydrodynamics in 3 Dimensions - Waterways Experiment Station) (Johnson et al, 1991). As illustrated by the numerical grids presented in Figures 2 (a) and (b), computations are made on a curvilinear or boundary-fitted grid. The vertical coordinate is the Cartesian coordinate. Physical processes impacting bay-wide circulation and vertical mixing that are modeled include tides, wind, density effects, freshwater inflows, turbulence, and the effect of the earth's rotation. Adequately representing the vertical turbulence is crucial to a successful simulation of stratification/destratification. A K-E turbulence closure model is employed. The boundary-fitted coordinates feature of the model provides enhancement to fit the deep navigation channel and irregular shoreline of the Bay and permits adoption of an accurate and economical grid schematization.

Several model developments/modifications were required in the latest modeling effort. These included allowing for part of the grid to only contain one layer, i.e., a mixture of 3D and vertically-averaged computations, implementation of a K-E turbulence closure scheme to replace the old scheme which required smoothing over the water depth to simulate diffusion of the turbulence, and linking with the watershed model to obtain lateral inflows on the tributaries.

STUDY STRATEGY

With the final calibration/verification of the suite of models, the integrated modeling system will provide the foundation for developing nutrient reduction goals, e.g., a 40 percent reduction by the year 2000. These nutrient reduction scenarios will be run in the water quality by using 10 years of hydrodynamics generated for 1985-94 to drive the transport and with 10 years of nutrient loads coming from the watershed and airshed models. It is anticipated that runs longer than 10 years may be required. If so, the loads and hydrodynamics will be recycled to drive water quality runs of up to 30 years.

To ensure technical success and acceptance by the public, throughout the study monthly meetings that are completely open to the public have been held. In addition, a team of international experts meets four times a year for a complete technical review of the various modeling components. Finally, once a year the entire scientific Bay community comes together for an annual review of the study. Again, all meetings are completely open to the general public. Much effort has also been invested by environmental reporters to make the findings of the integrated modeling system understandable to the public.

HYDRODYNAMIC MODEL RESULTS

Model verification has proceeded by demonstrating the model's ability to reproduce both short-term events such as wind mixing and flood events, as well as long-term trends (e.g., water column stratification and residual currents). Initial model verification was conducted to reproduce the wind-mixing event that occurred in September 1983. Reproduction of this event in the previous modeling effort has been discussed by Johnson et al, (1991). Velocity verification results on the York River at a location about 20 km from its mouth are presented in Figure 4. Results from the NOAA tide tables have been used to verify the model's ability to accurately reproduce mean tide ranges on the York, Pamunkey, and Mattaponi Rivers to within 5-10 cm. Times to low and high water were reproduced to within 15 minutes of the NOAA Tide Table values.

The importance of including the watershed flows in the tributaries is demonstrated in Figure 5. With the proper watershed inflow specified, the model reproduces salinity on the Rappahannock River during August 1987 to within a couple of tenths of a part per thousand. Tide table data were also utilized to show the importance of including the littoral zone. This is the zone from the 2m depth contour to the shoreline and is important to the water quality model since that is where the submerged aquatic vegetation grows. With the littoral zone cells included, the mean tide range is reproduced well. However, without the littoral zone the range in the mid to upper part of the river is computed to be much too large and the times to high and low water are computed to be too small.

Salinity data during 1985-87 are available at two week intervals throughout the Bay and its tributaries. Figure 6 shows a comparison of model and observed salinities at a location near the Bay bridge at Annapolis, MD. It can be seen that the model reproduces mixing events and captures the large variations in salinity quite well. Similar results were obtained throughout the Bay and its tributaries. Figure 7 illustrates that the model does an excellent job of reproducing seasonally averaged salinities along the main Bay channel. Both the extent of salt intrusion as well as the stratification are correctly computed.

It is important for a hydrodynamic model to reproduce correctly the residual currents resulting from gravitational circulation due to density effects. Figure 8 shows near surface and near bottom residual currents computed from model results averaged over the month of April during 1986. An averaging of observed data collected over several years at several stations throughout the Bay (Goodrich and Blumberg, 1991) are presented in Figure 9. Since the magnitude of residual currents varies with fresh water inflow, it should be noted that residual currents can differ greatly from month to month. Thus, one should not place too much importance on a comparison of Figures 8 and 9 other than they show the same relative magnitude.

Finally, it is important to demonstrate that the proper flux through the Bay's mouth is being computed. Observed data are rather limited. However, results published by Boicourt (1973) indicate that the inward flux during July and November of 1971 was about 6.1 and 5.8 x 10^3 m^3/sec, respectively. Computed values from the model show for these two months during 1985 are 6.27 and 6.01 x 10^3 m^3 /sec, which are reasonably close to the observations.

Although only a portion of model results have been presented, they are representative of verification results that have been obtained. The general conclusion is that the 3D hydrodynamic model is a good representation of the Bay and its tributaries. The model is currently being used to generate ten years (1985-94) of hydrodynamic data to drive the water quality model of the Bay and its tributaries.

SUMMARY

The integrated model presented in this paper illustrated the importance of the interaction between the land, air and the water surrounding the Chesapeake Bay. Ten years of loads and hydrodynamics are being generated for input to the water quality model. Various nutrient control scenarios including 40% reduction scenario will then be simulated in the water quality model for periods up to perhaps 30 years. For these simulations, 10 year loads and hydrodynamic will be recycled.

The integrated models are an example of approaching an environmental problem in a holistic manner such that problems related to the land, the water and the air are examined together. This approach is expected to provides solutions that are most cost effective and protective of the environment. The Chesapeake Bay integrated models are the first and most extensive application of this technology. The integrated systems approach enables environmental managers to examine "what if" questions to allow the simultaneous growth of the environment and

ACKNOWLEDGMENT

The results presented were obtained from work conducted for the US Army Engineer District, Baltimore and the EPA. Permission to publish this paper was granted by the Chief of Engineers.

REFERENCES

Boicourt, William C. 1973. "Circulation of Water on the Continental Shelf from Chesapeake Bay to Cape Hatteras. Ph.D thesis", The Johns Hopkins University.

Cerco, C. F. And Cole, T. 1993. "Three-Dimensional Eutrophication Model of Chesapeake Bay", Journal of Environmental Engineering, ASCE, Vol. 119, No. 6, 1993.

Dennis, R. L. 1994. "The Linking of RADM to Water Quality and Watershed Models. Presentation at the Watershed, Estuarine, and Large Lakes Modeling Workshop, Bay City, MI, April 19, 19.

Donigan, A., Bicknell, B., Patwardhan, A., Linker, L., Alegre, D., Chang, C., and Reynolds, R. 1991. "Watershed Model Application to Calculate Bay Nutrient Loadings", Chesapeake Bay Program Office, US Environmental Protection Agency, Annapolis, MD.

Goodrich, D. M. and Blumberg, A. F. 1991. The Fortnightly Mean Circulation of Chesapeake Bay. Estuarine, Coastal ans Shelf Sciences, Vol 32, p451-462.

HydroQual, Inc. 1987. "A Steady State Coupled Hydrodynamic/Water Quality Model of Eutrophication and Anoxia Processed in Chesapeake Bay." EPA Contract 68-03-3319, EPA,Chesapeake Bay Program, Annapolis, MD.

Johnson, B. H., Kim, K. W., Heath, R., Hsieh, B., and Butler L. 1991. "Validation of A Three-Dimensional Hydrodynamic Model of Chesapeake Bay", Journal of Hydraulic Engineering, ASCE, 1991, 2-20.

Linker, L. 1996. "Models of the Chesapeake Bay", Sea Technology, September, 1996, pp 49-55.

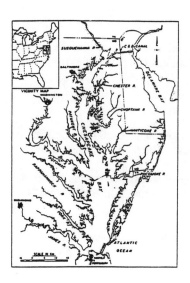

Figure 1. Chesapeake Bay

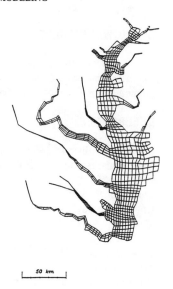

Figure 2(a). Chesapeake Bay model 68x35 coarse grid

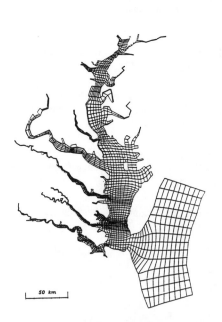

Figure 2(b). Chesapeake Bay model 105x88 fine grid

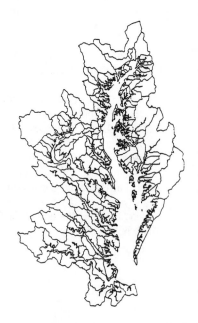

Figure 3. Chesapeake Bay watershed model subsegments

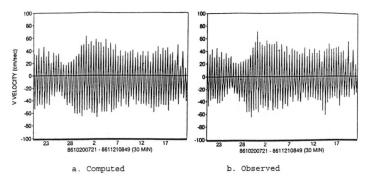

a. Computed   b. Observed

Figure 4. Velocity comparison on York River 20 km above the mouth

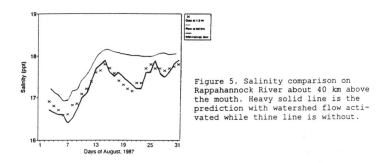

Figure 5. Salinity comparison on Rappahannock River about 40 km above the mouth. Heavy solid line is the prediction with watershed flow activated while thine line is without.

Figure 6. Salinity comparison at a location in the main bay near the Bay Bridge at Annopolis, MD.

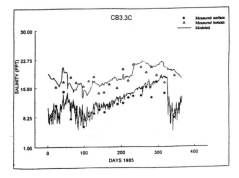

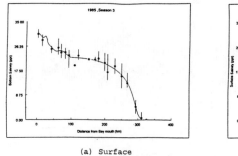 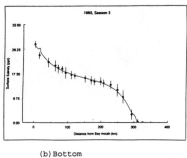

(a) Surface   (b) Bottom

Figure 7. Seasonally averaged salinities along the main bay channel,
Season 3 (summer), 1985

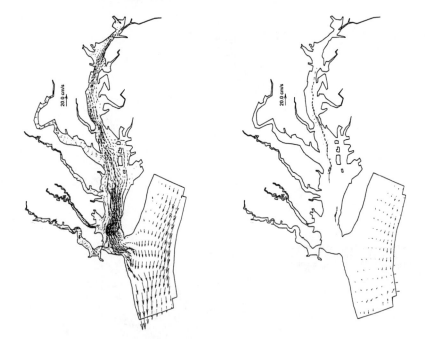

(a) near surface   (b) near bottom at 60 ft
Figure 8. Computed residual currents from model results during 1985

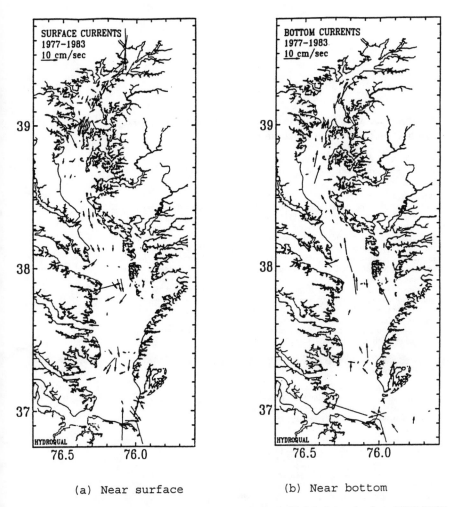

(a) Near surface   (b) Near bottom

Figure 9.   Computed residual currents from field data during 1977-1983
(Goodrich and Blumberg, 1991)

Improvement of Estuarine and Coastal Modeling Using High-Order Difference Schemes

Peter C. Chu, and Chenwu Fan

Department of Oceanography, Naval Postgraduate School, Monterey, California

Abstract.
How to reduce the computational error is a key issue of using σ-coordinate coastal ocean models. Due to the fact that the higher the order of the difference scheme, the less the truncation error and the more complicated the computation, we introduce three sixth-order difference schemes (ordinary, compact, and combined compact) for the σ-coordinate coastal models in order to reduce error without increasing much complexity of the computation. After the analytical error estimation, the Semi-Spectral Primitive Equation Model (SPEM) is used to demonstrate the benefit of using these schemes and to compare the difference among the three six-order schemes. Over a wide range of parameter space as well as a great parametric domain of numerical stability, the ordinary sixth-order scheme is shown to have error reductions by factors of 50 comparing to the second-order difference scheme. Among the sixth-order schemes, the compact scheme reduces error by more than 50-55% comparing to the ordinary scheme, and the combined compact scheme reduces error by more than 60% comparing to the compact scheme.

1. Introduction

Most estuarine and coastal models use second-order difference schemes (such as second-order staggered C-grid scheme) to approximate first-order derivative (Blumberg and Mellor, 1987; Hadivogel et al., 1991)

$$\left(\frac{\partial p}{\partial x}\right)_i \simeq \frac{p_{i+1/2} - p_{i-1/2}}{\Delta} - \frac{1}{24}\left(\frac{\partial^3 p}{\partial x^3}\right)_i \Delta^2, \quad (1)$$

where p, Δ represent pressure and grid spacing. Such a difference scheme was proposed by numerical modelers in early 50's as the first generation computers came into place. Since then the computer updates rapidly with several orders of magnitude increase in computational power. However, the difference schemes used by most modelers now are still staying at the 50's level (second-order schemes).

Besides, the current scheme uses the local Lagrangian Polynomials whose derivatives are discontinuous. Figure 1 shows the process of computing first-order derivative of function $\phi(x)$ at five neighboring grid points. At the grid x_i, $\phi'(x_i)$ is the tangential of the Lagrangian Polynomial L_i at x_i. As i increases, three neighboring L_{i-1}, L_i, L_{i+1} have different tangential at the point x_i. Such a hidden problem in discretization might distort the physical processes.

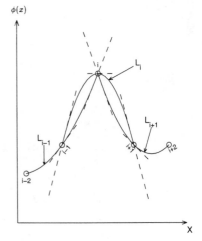

Figure 1. Discontinuity of the first derivatives of the Lagrangian Polynomials at each grid point.

$$= \frac{1}{240} \frac{-17p_{i-3/2} - 189p_{i-1/2} + 189p_{i+1/2} + 17p_{i+3/2}}{\Delta}$$
$$- \frac{117}{358,400} \left(\frac{\partial^7 p}{\partial x^7}\right)_i \Delta^6, \quad (5a)$$

The "left" boundary point is computed by

$$\left(\frac{\partial p}{\partial x}\right)_1 = \frac{1,627}{1,920}p_{1/2} - \frac{211}{640}p_{3/2} - \frac{59}{48}p_{5/2}$$
$$+ \frac{235}{192}p_{7/2} - \frac{91}{128}p_{9/2} + \frac{443}{1,920}p_{11/2} - \frac{31}{960}p_{13/2}$$
$$+ \frac{3,043}{107,520} \left(\frac{\partial^7 p}{\partial x^7}\right)_i \Delta^6 \quad (5b)$$

The "right" boundary point has the similar formulation. The last terms in (2) and (5) suggest that the compact scheme can provide more accuracy. The error ratio between the sixth-order compact and ordinary schemes is

$$r_6 = \frac{\left|-\frac{117}{358,400}\left(\frac{\partial^7 p}{\partial x^7}\right)_i \Delta^6\right|}{\left|\frac{5}{7,168}\left(\frac{\partial^7 p}{\partial x^7}\right)_i \Delta^6\right|} = 46.8 \quad (6)$$

which means a more than 50% reduction of the truncation error when we use the compact scheme for the 6th-order accuracy. We use the pressure gradient error reduction in the σ-coordinate system as an example to illustrate the benefit of using high-order compact difference schemes.

4. Sixth-Order Combined Compact Difference (CCD) Scheme

Recently, Chu and Fan (1998) proposed a three-point combined compact scheme

$$\left(\frac{\delta p}{\delta x}\right)_i + \alpha_1 \left(\left(\frac{\delta p}{\delta x}\right)_{i+1} + \left(\frac{\delta p}{\delta x}\right)_{i-1}\right)$$
$$+ \beta_1 \Delta \left(\left(\frac{\delta^2 p}{\delta x^2}\right)_{i+1} - \left(\frac{\delta^2 p}{\delta x^2}\right)_{i-1}\right) + \cdots$$
$$= \frac{a_1}{2\Delta}(f_{i+1} - f_{i-1}) \quad (7)$$

$$\left(\frac{\delta^2 p}{\delta x^2}\right)_i + \alpha_2 \left(\left(\frac{\delta^2 p}{\delta x^2}\right)_{i+1} + \left(\frac{\delta^2 p}{\delta x^2}\right)_{i-1}\right)$$
$$+ \beta_2 \frac{1}{2\Delta} \left(\left(\frac{\delta p}{\delta x}\right)_{i+1} - \left(\frac{\delta p}{\delta x}\right)_{i-1}\right) + \cdots$$
$$= \frac{a_2}{\Delta^2}(f_{i+1} - 2f_i + f_{i-1}) \quad (8)$$

2. Six-Order Ordinary Difference (OD) Scheme

Since the truncation error decreases with the increase of the order of the difference scheme, it might be benefited to use a higher order difference schemes. For example, Chu and Fan [1997a] proposed to use a six-order ordinary difference scheme

$$\left(\frac{\partial p}{\partial x}\right)_i \simeq \left(\frac{-9p_{i-5/2} + 125p_{i-3/2} - 2,250p_{i-1/2}}{1920\Delta}\right)$$
$$+ \left(\frac{2,250p_{i+1/2} - 125p_{i+3/2} + 9p_{i+5/2}}{1920\Delta}\right)$$
$$- \frac{5}{7,168} \left(\frac{\partial^7 p}{\partial x^7}\right)_i \Delta^6, \quad (2)$$

to compute the horizontal pressure gradient. Comparing (2) with (1), the error ratio between the sixth-order and second-order schemes is estimated by

$$r_{6,2} = \left|\frac{\frac{5}{7168}\left(\frac{\partial^7 p}{\partial x^7}\right)_i}{\frac{1}{24}\left(\frac{\partial^3 p}{\partial x^3}\right)_i}\right| \Delta^4 = 0.0167 \left|\frac{\left(\frac{\partial^7 p}{\partial x^7}\right)_i}{\left(\frac{\partial^3 p}{\partial x^3}\right)_i}\right| \Delta^4. \quad (3)$$

which is proportional to Δ^4.

There are two weaknesses of using ordinary high-order difference schemes. The first one is more grid points (6 points) needed for the computation. The second one is using lower-order schemes at two boundaries.

3. Sixth-Order Compact Difference (CD) Scheme

The compact scheme is used to overcome such a weakness by connecting the derivative at the grid with that at the two neighboring grids, e.g.,

$$\alpha^+ \left(\frac{\partial p}{\partial x}\right)_{i+1} + \alpha^0 \left(\frac{\partial p}{\partial x}\right)_i + \alpha^- \left(\frac{\partial p}{\partial x}\right)_{i-1}$$
$$= \beta_1 p_{i+3/2} + \beta_2 p_{i+1/2} + \beta_3 p_{i-1/2} + \beta_4 p_{i-3/2}. \quad (4)$$

The parameters $\alpha^0, \alpha^+, \alpha^-, \beta_1, \beta_2, \beta_3, \beta_4$ are determined by requiring the minimum error (Chang and Shirer, 1985). Such compact schemes guarantee continuity of the first derivative at each grid point.

By requiring the minimum truncation error for the sixth-order difference scheme we obtain the sixth-order compact scheme (Chu and Fan, 1997b)

$$\frac{1}{80} \left[9 \left(\frac{\partial p}{\partial x}\right)_{i-1} + 62 \left(\frac{\partial p}{\partial x}\right)_i + 9 \left(\frac{\partial p}{\partial x}\right)_{i+1}\right]$$

to compute p'_i, $p''_i, \ldots p_i^{(k)}$ by means of the values and derivatives at the two neighboring points. Moving from the one boundary to the other, CCD forms a global algorithm to compute various derivatives at all grid points, and guarantees continuity of all derivatives at each grid point.

5. Pressure Gradient Error in the σ-Coordinate System

Coastal ocean models usually use a terrain-following σ-coordinate system to handle the effects of bottom topography. Here the water column is divided into the same number of grid cells independence of depth . Let (x_*, y_*, z) denote Cartesian coordinates and (x, y, σ) sigma coordinates. In most sigma coordinate ocean models the relationship between the two coordinate systems are:

$$x = x_*, \qquad y = y_*, \qquad \sigma = 1 + 2\frac{z}{H(x,y)} \qquad (9)$$

where z and σ increase vertically upward such that $z = 0, \sigma = 1$ at the surface and $z = -H$ and $\sigma = -1$ at the bottom. $H = H(x, y)$ is the bottom topography.

A problem has long been recognized in computing the horizontal pressure gradient in the σ-coordinate system (e.g., Gary, 1973; Haney, 1991; and Mellor et al., 1994, McCalpin, 1994): the horizontal pressure gradient becomes a difference between two terms, which leads to a large truncation error at a steep topography.

6. Seamount Test Case

6.1 Model Description

Suppose a seamount located inside a periodic f-plane ($f_0 = 10^{-4}s^{-1}$) channel with two solid, free-slip boundaries along constant y. Unforced flow over seamount in the presence of resting, level isopycnals is an idea test case for the assessment of pressure gradient errors in simulating stratified flow over topography. The flow is assumed to be reentrant (periodic) in the along channel coordinate (i.e., x-axis). We use this seamount case of the Semi-spectral Primitive Equation Model (SPEM) version 3.9 to test the new difference scheme. The reader is referred to the original reference (Haidvogel et al., 1991) and the SPEM 3.9 User's Manuel (Hedstrom, 1995) for detail information. In the horizontal directions the model uses the C-grid and the second-order finite difference discretization except for the horizontal pressure gradient, which the user has choice of

either second-order or fourth-order difference discretization (McCalpin, 1994). In the vertical direction the model uses a boundary fitted σ-coordinate system. The discretization is by spectral collocation using Chebyshev polynomials. Our model configuration is similar to that of Beckmann and Haidvogel (1993), McCalpin (1994), and Chu and Fan (1997a). The time step and grid size used here are,

$$\Delta t = 675s, \Delta x = \Delta y = 5km. \qquad (10)$$

6.2 Topography

The domain is a periodic channel, 320 km long and 320 km wide. The channel walls are solid (no normal flow) with free-slip viscous boundary conditions. The channel has a far-field depth h_{max} and in the center includes an isolated Gaussian-shape seamount with a width L and an amplitude h_s (Figure 2),

$$h(x,y) = h_{\max} - h_s \exp\left[-\frac{(x - x_0)^2 + (y - y_0)^2}{L^2}\right] \qquad (11)$$

where (x_0, y_0) are the longitude and latitude of the seamount center. The far-field depth (h_{max}) is fixed as 5,000 m. But the seamount amplitude (h_s) changes from 500 to 4,500 m, and the lateral scale of the seamount (L) varies from 10 to 40 km. for the study.

6.3 Density Field

The fluid is exponentially stratified and initially at rest. The initial density field has the form,

$$\rho_i = \overline{\rho}(z) + \widehat{\rho}\exp(\frac{z}{H_\rho}) \qquad (12)$$

where z is the vertical coordinate, and $H_\rho = 1000$ m, and

$$\overline{\rho}(z) = 28 - 2.\exp(\frac{z}{H_\rho}) \qquad (13)$$

is a reference density field. Here a constant density, 1000 kg m^{-3}, has been subtracted for the error reduction. Following Beckmann and Haidvogel (1993) and McCalpin (1994), we subtract the mean density field $\overline{\rho}(z)$ before integrating the density field to obtain pressure from the hydrostatic equation.

The density anomaly, $\widehat{\rho}$, indicates that the initial condition of the model was slightly less stably stratified than the reference field for each computation. In this study, the density anomaly varies from 0.1-1 kg m^{-3}.

6.4 Lateral Viscosity

Ideally, the new difference scheme should be tested with no lateral diffusion of density. This is due to the fact that the density diffusion along σ surfaces generates horizontal gradients wherever the σ surfaces are not flat, and then produces horizontal pressure gradients which drive currents in much the same way as the pressure gradient errors (McCalpin, 1994). Unfortunately, the absence of the horizontal diffusion keeps the small-scale pressure disturbances generated by topographic scale density advection, and the induced small-scale velocity fields, which in turn cause the computational instability problem. Thus, some lateral viscosity on σ surfaces is required in the momentum equations to maintain stability. A constant coefficient (A_H) biharmonic formulation is used here for the lateral viscosity, which varies from $10^{10} - 10^8$ m^{4}s^{-1} in this study.

7. Standard Case

7.1 Model Parameters

At first, we set up a standard test case to compare errors among the ordinary, compact, and combined compact difference schemes: $L = 40$ km, $h_s = 4{,}500$ m,

$$\widehat{\rho} = 0.2 kgm^{-3}, \ A_H = 10^{10} m^4 s^{-1} \qquad (14)$$

Figure 3 displays results from the sixth-order compact case: the error in the streamfunction and velocity fields after performing 5 days of integration. The mass transport streamfunction has a large-scale eight-lobe pattern centered on the seamount. This symmetric structure can be found in all the fields. After 5 days of integration, the model generates spurious currents of O(0.03 cm/s), much smaller than that of O(6 cm/s) generated with the second order scheme and O(0.2 cm/s) generated with the fourth order ordinary scheme (Figures 1 and 2 in McCalpin, 1994).

7.2 Temporal Variations of Peak Error Velocity

Owing to a very large number of calculations performed, we discuss the results exclusively in terms of the maximum absolute value the spurious velocity (called peak error velocity) generated by the pressure gradient errors. Figure 4 shows the time evolution of the peak error velocity for the first 20 days of integration with the second-, fourth-, and sixth-order ordinary schemes. Figure 5 shows the time evolution of the peak error velocity with the sixth-order ordinary, compact, and combined compact schemes. The peak error velocity fluctuates rapidly during the first few days integration. After the

5 days of integration, the peak error velocity show the decaying inertial oscillation superimposed into asymptotic values. The asymptotic value is near 10^{-4} m/s for the ordinary scheme, 0.44×10^{-4} m/s for the compact scheme, and 0.16×10^{-4} m/s for the combined compact scheme. Thus, the compact scheme reduces error by more than 50-55% comparing to the ordinary scheme, and the combined compact scheme reduces error by more than 60% comparing to the compact scheme.

8. Conclusions

(1) The σ-coordinate, pressure gradient error depends on the choice of difference schemes. By choosing an optimal scheme, we may reduce the error in a great deal without increasing the horizontal resolution. Analytical analysis shows that the truncation error ratio between the fourth-order scheme and the second-order scheme is proportional to Δ^2, and the truncation error ratio between the sixth-order scheme and the second-order scheme is proportional to Δ^4. The compact scheme may reduce near 30% error for the fourth-order difference and more than 50% error for the sixth order difference.

(2) The SPEM Version 3.9 is used to demonstrate the benefit of using the sixth-order scheme. A series of calculations of unforced flow in the vicinity of an isolated seamount are performed. Over a wide range of parameter space as well as a great parametric domain of numerical stability, the sixth-order scheme has error reductions by factors of 50 comparing to the second-order difference scheme. Furthermore, the compact scheme reduces error by more than 50-55% for the sixth-order difference scheme, and the combined compact scheme reduces error by more than 60% comparing to the compact scheme.

(3) Using the sixth-order scheme does not require much more CPU time. Taking SPEM3.9 as an example, the CPU time for the sixth-order scheme is almost the same as for the fourth-order scheme, and 10% more than for the second-order scheme, and the CPU time for the compact scheme (both fourth- and sixth-order) is only 5% more than the ordinary scheme.

(4) Since the fourth-order different scheme has error reductions by factors of 10 comparing to the second-order difference scheme, there is no real advantage to going to a higher order scheme if the bottom topography is not too complicated. The need for a lot of accuracy will go up with increasingly complex bottom topography on small scales, so one might expect that future demand of the accuracy will increase as models strive for more realism.

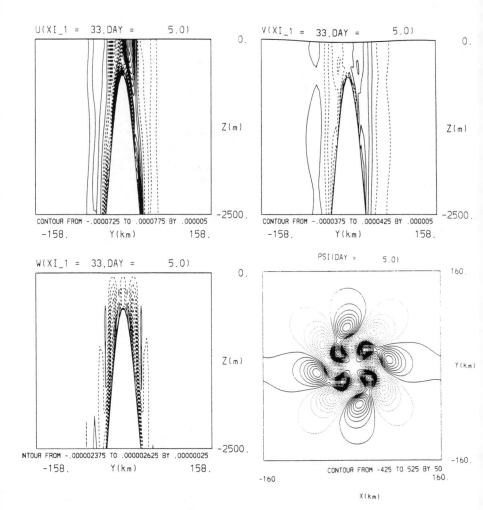

Figure 3. Instantaneous error pattern after 5 days of integration for the unforced experiment with sixth-order compact scheme: (a) u, (b) v, (c) w, and (d) mass transport streamfunction. Here u, v, w are evaluated at the slice (facing upchannel) through the center of the seamount (after Chu and Fan, 1997a.)

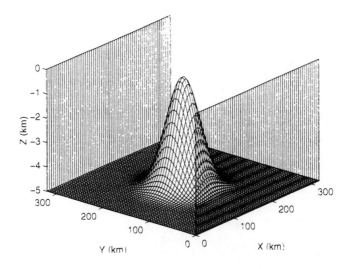

Figure 2. The seamount geometry.

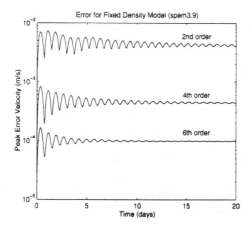

Figure 4. Peak error velocity for the second-, fourth-, and sixth-order ordinadry schemes (after Chu and Fan, 1997a.)

Acknowledgments

The authors wish to thank Dale Haidvogel and Kate Hedstrom of the Rutgers University for most kindly proving us with a copy of the SPEM code. This work was funded by the Office of Naval Research NOMP Program, the Naval Oceanographic Office, and the Naval Postgraduate School.

References

Beckmann, A., and D. B. Haidvogel, 1993: Numerical simulation of flow around a tall isolated seamount. Part 1: Problem formulation and model accuracy. *J. Phys. Oceanogr.*, **23**, 1736-1753.

Chang, H.R., and H.N. Shirer, 1985: Compact spatial differencing techniques in numerical modeling. *Mon. Wea. Rev.*, **113**, 409-423.

Chu, P.C., and C.W. Fan, 1997a: A sixth-order difference scheme in sigma coordinate ocean models. *J. Phys. Oceanogr.*, **27**, 2064-2071.

Chu, P.C., and C.W. Fan, 1997b: Compact difference schemes sigma coordinate ocean models. *Internat. J. Num. Methods in Fluids*, submitted.

Chu, P.C., and C.W. Fan, 1998: A three-point combined compact difference scheme. *J. Compt. Phys.*, **140**, 1-30.

Gary, J.M., 1973: Estimate of truncation error in transformed coordinate primitive equation atmospheric models, *J. Atmos. Sci.*, **30**, 223-233.

Haidvogel, D.B., J.L. Wikin, and R. Young, 1991: A semi-spectral primitive equation model using vertical sigma and orthogonal curvilinear coordinates. *J. Comput. Phys.*, **94**, 151-185.

Haney ,1991: On the pressure gradient force over steep topography in sigma coordinate ocean models. *J. Phys. Oceanogr.*, **21**, 610-619.

Hedstrom, K., 1994: User's Manual for a Semi-Spectral Primitive Equation Ocean Circulation Model Version 3.9, Rutgers University, New Jersey.

McCalpin, J.D., 1994: A comparison of second-order and fourth-order pressure gradient algorithms in a σ-coordinate ocean model. *Internat. J. Num. Methods in Fluids*, **18**, 361-383.

Mellor, G.L., T. Ezer, and L.-Y. Oey, 1994: The pressure gradient conundrum of sigma coordinate ocean models. *J. Atmos. Oceanic Technol.*, **11**, 1126-1134.

Sunsqvist, H., 1976: On vertical interpolation and truncation in connection with use of sigma system models. *Atmosphere*, **14**, 37-52.

Robinson, A.R., 1993: Physical precesses, field estimation and interdisciplinary ocean modeling. *Harvard Open Ocean Model Reports* **Harvard University**, Cambridge, 71 pp.

P. Chu, and C. Fan, Department of Oceanography, Naval Postgraduate School, Monterey, California. (e-mail: chu@nps.navy.mil)

(Received August 19, 1997; revised February 10, 1998; accepted February 25, 1998.)

MODELING DISSOLVED OXYGEN DYNAMICS OF TAMPA BAY DURING SUMMER OF 1991

Eduardo A. Yassuda[1] and Y.Peter Sheng[2]

ABSTRACT

An integrated modeling study of the Tampa Bay Estuarine System (Florida) was conducted to further our understanding of the estuary as an integrated system, and to provide a tool for quantitative assessment of various management practices. By integrating several models and synthesizing field data, we produce a detailed characterization of the circulation and water quality dynamics within the Bay. The models integrated in this study included a hydrodynamic model (Sheng, 1989; Sheng and Yassuda, 1995), a nitrogen cycle model (Chen and Sheng, 1995), a dissolved oxygen model (Yassuda, 1996), and a seagrass model (Fong and Harwell, 1994; Sheng et al., 1995). In this paper, the results of the water quality simulations are presented in terms of the dissolved oxygen dynamics and nutrient budget in the summer of 1991. Model simulations successfully reproduced the measured water surface elevation, currents, and salinity dynamics. In addition, the water quality model successfully simulated the development of hypoxia conditions and subsequent recovery during the summer of 1991.

INTRODUCTION

Competitive demands for natural resources in estuarine systems can lead to a serious deterioration of the environment. In order to obtain solutions to environmental problems, resources management agencies are supporting a holistic approach to environmental management. For example, the Florida Department of Environmental Protection has recently emphasized ecosystem management as an integrated, flexible approach to manage Florida's biological and physical environment.

An efficient strategy to prevent or reverse the degradation of important estuarine systems is to make use of numerical models in conjunction with monitoring programs. Through monitoring, not only the present state of the system can be

---

[1] Formerly Ph.D. Candidate, Coastal and Oceanographic Engineering Dept., University of Florida. Presently, Water Resources Engineer, Applied Technology & Management, Inc., Gainesville, FL. (yassuda@worldnet.att.net).
[2] Professor, Coastal and Oceanographic Engineering Dept., University of Florida, Gainesville, FL. (pete@coastal.ufl.edu).

35

obtained, but it is also possible to evaluate the effectiveness of past and proposed management efforts. Numerical models can be used to study the response of the system to various management options. Numerical models can be applied to study the hydrodynamics, sediment dynamics, water quality dynamics and system ecology in estuarine systems.

Recognizing the important relationships among the various ecological components (hydrodynamics, sediments, water quality, aquatic vegetation, etc.) and advancement in computer resources and scientific understanding, it is now appropriate to integrate the various component models with multiple dimensions and coupled processes. An integrated model can be used to further the understanding of an estuary as an integrated system, and to provide a quantitative evaluation of various management practices.

Previous hydrodynamics modeling studies on Tampa Bay have been conducted by Goodwin (1987) who used a vertically-integrated two-dimensional model, Galperin et al. (1991) and Hess (1994) who used an three-dimensional curvilinear orthogonal-grid model, and Sheng and Yassuda (1995) who used a three-dimensional boundary-fitted-grid model.

Previous water quality modeling studies on Tampa Bay have been conducted by Ross et al. (1984) who developed a vertically-integrated two-dimensional hydrodynamics model coupled to a water quality box model, and AScI (1995) who applied the WASP box model to Tampa Bay. Ross et al. (1984) did not have sufficient data to calibrate and verify their water quality model, while AScI (1995) used a daily time step to simulate the daily-averaged water quality dynamics.

For the present study, the hydrodynamics model is based on "CH3D" developed by Sheng (1989) and Sheng and Yassuda (1995). The water quality model is based on that developed by Chen and Sheng (1995). It includes the biogeochemical processes related to: sediment dynamics (advection, mixing, deposition, resuspension, and settling), nitrogen cycle (mineralization, hydrolysis, nitrification, denitrification, volatilization, sorption/desorption, and uptake), phytoplankton (uptake, growth, grazing, respiration, and settling), and dissolved oxygen (reaeration, photosynthesis, oxidation, and nitrification). In addition, a conceptual seagrass model (Fong and Harwell, 1994) has been incorporated into the integrated Tampa Bay model.

The hydrodynamics component of this integrated Tampa Bay model has been successfully calibrated and verified using Tampa Bay data provided by the National Oceanic and Atmospheric Administration (NOAA) and the United States Geological Survey (USGS). The hydrodynamics have been directly incorporated into the water quality model by using the same grid spacing and time step for both the hydrodynamics and water quality models. If the two models had used different grids and time steps, it would have been necessary to conduct ad-hoc tuning of advective fluxes and dispersion coefficients. The water quality component of the integrated Tampa Bay model has been tested using monthly water quality data provided by the Hillsborough County Environmental Protection Commission (EPC), although a more comprehensive data set is needed to fully validate the water quality model. Results of previous statistical and mass-balance models were used to determine the relevant biogeochemical processes and to test causal relationships among state variables. These simple models also proved to be useful tools for calibration of the water quality model coefficients in the absence of process-specific data (e.g., remineralization, nitrification, and denitrification).

In the following, we will mainly present results of hydrodynamic and water quality model simulations for the summer of 1991, with a detailed comparison between simulated dissolved oxygen and EPC's data.

SIMULATION OF THE SUMMER 1991 CONDITION

The integrated model of the Tampa Bay Estuarine System was used to perform a four-month simulation of the hydrodynamics and water quality dynamics during the summer of 1991, when sufficient hydrodynamics data (Hess, 1994) and water quality data (Boler, 1992) are available for model calibration. The water quality dynamics in Tampa Bay were simulated with real tidal, river discharge, rainfall, and wind forcing.

Previous hydrodynamic studies (Goodwin, 1987; Galperin et al., 1991; Hess, 1994; Sheng and Yassuda, 1995) demonstrated that complex geometrical and bathymetrical features in the Tampa Bay Estuarine System have a profound influence on the circulation and transport within the Bay. Thus, to successfully apply the integrated model of Tampa Bay, it is essential to design a numerical grid which accurately represents the dominant geometric and bathymetric features. The computational grid used for the integrated model of Tampa Bay Estuarine System is shown in Figure 1, and contains 45 by 85 horizontal cells and 8 vertical "sigma" layers, with a total of 30600 grid cells. Horizontal grid spacing varies from 100 to 1500 meters.

Hydrodynamic Simulation

Calibration of the hydrodynamics model component was conducted with a 30-day simulation, using data for tides, river discharge, and wind. For model verification, a four-month simulation was performed using data for the summer of 1991, during June through September. Model skill assessment demonstrated the model's ability to simulate surface elevation, current velocity, and salinity with 5-10%, 10-15%, and 10-15% error, respectively (Yassuda, 1996). The Eulerian residual currents and salinity for the summer of 1991 condition revealed the traditional estuarine circulation, with the bottom water flowing into the Bay, mainly along the navigation channel (Figure 2), while surface water flowing out, mainly along the shallows of the bay (Figure 3). The salinity distributions shown in Figures 2 and 3 do not indicate a strong stratification in the vertical direction, but a distinct horizontal gradient with salinity ranging from 20 psu in the upper reaches of Hillsborough Bay to 36 psu in the Gulf of Mexico.

Water Quality Simulation

The water quality simulation started on May 26 and ended on September 30, 1991. The initial conditions for the water quality parameters were obtained from monthly-averaged data provided by EPC. Figure 4 shows the EPC monitoring network, and the assumption implicit in their sampling methodology is that the 52 EPC stations inside the four major sub-basins (Hillsborough Bay, Old Tampa Bay, Middle Tampa Bay, and Lower Tampa Bay) are representative of each area.

The water quality model used for this study included the following parameters: ammonia nitrogen, represented by the state-variable NH3; ammonium nitrogen (NH4); nitrate+nitrite (NO3); particulate organic nitrogen (PON); particulate inorganic nitrogen (PIN); dissolved oxygen (DO); biological and chemical oxygen demand (CBOD); phytoplankton (ALG); zooplankton (ZOO); and suspended sediments. The water quality equations are derived from an Eulerian approach, using a control volume formulation. In this method, the time rate of change of the concentration of any substance within this control volume is the net result of (1) concentration fluxes through the sides of the control volume, and (2) production and sink inside the control volume. The conservation equation for any water quality parameter (C) is given by:

$$\frac{\partial C}{\partial t} + \nabla \cdot (C\,\bar{u}) \;=\; \nabla \cdot \left[D_H \, \nabla(C\,\bar{u})\right] + Q$$
$$(i) \qquad (ii) \qquad\qquad (iii) \qquad (iv)$$

where *(i)* is the evolution term (rate of change of concentration in the control volume), *(ii)* is the advection term (fluxes into/out of the control volume due to advection of the flow field), *(iii)* is the dispersion term (fluxes into/out of the control volume due to turbulent diffusion of the flow field), and *(iv)* is the sink/source term, representing the kinetics and transformations due to sorption/desorption, oxidation, excretion, decay, growth, biodegradation, etc. The bottom sediments included a thin aerobic layer on top of an anaerobic layer. Details of the water quality model can be found in Chen and Sheng (1995) Yassuda (1996).

In the following, model results will be presented in terms of dissolved oxygen dynamics in the summer of 1991. Time series of DO at specific grid cells that coincide with EPC stations are compared with data. Contour plots of DO representing snapshots of the Bay at the end of four 30-day periods (June to September of 1991) are also presented.

Dissolved Oxygen Simulation

The dissolved oxygen balance is one of the most important ecological processes in an estuary because living organisms depend on oxygen in one form or another to maintain their metabolic processes. In the present water quality model, dissolved oxygen is a function of photosynthesis and respiration by planktonic organisms, reaeration, nitrification and denitrification, decomposition of organic matter, tidal and wind-mixing, and pollutant loading. In this regard, the model's ability to accurately simulate the DO dynamics can be considered an indicator of its ability to synthesize the overall water quality dynamics within the estuary.

Data during the summer of 1991 showed the development and later recovery of hypoxia in Hillsborough Bay, while DO in other parts did not change significantly. Simulated DO concentrations through the summer months are shown in Figures 5 through 8. Figure 5 shows the near-bottom DO distribution in Tampa Bay on June 26, after 30 days of simulation. Figure 6 shows the snapshot on July 26, after 60 days of simulation, Figure 7 shows the snapshot on August 25, after 90 days of simulation, and Figure 8 shows the result on September 24, after 120 days of simulation.

Due to its shallowness and significant wind and tidal mixing, Tampa Bay generally exhibits a vertically well-mixed distribution of DO. In Hillsborough Bay, which usually presents the lowest levels of DO, some stratification may occur due to high consumption near the bottom and super-saturation near the surface. High oscillations in DO concentration that are commonly seen in Hillsborough Bay are characteristic of a eutrophic water body. The upper reaches of Old Tampa Bay also exhibited some low levels of DO (between 5 and 6 mg/L), but Figures 5 to 8 show that it does not seem to evolve throughout the summer. Model results showed that DO variations in Middle and Lower Tampa Bay are small, without a significant trend from June to September of 1991.

EPC data and model results were plotted for comparison. Figure 9 shows the time series of model results for the near-bottom segment-averaged DO concentration and the EPC data inside Hillsborough Bay. The simulated segment-averaged DO represented by the solid line in Figure 9 appears to capture the EPC averages represented by the four diamond symbols. The maximum and minimum values inside Hillsborough Bay obtained from model simulations, which are represented by the upper and lower dashed lines, are within the range of the measured data at all EPC stations. Examination of model results and data indicates that organic matter decomposition and nitrification/ denitrification processes are the major sinks for DO in the bottom layers of this bay. The maximum DO values in Hillsborough Bay exhibit a very dynamic fluctuation. The super-saturation levels of DO suggest high phytoplankton activity which is typical of eutrophic conditions.

A more direct assessment of model's accuracy was performed by comparing model results at specific grid cells corresponding to the locations of the EPC stations. Figure 10 shows the model results and measured data for near-bottom dissolved oxygen at EPC stations 8, 70, 73, and 80 in Hillsborough Bay. The figure shows that the model was able to capture the different dissolved oxygen trends occurring in different areas of Hillsborough Bay. The near-bottom dissolved oxygen balance in the upper reaches of Hillsborough Bay (Figure 7) shows a very dynamic environment, with concentrations varying from super-saturation values to hypoxic and even anoxic conditions. On the other hand,

Figure 10 shows that, close to the mouth of Hillsborough Bay, the near-bottom dissolved oxygen dynamics exhibit a more stable trend, similar to the average conditions found in the other parts of Tampa Bay. Moreover, these figures show the significance of using a three-dimensional fine-resolution model to fully understand the water quality dynamics in Hillsborough Bay. The detailed spatial and temporal dynamics of DO could not be simulated with the box model of AScI (1995) or Ross et al. (1984).

A Conceptual Tampa Bay Water Quality Framework

Results of the simulation of the summer of 1991 condition suggested a conceptual framework for the water quality dynamics in the Tampa Bay Estuarine System, as summarized in Figures 11 and 12. The water quality parameters included nitrate+nitrite (NO3), ammonium nitrogen (NH4), ammonia nitrogen (NH3), soluble organic nitrogen (SON), phytoplankton (ALG), zooplankton (ZOO), particulate organic nitrogen (PON), particulate inorganic nitrogen (PIN), and gaseous nitrogen (N2). For illustrative purposes, the nitrogen cycle is divided into three separate steps, corresponding to (a) loading from point and non-point sources and atmospheric deposition, (b) water-column biogeochemical processes, and (c) sediment-column biogeochemical processes.

For the loading step, Figure 11 shows organic and inorganic nitrogen species and phytoplankton being released into the Bay through point sources, non-point sources, and atmospheric deposition (collectively combined into point discharges during model simulations). The water-column and sediment-column biogeochemical processes shown in Figure 12 demonstrate the central role played by soluble organic nitrogen (SON). The rate at which it is mineralized dictates both the formation of nitrate+nitrite, through nitrification, and the availability of dissolved inorganic nitrogen to phytoplankton uptake. The high levels of soluble organic nitrogen and phytoplankton inside the Bay produce a net export of these water quality parameters to the Gulf of Mexico. On the other hand, the nitrogen limiting condition of the Bay causes a depletion of dissolved inorganic nitrogen species, and the net transport generated is from the Gulf of Mexico into the Bay.

Figure 12 also shows that particulate organic and inorganic nitrogen have a net depositional flux. Two sources were considered in the simulations: the settling and deposition of particulate species following the suspended sediment dynamics, and the burial of algal cells from the first vertical layer next to the bottom. Particulate inorganic nitrogen (adsorbed ammonium nitrogen) and interstitial ammonium interchange according to the sorption/ desorption reaction. Particulate organic nitrogen is converted to soluble organic nitrogen at a constant hydrolysis rate. Since both organic an inorganic dissolved species exhibit higher concentrations in the sediment layer than in the water column, the net diffusive flux is from the sediment into the water column. Resuspension of particulate nutrients was not significant during the summer of 1991.

CONCLUSIONS

An integrated model which combines the enhanced versions of a 3-D hydrodynamics model (Sheng, 1989; Sheng and Yassuda, 1995) and a 3-D water quality model (Chen and Sheng, 1994; Sheng et al., 1995), has been developed for the Tampa Bay Estuarine System (Yassuda, 1996). This paper presents results of model simulations of DO dynamics in Tampa Bay during the summer of 1991, and a detailed comparison with data. Major conclusions are summarized in the following:

• A fine-resolution and smooth boundary-fitted numerical grid, which accurately represents the complex geometrical and bathymetrical features in Tampa Bay, was generated. Circulation patterns produced by the hydrodynamics model in this grid revealed flow features, which agree well with existing information on Tampa Bay circulation, available from past modeling and field studies.

- Results presented in this paper were obtained with a coupled hydrodynamics-sediment-water quality model with the same time step and grid spacing for all models. Hence, the effects of hydrodynamics were incorporated into the water quality model without any ad-hoc tuning of advective fluxes and dispersion coefficients.

- Information obtained from simple water quality models such as regression models and box models were used for determination of relevant processes, relationships between state variables, and for calibration of water quality model coefficients in the absence of process-specific data (e.g., nitrification, denitrification, etc). However, these simple models could not be used for prediction of such important dynamic events as the development and recovery of hypoxia, simulated by the time-dependent and fine-resolution three-dimensional model.

- The summer of 1991 simulation demonstrated the central role that organic nitrogen and mineralization of organic matter have on the water quality dynamics of the Tampa Bay Estuarine System. Simulation results showed that high levels of water column soluble organic nitrogen and phytoplankton inside the Bay produce a net export of these water quality parameters to the Gulf of Mexico. On the other hand, the nitrogen limiting condition of the Bay causes a depletion of dissolved inorganic nitrogen species, with a net transport from the Gulf of Mexico into the Bay. Model results also showed that during the summer of 1991, particulate organic and inorganic nitrogen have a net depositional flux. Since both organic and inorganic dissolved species exhibited higher concentrations in the sediment layer than in the water column, the net diffusive flux was from the sediment into the water column.

- Hillsborough Bay had the poorest water quality in the Bay. Although the EPC water quality index has consistently improved since 1987 (Boler, 1992), eutrophic conditions and hypoxia events still occurred in the upper reaches of the Bay. Model results demonstrated that external loading, nutrient-enriched sediments and limited flushing capacity are the primary causes of these characteristics.

ACKNOWLEDGMENT

The authors wish to thank the financial support of University of Florida, CNPq (Brazil), and Tampa Bay National Estuary Program for making this study possible. They would also like to thank NOAA/NOS, TBNEP, SWFWMD, and EPC for providing the data.

REFERENCES

AScI, 1994: "Tampa Bay Water Quality Model," Draft Final Report for the Southwest Florida Water Management District.

Boler, R.N. (ed.), 1992: "Surface water quality, Hillsborough County, Florida: 1990-1991." Hillsborough County Environmental Protection Commission. Tampa, FL.

Chen, X., and Y.P. Sheng, 1995: "Application of a Coupled 3-D Hydrodynamics-Sediment-Water Quality Model." In Estuarine and Coastal Modeling IV, ASCE, pp. 315-339.

Fong, P., and M.A. Harwell, 1994: "Modeling seagrass communities in tropical and sub-tropical bays and estuaries: a mathematical model synthesis of current hypothesis," Bull.Mar.Sci., 54, 3, 757-781.

Galperin, B., A.F. Blumberg, R.H. Weisberg, 1991: "The Importance of Density Driven Circulation in Well-mixed Estuaries: The Tampa Bay Experience." In Estuarine and Coastal Modeling II, ASCE, pp. 332-356.

Goodwin, C.R., 1987: "Tidal-flow, circulation, and flushing changes caused by dredge and fill in Tampa Bay, Florida." U.S. Geological Survey Water-Supply Paper 2282. Tampa, FL.

Hess, K.W., 1994: "Tampa Bay oceanographic project: development and application of the numerical circulation model." *NOAA Technical Report NOS OES 005.* Silver Spring, MD.

Sheng, Y.P., 1989: "Evolution of a 3-D curvilinear grid hydrodynamic model: CH3D." In Estuarine and Coastal Modeling I, ASCE, pp. 40-49.

Sheng, Y.P. and E.A. Yassuda, 1995: "Application of a three-dimensional circulation model to Tampa Bay to support water quality modeling." *Coastal and Oceanographic Eng. Dept,* University of Florida.

Sheng, Y.P., E.A. Yassuda, and C. Yang, 1995: "Modeling the Effects of Reduced Nitrogen Loading on Water Quality." In Estuarine and Coastal Modeling IV, ASCE, pp. 644-658.

Ross, B.E., M.A. Ross, and P.P. Jenkins, 1984: "Waste Load Allocation Study, Vol. I – Hydraulic Model, Vol. II – Transport Model, Vol. III – Model Validation, Vol. IV – Nutrient Box Model." University of South Florida.

Yassuda, E.A., 1996: "Integrated Modeling of the Tampa Bay Estuarine System." Ph.D. dissertation. University of Florida.

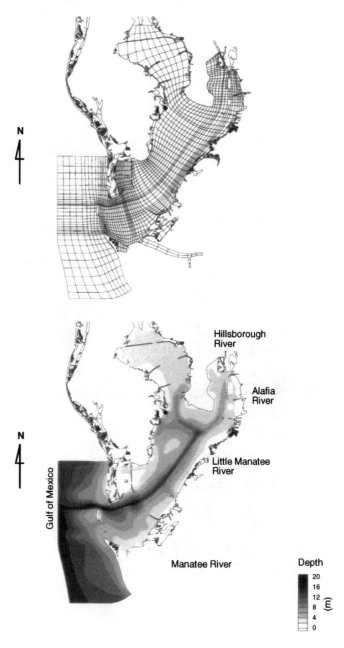

Figure 1 - Tampa Bay Grid and Bathymetry.

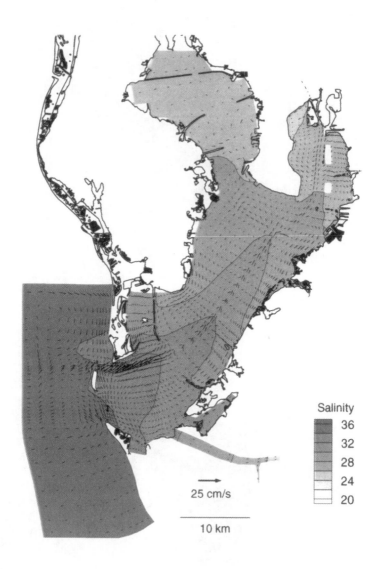

Figure 2 - Residual near-bottom currents and salinity for the summer of 1991.

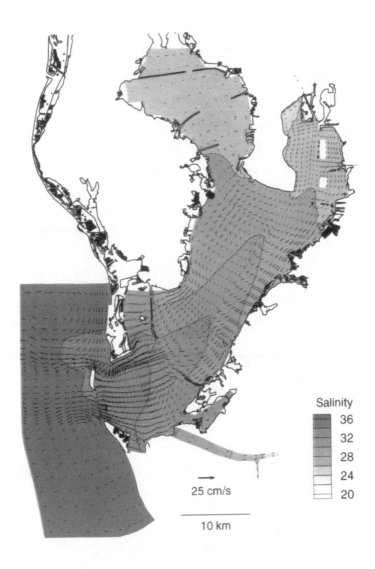

Figure 3 - Residual near-surface currents and salinity for the summer of 1991.

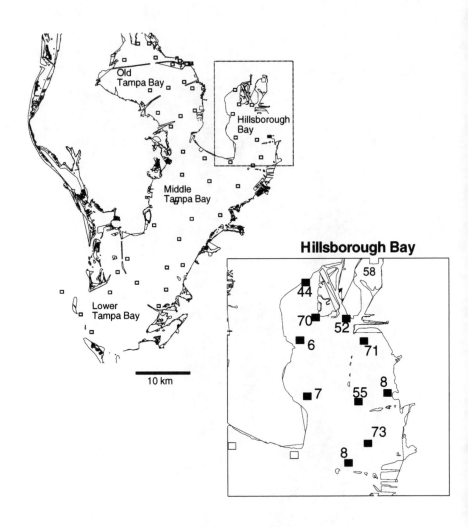

Figure 4 - EPC Water Monitoring Network in Tampa Bay.

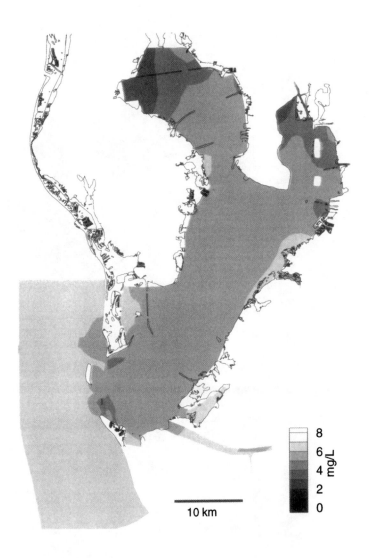

Figure 5 - Dissolved Oxygen Concentration after 30 days of simulation (June 26, 1991).

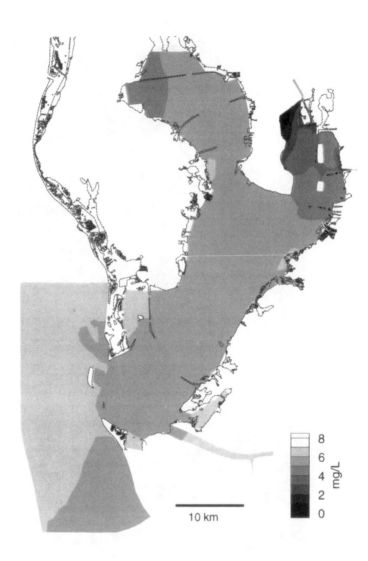

Figure 6 - Dissolved Oxygen Concentration after 60 days of simulation (July 26, 1991).

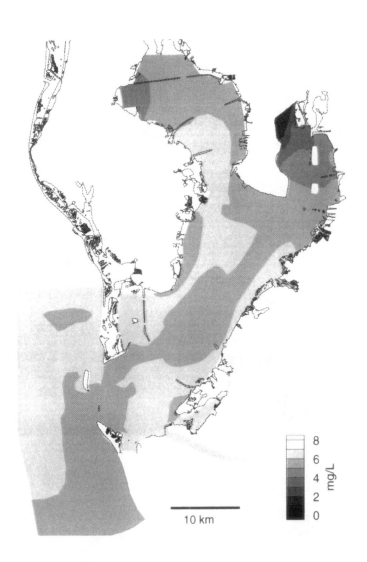

Figure 7 - Dissolved Oxygen Concentration after 90 days of simulation (August 26, 1991).

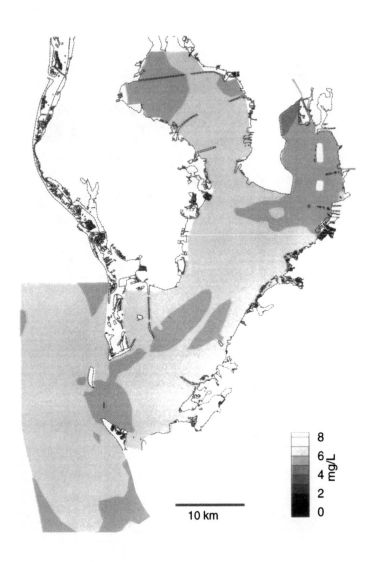

Figure 8 - Dissolved Oxygen Concentration after 120 days of simulation (September 24, 1991).

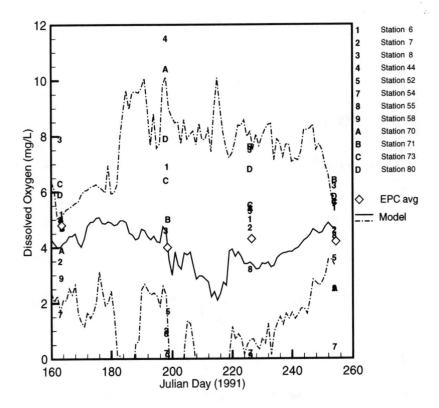

Figure 9 - Measured and simulated near-bottom dissolved oxygen concentration in Hillsborough Bay during the summer of 1991.

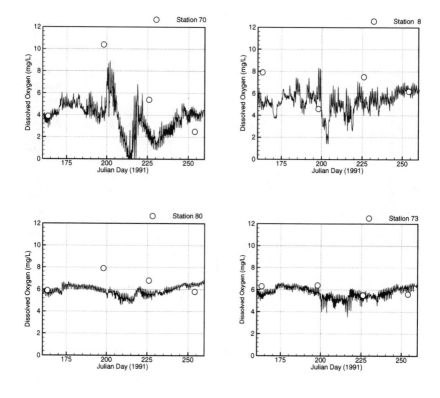

Figure 10 - Measured and simulated dissolved oxygen concentration in Hillsborugh Bay
(summer of 1991).

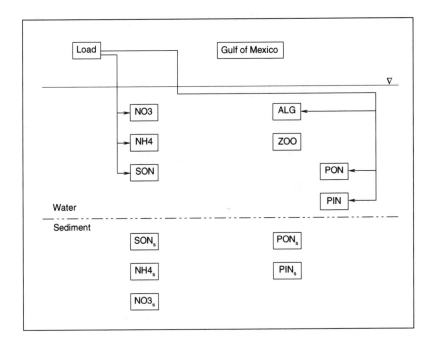

Figure 11 - Water quality dynamics in Tampa Bay - Loading.

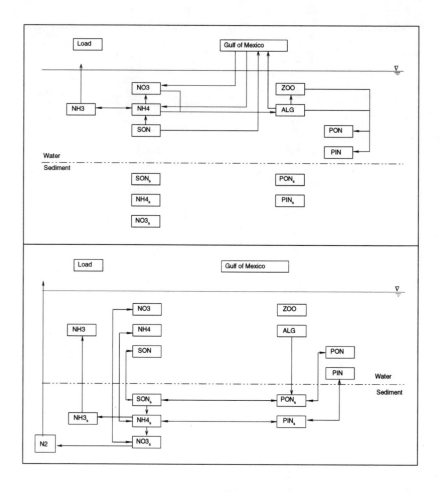

Figure 12 - Water quality dynamics in Tampa Bay - Water and sediment column biogeochemical
            related processes.

A 3-D MODEL OF FLORIDA'S SEBASTIAN RIVER ESTUARY

Peter V. Sucsy, Frederick W. Morris, Martien J. Bergman, and Lou J. Donnangelo

ABSTRACT

This study describes the application of a 3-D hydrodynamic and salinity model to Florida's Sebastian River estuary. The goal of the modeling project is to define freshwater reduction goals that meet resource-based salinity targets. Freshwater discharge to the estuary has increased since the 1920's as a result of flood control projects and farming practices. The increased discharges have harmed estuarine biota in both the Sebastian River and adjacent Indian River Lagoon.

Simulated salinity compared well with observed salinity for both summers of 1994 and 1995, demonstrating that this modeling approach is useful for determining freshwater reduction goals for Sebastian River. Model results indicate that some modifications of the salinity targets may be necessary because of a significant bottom inflow of saline water from the Indian River Lagoon into Sebastian River under moderate discharge conditions.

INTRODUCTION

The Sebastian River is a tidal estuary that discharges to the Indian River Lagoon (IRL) (Figure 1). The historic watershed of the Sebastian River was restricted by a system of relict dunes to the west of the IRL (Steward and VanArman, 1989) and was characterized by flat, poorly-drained and swampy areas. Beginning in the 1920's, a network of canals was constructed within the watershed to drain the land for farming and flood control. Two large canals, Fellsmere and C54, were constructed to divert water from the St. Johns River basin to the Sebastian River. Presently, freshwater discharge to the Sebastian River occurs from 3 gauged

[1]Department of Water Resources
St. Johns River Water Management District
Palatka, Florida 32178

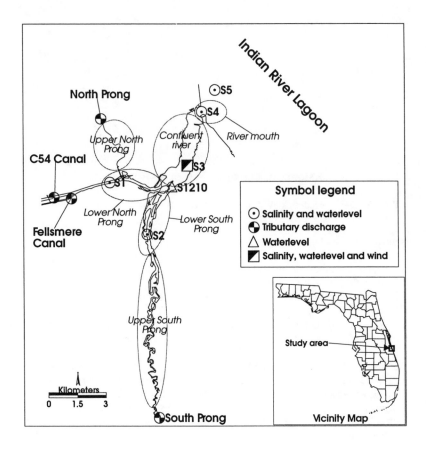

Figure 1. Sebastian River estuary with location of sampling stations. *Italicized names identify river segments with differing salinity targets (see text, Table 1)*

watersheds: the North Prong (95 km^2), the South Prong (145 km^2), Fellsmere Canal (80 km^2), and from 9 ungauged watersheds (60 km^2). Controlled discharge through C54 Canal is used intermittently to prevent flooding in the upper St. Johns River basin (Garver et al., 1996).

The alteration and expansion of the Sebastian River watershed has increased both the volume and peak of freshwater discharges, which has altered the salinity regime within the river and adjacent IRL. The present salinity regime is sub-optimal for the growth and reproduction of seagrass, clams, and many other important estuarine biota (Estevez and Marshall, 1994). Seagrass beds, in particular, are essential to the ecology of the lagoon estuary, because they provide the habitat and primary food source for most of the lagoon biota. Approximately 90% of seagrass beds in the area have been lost since 1943 (Woodward-Clyde, 1994).

The objective of this study is to develop a hydrodynamic model of the Sebastian River that can be used to determine the range of discharges required to maintain salinity within established salinity targets. Ultimately, the recommended discharge limits will be used as design parameters for controlling discharge to the river.

This study required a 3-D hydrodynamic model capable of simulating stratification and two-layer flow, and we chose EFDC (Environmental Fluid Dynamics Code), written by John Hamrick (1992a). EFDC was previously used to model salinity in and near Turkey Creek, about 26 km north of Sebastian River (Moustafa and Hamrick, 1993).

SALINITY TARGETS FOR SEBASTIAN RIVER

Salinity targets for Sebastian River are resource-based, and incorporate the salinity requirements of specific organisms and habitats within different river segments (Estevez and Marshall, 1994). General resource goals for Sebastian River and the adjacent IRL are that (a) primary productivity should be macrophyte-based, (b) species at risk should be protected, and (c) economically important species should be protected. For Sebastian River these goals specifically target seagrass, 4 species of threatened fish (Gilbert, 1992), the endangered West Indian Manatee (*Trichechus manatus*), hard clams (*Mercenaria mercenaria*), oysters (*Crassostrea virginica*), snook (*Centropomus undecimalis*) and spotted seatrout (*Cynoscion nebulosus*).

The general salinity targets for Sebastian River are to maintain low surface salinity variability at the mouth while providing permanent oligohaline and low-salinity estuarine habitat in the upper reaches of the river. Table 1 shows specific targets for surface salinity (Practical Salinity Scale 1978, PSS78) within different river segments (see Figure 1).

Table 1. Recommended surface salinity targets for Sebastian River.

River Segment	Mean	Standard Deviation	Minimum	Maximum
River mouth	20	10	10	---
Confluent river	15	15	5	30
Lower Prongs	6	8	0	20
Upper Prongs	1	2	0	4

Source: Estevez and Marshall (1994)

Bottom salinity targets for the confluent river are based on the salinity requirements of oysters:

- A preferred bottom salinity range of 10–28 (PSS78)
- Bottom salinity < 6 (PSS78) for no more than 2 consecutive weeks
- Bottom salinity < 2 (PSS78) for no more than 1 week

REVIEW OF AVAILABLE DATA

The principle data set used for this study was obtained from a network of stations operated by the U.S. Geological Survey (USGS) from April 1992 through March 1996 (Figure 1). All variables were measured continuously at 15 minute intervals. USGS also collected vertical salinity profiles at approximate two-week intervals at four sites: S1, S2, S3, and S4.

The St. Johns River Water Management District (SJRWMD) monitored rainfall throughout the watershed during this same period. The rainfall data were used to estimate the daily contribution of surface flow from 9 ungauged sub-basins.

Present Compliance with Salinity Targets

Descriptive statistics for observed surface salinity were calculated at S2, S3, and S4 for each dry- and wet-season (Table 2) to evaluate compliance with recommended salinity targets. A wet-season was defined as the period from 1 June to 31 October, and a dry-season from 1 November to 31 May. The dry-season year is defined as the calendar year beginning the dry-season.

Surface salinity at S2 generally met the recommended salinity targets for the lower South Prong. Surface salinity greater than 20 (PSS78) occurred infrequently. 0.9% of surface salinity observations exceeded 20 (PSS78) during the 1994 dry-season and 2.6% exceeded 20 (PSS78) during the 1995 wet-season.

Table 2. Descriptive statistics by wet- and dry-season for observed surface salinity (PSS78)

	S2		S3		S4	
Season	Mean±SD	Min-Max	Mean±SD	Min-Max	Mean±SD	Min-Max
Wet 1992	No Data	No Data	15.7±9.3	0.2–39.5	18.5±10.3	0.0–40.0
Dry 1992	No Data	No Data	13.0±8.7	0.0–36.5	17.5±10.0	0.0–39.5
Wet 1993	5.8±4.3	0.0–19.7	14.3±7.8	0.0–40.0	25.9±7.5	3.2–40.0
Dry 1993	3.8±3.0	0.3–18.8	13.6±7.8	0.0–40.0	23.6±8.0	1.3–40.0
Wet 1994	0.8±1.1	0.0–8.1	3.6±3.0	0.2–18.6	11.6±9.5	0.3–40.0
Dry 1994	4.7±6.0	0.0–40.1	6.4±5.4	0.0–30.5	19.9±9.8	0.0–40.0
Wet 1995	4.6±5.8	0.0–22.8	6.3±10.2	0.0–39.5	17.9±12.0	0.0–40.0
Dry 1995	3.5±2.5	0.2–12.3	13.5±6.7	0.0–39.9	20.8±6.9	0.0–39.9

Source of data: USGS

Mean surface salinity at S3 generally did not meet the recommended salinity targets for the confluent river. 72.9% of the observed values of surface salinity were below 5 (PSS78) during the wet-season of 1994.

Bottom salinity at S3 (not shown in Table 2) was often outside the optimal range for oysters. 39.9% of bottom salinity observations at S3 were below 10 (PSS78) during the 1994 wet-season and 61.7% of observations exceeded 28 (PSS78) during the 1993 wet-season. Bottom salinity was below 2 (PSS78) for 10 consecutive days in March 1993.

Mean surface salinity at S4 was below 20 (PSS78) for 3 of 4 wet-seasons. This target was met for 3 of 4 dry-seasons, however.

WATERSHED SIMULATIONS

Due to tidal influence and multiple, shifting drainage outlets, downstream drainage areas of the Sebastian River could not be gauged. Collectively, these areas contribute about 16% of the total drainage area. Discharges from these ungauged areas were simulated using the Hydrologic Simulation Program Fortran (HSPF; Bicknell et al., 1993). Because the majority of watersheds in the region can be characterized by similar physiography and hydrology, a set of regional parameters was developed, reflecting the hydrologic and physiographic characteristics of the region. The resulting parameter set was tested on 11 gauged watersheds in the region. In all applications the regional parameters met well defined calibration criteria (Bergman and Donnangelo, 1996). Because the ungauged drainage basins reflect the hydrologic and physiographic characteristics of the region, it is expected that simulated freshwater discharges using the developed regional parameters are reliable.

MODEL CALIBRATION AND VERIFICATION

We simulated salinity in the Sebastian River using the 3-D hydrodynamic model EFDC (Hamrick, 1992a; 1992b), which solves finite-differenced forms of the hydrostatic Navier-Stokes equations, together with transport equations, for salinity, temperature, turbulent kinetic energy and turbulent macroscale. The equations are solved horizontally on a curvilinear, orthogonal grid and vertically on a stretched, sigma grid. Vertical diffusion coefficients for momentum, mass, and temperature are determined by the second-momentum closure scheme of Mellor and Yamada (1982) and Galperin et al. (1988).

An irregular, orthogonal model grid (Figure 2) was applied to Sebastian River and a portion of the IRL from Wabasso to Grant. The model grid contained 2905 cells (581 horizontal cells and 5 vertical cells). Horizontal cell widths ranged from 21 m within the upper reaches of Sebastian River to 1193 m within the IRL. Vertical cell thickness ranged from 5 to 160 cm.

Model bottom elevations ranged from -0.8 to -8.7 m NAVD88, with the lowest elevations occurring in the dredged C54 Canal, near the S157 structure. Water surface elevation ranged from -0.54 to 0.32 m during the simulation periods, so that model depths ranged from 0.27 to 9.02 m.

Water elevation and salinity were specified near Grant, near Wabasso, and at the ocean inlet (Figure 3). Boundary data were obtained from several sources (Table 3) for the summers of 1994 and 1995.

At both the northern and ocean inlet boundaries, time series of water surface elevation were created from the superposition of long-period (tidally-filtered) water surface elevation data from a nearby gauge with the harmonic tide predicted using tidal constituents at the boundary. This superposition was necessary because water level data were not available at these locations for the required simulation periods. A constant salinity of 35 was applied at the ocean inlet.

Observed hourly discharges were specified for North Prong, South Prong, Fellsmere Canal, and C54 Canal. Simulated daily discharges were used to represent the freshwater contribution from 9 ungauged sub-basins.

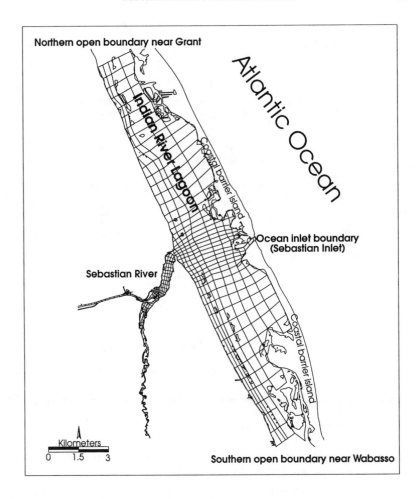

Figure 2. Sebastian River model grid covering Sebastian River and a portion of the Indian River Lagoon between Grant and Wabasso. *Water level and salinity were specified at the 3 locations indicated.*

Table 3. Sources of data used to estimate water surface elevation and salinity at the open model boundaries for the summers (June–September) of 1994 and 1995

Station Location	Measured variable used in model	Open Model Boundary	Collecting Agency
IRL nr Eau Gallie River[1]	Long-period (tidally-filtered) water elevation	Grant	USGS
IRL nr Grant	Tide	Grant	F.I.T.
IRL nr Wabasso	Water elevation	Wabasso	USGS
Atlantic Ocean nr Cocoa[2]	Long-period (tidally-filtered) water elevation	Ocean inlet	NOAA
Atlantic Ocean nr Sebastian Inlet	Long-period (tidally-filtered) water elevation	Ocean inlet	F.I.T.
Sebastian Inlet	Tide	Ocean inlet	FDEP
IRL nr Grant	Salinity	Grant	F.I.T.
IRL nr Wabasso	Salinity	Wabasso	F.I.T.

F.I.T.=Florida Institute of Technology, Melbourne, Fla.
FDEP=Florida Dept. of Environmental Protection, Tallahassee, Fla.
[1] Eau Gallie River is located approximately 20 miles north of Grant
[2] Cocoa is located approximately 27 miles north of Sebastian Inlet

Tidal Calibration

The model was first calibrated for tide. Simulated water level at 8 locations was analyzed for tidal harmonics and compared to data. Depths in the AIWW (Atlantic Intracoastal Waterway) were decreased 0.5 m to produce a good match of observed and simulated tidal harmonics for the M2, N2, S2, K1, and O1 components. Simulated and observed M2 amplitudes and phases agreed to within 0.7 cm and 10 degrees, respectively (Table 4). For the other tidal components, amplitudes and phases were within 0.2 cm and 15 degrees of observed values.

Table 4. Comparisons of observed and simulated M2 amplitude and local phase angle at selected stations

Station	Source of harmonic data[1]	Observed M2 Amplitude (cm)	Simulated M2 Amplitude (cm)	Observed M2 Phase (Degrees)	Simulated M2 Phase (Degrees)
Micco	NOS	3.2	3.6	288	291
Sebastian-IRL	NOS	3.9	4.6	297	290
S1	SJRWMD	4.3	4.3	288	288
S2	SJRWMD	4.6	4.5	292	293
S3	SJRWMD	4.2	4.4	286	288
S4	SJRWMD	4.2	4.2	287	283
S5	SJRWMD	4.9	4.2	287	284
S1210	SJRWMD	4.5	4.4	309	289

[1] SJRWMD harmonic data was obtained using a least squares harmonic analysis program by Foreman (1977)

1995 Salinity Calibration

Simulated salinity compared well with observed salinity throughout the river for the period of 1 July–1 September 1995. The model correctly simulated two low-salinity events during this period (Figure 3). The first low-salinity event followed Hurricane Erin on 31 August when combined discharge to the river exceeded 62 m^3s^{-1} (Figure 4). The second low-salinity event began about 25 August when combined discharge exceeded 65 m^3s^{-1}.

Observed and simulated mean salinity were within 1.3 (PSS78) at stations S1, S2 and S4 bottom (Table 5). Standard deviations of observed and simulated salinity were within 1.6 (PSS78) at stations S1, S2, and S4 surface. Observed and simulated values of daily average salinity were compared by calculating the mean error, and correlation coefficient at each station (Table 5). Correlation coefficients were greater than 0.8 at all stations except the bottom of S4.

Table 5. **Comparison of observed and simulated salinity (PSS78), 1 July to 1 September 1995**

Station	S1 Bottom		S2 Surface		S3 Bottom		S3 Surface		S4 Bottom	
	Obs.	Sim.	Obs.	Sim.	Obs.[1]	Sim.	Obs.[1]	Sim.	Obs.	Sim.
Mean	20.0	21.0	4.2	3.1	---	8.8	---	13.8	22.1	23.4
S.D.	10.6	10.5	4.6	3.0	---	5.4	---	7.7	10.1	6.8
MIN	0.3	0.0	0.2	0.0	---	0.0	---	0.0	1.0	0.8
MAX	33.3	32.0	18.1	9.3	---	17.6	---	27.7	38.3	36.4
Mean Error[2]	0.2		-1.7		5.6		2.0		-2.6	
Correlation	0.82		0.88		0.87		0.86		0.46	

[1]Insufficient number of observations for representative statistics over period of record
[2]Mean error is the mean of the time series of observed and simulated daily averaged salinity differences

1994 Salinity Verification

The period of 15 June–1 September 1994 was used for model verification. Rainfall totals were greater during this wet-season (108 cm) than for the 1995 wet-season (96 cm) and combined discharge (14–40 m^3s^{-1}) during August 1994 was also higher (Figure 5). Simulated and observed salinity again compared well (Figure 6). Simulated bottom salinity at S1 responded more slowly to discharge than observed salinity because the model depth (4 m) is greater than the station depth (1.2 m) at this site. S1 was located on the side of the channel that forms the C54 Canal, but the model grid does not resolve the lateral variability in this area.

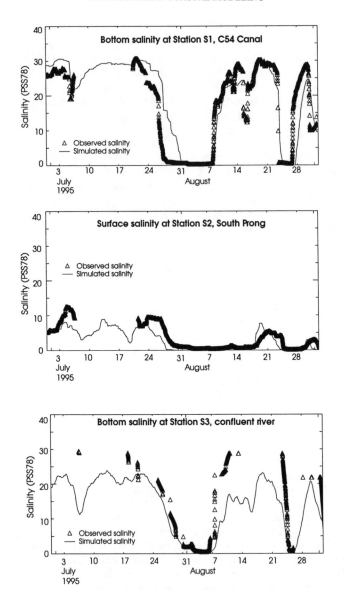

Figure 3. Comparison of observed and simulated salinity in Sebastian River, 1
        July–1 September 1995.

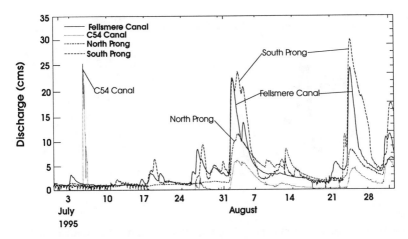

Figure 4. Hourly discharge into Sebastian River estuary from 4 main tributaries during model calibration of 1 July-1 September 1995. *Source: USGS and SJRWMD.*

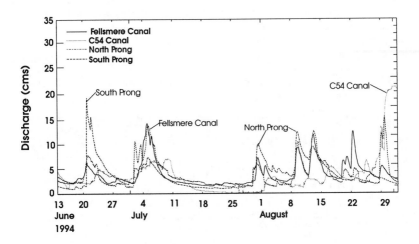

Figure 5. Hourly discharge into Sebastian River estuary from 4 main tributaries during model validation of 15 June–1 September 1994. *Source: USGS and SJRWMD.*

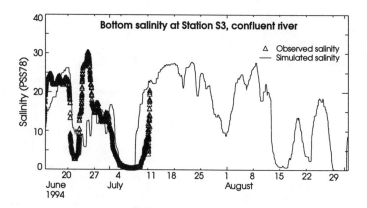

Figure 6. **Comparison of observed and simulated salinity in Sebastian River, 15 June–1 September 1994.**

Observed and simulated mean salinity were within 2.5 (PSS78) and standard deviations of salinity were within 1.6 (PSS78) at all stations (Table 6). The correlation coefficient for observed and simulated daily-average surface salinity at S2 was poor (-0.05), but the model correctly simulated very low salinity at this site.

Table 6. **Comparison of observed and simulated salinity (PSS78), 15 June–1 September 1994**

Station	S1 Bottom		S2 Surface		S3 Bottom[1]		S3 Surface[1]		S4 Bottom	
	Obs.	Sim.	Obs.	Sim.	Obs.[1]	Sim.	Obs.[1]	Sim.	Obs.	Sim.
Mean	5.7	6.9	0.8	0.4	13.9	15.1	3.6	3.2	30.3	32.2
S.D.	7.1	7.1	1.0	1.1	10.0	8.4	2.8	3.0	3.9	3.1
MIN	0.3	0.0	0.3	0.0	0.4	0.0	0.3	0.0	19.1	21.4
MAX	27.9	29.7	5.5	7.2	35.5	31.1	18.6	12.1	41.8	36.0
Mean Error	-0.7		0.4		-1.0		-0.3		-2.5	
Correlation	0.65		-0.05		0.64		0.82		0.43	

[1]Observed data from 15 June–11 July 1994 only

RESULTS AND DISCUSSION

Two periods from the 1995 model calibration were used to characterize circulation within the Sebastian River: (a) 16–22 July, when combined discharge was moderate (4–10 m^3s^{-1}) and (b) 31 July–6 August, when combined discharge was high (30–65 m^3s^{-1}). Residual salinity and eulerian velocity at both surface and bottom were calculated for each of these two periods.

Under moderate discharge, a strong two-layer flow advects saline bottom water well into the C54 Canal (Figure 7). As a result, residual bottom salinity throughout much of the confluent river and C54 Canal was near ambient lagoon salinity (30, PSS78). By contrast, the gradient of residual surface salinity (not shown) essentially followed the thalweg.

The bottom inflow was disrupted during high discharge (Figure 8). However, residual bottom salinity in C54 Canal increased slightly in the upstream direction, indicating that C54 Canal acts as a reservoir of saline water.

During periods of moderate discharge bottom salinity in much of the confluent river is governed by the ambient lagoon salinity, which often reaches oceanic levels. It is probably not possible to maintain bottom salinity below 28 (PSS78) under these conditions to provide optimal salinity for oysters. The bottom inflow also produces pronounced stratification in the lower and northern conflent river that is less pronounced nearer South Prong. The segmentation of the river (Table 1) may need to be modified to account for these vertical salinity differences.

Preliminary results indicate that under moderate discharge conditions it is possible to meet the recommended targets for mean surface salinity by controlling large peak discharges. In addition, model tests showed that increasing baseflow into C54 Canal by 1.5 m^3s^{-1} had no effect on either the surface or bottom salinity at S1. Small increases of baseflow, then, could probably be used to enhance oligohaline conditions in the upper reaches of the river without compromising the estuarine salinity regime downstream.

CONCLUSIONS

A 3-D salinity model (EFDC) has been applied to the Sebastian River and calibrated for a period containing two significant low-salinity events. The model accurately simulates the effect of freshwater discharge on salinity. The calibrated model can now be used to define reduction goals for freshwater discharge that meet recommended salinity targets within different segments of the Sebastian River. In particular, the model can be used to determine under what conditions the two-layer bottom inflow into Sebastian River is disrupted. This calculation would provide an

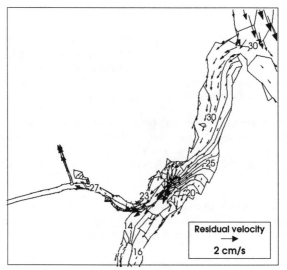

Figure 7. Contours of residual bottom salinity and vectors of eulerian residual
bottom velocity over 16–22 July 1995.

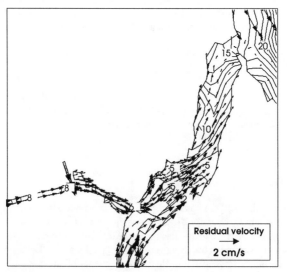

Figure 8. Contours of residual bottom salinity and vectors of eulerian residual
bottom velocity over 31 July–6 August 1995.

upper bounds for acceptable combined discharge to the river. Likewise, the model could be modified to calculate the minimum flow required to prevent intrusion of salt into the river's upper prongs.

Preliminary results indicate that the resource-based salinity targets for Sebastian River are also physically realistic. Salinity targets are generally met under moderate discharges. However, bottom salinity throughout much of the estuary is controlled by ambient lagoon salinity under moderate discharges and could probably not be maintained below 28 (PSS78) to accommodate oysters.

REFERENCES

Bergman, M. and L. Donnangelo. 1996. Development of HSPF hydrologic simulation model for the North Prong drainage basin of the Sebastian River. Tech. Mem. No. 18. Dept. Water Resources, St. Johns River Water Management District, Palatka, Fla.

Bicknell, B.R., J.C. Imhoff, J.L. Kittle Jr., A.S. Donigian, and R.C. Johanson. 1993. Hydrologic Simulation Program -- Fortran. User's Manual for Release 10. Environmental Research Laboratory, Office of Research and Development, U.S. Environmental Protection Agency, Athens, Ga.

Estevez, E.D. and M.J. Marshall. 1994. Sebastian River salinity regime. Mote Marine Lab, Sarasota, Fla. Special Publ. SJ94-SP1, St. Johns River Water Management District, Palatka, Fla.

Foreman, M.G.G. 1977. Manual for tidal height analysis and prediction. Pacific Marine Science Report 77-10. Inst. of Ocean Sciences, Patricia Bay, Victoria, B.C.

Galperin, B., L.H. Kantha, S. Hassid, and A. Rosati. 1988. A quasi-equilibrium turbulent energy model for geophysical flows. *J. Atmos. Sci.*, 45: 55–62.

Garver, R., M. Ritter, D. Dycus, T. Ziegler, and D. Clapp. 1996. Tropical Storm Gordon, November 14–21, 1994. Tech. Mem. No. 16. Dept. Water Resources, St. Johns River Water Management District, Palatka, Fla.

Gilbert, C.R. 1992. Rare and endangered biota of Florida, Volume 2. Fishes. University Press of Florida. 247 pp.

Hamrick, J.M. 1992a. A three-dimensional environmental fluid dynamics computer code: Theoretical and computational aspects. Special Report 317. The College of William and Mary, Virginia Institute of Marine Sciences, Va.

———— 1992b. Estuarine environmental impact assessment using a three-dimensional circulation and transport model. In *Estuarine and Coastal Modeling, Proceedings of the 2nd International Conference*, M.L. Spaulding, ed. American Society of Civil Engineers. N.Y.

Mellor, G.L. and T. Yamada. 1982. Development of a turbulence closure model for geophysical fluid problems. *Rev. Geophys. Space Phys.*, 20(6): 851–875.

Moustafa, M.Z. and J.M. Hamrick. 1993. Modeling circulation and salinity transport in the Indian River Lagoon. In *Estuarine and Coastal Modeling, Proceedings of the 3rd International Conference*, M.L. Spaulding, ed. American Society of Civil Engineers. N.Y. pp. 381–395.

Steward, J. and J.A. VanArman, eds. 1989. Indian River Lagoon joint reconnaissance report. St. Johns River Water Management District, Palatka, Fla.

Woodward-Clyde Consultants. 1994. Physical features of the Indian River Lagoon. Indian River Lagoon National Estuary Program. Melbourne, Fla.

Three-Dimensional Contaminant Transport/Fate Model

Mark Dortch[1], Fellow ASCE, Carlos Ruiz[1], Terry Gerald[2], and Ross Hall[1]

Abstract

Management of estuarine and coastal waters requires prediction of contaminant transport and fate with relatively high spatial resolution and complete process descriptions. A recently developed model, referred to as CE-QUAL-ICM/TOXI, combines parts of two previously existing models to satisfy these needs. The transport framework of the eutrophication model, CE-QUAL-ICM, was combined with the toxic substance fate routines of EPA's WASP model. The result is an improved contaminant model that exhibits accurate transport capabilities with widely used fate process descriptions. Additionally, a benthic sediment module was formulated using variable numbers of layers and layer thicknesses to provide improved capabilities for simulating long-term benthic contaminant fate. The contaminant model is indirectly linked to a three-dimensional, hydrodynamic model, CH3D-WES, to drive the water column transport terms of the contaminant model. A dynamic solids balance for the water column and bed can be simulated within the model or read in from a sediment transport model, such as CH3D-COSED. For validation purposes, the model was applied to a flooded limestone quarry dosed with DDE. The model compared favorably against observed data and results from another simpler contaminant model. An application of the model to the New York-New Jersey harbor-estuary system for total PCBs is presented.

[1] U.S. Army Engineer Waterways Experiment Station, Vicksburg, MS 39180

[2] AScI, Inc., 3402 Wisconsin Ave., Vicksburg, MS 39180

Introduction

Organic chemicals and trace metals tend to concentrate in sediments and persist in aquatic systems for many years, especially in waterways adjacent to large industrial areas. Waterways in industrial areas are often navigable to allow waterborne transportation and port access. Dredging of navigation channels for maintenance and channel improvements can result in redistribution of contaminated sediments. Estimates of present and future contaminant exposure concentrations are required to assess environmental impacts associated with channel dredging. Mathematical transport/fate models provide a means of computing future contaminant exposure concentrations in both the water column and bottom sediments.

Ports and navigable waterways are often in estuarine and coastal zones where complex geometry and currents may exist. Thus, a good general purpose contaminant model must provide sufficient spatial and temporal detail to capture accurate hydrodynamics and transport processes for these types of systems. Although contaminant models have been previously developed and applied for estuarine and coastal systems (e.g., Ambrose 1987 and Thomann et al. 1991), these models used rather coarse spatial and temporal scales typical of the box modeling technology of the time. Additionally, simplifying assumptions (e.g., steady-state total suspended solids balance) are often made for computing sediment transport. Such approaches may not be adequate for the increasing scrutiny placed on contaminant issues today. To satisfy the need for higher resolution contaminant transport/fate models, we have developed a new model by coupling recent approaches for three-dimensional (3D) hydrodynamic and constituent transport modeling with existing, generally accepted contaminant fate process descriptions. This model is described below along with validation results and example application for the New York and New Jersey harbors and estuaries.

Model Description

The 3D eutrophication model, CE-QUAL-ICM (Cerco and Cole 1995), which was originally developed during a study of Chesapeake Bay (Cerco and Cole 1993), was modified (Wang et al. 1997) for application to trace contaminants. The contaminant version is referred to as CE-QUAL-ICM/TOXI, or simply ICM/TOXI. The chemical kinetic algorithms included in ICM/TOXI were based on those of the Water Quality Analysis Program, WASP, (Ambrose, Wool and Martin 1993). ICM/TOXI solves the following mass balance equation for each water column control volume and for each water quality constituent state variable:

$$\frac{\delta \left(V_j \, C_j \right)}{\delta t} = \sum_{k=1}^{n} Q_{jk} \, C_k^* + \sum_{k=1}^{n} A_k \, D_k \, \frac{\delta C}{\delta x_k} + \sum S_j \qquad (1)$$

where

V_j = volume of the jth control volume (m^3)
C_j = concentration in jth control volume (g m^{-3})
Q_{jk} = volumetric flow across flow face k of jth control volume (m^3 sec^{-1})
C_k^* = concentration in flow across flow face k (g m^{-3})
A_k = area of flow face k (m^2)
D_k = diffusion coefficient at flow face k (m^2 sec^{-1})
S_j = external loads and kinetic sources and sinks in the jth control volume (g sec^{-1})
n = number of flow faces surrounding jth control volume
t = time, sec
x_k = coordinate distance between jth control volume and adjacent cell across flow face k, m

The summations are applied to all faces of each control volume. The transport solution scheme employed is common to both the eutrophication and contaminant models. The primary difference in the two models is in the number of state variables simulated (22 in the eutrophication model versus 8 in the contaminant model) and in the form of the kinetic sources and sinks incorporated into the term S_j, since the processes controlling trace chemicals differ from those simulated in the eutrophication model.

ICM/TOXI allows the simulation of a variety of processes that may affect toxic chemicals in the aquatic environment. These include physical processes such as hydrophobic sorption, volatilization, and sedimentation; chemical processes such as ionization, hydrolysis, photolysis, and oxidation; and biological processes such as biodegradation. The eight state variables included in the model are temperature, salinity, three types of solids classes (e.g., silts and clays), and three chemicals. The three chemicals may be independent or they may be linked with reaction yields, such as a parent compound-daughter product sequence. Salinity is treated as a conservative substance. Temperature is simulated using the equilibrium temperature approach to describe surface heat exchange. Solids, which can settle and be resuspended, are included in the model due to their importance as a sorbing surface for contaminants. Many water-borne contaminants exist both in dissolved form and forms sorbed to solids and other surfactants, depending on characteristics of the contaminants and the solids. The distribution between dissolved and sorbed contaminants impacts the transport and fate of the contaminants in the water column and sediment bed. A dynamic solids balance is provided in the sediment bed as described later.

The simulation of organic chemicals is the primary purpose of this model. The chemicals themselves are arbitrary in that the specific chemical to be simulated is defined through the specification of kinetic constants. Although the model is best suited for simulating organic chemicals, it can also be used for trace metals if care is exercised.

Each chemical may exist as a neutral compound and up to four ionic species. The neutral and ionic species can exist in five phases: dissolved, sorbed to dissolved organic carbon (DOC), and sorbed to each of the three types of solids. Local chemical and sorption equilibrium are assumed so that the distribution of the chemical between each of the ionic species and the phases is defined by chemical equilibrium constants (and pH) and partition coefficients, respectively. In this fashion, the concentration of any specie in any phase can be calculated from the total chemical concentration. Therefore, only a single state variable representing total concentration is required for each chemical. A description of the equilibrium distribution methods and the kinetic processes affecting trace chemicals is provided in the model documentation and user guide (Wang et al. 1997).

Solids and contaminants in benthic sediments are simulated using a benthic submodel, which includes such processes as accretion due to settling from the water column, resuspension, burial, pore water diffusion, and consolidation with pore water extrusion, in addition to partitioning among phases and biochemical kinetic processes. All benthic transport processes are in the vertical direction only, thus, horizontal advection and diffusion of solids and water are not allowed in the sediments. Contaminant mass in the water column and bed are dynamically linked through deposition, resuspension, and water diffusion across the sediment-water interface.

Solids transport and fate can be dynamically simulated within ICM/TOXI or read from output of another model, such as the CH3D-COSED hydrodynamic and sediment transport model (Spasojevic and Holly 1994 and 1997). For the latter option, water column concentrations and fluxes between the water column and bed are read from CH3D-COSED output. It is not necessary to transport solids within the water column of ICM/TOXI with the equilibrium assumption and known (computed) solids concentrations. However, solids transport calculations are still required within the ICM/TOXI bed. When water column solids transport are simulated within ICM/TOXI, each solids variable is treated as independent with no interactions, such as blocking or flocculation. Settling rates are either specified or computed from Stokes law. Sediment resuspension is either specified or computed from bottom shear stresses that must be provided from hydrodynamic model output. Various resuspension formulae for cohesive sediments are available.

The bed is modeled with multiple layers that can vary both spatially and temporally. A Lagrangian approach is used for sediment mass in the benthic layers. Sediment mass in surficial benthic layers increases or decreases during sediment deposition and resuspension, respectively. Sorting and armoring are not considered. Sediment mass accumulation is dependent on the deposition of each solids class at its respective density. As surficial layer mass changes, the surficial layer thickness varies. Erosion and deposition can alter the thickness of the surficial and sub-surficial layers, but these layer thicknesses are constrained to remain within a user specified range of the initial layer thicknesses; thus, these layers are split or combined if necessary, altering the number of layers. This semi-Lagrangian approach for simulating benthic layers is physically more realistic than using an Eulerian approach of fixed numbers of layers that require solids advection as deposition and erosion occur.

The porosity (ϕ) and bulk density (ρ_b), i.e., benthic sediment solids concentration, for each solids type of the sediment bed can vary from layer to layer and over time. Only the dry sediment density of each solids type is constant over space and time. Porosity is assumed to decrease exponentially with depth, where the exponential shape coefficient and the porosities of the surficial and bottom benthic layers are specified as constants, and the porosities for layers in between are computed according to the exponential function. As new layers are added for depositional areas, layer porosities are re-computed using the exponential equation. Since porosity is a function of depth, and layer thicknesses (thus depth) depend on bulk density which depends on porosity, an iterative solution for porosity is required. This approach imitates consolidation since the porosities of layers below the surficial layer decrease as layers are added. Extruded pore water is computed based on the amount of consolidation, and the amount of extrusion is used for pore water advective contaminant transport. For erosional segments, it is assumed that the removal of overburden does not cause un-consolidation, thus, there is no need to re-compute layer porosities as layers are removed.

Benthic sorbed, particulate-phase contaminant mass migrates according to the migration of the solids classes as described above. The dissolved, water-phase and the DOC-sorbed phase are transported vertically within the bed according to a Crank-Nicholson, finite difference, solution of the advection-diffusion equation. Upward advection occurs due to pore water extrusion during compaction. Additionally, groundwater flow associated with recharge or discharge can be specified. Pore water diffusion occurs between sediment layers and across the sediment-water interface. Pore water contaminants are also transported from the surficial sediment layer into the water column during resuspension.

Model Validation

A partial validation of the model was conducted by applying the model to a field-scale experiment in which a flooded limestone quarry near Oolitic, Indiana, was dosed with the insecticide DDE (dichloro-diphenyldichloro-ethylene) as reported by Waybrant (1973). ICM/TOXI was compared with observed data and results from the screening-level model, RECOVERY, which had been previously applied to this data set (Boyer et al. 1994). The same experiment was also previously analyzed with a time variable model by DiToro and Paquin (1983).

The RECOVERY model is time-varying with a well-mixed, zero-dimensional water column underlain by a vertically stratified, one-dimensional (1D) sediment column. RECOVERY contaminant fate processes include: water column and sediment sorption and decay; water column volatilization; sediment burial and resuspension; water column settling; sediment and pore water advection; and pore water diffusion among sediment layers and across the sediment-water interface. RECOVERY assumes a steady-state solids balance for one class of solids where the user specifies the suspended solids concentration and two of the three rates for burial, settling, and resuspension, and the model computes the third rate. Sediment porosity can vary over depth but is constant over time. RECOVERY uses an Eulerian framework with a fixed number of layers. Thus, burial and resuspension result in vertical advection of sediment-bound and pore water contaminant relative to the fixed surficial sediment reference. The 1D, total contaminant concentration equation for the sediment bed is solved with a Crank-Nicholson finite difference representation.

The quarry was approximately 91 m long, 41 m wide, and 14 m deep. The quarry had little groundwater interaction and contained primarily rainwater and local runoff. The bottom sediment consisted of approximately 1 percent sand, 42 percent silt, and 57 percent clay. The quarry water was dosed with DDE at a level equivalent to 50 parts per trillion (pptr) overall throughout the water column (Waybrant 1973). However, within a few days following the dose, a significant runoff event occurred, washing sediment into the quarry. The suspended sediment load caused the DDE concentration in the water column to decrease rapidly as DDE sorbed to solids and settled to the bottom. Within about two months following the dose, the water column and bottom sediments appeared to have reached a new initial condition with observed concentrations of 2.0 pptr and 35 parts per billion (ppb) throughout the water column and in the top 1.5 cm of sediments, respectively. These initial conditions for water column and sediment DDE concentrations were used for the model runs. Both the water column and bottom sediments were periodically analyzed for DDE concentrations over a course of eight months after dosage. Bottom sediments were sampled and analyzed for DDE again after five years (DiToro and Paquin 1983).

For simulation of the quarry, it was assumed that inflow was insignificant relative to the quarry volume. For the observed water depth of 13.9 m, the volume was 5.23×10^4 m^3. The suspended solids concentration was set to 5 mg/L based on an analysis by DiToro and Paquin (1983) for Secchi disc readings and sediment trap data. Resuspension was set to zero, and the settling rate was set to 0.1 m/day (36.5 m/yr), based on Stokes law for clay particles of about 1 μm diameter. The burial velocity was computed within RECOVERY from a steady-state solids balance to be 3.65×10^{-4} m/yr (0.0365 cm/yr). The partition coefficient (K$_d$) for DDE was computed as 1.54×10^{-3} m^3/g (1.54×10^3 L/kg) based on an assumed organic carbon content (f$_{oc}$) of 0.005 (by weight), an octanol/water partition coefficient (K$_{ow}$) of 5×10^4 L/kg, and the relation K$_d$ = 0.617 f$_{oc}$ K$_{ow}$. The relatively low organic carbon content is expected for inorganic sediments associated with quarries in the absence of any significant allochthonous influent. The molecular diffusivity was set to 1×10^{-5} cm^2/sec. Henry's constant and the molecular weight for DDE are 2.34×10^{-5} atm-m^3/mole and 319, respectively. With Henry's constant, the molecular weight, and a wind speed of 2 m/sec, the volatilization rate was computed as 38.6 m/yr using the Whitman two-film model (Boyer et al. 1994). The degradation rate for DDE was set to 10.0 yr^{-1} to account for photolysis, which is within the values of photolysis rates reported by Howard et al. (1991) corrected for the quarry water depth. Sediment porosity (ϕ) was set to 0.8 for the surficial layer and 0.65 for other layers. Dry sediment density was set to 2.5×10^6 g/m^3.

For hydrophobic contaminants, such as DDE, contaminant mass concentrates in bottom sediments. It is important to properly specify the mass transfer rate across the sediment-water interface. The sediment-water mass transfer rate (V$_d$) was computed in both models from

$$V_d = \frac{\phi D_s}{z'} \tag{2}$$

where D$_s$ is the effective sediment pore water diffusion coefficient (m^2/yr), and z' is the diffusive sublayer thickness, which is assumed to be 0.5 cm, a reasonable value for quiescent water (Thibodeaux 1979). D$_s$ is estimated from D$_s$ = D$_m$ ϕ^2, where D$_m$ is molecular diffusivity in water.

The same parameters discussed above were used for RECOVERY and ICM/TOXI. RECOVERY uses a bed layer thickness of 1.0 cm. The total depth of the contaminated sediment was specified as 5.0 cm, and the total bed depth modeled was 3.0 m. The same bed thicknesses and depths were used for ICM/TOXI, thus, 300 bed layers were employed. Normally this many bed layers would not be used with ICM/TOXI, but for this application, we wanted to force the model to have the same conditions as RECOVERY. The ICM/TOXI water column was modeled with four computational cells in the plan view with two layers, thus eight cells total. Relatively high horizontal and vertical water diffusion coefficients were used to maintain fully mixed conditions with uniform DDE concentrations, consistent with the zero dimensional, fully mixed assumption for the RECOVERY water column. Additionally, the suspended solids concentration was fixed as a constant within the ICM/TOXI code, rather than simulated, to follow the RECOVERY assumption of constant suspended solids concentration.

ICM/TOXI results for a ten year simulation of the quarry are compared with results from RECOVERY and observed data as shown in Figure 1 for the water column and the surficial sediment layer. All results and the observed data are presented in terms of total chemical concentration on a total volume basis. Both models agree well with the observations for DDE. ICM/TOXI produced results that are almost identical with those obtained with RECOVERY. This close comparison with another model helps to validate that the ICM/TOXI algorithms utilized in this test case are correctly implemented.

Example Application

ICM/TOXI was applied for total polychlorinated biphenyls (PCBs) in the NY-NJ harbors and estuary system as part of a demonstration project. The 3D hydrodynamic model, CH3D-WES (Chapman et al. 1996), had been applied to the harbor-estuary system as part of another effort, thus, providing hydrodynamics to drive the contaminant model transport. The model domain included the inner harbors and estuaries, the tributaries to the head of tide, Long Island Sound, and the apex of the NY Bight (Figure 2). This model has been referred to as the harbor-apex model, or HAM. The HAM hydrodynamic model (HM) employed a 241 x 55 curvilinear, boundary-fitted, plan-form grid (Figure 2) and stretched (sigma) vertical coordinates with five layers. The HM grid consisted of 2,953 active surface cells for a total of 14,765 active water column cells. The ICM/TOXI grid was an approximate 4:1 plan-form overlay of the HM grid with five layers, resulting in 3990 water column cells with an average cell horizontal dimension of 1865 x 1500 m. Three sediment layers were used throughout the ICM/TOXI grid, with thicknesses of 10 cm, 90 cm, and 100cm from top to bottom, respectively, for a total sediment bed depth of 2.0 m. The HAM consisted of two ocean boundaries (Long Island Sound and apex) and four tributary boundaries (Hudson, Raritan, Passaic, and Hackensack Rivers).

Total suspended solids (TSS) and total PCBs were modeled for the HAM application. The period January through September 1991 was chosen for the application since this period contained some of the most abundant data for both TSS and PCBs. Data sources for TSS were obtained from Farrow et al. (1986),

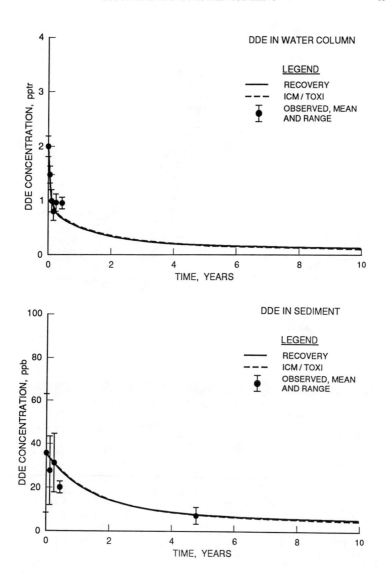

Figure 1. Computed and observed DDE concentrations versus time in quarry water column and sediment

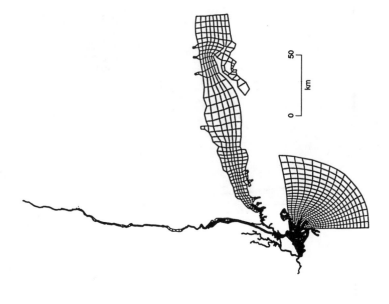

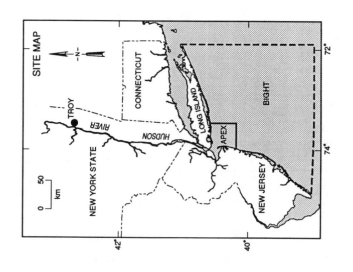

Figure 2. HAM Site map and numerical model grid

Hydroscience (1978), HydroQual (1989 and 1995), Mueller et al. (1982), and NYC-DEP (1991). PCB data were obtained from HydroQual (1989 and 1995), Mueller et al. (1982), Achman et al. (1994), and Stevens Institute of Technology (1995).

The model initial conditions were based on observed data. Initial conditions for water column concentrations of TSS and PCB were 25.0 g/m^3 and 20 pptr, respectively. Initial conditions for sediment bed TSS and PCB were 10^6 g/m^3 (1,000 g/L) and 2.0 parts per million (ppm), respectively. Estimates from the above referenced data sources for TSS and PCB loadings introduced by tributaries, wastewater discharges, and surface runoff were provided to the model. DOC and fraction of organic carbon in sediments were set everywhere to 2.0 g/m^3 and 0.04, respectively. K_{ow}, Henry's constant, and the molecular weight were specified to 1 x 10^5 L/kg, 5 x 10^{-3}, and 258, respectively. Volatilization was modeled with the Whitman two-film model using Option 2 (Wang et al. 1997), where the reaeration rate is specified, corrected for molecular weight, and used to estimate the liquid phase transfer coefficient; the gas phase transfer coefficient is assumed constant at 100 m/day for a flowing system (Wang et al. 1997). A value of 1.0 m/day was used for the reaeration rate. No degradation through other pathways, such as biodegradation and photolysis, was assumed. TSS settling velocity was specified globally as 1.0 x 10^{-6} m/sec (0.086 m/day). The resuspension rate was varied during model calibration and was finally set to a value of 1.0 x 10^{-11} m/sec, or essentially zero. The molecular diffusion coefficient was set to 6.0 x 10^{-6} cm^2/sec.

Results for TSS and PCB are plotted for a transect extending along the Hudson River into the Apex. River mile 0.0 corresponds to the Battery at the southern tip of Manhattan Island, negative values extend into the bays toward the Apex, and positive values are upstream in the Hudson River. Simulation means and minimum and maximum concentrations are plotted along with the means and ranges of the observed data. Figure 3 shows computed and observed TSS. Computed values generally compare favorably with observed values, except for some over-prediction in the surface layer and under-prediction in the bottom layer, possibly indicative of too much vertical mixing. Figure 4 shows computed and observed water column PCB. There was little or no difference in computed values for the surface and bottom layers. The model compares remarkably well with the few available observations. With a nine month simulation, the initial conditions are flushed out, and the computed water column PCB concentrations are influenced by bottom sediment PCB concentrations and boundary loadings. Imposing the proper sediment influence on the water column is highly dependent on accurately representing the PCB partitioning to solids and the sediment-water mass transfer rate. Computed benthic sediment PCB concentrations changed very little from the initial condition concentrations during the nine-month simulation, thus, a comparison of observed and computed sediment PCBs is not useful.

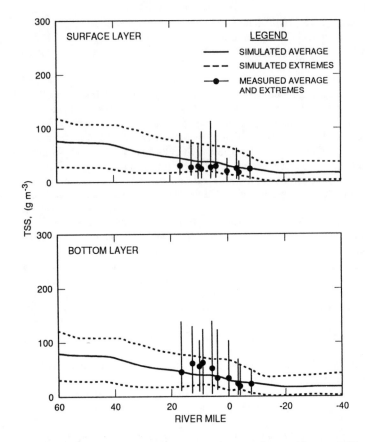

Figure 3. Computed and observed water column surface and bottom TSS along
HAM Hudson River transect

Conclusions

Hydrophobic organic chemicals, such as DDE and PCB, strongly sorb to solids
and tend to concentrate in bottom sediments. Thus, it is important for contaminant
models to accurately simulate conditions within the sediment bed and exchange
across the sediment-water interface. ICM/TOXI compared favorably with
RECOVERY and observations for the quarry water column and sediment DDE.
ICM/TOXI also compared closely with water column PCB observations for the
HAM simulation, indicating that the bottom sediment influence and boundary

ESTUARINE AND COASTAL MODELING 87

loadings are represented reasonably well. There was an observed sediment DDE data point observed after five years for the quarry. However, the HAM observations and simulation extended over only nine months, which is not really long enough to experience significant changes in sediment conditions. Observations over much longer time periods (e.g., decades) are required to more fully evaluate contaminant transport/fate models. With the ability to link to output from the 3D hydrodynamic model, CH3D-WES, ICM/TOXI can be applied with relatively high spatial resolution to a wide variety of coastal and estuarine systems, as well as other types of water bodies.

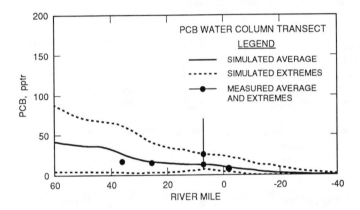

Figure 4. Computed and observed water column PCB along HAM Hudson River transect

Acknowledgment

This study was funded by the Corps of Engineers Water Quality Research Program. The Chief of Engineers has granted permission to publish these results.

References

Achman, D. R., Brownawell, B. J., and Zhang, L. (1996). "Exchange of polychlorinated biphenyls between sediment and water in the Hudson River Estuary," *Estuaries,* 19 (4), 950.

Ambrose, R. B., Jr. (1987). "Modeling volatile organics in the Delaware Estuary," *J. Environmental Engineering*, Am. Soc. Civil Eng., 113 (4), 703-721.

Ambrose, R. B., Wool, T. A., and Martin, J. L. (1993). "The Water Quality Analysis Simulation Program, WASP5; Part A: Model documentation," U.S. Environmental Protection Agency, Center for Exposure Assessment Modeling, Athens, GA.

Boyer, J. M., Chapra, S. C., Ruiz, C. E., and Dortch, M. S. (1994). "RECOVERY, a mathematical model to predict the temporal response of surface water to contaminated sediments," Technical Report W-94-4, U.S. Army Engineer Waterways Experiment Station, Vicksburg, MS.

Cerco, C. F., and Cole, T. (1993). "Three-dimensional eutrophication model of Chesapeake Bay, *J. Environ. Eng.*, Am. Soc. Civil Eng., 119(6), 1006-1025.

Cerco, C. F., and Cole, T. M. (1995). "User's guide to the CE-QUAL-ICM three-dimensional eutrophication model", Technical Report EL-95-15, U.S. Army Engineer Waterways Experiment Station, Vicksburg, MS.

Chapman, R. S., Johnson, B. H., and Vemulakonda, S. R. (1996). User's guide for the sigma stretched version of CH3D-WES," Technical Report HL-96-21, U.S. Army Engineer Waterways Experiment Station, Vicksburg, MS.

DiToro, D. M., and Paquin, P. R. (1983). "Time variable model of the fate of DDE and lindane in a quarry," Presented at the Society of Environmental Toxicology and Chemistry 3rd Annual Meeting (November 1983), Arlington, Va.

Farrow, D. R. G., Arnold, F. D., Lombardi, M. L., Main, M. B., and Eichelberger, P. D. (1986). "The national pollutant discharge inventory, estimates for Long Island Sound," NOAA, Rockville, MD.

Howard, P. H., Boethling, R. S., Jarvis, W. F., Meylan, W. M., and Michalenko, E. M. (1991). *Handbook of environmental degradation rates*, Lewis Publ., Chelsea, MI.

HydroQual, Inc. (1989). "Assessment of pollutant inputs to New York Bight," prepared for Dynmac Corp., Rockville, MD.

HydroQual, Inc. (1995). Personal transmission of Battelle and NYC-DEP suspended solids and sediment data for 1991.

Hydroscience, Inc. (1978). "Rainfall-runoff and statistical receiving water models, NYC 208 Task Report 225, prepared for Hazen & Sawyer, Eng.

Mueller, J. A., Gerrish, T. A., and Casey, M. C. (1982). "Contaminant inputs to the Hudson-Raritan estuary," NOAA Technical Memorandum OMPA-21, Office of Marine Pollution Assessment, Boulder, CO.

NYC-DEP (1991). "New York Harbor water quality survey, 1988-1990," City of New York, Department of Environmental Protection, New York, NY.

Spasojevic, M., and Holly, Jr., F. M. (1994). "Three-dimensional numerical simulation of mobile-bed hydrodynamics," Contract Report HL-92-2, U.S. Army Waterways Experiment Station, Vicksburg, MS.

Spasojevic, M., and Holly, Jr., F. M. (1997). "Cohesive sediment capabilities in CH3D: formulation and implementation," IIHR Report No. 386, Iowa Institute of Hydraulic Research, Univ. of Iowa, Iowa City, IA.

Stevens Institute of Technology (1995). Personal transmission of PCB data assembled for 1991 from various sources by Stevens Institute.

Thibodeaux, L. J. (1979). *Chemodynamics, environmental movement of chemicals in air, water, and soil*, John Wiley & Sons, New York, NY.

Thomann, R. V., Mueller, J. A., Winfield, R. P., Huang, C. R. (1991). "Model of fate and accumulation of PCB homologues in Hudson Estuary," *J. Environmental Engineering*, Am. Soc. Civil Eng., 117 (2), 161-178.

Wang, P. F., Martin, J. L., Wool, T., and Dortch, M. S. (1997). "CE-QUAL-ICM/TOXI, a three-dimensional contaminant model for surface water: Model theory and user guide," draft Instruction Report, U.S. Army Engineer Waterways Experiment Station, Vicksburg, MS.

Waybrant, R. C. (1973). "Factors controlling the distribution and persistence of lindane and DDE in lentic environments," Ph.D. diss., Purdue Univ., Univ. Microfilms, Order No. 74-15, 256, Ann Arbor, MI.

Modeling the Baroclinic Residual Circulation in Coastal Seas under Freshwater Influence

Y. Sugiyama[1], F. Takagi[2], K. Nakatsuji[3] and T. Fujiwara[4]

Abstract

Baroclinic residual circulation processes are examined in gulf-type coastal seas with freshwater influence. Such influence includes rivers discharging into a rounded head, which is wider than the Rossby internal deformation radius. Field surveys and three-dimensional numerical experiments concentrate on Ise Bay. Ise Bay is a gulf-type coastal sea like Tokyo Bay and Osaka Bay in Japan. On the other hand, all the three bays are also estuaries in a broad sense. There are three key points to understanding the circulation in Ise Bay. First, we divide the water body into three main water masses : the river plume, the low salinity upper layer , and the high salinity lower layer. Even the river plume eventually cause the low salinity of the upper layer through mixing, initially it is distinct from the upper layer and follows the plume dynamics. Second, baroclinic processes in Ise Bay are influenced by Earth's rotation firstly, since the Rossby internal deformation radius (2 - 7 km) is smaller than the horizontal dimensions of Ise Bay (20 - 40 km). Residual currents are quasi-geostrophic and the potential vorticity is approximately conserved. Third, the combined effects of classical longitudinal estuarine circulation and horizontal anti-cyclonic circulation are necessary to produce the resulting circulation. The classical estuarine circulation induces upward entrainment and horizontal divergence that leads to the increasing of seaward upper layer transport and the formation of an anti-cyclonic gyre. In the present study, the adaptability of a 3-D baroclinic flow model to such complicated circulation is examined by comparison of the model simulation results and the three-dimensional field observation data. As a result of numerical experiments using ODEM, the seasonal variation of the residual current system in Ise Bay is also discussed with the consideration of the fresh water influence in Ise Bay.

Introduction

Ise Bay is one of the representative bays of Japan, which is located in the central area of the Honshu Island. Ise Bay, togather with Tokyo Bay & Osaka Bay are gulf-type coastal seas, in addition, are also a kind of estuaries from the physical viewpoint. Classical drowned river valley estuaries, such as Chesapeake Bay, in the U.S.A., are wider and deeper at their mouths but become progressively narrower and shallower in up-estuary. In

[1] Electric Technology Research Department, Chubu Electric Power Co. Inc., Nagoya, Aichi 459, Japan.

[2] Dept. of Civil Engineering, Graduate School of Engineering, Nagoya Univ., Nagoya, Aichi 464-01, Japan

[3] Dept. of Civil Engineering, Graduate School of Engineering, Osaka Univ., Suita, Osaka 565, Japan

[4] Dept. of Fisheries, Graduate School of Agriculture, Kyoto Univ., Kyoto, Kyoto 606-01, Japan

contrast, gulf-type coastal seas remain wide enough so that the width of the estuary head is greater than the Rossby internal deformation radius. The density stratification in these bays is governed by the freshwater discharge from rivers and the seasonal heating and cooling at water surface.

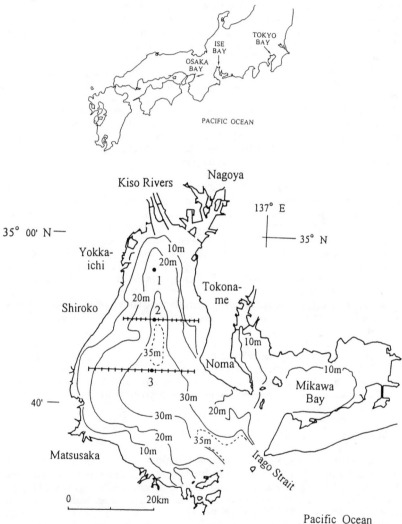

Figure 1 : Map of Japan showing Osaka Bay, Ise Bay and Tokyo Bay (upper panel), and Bathymetric map of Ise Bay with observation station locations (lower panel).

Figure 1 shows the topography of Ise Bay with the locations of the observations described in Chapter 2. The Bay is open to the Pacific Ocean in the south, and connects with the relatively shallow Mikawa Bay (average depth: 9.2 m) in the southeast. Ise Bay has a mean depth of 19 m, surface area of 1,738 km², and a volume of 33.9 km³. The Rivers of Kiso, Ibi, and Nagara, (together known as the Kiso Rivers), flow into the bay head in the north. In addition, other rivers also flow into the west side of the bay. The discharge of the Kiso Rivers is about 230 m³/s in minimum in winter to a peak value of 1000 m³/s in summer, with an annual mean value of 530 m³/s. The density structure is dominated by the salinity in Ise Bay, as well as in most estuaries, and in fact it remain as a typical salt-wedge estuary for three out of the four seasons. From spring to fall the water is strongly stratified, while it is weakly stratified in winter (note the unstable winter temperature distribution). The Rossby internal deformation radius calculated on the basis of 10-year monthly mean density structure is about 4.5 km to 6.9 km during April to October, and about 1.4km to 2.5 km during November to March. The prevailing winds across in the bay are southeasterly breezes in the summer and relatively strong northwesterly monsoon winds in the winter. These winter winds presumably provide the energy to the bay for the mixing the water column.

Gravitational estuarine circulation dominates the residual current system in gulf-type bays which influenced by freshwater discharge, while less saline water flowing seaward in the upper layer and saltier water flowing landward in the lower layer. The dominance of the gravitational circulation is, however, only in a laterally averaged sense. Anti-cyclonic circulation is found to be generated in the upper layer, caused by the horizontal divergence associated with upward entrainment, which is part of the estuarine circulation (Fujiwara et al.; 1997). The combined effect of classical longitudinal estuarine circulation and horizontal anti-cyclonic circulation is necessary to produce the resulting residual circulation in this type of bay. The baroclinic flow processes in coastal seas under freshwater influence are affected not only by stratification but also by the Earth's rotation. Thus, in this case, residual currents are quasi-geostrophic and the potential vorticity is approximately conserved.

In the present study, the characteristics of the residual circulation in Ise Bay are discussed upon field surveys, and the adaptability of a 3-D baroclinic flow model. The complicated circulation is examined in comparison with three-dimensional field data. Finally, the seasonal variation of the residual current system in Ise Bay is examined by numerical experiments.

Field surveys

Detailed observations of the currents and temperature and salinity in Ise Bay were carried out in October, 1994. The current distributions were measured on two east-west transects by a shuttling operation with two separate shipboard ADCP (Acoustic Doppler Currents Profiler), while the third vessel carried out STD observations in the neighbor aria (as shown in Figure 1). The northern transect was conducted on the 29th October and the southern transect on the 30th October. These observations were set during a period of neap tide and low diurnal inequality in order to minimize the effects of tidal currents and maximize the accuracy of the residual current estimates. In observation, we used The two ADCPs (RD Instruments, 600 kHz, Broad band). In addition, two ADCP-equipped vessels ran back and forth at the same transect for 12 hours, measuring the current velocities at 1 m depth intervals starting at 2 m below the surface. The measurement stations were set to be separated by 1 km on the northern transect and 1.5 km on the southern transect. Between these stations the ADCPs were raised and the vessels moved at full speed, while in the vicinity of each station the ADCPs were set for measurements and the velocities were measured for 90 seconds while proceeding at 2-3 knots. In this way, each station was visited 11 and 14 times on the northern and southern transects, respectively. The data were analyzed by a harmonic method to obtain residual current estimates.

The vertical profiles of temperature and salinity are shown in Figure 2. The figure shows a clear separation of the river plume from the upper layer. The measured stations are 1, 2 and 3 as shown in Figure 1. Three water masses are apparently distinguished by the salinity and temperature: the river plume, the upper layer, and the lower layer. The river plume is characterized by salinity of less than 30 ‰ and by slightly lower temperatures, and it stays in the water layer shallower than 2 m. The upper and lower layers have salinities of 31 ‰ and 33 ‰, respectively, with corresponding temperatures of 21 °C and 23.5 °C. Between the upper and lower layers, there is a sharp pycnocline at the 12 m to 20 m depth represented 32.5 ‰ in salinity and 22.5 °C in temperature during the observation period. Salinity affects the density profiles, while the gradient of temperature increases with an increase of depth to result in densimetrically unstable situation.

Cross-section views of the residual current and salinity distributions in October, 1994 are shown in Figure 3. In the northern cross-section, the river plume is apparent in water surface, which corresponds to a thin surface layer of the water body. The resulting residual currents are indicated only by the uppermost current vectors at 2 m depth. The flow directions in the river plume differ from those in the upper layer, tending more toward the west and northwest. The halocline separating the upper and lower layers (S = 32.5 ‰) seems to be located at the depth of 9 m to15 m, with slopes upward to the right facing upstream. In the upper layer, strong southward currents turn to the east, eastward currents prevail at the center of the bay, and weak northeastward currents are present in the west, with a general clockwise circulation. In the lower layer, there is a northward current above the deepest point of the cross section, with a landward estuarine circulation. What is more, we find that the velocity at this deep sea bottom reaches 20cm/s, which is surprisingly high.

In the southern cross-section, the signature of the river plume disappeared. The halocline slopes upward to the east, creating a front off the eastern coast where it intersects the surface. The main features of the upper layer flow are similar to those of the northern section: a weak northward flow in the west and a strong southward flow in the eastern frontal zone. A distinguishing feature of this flow pattern is the sharp boundary of the frontal zone, with large shears and direction changes across the sharpest salinity (density) gradients. The southward flow is parallel to the frontal boundary in the upper layer. It shows a geostrophic balance in the frontal zone. In a view of geostrophic balance, upward bending (concave curvature) of the pycnocline implies negative vorticity in the upper

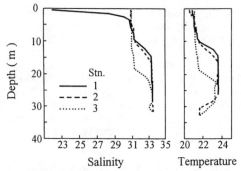

Figure 2 : Vertical profiles of temperature and salinity in Ise Bay observed in October, 1994 at the station shown in Figuer 1.

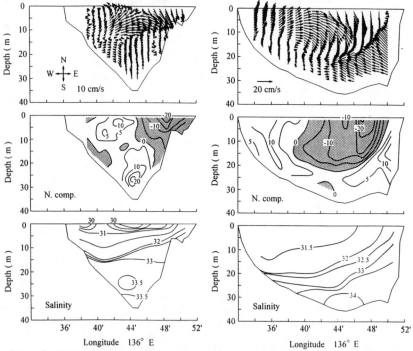

Figure 3 : Distributions of residual horizontal current vectors (upper panels), north-south components of the same currents (middle panels : northward velocity is positive), and salinity (lower panels) observed in Ise Bay during October, 1994 along the northern transect (left) and southern transect(right) shown in Figure 1.

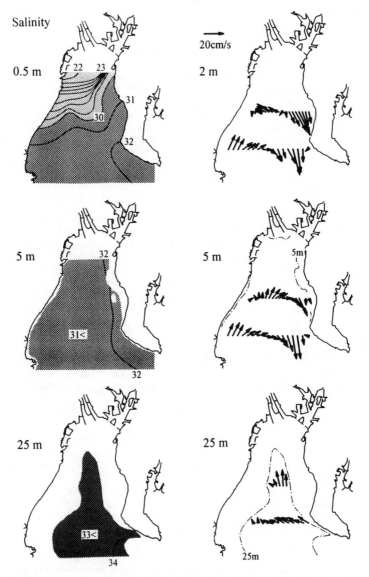

Figure 4 : Plan view of salinity and velocity vectors in the same observational period as in Figure 3 at the three depth : 0.5 m (upper), 5 m (middle) and 25 m (lower).

layer, which is also consistent with the larger scale flow pattern. Currents in the lower layer are, again, quite different from that in the upper layer, with westward to northwestward velocities prevailing.

Figure 4 presents the plan view distributions of the salinity and residual currents at three depths. At the depth of 2 m from water surface, the western half of the northern current section is covered by the river plume and the residual currents are directed towards the west and northwest. The remainder of the northern section at 2 m and the entire southern section at 2 m are contained in the upper layer, and here a clockwise circulation is prominent. This clockwise circulation is also obvious at 5 m depth, which is entirely contained in the upper layer. The upper layer vorticity calculated from the residual currents at 5 m depth is -10.1×10^{-6} s^{-1}, which is 24 % of $-f/2$ ($f=84. \times 10^{-6}$ s^{-1}), and the horizontal divergence in the upper layer is 5.5×10^{-6} s^{-1}. It seems that the clockwise circulation shown in the upper layer is an anti-cyclonic gyre. Both sections at 25 m depth are in the lower layer, water with high salinity flowing into the bay along the axis of the deep sea bottom.

Based on the results of the ADCP observations during October, 1994, the volume fluxes were estimated for Ise Bay using a box model by Fujiwara et al, (1997). Upwelling velocities of 0.9 m/day were found at 12 m depth. The estuarine circulation is apparent, with lower layer water flowing landward and upper layer water flowing seaward. We note that, while upwelling always implies horizontal divergence in the upper layer, it does not always imply horizontal convergence in the lower layer. In the coastal areas of a landward shallowing of the layer is present, upwelling from the lower layer can be achieved without horizontal convergence. For example, a horizontal current of 2.0 cm/s in the case of the bottom slope of 0.3×10^{-3} yields an upwelling velocity of 0.5 m/day.

Numerical Simulation Results

The above-mentioned observation results suggested that the density stratification and the Earth's rotation play significant roles in determining the distributions of residual currents and density structure as shown in Figure 3. They also recommended that a three-dimensional treatment should be required for simulating the complicated current and density fields observed in Ise Bay. Therefore, a three-dimensional baroclinic flow model, ODEM (Osaka Daigaku Estuarine Model) is applied in the present study, which was developed by Murota et al.(1988). The influence of stratification due to density difference and the Earth's rotation can be account into the model. The basic equations of continuity, motion in three-dimensional directions, and transport of temperature and salinity are solved under the hydrostatic and Boussinesq assumption The details of ODEM and computation conditions are described in the paper of Nakatsuji and Fujiwara (1997). It is worthy that the coefficients of the vertical eddy viscosity and diffusion in the model are adapted Webb's function and Munk-Anderson's function according to the primitive hydraulic experiments.

The diagnostic computation is done by solving the steady flow except the density calculation from the ordinary ODEM algorithm. The primary density distribution adapted to the diagnostic calculation is made from the temperature and salinity data which was

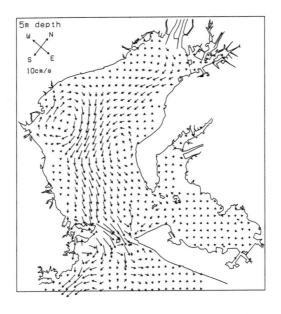

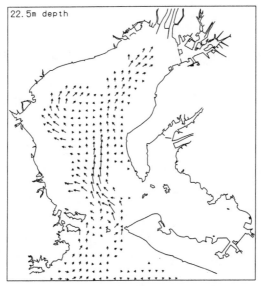

Figure 5 : Horizontal distribution of computed residual current at the depth of 5 m (upper) and 22.5 m (lower) in which the results of diagonistic computation used as initial condition.

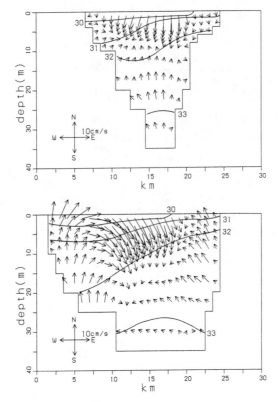

Figure 6 : Vertical distributions of computed current and one-tidal field integrating salinity corresponding to Figure 5; (a)upper: northern cross-section and (b) lower: southern one shown in Figure1.

Table 1 : Comparison of river discharges flowing into Ise Bay in August and October.

River	August	October
Shonai	12. 6	17. 7
Kiso	139. 6	122. 9
Ibi	13. 5	49. 0
Nagara	41. 7	102. 9
Suzuka	1. 4	7. 7
Kumozu	3. 5	15. 8
Kusida	5. 6	25. 0
Miya	3. 2	77. 9

unit : m^3/s

collected at 34 stations around Ise Bay during the ADCP observations. Firstly, the value at the vertical level of computation control volume are determined to be in an inverse proportion to the distance between its depth and each observation depth. Secondly, the values of the horizontal grids of control volumes are determined in the same way.

The computation domain is the area containing Ise Bay and Mikawa Bay. The model's resolution is 1 km in the horizontal plane, and has 10 vertical layers, the thicknesses of which are 2, 2, 2, 2, 2, 5, 5, 5, 10 and 15 m in respectively from the sea surface. Firstly, the computation was carried out diagnostically for 72 hours. And next, in the 48 hours computation, the computed results of diagnostic computation are used as an initial condition. The tidal elevation is with an amplitude of 44 cm, and a cycle of 12 hours at the open boundary near the bay mouth, in order to add the tide influence. The discharge from rivers and the heating and cooling at the water surface has been taken into account. The external wind force is, however neglected in this computation.

The computed results averaged for the last tide are shown in Figure 5 and Figure 6. Figure 5 presents the plan view distributions of the one-tidal averaging baroclinic residual currents at two depths. At the depth of 5 m, the clockwise vorticities are prominent at the head and the central area of the bay, which are regarded as anti-cyclonic vorticities. Roughly averaged currents are directed seaward at the 5 m depth, while at 22.5m depth, they are directed landward. However, river plume currents which are assumed to exist in the upper 2 m depth according to the observation cannot be reproduced clearly. Figure 6 shows the computed results along the same two transections as shown in Figure 3. In the northern transection, the upper layer currents are clearly separated from the lower layer currents. The computed upper layer currents are entirely directed southward, not like the observed currents which are directed eastward. It is probably because of other external forces such as the wind effects. In the southern transection, a strongly geostrophic southerly flow at the frontal zone of the density in the upper layer appears similar to the observed currents.

The computed results presented here show that the observed currents is mainly determined by its density distribution. It is also known that the numerical model utilized in the present study was mounted to allow for the necessary dynamic conditions to compute the baroclinic processes in gulf-type coastal seas under a freshwater influence.

<u>Computed Baloclinic Residual Current System in Ise Bay in Summer and Fall</u>

The current system of Ise Bay is significantly affected by its density distribution, which is governed by the freshwater discharge from rivers and the heat exchange between the sea and the atmosphere. Especially, the fresh water discharges are important to form the spatial characteristics of the density field because of the uneven distribution of the river mouths and the seasonal variation of the flow rate. Case studies are carried out to clarify the current system under the influence of fresh water from rivers, especially in summer and fall. It is because the density stratification becomes weak and the wind-driven current becomes dominated in the water. On the other hand, physical situation in the spring is similar to that in the fall.

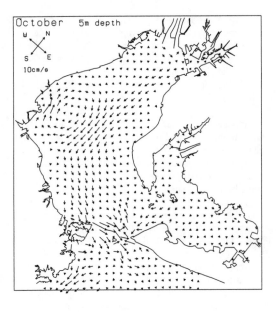

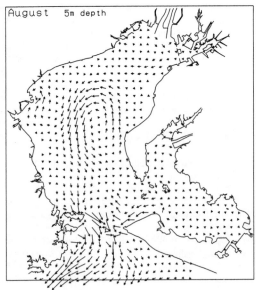

Figure 7 : Horizontal distributions of baroclinic residual current computed by forecasting computation. Fall (upper), Summer (lower).

The flow rates of the main rivers flowing into Ise Bay are shown in Table 1, There is the distinguished difference between summer (August) and fall (October). The Rivers of Shonai, Kiso, Ibi and Nagara flowing into the bay head are regarded as group A, and the rest of the 4 rivers flowing into the west side of Ise Bay are regarded as group B in the computation. The flow rates of the rivers belonging to group B increase remarkably in the fall under the influence of typhoon, especially, the Miya River has as much discharge as the Kiso River in the fall.

The above-mentioned same model as shown in Chapter 3 with the same grids is adapted in these forecasting computations. The computations have been carried out for 40 tides (480 hours). The initial conditions are given in a uniform value for the area, which have been selected in reference to past records. The external wind force is neglected again in these computations in order to examine the effect of the fresh water discharge.

Figure 7 shows the computed baroclinic residual currents at 5 m depth in the case of the fall and the summer seasons respectively. The distinct currents in Ise Bay such as the anti-cyclonic vorticities and the strongly geostrophic flow at the frontal zone appear in both cases, but their scale and location are quite different from each other. In the case of the fall season, the current distribution is similar to the observed situation. But the clockwise vorticity which is shown at the central area of Ise Bay in the fall disappears in the summer season. Instead of a clockwise vorticity, we observe an anti-clockwise vorticity at the bay center. We then assume that the small flow rates of the group B rivers cause the frontal zone moving westward and the anti-clockwise vorticity advance as a counter-current on the eastern bay across the frontal zone. The results computed for the summer show good agreement with the distribution of the residual currents described in the "Chart of Tidal Streams in Ise Bay" which was published by the Maritime Safety Agency of Japan (1995). The seasonal variation of the baroclinic residual circulation is mainly estimated by the ratio of the river discharge of those in group A to those in group B.

Conclusion

There are three key points to understand the circulation in Ise Bay as follows. (1) There are mainly three distinct water masses: the river plume, a low salinity upper layer, and a higher salinity lower layer. Though the river plume is eventually responsible for the low salinity of the upper layer through mixing, initially it is distinct from the upper layer and displays characteristic plume dynamics. (2) The baroclinic processes in Ise Bay are influenced by the Earth's rotation in the first order, since the Rossby internal deformation radius (2 - 7 km) is rather smaller than the horizontal dimensions of Ise Bay (20 - 40 km). The baroclinic residual currents are quasi-geostrophic and the potential vorticity is approximately conserved. (3) The combined effects of a vertical estuarine circulation and horizontal anti-cyclonic circulation are both necessary to produce the resulting circulation. The classical estuarine circulation induces upward entrainment and horizontal divergence that leads to seaward   increases in the upper layer transport and to the formation of an anti-cyclonic gyre.

A numerical model which is adapted to these circulations must considere two conditions. First, accurate models of the vertical eddy viscosity and diffusion are needed to reproduce the clearly separated density stratification and shear flow. Second, the term of the Coriolis force must be added to the momentum equations. Such conditions though are not a new concept, as most 3-D baroclinic flow models applied them in several ways recently to forecast sea water. The ODEM utilized in the present study is also one of such type. Since the observed circulation is reproduced with good agreement in the diagnostic computations, we think that the model has satisfied the necessary dynamic conditions to simulate the baroclinic processes in gulf-type coastal seas under a freshwater influence. It is, however, difficult to simulate the dynamics of river plumes, because the thickness of surface layer must become 2m and more in order to simulate the tidal elevation only in the surface layer.

It was mentioned in Chapter 4 that there is a large clockwise vorticity at the bay center in October, but this changes to an anti-clockwise vorticity in August. Of course, the wind external force causes the moving of the frontal zone and it fortifies the estuarine circulation. But, the remarkable variations of the current distribution in Ise Bay are mainly dominated by the disposition of the river mouths and their flow rates.

Acknowledgment

The founds for this research effort were provided by the Grant in Aid for Science Research (c), Ministry of Education, Science, Courtier and Sports (09660200 and 09650565) and Grant of Nippon Live Insurance Grant.

References

Chao, S.-Y., 1988 : River-Forced Estuarine Plumes. J.Phys. Ocesn., 18:72-88.
Fujiwara, T.,Y.Sawada, K.Nakatuji and S.Kuramoto, 1994 : Water exchange time and flow characteristics of the upper water in the eastern Osaka Bay. Bullenin on Coastal Oseanography, 31, 227-238. (in Japanese with English abstruct)
Fujiwara,T., Sanford, L. P., Nakatuji, K., and Sugiyama, Y., 1997 : Anti-cyclonic Circulation Driven by the Estuarine Circulation in Gulf Type ROFI. J. Marine Systems, 12
Mritime Sfety Agency, 1995 : Chart of tidal stream in Ise Bay. Maritime Safety Agency, Tokyo, 20pp.
Murota, A., Nakatsuji, K. and Hoh, H. Y., 1988 : A three-dimensional computer simulation model of river plumes, Rifined Flow Modeling and Turbulence Measurement, University Academic Press, pp.539-547.
Nakatuji,K., T. Fujiwara and H.Kurita, 1995 : An estuarine system in semi-enclosed Osaka Bay in Japan. In: K.. R. Dyer and R. J. Orth (Editors), Changes in fluxes in estuaries. Olsen and Olsen, Fredensborg, pp. 79-84.
Nakatsuji, K. and Fujiwara, T., 1997 : Residual baroclinic circullation in semienclosed coastal Seas, J. Hydraulic Engineering, ASCE, Vol.123, No.4, pp.362-377.

Modeling the Mighty Burdekin River in Flood

Brian King, Simon Spagnol, Eric Wolanski and Terry Done

Australian Institute of Marine Science, PMB #3 Townsville MC, Townsville Qld 4810, Australia
Phone: +61-7-4753 4444   Fax: +61-7 4772 5852

Abstract

A 3-dimensional hydrodynamic model was employed to simulate a flood of freshwater from the Burdekin River into the coastal waters of the Great Barrier Reef (GBR). The model was verified for the entire 1981 flood event using actual discharge data and actual wind data for the river, against the historical field data. This flood was chosen for the validation of the model as it was the only flood event for which an extensive survey and data set of the river plume existed. The river delivered almost 19 billion tonnes of freshwater to the GBR lagoon from 3 separate flood events over the period, with peak discharges exceeding 12,000 tonnes of water per second. Comparisons between model results and field data for 3 different days of field surveys shows very good agreement between the observed and predicted salinity distribution in coastal waters at corresponding times.

Sensitivity analysis on the model runs showed that the main driving influences on the fate of the plume water were the discharge volume of the river and the local wind forcing. Thus each year, one would expect different plume trajectories depending on the time-varying nature of both the wind and the rainfall/catchment. The model simulations also showed that patchiness in the far-field salinity field was enhanced by discharges from neighbouring rivers and by tidal interactions with headlands along the coast.

From a management perspective, the simulations can identify the fate of the Burdekin River plume in isolation to other freshwater discharges which cannot be done with field observations alone. This information should provide useful information on catchment management implications of the Burdekin region and its impact on shelf and Great Barrier Reef waters.

1. Introduction

The Burdekin River, in tropical north-east Australia, has a catchment area of 130,000 km^2 and has the largest recorded mean annual flow (approximately 9,700,000,000 m^3 per year) for any river adjacent to the world heritage listed Great Barrier Reef (Wolanski, 1994). The river mouth is located between Cape Upstart to the south and Cape Bowling Green to the north (see figure 1). The run-off, while extensive, is also highly variable, limited to the occasional flood event (0-3 per year) usually occurring during the Austral summer months of December to March.

Information on the fate of flood waters during and after the massive discharge events of the Burdekin River is limited. Wolanski and Van Senden (1983) reported the most detailed survey todate which covered the 1981 flood events. This dataset provided a unique opportunity to calibrate and verify a 3-dimensional hydrodynamic model of the Burdekin River in flood.

2. Methods

Model Description

The "MECCA" 3-dimensional hydrodynamic model (see Hess, 1989) from NOAA, which incorporated river plume dynamics into the governing equations, was used for this exercise. The model was designed to predict tidal, wind and density driven flows in bay and on continental shelves. MECCA has been applied to study the salinity and temperature distribution in Chesapeake Bay and surrounding shelf areas (Hess, 1986).

The model utilizes a three time level, semi-implicit alternating direction numerical scheme for the external mode calculation. The internal mode is computed implicitly. The model offers an ability to split the external mode and internal calculations and run each mode at different time-steps.

Model Configuration

Wind data (a 3 hour intervals) was obtained from the nearby Mackay weather station for years 1966 to 1991, and from the nearby AIMS weather station for years 1992 to 1996, for the purposes of simulating historical Burdekin River flood events. Daily discharge data from the Burdekin River since 1951 was also obtained from the Queensland Water Resources Commission. Tidal influences were included implicitly via the mixing parameters based on the work of King and Wolanski (1996). Thus these data can be applied to the model to simulate any of the major flood events of the Burdekin River of the last 30 years.

The MECCA model uses a 3-dimensional grid to mathematically represent the study area domain and covered the entire shelf of the central section of the GBR from Cairns to Bowen.

The numerical grid representing the domain had over 100,000 computational points, that is, 5 layers in the vertical, 211 points in the along-shelf direction and 95 in the across-shelf direction. The grid spacing in the horizontal plane were 2 km x 2 km throughout, while the vertical grid spacing varied according to the depth (sigma representation). For example, depths near the coast were of the order of 5 - 10 m, thus the vertical grid spacing would be 1 - 2 m respectively.

The time-step of the model was set to 30 seconds for both the external mode and internal mode due to the limitations of stability of the numerical representation of the buoyancy terms, since discharge rates from the Burdekin can exceed 25,000 m^3/s.

3. Results and Discussion

Model Results

The entire 1981 flood event period was simulated by the model for the purposes of calibration and verification against the field data of Wolanski and Van Senden (1983), covering the 90 day period from 0000 hrs on the 1st January, 1981 until 0000 hrs on the 31st March, 1981. Figure 1 shows the occasional snapshot of the surface salinity at times throughout the flood. From the discharge data in 1981 (see "Flow rate" Insert in Figure 1), we see 3 separate flood events from rainfall in the discharge data. The total volume of freshwater discharged into the GBR from the Burdekin River catchment area at this time was a massive 18.95 billion tonnes of water.

Figure 1a (top) shows that surface distribution of salinity over a 3-dimensional bathymetric map of the GBR at midnight on the 24th January 1981 when the river was in peak flood (12,000 tonnes of river water per second). The wind vector in Figure 1(a) showed that the wind was strong and from the SE at 10 ms^{-1} or about 20 knots. As a result of these conditions, freshwater filled the entire Upstart Bay and a tongue of the brackish water (<18 ppt) stretched 100 km northward along the coast. Figure 1a (bottom) shows the surface salinity distribution 5 days later at the end of the first and major discharge event. The SE winds weakened at this time, though had advected the plume and the 30 ppt contour almost 200 km from the mouth of the river to surround the continental islands of Cleveland Bay and Halifax Bay.

Figure 1b (top) shows the fate of the plume water from the second peak in the flood which occurred during north-east winds. At midday on the 8th February, 1981, Upstart Bay was almost completely freshwater at the surface and the winds had pushed plume waters into the bay to the south of the mouth of the Burdekin River. Note that the plume waters to the north from the previous peak had mixed with continental shelf water and were diluted further. The second flood event subsided after 10 days and a wind change to the south-east pushed the plume waters northward again as shown Figure 1b (botton). Note that the bay to the south of the river had been flushed and no longer was exposed to any river water under these wind conditions.

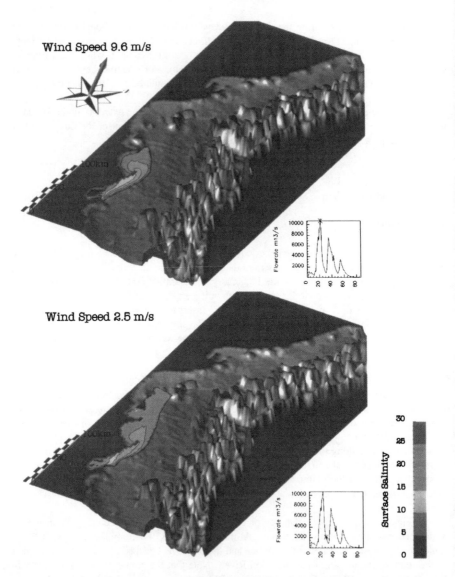

Figure 1a: Surface salinity distribution of Burdekin River plume waters on the shelf of the Great Barrier Reef at times: (top) midnight January 24, 1981 and (bottom) midnight January 29, 1981. The insert graphs the discharge rates over the 1981 wet season (m³/s) starting from January 1, 1981 and the asterisk indicates the flow rate at the time of output. The vector on the compass represents the wind speed and direction at each time.

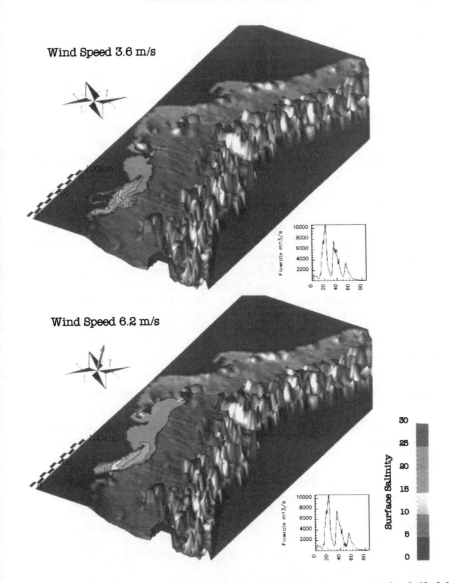

Figure 1b: Surface salinity distribution of Burdekin River plume waters on the shelf of the Great Barrier Reef at times: (top) midday February 8, 1981 and (bottom) midday February 18, 1981. The insert graphs the discharge rates over the 1981 wet season (m³/s) starting from January 1, 1981 and the asterisk indicates the flow rate at the time of output. The vector on the compass represents the wind speed and direction at each time.

Computational Demands

Given the huge discharges and rates simulated by this model, the model required a small model time-step. Thus 3-dimensional calculations were needed every 30 seconds for a 90 days period. This simulation of the 1981 flood event took 22 actual days of CPU time on an IBM RS6000/590 supercomputer and resulted in 250 trillion computations for the 3 month simulation. With such large amounts of data being generated by the model, only snapshots (every 6 hours) of salinity and velocity distributions were stored for post-processing. The output of the model run was visualised using a specialised animation program developed using IBM's Data Explorer (DX). A full animation from the DX program can be seen via **http://ibm590.aims.gov.au/**.

Model Verification

The model was verified for the 1981 flood using actual discharge data and actual wind data for the river, against the historical data of Wolanski and Van Senden (1983). This flood was chosen for validation of the model as it was the only flood event for which an extensive map and data set of the river plume exists. Comparisons between model results and field data for 3 different days of field surveys are shown in Figure 2 - 4. Figure 2 - 4 show snapshots of the predicted plume and the corresponding field measurements around that time. Given that the field data were collected over a 2 day period, there is a good agreement between the observed and predicted salinity distribution in coastal waters at corresponding times (see figure 2 - 4). Note also that the measured distribution of surface salinity may also contain freshwater from smaller rivers along the coast in the north.

Sensitivity Analysis

Having verified the model for the 1981 flood event, sensitivity analysis on the model was undertaken to determine the key forcings controlling the fate of the plume. The sensitivity runs showed that the main driving influences on the fate of the plume water were the discharge volume of the river and the local wind forcing (see Figure 5 for example). In particular, since the discharge rates and volumes from this river were so significant at times, the water thunders into the coastal waters and flows northward along the coast, driven by its own huge momentum and the effect of Coriolis. This momentum enables the river water to rapidly mix with shelf waters. This plume behaviour is almost always evident except under the strongest of winds. Further away from the river, the wind effects dominate the far-field movement of the plume. Thus each year, one would expect different plume trajectories depending on the time-varying nature of both the wind and the rainfall/catchment.

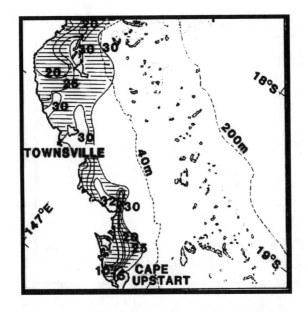

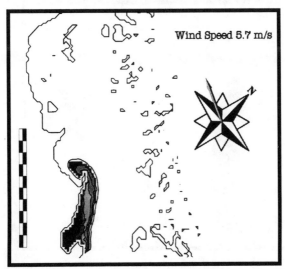

Figure 2: Predicted surface salinity distribution in the Great Barrier Reef during the Burdekin River flood at midday 21st January, 1981. The insert shows the observed distribution of salinity from Wolanski and Van Senden (1983) at this time for validation of the model simulation. Checked distance scale represents 100 km.

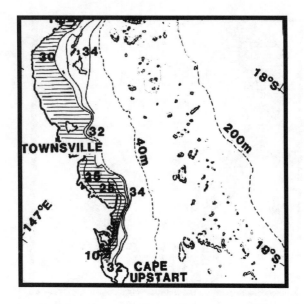

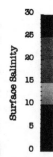

Figure 3: Predicted surface salinity distribution in the Great Barrier Reef during the Burdekin River flood at midday 27th January, 1981. The insert shows the observed distribution of salinity from Wolanski and Van Senden (1983) at this time for validation of the model simulation.

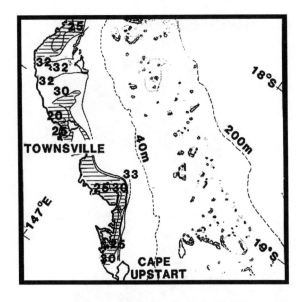

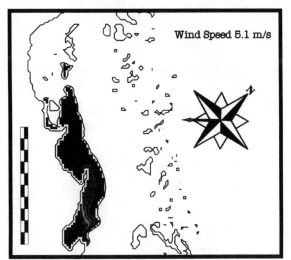

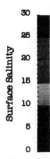

Figure 4:   Predicted surface salinity distribution in the Great Barrier Reef during the Burdekin River flood at midday 4th February, 1981.   The insert shows the observed distribution of salinity from Wolanski and Van Senden (1983) at this time for validation of the model simulation.

Figure 5: Comparison of surface salinity distribution using identical river discharge, but typical wind patterns for the region; that is: no winds, a 10 knot south-east wind and a 1 knot north-east wind.

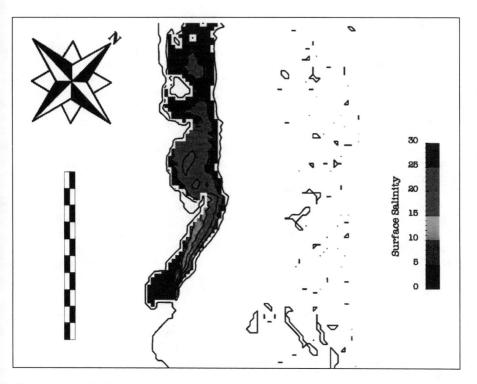

Figure 6. Predicted surface salinity distribution with enhanced vertical mixing to create observed patchiness in the plume near headlands and capes. Checked distance scale represents 100 km in length

Tidally Induced Patchiness

Field observations (Wolanski, 1994) revealed that about a week to ten days after the flood subsides the plume breaks up in patches spread along the coast which is very rugged. This patchiness was attributed to enhanced vertical mixing and flow separation at the numerous islands and headlands along the coast. The model grid was too coarse to implicitly calculate these secondary flows. Instead their effects were explicitly parameterised in the model by enhancing vertical mixing along the observed separation lines at the headlands. This had the effect of bringing deep, saltwater to the surface and breaking the plume into patches. It was possible to fine-tune this mixing enhancement factor so as to generate patchiness after the flood subsided, and not during the flood, in agreement with observations. The distribution of the patches also matched the observations.

4. Conclusion

The "MECCA" 3-dimensional hydrodynamic model (see Hess, 1989) from NOAA, was employed to simulate a flood of freshwater from the Burdekin River into the coastal waters of the Great Barrier Reef (GBR). The model domain was represented by a sigma grid and covered the entire shelf of the central section of the GBR from Cairns to Bowen. The grid comprised of over 100,000 computational points (5 layers in the vertical, 211 points in the along-shelf direction and 95 across-shelf).

By comparing model predicted results and observed surface salinity distributions over several days in 1981, the model was verified for the flood when forced by actual discharge data and actual wind data for the river. The entire 1981 flood event period was simulated and covered the period of 0000 hrs on the 1st January, 1981 until 0000 hrs on the 31st March, 1981; a 90 day period. This flood was chosen for the validation of the model as it was the only flood event for which an extensive survey and data set of the river plume existed. The river delivered a total of almost 19 billion tonne of freshwater into the GBR lagoon from 3 separate flood events over this period. The peak discharge rate of these floods in 1981 was over 12,000 tonnes of water per second at times. Comparisons between model results and field data for 3 different days of field surveys shows very good agreement between the observed and predicted salinity distribution in coastal waters at corresponding times.

Given the huge discharge from the river at this time, it was necessary to set the 100,000 node 3D model to a 30 second time step for both the external mode and internal mode. Consequently, this simulation for the 1981 flood event took 22 days of CPU time on the IBM RS6000/590 supercomputer at AIMS and resulted in 250 trillion computations for the 3 month simulation.

Sensitivity analysis on the model runs showed that the main driving influence on the fate of the plume water was the discharge volume of the river and the local wind forcing. Thus each year, one would expect different plume trajectories depending on the time-varying nature of both the wind and the rainfall/catchment. The model simulations also showed that patchiness in the far-field salinity field was enhanced by discharges from neighbouring rivers and by tidal interactions with headlands along the coast.

From a management perspective, the simulations can identify the fate of the Burdekin River plume in isolation to other freshwater discharges which cannot be done with field observations. This information should provide useful information on catchment management implications of the Burdekin region and its impact on shelf and Great Barrier Reef waters. Given that the fate of the river waters was found to be highly dependent on the ever changing discharge strength and the local wind field, a risk assessment analysis using the 25 years of wind and discharge data is needed to quantify the impact of river floods on the Great Barrier Reef. In addition, the model should find applications for pollutant transport studies to understand the fate of land-derived suspended sediments and dissolved nutrients and pesticides which are ejected into the coastal zones during these major floods.

Acknowledgments

This work was funded by the Cooperative Research Center for the Ecologically Sustainable Development of the Great Barrier Reef (CRC Reef), the Australian Institute of Marine Science (AIMS) and the IBM International Foundation. The authors acknowledge with thanks the contributions of Dr Chris Crossland, Dr Ian Gardiner, Felicity McAllister, Kris Summerhayes and Tim Waterhouse.

References

Hess, K.W. (1986). "Numerical model of circulation in Chesapeake Bay and the continental shelf". NOAA Technical Memorandum NESDIS AISC 6, National Environmental Satellite, Data, and Information Service, N.O.A.A., U.S. Department of Commerce, 47 pp.

Hess, K.W. (1989). "MECCA Program Documentation". NOAA Technical Report NESDIS 46, NOAA, U.S. Department of Commerce, 156 pp.

King, B. and E. Wolanski (1996). "Tidal current variability in the Central Great Barrier Reef". *Journal of Marine Systems*, Volume 9, pp187-202.

Wolanski E and D. Van Senden (1983) "Mixing of Burdekin River Flood Waters in the Great Barrier Reef", *Australian Journal of Marine and Freshwater Research*, Volume 34, pp 49-63.

Wolanski, E. (1994). Physical oceanographic processes of the Great Barrier Reef. CRC Press (Florida), Marine Science Series, pp. 194.

Modeling Realistic Events of the Maine Coastal Current

Monica J. Holboke[1]

Abstract

The Maine Coastal Current (MCC) is an along-shore current off the coast of New England in the Gulf of Maine. It plays an important role in transporting nutrients, biological species, and pollutants. This is part of a larger piece of work to determine the dynamics that influence the MCC variability which has direct importance for environmental managers. The purpose here is to investigate the use of realistic forcing and initial conditions with a three-dimensional fully nonlinear hydrodynamic model.

The model is a state-of-the-art prognostic finite element code that incorporates heat and salt transport and turbulence closure in tidal time. The computational domain encompasses the Gulf of Maine with a cross-shelf upstream boundary at Halifax and an along-shelf boundary offshore past the shelf break. The boundary conditions are forced with data on the upstream while the downstream allows for internally generated waves to radiate out. Initial conditions of temperature and salinity are obtained through optimally estimating observations taken from Brown and Irish (1992, 1993) during April 1987.

Comparison of the April 1987 realization with previously computed climatology shows:

- More localized gyres in the deep basins, especially Jordan Basin,

- A more pronounced meander around Penobscot,

- Increased transport along the western Gulf.

A prognostic calculation of the realization under realistic forcing displays the Gulf scale response. Transects taken at cross-shore points are used to calculate transport through the various segments of the MCC. The simulation additionally highlights the use of a second-order radiating boundary condition.

---

[1]Dartmouth College, 8000 Cummings Hall, Hanover, NH 03755

Introduction

The Gulf of Maine is a semi-enclosed coastal sea off the coast of New England, partially bounded offshore by Georges Bank (Figure 1 left). The Maine Coastal Current is at the confluence of the Gulf of Maine cyclonic flow and the coastal riverine induced circulation. It is roughly centered over the 100m isobath with the Gulf of Maine flow dominant seaward and the coastal circulation dominant land-ward. The large-scale cyclonic circulation is modulated by throughflow from the Scotian Shelf and dense slope water intrusion through the Northeast Channel that spills into the deep basins (Brooks 1985; Brown and Irish 1992, 1993). The coastal flow is modulated by the along-coast frontal structure and buoyancy inputs from freshwater runoff (Lynch et al, 1997).

In previous work (Lynch et al 1997, Holboke and Lynch 1995) computational results focused on understanding the spring climatological circulation. Herein the work departs from climatology with the modeling of a realistic event, the period assessed is April 1987 (Brown and Irish 1992, 1993). The data available includes a series of CTD scans that provides wide coverage of the Gulf, sub-tidal synthetic subsurface pressure (SSP) measurements along the coast, bottom pressure measurements in the deep basins, and wind speed over the Gulf on hourly intervals. [The CTD scans, SSP and BP measurements are from Brown and Irish 1992 and 1993. The wind fields were optimally estimated onto the Gulf of Maine from NDBC moorings; these fields were produced by Hui Feng and Wendell Brown at UNH (Feng and Brown 1997, Brown 1997).] The goal of this study is to examine the role of Gulf-scale forcing on MCC variability and compare this variability to climatology.

Prognostic Time-Domain Model

QUODDY, a state-of-the-art finite-element circulation model is employed. It is described in detail in Lynch et al 1996, 1997. The governing equations are the canonical set: continuity and transport equations for horizontal momentum, heat, salt, and turbulence. The model is three-dimensional with a free surface, partially mixed vertically, hydrostatic, and fully nonlinear. All terms are resolved in tidal time. A level 2.5 turbulence closure scheme (Mellor and Yamada, 1982; Galperin et al., 1988; Blumberg et al. 1992) determines the vertical mixing and the horizontal mixing is represented by a mesh- and shear-dependent eddy viscosity similar to Smagorinsky (1963).

Variable horizontal resolution is achieved with an unstructured mesh of conventional linear triangles. In the vertical, a general terrain-following coordinate system is used, that incorporates flexible, nonuniform vertical discretization. Therefore continuous tracking of the free surface, including proper resolution of surface and bottom boundary layers, is accomplished.

Procedure

The CTD observation locations are shown in Figure 1 right. The 145 scans were obtained on a cruise from April 4-18 1987. While the data had significant coverage of the interior Gulf of Maine, it did not provide information near the coast, on Georges Bank, the Scotian Shelf, or past the shelf break. Therefore the prognostic March-April climatology from Lynch *et al* 1997 is used as the spatially variable mean upon which to objectively analyze(OA) the data and results in the April realization. The correlation base scale is 25km with locally anisotropic correlation scales aligned with the bathymetry (Loder *et al* 1997). The initial field is dated at April 12th 00hrs 1987, the midpoint of the CTD observations. The initial pressure and velocity is computed with FUNDY5, a harmonic linear finite element hydrodynamic model, see Lynch *et al* 1987 and 1992. The forcing includes tidal (M_2 only) elevations, residual elevations, zero wind stress, and the initial density field. The residual elevations are a combination of interpolated values from the March-April prognostic climatology and a steric height calculation (see Loder *et al* 1997 and Hannah *et al* 1996).

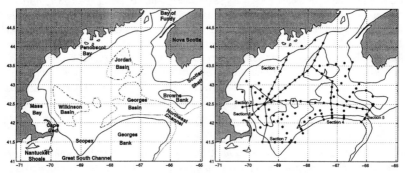

Figure 1: The contours in each image are the 100m and 200m isobaths. Left is a location map of distinct geographical features in the Gulf of Maine. Right displays the CTD scan locations for April 4-18 1987 cruise, sections marked are the same sections used in Brown and Irish 1993.

The diagnostic calculation provides a dynamically consistent evaluation of the April realization. However, within the April realization aliasing of the data necessarily exists. Also nonlinearities and additional dynamics within QUODDY vs. FUNDY5 require spin up time. Therefore to obtain an optimal evaluation of the April 12th 00hrs 1987 realization, barotropic and baroclinic adjustment must take place. This can be accomplished through a prognostic calculation with QUODDY. Baroclinic adjustment in the Gulf of Maine requires about 3 days in April (personal communication from Wendell Brown), barotropic adjustment requires less time. Three days is also the amount of time for the nonlinearities and additional physics to spin up. Therefore this adjustment simulation is run for 6 M_2 tidal periods. The first two tidal periods linearly ramp up the Halifax sea level and the wind field for April 12th 00hrs. For the remaining 4 tidal cycles these quantities are held constant. A restoring surface boundary condition is used on temperature with a heating rate of $2.3e^{-5}m/s$, as in Naimie 1996, Lynch *et al* 1996, 1997.

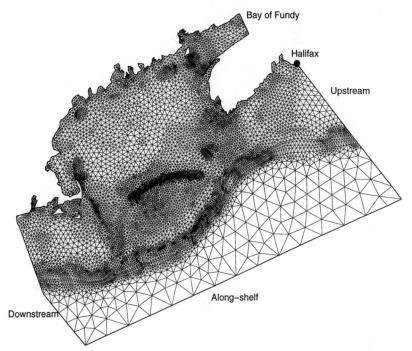

Figure 2: GHSD Mesh. There are 8787 nodes and 17019 elements. This mesh has the special feature of discretization along the shelf break that satisfies the $(h_{max} - h_{min})/h_{min} < 0.5$ criterion (Naimie 1996)

In Figure 2 the mesh displays four open boundaries; each boundary is treated differently in the prognostic simulation. Halifax is a specified elevation boundary node with M_2 and residual components as in the FUNDY5 run. The residual elevation is modulated by the Halifax SSP measurements. The rest of the upstream boundary, between Halifax and the along-shelf boundary, is allowed to radiate. The radiating boundary condition implemented is a second-order RBC that drives tides externally while internally generated waves can radiate out. The formulation implemented here is similar to Johnsen and Lynch 1995. The external elevation used along this boundary are the tidal and residual elevations as in the FUNDY5 simulation with the residual component modulated by an extrapolation of the Halifax SSP. The along-shelf boundary is clamped with the tidal and residual elevations as in the FUNDY5 run. The downstream boundary is another radiating boundary. It is externally forced with the tidal and residual elevations as in the FUNDY5 run. The Bay of Fundy boundary is forced by periodic normal flow boundary conditions (Lynch and Holboke 1996), that allow the elevation there to respond freely to the wind.

After the simulation has undergone the prognostic adjustment it is additionally run for 14 M_2 tidal periods, $\sim$ one week. In this 'realistic' simulation the boundary conditions and wind fields are varying with time, yielding a synoptic forcing of the Gulf. The boundary condition on temperature is set to a April climatological heat flux value of $100W/m^2$. A one week time period was chosen to observe the variability of the MCC while the information in the density structure was still valid. These results are compared to spring climatology. The important dynamical features compared are basin gyre strengths and locations, coastal current branch points, and transport in each leg of the coastal current.

Results
Initial Conditions

Previously illustrated in Figure 1 are the CTD measurement locations. A line joining a subset of these measurements is labeled section 2. This section is chosen to compare the data, initial fields, and climatology. These comparisons are shown in table 1 and Figure 3 for temperature, salinity, and σ_t fields. Section 2 generally demonstrates that the objective analysis procedure was successful in capturing the prominent features and smoothing the small scale features in the data.

	Temperature(deg C)		Salinity(psu)		σ_t	
	mean	std	mean	std	mean	std
OA	5.31	1.50	33.16	0.95	26.16	0.58
Data	5.28	1.53	33.14	0.95	26.15	0.58
Climatology	4.95	1.43	33.31	0.81	26.33	0.48
OA - Data	0.03	0.23	0.02	0.09	0.01	0.05
OA-Climatology	0.36	0.38	-0.16	0.21	-0.17	0.14

Table 1: Section 2 mean values and standard deviations.

The mean and standard deviations of the OA, data, climatology, OA-data difference, and OA-climatology difference for section 2 fields are listed in table 1. The statistics for the difference OA-data shows small errors, therefore demonstrating the success at capturing the data in the realization. Compared to the climatology the OA fields are fresher, warmer, and lighter. These differences are caused by increased Scotian Shelf inflow, increased heat flux, increased river discharge, and slope water inflows, all of which were noted in Brown and Irish 1992, 1993. Scotian Shelf water is characterized as fresh cold water coming down the Scotian

Shelf into the Gulf of Maine, hugging the coast of Nova Scotia. Slope water is characterized as warm salty water that enters through the Northeast Channel at depth.

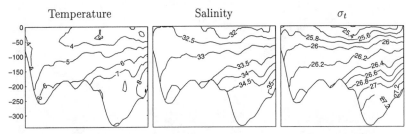

Figure 3: Section 2: Objectively analyzed temperature, salinity and σ_t fields.

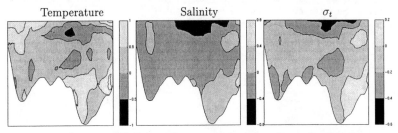

Figure 4: Section 2 difference fields, OA - climatology.

Section 2 (Figures 3, 4) extends from the left at Massachusetts Bay across the Gulf of Maine through Georges Basin to the Northeast Channel. The transect is generally fresher than the climatology, especially near the surface. Additionally, compared to the climatology it is cooler near the surface and warmer at depth in Georges Basin. Consequently, there is an increase in stratification from the climatology, mainly over Georges Basin. Again these differences are directly related to the significant Scotian Shelf inflow and the episodic slope water intrusions.

The surface OA mass fields and OA - Climatology difference fields are shown in Figure 5. Outside the data region up on the Scotian Shelf, offshore from the shelf break, and downstream by Nantucket Shoals the difference fields are close to zero; therefore, these regions are very similar to the climatology. The Gulf is warmer in the west and cooler in the east compared to climatology, caused by increased warming and increased Scotian Shelf inflow. The surface of the Gulf is fresher overall with notable freshness over the deep basins and in Massachusetts Bay. The combination of these yields lighter water at the surface than in the climatology. Different features from the climatology seen in the OA are the Scotian Shelf inflow spreading over the eastern Gulf and an intensified front along the western coast of the Gulf.

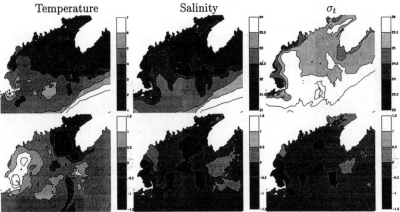

Figure 5: Surface values of mass field approximation for April 12th 00hrs 1987 -
Top. OA-Climatology difference fields - Bottom.

Diagnostic April Circulation

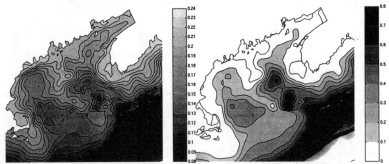

Figure 6: FUNDY5 results: Left, residual surface elevation(m). Right, residual
stream function(Sv).

   Figure 6 shows the FUNDY5 computational result for the residual surface elevation and stream function. The elevation establishes high values at the coast
and low values over the basins. Subsequently, flow circuits the Gulf with pathways along the coast and cyclonic gyres over the deep basins, consistent with
the general circulation observed in the Gulf. In Brown and Irish 1992, surface
dynamic topography was calculated. While elevation estimates the surface flow,
surface dynamic topography estimates the surface geostrophic flow, therefore it
is possible to compare the two calculations. The surface geostrophic flow crosses
Jordan basin to the southwest then splits with one arm returning north toward
the coast and the other directed south toward Georges Bank. The arm heading

towards Georges Bank diverts away from it, heads west through Wilkinson Basin and is rerouted south to exit the Gulf. These features are also visible in the elevation and stream function. Therefore the FUNDY5 calculation on the mass fields approximation is consistent with the dynamic height.

Prognostic April 12th 00hrs 1987 Circulation

In Figure 7 the April 12th 00hrs 1987 surface temperature, salinity, σ_t fields and their associated difference to climatology fields are shown which have been dynamically adjusted compared to Figure 5. The dynamic adjustment has smoothed the mass fields by mixing any dynamically inconsistent features.

Temperature   Salinity σ_t

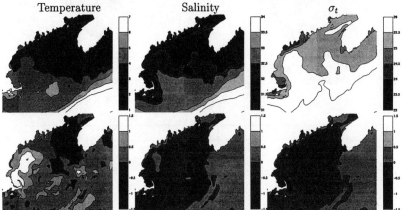

Figure 7: Top: April 12th 00hrs 1987 Surface mass fields after adjustment with QUODDY. Bottom: Difference to climatology mass fields.

The residual stream-function and transports through key sections in the coastal current for the April 12th 00hrs 1987 realization and climatology are compared. In Figure 8 left, is the Gulf circulation for the March-April climatology. The cyclonic flow around the Gulf with cyclonic basin gyres and anti-cyclonic flow around Georges and Browns banks are typical features in the Gulf of Maine. In the April 12th 00hrs realization (Figure 8 right) the general circulation is similar, however the details are different. The Jordan basin gyre is more localized, the Georges basin gyre has intensified and moved closer to the Northeast Channel. The flow around Wilkinson basin has spread to incorporate more of the SCOPEX region with increased circulation. Georges and Browns banks anti-cyclonic flows are similar to the climatology, however depicts increased flow around Georges Bank. More localized flow around Jordan Basin along with the temperature and salinity structure generates a meander around Penobscot Bay that extends further offshore than the climatology. However, as the surface temperature and salinity fields return to shore and merge with the sharp front along the western coast of the Gulf, the coastal current follows this structure. Light water and Ekman transport from the April 12th 00hrs northward wind pushes the coastal current offshore, inhibiting the formation of a Mass Bay leg.

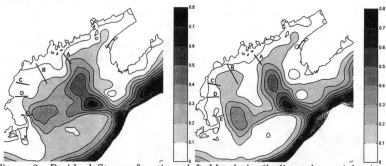

Figure 8: Residual Stream-function: left March-April climatology, right April 1987. Marked are transects A, B, C, D, E, where transports through are listed in table 2.

Transect	A	B	C	D	E
Climatology	0.225Sv	0.139Sv	0.046Sv	0.059Sv	0.293Sv
April 1987 Prognostic	0.178Sv	0.116Sv	0.129Sv	0.082Sv	0.129Sv

Table 2: Net transports through the transects marked on Figure 8.

In table 2 are the transports through each transect as marked in Figure 8. The climatology shows a stronger eastern leg that departs the coast around Penobscot Bay with only half of the current returning to the western Gulf and only half of that amount staying near shore. The current increases again after Mass Bay with flow from Wilkinson Basin. In the April 1987 coastal current the eastern leg is smaller, however it maintains more of its strength around Penobscot Bay than in the climatology and along the western coast of the Gulf. Around Mass Bay it is deflected offshore and returns with increased flow from Wilkinson Basin to exit towards Georges Bank and Nantucket Shoals.

The effects of the new baroclinic forcing from the realistic mass fields on the Gulf-scale circulation are: Jordan Basin gyre is more localized, Georges Basin gyre is more localized, intense, and closer to the Northeast Channel, Wilkinson Basin gyre is physically larger encompassing more of the SCOPEX region and is connected to the other basins. The amount of transport through the Gulf is similar to the climatological amounts. The influence of the realistic mass fields on the MCC are: a reduced eastern leg, more retention through the Penobscot branch point to the western leg, a larger excursion around Penobscot Bay and deflection offshore from Mass Bay.

Prognostic April 12th 00hrs-19th 5.88hrs 1987 Circulation

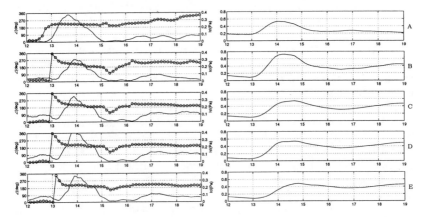

Figure 9: Left: Wind Stress (Pa, solid line right axis) at transects averaged over tidal periods and centered in time (days) for specific tidal period. Line with circles are the wind stress angle from true north (left axis, ie. 0=N, 90=E, 180=S, 270=W, 360=N). Right: Transport through transects A-E in Sv., centered in time for the specific tidal average.

The transport time series in Figure 9 shows significant coherence with the wind fields. The peak in wind stress occurs $\sim$ 3hrs before the 14th at transects B-E, in the western Gulf. At transect A it occurs about 8 hours before the 14th. In Brown 1997 the peak pressure response was for wind 260° T during most of the year and the western Gulf response was the highest from Portland southward past Cape Cod. Although the peak wind stress here is $\sim$ 230° T the maximum response is in the western Gulf, which corresponds with Brown's analysis.

At transect A the computed transport is generally constant until the wind event, it then responds with a maximum of 0.52Sv lagging the peak wind stress by about 8hrs. After the wind event the response gradually decreases back to approximately 0.2Sv. Transects B-E initially decrease under the northward wind consistent with the deflection offshore of the MCC and its reduced strength. Transect B responds to the large wind event with approximately 0.7Sv, the largest response in any of the legs, and lags the local wind stress maximum by about 6 hours. Transect C responds with a maximum of 0.55Sv that lags the maximum wind stress by about 1 day. Transect D increases to 0.55Sv for the large wind event and lags the wind $\sim$ 6 hours. This maximum transport plateaus for most of the 14th caused by the local wind stress being largest at this transect. Transect E shows a delayed and reduced response to the wind event with a maximum of 0.5Sv and a lag time of approximately 1 day. Transects B-E decrease after the large wind event hitting a minimum inbetween the 16th and 17th of $\sim$ 0.3Sv and then increasing again under the southward wind to 0.45Sv. This spatially cohesive response from the 15th on demonstrates that the large scale dynamic of the wind has set up the coast uniformly through this region. Transect A has contrary results caused by the different local wind field in the beginning and end of the simulation. The lag times for transport response in transect B is about two hours slower than the transport response time for section 7, similar in location to transect B, of Greenberg *et al* 1997. Therefore, demonstrating a slight time lag error in these results.

In Figure 10 are the 2-D residual stream function results with vertically averaged residual velocities overlaid. In the first panel (IC) is the April 12th 00hrs result a variation of Figure 8. The MCC eastern leg travels along-shore from Grand Manan towards Penobscot Bay it branches offshore entrained in the gyre in Jordan Basin. It travels along with the Gulf-scale circulation towards Georges Bank before returning to the coast and joining the western leg. The western leg continues on with some deflection offshore from the local northward wind stress. A branch point off Cape Ann has a small leg entering Mass Bay, while most of the current remains in the Stellwagen leg, the leg that bypasses Mass Bay to the east. It continues on toward the third branch point, southeast of Cape Cod, where flow splits either exiting the Gulf toward Nantucket Shoals or returning to the Gulf circulation via the northern flank of Georges Bank.

In the second panel, the April 13th 00hrs result, the wind is generally northward in the western Gulf and weak in the eastern Gulf. The Scotian Shelf flow has decreased, therefore the strength of the gyres in Jordan and Georges Basin have decreased. The northward wind in combination with the decrease of throughflow from the Scotian Shelf has weakened the cyclonic circulation in the western Gulf. At the Penobscot branch point the flow that exits the MCC does not return and instead is incorporated into more localized flow around southern Wilkinson Basin and into an anti-cyclonic gyre that has formed offshore from Penobscot Bay. Flow that continues on from the eastern leg past Penobscot Bay defines the MCC western leg. The western leg is then pushed offshore by increased upwelling from the local northward wind and inhibits the formation of a Mass Bay leg. The Stellwagen leg reaches the third branch point and is mostly deflected towards Georges Bank.

On April 14th 00hrs, the middle of the wind event, the Gulf experiences significant throughflow. This throughflow travels down the Scotian Shelf circuits through the Gulf and overwhelms any structure in the coastal current. The only distinctive features to remain are the anti-cyclonic flow around Georges Bank and the gyre in Georges Basin. Ekman transport pushes water off to the east-northeast, thereby causing the significant increase in the western leg noted in the transport time series. On the 15th the wind has started to decrease and the inflow from the Scotian Shelf has also decreased so that local structure in the Gulf is more apparent. A gyre inside Georges and Jordan basin has formed. Subsequently, a meander in the MCC at the Penobscot branch point is noticeable. The meandered

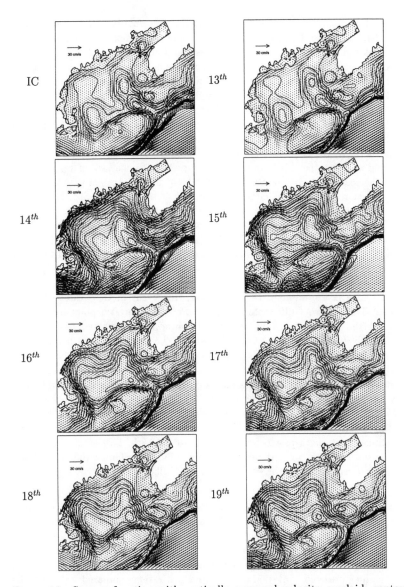

Figure 10: Stream function with vertically averaged velocity overlaid, contour interval = 0.1Sv, thicker lines are even values (ie. 0Sv 0.2Sv. etc.). Images are tidally averaged and centered at 13th 00hrs, 14th 00hrs, 15th 00hrs, 16th 00hrs, 17th 00hrs, 18th 00hrs and 19th 00hrs.

flow returns augmented to the coast in a narrow current that stays along-shore. The 14th had shown flow entering Mass Bay and this remains through this time period as well. The circulation in Georges and Jordan basins also extends into Wilkinson Basin. Increased anti-cyclonic flow around Georges Bank is also noted. As the coastal current travels towards the third branch point it splits either exiting the Gulf toward Nantucket Shoals or heading towards Georges Bank.

On the 16th the wind has receded, however the Gulf is still responding to the wind event. The Gulf circulation contributes to the western leg of the MCC from the Penobscot branch point. This contribution is retained as the current travels along-shore. At the Cape Ann branch point the current is denied access into Mass Bay and splits at the third branch point. On the 17th the lack of recirculation along the coast in Penobscot Bay allows the MCC to flow straight past with little augmentation from the Gulf circulation entering at this branch point. Essentially the eastern leg of the MCC determines the western leg by inhibiting flow from the Gulf into the western leg. A Mass Bay leg has reformed and the current splits at the third branch point. The Gulf circulation is gaining more structure as the gyre in Georges Basin encompasses a larger region. The circulation that was present in Wilkinson Basin on the 15th has moved southward to include part of the SCOPEX region as well.

The 18th is the onset of the southward wind event in the western Gulf of Maine. The eastern Gulf, eastern leg, and Penobscot branch point are similar to the 17th. The western Gulf responds by forming a larger and more intense gyre in the Wilkinson Basin/SCOPEX region. The coastal current becomes the shore-ward manifestation of this circulation. At the Cape Ann branch point, a Mass Bay leg has formed in part by increased Ekman transport. At the third branch point the current separates into two legs. On the 19th the eastern Gulf is now forced by a northward wind, therefore reducing the strength of the eastern leg slightly. Recirculation in the Penobscot Bay region increases the meander around it. The flow in the western leg is increased relative to the eastern leg and maintains its strength along-shore. Similar to the 18th, flow enters Mass Bay. At the third branch point the current again splits. The Gulf circulation shows increased circulation in Georges Basin and the Wilkinson/SCOPEX region.

Conclusion

Creation of an initial condition field by objective analysis(OA) of the data with climatology as the spatially variable mean was successful. It yielded not only a fair representation of the data, but also a smooth field outside of the data. The radiating boundary condition also proved successful by allowing the pressure response to propagate through the boundary.

The diagnostic computation on the April realization compared well with the Brown and Irish 1992 analysis of the geostrophic flow. The prognostic adjustment simulation compared to climatology showed more localized and intensified gyres in Jordan and Georges Basins and spatially larger circulation in Wilkinson Basin/SCOPEX region. The MCC has a reduced eastern leg, more retention through the Penobscot branch point, a larger excursion offshore in this branch point, and deflection offshore from Mass Bay.

The synoptic forcing of the wind and elevation combined with the OA initial conditions produces a 'realistic' simulation to observe variability of the MCC. The details to note are: Scotian Shelf inflow affects the gyres in Georges and Jordan Basins, the gyre in Jordan Basin and recirculation in Penobscot Bay influence the Penobscot branch point, upwelling in the western Gulf will deflect the MCC offshore, and intensified circulation in Wilkinson Basin will dominate the western leg of the MCC. The gyre in Georges Basin and the anti-cyclonic gyre around Georges Bank remain structured through the wind event. This robustness highlights the importance of baroclinic structure in the deep basin and the tidal nonlinearities around Georges Bank.

Acknowledgments

Wendell Brown provided the CTD data and SSP measurements; Christopher Naimie provided the Gulf-wide climatological solutions; John Loder and Peter Smith provided the climatological wind and (T,S) fields; David Greenberg provided the original finite element bathymetry and tidal boundary conditions; OAX software was provided by BIO; and Wendell Brown and Hui Feng provided the objectively analyzed wind fields. This work is sponsored by the Gulf of Maine Regional Marine Research Program and the New Hampshire Sea Grant College Program.

References

Blumberg, A. F., B. Galperin, and D. J. O'Connor, "Modeling vertical structure of open-channel flows", *ASCE, J. Hyd. Engg.*, **118**, 1119–1134, 1992.

Brooks, D.A., "The vernal circulation in the Gulf of Maine", *J. Geo. Res.* **90**, C5, 4687-4705, 1985.

Brown, W.S., "The wind-forced pressure response of the Gulf of Maine", *J. Geo. Res.* 1997 in press.

Brown, W.S. and J.D. Irish, "The annual variation of water mass structure in the Gulf of Maine: 1986-1987", *J. Mar. Res.* **51**, 53-107, 1993.

Brown, W.S. and J.D. Irish, "The annual evolution of geostrophic flow in the Gulf of Maine: 1986-1987", *J. Phys. Ocean.* **22**, 445-473, 1992.

Feng, H. and W.S. Brown, "The wind-forced response of the western Gulf of Maine coastal ocean during Spring and Summer 1994", *J. Geo. Res.* submitted 1997.

Galperin, B., L.H. Kantha, S. Hassid, and A. Rosati, "A quasi-equilibrium turbulent energy model for geophysical flows", *J. Atmos. Sci.*, **45**, 55-62, 1988.

Greenberg, D.A., J.W. Loder, Y. Shen, D.R. Lynch, and C.E. Naimie, "Spatial and temporal structure of the barotropic response of the Scotian Shelf and Gulf of Maine to surface wind stress: A model-based study", *J. Geo. Res.*, **102**, C9, 20897-20915, 1997.

Hannah, C.G., J.W. Loder, and D.G. Wright, "Seasonal Variation of the Baroclinic Circulation in the Scotia Maine Region". In D.G. Aubry and C.T. Friedrichs, editors, *Buoyancy Effects on Coastal and Estuarine Dynamics.* AGU, 1996.

Holboke, M.J. and D.R. Lynch, "Simulations of the Maine Coastal Current", Proc. ASCE Conf. on Estuarine and Coastal Modeling, San Diego, Oct. 1995.

Johnsen, M. and D.R. Lynch, "Assessment of a Second-order Radiation Boundary Condition for Tidal and Wind Driven Flows". In D.R. Lynch and A.M. Davies, editors, *Quantitative Skill Assessment for Coastal Ocean Models.* AGU, 1995.

Loder, J.W., G. Han, C.G. Hannah, D.A. Greenberg and P.C. Smith, "Hydrography and baroclinic circulation in the Scotian Shelf Region: winter vs summer", *Can. J. of Fish. and Aq. Sc. Sup.*, **54**, 40-56, 1997.

Lynch, D.R. and M.J. Holboke, "Normal Flow Boundary Conditions in 3D Circulation Models", *Int. J. Num. Meth. Fluids*, **25**, 1-21, 1997.

Lynch, D.R., M.J. Holboke, and C.E. Naimie, "The Maine Coastal Current: Spring Climatological Circulation", *Cont. Shelf Res.*, **17**, 6, 605-634, 1997.

Lynch, D.R., J.T.C. Ip, C.E. Naimie, and F.E. Werner, "Comprehensive coastal circulation model with application to the Gulf of Maine", *Cont. Shelf Res.*, **16**, 7, 875-906, 1996.

Lynch, D.R., F.E. Werner, D.A. Greenberg, and J.W. Loder, "Diagnostic Model for Baroclinic, Wind-driven and Tidal Circulation in Shallow Seas", *Cont. Shelf Res.*, **12**, 37-64, 1992.

Lynch, D.R. and F.E. Werner, "Three-Dimensional Hydrodynamics on Finite Elements. Part I: Linearized Harmonic Model.", *Int. J. Num. Meth. Fluids*, **7**, 871-90, 1987.

Mellor, G. L., and T. Yamada, "Development of a turbulence closure model for geophysical fluid problems", *Rev. of Geophys. Space Phys.*, **20**, 851–875, 1982.

Naimie, C.E. "Georges Bank Residual Circulation during weak and strong stratification periods - Prognostic numerical model results", *J. Geo. Res.*, **101**, 6469-6486, 1996.

Smagorinsky, J., "General circulation experiments with the primitive equations I. The basic experiment", *Monthly Weather Review*, **91**, 99–164, 1963.

Assessment of the Random Choice Method for Modeling
Tidal Rectification in Coastal Water Bodies

Eugenio Gomez-Reyes[1]

ABSTRACT

Much of the problem to simulate tidal rectification processes resides in the
numerical treatment to the nonlinear advective terms of the momentum equation. In
the Random Choice Method (RCM) these advected terms are treated as part of the
Riemann problem, resolving accurately the non-linear dynamics. Here, RCM is
extended and assessed its capability for simulating tidal rectification processes in
coastal systems. The assessment is carried out through analysis of the induced
overtide and residual currents, as generated by the solution of the one-dimensional
shallow water equation. Results indicate that the RCM is capable of accurately
reproducing overtides and residual currents, with as much resolution as obtained with
centered differences schemes. Although further numerical assessments are required
before applying RCM to real coastal water bodies.

INTRODUCTION

The non-linearities of the dynamics of the tidal flow is one of the generators of
overtides and residual currents in coastal water bodies. This tidal rectification
requires that numerical schemes, for the non-linear momentum terms, conserve
squared properties to allow the transfer of energy from fluctuating tidal oscillations to
both overtide and mean flows (Grammeltvedt, 1969). Hence, much of the problem to
simulate tidal rectification resides in the numerical treatment to the nonlinear
advective terms of the equations. Assessments of the numerical schemes for
simulating tidal rectification processes may carried out through direct analysis of the

---

[1] Departamento de Ingenieria de Procesos e Hidraulica. Universidad Autonoma
Metropolitana. Av. Michoacan y la Purisima, Col. Vicentina Iztapalapa, Mexico D.F.
09340.

scheme for conserving squared properties (*i.e.*, truncation error analysis), numerical calculations of squared properties (for instance, kinetic energy, square of vorticity) over the simulated velocity field, or through analysis of the induced overtide and residual currents.

Since the last decade, the Random Choice Method (RCM) have been applied in gas dynamics (Chorin and Marsden, 1979) and hydraulics (Marshall and Menendez, 1981) to successfully simulate discontinuities, without introducing numerical diffusion. In the RCM the advected terms of the momentum equations are treated as part of the Riemann problem, resolving accurately the non-linear dynamics. Marshall and Menendez (1981) extended the method to allow solutions for the nonconservation forms of the shallow water equations, opening an avenue for research of the RCM to simulate tidal rectification in coastal systems, not undertaken since.

The present study assesses the ability of the RCM to simulate tidal rectification processes in coastal water bodies through analysis of overtide and residual currents, as generated by the solution of the one-dimensional shallow water equation. In the following section the RCM is described briefly. Performance of the one-dimensional shallow water equation is shown in the subsequent section. Conclusions are presented in the final section.

THE RANDOM CHOICE METHOD

The RCM was first introduced by Glimm (1965) as part of a technique to prove the existence of solutions for hyperbolic equations. Chorin (1976) developed RCM into a numerical method. Further developments and applications of the method have been documented elsewhere (see, for instance, Jazcilevich-Diamant *et al.* 1995).

In the RCM, the solution (u) of the momentum equation is constructed as a superposition of local analytical solutions of Riemann problems, which are evaluated at a randomly chosen point. At time $t = n\Delta t$, the solution is first approximated by a piecewise constant function U, on intervals of size Δx, at the point $x = i\Delta x$, *i.e.*,

$$u((i+a)\Delta x, n\Delta t) = U_i^n \qquad for \qquad -\frac{1}{2} \le a < \frac{1}{2} \tag{1}$$

Each discontinuity between adjacent piecewise approximation give rise to a Riemann problem with the following initial conditions:

$$u(x,0) = \begin{cases} U_R = U_{i+1}^n \\ U_L = U_i^n \end{cases} \qquad for \qquad \begin{matrix} x \ge 0 \\ x < 0 \end{matrix} \tag{2}$$

From where an exact solution can be obtained for u along the characteristics, the Riemann problem solution, provided that the Courant-Friedrichs-Lewy condition is satisfied to avoid intersection of waves propagating from adjacent discontinuities. Finally, new values for the u are sampled from the Riemann problem solution at each interval [$-\Delta x/2$, $\Delta x/2$]. This sampling is carried out through an equidistributed random variable (ξ) which has a probability density function that takes the value of 1 in [$-1/2$, $1/2$] and 0 elsewhere. In this way a single value of u (*i.e.*, U_R or U_L) is assigned at time $(n+1)\Delta t$.

Because the solution is either U_R or U_L, the region in which the velocity is different from zero remains sharply defined if it is sharply defined initially. That is, if the velocity at the initial time is a step function, it remains so at all times. The method therefore has no numerical diffusion.

PERFORMANCE COMPARISONS

Here, the RCM is extended to simulate tidal rectification by solving the one-dimensional shallow water equation:

$$\frac{\partial u}{\partial t} + u \frac{\partial u}{\partial x} = -g \frac{\partial \eta}{\partial x} + R_{(u)}$$

$$\frac{\partial \eta}{\partial t} + \frac{\partial}{\partial x}(Hu) = 0$$

(3)

Together with its boundary conditions

$$\eta = A_o \cos(\omega t) \qquad \qquad at \ \ x = 0$$

$$\eta = \begin{cases} \dfrac{\partial \eta}{\partial t} - c \dfrac{\partial \eta}{\partial x} = 0 & for \quad c > 0 \\[2mm] \dfrac{\partial \eta}{\partial t} = 0 & for \quad c \leq 0 \end{cases} \qquad at \ \ x = L$$

(4)

Where η is the water surface elevation, A_o is the amplitude and ω is the frequency of the tidal forcing, L is a distance of several wave lengths, c is the phase speed of the radiating wave (*i.e.*, $c = (gH)^{1/2}$), g is the gravity, H is the total water depth, and $R_{(u)}$ is a linear friction term (function of u, *i.e.*, $R_{(u)} = k \bullet u$, where k is the friction coefficient).

The velocity field described by Equations (3) and (4), for the case of a flat bottom, is characterized by a progressive wave traveling from the forcing to the radiating boundary. This wave is damped by friction and modified by the rectification mechanism of the nonlinear advective term of the momentum equation, which generates higher harmonics and residuals from the fundamental frequency forcing. For instance, when the forcing is the M2 (lunar principal tidal harmonic with period of 12.42 hours), the generated overtide is the M4 at twice the M2 frequency.

The RCM was then applied to the inhomogeneous momentum equation in (3). Such equation was treated using a fractional two-step method, as described by Sod (1977). In the first step the inhomogeneous term is removed and the resulting conservation form equation is solved be the RCM. In the second step a system of ordinary differential equations is solved by finite differences, using as initial conditions the solution obtained in the first step, *i.e.*,

$$1st \quad step \qquad \frac{\partial u'}{\partial t} + u' \frac{\partial u'}{\partial x} = 0 \qquad\qquad from \quad n-1 \quad to \quad n \quad time \quad level$$

$$U_i^n = u\left(\left(i + a \pm \tfrac{1}{2} \xi^{n-1} \right) \Delta x, n\Delta t \right) \qquad for \qquad -\frac{1}{2} \le a < \frac{1}{2}$$

$$2nd \quad step \qquad \frac{\partial u}{\partial t} = -2g \frac{\partial \eta}{\partial x} + 2R(u) \qquad\qquad from \quad n \quad to \quad n+1 \quad time \quad level$$

$$U_i^{n+1} = U_i^n - 2g \frac{\Delta t}{\Delta x} \left(\Delta \eta^n \right) + 2 \bullet \Delta t \bullet R(U_i^n)$$

where $u' = 2u$.

The Riemann problem for the homogeneous system of the first step was solved using a numerical method to integrate the characteristic curves. The sampling for the Riemann problem solution was carried out using the Van der Corput random sequence (Hammersley and Handscomb, 1965). The Asslin (1972) filter was applied to remove the tendency for the time to split, due to the use of the fractional step method with leap-frog differencing.

The solution for the one-dimensional shallow water equation using the RCM, as described above, is shown in Figure 1. The solution using the centered difference scheme is also shown for comparison. Their corresponding amplitudes and phases are shown in Table 1. These amplitudes and phases were calculated by extracting, during a tidal cycle, the Fourier coefficients for M2 and M4 oscillations from the simulated velocity field.

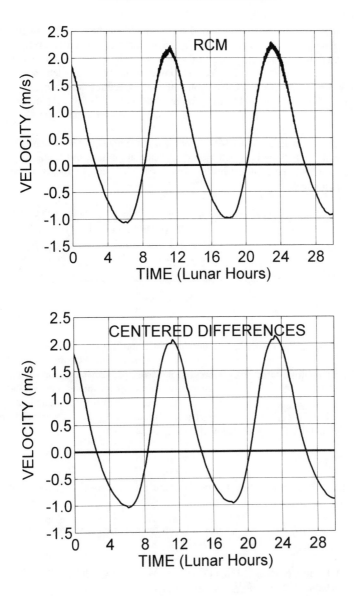

Figure 1. Solutions for the one-dimensional shallow water equation

Solutions in Figure 1 are displayed for the midpoint between the forcing and radiating boundaries, and were obtained by driven the system with 12 cycles of M2 tide (1 m amplitude), over an hypothetical coastal water body of constant depth (5 m). The RCM solution shows that the method reproduces the asymmetric character of the non-linear traveling wave (*i.e.*, interaction with the induced residual velocity). While the values of amplitudes and phases for the M2 and M4 (Table 1) indicate the presence of distortion in the solution (*i.e.*, non-linear character of the signal).

Further assessment of the RCM for reproducing the non-linearities were provided by additional simulations in which the tidal forcing amplitude was changed. These simulations shown that tidal rectification velocity fields become stronger as the tidal forcing amplitude increases.

Table 1. Computed amplitudes and phases for the tidal rectification velocity field.

SCHEME	AMPLITUDE (m/s)			PHASE (°)	
	M2	M4	RESIDUAL	M2	M4
RCM	1.51	0.26	0.65	18.94	71.34
CENTER DIFFERENCES	1.53	0.26	0.60	18.60	66.82

CONCLUSIONS

The RCM is been extended to simulate tidal rectification processes in coastal water bodies. This method is capable of reproducing the non-linearities of the advective momentum term. It is necessary, however, to further develop the method before applying RCM to a real coastal water body, *i.e.*, evaluate conservation of squared properties, extension to two and three-dimensional simulations, and comparisons with well suited schemes to the modeling of vorticity transfer from fluctuating tidal oscillations to the overtide and residual fields.

REFERENCES

Asslin, R. (1972). "Frequency Filters for Time Integrations." Monthly Weather Review, 100: 487- 490.

Chorin, A.J. (1976). "Random Choice Solution for Hyperbolic Systems." Journal of Computational Physics, 22: 517-533.

Chorin, A.J., and Marsden, J.E. (1979). "A Mathematical Introduction to Fluid Mechanics." Springer-Verlag, New York.

Glimm, J. (1965). "Solutions in the Large for Nonlinear Hyperbolic Systems of Conservation Laws." Communications of Pure and Applied Mathematics, 18: 697-715.

Grammeltvedt, A. (1969). "A Survey of Finite-Difference Schemes for the Primitive Equations for a Barotropic Fluid." Monthly Weather Review, 97(5): 384-404.

Jazcilevich-Diamant, A., Gomez-Reyes, E., Valle-Levinson, A. and Fuentes-Gea, V. (1995). "Application of the Random Choice Method to the Estuarine Mass Transport." Proceedings of the Third Water Pollution Conference, L.C. Wrobel and P. Latinopoulos (editor), Computational Mechanics Publications, Porto Carras Greece, pp. 253-259.

Hammersley, J.M. and Handscomb, D.C. (1965). "Monte Carlo Methdos." Methuen, London.

Marshall, G. and Menendez, A.N. (1981). "Numerical Treatment of Nonconservation Forms of the Equations of Shallow Water Theory." Journal of Computational Physics, 44: 167-188.

Sod, G.A. (1977). "A Numerical Solution of a Converging Cylindrical Shock." Journal of Fluid Mechanics, 83: 785-794.

Analysis and application of Eulerian
finite element methods for the transport equation

James L. Hench[1] and Richard A. Luettich, Jr.[1]

Abstract

A comparison of various Eulerian finite element schemes for the transport equation is made using Fourier analysis. The relative merits of each scheme are discussed and one scheme is selected for a two-dimensional field application. A series of runs are made to test global scalar mass conservation which is found to depend on the complexity of the flow field. Difficulties with global scalar mass conservation are related to local fluid mass errors. The results stress the importance to transport modeling of using hydrodynamic schemes which conserve fluid mass on both a local and global basis.

Introduction

Circulation and transport in coastal and estuarine waters presents many challenges to accurate and efficient numerical modeling due to strong spatial and temporal gradients in velocity, salinity, temperature, and other scalars. Simple scaling of the governing equations of mass, momentum, and scalar transport shows that in contrast to flows on the outer continental shelf or open ocean, coastal and estuarine flows are often advection dominant -- a condition well known to cause numerical difficulties (e. g. Hughes (ed.), 1979). Moreover, small scale flow features (<1 km) are often of interest and require correspondingly small grid spacing (<100 m) which places a high demand on computer storage and computational speed. Numerical schemes which perform satisfactorily on the relatively quiescent waters of the outer continental shelf may not be robust enough for inshore applications and more accurate but potentially more expensive alternatives are required.

[1]Institute of Marine Sciences, University of North Carolina at Chapel Hill
3431 Arendell Street, Morehead City, NC 28557

In this paper we use Fourier analysis to compare various Eulerian finite element schemes for the transport equation with the goal of identifying schemes which will be the most useful for solving difficult coastal and estuarine applications. We select one scheme for a field application along the North Carolina coast. The results of this application are analyzed for global scalar mass conservation. Some of the difficulties with scalar mass conservation appear to be explained by errors in local fluid mass conservation.

Fourier analysis of Eulerian finite element methods

Within the context of Eulerian finite element methods there are a wide range of discretization schemes available in the numerical methods literature. The majority of the schemes remain untested on field applications, often due to their complexity, inefficiency on large problems, or from assumptions in their derivation which confine them to regular grids and uniform flow fields. Given a wide range of possible numerical methods, a natural question to ask is what is the comparative benefit of using a more "exotic" numerical scheme versus a simpler but less accurate scheme. To begin to address this question, we compare five schemes of varying complexity and accuracy using Fourier analysis: Crank-Nicolson Galerkin (CNG), Streamline Crank-Nicolson Galerkin (SCNG), Streamline Upwind / Petrov Galerkin (SUPG), N+2 Petrov Galerkin (N+2PG), and Crank-Nicolson Quadratic Galerkin (CNG-Q).

The first is a conventional Bubnov Galerkin scheme which uses linear elements for the spatial discretization. Time derivatives are treated with a Crank-Nicolson approach in which the convective and diffusive terms are centered between two time levels. We designate this scheme as Crank-Nicolson Galerkin (CNG). The scheme is second order in time but is well known to be prone to oscillations in convection dominated flows.

The second scheme is the same as CNG but with a modification that adds artificial diffusion. Recent experimental (Noorishad et al., 1992) and theoretical work (Perrochet and Berod, 1993) has shown that non-oscillatory solutions to the transport equation using a modified CNG scheme can be obtained by adding an optimal amount of diffusion using criteria related to the local Courant (Cr) and Peclet (Pe) numbers. Specifically, it can be shown that for linear elements, numerical oscillations can be avoided when $PeCr \leq \gamma$ is satisfied everywhere. The parameter γ is a performance index and when $\gamma \leq 2$ the scheme will be non-oscillatory. Our experience indicates that acceptable results can be obtained with γ as high as ten. With some manipulation, the horizontal diffusion tensor can be specified such that a minimal amount of diffusion can be introduced to ensure numerical stability. Since diffusion is added to the solution only in the direction of the flow, we designate this scheme as Streamline CNG (SCNG).

Third is the Streamline Upwind / Petrov-Galerkin (SUPG) scheme (Brooks and Hughes, 1982; Hughes and Brooks, 1982) which uses linear basis functions and weight functions which are perturbed by a polynomial which is a derivative of the basis functions. In this way, artificial diffusion is added only in the direction of the flow, much in the same way as for the SCNG scheme. However, we expect the SUPG scheme to produce more accurate results than SCNG since it also introduces a modification to the mass matrix and thereby improves the time discretization.

Fourth is the N+2 Petrov Galerkin scheme developed by Westerink and Shea (1989) and extended to bilinear quadrilaterals by Cantekin and Westerink (1990). Unlike other Petrov Galerkin schemes which introduce numerical diffusion (such as SUPG), this scheme is completely nondiffusive. The method uses weight functions perturbed by a polynomial two degrees higher than the basis functions and as such is designated a N+2 Petrov Galerkin scheme (N+2PG).

Finally, we analyze a conventional CNG scheme but with quadradic elements (CNG-Q) to use as an upper bound of the capability of a conventional, albeit expensive, Eulerian method.

We compare the amplitude and phase characteristic of the five schemes using Fourier analysis (e.g. Pinder and Gray, 1977). Figures 1a,b show comparative phase and amplitude portraits at Courant number equal to 0.1 and infinite Peclet number. [The amplitude and phase portraits at smaller Peclet numbers are quite similar as long as the flow is in the convection dominated range.] The phase portrait, Figure 1a, shows that all schemes propagate long wavelength components of the solution with very little phase error. However, none of the schemes accurately propagate wavelengths near $2\Delta x$. At intermediate wavelengths the schemes exhibit differing phase errors. The CNG-Q exhibits the best phase characteristics with significant phase errors only at the shortest wavelengths. The N+2PG scheme has the next smallest phase error, followed by the CNG, SCNG, and SUPG schemes which exhibit nearly identical phase portraits.

Amplitude portraits give a measure of the damping a numerical scheme introduces into the solution. Figure 1b shows the comparison for the low Cr case. The CNG, N+2PG, and CNG-Q all exhibit perfect damping (i.e. they do not introduce numerical diffusion into the solution). SUPG shows selective damping of short wavelength components and very little damping at long wavelengths. The SCNG scheme shows similar damping to SUPG for short wavelengths, but is less selective since it shows a small amount of damping at long wavelengths as well.

The comparison is somewhat different at higher Cr as shown in Figures 1c,d. The N+2PG scheme exhibits the best phase characteristics and in fact has actually improved over the low Cr case. In contrast the other four schemes show increased phase errors with increased Cr. The CNG-Q shows the next smallest phase errors, but at higher Cr its relative benefit over CNG, SCNG, and SUPG is reduced. The high Cr amplitude portrait (Figure 1d) again shows no damping for the CNG, N+2PG and CNG-Q schemes. The SUPG scheme exhibits increased damping at the short and intermediate wavelengths but still very little at long wavelengths. The SCNG also exhibits increased damping but is much less selective with even the $100\Delta x$ wave being somewhat damped.

Based on the results of the Fourier analysis, one would intuitively select the CNG-Q at low Cr and the N+2PG at high Cr since they have the smallest phase errors and no artificial diffusion. However in field applications we have found these two schemes to be unsatisfactorily oscillatory. Despite the much better phase characteristics, oscillations persist because of the presence of short wavelength components that are not propagated correctly. Since the short wavelength components are undamped, they continue to propagate at incorrect phase speeds. With constructive and destructive superposition, oscillations are seen following the signal. Gresho and Lee (1981) suggest that these oscillations are indicative of poor grid resolution and identify places where grid refinement is needed.

We concur with Gresho and Lee that heavy-handed application of artificial diffusion to dampen oscillations can lead to misleading or even useless results. However in estuarine and coastal field applications with sharp concentration gradients, we find it most practical to apply a small amount of selective damping to quell oscillations from the shortest wavelengths, but not damp intermediate and long wavelength components which are of primary interest. By taking the fall-back position of using some selective damping, we see that SUPG and SCNG provide this capability. SUPG is particularly attractive since it exhibits a close match between phase errors and damping. That is, wavenumbers with large phase errors are removed from the solution by damping before they can cause troublesome oscillations. At low Cr (i.e. $O(0.1)$), SCNG is also a reasonable choice.

Two-dimensional implementation

With a knowledge of how the various schemes can be expected to perform from Fourier analysis, we now move to a two-dimensional model in order to study transport processes in the coastal ocean. We solve the depth-integrated form of the transport equation expressed here in conservative form (e.g. Pinder and Gray, 1977):

$$\frac{\partial HC}{\partial t} + \frac{\partial}{\partial x}(UHC) + \frac{\partial}{\partial y}(VHC) = \frac{\partial}{\partial x}\left(D_{xx}H\frac{\partial C}{\partial x}\right) + \frac{\partial}{\partial x}\left(D_{xy}H\frac{\partial C}{\partial y}\right) + \frac{\partial}{\partial y}\left(D_{yx}H\frac{\partial C}{\partial x}\right) + \frac{\partial}{\partial y}\left(D_{yy}H\frac{\partial C}{\partial y}\right) \quad [1]$$

Where $H = \zeta + h$ is the total water column, U and V are depth-integrated velocities, C is the depth-integrated scalar concentration, and D_{ij} is the diffusivity tensor. Flow fields are obtained using the two-dimensional depth-integrated version of ADCIRC (Luettich, et al., 1992; Grenier, et al., 1995) which solves the fully nonlinear shallow water equations using a wave-equation formulation and a Galerkin finite element method. ADCIRC flow fields are harmonically analyzed to extract primary tidal constituents, overtides, and tidal residuals. The transport model uses these precomputed harmonic constituents to specify the flow field. Decoupling the transport model from the hydrodynamic model in this way is convenient since it saves the computational overhead of recomputing the hydrodynamics each time a scalar transport simulation is run.

We have selected the SCNG method for our two-dimensional implementation based on the results of our Fourier analysis. As noted above, this scheme is a reasonable choice at low Courant numbers since it has similar phase characteristics as CNG and SUPG and includes some selective damping. We confine our simulations with this method to low Cr, $O(0.1)$ or less.

The SCNG scheme uses simple C^0 basis and weight functions which can be integrated efficiently with analytic formulae. The amount of artificial diffusion to be added is computed from local flow variables and added into the diffusivity tensor as:

$$D_{ij} = \frac{|u_i u_j|\Delta t}{\gamma} \quad [2]$$

The SCNG scheme yields a sparse nonsymmetric system matrix which requires re-assembly each time step. Solving this global system of equations with a direct solver would require costly LU decomposition and back substitution at each time step which would be prohibitively expensive for large field applications. Instead, we solve the system of equations using a preconditioned Krylov subspace iterative solver, Bi-CGSTAB (van der Vorst, 1992) in conjunction with a compact storage scheme. We have found this solver to be very efficient for a system of over 30,000 equations with convergence typically in 2-6 iterations.

Application to inner shelf and tidal inlet

In this section we apply our two-dimensional transport code to a convection dominated field application. The model domain comprises the contiguous shelf, coastal and estuarine waters of North Carolina (Figure 2a). The grid has been refined in Beaufort Inlet (with resolutions down to 25 meters) while still including a realistic representation of the surrounding sounds, barrier islands, as well as the coastal ocean (Figure 2b). Model bathymetry (Figures 2c,d) varies from 2 meters in the estuarine areas to over 5000 meters off the continental shelf. The inlet has steep bathymetric gradients primarily due to a 12 meter deep navigation channel which splits into several shallower sloughs. The flow field was computed from a barotropic simulation with M_2 tidal forcing and has been has been verified for water level and velocity (see Luettich, et al., 1998). Maximum flow speeds in the vicinity of Beaufort Inlet are in excess of 1 m/s.

Figure 3 shows the transport of a passive scalar Gauss plume released in the vicinity of the inlet. The rapid and severe distortion of the plume is due to the strong horizontal velocity gradients near the inlet throat. Even under these difficult flow conditions, the concentration field does not exhibit the oscillatory behavior of nondiffusive schemes. Using the SCNG scheme, diffusivities are computed as a function of time and space and act only in the direction of the flow. For this application, diffusivity values range from near zero to 5 m^2/sec. The SCNG scheme appears to be effective in suppressing numerical oscillations, albeit at the cost of artificially smearing some of the fronts.

Scalar mass conservation

Global scalar mass conservation is of utmost importance in long-term simulations of transport processes. We conducted a set of model runs with different model formulations and initial conditions to test this property. Initial conditions were specified as a Gaussian plume starting on the continental shelf or in front of Beaufort Inlet (Figure 4). Although the spatial extent of the plumes is quite different, they both have about the same grid resolution (since the mesh is graded) with approximately 30 nodes across the breadth of each plume.

To test our code's mass conservation properties we calculate the total mass in the domain at selected times:

$$Mass_{global} = \sum_{ne} \left(\int_{\Omega_e} HC d\Omega_e \right) \qquad [3]$$

Figure 5a shows global scalar mass time series for the first two tidal cycles of the simulation for the continental shelf plume. Both the conservative and nonconservative forms of the transport equation conserved mass within one percent. Figure 5b shows a similar plot for the inlet plume. Here mass is conserved within only 30 to 60 percent and there is a marked difference between the conservative and nonconservative forms.

Relationship between fluid and scalar mass conservation

We suspect that much of the problem conserving scalar mass is tied to difficulties in conserving fluid mass. By applying the product rule to the transient and convective terms in equation [1], the conservative form of the transport equation can be equivalently rewritten as:

$$\frac{\partial C}{\partial t} + U\frac{\partial C}{\partial x} + V\frac{\partial C}{\partial y} = \frac{1}{H}\left[\frac{\partial}{\partial x}\left(D_{xx}H\frac{\partial C}{\partial x}\right) + \frac{\partial}{\partial x}\left(D_{xy}H\frac{\partial C}{\partial y}\right) + \frac{\partial}{\partial y}\left(D_{yx}H\frac{\partial C}{\partial x}\right) + \frac{\partial}{\partial y}\left(D_{yy}H\frac{\partial C}{\partial y}\right)\right] - RC \qquad [4]$$

where:

$$R = \frac{1}{H}\left(\frac{\partial H}{\partial t} + \frac{\partial UH}{\partial x} + \frac{\partial VH}{\partial y}\right) \qquad [5]$$

The right hand side of [5] is immediately recognizable as the primitive depth-integrated continuity equation multiplied by $1/H$. For a truly conservative flow field, this term will be identically zero. However, in field applications the flow field is typically specified by a hydrodynamic model with imperfect fluid mass conservation. These fluid mass errors are manifest as a residual fluid mass R.

From equations [4] and [5], it can be seen that errors in the hydrodynamic model are passed directly down to the transport model in the form of a first order reaction term. In equation [4] the residual fluid mass, R can be interpreted as a "reaction rate" coefficient in the source/sink term. When there is a deficit in fluid mass, R is negative and scalar mass is added to compensate due to a source-like term. Conversely, a surplus in fluid mass continuity yields a positive R, and a sink-like term for scalar mass. Of course given a flow field with perfect fluid mass conservation the conservative transport equation will reduce to the familiar nonconservative form.

When a wave equation formulation finite element method is applied to solve the shallow water equations, it can be shown that on a global basis mass is conserved to a very high degree of precision, provided proper boundary conditions are used (Lynch, 1985; Kolar et al., 1994; Lynch and Holboke, 1997). However on a local basis there is no such guarantee for these methods.

Local fluid mass conservation

The results from the previous section suggest that errors in scalar mass conservation can arise from errors in fluid mass conservation. To examine errors in local fluid mass conservation we integrate equation [5] over each element in the grid and in time:

$$R_e = \int_{t_1}^{t_2} \left\{ \int_{\Omega_e} \left[\frac{1}{H} \left(\frac{\partial H}{\partial t} + \frac{\partial UH}{\partial x} + \frac{\partial VH}{\partial y} \right) \right] d\Omega_e \right\} dt \qquad [6]$$

Known values of U, V, and H are expanded over each element using standard linear basis functions. The transient term is approximated using a forward finite difference. To complete the integration a time scale must be selected to integrate over. Here, we integrate over a single model time-step which is five seconds. The residual from this integration gives a measure of the local error in fluid mass conservation per time-step. The elemental fluid residual is then normalized by the element area, giving a nondimensional measure of fluid mass error:

$$E_e = R_e / Area_e \qquad [7]$$

Figure 6 shows fluid mass error contours during maximum ebb in the vicinity of Beaufort Inlet. The largest errors coincide with areas of abrupt changes in shoreline geometry, and areas with steep bathymetric gradients. Large errors are also evident in flow separation regions near headlands. These findings are consistent with those of Oliveira et al. (in review). The spatial pattern of errors shows local regions of excess fluid mass immediately adjacent to local regions of a deficit in fluid mass. The scale of these adjacent areas is roughly that of the length scale of the topographic features and span several elements; the errors do not appear to be $2\Delta x$ type oscillations. The magnitude of the fluid mass errors is of the order 0.1 percent near the inlet with values up to one percent in localized areas. Local errors offshore are substantially smaller. The magnitude of these errors should be considered in the context of a Lagrangian water parcel advecting through a series of elements. That is, the error that a fluid parcel experiences is proportional to the amount of time it spends within each element and is accumulative as it moves through the domain. Figure 7 shows a similar plot at maximum flood. By comparing Figures 6 and 7 one can see that local fluid mass errors are highly transient with the same element having a mass surplus at part of the tidal cycle and a mass deficit later in the tidal cycle.

Discussion and conclusion

We have compared various finite element schemes using Fourier analysis. Schemes with no damping appear to be too oscillatory for convection dominated field applications. For field applications optimal schemes are those with a close match between phase errors and selective damping. Of the schemes analyzed, SUPG and SCNG (only for low Cr) exhibit these qualities.

Our field application has shown that global scalar mass conservation depends strongly on the complexity of the flow. On the continental shelf where bathymetric and velocity gradients are small we find that mass conservation is acceptable for many transport calculations. Near a tidal inlet where bathymetric and velocity gradients are large we obtain poor global scalar mass conservation. The large discrepancy between the conservative and nonconservative forms of the transport equation points to errors in fluid mass conservation. While global fluid mass conservation is excellent, our analysis of local fluid mass errors shows these errors can be significant.

These results suggest that substantial improvements in transport modeling will be gained by using mass conserving schemes for the hydrodynamics. Several new techniques have emerged for enforcing local fluid mass conservation including finite volume methods (e.g. Chippada, et al., 1998) as well as post-processing corrections to finite element methods (e.g. Carey, et al., 1997).

Acknowledgments

The simulations in this paper were performed using the IBM SP2 at the Department of Defense MSRC at CEWES. Funding was provided by NOAA, National Marine Fisheries Service; Office of Naval Research, Naval Research Laboratory; and U.S. Army Corps of Engineers, Coastal Hydraulics Laboratory.

References

Brooks, A. N., and T. J. R. Hughes, 1982. Streamline upwind / Petrov-Galerkin formulations for convection dominated flows with particular emphasis on the incompressible Navier-Stokes equations. *Computer Methods in Applied Mechanics and Engineering*, 32: 199-259.

Cantekin, M. E., and J. J. Westerink, 1990. Non-diffusive N+2 degree Petrov-Galerkin methods for two-dimensional transient transport computations. *International Journal for Numerical Methods in Engineering*, 30: 397-418.

Carey, G. F., G. Bicken, C. Berger, and V. Carey, 1997. Constrained finite element projections for environmental flows. *Fourth SIAM Conference on Mathematical and Computational Issues in the Geosciences, Final Program and Abstracts*, Albuquerque, NM, 69-70.

Chippada, S., C. N. Dawson, M. L. Martinez, and M. F. Wheeler, 1998. A Godunov-type finite volume method for the system of shallow water equations. *Computer Methods in Applied Mechanics and Engineering*, 151: 105-129.

Grenier, R. R., R. A. Luettich, and J. J. Westerink, 1995. A comparison of the nonlinear frictional characteristics of two-dimensional and three-dimensional models of shallow tidal embayments. *Journal of Geophysical Research*, 100: 13,719-13,735.

Gresho, P. M., and R. L. Lee, 1981. Don't suppress the wiggles - they're telling you something! *Computers and Fluids*, 9: 223-253.

Hughes, T. J. R. (ed.), 1979. *Finite Element Methods for Convection Dominated Flows*. AMD - Vol. 34, American Society of Mechanical Engineers, New York, 227 pages.

Hughes, T. J. R., and A. N. Brooks, 1982. A theoretical framework for Petrov-Galerkin methods with discontinuous weighting functions: application to the streamline upwind procedure. In: *Finite Elements in Fluids, Vol. IV*, R. H. Gallagher, et al. (eds.), Wiley, Chichester, 47-65.

Kolar, R. L., J. J. Westerink, M. E. Cantekin, and C. A. Blain, 1994. Aspects of nonlinear simulations using shallow water models based on the wave continuity equation. *Computers and Fluids*, 23: 523-538.

Luettich, R. A., J. L. Hench, C. D. Williams, B. O. Blanton, and F. E. Werner, 1998. Modeling circulation and transport through a barrier island inlet. *Proceedings of the 5th International Conference on Estuarine and Coastal Modeling*, A. F. Blumberg and M. L. Spaulding (eds.), American Society of Civil Engineers.

Luettich, R. A., J. J. Westerink, N. W. Scheffner, 1992. ADCIRC: An advanced three-dimensional model for shelves, coasts, and estuaries, Report 1: Theory and methodology of ADCIRC 2DDI and ADCIRC 3-DL. *Dredging Research Program Technical Report DRP-92-6*, Coastal Engineering Research Center, U. S. Army Corps of Engineers, Waterways Experiment Station, Vicksburg, MS, 141 pages.

Lynch, D. R., 1985. Mass balance in shallow water simulations. *Communications in Applied Numerical Methods*, 1: 153-159.

Lynch, D. R., and M. J. Holboke, 1997. Normal flow boundary conditions in 3D circulation models. *International Journal for Numerical Methods in Fluids*, 25: 1-21.

Noorishad, J., C. F. Tsang, P. Perrochet, and A. Musy, 1992. A perspective on the numerical solution of convection-dominated transport problems: A price to pay for the easy way out. *Water Resources Research*, 28: 551-561.

Oliveira, A., A. M. Baptista, and A. B. Fortunato, in review. Local mass conservation in finite element shallow water models. *Communications in Numerical Methods in Engineering*.

Pinder, G. F., and W. G. Gray, 1977. *Finite Element Simulation in Surface and Subsurface Hydrology*, Academic Press, San Diego, 295 pages.

Perrochet, P., and D. Berod, 1993. Stability of the standard Crank-Nicolson-Galerkin scheme applied to the diffusion-convection equation: Some new insights. *Water Resources Research*, 29: 3291-3297.

van der Vorst, H. A., 1992. Bi-CGSTAB: a fast and smoothly converging variant of Bi-CG for the solution of nonsymmetric linear systems. *SIAM Journal on Scientific and Statistical Computing*, 13: 631-644.

Westerink, J. J., and D. Shea, 1989. Consistent higher degree Petrov-Galerkin methods for the solution of the transient convection-diffusion equation. *International Journal for Numerical Methods in Engineering*, 28: 1077-1101.

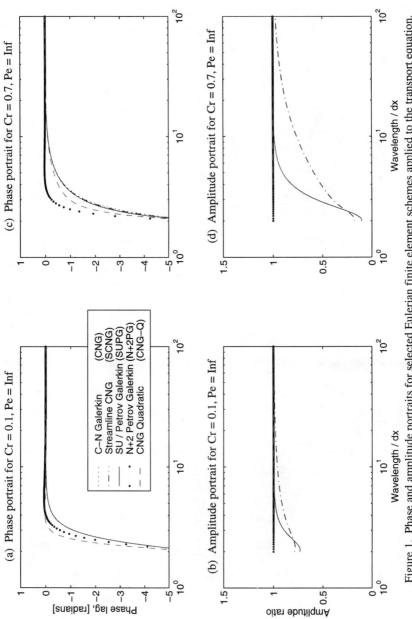

Figure 1. Phase and amplitude portraits for selected Eulerian finite element schemes applied to the transport equation.

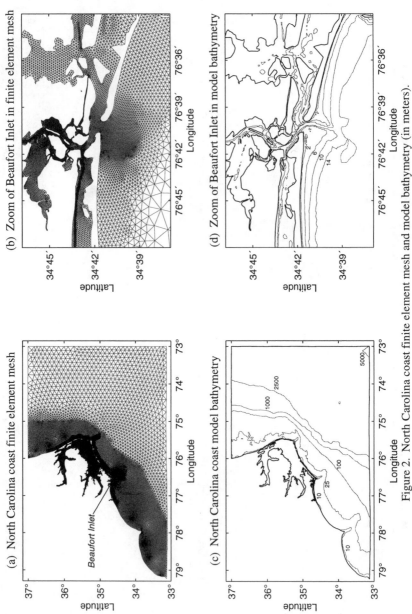

(a) North Carolina coast finite element mesh

(b) Zoom of Beaufort Inlet in finite element mesh

(c) North Carolina coast model bathymetry

(d) Zoom of Beaufort Inlet in model bathymetry

Figure 2. North Carolina coast finite element mesh and model bathymetry (in meters).

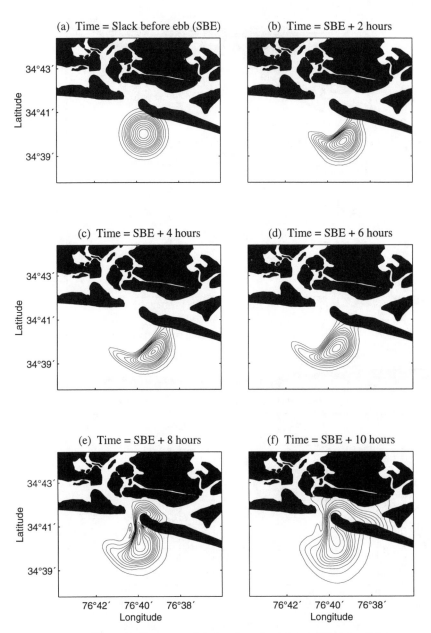

Figure 3. Transport of scalar plume in vicinity of Beaufort Inlet.

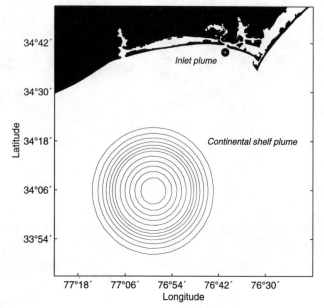

Figure 4. Initial conditions for scalar mass conservation experiments.

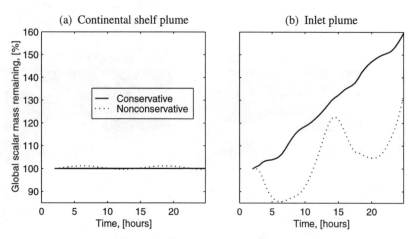

Figure 5. Global scalar mass conservation.

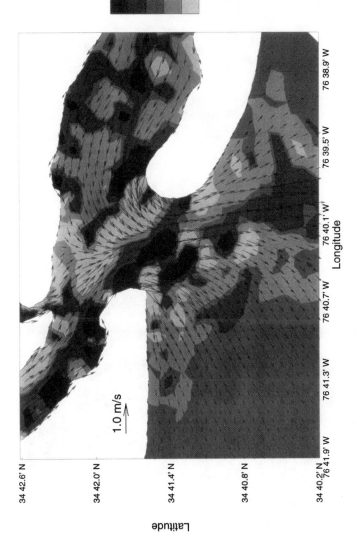

Figure 6. Local fluid mass errors in vicinity of Beaufort Inlet at maximum ebb.

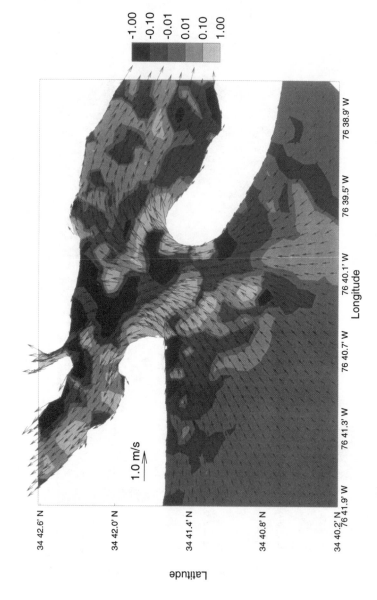

Figure 7. Local fluid mass errors in vicinity of Beaufort Inlet at maximum flood.

Interfacing Hydrodynamic and Water Quality Models with the Eulerian-Lagrangian Method

Ling Tang and E. Eric Adams

R.M. Parsons Laboratory
Dept. Civil and Environmental Engineering
Massachusetts Institute of Technology
Cambridge, MA 02139

Abstract

For a coastal water quality modeling project, we need two kinds of models, the water quality model and its driving hydrodynamic model. Because of the significant difference between time steps, and possibly space steps of these two models, an interfacing strategy has an important role for the efficiency of a water quality evaluation project. However, most dominant water quality models used to-date are Eulerian models requiring relatively small time steps, which detracts from their efficiency for long term simulations. In this paper, we present an alternative interfacing approach based on an Eulerian-Lagrangian method, which admits fairly large time steps. The method is applied to Massachusetts Bay, where 3-D calculations of a conservative tracer simulated by both the hydrodynamic model and the water quality model are compared. The results show that the ELM is promising for long term water quality models if combined with an appropriate mass correction technique. The concerns of mass conservation are briefly discussed and a total mass scaling technique is implemented.

1 Introduction

In coastal areas, we generally need two kinds of models for water quality evaluation. One is the water quality model (WQM), and the other is the driving hydrodynamic model (HM). For most hydrodynamic models, the Courant condition restricts the numerical time step in the range of several minutes to resolve free surface gravity wave motion. However, for a water quality model, the limiting physical, chemical or biological time scales may be much larger: e.g., a tidal period or a day. This suggests that the ideal time step of a water quality model should be around the same order. The significant difference of time scales between a HM and a WQM calls for an interfacing strategy that optimizes the use of the somewhat more accurate physical transport description from a HM. Furthermore, in long term water quality simulations, mixing tends to smooth the distribution of concentrations away from contaminant sources. Thus, it may be possible for a WQM to use a much coarser numerical grid than that in a HM in those areas. Therefore, a flexible interfacing strategy should allow a large time step for WQM (up to semi-diurnal or diurnal tidal periods in coastal areas) and coarser grids to reduce the cost of long term simulation both in computing time and disk storage space (Hamrick, 1993). The other important criterion is that the interfacing should be universally applicable, i.e., independent of the degree of nonlinearity of a tidal system. Since a WQM must be driven by a HM, then how to couple them efficiently is critical. Here we consider two general classes of WQMs, Eulerian models (EM) and Eulerian-Lagrangian models (ELM), with corresponding interfacing strategies, EM coupling or ELM coupling respectively.

2 Eulerian Coupling

The advantage of EM coupling, based on control volume grid, is its guaranteed mass conservation. This is the major reason for the prevalence of this class of models in real water quality applications. Unfortunately, the advective Courant number restriction may limit the time step to as little as several minutes depending on the magnitude of instantaneous current velocities. A small time step of a WQM reduces the efficiency of long term simulations, especially for multiple constituents. In the framework of Eulerian Methods, one way to increase the time step is to reduce the magnitude of the transport velocity u through inter-tidal average.

In inter-tidal averaging, any instantaneous variable ϕ can be decomposed into the tidal-averaged and the tidally varying components $\overline{\phi}$ and ϕ', i.e.,

$$\phi = \overline{\phi} + \phi' \tag{1}$$

where

$$\overline{\phi} = \frac{1}{T} \int_{t_0}^{t_0+T} \phi dt \tag{2}$$

$$\overline{\phi'} = 0 \tag{3}$$

Here, T is the averaging period, usually the tidal period. The transport equation then becomes

$$\frac{\partial \bar{c}}{\partial t} + \frac{\partial \bar{u}_i \bar{c}}{\partial x_i} + \frac{\partial \overline{u'_i c'}}{\partial x_i} = \frac{\partial}{\partial x_i}(\overline{D_{ij}} \frac{\partial \bar{c}}{\partial x_j}) \tag{4}$$

where $\overline{D_{ij}}$ is the tidally averaged turbulent eddy/sub-grid scale diffusion coefficient. After averaging, the tidally averaged velocity $\bar{u}_i$ (actually Eulerian averaged velocity $\overline{u_{Ei}}$) is much smaller than the instantaneous one. Therefore, the time step can be greatly enhanced in WQM. Early models used the idea of tidal dispersion stemming from Tayler's shear dispersion theory, where the term $\overline{u'_i c'}$ was defined as $-D_{ij}^T \frac{\partial \bar{c}}{\partial x_j}$, where D_{ij}^T is called the tidal dispersion coefficient. The averaged transport equation thus has the form

$$\frac{\partial \bar{c}}{\partial t} + \frac{\partial \bar{u}_{Ei} \bar{c}}{\partial x_i} = \frac{\partial}{\partial x_i}[(\overline{D_{ij}} + D_{ij}^T)\frac{\partial \bar{c}}{\partial x_j}] \tag{5}$$

where $\overline{u_{Ei}}$ is Eulerian residual velocity.

This method was extensively used in early WQMs. Its drawback, however, is the large variability in the range of observed or fitted tidal dispersion coefficients which can mask the more physically based turbulence diffusion and shear dispersion.

It was later recognized to be more natural to introduce the concept of the Lagrangian tidal residual velocity (Cheng and Casulli 1982, Feng et al 1986a,b), or

$$\frac{\partial \bar{c}}{\partial t} + \frac{\partial \bar{u}_L \bar{c}}{\partial x_i} = \frac{\partial}{\partial x_i}(\overline{D_{ij}} \frac{\partial \bar{c}}{\partial x_j}) \tag{6}$$

where $\overline{u_L}$ is Lagrangian residual velocity.

In regions where the tidal excursion is small compared to the scale of spatial variability in the flow, i.e., $\kappa << 1$, there exists an asymptotic relation between Eulerian residual and the Lagrangian motion. Feng (1986a,b) showed that for $\kappa << 1$ ($\kappa = \frac{\zeta_c}{h_c} = \frac{l_c}{L_c}$), the Lagrangian residual velocity u_L can be expressed as the sum of a series of perturbation terms

$$u_L = u_E + u_S + \kappa u_{LD} + H.O.T \tag{7}$$

where ζ_c is the tidal amplitude, h_c is the mean water depth, l_c is the tidal excursion and L_c is the scale of spatial variability. u_E is the Eulerian residual velocity, u_S is the Stokes drift and u_{LD} is the Lagrangian drift velocity.

It is clear that the method of Stokes drift correction allows a significantly enhanced WQM time step and the approach has been used successfully on the Chesepeake Bay simulations (Dortch, 1990). However, because it is based on a weakly non-linear tidal system, it cannot be applied to more strongly non-linear systems such as San Francisco Bay ($\kappa \approx 0.3$), and the English Channel ($\kappa \approx 0.25$). In general, it is quite complicated to explicitly relate the Lagrangian residual velocity to the corresponding Eulerian residual velocity, because of the non-linear relation between them. The Lagrangian velocity is a function of the particle trajectory and the trajectory is, in turn, a function of the Lagrangian velocity. This can be shown by the following expression:

$$u_L(\mathbf{x_0}, t) = u_E[\mathbf{x'}(\mathbf{x_0}, t)] \tag{8}$$

where $u_E(\mathbf{X}, t)$ is the Eulerian velocity, $u_L(\mathbf{x_0}, t)$ is the Lagrangian velocity of a particle moving from $\mathbf{x_0}$, and $\mathbf{x'}$ is the trajectory defined by

$$\mathbf{x'}(\mathbf{x_0}, t) = \mathbf{x_0} + \int_0^t u_L(\mathbf{x_0}, t') dt' \tag{9}$$

3 Eulerian-Lagrangian Coupling and Mass Conservation

ELMs split the non-conservative transport equation into two parts, the pure advection problem and the dispersion problem, governed by

$$\frac{\partial c}{\partial t} + \mathbf{u} \cdot \nabla c = 0 \tag{10}$$

and

$$\frac{\partial c}{\partial t} = \nabla \cdot (\mathbf{D} \nabla c) \tag{11}$$

For the dispersion part (11), a time implicit numerical scheme is used so that the time step Δt is not restricted by stability consideration (Baptista et al, 1984). For example, the tidal period, $\Delta t = T$, could be used. This will significantly reduce the computation expense, making the cost of long term simulation with multiple constituents more reasonable.

In typical ELM applications, the advection step is accomplished by backward tracking of characteristic lines emanating from each node, followed by concentration interpolation. The tracking part is defined by an ODE

$$\frac{d\mathbf{x}}{dt} = \mathbf{u} \tag{12}$$

and there are many standard numerical methods to solve it, such as Runge-Kutta and Euler method. Once the feet are located, the concentration can be interpolated from values at the previous time using a choice of interpolators. Considering the large time step in the dispersion problem, the tracking is naturally divided into several smaller steps δt associated with the hydrodynamic model. The backward tracking can be done after each time step Δt, by a subroutine in the hydrodynamic model. Of course, this means that the velocity field during the period Δt needs to be stored. This requires extra storage space, but the disk quota for this part is small compared with the space needed to store output for the whole simulation process. Furthermore, the space can be reused during the next Δt. See Figure 1.

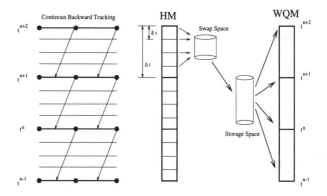

Figure 1: A Sketch for Backward Tracking and Storage Swapping

In ELM, there is no approximation to get a Lagrangian description of the transport. The backward particle tracking, which is independent of the degree of the nonlinearity of the tidal system, directly records this information without temporal averaging. Furthermore, the tracking needs to be done only once. The concentrations of different constituents can be interpolated from a single corresponding foot.

If we wish to aggregate a fine HM grid to get a coarser one for the WQM, allowing for further saving on CPU time and disk space, this can be easily implemented in ELM. We can simply pick the coarser nodal points and do tracking to get the corresponding feet. By contrast, in an EM, we need to at least average the fluxes on the basis of the fine grid to get the equivalent flux for the coarser grid.

ELM coupling is obviously quite flexible. However, the major drawback in conventional practice is the lack of absolute mass-conservation. This is not because of the scheme itself, but because most ELMs uses non-conservative transport equations. The following 1-D example shows that an ELM can actually conserve mass identically if it solves a conservative transport equation.

Assume a pure advection model as

$$\frac{\partial c}{\partial t} + \frac{\partial uc}{\partial x} = 0 \tag{13}$$

and its non-conservative form

$$\frac{\partial c}{\partial t} + u\frac{\partial c}{\partial x} = 0 \tag{14}$$

Using a staggered grid first-order upwind explicit Eulerian scheme based on (13) gives

$$c_i^{n+1} = (1 - \alpha_i)c_i^n + \alpha_{i-1}c_{i-1}^n \tag{15}$$

while its ELM counterpart based on (14) gives

$$c_i^{n+1} = (1 - \alpha_i)c_i^n + \alpha_i c_{i-1}^n \tag{16}$$

where $\alpha_i = u_i^n \Delta t/\Delta x$, $\alpha_{i-1} = u_{i-1}^n \Delta t/\Delta x$. The expression (16) implicitly assumes that the Courant number α is less than 1 just for purposes of comparing with the EM. (In general, the interpolation in an ELM should be taken between the nodes that bound the foot of each characteristic line, so that the restriction of CFL condition is alleviated.)

It is well-known that (15) conserves mass and (16) doesn't if the divergence of the flow field is not zero which, in this case, means u is not a constant. The difference between (15) and (16) is that the former one is consistent with the flux advection, i.e., the velocity nodal index is always the same as its concentration index, while the latter is not, which allows a hole for mass imbalance. The fractional mass change can be measured by

$$\frac{u_i^n - u_{i-1}^n}{\Delta x}\Delta t$$

which is the discretized form of flow divergence during period Δt period.

If the flow field u reaches steady state, so that

$$\frac{\partial u}{\partial t} = 0$$

we can multiply (13) by u and get

$$u\frac{\partial c}{\partial t} + u\frac{\partial uc}{\partial x} = 0 \tag{17}$$

That is

$$\frac{D\,u\,c}{Dt} = 0 \tag{18}$$

Then following the ELM strategy, we have

$$u_i c_i^{n+1} = (1 - \alpha_i)u_i c_i^n + \alpha_i u_{i-1} c_{i-1}^n \tag{19}$$

This gives

$$c_i^{n+1} = (1 - \alpha_i)c_i^n + \alpha_{i-1} c_{i-1}^n \tag{20}$$

which is exactly the EM scheme (15). This shows that an ELM can conserve mass if a conservative form of the transport equation is used. However, an ELM has to use a non-conservative form of the transport equation in most cases.

The trade-off of accuracy between mass and concentration are discussed in King and DeGeorge (1995) and Tang and Adams (1995). In this study, the technique of total mass scaling is implemented to correct for any mass imbalance. That is

$$c(x, t^{n+1}) = \frac{\int_{\partial V} c(x, t^n)dV}{\int_{\partial V} c^*(x, t^{n+1})dV} c^*(x, t^{n+1}) \tag{21}$$

where $c^*(x, t^{n+1})$ is the value obtained from the traditional ELM.

4 Case Study using Scaling ELM

The scaling ELM strategy was applied to Massachusetts Bay where previous HM (Signell et al, 1996) and WQM simulations (HydroQual,Inc., 1995) were performed to assess the potential impacts of Boston's re-located wastewater outfall. Our example calculations use the month of January 1992, and Table 1 summarizes the simulation parameters for the two models.

The 3-D HM is ECOMsi (Blumberg and Mellor, 1987) which is applied to the HM grid shown in Figure 2. The model uses a curvi-linear coordinate system with 68x68 cells in the horizontal and 12 sigma layers in the vertical. Meteorological conditions on the water surface, plus tidal and hydrographic conditions on the outer boundaries, were obtained from measurements and the output of a larger scale model, and were provided to us by R. Signell

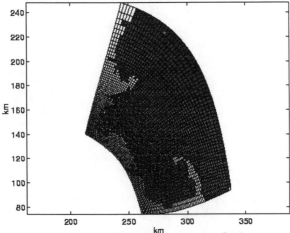
Figure 2: ECOMsi Computation Grid

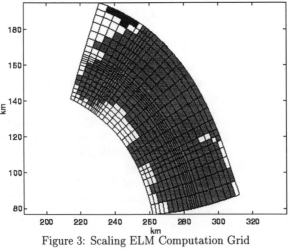
Figure 3: Scaling ELM Computation Grid

(personal communication). River water enters into Boston Harbor on the western side of the domain, and wastewater enters the domain, at an average rate of $18m^3/s$, at the outfall location approximately 15 km NE of Boston (HM grid location 30,47). We inserted a backward tracking routing into ECOMsi to derive the feet of the characteristic lines emanating from user-specified nodes used for subsequent WQM calculations. The tracking routine uses second order Runge-Kutta integration with the 7 minute HM velocities. We also save tidal-averaged horizontal and vertical diffusivities from the HM for use in the WQM. In order to provide a reference concentration for comparison with the WQM simulations, a conservative tracer, with arbitrary initial concentration of 100, was added to the wastewater inflow.

Our companion WQM model was applied to the 3-D grid shown in Figure 3, which has 24x29 grids in the horizontal. The grid is the same as used by HydroQual(1995) and was developed from the HM grid by eliminating the eastern portion of the domain, collapsing most horizontal grids by a factor of three in both directions (but retaining the same HM resolution in the vicinity of the outfall in our grid) and combining the three sigma layers nearest the surface into one layer. The WQM uses tri-linear concentration interpolation on the transformed grid to complete the advection calculations, and an iterative solver to perform the implicit diffusion calculations. After each time step, all concentrations are scaled uniformly to correct for any mass loss or gain of our "conservative" tracer. While we use the term WQM, we should note that the model is really more of a transport model, since this application codes only for chemically uncoupled contaminants. For our present calculations, the contaminant is assumed to be conservative in order to compare results with ECOMsi and to assess mass conservation before corrections.

The HM model was run for one month (January 1992) using a time step of about 7 minutes, with feet of the characteristic lines saved every tidal cycle (12.42 hours). The CPU time, based on a low-end Sun Sparc-4/110 workstation with 24 MB RAM, was approximately 44 hours, which includes 2 hours for calculating the feet of characteristic lines. The WQM was run with the 12.42 hour time step to simulate concentrations of freshwater and wastewater tracer over the month. The additional CPU time was 20 minutes, mainly for the diffusion calculations, making the total CPU time associated with WQM calculations about 2.3 hours, or about 5% of the HM CPU time.

Although neither the HM nor the WQM can be expected to fully reproduce the sub-grid features of concentrations within the outfall plume, we focus on this region because it produces the sharpest gradients, and hence provides the most difficult test of model interfacing. Tracer concentrations at a depth of 2 meters after 25 days are shown in Figure 4 for ECOMsi and in Figure 5 for our WQM. The WQM simulations shown are following mass corrections at each time step; the average correction is about 4% with

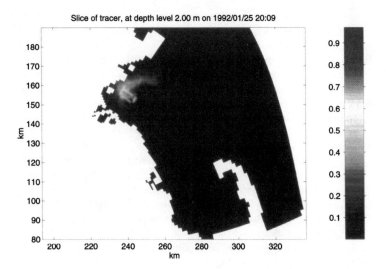

Figure 4: ECOMsi Result of Conservative Tracer in Mass and Cap Cod Bays

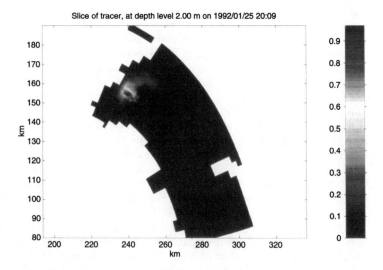

Figure 5: SELM Result of Conservative Tracer in Mass and Cap Cod Bays

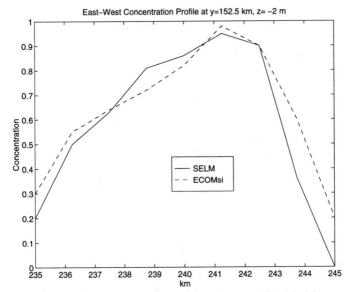

Figure 6: Concentration Comparison between Two Models

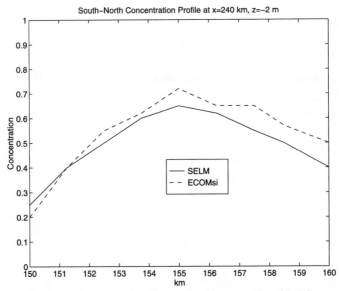

Figure 7: Concentration Comparison between Two Models

a maximum correction of 7%. A detailed comparison of concentration in an East-West and South-North sections across the surface of the plume is shown in Figure 6 and 7, indicating that concentrations with both models agree to within about 10%, which appears reasonable to support long term simulations.

Table 1: Simulation Parameters for ECOMsi and ELM WQM

	ECOMsi	ELM WQM
Grid Size	68x68x12	24x29x10
Outfall Location Index	$(30, 47)$	$(12, 22)$
Time Step	414 sec.	12.42 hours
CPU time	44 hours	20 mins.

5   Acknowledgements

We are grateful to Rich Signell (USGS) and Rich Isleib, Jim Fitzpatrick and Alan Blumberg (HydroQual, Inc.) for sharing their models with us.

References

[1] Baptista, A. E., Adams, E. E., and Stolzenbach, K. D., Eulerian-Lagrangian Analysis of Pollutant Transport in Shallow Water, Dept. of Civil and Environmental Engineering, MIT report No. 296, 1984

[2] Blumberg, A. F., and Mellor, G. L., A Description of a Three-Dimensional Coastal Circulation Model, Three-Dimensional Coastal Ocean Models. Ed. by Heaps. AGU. Washington D.C. 1-16.,1987

[3] Cheng, R. T., Casulli, V., On Lagrangian Residual Currents with Application in South San Fransico Bay, California, Water Resources Research, 18(6): 1652-1662, 1982

[4] Dortch, M. S., Three-Dimensional Lagrangian Residual Transport computed from an Intratidal Hydrodynamic Model, Technical Report El-90-11, Environmental Lab., Dept. of The Army, Waterways Experiment Station, Corps of Engineers, 1990

[5] Feng, S., Cheng, R. T. , On Tidal Induced Residual Current and Residual Transport, 1. Lagrangian Residual Current, Water Resources Research, Vol. 22, No. 12, pp. 1623-1634, 1986a

[6] Feng, S., Cheng, R. T. , On Tidal Induced Residual Current and Residual Transport, 2. Residual Transport with Application in South San Francisco Bay, California, Water Resources Research, Vol. 22, No. 12, pp. 1635-1646, 1986b

[7] Hamrick, J. M., Linking Hydrodynamic and Biogeochemical Transport Models for Estuarine and Coastal Waters, Proceedings of 3rd Intl. Conf. on Estuarine and Coastal Modeling III, pp. 591-608, 1993

[8] HydroQual Inc. , A Water Quality Model for Massachusetts and Cape Cod Bays: Calibration of the Bays Eutrophication Model, 1995

[9] King, I., DeGeorge, J. F., Multi-Dimensional Modeling of Water Quality Using the Finite Element Method, Proceedings of 4th International Conference on Estuarine and Coastal Modeling, pp. 340-354, San Diego, California, Oct. 26-28, 1995

[10] Signell, R. P.,Jenter, H.L.,and A.F.,Blumberg, Circulation and Effluent Dilution Modeling in Massachusetts Bay: Model Implemenetation, Verification and Results, USGS Open File Report 96-015, 1996

[11] Tang, L., Adams,E. E., Effect of Divergent Flow on Mass Conservation in Eulerian-Lagrangian Schemes, 4th Intl Conf. on Estuarine and Coastal Modeling, 1995

Barotropic Tidal and Residual Circulation in the Arabian Gulf

Cheryl Ann Blain[1]

Abstract

The tidal response in the Arabian Gulf is a mixed semi-diurnal and diurnal signal controlled through the Strait of Hormuz. Tidal elevations range between 0.5 -1.5 m and tidal velocities are on the order of 0.2 - 0.5 m/s. The tidal residual circulation is most prevalent over a gradually sloping bottom in depths less than 15-20 m (i.e. off the U.A.E coast, in the northern Gulf, and near the cape at Ras-e Jabrin in Iraq). Numerous islands dotting the eastern Gulf and ragged coastlines through the Strait induce residual currents having magnitudes nearly 10% of the semi-diurnal and 25% of the diurnal frequencies present. Generally, Coriolis forcing acts as a constraint on generated residuals, reducing their magnitudes and confining residual circulation to follow along topographic gradients. At the sea bed a decrease in the frictional force enhances the tidal residuals. Despite the role of these nonlinear processes, advection remains the primary mechanism for the generation of residual currents.

Introduction

Contained within the region defined by the Arabian Gulf, Strait of Hormuz and the Gulf of Oman is one of the busiest and most important shipping lanes in the world. A vessel passes through the Strait of Hormuz every 6 min and approximately 60% of the world's marine transport of oil comes from this region (Reynolds, 1993). For a region of such economic importance, comprehensive studies investigating the circulation of the region have been sparse, in part, due to the scarcity of observations.

A diagnostic study of the tidal-induced circulation in the Arabian Gulf is undertaken here to establish the barotropic response of the basin prior to the inclusion of more complex baroclinic dynamics. Conflicting opinions on the importance of the

---

[1]Oceanographer, Oceanography Division, Naval Research Laboratory, Code 7322, Stennis Space Center, MS 39529

tidal circulation are evident in the literature (e.g. Proctor et al, 1994; Reynolds, 1993; Chao et al, 1992; LeProvost, 1985). Of particular interest is the tidal residual circulation, the component of the current which remains after removing the oscillatory response at all significant tidal frequencies. This paper identifies those mechanisms that control the tidal residual circulation and the regions in which the tidal residual may be significant.

Numerical Modeling Strategy

The tidal computations for this paper are made using the fully nonlinear, two-dimensional, barotropic hydrodynamic model ADCIRC-2DDI (Luettich et al., 1992). The ADCIRC model has an extensive and successful history of tidal prediction in coastal waters and marginal seas (e.g. Mark and Scheffner, 1996; Grenier et al, 1995; Luettich and Westerink, 1995; Westerink et al, 1994). The model is well suited to tidal studies because of its computational efficiency which is due in part to implementation of the finite element method, vectorization of the code, and the use of an iterative sparse matrix solver. One advantage of the model for tidal analyses is the facility for timely and complete run-time harmonic decomposition for astronomical and nonlinear compound and over-tide components of the tidal signal. One can expect to complete simulations which extend from months to a year and employ 30 second time step increments expediently on both workstation and supercomputer platforms without unreasonable storage requirements.

The ADCIRC-2DDI model is based on the well-known shallow water equations which are derived through a vertical integration over the water column of the three-dimensional mass and momentum balance equations subject to the hydrostatic assumption and the Boussinesq approximation. Surface heating and evaporative fluxes together with strong wind forcing are known to be dominant mechanisms for inducing circulation in the Arabian Gulf (e.g. Chao et al, 1992). However, baroclinic processes, wind stress, and surface pressure effects are neglected so that the tidal component of the circulation can be examined independently. Bottom stress terms are parameterized using the standard quadratic friction law and no lateral mixing due to diffusion or dispersion is considered. Application of these simplifications lead to the following set of conservation statements in primitive, non-conservative form expressed in a spherical coordinate system (Kolar et al., 1994):

$$\frac{\partial \zeta}{\partial t} + \frac{1}{R\cos\phi}\left[\frac{\partial u_\lambda H}{\partial \lambda} + \frac{\partial(v_\phi H\cos\phi)}{\partial \phi}\right] = 0 \tag{1}$$

$$\frac{\partial u_\lambda}{\partial t} + \frac{1}{R\cos\phi}u_\lambda\frac{\partial u_\lambda}{\partial \lambda} + \frac{1}{R}v_\phi\frac{\partial u_\lambda}{\partial \phi} - \left[\frac{\tan\phi}{R}u_\lambda + f\right]v_\phi = -\frac{1}{R\cos\phi}\frac{\partial}{\partial \lambda}[g(\zeta-\eta)] - \tau_* u_\lambda \tag{2}$$

$$\frac{\partial v_\phi}{\partial t} + \frac{1}{R\cos\phi}u_\lambda\frac{\partial v_\phi}{\partial \lambda} + \frac{1}{R}v_\phi\frac{\partial v_\phi}{\partial \phi} + \left[\frac{\tan\phi}{R}u_\lambda + f\right]u_\lambda = -\frac{1}{R}\frac{\partial}{\partial \phi}[g(\zeta-\eta)] - \tau_* v_\phi \tag{3}$$

where t represents time, λ, ϕ are degrees longitude (east of Greenwich positive) and degrees latitude (north of the equator positive), ζ is the free surface elevation relative to the geoid, u_λ, v_ϕ are the depth averaged horizontal velocities, R is the

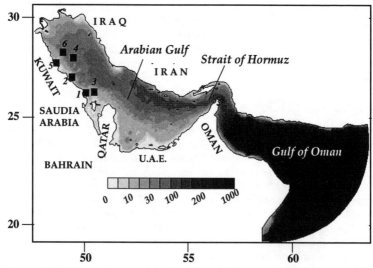

Figure 1: Bathymetric map of the Arabian Gulf and surrounding waters. The six tidal stations shown (1-6) correspond to Ras Tanura, Abu Ali, Abu Safah, Lawah, Safaniya, and Zuluf, respectively.

radius of the Earth, $H = \zeta + h$ is the total water column depth, h is the bathymetric depth relative to the geoid, $f = 2\Omega\sin\phi$ is the Coriolis parameter, Ω is the angular speed of the Earth, g is the acceleration due to gravity, η is the effective Newtonian equilibrium tide potential, and τ_* is given by the expression

$$c_f[u_\lambda^2 + v_\phi^2]^{1/2}/H \qquad (4)$$

where c_f equals the bottom friction coefficient.

Numerical solution of the governing equations (1)-(3) is achieved by re-casting the continuity equation into a generalized wave continuity equation (Lynch and Gray, 1979; Kinnmark, 1984). The discrete problem in space is approached through an application of the finite element method. The wetting and drying of computational elements is possible within ADCIRC-2DDI but is not activated for computations in this paper. As a consequence, near drying elements require the finite amplitude term linearizations which are applied to all of the simulations described in this paper.

The Arabian Gulf Model

The Arabian Gulf is a semi-enclosed marginal sea located at the south eastern end of the Arabian Peninsula. Moving from the SW coast counterclockwise along the SE coast, Gulf waters are bordered by the United Arab Emirates, Qatar, Bahrain, Saudi Arabia, Kuwait, Iraq and Iran. Depths in the Arabian Gulf are quite shallow averaging approximately 50 m. Connection to the Gulf of Oman and the eastern Indian Ocean is through the Strait of Hormuz whose narrow constriction controls the tidal character of the basin, separating it from the Indian Ocean co-tidal system. The bathymetry of the Arabian Gulf is markedly asymmetrical about the NW-SE axis of the basin (figure 1). A deep channel off the Iranian coast contrasts the shallow broad shelf on the Arabian side of the Gulf. The source of bathymetry is the DBDB5 5 minute data base (Naval Oceanographic Office, 1987) on which a minimum depth of 3 m is imposed to eliminate drying of computational points along the shoreline. The domain selected for the majority of model experiments includes the Arabian Gulf, Strait of Hormuz, Gulf of Oman, and extends into deep Indian Ocean waters (figure 1). The finite element mesh for this domain contains 8550 computational points whose spacing ranges from 0.5-1.0 km near the Shatt al Arab river inflow in the NW to 46 km in the deep waters of the Indian Ocean.

The initial set of experiments consists of a series of 20 day simulations each forced with a single tidal constituent, either M_2, S_2, K_1, or O_1. An expanded series of model runs forced by the M_2 tide only is used to investigate sensitivity of the bottom friction coefficient (BF#), the influence of topography (TOPO) and the importance of Coriolis forcing (NCOR) and advective term (NADV) contributions to the tidal residual circulation. Four long simulations of 215 days, a length appropriate for complete harmonic decomposition of 42 astronomical and nonlinear tides, have the following forcing configurations: semi-diurnal tides M_2 and S_2 (SD), diurnal tides K_1 and O_1 (D), all primary tides M_2, S_2, K_1 and O_1 (MAIN), and a run forced with the ten constituents listed in Table 1 (ALL). Tidal forcing is applied both at the open boundary and internal to the domain through the tidal potential. Values for tidal constituents prescribed at the open boundary are obtained from results of the Grenoble global tide model (LeProvost et al., 1994). A final set of model runs compares two open boundary placements and varied forcing at these boundaries. These simulations are described under a separate heading.

Model parameters are identical for all experiments just described so that comparisons between model simulations are possible. The bottom friction coefficient is uniform and set equal to 0.0015 for all model runs except those investigating sensitivity of the bottom friction coefficient. Simulations are spun up from a homogeneous initial condition using a ramp in time equivalent to five days for short period runs (20 days) and 12 days for the long term 215 day runs. A time step of 30 seconds is more than sufficient to satisfy the Courant criterion over the entire domain.

Description of the Tidal Dynamics

The tidal response of the Arabian Gulf is a mixed semi-diurnal and diurnal signal. Classification of tidal elevations according to the expression $(M_2 + S_2)/(K_1 + O_1)$ is shown in figure 2a where predominantly semi-diurnal areas have values < 0.25, mixed semi-diurnal and diurnal are in the range 0.25 - 3.0, and diurnal

Table 1: Tidal Forcing and Model Configuration for Arabian Gulf Simulations

Run	M_2	S_2	K_1	O_1	$Q_1\ P_1\ \mu_2$ $N_2\ K_2\ L_2$	Bottom Friction	Coriolis Term	Convective Terms
M_2	x					0.0015	x	x
BF1	x					0.00375	x	x
BF2	x					0.00075	x	x
BF3	x					0.003	x	x
BF4	x					0.006	x	x
NCOR	x					0.0015		x
NADV	x					0.0015	x	
TOPO	x					0.0015	x	x
S_2		x				0.0015	x	x
K_1			x			0.0015	x	x
O_1				x		0.0015	x	x
SD	x	x				0.0015	x	x
D		x	x			0.0015	x	x
MAIN	x	x	x	x		0.0015	x	x
ALL	x	x	x	x	x	0.0015	x	x

tides are > 3.0 (LeProvost, 1985). Note that the diurnal character of the tide appear-saround the amphidromic points of the semi-diurnal tide. An identical classification relative to the currents is shown in figure 2b. Most often, tidal elevations and currents correspond in frequency content, however anomalous ratios between the elevation and associated velocity are produced in the center of the Arabian Gulf at the diurnal amphidrome. Currents at this location are largely diurnal whereas the elevations are semi-diurnal. Generally, the tidal currents, too, are mixed, semi-diurnal and diurnal, except in the narrow channel north of the Strait of Hormuz where currents remain semi-diurnal.

With the natural oscillation period of the basin estimated to be approximately 21 hours ($2L/\sqrt{gh} = 2x850km/\sqrt{9.81x50m}$), the barotropic or external Rossby radius for the basin (c/f) at a latitude of 27° is just over 330 m, an order of magnitude similar to that of the basin width. As a result, energy enters the Arabian Gulf through the Strait of Hormuz and propagates for each constituent as a Kelvin wave sloping up towards the Iranian coast. These waves are reflected at the northern head of the Gulf and return along the southern side of the main channel. For semi-diurnal forcing, these opposing Kelvin waves produce a node in the center of the basin and two anti-cyclonic amphidromes, the first being 1/4 wavelength from the point of reflection. The broad shelf on the Arabian coast leads to increased frictional damping of the tidal amplitudes and so shifts the semi-diurnal amphidrome closer to the Arabian and U.A.E coasts SW of the basin axis (see the co-tidal chart for M_2 in figure 3a). Diurnal forcing results in a single half-wave basin oscillation which creates a single amphidrome in the central Gulf (figure 3b).

Resonant amplification of the semi-diurnal tides is evident in the northern

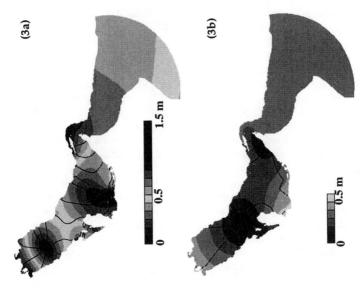

Figure 3: Co-tidal charts for the a) M_2 and b) K_1 tides computed by the Arabian Gulf model.

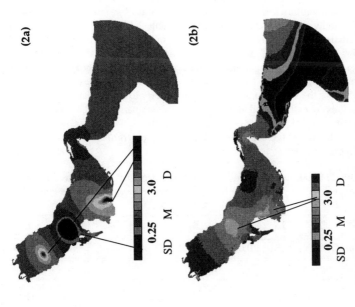

Figure 2: Tidal classification for a) elevation and b) currents in the Arabian Gulf. Semi-diurnal < 0.25, mixed $0.25 - 3.0$, and diurnal > 3.0.

Arabian Gulf and in the Strait of Hormuz (i.e. M_2 in figure 3a). Diurnal constituents are also amplified in the northern Gulf and additionally in the region off the Arabian coast (i.e K_1 in figure 3b). Through visual inspection, computed co-tidal charts for M_2 and K_1 (figures 3a and 3b) as well as the S_2 co-tidal chart (not shown) compare very favorably with empirical co-tidal charts by the Hydrographic Department of the British Admiralty generated from a set of observational data (reprinted in LeProvost, 1985) and the co-tidal charts of the U.S. Hydrographic Office (reprinted in Lardner et al., 1986). Tidal elevations computed for O_1 forcing tend to be higher in the northern Arabian Gulf when compared to the published British Admiralty co-tidal chart. Also evident are some notable differences in the amplitude pattern through the Strait of Hormuz and off the Arabian coast. Numerical comparisons of the predicted tidal elevation at individual stations follows.

ADCIRC model predictions of elevation for four tidal constituents (M_2, S_2, K_1 and O_1) are evaluated by comparison to tidal amplitudes and phases at six stations located in the northwestern section of the Gulf off the Saudi Arabian coast (identified in figure 1). The station data is obtained from the published comparisons of Al-Rabeh et al. (1990). Two error measures are presented in Tables 2 and 3, a root mean square (RMS) error and a total error computed as a distance, D, in the complex plane (Foreman, 1993):

$$D = \left\{ (A_o \cos P_o - A_m \cos P_m)^2 + (A_o \sin P_o - A_m \sin P_m)^2 \right\}^{1/2} \qquad (5)$$

where A_o, A_m, P_o, P_m are the observed and modeled amplitudes and phases, respectively. For all stations, the mean elevation errors are 6.8 cm and 4.6 cm when comparing individual tidal constituents. From Table 2, model computations for the semi-diurnal phases comprise the largest source of error. Simulations of the M_2 tide examining the sensitivity of the frictional coefficient indicate that increased bottom stress shifts amphidrome locations further toward the Arabian coast (not shown). LeProvost (1985) has noted this behavior as well. The large phase error at Safaniya is likely due to the station locations slightly exterior to the model domain (model computations are extrapolated to that location for comparison). The best agreement is at Lawah and Zuluf, stations both located in the basin's central waters, away from the influence of the coastline and unmarked underground structures associated with the oil fields.

Table 2: Elevation Amplitude and Phase Errors by Tidal Constituent

Error Measure	AMPLITUDE				PHASE			
	O_1	K_1	M_2	S_2	O_1	K_1	M_2	S_2
RMS, m	0.0606	0.0816	0.0584	0.0222	31.66	42.46	116.62	156.08
	O_1		K_1		M_2		S_2	
D, m	0.0958		0.1659		0.5965		0.1948	

Table 3: Elevation Amplitude and Phase Errors by Tidal Station

Stations	RMS Error		D, m
	Amplitude, m	Phase, deg	
Ras Tanura	0.0518	84.92	0.4041
Abu Ali	0.0323	78.96	0.3053
Abu Safah	0.0474	80.30	0.3374
Lawah	0.0685	58.57	0.1249
Safaniya	0.0764	104.59	0.2354
Zuluf	0.0697	81.74	0.1515

Influence of Boundary Location and Forcing

A domain which includes only the Arabian Gulf is constructed to have an open ocean boundary situated in the Strait of Hormuz on a line from 56°E, 26.73°N to 56.37°E, 26.38°N. Three sources of tidal elevation forcing are applied separately at this open boundary of the smallest model domain. The first set of elevations is obtained from the Grenoble global tidal database. A second and third set of boundary elevations are taken from Al-Rabeh et al. (1990) with the third set simply excluding the nonlinear tidal constituents, M_4 and MS_4. Tidal elevation profiles across the Strait of Hormuz are compared in figure 4 for the three sources of forcing just described and elevations computed by the model whose domain has an open boundary in the deep Indian Ocean. Notable differences are evident between the tidal data sources and model computations derived from the large model domain. Inaccuracies in the Grenoble database at the Strait of Hormuz are not surpris-

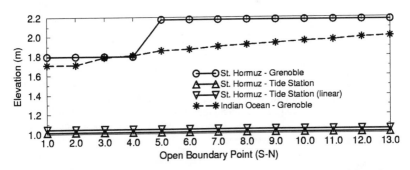

Figure 4: A comparison of tidal elevations across the Strait of Hormuz either specified as forcing for an Arabian Gulf domain or computed from a larger model domain which has an open boundary in the deep Indian Ocean.

ing since the global tide model does not contain a high degree of refinement relative to the bathymetry or shoreline in the Strait. Elevations from Al-Rabeh et al. (1990) extrapolated from coastal tidal stations are quite diminished compared to elevations computed using the largest Arabian Gulf-Gulf of Oman model domain.

Differences between tidal elevations predicted by the four model configurations at six stations located off the Saudi Arabian coast are represented by the error measure, D, in Table 4. At half of the stations, the Arabian Gulf-Gulf of Oman model yields the lowest errors. For each of the stations Abu Ali and Abu Safah, two physical locations can be identified with each name (i.e. an island and a bay and an oil field and a reef, respectively). It is possible that the coordinates defining the stations herein do not correspond to the unpublished values used by Al-Rabeh et al. (1990). Furthermore, an abundance of submerged structures associated with the oil industry are not represented in the bathymetry may account for the mixed results in these data comparisons.

Table 4: Average Error, D, at Six Tidal Stations for Four Model Configurations

Station Names	D, m			
	Boundary Location Source of Boundary Forcing Data			
	St. Hormuz Grenoble	*St. Hormuz Al-Rabeh et al. (1990)*	*St. Hormuz Al-Rabeh et al. (1990) (exclude M_4 and MS_4)*	*Indian Ocean Grenoble*
Ras Tanura	0.3730	0.2746	0.2748	0.4041
Abu Ali	0.2826	0.2025	0.2027	0.3053
Abu Safah	0.3081	0.2236	0.2237	0.3374
Lawah	0.1486	0.1400	0.1400	0.1249
Safaniya	0.2407	0.2641	0.2642	0.2354
Zuluf	0.1554	0.2326	0.1762	0.1515

The domain encompassing the Arabian Gulf-Gulf of Oman appears a better choice since open boundary forcing is far from the region of interest and errors tend to be lower. Even though the Arabian Gulf is tidally isolated from the Gulf of Oman, exact placement of an open boundary within the Strait of Hormuz is not obvious. Furthermore, such a boundary may become problematic once wind and baroclinic dynamics are considered.

Tidal Residual Circulation

The harmonic decomposition of the tidal time series generated by simulation ALL into 42 tidal constituents is used to assess the significance of nonlinear tidal interactions to the overall tidal signal. Ratios of the mean elevation (figure 5a) and mean current amplitudes (figure 5b) for all frequencies relative to the mean M_2 tide indicate that a portion of the tidal energy is manifest in nonlinear compound and over-tide frequencies (i.e. greater than 2.0 cycles/day) and in the tidal residual (i.e. zero cycles/day)

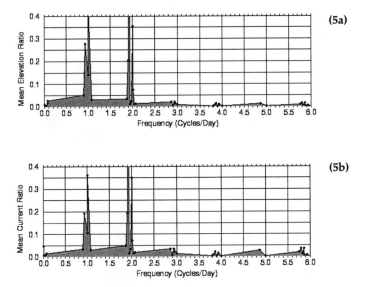

Figure 5: Ratios of the mean amplitude for 42 frequencies relative to the M_2 tide for (a) elevation and (b) currents.

For the majority of the Arabian Gulf, tidal residual currents are on the order of 2 cm/s or less. Though small, this residual is in fact nearly 25% of the mean K_1 tidal current and more than 10% of the mean M_2 tidal current. If one considers that the tides over the domain are of mixed semi-diurnal and diurnal frequency, the importance of the tidal residual only increases. For run MAIN, the RMS value of the residual velocity east of 56.4° (an arbitrary boundary placed in the St. of Hormuz to delineate the Arabian Gulf from the Gulf of Oman) is 2.67 cm/s. This value differs from the RMS residual current generated in the simulation ALL by a mere 0.5%. Recall that the simulation ALL has Q_1, P_1, μ_2, N_2, K_2, and L_2 as additional forcing at the open boundary. While forcing from the additional constituents may be important to the tidally-driven currents and elevation (i.e. Proctor et al., 1991), they contribute little to the residual circulation.

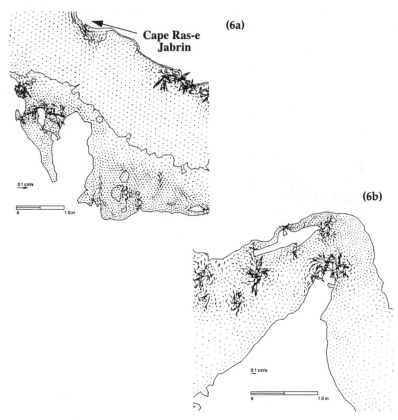

Figure 6: Tidal residual circulation from Run MAIN in the a) lower Arabian Gulf and b) Strait of Hormuz.

The residual circulation is largely controlled by topographic changes and shoreline tortuosity. Residual currents are clearly evident off the coast of Iraq in the vicinity of cape Ras-e Jabrin (figure 6a), off the U.A.E. coast (figure 6a), and around numerous islands scattered throughout the basin and in the Strait of Hormuz (figure 6b). At each of the locations in figure 6a, depths are less than 20 m, as shown by the single contour line, and topographic slopes are broad. In figure 6b, the irregular land boundaries of islands located in the midst of the primary axis of basin flow and the ragged coastline at the southern shore in the Strait of Hormuz contribute to the generation of residual velocities. The augmented magnitude of these residuals, however, is likely a non-physical consequence of insufficient mesh resolution. This issue is under investigation through use of a doubly refined mesh.

Since the tidal structure over the basin is of mixed character, contributions from both diurnal and semi-diurnal signals are prevalent. RMS values representing the diurnal and semi-diurnal contribution to the residual are derived from differences between tidal residuals generated for simulations MAIN and D and simulations MAIN and SD. Overall, the RMS values associated with the residual circulation relative to run D (diurnal forcing only) is 1.7 times smaller than that computed relative to run SD (semi-diurnal forcing) indicating that diurnal forcing may be more influential in the generation of significant residual currents. If one sums residual current contributions from each of simulations M_2, S_2, K_1, and O_1, the total residual is twice the actual residual computed in run MAIN. Clearly all of the additional energy generated by the semi-diurnal and diurnal forcing in run MAIN is not transferred to the mean residual circulation but instead is manifest in both higher and lower frequencies as evident in figure 5b.

In the following sensitivity analyses, only M_2 tidal forcing is considered. For a series of five experiments in which the bottom drag coefficient is varied, domain-wide RMS residual current values are compared to determine the influence of bottom friction on the generation of residual circulation (see Table 5). As expected, RMS values for the residual velocity decrease with increasing drag coefficients. When the drag coefficient is doubled between runs BF1, BF2, and M_2, the relative decrease in the RMS of the residual remains at a constant factor of 1.8. A significant reduction (i.e. fourfold) in the RMS residual occurs when moving from a drag coefficient of 0.0015 to 0.003. A slight increase in the residual occurs for the final doubling of the friction coefficient. The spatial pattern of the residual circulation throughout the basin remains largely intact except at high drag coefficients (i.e. 0.003 and 0.006). For these cases, residual currents essentially disappear as a consequence of the severe damping created by the large bottom friction force. Accounting for the experience of others (i.e. Proctor et al., 1991) and the reasonably good correspondence to published empirical co-tidal charts, a value for the bottom drag coefficient equal to 0.0015 is a realistic value. Note however that the tidal elevations, currents, and residuals exhibit marked sensitivity to the bottom friction particularly for drag coefficients in the vicinity of 0.0015.

Table 5: Sensitivity of the Residual Current to the Bottom Drag Coefficient

Bottom Drag Coefficient	Domain-Avg. RMS Residual Current
0.000375	0.0778974
0.00075	0.0422771
0.0015	0.0226743
0.003	0.0053030
0.006	0.0065843

Further continuing the investigation of the mechanisms driving generation of the tidal residual, model experiments removing the effects of the earth's rotation and advection are examined independently. The inclusion of Coriolis terms within the model decreases the mean RMS residual currents (labeled in figures 7a and 7b) by nearly 17.7%. Residual circulation off the U.A.E. coast produced without the influence of Coriolis "forcing" is shown in figure 7a. A comparison of the residual circulation in figures 7a and figure 7b (which includes Coriolis effects) indicates that the earth's rotation acts to damp and constrain the tidal residual. So in this case, residuals generated by topographic anomalies are opposed by the effect of Coriolis forcing and are thus confined to topographic contours. When advection is omitted from the model, domain-averaged RMS residual currents are three orders of magnitude smaller than residuals produced under the influence of advection and are not shown. Nonlinear advection is clearly the primary means for tidal residual generation in the Arabian Gulf.

To ascertain the importance of the tidal flats off the U.A.E. coast to the residual circulation, the flats and surrounding bathymetric features are modified by a factor of 0.75 for depths less than 50 meters in a region contained by 52.5°E, 56.2°E, and 26°N. Essentially, near coastal depths are deepened from less than 5 m to nearly 30 m. Removal of the very shallow water all but eliminates tidal residual currents in the region. The tidal velocities are significantly reduced in the deeper water which in turn minimizes the influence of the advective terms in the model, the primary mechanism for residual current generation.

Summary and Conclusions

The mixed frequency content of the tides in the Arabian Gulf results in a residual circulation that has contributions from both diurnal and semi-diurnal tidal constituents. Nonlinear advection is found to be the primary mechanism for the generation of tidal residual currents. In gradually sloped, shallow nearshore waters, tidal velocities are large enough in magnitude (up to 0.5 m/s) that advective contributions to the flow are important. In a single experiment investigating sensitivity to the bottom topography, deepened waters off the U.A.E. coast demonstrate the necessity of significant tidal velocities for advective transport and residual circulation. The rotation of the earth, represented through the Coriolis "force", counterbalances the effects of advection, reducing the residual magnitude and confining residual currents to flow along topographic contours.

The magnitude of the bottom drag coefficient obviously affects elevation amplitudes but it also controls the location of tidal amphidromes. Considerable sensitivity of the tidal response is noted between bottom drag coefficients of 0.0015 and 0.00075. With regard to the residual circulation, bottom friction does not affect the structure only the relative damping of the current magnitude. Generally, tidal residual currents are quite small (a few cm/s) and their influence is limited to advection-dominated localized areas having shallow, broad shelves.

Residual currents evident around islands and the ragged coastline in the Strait of Hormuz warrant further investigation at a finer mesh resolution. Future directions include a more thorough comparison of model predictions to tidal station data and characterization of tidal mixing over the water column. Identified regions of mixing may result in interactions between the tidal residual circulation and the

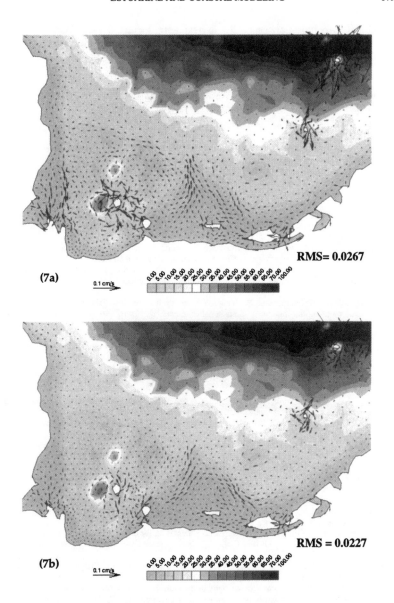

Figure 7: Tidal residual currents due to M_2 forcing a) without Coriolis terms and b) with Coriolis and advection effects included.

dominant baroclinic flow especially in nearshore, localized areas. While perhaps not important to the overall large scale Gulf circulation, these smaller scale deviations have importance to the oil drilling and shipping communities.

Acknowledgments. This work was funded through the Office of Naval Research's Navy Ocean Modeling and Prediction Program (Program element 62435N). This paper, NRL contribution PP-7322-97-0038, is approved for public release; distribution is unlimited.

References

Al-Rabeh, A. H., N. Gunay, H. M. Cekirge, 1990: A hydrodynamic model for wind-driven and tidal circulation in the Arabian Gulf, *Appl. Math. Modelling.*, 14, 410-419.

Chao, S-Y., T. W. Kao, and K. R. Al-Hajri, 1992: A numerical investigation of circulation in the Arabian Gulf, *J. Geophys. Res.*, 97(C7), 11219-11236.

Foreman, M. G. G. and R. F. Henry, 1993: A finite element model for tides and resonance along the north coast of British Columbia, *J. Geophys. Res.*, 98(C2), 2509-2531.

Grenier, R. R., R. A. Luettich, and J. J. Westerink, 1995: A comparison of the nonlinear frictional characteristics of two-dimensional and three-dimensional models of a shallow tidal embayment, *J. Geophys. Res.*, 100(C7), 13719-13735.

Kinnmark, I.P.E. 1984. *The Shallow Water Wave Equations: Formulation, Analysis and Application*, Ph.D. Dissertation, Department of Civil Engineering, Princeton University.

Kolar, R.L., W.G. Gray, J.J. Westerink, and R.A. Luettich. 1994. Shallow water modeling in spherical coordinates: Equation formulation, numerical implementation, and application, *J. Hydraul. Res.*, 32, 3-24.

Lardner, R. W., H. M. Cekirge, and N. Gunay, 1986: Numerical solution of the two-dimensional tidal equations using the method of characteristics, *Comp. & Maths. with Appls.*, 12A(10), 1065-1080.

LeProvost, C., 1985: Models for tides in the Kuwait Action Plan (KAP) region, *Proceedings of the Symposium/Workshop on oceanographic modelling of the Kuwait Action Plan (KAP) Region, UNEP Regional Seas Reports and Studies No. 70*, UNEP, 209-225.

LeProvost, C., M. L. Genco, F. Lyard, P. Vincent, and P. Canceil, 1994: Spectroscopy of the world ocean tides from a finite element hydrodynamic model, *J. Geophys. Res.*, 99(C12), 24777-24798.

Luettich, R. A., J. J. Westerink, and N. W. Scheffner, 1992: ADCIRC: An advanced three-dimensional circulation model for shelves, coasts, and estuaries, Report 1: Theory and methodology of ADCIRC-2DDI and ADCIRC-3DL, *Tech. Rep. DRP-92-6*, Department of the Army, Washington., D. C.

Luettich, R.A. and J.J. Westerink, 1995: Continental shelf scale convergence studies with a barotropic model, *Quantitative Skill Assessment for Coastal Ocean Models, Coastal and Estuarine Studies Vol. 47*, American Geophysical Union, 349-371.

Lynch, D.R. and W.G. Gray, 1979: A Wave equation model for finite element tidal computations, *Comp. Fluids, 7*, 207-228.

Mark, D. J. and N. W. Scheffner, 1996: Development of a large scale tidal circulation model for the Mediterranean, Adriatic, and Aegean Seas, *Estuarine and Coastal Modeling, Proceedings of the 4th International Conference*, ASCE, 168-179, 1996.

Naval Oceanographic Office, 1987: Data base documentation for digital bathymetric data base confidential (DBDBC), Doc. OMAL-DBD-17A, Naval Oceanographic Office, Stennis Space Ctr. MS.

Proctor, R., R. A. Flather, A. J. Elliott, 1994: Modeling tides and surface drift in the Arabian Gulf-application to the Gulf oil spill, *Cont. Shelf. Res.*, 14(5), 531-545.

Reynolds, R. M., 1993: Physical oceanography of the Gulf, Strait of Hormuz, Gulf of Oman-Results from the *Mt. Mitchell* Expedition, *Marine Pollution Bulletin*, 100, 35-59.

Westerink, J. J., R. A. Luettich, J. Muccino, 1994: Modeling tides in the Western North Atlantic using unstructured graded grids, *Tellus*, 46a(2), 178-199.

Seasonal Characteristics of the Tidal Current Circulation in Omura Bay

Tadashi Fukumoto[1], Akihide Tada[1],
Takehiro Nakamura[2] and Hiroyoshi Togashi[3]

Abstract

A three-dimensional hydrodynamic numerical model is applied to simulate the seasonal tidal stratification and circulation in Omura Bay in Japan. In addition, field measurements have been carried out to identify the characteristics of the currents in the entire bay. The major axis of the M_2 tidal ellipse near the bottom layer has been found to be shifted from that of the surface layer in summer. On the other hand, both layer's tidal ellipses in winter are the same as the ellipse near the bottom in summer. These phenomena affected by thermal stratification and vertical current profiles are apparent in the numerical simulations. Current observations using ADCP were carried out near the bay mouth of Omura Bay. An anti-clockwise current circulation was observed in this area. In particular, complicated flow patterns caused by the effects of bottom topography and bay configuration have been observed and also simulated numerically. Moreover, the tidal residual circulation seems to meander in the western part of Omura Bay.

1. Introduction

Omura Bay, which is located in the western part of Japan, is a complex coastal plain estuary with 320 km^2 surface area and 16 m average water depth. It is connected with Sasebo Bay by two narrow channels as shown in **Figure 1**. The inflow of ocean water is small because of these topographic features. The mouth of Omura Bay, which is called Hario-Seto, is very narrow (200 m wide and 40 m deep). The currents in Omura Bay, which is typical of enclosed bays in Japan, are caused mostly by tidal motion. The water near the bay mouth is approximately 30 m deep and vertically well mixed. Currents in this region are primarily controlled by the tides. The peak magnitude of the spring flood currents exceeds 5.0 m/s. However, in the center of Omura Bay the characteristics of the tidal circulation pattern in summer are different

[1] Research Engineer, Nishimatsu Construction Co., Ltd., Technical Research Institute
2570-4 Shimotsuruma, Yamato, Kanagawa 242-8520, Japan
[2] Associate Professor, Faculty of Environmental Studies, Nagasaki University
1-14 Bunkyo-machi, Nagasaki, Nagasaki 852-8131, Japan
[3] Professor, Department of Civil Engineering, Nagasaki University
1-14 Bunkyo-machi, Nagasaki, Nagasaki 852-8131, Japan

from those in winter, due to the significant density current associated with the
formation of strong thermal stratification. This has led to generations of red tide
formation. Damages caused by the red tide have often been reported (Iizuka et al. 1989,
Iizuka and Min 1989). Lately, the developments along the bay, such as reclamation for
housing, have taken place and the inflow pollutant loads exceed the self-purification
rate of Omura Bay. Thus, in order to study the long-term transport and baroclinic flow,
field observations are very important and numerical models are need to simulate not
only tidal time-scale variables but also the long-term residual current and transport.
The authors have been carrying out both field observations and numerical simulations
on the currents, on water temperature and on salinity in order to comprehend the
behaviors of the currents and water quality in Omura Bay. These include the current
measurements using ADCP and three-dimensional baroclinic flow modeling.

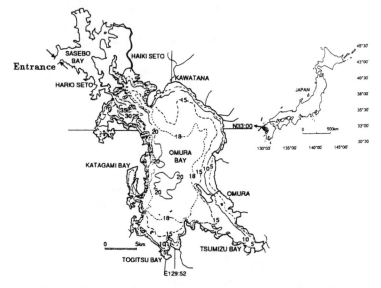

Figure 1. Location and contour map of bottom topography in Omura Bay.

2. Methods of study on the currents

To predict the generation of red tide or blue tide and to conserve the ecosystem in
enclosed bays, it is necessary to understand the real behavior of the currents and water
quality by means of field observations and numerical simulations.

2.1 Field observations

Field observations on the currents in the entire area of Omura Bay have been
carried out over 15 days every year since 1989. The measurements include water
temperature and tide level as well as current velocity and direction using the electro-
magnetic current meters (ALEC ACM16M and ACM8M). Figure 2 shows the
locations of measurement stations. These instruments were attached at two or three

elevations and data was stored for 30 seconds (sampling time is 1 s) every 10 minutes. An example of such time-series data collected at Stn.P6 is shown in **Figure 3**. Water depth at Stn.P6 is about 16 m. Tidal current velocities were recorded at two elevations. The tidal currents are typically mixed semi-diurnal. The most dominant tidal constituent is M_2 tide.

An anti-clockwise horizontal circulation pattern of residual currents in the northern part of Omura Bay was observed during winter (Fukumoto et al. 1992). The correlation between wind-stress and currents was also found in these studies (Nakamura et al. 1992, Fukumoto et al. 1996).

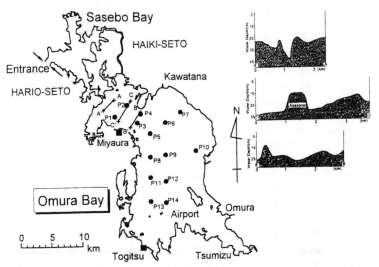

Figure 2. Stations and lines for the field observation, showing the locations of the tide gauges (squares), current meters (circles) and three lines (A-A', B-B' and C-C') for ADCP. Also shown is the bottom topography along the three lines.

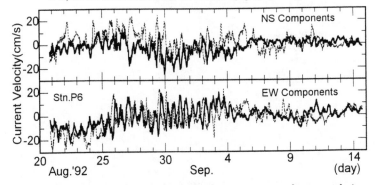

Figure 3. Time-series of NS (top) and EW (bottom) components of current velocity were collected at Stn.P6 in 1992. The current meters were located at 10 m (dotted) and 5 m (solid) above the sea bed.

It however was very difficult to get consistent field measurements. by moored current meters, near the mouth of the bay. The inflow velocities during flood tide are very large and the bottom topography in this region is very complex. The collection of data using Acoustic Doppler Current Profilers (RD Instruments 600 kHz ADCP) was carried out along three lines shown in Figure 2. Recently. ADCP has been mounted on moving ships and used for routine tidal current measurements. High data recovery rate has been achieved in tidal current measurements employing the Real Time Kinematics Global Positioning System (RTK-GPS, Tsuboi and Okamoto 1996). Current measurements were collected 13 times during one tidal cycle beginning from 3 m below the water surface to 2 m above the sea bed at 1 m intervals. The ship's speed was almost constant at about 2.5 m/s. The sampling time was about 7s. The data obtained using ADCP revealed a three dimensional tidal current structures.

2.2 Numerical simulations

A two-dimensional (2-D) flow model for the tidal currents (tide level, horizontal velocity) in Omura Bay was shown to be in good agreement with limited field observations (Fukumoto et al. 1992), but it could not predict the vertical velocity profile. To get an accurate reading of the tidal current circulation, vertical velocity in a three dimensional (3-D) model needed to be measured. Numerical simulations for estuarine flows in Omura Bay were therefore performed to reproduce the hydrodynamic properties of tide, tidal residual current and density flow. The numerical results of the tidal current and the thermocline agree with the observed one. We therefore concluded that the 3-D baroclinic flow model can accurately reproduce the realistic behavior of the current qualitatively in Omura Bay (Fukumoto et al. 1995).

The 3-D shallow water equations are solved here by a finite difference scheme. The estuarine system is sufficiently large so that Coriolis acceleration effects must be included in the momentum equations. To simplify the governing equations based on incompressible Navier-Stokes equations, the hydrostatic pressure is assumed and the Boussinesq approximation is applied. In Cartesian coordinates, the governing equations are written as follows (Matsuo et al 1993);

· Continuity Equation

$$\frac{\partial u}{\partial x} + \frac{\partial v}{\partial y} + \frac{\partial w}{\partial z} = 0 \tag{1}$$

· Momentum Equations

$$\frac{Du}{Dt} - f v = -\frac{1}{\rho_0}\frac{\partial p}{\partial x} + \frac{\partial}{\partial x}\left(A_h \frac{\partial u}{\partial x}\right) + \frac{\partial}{\partial y}\left(A_h \frac{\partial u}{\partial y}\right) + A_v \frac{\partial^2 u}{\partial z^2} \tag{2}$$

$$\frac{Dv}{Dt} + f u = -\frac{1}{\rho_0}\frac{\partial p}{\partial y} + \frac{\partial}{\partial x}\left(A_h \frac{\partial v}{\partial x}\right) + \frac{\partial}{\partial y}\left(A_h \frac{\partial v}{\partial y}\right) + A_v \frac{\partial^2 v}{\partial z^2} \tag{3}$$

$$0 = -g - \frac{1}{\rho}\frac{\partial p}{\partial z} \tag{4}$$

· Diffusion Equation of Water Temperature

$$\frac{DT}{Dt} = \frac{\partial}{\partial x}\left(K_h \frac{\partial T}{\partial x}\right) + \frac{\partial}{\partial y}\left(K_h \frac{\partial T}{\partial y}\right) + K_v \frac{\partial^2 T}{\partial z^2} + q \tag{5}$$

- Diffusion Equation of Salinity

$$\frac{DS}{Dt} = \frac{\partial}{\partial x}\left(D_h\frac{\partial S}{\partial x}\right) + \frac{\partial}{\partial y}\left(D_h\frac{\partial S}{\partial y}\right) + D_v\frac{\partial^2 S}{\partial z^2} \tag{6}$$

- Equation of State of Density

$$\rho = 999.84 + 6.79 \times 10^{-2}T - 9.10 \times 10^{-3}T^2 - 1.00 \times 10^{-4}T^3 - 1.12 \times 10^{-6}T^4 + 6.54 \times 10^{-9}T^5$$
$$+ (8.24 \times 10^{-1} - 4.09 \times 10^{-3}T + 7.64 \times 10^{-5}T^2 - 8.25 \times 10^{-7}T^3 + 5.39 \times 10^{-9}T^4)S \tag{7}$$
$$- (5.72 \times 10^{-3} - 1.02 \times 10^{-4}T + 1.65 \times 10^{-6}T^2)S^{3/2} + 4.83 \times 10^{-4}S^2$$

where x, y, z = Cartesian coordinates; t = time: $u, v, w = x, y, z$ velocity components; ρ = density of seawater, which is based on the international equation of state of seawater (Millero and Poisson 1981); ρ_0 = reference density (= 1026.64 kg/m³); p = pressure; g = gravitational acceleration (= 9.8 m/s²); f = Coriolis parameter (= 7.92×10^{-5} rad/s); T = water temperature; S = salinity; q = water temperature generation term by radiation; A = eddy viscosity and K and D = eddy diffusivities of T and S, respectively. The subscripts, h and v, denote the horizontal and vertical direction, respectively.

These equations integrated over a control volume are discretized by means of a space-staggered grid system (Iwasa et al. 1983). A Leap-Frog and Adams-Bashforth method for the temporal differences and an upwind scheme (Doner-Cell) for the convective terms are also adopted (Shen 1991). For the initial conditions, the water temperature and salinity are assumed to be uniform and the still water state is given. The water temperature and salinity of the ocean water and tidal level are given as the boundary conditions at the grid of the bay mouth, Hario-Seto. The discharge and water temperature of 24 rivers flowing into Omura Bay are also considered. Meteorological heat exchanges are modeled at the water surface (Schertzer and Lam 1991). The vertical eddy viscosity and eddy diffusivity are given as a function of Richardson Number (Nakatsuji 1994). The Sub-Grid Scale (SGS) model is used to set the coefficients of eddy viscosity and eddy diffusivity in the horizontal plane (Smagorinsky 1963, Nakatsuji and Fujiwara 1995).

In this paper, three horizontal grid spacings are chosen for the simulation of real behaviors of the currents near the bay mouth from winter to summer. Note the explicit integration of the shallow water equations is known to allow for a time step limited by the Courant condition based on fast gravity wave. These computational conditions and layer thicknesses are tabulated in Table 1. The region calculated using 100m horizontal grid spacing is limited near the bay mouth to the northern one-third of Omura Bay. Table 2 lists the typical data at the surface and open boundary conditions and the initial conditions of inner water in Omura Bay obtained by field observations.

Table 1. Parameters for Calculation

	Grid spacing $\Delta x, \Delta y$ (m)		
	1000	500	100
Time step Δt (s)	20.0	10.0	2.0
Number of grid (x×y)	39×26	74×45	140×152
15 thickness of layers	11×2 m + 3×4 m + 8 m (bottom)		

Table 2. Typical atmospheric and oceanographic conditions

Initial conditions		in winter	in summer
Water temperature of Omura Bay	(°C)	10.0	21.0
Salinity of Omura bay	(‰)	32.5	31.5
Water temperature of ocean	(°C)	11.0	25.0
Salinity of ocean	(‰)	33.0	33.0
Air temperature	(°C)	8.0	27.0
Cloudiness	(%)	50.0	50.0
Relative humidity	(%)	70.0	80.0
Wind direction		NW	SWS
Wind velocity	(m/s)	3.5	3.0
Solar radiation	(MJ/day)	7.0	27.0

All calculations are based on the sequences as follows: first, initial calculation of surface level and current caused only by tide is carried out during 10 tidal cycles using a 3-D barotropic flow model; second, a 3-D baroclinic model using the initial conditions based on the output of the first model is carried out during 20 tidal cycles. The calculated results during the last tidal cycle of the second model is presented in the following.

3. Seasonal characteristics of tidal current ellipses

The currents in Omura Bay are dominated by tidal motion; however, tidal currents at the head of bay are relatively small. Thus, the effect of wind-induced current can not be neglected in this region far from of the bay mouth. In the eastern part of Omura Bay, the currents are complex and affected by both tidal and density flows due to river inflow along the eastern coastline of Omura Bay. Particularly, the stratification becomes pronounced in summer.

To show the effects of stratification in summer, **Figure 4** shows the M_2 tidal ellipses at Stn.P6 obtained by harmonic analyses of the measured currents both in winter and summer corresponding to the data shown in **Figure 3**. The residual currents obtained by time-averaging of measured current are also shown in **Figure 4**. The M_2 tidal ellipse at the both layers are similar in the absence of stratification in winter, whereas the oscillatory tidal currents at surface and bottom layers are noticeably different due to stratification in summer. Relatedly, the residual currents at the surface and bottom layers are in the same direction in winter but in the opposite direction in summer. As a result, the effects of summer stratification are significant in the bottom layer as expected.

The computed M_2 tidal current ellipses near the bottom layer at Stn.P6 in summer, based on horizontal grid spacing of 1000 m, are shown in **Figure 5**. The conditions of three computed cases are listed in **Table 3**. The M_2 ellipse for case-1 (3-D baroclinic model) is similar to the measured ellipse shown in **Figure 4(b)**. Comparing the M_2 ellipse for case-2 and case-3 (2-D model), which includes the vertical velocity variation, shows increased data correlation. However, the vertical density variation needs to be included to obtain more realistic tidal ellipses near the bottom in summer.

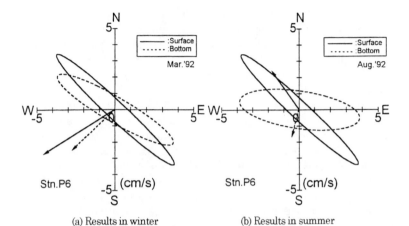

(a) Results in winter   (b) Results in summer

Figure 4. M₂ tidal current ellipses and mean flow velocities at Stn.P6
near the surface (solid) and bottom (dotted) layers.

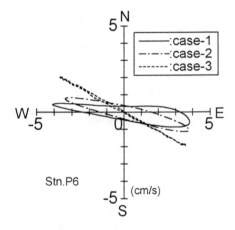

Figure 5. Results of the calculated tidal current ellipses at Stn. P6
near the bottom layer (see Table 3).

Table 3. Case of calculations

	Model Name	Initial water conditions	Solar radiation
case-1	3-D Baroclinic Model	stratified	yes
case-2	3-D Barotropic Model	uniform	no
case-3	2-D Model	uniform	no

4. Large scale residual circulation pattern in Omura Bay

Fukumoto et al. (1992) found an anti-clockwise residual current circulation near the bottom layer during winter in the northern part of Omura Bay. The mean current velocity in summer was reproduced by a 3-D baroclinic flow model (Fukumoto et al. 1995). The characteristics of residual current circulation are described in detail using both observed and calculated results.

Figure 6 shows the residual current based on all the observed records from 1989 to 1992. The data near the bay mouth, where tidal currents are strong, are not shown due to noise caused by oscillations of the moored current meters. Observed residual currents vary seasonally in each layer. The circulation pattern could not be identified near the surface layer because the data points were sparse. In the northern part of the bay, an anti-clockwise current circulation pattern was apparent in the bottom layer both in winter and summer, probably caused by the topography of the bay. In the western part of Omura Bay, a small clockwise current seemed to exist. The residual circulation might meander in the western part of Omura Bay.

The similar circulation pattern is calculated by the 3-D baroclinic model as shown in **Figure 7** and **Figure 8**. Top (a), middle (b) and bottom (c) panels are at 1m, 9m and 17 m below the still water level (SWL), respectively. These figures show the results based on a 500 m grid spacing in winter and summer without the influence of wind. The residual current circulation pattern in winter is similar in all layers and anti-clockwise. There is however a density flow caused by freshwater inflow from rivers located along the north and east coastlines of Omura Bay. This effect becomes strong in summer due to thermal stratification. Thus, the outflow of the inner water of Omura Bay toward the bay mouth is dominated near the surface layer in summer and the inflow of ocean water near the bottom layer is larger in summer than in winter.

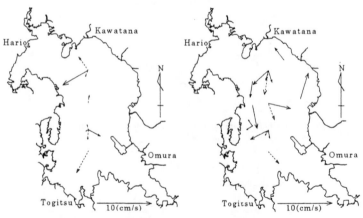

(a) Results near surface layer   (b) Results near bottom layer
Figure 6. Observed residual current in winter (solid arrow)
and in summer (dotted arrow).

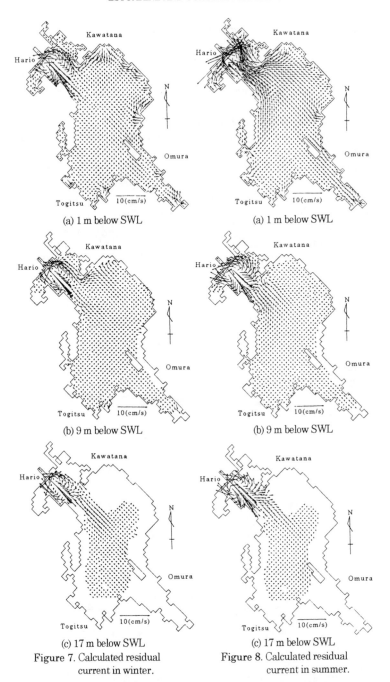

(a) 1 m below SWL

(a) 1 m below SWL

(b) 9 m below SWL

(b) 9 m below SWL

(c) 17 m below SWL
Figure 7. Calculated residual
current in winter.

(c) 17 m below SWL
Figure 8. Calculated residual
current in summer.

An anti-clockwise circulation pattern near the middle layer is similar in winter and summer.

Consequently, the residual current in Omura Bay is formed by both tidal and density effects in summer. The anti-clockwise current circulation pattern is caused by tidal currents affected by the particular topography of Omura Bay.

5. Characteristics of the currents near the bay mouth

The currents near the mouth of Omura Bay are very strong and complex. The magnitude of inflow velocity can be more than 5 m/s. The current meters were washed away and broken near this region in the field experiment conducted in 1994. The measurements using ADCP attached to a moving ship were carried out from July 25 to 27 in 1996 during neap tides. The observed periods of the three lines (see **Figure 2**) are shown in **Figure 9** relative to the records of tide level at Miyaura port. In this figure, the subscripts of " e " and " f " of A, B and C indicate the times of ebb and flood tide. These observations were carried out within one tidal cycle from high tides. The measured velocity magnitude was very large along the A-A' Line. The pattern of currents along the C-C' Line is very complex due to Asasone shoal, where the water depth is about 15 m in contrast to the water depth of about 30 m outside this region.

The horizontal current velocity vectors at elevations of 5 m intervals along the A-A' Line in flood tide are shown in **Figure 10**. In front of Hario-Seto (bay mouth) where

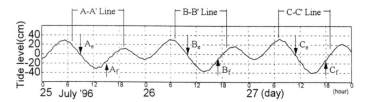

Figure 9. Time series of tide level at Miyaura port, observation periods using ADCP and the cross sections along the three lines.

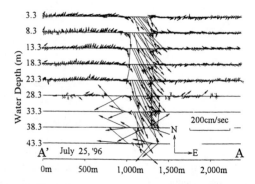

Figure 10. Vertical structure of flood current along the A-A' Line.

the water depth is about 40 m, current velocity magnitude was large (over 2.0 m/s) and the bottom current was larger than the surface current. To compensate for this inflow rate, the surrounding currents were toward the opposite direction. **Figure 11** shows the vectors of flood current at elevations of 5 m intervals along the C-C' Line. The currents on the west side of Asasone were to the south away from the bay mouth. The maximum velocity magnitude was about 1.5 m/s and currents above the shoal, Asasone, were smaller. On the east side of Asasone, the currents were to the north toward the bay mouth. This phenomenon occurred on the east side during flood tide. The western side is deeper as may be inferred from **Figure 1** and flood tides flow along this deeper channel.

To investigate the spatial variations of the currents near the bay mouth, the records were selected for the flood tide data of A_f, B_f and C_f and ebb tide data of A_e, B_e and C_e as shown in **Figure 9**. **Figure 12** shows the currents at about 3 m and 19 m below the water surface along three lines for the flood tide. **Figure 13** shows the corresponding currents for the ebb tide. Ocean water passing through the bottom layer in front of the bay mouth is clearly seen along the A-A' Line in **Figure 12**. There are two types of current directions. One is rotated counterclockwise about the shoal,

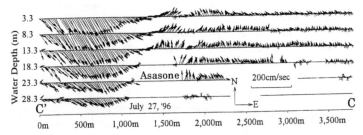

Figure 11. Vertical structure of flood current along the C-C' Line.

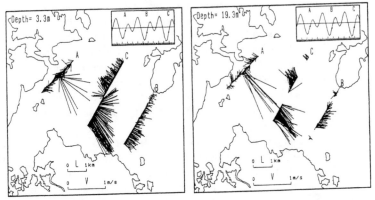

(a) 3 m below the water surface   (b) 19 m below the water surface
Figure 12. Observed tidal currents during flood tide.

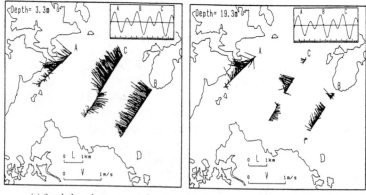

(a) 3 m below the water surface   (b) 19 m below the water surface
Figure 13. Observed tidal currents during ebb tide.

Asasone and the other is toward the bay head along the south-west coastline. Since the thermal stratification is formed from about the B-B' Line to the bay head and the water temperature near the bay mouth is similar to that at middle layer (between 5 m and 15 m below the water surface) in the bay center (Fukumoto et al. 1995). The ocean water seems to flow upward along the sloping bottom into the middle layer of the B-B' Line.

For the ebb tide, the currents in Omura Bay are concentrated near the B-B' Line and flow toward Hario-Seto by way of the north-eastern part of this region. The currents on the south-west side of the C-C' Line are in the opposite direction and there is a counterclockwise circulation during the ebb tide as well. The counterclockwise current formed near the bay mouth is caused by the interaction between the tidal currents, the bottom topography and coastline. The thermocline was absent during this period in this region.

Moreover, to clarify the characteristics of the vertical structure of the currents near the bay mouth, the inflow and outflow along the B-B' Line are analyzed. Figure 14 shows the temporal variations of the average velocity in upper (O), middle (△) and lower (□) layers in the northern part of the B-B' Line. The vertical cross section along the B-B' Line is divided into two parts about the center of this line. The upper, middle and lower layers are separated at 6 m and 16 m below the water surface. The upper layer is a region between the surface and 5 m below the water surface. The middle layer is a region between 6 m and 15 m below the water surface. The lower layer is a region between 16 m below the water surface and bottom. The average velocity in each layer is calculated from 13 ADCP data sets.

The temporal variations of the velocity in lower layer (□) is similar to that in the middle layer (△). The current in the upper layer (O) is directed toward the bay mouth except for a few hours when flood current becomes maximum. On the other hand, the current in the middle layer (△) is directed mostly towards the bay center. This tendency in the middle layer also occurs in the southern part of the B-B' Line. The

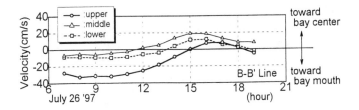

Figure 14. Temporal variations of average velocity in upper (~5 m;○),
middle (6 m~15 m;△) and lower (16 m~:□) layers
in the northern part of the B-B' Line.

ocean water appears to invade in the middle layer near the B-B' Line and the inner water of Omura Bay outflow through the upper layer in the northern part of the B-B' Line while rotating counterclockwise in this region. These current patterns in the upper and middle layers do not appear along the other lines. The currents along the A-A' and the C-C' Line are almost uniform vertically.

In order to reproduce these observed results and clarify the characteristics of the currents near the bay mouth, a numerical simulation using a 100 m grid spacing in this area is carried out. The boundary conditions shown in Table 2 are adopted. The initial and open boundary conditions at the center of the bay are based on the computed results using a 500 m grid spacing. Figure 15 shows the calculated residual currents on the vertical section along the solid line shown in this figure. The surface currents in the northern part near the B-B' Line tend to flow toward the bay mouth and bottom current flows toward the center of bay in a manner consistent with the data shown in Figure 14.

In addition, the calculated currents at 3 m (2nd layer) and 19 m (10th layer) below the water surface near the bay mouth at the flood tide in summer are shown in

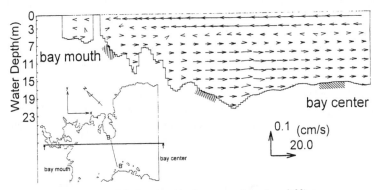

Figure 15. Calculated residual currents along the solid line
shown on the left using a 100 m grid spacing.

Figure 16. The current patterns near the A-A' Line shown in **Figure 12** is reproduced. A small counterclockwise current in the north-east part of bay mouth is noticed in **Figure 16**. It is similar to the observed current pattern along the A-A' Line shown in **Figure 12**, that is, the surrounding currents in front of bay mouth near the A-A' Line are toward the opposite direction to compensate for this inflow rate.

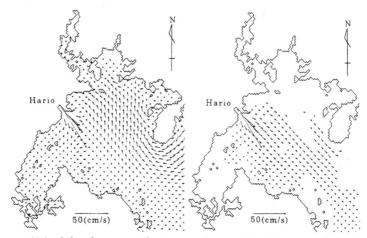

(a) 3 m below the water surface   (b) 19 m below the water surface

Figure 16. Calculated residual currents using a 100 m grid spacing

6. Conclusions

Many current observations including ADCP and a high resolution numerical model were used in this investigation of Omura Bay. The numerical model has been calibrated against observed data for tidal currents. These results suggest that the tidal residual circulation seems to meander in the western part of Omura Bay. The seasonal characteristics of tidal ellipses are affected by the vertical velocity variation and stratification in summer. The counterclockwise current formed near the bay mouth is caused by the interaction between the tidal currents and the bottom topography and coastline. As a result, the tidal residual currents in Omura Bay are strongly influenced by its topography and stratification.

Future work will include the investigation that the wind-induced current has an effect on the residual current in Omura Bay. Additionally, the horizontal and vertical mixing current with the large tidal current near the bay mouth will be studied using the results of the 3-D baroclinic model and field observations.

Acknowledgments

The authors would like to thank the people in the 15 fisheries co-operative associations around Omura Bay and Mr. Syunjiro Haraguchi, who is the captain of fishing boat "AJIROMARU", for their assistance in the field observations, and the students in the Department of Civil Engineering. Nagasaki University for the data analysis. The authors also appreciate the useful advice by Prof. Nobuhisa Kobayashi. Center for Applied Coastal Research. University of Delaware.

References

Fukumoto,T., T.Nakamura, and H.Togashi (1992), " Tidal Residual Current in Omura Bay." *Proc. of 2nd Int. Conf. on Hydr. and Envir. Modeling of Coast. Engrg. and River Waters*, Bradford, 79-87.

Fukumoto,T., A.Tada. T.Nakamura and H.Togashi (1995), " Three-Dimensional Tidal Currents and Water Quality in Stratified Omura Bay." *Proc. of the 4th Int. Conf. on Estuarine and Coast. Modeling*, ASCE, San Diego, 707-721.

Fukumoto,T.. A.Tada, T.Nakamura and H.Togashi (1996), " Study on the Variations between Wind-Induced Currents and Thermocline in Omura Bay." *Proc. of Coast. Engrg, Vol. 43*, JSCE, 396-400 (in Japanese).

Iizuka,S., H.Sugiyama and K.Hirayama (1989), " Population Growth of Gymnodinium Nagasakiense Red Tide in Omura Bay." in T.Okaichi et al. (eds.): Red Tide: *Biology Environmental Science and Toxicology, Elsevier Science Publishing Co. Inc.*, 269-272.

Iizuka,S. and Si Hong Min (1989), " Formation of Anoxic Bottom Waters in Omura Bay." *Bulletin on Coast. Oceanogr. Vol.26, No.2*, Coast. Oceanogr. Res. Committee, The Oceanographical Society of Japan, 75-86 (in Japanese).

Iwasa,Y., K.Inoue, S.Liu and T.Abe (1983), " Numerical Simulation of Flows in Lake Biwa by Means of a Three-Dimensional Mathematical Model." *Annuals, Disas. Prev. Res. Inst., Kyoto University, No.26 B-2*, 531-542 (in Japanese).

Millero,F.J. and A.Poisson (1981), " International One-Atmosphere Equation of State of Seawater." *Deep Sea Res., Part A, Vol.28, No.6*, 625-629.

Matsuo,N., K.Inoue and N.Nagata (1993), " Prediction of Algae Blooming and its Control by Bubble Plume in a Eutrophicated Reservoir." *Proc. of 25th IAHR Congr., Vol.5*, Tokyo, 101-108.

Nakamura,T.. H.Togashi, T.Fukumoto and S.Mikuriya (1992), " Study on Wind-Induced Currents in Omura Bay." *Proc. of Coast. Engrg, Vol.39*, JSCE, 246-250 (in Japanese).

Nakatsuji,K. (1994), " Estuarine Circulation and Mass Transport in Osaka Bay." *Lecture Notes of the 30th Summer Seminar on Hydr. Engrg., Committee on Hydr., Course A*. JSCE, 1-28 (in Japanese).

Nakatsuji,K. and T.Fujiwara (1995), " Anticyclonic Upper Layer Residual Circulation and Estuarine Circulation in Osaka Bay." *Proc. of the 4th Int. Conf. on Estuarine and Coast. Modeling*, ASCE, San Diego, 128-142.

Schertzer,W.M. and D.C.L.Lam (1991), " Modeling Lake Erie Water Quality - A Case Study." in Brian Henderson-Sellers (eds.): *Water Quality Modeling Vol.IV :CRC Press, Inc.*, 27-68.

Shen,H. (1991), " Numerical Analysis of Large-Scale Flows and Mass Transports in Lakes." *Doctoral Thesis*, Kyoto University, 1-197.

Smagorinsky,J. (1963), " General Circulation Experiments with Primitive Equations I ." *The Basic Experiment, Monthly Weather Review, Vol.91, No.3*, 99-164.

Tsuboi,H and O.Okamoto (1996), " Experimental Study on the Positioning Error in Real Time Kinematic GPS." *Nishimatsu Tech. Res. Report, Vol.19*, 33-40 (in Japanese).

FIELD DATA COLLECTION AND MODELING FOR VERIFICATION OF AN ECOSYSTEM MODEL IN OSAKA BAY, JAPAN

N.Yamane[1], K. Nakatsuji[2], H. Kurita[1] and K. Muraoka[2]

Abstract

The structure of water quality and density in Osaka Bay is examined on the basis of the field observation data collected during the past 9 to 20 years. From the density distribution, the density front can be identified along the 20-meter depth contour, which has some connection with the formation of the stratification that develops at the bay head throughout the year. This density structure indicates that Osaka Bay can be considered an estuary. Having found that the water quality has strong relation with the density structure, the authors set up a primary ecosystem model connected with a three-dimensional baroclinic current numerical model, in order to study water quality in Osaka Bay. The baroclinic residual current system and the estuarine circulation control the transformation of matter in Osaka Bay. The fresh water which flowing into the eastern bay contains a huge amount of organic matter and nutrients in the upper stratified layer. The pollution has been fast deteriorating in upper layer at the bay head combined with lots of development projects in coastal waters. The pollution level of the middle and lower layers of water is moderate: it is because the estuarine circulation brings in relatively clean water from the outer sea, maintaining better water quality there.

Introduction

Osaka Bay is an oval-shaped bay with 58 km length and 30 km in width, open to the Seto Inland Sea and The Pacific Ocean through the Akashi Strait and the Kitan Strait (see Figure 1). Since the the Yodo and Yamato Rivers flow in at the head of the bay, the soft clay and silt, therefore, form a sediment layer in the sea bottom of the eastern bay at a depth of 20 m or lower. The bed slopes down moderately from here toward the west, reaching a depth of 40 to 70 m at the east side of Awaji Island, forming a sea valley. The inflow and outflow of tidal-induced flow through both straits make tidal flow in the western bay very fast and

[1] CTI-Engineering Co., Osaka Branch, 1-2-15 Otemae, Chuo-ku, Osaka City, Osaka 540, Japan.
[2] Department of Civil Engineering, Osaka University, Suita City, Osaka 565, Japan

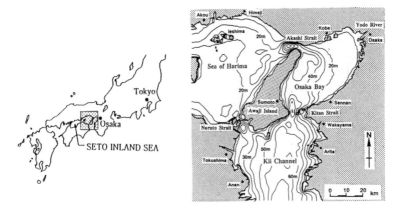

Fig. 1   Map of Japan and topographical features of Osaka Bay

result in a vertically well mixing section.   On the other hand, the tidal flow in the eastern bay is stagnant.   The inflow of fresh water from the rivers and the heating of the seawater surface during summer cause the formation of stratification, producing a tidal front along the 20-meter depth contour.   The tidal front marks the borderline between the stratified eastern bay and the well-mixed western bay.

Although water quality control regulations have been implemented since 1975 for the Seto Inland Sea and Osaka Bay, the effect of such regulations has not been significant: the algal blooms still appear on the sea surface, and anoxic water is produced near the sea bed particularly during summer.   In order to understand the pollution mechanism of this semi-enclosed bay, a number of contributing factors are under our study.   Such as the load inflow from the land, load release from the sea bed sediment, transportation of nutrients within the bay, three-dimensional structure of the production and elimination of pollutants in the course of water circulation, and also the seasonal variations.

The baroclinic residual current system is strongly related to long-term mass transport in Osaka Bay, it has been already clarified by three-dimensional numerical experiments and field observations (Nakatsuji and Fujiwara, 1997).   In the present study, the characteristics of water quality and density distribution are examined based on the field data and numerical simulation using a primary hydrodynamics-dominated ecosystem model.   The objectives of the present paper are as follows; (1) to collect field data of water qualities and density measured in Osaka Bay, (2) to establish a hydrodynamics-dominated ecosystem model including its verification, and (3) to clarify the transport processes of water qualities.

The water quality structure in Osaka Bay

The Osaka Prefecture Fisheries Experimental Station has been conducting regular field surveys on the water quality in Osaka Bay since 1972.   Figure 2(a) indicates the location of 20 observation stations.   The water temperature, salinity, and transparency are measured monthly at up to six different depths (0, 5, 10, 20, 30 m and the sea bottom).   Dissolved oxygen (DO) and inorganic phosphorus (IP) are measured in February, May, August and November at two different

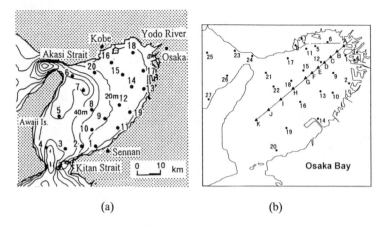

(a)   (b)

Fig. 2   Location of observation stations conducted by
          (a) Osaka Prefecture Fisheries Experimental Station;
          (b) The Environmental Agency of Osaka and Hyogo
              Prefectures

depths: namely, the surface and bottom layers.   In the present analysis using the
data collected from 1973 to 1992, twenty-year monthly averages are calculated for
each test station, and the density (σ_t) is calculated from the temperature and
salinity data. The Environmental Agencies of Osaka and Hyogo Prefectures also
have a surface water quality monitoring system (27 stations) in Osaka Bay (shown
in Fig. 2(b)).   From this set of data, we select those collected between 1984 and
1992 for our tests.   To identify the effect of the seasonal variation, we calculated
9 year averages of 9 parameters: water temperature, transparency, pH, DO, COD,
total nitrogen, total phosphorus, PO_4-P and chlorophyll-a.

Changes in density distribution throughout the year

        Figure 3 shows the horizontal distribution of the surface water density in
August and February.   Figure 4 shows the vertical density distribution at the
cross-section connecting line of observation stations 4, 5, 8, 12, 14 and 17 as
shown in Figure 2(a).   Representing the value of density(σ_t), these figures show
a significant difference between the western and eastern bays in August.   In
addition, the tidal front that runs from north to south along the 20-m depth contour
is clearly seen.   At the eastern area of the bay head, the inflow from the rivers
causes stratification in the top 5-meter layer.   The front of stratified area reaches
the 20-m contour between test locations 8 and 12.   The stratified area shrinks in
February, but can still be seen off the mouth of the Yodo River.   The tidal front
has long been known to be located along the 20 m contour, and remains at about
the same location in winter (Ueshima, 1987).   Figure 4, however, indicates that
the tidal front located on the line connecting observation station 12 and 14 recedes
from the 20-m contour.   In the western bay, the difference in densities between
surface and bottom water is minimal in both summer and winter because the
strong tidal flow through the Akashi Strait and the Kitan Strait mixes the water
well.

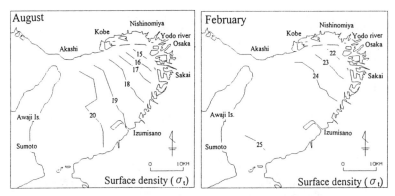

Fig. 3   Horizontal distribution of surface density in summer and winter

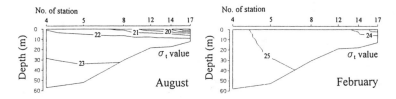

Fig.4   Vertical distribution of density at cross section connecting
observation stations from 4 to 17 as shown in Fig.2(a)
in summer and winter

Changes in water quality throughout the year

Figures 5 - 10 show the horizontal distributions of parameters which represent water quality in Osaka Bay during summer and winter. The figures mean 20-year average values, and the solid and dash lines indicate the data in August and February, respectively. Shown in Figure 5, the transparency in the western bay is 5 to 7 m throughout the year, but that in the eastern bay falls significantly toward the inner bay. The lower transparency caused by the large amount of pollutant load, the suspended solids coming from the rivers, and the algal blooms often occurred in this area. Figures 6 and 7 show the surface distributions of chlorophyll-a and COD. The chlorophyll-a distribution in August shows significant change across the 20-m contour, which is 10 mg/l in the western bay and as high as 20 - 60 mg/l in the eastern bay. In February, the amount of chlorophyll-a is approximately 10 mg/l throughout the bay. The similar distribution of COD and chlorophyll-a in August can be considered as the index of phyto-plankton growth.

Figure 8 shows the horizontal distribution of oxygen saturation levels at the sea bottom. In February, these are 90% or higher in most parts of the bay. In August, however, the saturation levels drop to below 30% in a broad area at the head of the bay. In May and November, although the data is not shown in this figure, the saturation levels are 60% or higher at the head of the bay, the levels which are not life-threatening for marine organisms. The oxygen saturation level

Fig. 5   Horizontal distribution of
transparency(m)

Fig.6   Surface chlorophyll-a
(μ g/ l)

Fig. 7   Surface COD (mg/ l)
distribution

Fig. 8   DO saturation (%)
at sea bottom

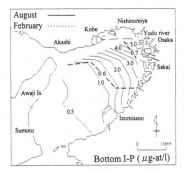

Fig. 9   Inorganic phosphorus
(mg/ l) at surface

Fig. 10   Inorganic phosphorus
(mg/ l) at sea bottom

at the sea bottom has high correlation with the differences in temperature and salinity between the sea surface and the sea bottom (Joh, 1986; Yamane et al., 1996). Factors exacerbating the diminished oxygen level at the sea bottom include oxygen consumption by bottom-dwelling organisms, temperatures much lower than at the surface, and the enhancement of stratification between the surface and the sea bottom, which presents the oxygen supply from the surface. Figures 9 and 10 show the horizontal distributions of inorganic phosphorus in the surface water and at the sea bottom, respectively. At the surface water, the inorganic phosphorus is always high around the mouth of Yodo River, with the same level in both February and August. The inorganic phosphorus level at the sea bottom is high throughout the year, being much higher in the eastern bay than the western bay in August, but having little difference in February. The level differs significantly between the surface and the sea bottom in the eastern bay, in particular, at the bay head in August, while the difference is much reduced in February. A well-established stratification makes the phosphorus level on the sea bottom much higher than that in the surface water. The sea bottom sediments continually release the inorganic phosphorus, so accelerates the consumption of nutrients by phyto-plankton at the surface. The inorganic phosphorus level at the sea bottom has close relationship with the surface-seabed temperature and salinity differences as well as sea bottom oxygen saturation. Figures 9 and 10 indicate that the existing stratification prevents the surface-to-seabed water exchange, and the oxygen consumption near the sea bottom facilitates the release of inorganic phosphorus. In the western bay, however, the difference of inorganic phosphorus release between the surface and the sea bottom is not obvious all the year round because the seawater is well mixed.

Numerical simulation model

For understanding the water quality distribution in estuaries, the numerical simulation model used in this study is an ecosystem model with a three-dimensional baroclinic flow model that taking biochemical primary production into account.

Basic equations of three-dimensional baroclinic flow model
The three-dimensional baroclinic equations are derived from the conservation of volume, momentum in three dimensions, scalar quantities such as temperature and salinity simplified under the Boussinesq assumption and the hydrostatic approximation. Basic equations as follows.

$$\frac{\partial U}{\partial x} + \frac{\partial V}{\partial y} + \frac{\partial W}{\partial z} = 0 \quad\quad\quad\quad\quad\quad \cdots\cdots (1)$$

$$\frac{\partial U}{\partial t} + U\frac{\partial U}{\partial x} + V\frac{\partial U}{\partial y} + W\frac{\partial U}{\partial z} = fV - \frac{1}{\rho_a}\frac{\partial P}{\partial x}$$

$$+ \frac{\partial}{\partial x}\left(A_H\frac{\partial U}{\partial x}\right) + \frac{\partial}{\partial y}\left(A_H\frac{\partial U}{\partial y}\right) + \frac{\partial}{\partial z}\left(A_V\frac{\partial U}{\partial z}\right) \quad \cdots\cdots (2)$$

$$\frac{\partial V}{\partial t} + U\frac{\partial V}{\partial x} + V\frac{\partial V}{\partial y} + W\frac{\partial V}{\partial z} = -fU - \frac{1}{\rho_a}\frac{\partial P}{\partial y}$$

$$+ \frac{\partial}{\partial x}\left(A_H\frac{\partial V}{\partial x}\right) + \frac{\partial}{\partial y}\left(A_H\frac{\partial V}{\partial y}\right) + \frac{\partial}{\partial z}\left(A_V\frac{\partial V}{\partial z}\right) \quad \cdots\cdots (3)$$

$$0 = -g - \frac{1}{\rho}\frac{\partial P}{\partial z} \qquad \qquad \cdots\cdots\cdots\cdots\cdots(4)$$

$$\frac{\partial \Delta T}{\partial t} + U\frac{\partial \Delta T}{\partial x} + V\frac{\partial \Delta T}{\partial y} + W\frac{\partial \Delta T}{\partial z} = \frac{\partial}{\partial x}\left(K_H\frac{\partial \Delta T}{\partial x}\right)$$

$$+ \frac{\partial}{\partial y}\left(K_H\frac{\partial \Delta T}{\partial y}\right) + \frac{\partial}{\partial z}\left(K_V\frac{\partial \Delta T}{\partial z}\right) + \frac{1}{\rho_a C_P}\frac{dq}{dz} \qquad \cdots\cdots(5)$$

$$\frac{\partial \Delta S}{\partial t} + U\frac{\partial \Delta S}{\partial x} + V\frac{\partial \Delta S}{\partial y} + W\frac{\partial \Delta S}{\partial z} = \frac{\partial}{\partial x}\left(K_H\frac{\partial \Delta S}{\partial x}\right)$$

$$+ \frac{\partial}{\partial y}\left(K_H\frac{\partial \Delta S}{\partial y}\right) + \frac{\partial}{\partial z}\left(Kv\frac{\partial \Delta S}{\partial z}\right) \qquad \cdots\cdots\cdots\cdots(6)$$

where, U, V, W; velocities in the x, y and z directions, P; the pressure, ρ; density, ρ_a; the sea water density, $\Delta\rho$; the density difference of fluid from the sea water ($= \rho_a - \rho$), ΔT; water temperature deviation from that of sea water, ΔS; salinity deviation from that of sea water, g; gravity acceleration (= 9.80 m²/s), f; the Coriolis parameter, A_H, A_V; horizontal and vertical eddy viscosity coefficients, K_H, K_V; horizontal and vertical eddy diffusion coefficients, C_b; weight specific heat at 20°C (= 4.18 x 10⁻⁵ J/ton/ °C), q; heat source function for short-wave radiation amount (W/m²)

The SGS model concept is introduced to represent the horizontal eddy viscosity and diffusion coefficients, A_H and K_H, which was proposed by Smagorinsky (1963). As a strong stratification forms in Osaka Bay, the Webb function (1970) is used for the vertical eddy viscosity coefficient, A_V; the Munk and Anderson function (1970) is for the vertical eddy diffusion coefficient, K_V.

$$A_H = (C\Delta x)^2\left[2\left(\frac{\partial U}{\partial x}\right)^2 + \left(\frac{\partial U}{\partial y} + \frac{\partial V}{\partial x}\right)^2 + 2\left(\frac{\partial V}{\partial y}\right)^2\right]^{1/2}, \qquad K_H = A_H \quad , \qquad \cdots\cdots\cdots(7)$$

$$A_V = A_{VO}\left(1 + 5.2 Ri\right)^{-1} \quad , \qquad K_V = A_V\frac{\left(1 + 10Ri/3\right)^{-3/2}}{\left(1 + 10Ri\right)^{-1/2}} \quad , \qquad \cdots\cdots\cdots\cdots\cdots(8)$$

where C is a turbulence constant (= 0.12), Δx is a horizontal grid interval, A_{vo} is a neutral value of vertical eddy viscosity coefficient, and R_1 is Richardson number.

Basic equations of ecosystem model

In the present study, an ecosystem model is designed to estimate the biochemical reaction of nitrogen, phosphorus, COD and DO in conjunction with the three-dimensional baroclinic flow model.   The predicted substances are: chlorophyll-a (Chl-a) as phyto-plankton; nitrogen (ON), phosphorus (OP) and COD as non-living organic matter; and inorganic nitrogen (IN), inorganic phosphorus (IP) as nutrients and dissolved oxygen (DO).   The outline of the proposed ecosystem is shown in Figure 11.   The circulation of each nutrient is separated into two processes: the advection-diffusion process (hydrodynamics) which is common to all matters, and the internal conversion process that consists

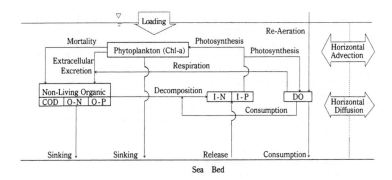

Fig. 11   Schematic view of proposed ecosystem model

mainly of biochemical reactions.

a) Advection-diffusion process

$$\frac{\partial Ci}{\partial t} = -\frac{\partial U Ci}{\partial x} - \frac{\partial V Ci}{\partial y} - \frac{\partial W Ci}{\partial z} + K_x \frac{\partial^2 Ci}{\partial x^2} + K_y \frac{\partial^2 Ci}{\partial y^2} + K_x \frac{\partial^2 Ci}{\partial z^2} + \Sigma Si(Cj) \qquad \cdots\cdots\cdots(9)$$

b) Internal change process

$$\frac{\partial PP}{\partial t} = \{G_P - R_P - D_P\} PP + U_{PP} \frac{\partial PP}{\partial z} \qquad \cdots\cdots\cdots\cdots\cdots(10)$$

$$G_P = \mu_{max} \cdot F_N \cdot F_i \cdot F_T$$

$$F_N = Min\left[\frac{IP}{K_{IP} + IP}, \ \frac{IN}{K_{IN} + IN}\right] \ , \quad F_I = \frac{I}{I_S}\exp\left[1 - \frac{I}{Is}\right], \quad F_T = \frac{T}{T_S}\exp\left[1 - \frac{T}{T_S}\right]$$

$$\frac{\partial IP}{\partial t} = -F_P G_P PP + K_P OP + \frac{W_{IP}}{H_b} \qquad \cdots\cdots\cdots\cdots\cdots(11)$$

$$\frac{\partial IN}{\partial t} = -F_N G_P PP + K_N ON + \frac{W_{IN}}{H_b} \qquad \cdots\cdots\cdots\cdots\cdots(12)$$

$$\frac{\partial OP}{\partial t} = F_P\{R_P + D_P\} PP - K_P OP + U_{OP} \frac{\partial OP}{\partial z} \qquad \cdots\cdots\cdots\cdots\cdots(13)$$

$$\frac{\partial ON}{\partial t} = F_N\{R_P + D_P\} PP - K_N ON + U_{ON} \frac{\partial ON}{\partial z} \qquad \cdots\cdots\cdots\cdots\cdots(14)$$

$$\frac{\partial COD}{\partial t} = F_C\{R_P + D_P\} \cdot PP - K_C COD + U_{COD} \frac{\partial COD}{\partial z} + \frac{W_C}{H_b} \qquad \cdots\cdots\cdots\cdots\cdots(15)$$

$$\frac{\partial DO}{\partial t} = F_{PDO}\{G_P - R_P\}PP - F_{CDO} K_C COD K_S (DOS - DO_S) - \frac{W_{DO}}{H_b} \qquad \cdots\cdots\cdots\cdots (16)$$

Where Ci ; concentration of matters i (g/m³), PP ; concentration of chlorophyll-a, G_P ; phyto-plankton growth rate, R_P ; phyto-plankton extracellular excretion rate,

D_P ; phyto-plankton mortality rate, U_{PP} ; phyto-plankton sinking velocity, μ_{MAX} ; phyto-plankton maximum reproduction rate, F_N, F_I, F_T ; the influence of nutrients, light intensity, and water temperature on the growth rate, I, I_S ; light intensity in water and optimal light intensity, T, T_S ; water temperature and optimal water temperature, IP, IN ; concentrations of inorganic phosphorus and nitrogen, OP, ON ; concentrations of non-living organic phosphorus and nitrogen, K_P, K_N ; decomposition rates of non-living organic phosphorus and nitrogen, W_{IP}, W_{IN} ; releasing rates of inorganic phosphorus and nitrogen, H_b ; bottom layer thickness, U_{OP}, U_{ON} ; sinking velocities of non-living organic phosphorus and nitrogen, COD ; non-living COD concentration, K_C ; decomposition rate of non-living COD, W_C ; releasing rate of non-living COD, U_{COD} ; sinking velocity of non-living COD, DO ; dissolved oxygen concentration, DOS ; dissolved oxygen saturation concentration, DO_S ; dissolved oxygen concentration in surface water, K_S ; re-aeration coefficient, W_{DO} ; oxygen consumption rate by sea bottom dirt, F_P ; P/Chl-a ratio, F_N ; N/Chl-a ratio, F_C ; COD/Chl-a ratio, F_{PDO} ; DO /Chl-a ratio and F_{CDO} ; DO/COD ratio.

Boundary conditions for calculation
Figure 12 shows the 64 km x 64 km area around Osaka Bay subject to computation.   The model's resolution is 1 km in the horizontal plane with 19 vertical levels having thickness of fifteen 2-m, one 4-m, one 6-m and two 10-m. Based on our computation, these vertical levels are sufficient to accurately model the vertical distribution of the water circulation and density.   Table 1 shows the boundary conditions used for the computation.   For the flow computation, we use the temporal changes in the tidal elevation (M_2) for the outside of the Akashi and the Kitan Straits, the vertical distribution of the water temperatures and salinity as boundary conditions.   In addition the heat exchange at the water surface is also obtained from meteorological data.   The inflow rates, temperature and salinity of 21 major rivers are yielded at the corresponding boundary grids. These values are set up according to July and August averages of the monthly data. The computation of 50 tidal cycles is required to attain the steady state of

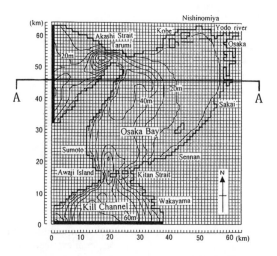

Fig. 12   Computation domain of Osaka Bay

Table 1   Boundary conditions of 3-D baroclinic flow and ecosystem model

3-D Baroclinic Flow Model

Sea Surface	Temperature	26.7	℃
	Cloudiness	6.7	
	Wind Velocity	3.0	m/s
	Water Vapor Pressure	25.8	hpa
	Light Intensity	193.5	W/m^2
River Mouth	Discharge (Yodo Riv.)	367.9	m^3/s
	(Yamato Riv.)	27.2	m^3/s
	Water Temperature	28.9	℃
	Salinity	20.0	psu
Bay Mouth	Tide	M$_2$	
	Water Temperature	28.9	℃
	Salinity	20.0	psu

3-D Ecosystem Model

River Mouth	Loading	I-P	10,800	
	Inputs	I-N	186,000	
	(Total)	O-P	6,200	
		O-N	67,000	
	(kg/day)	COD	423,000	
Bay Mouth	Water	I-P	0.015[1]	0.007[2]
	Quality	I-N	0.16[1]	0.016[2]
		O-P	0.015[1]	0.015[2]
	(mg/l)	O-N	0.04[1]	0.04[2]
		COD	1.8[1]	1.4[2]
		Chl-a	0.005[1]	0.005[2]
		DO	8.1[1]	8.1[2]

[1] Akashi Strait   [2] Kitan Strait

Table 2   Specific parameters used in ecosystem model

μ_{MAX}	Maximum Photosynthetic Rate (/day)	2.4
T_S	Optimum Temperature (℃)	25
I_S	Optimum Light Intensity (cal/cm^2/day)	200
K_{IP}	Half Saturation Coefficient for I-P (g/m3)	0.005
K_{IN}	Half Saturation Coefficient for I-N (g/m3)	0.025
R_p	Phytoplankton Extracellular Excretion Rate (/day)	$0.13*1.07^{T-20}$
D_P	Phytoplankton Mortality Rate (/day)	$0.10*1.07^{T-20}$
U_P	Phytoplankton Sinking Velocity (m/day)	0.0
K_P	O-P Decomposition Rate (/day)	$0.02*1.09^{T-20}$
U_{OP}	O-P Sinking Velocity (m/day)	0.3
W_{IP}	I-P Release Rate (gP/m^2/day)	$0.007\sim0.025*1.05^{T-20}$
K_N	O-P Decomposition Rate (/day)	$0.02*1.09^{T-20}$
U_{ON}	O-N Sinking Velocity (m/day)	0.3
W_{IN}	I-N Release Rate(gN/m^2/day)	$0.03\sim0.05*1.05^{T-20}$
K_C	COD Decomposition Rate (/day)	$0.02*1.09^{T-20}$
U_C	COD Sinking Velocity (m/day)	0.3
W_C	COD Release Rate (gCOD/m2/day)	$0.10\sim0.15*1.05^{T-20}$
Ks	Surface Aeration Coefficient (day-1)	0.1
W_{DO}	DO Consumption Rate in Sea Bed (gO/m^2/day)	$1.3\sim5.0*1.05^{T-20}$
F_P	P/Chl-a Ratio (gP/gChla)	1.8
F_N	N/Chl-a Ratio (gN/gChla)	36.0
F_C	COD/Chl-a Ratio (gCOD/gChla)	129.0
F_{PDO}	DO/Chl-a Ratio (gO^2/gChla)	113.0
F_{CDO}	DO/COD ratio(gO^2/gCOD)	0.88

baroclinic residual current, namely one-tidal integrated current.   The flow and density fields for the 50th tidal cycle are used in the computation of ecosystem as the advection and diffusion terms of   Eq.(9).   As the boundary conditions for water quality computations, we use observed values of each nutrient at the open sea boundary and also inflow loads of nutrients at the river mouth.   The model parameters for the internal conversion processes are chosen basing on the values conventionally used for the study of Osaka Bay (See Table 2).   The rates of IN and IP release and dissolved oxygen consumption of the sediments are set for various locations according to previously published data.

The details of difference schemes of basic equations and computational conditions are mentioned in the published papers (for example, Nakatsuji and Fujiwara; 1996).   The calculation of water surface elevation is differenced in an implicit scheme; while other equations for both baroclinic flow model and ecosystem model are in an explicit scheme.   The model's resolution is 1 km in the horizontal plane with 19 layers of 2m x 15 levels, 4m, 6m, 10m, 10m.   The time increment Δt is determined to be 30 seconds due to CFL condition. A hydrodynamics-dominated ecosystem model is also time-marching scheme and it requires the temporal change of three dimensional components of flow, water temperature, eddy diffusion coefficient of each control volume. The time increment of the water quality computation is 30 minutes ; the boundary conditions are yielded by the linear interpolation of the flow computation data taken at 30-minute intervals. The computations have been repeated for 30 tidal cycles.

Three-dimensional baroclinic flow and density structure

Figure 13 shows the horizontal and east-west cross-sectional distributions of computed baroclinic residual current and one-tidal integrated density in summer. In the present computation, the discussion is limited only to the summer case. The computation of the density distribution shows a well-mixed area in the western part of the bay which has a uniform vertical density distribution. It also suggests that stratification in the eastern part of the bay, which is formed by lots of inflow from the Yodo River and the Yamato River. The major residual currents seen in Osaka Bay are the Okinose Circulation in the western bay and the Off-Nishinomiya Circulation in the bay head. The former is a tidal-induced residual current, in which circulation appears independent on water depth. The latter is a anticyclonic upper layer residual current. The horizontal clockwise flow is intimately related with the stratification and the earth's rotation. The production mechanism of these particular residual currents were clarified by Nakatsuji and Fujiwara (1996, 1997). This computed result has good agreement with the observed result shown Figures 3 and 4. Upward entrainment can be seen in the stratified interface and the flow toward the bay head appears to compensate the water volume of entrainment. Such vertical circulation can be often observed and it is known as estuarine circulation.

Distribution of water quality in Osaka Bay

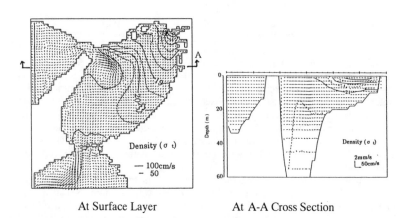

At Surface Layer   At A-A Cross Section

Fig. 13   Residual current flow and density distribution in surface water
and at the A - A vertical cross section in summer

Figure 2(b) shows the locations of 27 observation stations, where the Environmental Agencies of Osaka and Hyogo Prefectures conduct monthly regular field surveys on water quality.   We use 9-year average values of observed data collected in July from 1984 to 1992 to verify the accuracy of computational results.   Figure 14 shows the comparison of computed and observed chlorophyll-a and total phosphorus.   Figure 15 shows the horizontal distributions of surface chlorophyll-a, total phosphorus and inorganic phosphorus and their vertical distributions at the cross-section A-A.

The computed and observed levels of chlorophyll-a are about the same at 20

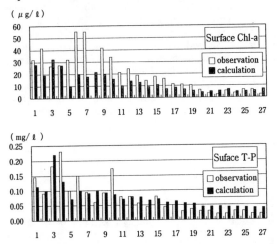

Fig. 14   Comparison of computed and observed Chlorophyll-a and TP in July

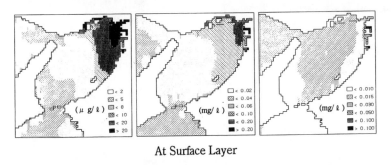

At Surface Layer

At A-A Cross Section

Fig. 15   Distribution of computed chlorophyll-a , total phosphorus and inorganic phosphorus of Osaka Bay in summer

μ g/l or higher at the head of bay covering the area off Kobe and Sen-nan. Both the values fall to 10 μ g/l or lower toward the western bay. However, the computation can't predict the value of chlorophyll-a of 30 μ g/l or higher at the stations of 6, 7, 9 and 10. Considering that stations 6 and 7 located within the Kobe Port and that stations 9 and 10 in the pass of Off-Nishinomiya Circulation, we can somehow guess the reason of the difference between computed and observed values. Regarding to total phosphorus, the distribution of the computed values matches that of the observed ones. The computed values, however, are slightly lower at the bay head where the chlorophyll-a level is generally high. The chlorophyll-a and total phosphorus distributions in the surface water are different between the eastern part and the western part of the bay, which is separated by the tidal front developing along the 20-m contour line. The load that comes from the rivers stratifies in the eastern part of the bay, contributing to the active internal production. The quality of the water at depth deeper than 5 m is influenced by the flow which is running from the western bay to the bay head due to the estuarine circulation, and the concentrations of chlorophyll-a and total phosphorus which are lower than those in the surface water. The inorganic phosphorus level in summer is low in the

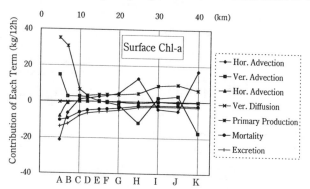

Fig. 16   Contribution of each term of chlorophyll-a transport equation
along the line of A to K shown Fig. 2(b)

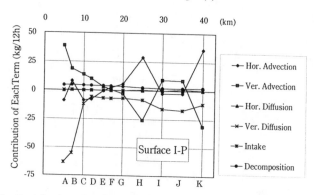

Fig. 17   Contribution of each term of inorganic phosphorus transport equation
along the line of A to K shown Fig. 2(b)

surface and high near the sea bottom. The high phosphorus level near the sea bottom is caused by the phosphorus released from the sediments at the head of bay. The low phosphorus level in the surface water can be explained by the intake of the phosphorus by plankton.

In order to understand the whole structure and the mutual interaction between nutrients determining the water quality distribution, the contribution of each term included in nutrients transport equations are examined. Figure 16 shows the comparison of each term of chlorophyll-a transport equation as shown in equations. 9 and 10. The alphabets of A to K in the figures indicate the stations as shown in Figure 2(b). Therefore, the connecting line from A to K corresponds to the expanding center-line of Yodo River. The season computed is summer. The average of the top three layers in the computation grids (approximately 6 m thickness) as the upper layer. Figure 16 shows that the chlorophyll-a production indicates significantly increasing tendency at the bay head, namely in the area of A, B and C. The chlorophyll-a produced in this area balances horizontal advection, excretion and mortality. The production of chlorophyll-a, however, becomes much lower in the area toward the sea beyond the station C. Figure 17 shows the contributions of each term in transport equation of inorganic phosphorus, which controls phyto-plankton reproduction. Around stations A and B, the terms of intake of inorganic phosphorus is extremely increasing. That is to say, IP decreases in this area where chlorophyll-a abounds. This decrease in IP is supplemented mainly by the transport of IP from lower layer, i.e. the upwelling (sometimes we call it entrainment). Estuarine circulation takes part in transport of water and nutrients as well. This results leads to the conclusion that the estuarine circulation is of importance for the production of phyto-plankton at the bay head.

Conclusions

The structure of water quality and density distribution in Osaka Bay is examined on the basis of field data collected during 9 and 20 years. A hydrodynamics-dominated ecosystem model was developed and applied to Osaka Bay. The model results were compared with the field data, and examined the adaptability. Main results are as follows:

(1) The seasonal variation of water quality and density in Osaka Bay can be simulated by the primary ecosystem model. More precise threshold settings, however, are required to simulate the water quality at the bay head of Osaka Bay.
(2) Water quality distribution is significantly influenced by the baroclinic residual current system and density field inside the bay throughout the year.
(3) Phyto-plankton production at the bay head is strongly supported by the upwelling of nutrients from the lower layer. This upwelling and mas transport in the lower layer is found to be intimately related with the estuarine circulation.

Acknowledgment

This research was partly supported by Grant-in-Aid for Scientific Research (c) of the Ministry of Education, Science, Culture and Sports (09650565 and 09660200), and Grant of Nippon Life Insurance Found.

References

Joh, H. (1986). Studies on the mechanism of eutrophication and effect of it on fisheries production in Osaka Bay, Bull. Osaka Pref. Fish. Exp. Stat., 1-174. (in Japanese)

Munk, W.H. and E.R. Anderson (1970). Notes on a theory of the thermocline, J. Marine Res., 7, 276-295.

Nakatsuji, K., T. Sueyoshi and K. Muraoka (1994). Environmental assessment of hypothetical large-scale reclamation in Osaka Bay, Japan, Proc. 24th ICCE, 3178-3192.

Nakatsuji, K., and T. Fujiwara (1996). Anticyclonic upper layer residual circulation and estuarine circulation in Osaka Bay, Proc. 4th Int'l Conf. on Estuarine and Coastal Modeling, ASCE, 128-142.

Nakatsuji, K., and T. Fujiwara (1997). Residual baroclinic circulation in semi-enclosed coastal seas, J. Hydraulic Engineering, ASCE, Vol.123, No.4, 362-373.

Osaka Prefectural Fisheries Experimental Station (1973-1992). Regular observation in shallow sea, Bull. Osaka Pref. Fish. Exp. Stat. (in Japanese)

Smagorinsky, J. (1963). General circulation experiments with primitive equations, I. The basic experiment, Monthly Weather Review, 91, 3, 99-164

The Environmental Agency of Osaka Prefecture (1984-1992). Water quality monitoring report in the public water, Bull. Osaka Prefecture (in Japanese)

The Environmental Agency of Hyogo Prefecture (1984-1992). Water quality monitoring report in the public water, Bull. Hyogo Prefecture (in Japanese)

Ueshima, H., I. Yuasa, M. Takarada, H. Hashimoto, M. Yamazaki, and H. Tanabe (1987). Flow and its structure in stagnant region of Osaka Bay, Proc. 34th, Japanese Conf. on Coastal Engineering, 661-665. (in Japanese)

Webb, W.K. (1970). Profile relationship. The log-linear range and extension to strong stability, Quarterly, J. Royal Meteorological Soc., 6, 67-90.

Yamane, N., K. Nakatsuji and K. Muraoka (1996). Seasonal variation of density structure in Osaka Bay, Proc. 2th Int'l Conf. On Hydrodynamics, 1087-1092.

DEVELOPMENT OF A WASTE LOAD ALLOCATION
MODEL WITHIN CHARLESTON HARBOR ESTUARY
PHASE I: BAROTROPIC CIRCULATION

Steven Peene[1], Eduardo Yassuda[1], and Daniel Mendelsohn[2]

ABSTRACT

A two-dimensional curvilinear grid model was used to simulate the barotropic circulation within the Charleston Harbor Estuary. This application is the first in a three phase effort to develop a comprehensive hydrodynamic and water quality model for wasteload allocation purposes. Model accuracy for barotropic circulation was assessed using hydrodynamic data collected over a 60-day period in the summer of 1996. Model skill assessment was accomplished through multiple analyses of the measured and simulated water level, velocites and fluxes. Data reduction methods included graphical comparisons, spectral analysis, RMS error analysis, and harmonic analysis.

Discharge curves were calculated from continuous water column velocity measurements and discrete discharge transects. The goal was to provide continuous flows for model comparison, as in-stream dilution of effluent discharge is primarily a function of the rate of flow past the outfall rather than local current velocities.

---

[1] Applied Technology and Management, Inc, 2770 NW 43rd Street, Suite B, Gainesville, Fl. 32606
[2] Applied Science Associates, Inc., 70 Dean Knauss Drive, Narragansett, RI. 02882

Comparison of simulated and measured water level showed normalized Root Mean Squared errors ranging from 5 to 15 percent, while flux errors ranged from 10 to 12 percent, and velocities ranged from 15 to 20 percent. The simulations were calibrated to provide the least error in the fluxes rather than the velocites; this is reflected in the percent errors. The lower errors in the fluxes identify that the calibrated model accurately simulates the tidal prism but is less accurate for the cross-sectional velocity distribution.

Spectral and harmonic analyses demonstrated that the model and observations exhibit similar characteristics in the distribution of the energy and damping of frequency components; this is most evident in the water surface elevation results.

INTRODUCTION

Since the passage of the Clean Water Act in 1972, increased awareness of the environmental impacts of point source and non-point source loads to estuarine systems has prompted numerous studies to quantify the assimilative capacity of these complex water bodies. Historically, hydrodynamic model limitations have led to oversimplification of the physical processes that drive circulation and transport. Often the hydrodynamic component within estuarine waste load allocation models is treated as secondary to the water quality kinetics, both by the model developers and regulatory personnel. Given that the hydrodynamics become the foundation for the water quality calibration and application this can lead to inaccurate or biased determination of the assimilative capacity of a system.

Recent advancements in computing technology and hydrodynamic models, along with successful linkage of more robust hydrodynamics to water quality, allow developers of waste load allocation models to more accurately project the localized transport of effluent loads. In addition, use of models to evaluate detailed spatial distribution of pollutants once they enter the system provides a clearer understanding of the extent or degree of impacts rather than simply providing system wide allowable loads.

This paper summarizes the results of the first phase in the development of a waste load allocation model for Charleston Harbor Estuary. The phases are as follows:

- Phase I: Barotropic Circulation
- Phase II: Mass Transport and Baroclinic Residual Circulation
- Phase III: Water Quality Kinetics

The overall project goal is to provide assessment and quantification of model accuracy and performance through each phase. The reasoning for this approach is that each phase is the foundation for the next. Minimization of errors at each step will reduce cumulative errors, and produce a more complete understanding of all of the processes that impact assimilative capacity.

The model used in this application is a 2-D, vertically averaged, curvilinear grid, finite difference, hydrodynamic model (BFHYDRO), developed by Applied Science Associates of Narragansett, Rhode Island. This model has undergone extensive testing and development over the last decade (Swanson, 1986; Muin and Spaulding, 1993).

The following sections present: discussion of the characteristics of the Charleston Harbor Estuary and associated model domain, a methodology for analyzing continuous tidal discharge data for model comparison, skills assessment of the model calibration, and results, summary and conclusions.

DESCRIPTION OF CHARLESTON HARBOR ESTUARY AND MODEL DOMAIN

Figure 1 presents a map of the Charleston Harbor Estuary with the data collection stations identified. The system is made up of the following components:

- the harbor entrance from the offshore jetties to the split of the Cooper and Ashley River,
- the Ashley River,
- the lower Cooper River (from the US-17 Bridge to Dean Hall),
- the upper Cooper River (from Dean Hall upstream to the West and East Branch up to Pinopolas Dam) and,
- the Wando River .

The Ashley and Wando Rivers are tidal sloughs that receive minimal freshwater inflow; the yearly average is on the order of 10 cubic meters per second (cms). The Cooper River was historically a tidal slough, but the United States Army Corps of Engineers (USACE) connected it to Lake Moultrie through the Pinopolas Dam providing freshwater inflow rates which presently average 100 to 200 cms. The Wando and Cooper Rivers are presently maintained for shipping traffic to depths averaging 15 meters.

Tidal range is 2 meters at the Harbor Entrance and this range initially amplifies moving upstream into the tributaries. The tides amplify slightly traveling up the Cooper River to near Mobay Chemical where damping of the wave and super elevation of mean water level is evident.

The main tributaries are bordered by extensive marsh areas, which flood and dry over each tidal cycle. These marsh areas are characterized by narrow interconnected tidal creeks, which provide the primary flow pathways for the interior marsh areas. Vegetation is dense and characterized by typical saltwater marsh species in the lower reaches of the tributaries changing to primarily freshwater species moving upstream.

Data utilized for this study were collected over a 60-day period in a cooperative effort between the United States Geologic Survey (USGS) and Applied Technology and Management, Inc (ATM). Hydrodynamic data collected included the following:

- water surface elevations relative to NGVD at 11 stations,
- continuous vertical velocity profiles at two locations using bottom mounted Accoustic Doppler Current Profilers (ADCP),
- continuous mid-depth velocities at three locations using Marsh McBirney current meters (MMB),
- discharge at 5 transects over 12 hour periods during both spring and neap tides. These data were collected using a towed ADCP with bottom tracking capability.

The location of the data collection stations is shown in Figure 1. These stations coincided with historic gauging stations operated by the USGS (Conrads et al., 1996).

Figure 2 presents the grid used in the simulations. For visual clarity, the extents of the grid are only shown from just above the tee at Dean Hall to the offshore boundary. In reality the grid extends up to the Pinopolas Dam and covers all primary tributaries within the system. Historic hydrodynamic model applications to the Charleston Harbor Estuary have characterized various portions (Conrads and Smith, 1996; Bower et al., 1993) but did not quantify the entire system. The model calibration process identified the need to accurately represent the entire system in order to simulate the damping of the tidal wave and the volume and phasing of the tidal prism. The extensive marsh areas were represented as side storage zones which drew water volume off of the primary tributaries through narrow connecting channels; velocities within these storage areas were essentially zero.

CALCULATION OF CONTINUOUS DISCHARGE CURVES

As described above, discharge measurements were taken at 5 transects within the estuary. These transects coincided with the locations of the continuous velocity measurements. The velocity station measurements include, bottom mounted ADCPs at the US-17 Bridge and Daniel Island Bend, mid-depth point velocities at Woods Point, the Wando River above 526, and the Ashley River .

The primary mechanism governing the in-stream dilution of point source discharges is the rate of flow passing the effluent outfall. As a result, the accurate simulation of the tidal prism is paramount in the development of the hydrodynamic base for a wasteload allocation model. Use of single point or even vertically averaged velocities from a single location within the tributary cross-section for model calibration will not necessarily insure accurate simulation of the flows if the cross-sectional velocity distribution is not simulated accurately. A methodology was developed to calculate continuous flows from the ADCP and MMB current meters and the discrete ADCP discharge transects. The equation used to calculate the discharge is of the form:

$$Q = V_p \times A_c \times C_d$$

Where: V_p = vertically averaged or mid-depth velocity

A_c = cross-sectional area

C_d = velocity to discharge correction factor

V_p and A_c were measured quantities. C_d was calculated for each of the discrete flow measurements by dividing the measured flow rate by the coincident velocity and cross-sectional area. C_d values were then averaged for the flood and ebb tides. Multiplying the average C_d values by the continuous velocities produced a continuous discharge curve. Figure 3 presents measured flow rates plotted against calculated flow rates for Daniel Island Bend and Woods Point. The normalized RMS errors from the measured versus calculated discharges ranged from 6 to 8 percent with no clear trend of over- or under-prediction.

HYDRODYNAMIC MODEL SKILL ASSESSMENT

To assess the accuracy of the model in simulating the components of the barotropic circulation, the following comparative analyses were performed:

- Graphical comparison of the measured versus simulated water surface elevation, velocities and fluxes.
- Comparison of normalized Root Mean Squared (RMS) errors between the measured versus simulated water surface elevation, velocities, and fluxes. The errors are normalized by dividing the RMS errors by the average range of the data signal (Hess and Bosley, 1992).
- Comparison of spectra calculated from measured and simulated water surface elevation and velocities.
- Comparison of the primary harmonic constituents calculated from the measured and simulated water surface elevation and velocities.

The use of multiple analyses to quantify the errors between the simulated and measured water surface elevation, velocities and fluxes isolates individual weaknesses in the model simulations. The graphical comparisons and the normalized RMS errors identify the overall performance of the simulations. Spectral comparisons isolate the ability of the model to simulate the energy in the various major frequencies (i.e. the sub-tidal, the diurnal, semi-diurnal and higher order harmonics). Harmonic analyses isolate the model's ability to replicate the primary astronomical forcing constituents and the phase lags. The following presents a model skills assessment for the barotropic circulation.

Water Surface Elevation

A total of 11 stations within the Charleston Harbor Estuary measured water surface elevations. These include: Fort Sumter, Customs House, US-17 Bridge, Daniel Island Bend, Goose Creek, Mobay Chemical, Dean Hall, Wando River at 526, Cainhoy, Ashley River at 526, and Magnolia Gardens.

Figures 4 through 6 present graphical and spectral comparisons of the water surface elevations along the Cooper River at three selected stations positioned at the lower, middle, and upper sections of the Cooper River. The graphical comparisons show amplification and damping of the tidal wave moving upstream; amplified between Customs House and Goose Creek, and damped on the order of 50 percent moving upstream from Goose Creek to Dean Hall. The authors note here that initial modeling efforts which did not include the upper reaches of the system, and the full marsh storage volumes, did not demonstrate the amount of damping seen in the upper reaches.

Comparison of the spectral plots shows similar energy levels in the various frequency domains. Both data and model show a shift in percent energy in various frequencies. The semi-diurnal components are more damped than the diurnal moving upstream. The energy of the higher harmonic components increases moving upstream.

Figure 7 shows the normalized RMS errors at all of the elevation stations. The results show increasing errors moving upstream in each of the tributaries. The normalized errors range from 2 to 12 percent on the Cooper River, 5 to 6 percent on the Wando River, and 6 to 16 percent on the Ashley River. Examination of the graphical comparisons indicates that much of the error in the upstream portions of the Cooper River may be due to inaccurate simulation of the super elevation of the mean water level rather than the tide range.

Table 1 presents the comparison of the harmonic constituents at seven of the 11 stations. The primary constituent in both the simulated and measured data is the M2 component. Percent errors for the M2 constituent are of the same order as the overall RMS errors and show increased percentages moving upstream. The one exception is Dean Hall which shows lesser error in the M2 component than the RMS. This further supports the statement made earlier that the larger errors at this station may be due primarily to errors in mean water level. Phase errors range from 15 minutes to 1 hour with larger phase errors in the upstream stations.

Velocities and Fluxes

Velocity was measured at 5 stations within the Charleston Harbor Estuary. The station locations were: US-17 Bridge, Daniel Island Bend, Woods Point, Wando River above 526, and Ashley River above US-17. Concurrent discrete flux measurements were taken at each of these locations over a spring and neap tidal cycle.

Model calibration attempted to minimize the errors in flux while not adversely affecting the resulting errors in velocity. Figures 8 and 9 present graphical and spectral comparisons of the measured versus simulated velocities along the principal axis at the Daniel Island and Woods Point stations. Figure 11 presents a graphical comparison of the observed and simulated fluxes at these same stations. The spectral distribution was not determined for the fluxes as they were generated directly from the velocities and there is no reason to believe their percent distribution would change. Examination of the velocities shows that the simulations under-predict the magnitude of the velocities while slightly over-predicting the fluxes. Discrete transect flux measurements are also plotted in Figure 11 to provide comparison with the simulations. The discrepancy between the flux and velocity comparisons identifies that the model, while simulating the overall discharge relatively well, misses the lateral distribution of the velocities in the cross-section. Were the model calibration performed solely to minimize the errors in velocity, the most likely outcome would have been the over-prediction of the overall flux magnitudes.

Comparison of the spectral plots identifies that the simulated velocities miss the energy within the sub-tidal, diurnal and higher harmonic energy bands while capturing the energy in the semi-diurnal band. The offset in the measured spectrum at Daniel Island Bend (within the regions outside of the peaks) is most likely due to noise in the velocity data at that station.

Figures 10 and 12 present normalized RMS errors for velocities and fluxes respectively. The results show errors in the flux comparisons ranging from 8 to 12 percent, while the velocity comparisons range from 15 to 22 percent. There is no clear pattern moving upstream as was seen in the water surface elevations, and the calibrations were most sensitive to the accurate representation of the upstream storage areas.

Table 2 provides a comparison of harmonic constituents for the velocity. As with water surface elevation, the percent errors in the M2 constituent closely mirror the RMS error results. The percent errors reflect normalization by the total signal rather than individual signals. The global under-simulation of the diurnal components that was found in the spectral comparisons are also.seen in the harmonic comparisons. Phase errors in the currents range from 10 to 40 minutes.

CONCLUSIONS AND RECOMMENDATIONS

A barotropic model for the Charleston Harbor Estuary has been developed. This constitutes the first of three phases in the development of a wasteload allocation model. Major conclusions from this study and recommendations for subsequent phases are summarized as follows:

- Utilizing an extensive grid, which incorporates all regions within the estuary which show impact from the tides, provided simulations of the water surface elevations, velocities and fluxes which ranged in error from 2 to 15 percent.

- The accuracy of the tidal simulations decreases moving upstream with much of the error in the upper reaches due to inaccurate simulation of the super elevation of mean water level.

- The use of calculated flows from continuous velocity and discrete discharge measurements provided a more appropriate calibration for the development of a wasteload allocation model. In addition, the comparisons identified the potential errors associated with calibrating to single velocity measurements in the development of accurate tidal prism simulations.

- The model was calibrated to within 8 to 12 percent error for tidally driven fluxes but was less accurate in simulating the cross-sectional velocity distribution.

- Subsequent modeling efforts will focus on the application of a 2-D or fully 3-D baroclinic hydrodynamic model to assess the levels of impact of the density driven circulation on residual flows, and quantification of the model's ability to accurately transport conservative substances. These results will be compared with measured values of salinity, temperature, and residual velocity at various stations within the Estuary.

REFERENCES

Bower, D.E., Sanders, C.L., Conrads, P.A., 1993. Retention time simulation for Bushy Park Reservoir near Charleston, South Carolina. U.S. Geological Survey Water-Resources Investigations Report 93-4079.

Conrads, P.A., Cooney, T.W., and Long, K.B., 1996. Hydrologic and water-quality data from selected sites in the Charleston Harbor Estuary and tributary rivers, South Carolina, water years 1992-1995. U.S. Geological Survey Open-File Report 96-418.

Conrads, P.A., and Smith, P.A., 1996. Simulation of water level, streamflow, and mass transport for the Cooper and Wando Rivers near Charleston, South Carolina, 1992-1995. U.S. Geological Survey Water-Resources Investigations Report 96-4237.Spaulding, M.L., 1984. A vertically averaged circulation model using boundary –fitted coordinates. Journal of Physical Oceanography, 14, p. 973-982.

Hess, K.W., and K.T. Bosley, 1992. Techniques for validation of a model for Tampa Bay. Proceedings, 2nd International Conference on Estuarine and Coastal Modeling, Tampa, Florida, November 11-13, 1991. 83-94.

Muin, M. and M.L. Spaulding, 1993. A vertically averaged boundary-fitted circulation model in spherical coordinates for coastal seas. Department of Ocean Engineering, University of Rhode Island, Narragansett, Rhode Island.

Swanson, J.C., 1986. A three-dimensional numerical model system of coastal circulation and water quality. Ph.D. Dissertation, Department of Ocean Engineering, University of Rhode Island, Kingston, Rhode Island.

		Data Amplitude (meters)	Model Amplitude (meters)	Error %	Data Phase (hours)	Model Phase (hours)	Error (hours)
Fort Sumter (2172100)	M2	0.72	0.73	-0.8	-2.06	-2.04	-0.02
	S2	0.13	0.13	0.0	-3.95	-3.85	-0.1
	N2	0.21	0.22	-0.8	6.16	6.24	-0.08
	K1	0.06	0.06	0.0	-9.03	-9	-0.03
	O1	0.07	0.07	0.0	8.26	8.26	0
Custom House (2172710)	M2	0.73	0.75	-1.7	-1.93	-1.91	-0.02
	S2	0.13	0.13	0.0	-3.7	-3.68	-0.02
	N2	0.22	0.23	-0.8	-6.27	-6.26	-0.01
	K1	0.06	0.06	0.0	-8.73	-8.86	0.13
	O1	0.07	0.07	0.0	8.4	8.36	0.04
US-17 Bridge (LT-02)	M2	0.73	0.76	-2.5	-1.93	-1.86	-0.07
	S2	0.13	0.13	0.0	-3.7	-3.61	-0.09
	N2	0.22	0.23	-0.8	-6.27	-6.21	-0.06
	K1	0.06	0.06	0.0	-8.73	-8.81	0.08
	O1	0.07	0.07	0.0	8.4	8.4	0
Daniel Island Bend (LT-03)	M2	0.74	0.77	-2.4	-1.44	-1.77	0.33
	S2	0.13	0.14	-0.8	-3.06	-3.5	0.44
	N2	0.23	0.24	-0.8	-5.65	-6.09	0.44
	K1	0.07	0.06	0.8	-7.73	-8.7	0.97
	O1	0.07	0.07	0.0	9.07	8.5	0.57
Cooper @ Goose Creek (21720675)	M2	0.74	0.77	-2.4	-1.44	-1.57	0.13
	S2	0.13	0.14	-0.8	-3.06	-3.26	0.2
	N2	0.23	0.24	-0.8	-5.65	-5.86	0.21
	K1	0.07	0.06	0.8	-7.74	-8.44	0.7
	O1	0.07	0.07	0.0	9.07	8.74	0.33
Cooper @ Mobay (2172053)	M2	0.65	0.71	-5.5	-1.04	-1.07	0.03
	S2	0.12	0.12	0.0	-2.54	-2.69	0.15
	N2	0.2	0.21	-0.9	-5.13	-5.27	0.14
	K1	0.07	0.06	0.9	-6.36	-7.74	1.38
	O1	0.06	0.07	-0.9	9.9	9.41	0.49
Cooper @ Dean Hall (2172050)	M2	0.48	0.53	-8.8	-0.03	0.64	-0.67
	S2	0.07	0.08	-1.3	-1.42	-0.58	-0.84
	N2	0.13	0.17	-5.3	-3.8	-3.15	-0.65
	K1	0.09	0.06	3.7	-4.25	-5.66	1.41
	O1	0.06	0.07	-1.3	11.73	11.42	0.31

Table 1: Comparison of Simulated Versus Measured Harmonic Constituents for Water Surface Elevation

		Data Major Axis Amplitude (cm/s)	Model Major Axis Amplitude (cm/s)	Error %	Data Angle (deg)	Model Angle (deg)	Error	Data Phase (hour)	Model Phase (hour)	Error
	M2	70.2	52.8	15.8	-100.7	-103.8	3.1	2.0	2.1	-0.1
Cooper @	S2	11.8	9.5	2.1	-105.3	-103.7	-1.6	0.4	0.3	0.1
US-17 Bridge	N2	18.0	15.5	2.2	-105.8	-103.6	-2.2	-2.0	-2.1	0.1
(LT-02)	K1	5.4	2.5	2.7	-99.2	-103.7	4.5	-1.5	-1.7	0.2
	O1	5.0	2.5	2.2	80.2	76.3	3.9	2.2	3.1	-1.0
	M2	62.1	47.9	14.2	-74.6	-76.1	1.5	2.7	2.8	-0.1
Cooper @	S2	9.6	7.8	1.7	-81.3	-76.1	-5.2	0.7	1.0	-0.3
Daniel Island	N2	17.6	13.6	4.1	-83.3	-76.2	-7.1	-1.6	-1.3	-0.3
Bend (LT-03)	K1	6.8	2.6	4.3	-80.6	-76.2	-4.4	0.7	-0.8	1.5
	O1	3.6	2.8	0.8	106.4	103.8	2.6	4.0	4.0	0.0
	M2	54.1	47.6	7.5	-68.4	-60.6	-7.8	3.5	2.8	0.7
Cooper @	S2	8.2	7.7	0.6	-58.6	-60.7	2.2	1.7	1.3	0.4
Woods Point	N2	14.1	14.1	0.0	-60.3	-60.7	0.4	-0.3	-1.0	0.7
(MMB-1)	K1	4.6	2.4	2.6	-62.4	-60.4	-2.0	-0.8	-0.8	0.0
	O1	4.5	2.7	2.2	-63.0	-60.4	-2.6	-7.0	-8.9	1.9
	M2	44.5	36.8	10.0	-112.2	-117.1	4.9	1.8	1.7	0.1
Wando @	S2	9.4	7.4	2.6	-109.0	-117.0	8.0	0.2	0.0	0.2
"N-16"	N2	17.5	11.8	7.4	-110.2	-117.0	6.8	-2.0	-2.4	0.4
(MMB-3)	K1	2.9	1.5	1.9	-115.1	-117.1	2.0	-2.2	-2.3	0.1
	O1	2.8	1.5	1.7	65.2	63.2	2.0	2.6	2.5	0.1
	M2	55.9	54.0	2.0	58.1	64.3	-6.2	-4.1	-4.3	0.2
Ashley @	S2	13.1	10.7	2.5	-117.3	-115.6	-1.7	0.3	0.4	-0.1
"C-13"	N2	20.7	17.7	3.1	-116.1	-115.4	-0.7	-2.2	-2.2	-0.1
(MMB-2)	K1	3.7	2.2	1.6	-115.2	-115.5	0.3	-2.2	-2.0	-0.1
	O1	3.6	2.3	1.4	62.1	64.6	-2.5	2.4	2.6	-0.2

Table 2: Comparison of Simulated Versus Measured Current Harmonics

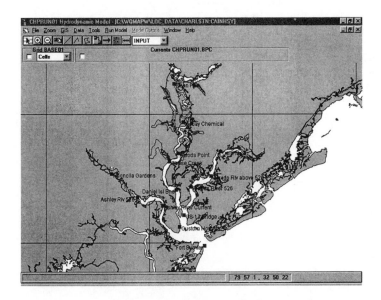

Figure 1: Charleston Harbor Estuary with Data Collection Stations

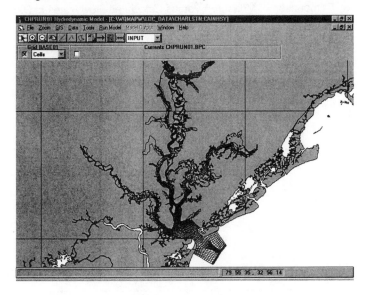

Figure 2:   Charleston Harbor Estuary with Model Grid

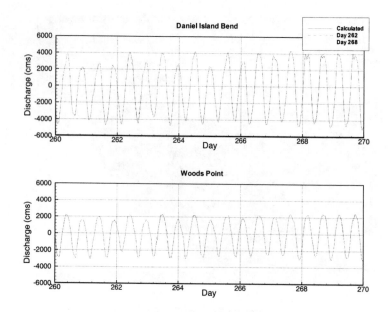

Figure 3: Comparison of Measured Versus Calculated Flow Rates at Daniel
Island Bend and Woods Point.

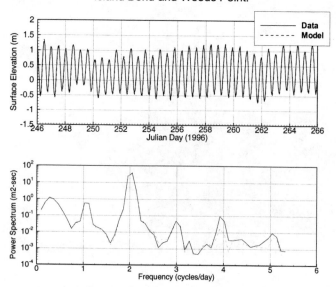

Figure 4:   Graphical and Spectral Comparison of Simulated Versus Measured
Water Surface Elevations at Customs House

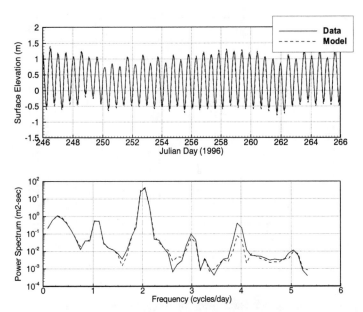

Figure 5:   Graphical and Spectral Comparison of Simulated Versus Measured
Water Surface Elevations at Goose Creek

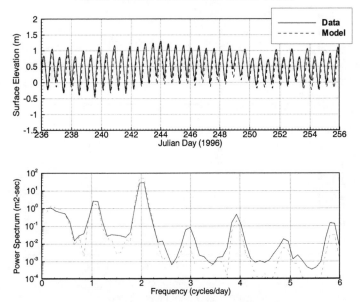

Figure 6:   Graphical and Spectral Comparison of Simulated Versus Measured
Water Surface Elevations at Dean Hall

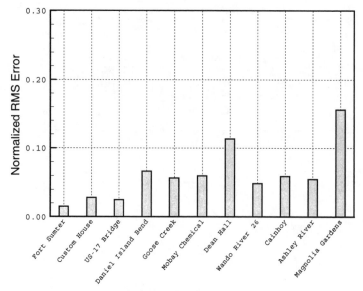

Figure 7:   Normalized RMS Error for Simulated Versus Measured Water
Surface Elevations at All Stations

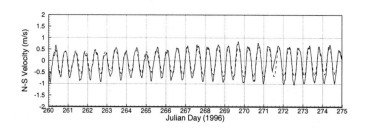

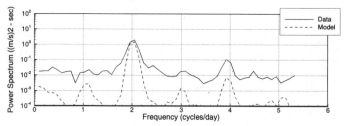

Figure 8:   Graphical and Spectral Comparison of Simulated Versus Measured
Currents at Daniel Island Bend

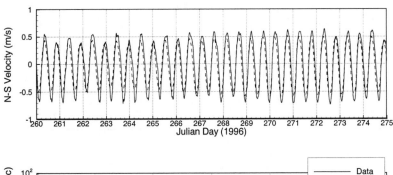

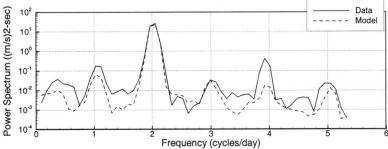

Figure 9:   Graphical and Spectral Comparison of Simulated Versus Measured
Currents at Woods Point

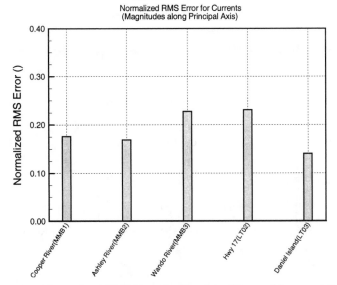

Figure 10:   Normalized RMS Error for Simulated Versus Measured Currents at
All Stations

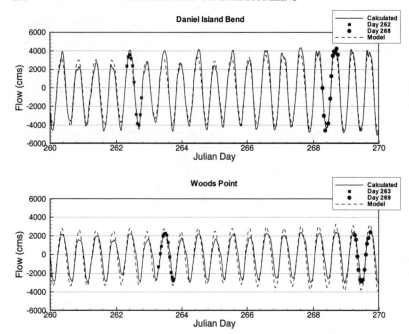

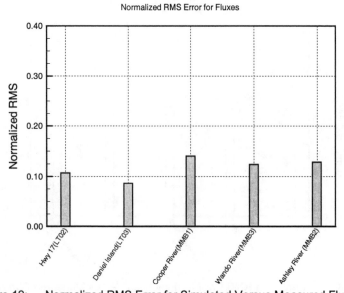

Figure 11:   Graphical Comparison of Simulated Versus Calculated Fluxes at
              Daniel Island Bend and Woods Point

Figure 12:   Normalized RMS Error for Simulated Versus Measured Fluxes at
              All Stations

COASTMAP, An integrated system for monitoring and modeling of coastal waters:
Application to Greenwich Bay

M. L. Spaulding[1], S. Sankaranarayanan[1], L. Erikson[1], T. Fake[1], and T. Opishinski[2]

Abstract

COASTMAP is an integrated environmental modeling and monitoring system designed
for application to estuarine and coastal seas. COASTMAP allows the user to collect,
manipulate, display, and archive environmental data through embedded environmental data
management tools (e.g. time series analysis including filtering, power spectral analysis, and
harmonic analysis) and a geographic information system. Data collection and transmission
can be by use of either conventional instrumentation or through a real time radio/satellite
telemetry based monitoring system. The system includes a suite of environmental models
(three dimensional hydrodynamics, particle based pollutant transport and runoff) to predict
the dynamics in the operational area and to assimilate real time data into the models to
allow hindcasting, nowcasting and forecasting. COASTMAP operates on a PC with a
Windows based user interface. It is controlled by pull down menus and makes extensive
use of color graphics to display model predictions (e.g. animations, plots, 3-D perspective
views) and the results of data analysis. The software is designed using a shell based
architecture making application to any geographic area simple and fast.

In the present paper the application of the system to monitor and model the circulation and
pollutant transport in Greenwich Bay, a small embayment in Narragansett Bay, is described.
Three real time water quality monitoring stations were established and collected data on
sea surface elevation, temperature, salinity, and dissolved oxygen. In addition wind speed
and direction were also collected at one location. At the offshore station an upward looking
ADCP collected current data. All data were sent to the base station via radio telemetry.
A boundary fitted coordinate hydrodynamic model and a particle based pollutant transport
model were employed to predict the tidal and wind driven circulation and dispersion of

---

[1] Ocean Engineering, University of Rhode Island, Narragansett, RI 02882
[2] Ocean Data Equipment Corporation, East Walpole, MA 02032-1155

material released from point sources in the system. Simulations of the circulation were in very good agreement with the surface elevations and currents. Simulations of two dye release experiments in small coves within the bay were in good agreement with field data. In addition simulations were performed to estimate the flushing time in the central section of the bay and were in good agreement with independent box model calculations.

Introduction

Application of numerical models to solve a wide variety of marine environmental and engineering problems is now commonplace as evidenced by the proceedings of the Estuarine and Coastal Modeling conference series. In a typical application the study area is discretized with a system of grid cells and the conservation equations governing the problem of interest are solved on the grid to predict the temporal evolution of the spatial distribution of variables (e.g. currents, sea surface elevation, pollutant concentrations). Data preparation, gridding, specification of model parameters, simulations, model calibration/validation, and visualization are normally done as separate steps or tasks in the analysis. Model predictions are generally displayed as spatial distributions at fixed times, as time series at fixed locations or as animations. The procedures to perform the various tasks are straightforward but often time consuming. This is particularly true for applications to new areas and even for different applications (higher grid resolution) in the same area.

Based on evolving work in oil spill and water quality modeling, Spaulding and Howlett (1995) demonstrated an integrated shell based approach to marine environmental problems. In this strategy the process models are integrated with a graphical user interface that allows the user to set up, run, and display model predictions. A geographic information system (GIS) is included to assist in the management of marine environmental data: to prepare data for input to the models and as a background on which model results can be presented. The shell based architecture allows the models to be applied rapidly and simply to any location in the world at a user specified resolution. Spaulding and Howlett (1995) summarized the application of this shell based approach to evaluate the water quality impacts of combined sewer overflows, to predict the dispersion of drilling fluids, to estimate shallow water wave transformations in a harbor, and to predict the movement of an oil spill.

Spaulding et al (1996) and Opishinski and Spaulding (1995) extended this basic approach to develop an integrated system for environmental monitoring and modeling. Application of the system for typical hydrodynamic and water quality problems was described in Opishinski and Spaulding (1995, 1996). Spaulding and Opishinski (1996) and

Spaulding et al (1996) described its application to oil spill response and in particular its use to nowcast and hindcast the movement of oil tracking buoys and the 1996 North Cape barge spill off the southern coast of Rhode Island.

The present paper begins with a brief overview of COASTMAP focusing on its major operational features and environmental models. Application of the system to Greenwich Bay, a small bay within Narragansett Bay, is then given. The application includes descriptions of the installation and operation of the real time sensing system, the analysis of the time series and the use of the resulting data and hydrodynamic and pollutant transport models to understand the circulation and movement of contaminants in the system. Conclusions and references are then provided.

The COASTMAP System

COASTMAP is an integrated system for environmental monitoring and modeling of fresh and marine water systems. The system allows the user to collect, manipulate, analyze, display, and archive environmental data; to perform simulations with a suite of environmental models (e.g. hydrodynamic, pollutant transport and fate, wave, oil spill) in order to predict the dynamics in the operational area, and to assimilate real time data into the models to allow hindcasting, nowcasting, and forecasting. COASTMAP operates on a personal computer, features a Windows based user interface, and is controlled by pull-down menus and point-and-click techniques. Color graphics are used extensively to present the results of data analysis and model predictions (e.g. graphs, tables, animations). The software is designed using a shell based architecture making application to any geographic area simple and fast. Some of the key features of the software are summarized here.

The location function within COASTMAP, and the core of the shell based architecture, allows the user to set up the system for a base area. The area, defined by a base map, can be at any resolution and at any location in the world. The base map can be of vector or raster form and provides the reference map for all model applications and GIS data display. Base maps are prepared by digitizing a map or by importing data bases that give the location of shoreline and other water/land features. All GIS data, monitoring data, and model input/output are linked to the location by the file structure. The user can rapidly switch from one location to another by simply "pointing" the system at the appropriate data set. The number of locations is limited only by the amount of available storage space on the operational computer.

The data function provides access to the real time monitoring module, the embedded GIS, environmental data management tools, and gridding modules for the environmental models.

The real time monitoring module allows the user to directly interface with the instruments and the associated communication system. The user, via the base communication system, can retrieve data, set the instrument sampling and storage intervals, turn sensors on and off, and alter the operation of the communication system. He can also add, update, and delete observation stations from the system. A status board is also provided to allow the user to view the output from each sensor in the system (e.g. last value, most recent time series) and the operational status of the power supply. The user can also link directly to a data processing module that allows standard time series analyses (e.g. filtering [high, low and band pass with selectable filter type and characteristics], harmonic composition and decomposition, and spectral analysis) and plotting of the resulting data. This analysis can be done either on data archived in the system or the most recently collected information. Output from this module can be linked directly to the environmental models. Data can be provided via radio telemetry, satellite, cell telephone or the Internet. The system can also be configured with an automated FTP module that can routinely distribute data to a selected distribution list.

The gridding system for the environmental models is typically located in the data section. The gridding system is selected to match the needs of the environmental models with square, rectangular, boundary conforming, triangular or other options. In all cases the gridding system features a split window view where the two windows show the grid system in physical and computational space. The gridding is largely automated but the user is given extensive hands-on control.

The embedded GIS allows the user to input, store, manipulate, analyze, and display geographically referenced information. The GIS is fast, user friendly, interactive, and simple to operate. The GIS is often used to provide input data to the models. It is also helpful in the presentation and interpretation of model predictions. Additional information can be linked to the GIS including charts, graphs, tables, tutorials, bibliographies, photographs, animations, text, or other software (e.g. spreadsheets, word processor).

COASTMAP is configured to incorporate a wide range of environmental models. These are accessed through a model directory. The models are setup through the user interface and supporting data is accessed through the location specific data sets and the GIS. For the present application hydrodynamic and pollutant transport models are included in the system. These are described in more detail below.

The hydrodynamic model included in the system solves the conservation of water mass, momentum, salt and heat equations on a spherical, non-orthogonal, boundary conforming grid system and is applicable for estuarine and coastal areas (Muin and Spaulding, 1996, 1997a,b). The eddy viscosities can be specified by the user or obtained via a one-equation turbulent kinetic energy model. The velocities are represented in their contravariant form.

A sigma stretching system is used to map the free surface and bottom to resolve bathymetric variations. The model employs a split mode solution methodology. In the exterior (vertically averaged) mode the Helmholtz equation, given in terms of the sea surface elevation, is solved by a semi-implicit algorithm to ease the time step restrictions normally imposed by gravity wave propagation. In the interior (vertical structure) mode the flow is predicted by an explicit finite difference method, except that the vertical diffusion term is treated implicitly. The time step generally remains the same for both exterior and interior modes. Computations are performed on a space-staggered grid system in the horizontal and a non-staggered system in the vertical. Time is discretized using a three level scheme. Muin and Spaulding (1996, 1997a) provide a detailed description of the governing equations, numerical solution methodology, and in-depth testing against analytic solutions for two and three dimensional flow problems.

The pollutant transport model is based on a random walk solution to the convective diffusion equation. In operation particles are released from the pollutant source with a specified mass for each particle. By appropriately selecting the particle mass and release rate time dependent releases can be modeled. Particles can be assigned rise/settling velocities to represent buoyant/settling material. They can also be given growth/decay rates so that the mass increases/decreases in time. The particles are advected by the current field and diffused using a random walk procedure. The currents are obtained by interpolation from the three dimensional boundary fitted hydrodynamic model results. The impact of sub-grid scale eddies and turbulence on the transport of pollutants are parameterized by user specified horizontal and vertical turbulent dispersion coefficients. The model predicts the location of each particle with time. Predictions are normally given in the form of animations of the time evolution of the spatial distribution of the particles. This particle data can also be presented in terms of concentration by overlaying a grid system and calculating the mass per unit volume.

The model output module of COASTMAP allows the user to present the results of any simulations performed. The user opens the scenario of interest and then selects the output desired. As an example for the hydrodynamics model, the user selects the variable of interest (currents, temperature, surface elevation, salinity) and the vertical level (sigma layer) and then animates the results (color contours or vectors). The animation controls allow forward, step forward, reverse, rewind, and stop options. The user can also select a section view window which allows the variable of interest to be visualized along the section simultaneously as the plan view is animated.

Application of COASTMAP to Greenwich Bay

Greenwich Bay (Figure 1) is a small bay located on the western side of Narragansett Bay. The bay is composed of a central basin and four smaller coves (Warwick, Brush Neck,

Apponaug, and Greenwich Coves). The mean low water depths in the coves are about 1 to 2.0 m while the central bay has a mean depth of 2.7 m. The bay has a deep (10 m) entry channel near its mouth which decreases in depth toward the west. The east-west oriented channel is bounded by broad shallow areas to the north and south. The circulation in the bay is driven mainly by the semi-diurnal tides with a mean tidal amplitude of about 0.55 m and maximum flood/ebb tidal current speed on the order of 15 cm/sec. Mean annual average freshwater inputs to the bay are approximately 1.3 m^3/sec. Most of the discharge is via Hardig Brook to the head of Apponaug Cove and from the East Greenwich waste water treatment facility (WWTF) to the center of Greenwich Cove. Because of the low freshwater input rates, the salinities are generally uniform across the central portion of the bay at 27 to 30 ppt and depend on salinities in the adjacent upper west passage of Narragansett Bay. Salinities in the coves, particularly Apponaug Cove, vary substantially with freshwater inputs with lower values on the order of 21 ppt. During high flow periods the water column in the coves can be highly stratified but vertical density differences are greatly reduced due to tidal mixing in the central portion of the main bay

The COASTMAP system was applied to Greenwich Bay. Three monitoring stations were installed: two land-based (Apponaug and Greenwich Cove) and the third buoy-based. The land-based stations indicate that the radio transmitters for those stations were located on land. The location of the three stations are shown in Figure 1. ENDECO/YSI 6000 water quality sensors were located at each of the stations and measured pressure, salinity, temperature, and dissolved oxygen approximately every 15 minutes. The sensors were located 25 cm above the sea bed at Greenwich Cove (depth of 3 m) and Apponaug Cove (depth of 2.0 m). At the buoy station (depth of 5 m) the sensor was located approximately 30 cm above the sea bed. In addition a Young meteorological station, with data logger, was placed at the Apponaug Cove station and recorded wind speed, direction, and barometric pressure. A bottom- mounted 75 kHz Sontek ADCP was located at the buoy station and configured to measure the vertically averaged current velocity. All stations were equipped with ENDECO/YSI 1240 radio transmitter/receivers to allow for radio communication. All systems were battery powered. Solar panels were used to recharge the batteries. The base communication station for the system was located at the University of Rhode Island, Narragansett Bay Campus, 20 km to the south on the western side of Narragansett Bay. Because of the distance and land-based obstructions between the Greenwich Bay sensors and the Bay Campus a repeater receiver/transmitter station was placed at Quonset Point (about mid-way between the two locations).

The base communication station was interfaced with a Pentium II personal computer. This PC acted as the server for the system and allowed control of the sensor system. The system was placed in operation in May 1996 and was removed in October 1997. Sampling was typically performed every 15 minutes. The water quality sensors were checked and cleaned approximately every two weeks. Data recovery was highly variable throughout the period due to problems with the power supply, repeated vandalism of the buoy sensors, partial ice coverage and its impact on the land-based stations in the winter, flooding of the sensor housings due to unexplained seal leaks, lightning strikes to the weather and telemetry stations, and repeated attempts to steal the offshore buoy and its three leg mooring system.

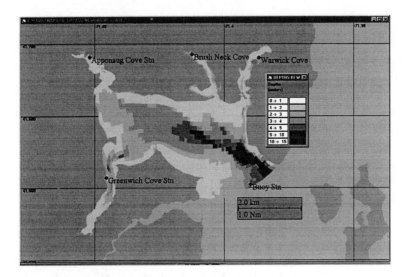

Figure 1   Greenwich Bay study area showing the location of the COASTMAP monitoring stations.

Given the modeling based nature of the conference the focus of the presentation given below is on the application of the models to understand the circulation and pollutant transport in the bay. Only a brief example of the data collection and analysis portion of the system is given.

Data Collection and Analysis

A typical output from the Greenwich Cove station for October 1996 is shown in Figure 2. The figure shows a stack plot of the temperature, salinity, dissolved oxygen, and sea surface elevation versus time. Data were sampled every 15 minutes.

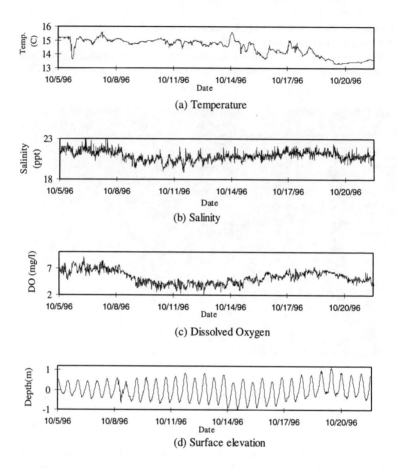

(a) Temperature

(b) Salinity

(c) Dissolved Oxygen

(d) Surface elevation

Figure 2   Example of real time output from the COASTMAP for Greenwich Cove. Stack plots of temperature, salinity, dissolved oxygen and surface elevation versus time.

COASTMAP was used to analyze the data. As an example Figure 3 shows the results of spectral analyses on the sea surface elevation and salinity time series shown in Figure 2. The surface elevation shows that the dominant energy is at the lunar, semi-diurnal forcing period (1.93 cycles/day). The M_4 and M_6, harmonics of the M_2, are also clearly evident. These results are consistent with a tidal harmonics analysis of the record given in Table 1 which show the amplitudes and phases for the major tidal constituents. The harmonic constants for the NOAA/NOS Newport, RI station are given for reference. The M_2 tide is slightly amplified (3.8 %) compared to the value at Newport and lagged by about 18.6 degrees or 0.64 hours. Harmonic analysis of the sea surface elevations from the buoy and Apponaug and the Greenwich Cove stations shows that they can't be distinguished from one another.

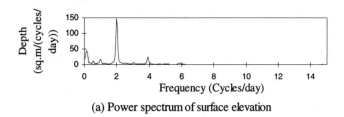

(a) Power spectrum of surface elevation

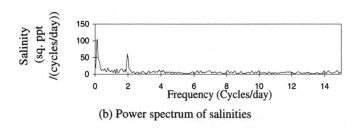

(b) Power spectrum of salinities

Figure 3   Power spectra of surface elevation and salinity for the Greenwich Cove station.

The lower portion of Figure 3 shows the results of the spectral analysis on the salinity data. While the M_2 tidal signal is clearly noted, the majority of the energy is at lower frequencies corresponding to variations in the freshwater input to the cove. Variations at the diurnal and the M_4 and M_6 tidal frequencies are masked by the natural variability in the system.

Hydrodynamics

COASTMAP's boundary fitted coordinate hydrodynamic model was applied in a two dimensional, vertically averaged mode to the Greenwich bay study area using the grid system shown in Figure 4. The grid system was designed to provide good resolution throughout the main body of the study area and higher resolution in the small coves on the northern and western sides of the bay. The ability of the boundary conforming system to allow variable aspect ratios in grid length in the along and cross channel directions is clearly seen in the grid design for these small coves and particularly in the narrow restrictions. Depth data for the area (Figure 1) were derived from the NOAA bathymetry (Chart # 13224) for Narragansett Bay. Data were mapped to provide a mean depth for each boundary fitted grid cell in the study area. The model was forced by the sea surface elevation data collected from the buoy based station and assumed to be spatially invariant across the open boundary of the model domain.

Table 1 Amplitudes and Greenwich phases at the Newport and Greenwich Cove station.

Station	Greenwich Cove		Newport	
Tidal constituent	Amplitude (m)	Phase (degrees)	Amplitude (m)	Phase (degrees)
M_2	0.549	235.15	0.529	216.4
N_2	0.133	223.26	0.127	202.6
S_2	0.127	254.68	0.123	232.9
M_4	0.073	152.12	0.056	110.8
M_6	0.10	261.81	0.000	128.9
K_1	0.068	101.43	0.064	93.1
SA	0.187	203.29	0.000	169.6
O_1	0.045	134.60	0.050	129.0
P_1	0.025	98.41	0.020	103.6

To calibrate the model predictions were performed varying the quadratic bottom drag coefficient from 0.001 to 0.005. Results showed that the predicted surface elevation at the measurement station was relatively insensitive to the drag coefficient. This result is consistent with the limited length of the system and the relatively uniform bathymetry. Based on a comparison between model predictions and the currents at the buoy the drag coefficient was set at 0.0035.

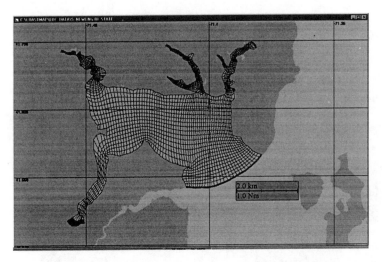

Figure 4   Boundary fitted coordinate hydrodynamic model grid system for
             Greenwich Bay.

Figure 5 shows the model predicted maximum flood tide current. The currents are
predicted to be relatively uniform across the central bay. The increase of currents in the
vicinity of headlands on the south side of the bay is clearly noted. Toward the western end
of the bay the current speeds show a sharp decrease as the flood tide water enters
Apponaug and Greenwich Coves. Current speeds generally decrease as one moves toward
the heads of the coves. The current speeds can increase quite substantially locally,
however, if the cove widths decrease markedly. This behavior is clearly observed in
several locations in Apponaug, Brush Neck, and Warwick Coves. The ebb tide currents
look very similar to those on flood, except in the opposite direction. The model predicts
essentially no tidal mean flows because of the smooth bathymetry. Animations of the sea
surface elevation contours show that they are in phase (within several minutes) throughout
the area. Sea surface elevations are predicted to be phase lagged about 90 degrees from
the velocity, consistent with simple standing wave theory.

Figure 6 shows a comparison between the model predicted and observed sea surface
elevations at Apponaug Cove for a five day period in mid-September 1996. Model
predictions are generally in good agreement with the observations accurately representing
the tidal phasing and correctly estimating the tidal amplitudes. Predictions of individual
tidal troughs can be in error as much as 15 to 20 cm. Note that the depths of individual
cells, near the heads of the coves were artificially increased to allow the model to run
without error from dry cells at low tide. Comparison to ADCP measurements were made

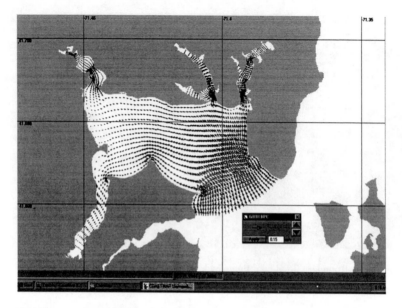

Figure 5   Hydrodynamic model predicted maximum flood tide currents for
               Greenwich Bay.

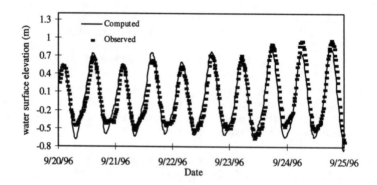

Figure 6   Comparison of model predicted and observed surface elevation in
               Apponaug Cove.

(not shown) and showed the model predictions were consistent with the observations. The observations displayed significant variability at wind forced time scales.

Pollutant Transport

To test the pollutant transport model in COASTMAP a simulation of a dye release event was performed. The experiment (Turner, 1986) consisted of a 5 day continuous release of Rhodamine WT dye through the East Greenwich's sewage treatment plant discharge line which ultimately discharges into Greenwich Cove. The dye was premixed with seawater to adjust its density so that it would be similar to that of the receiving water. The continuous dye release was made using a peristaltic pump so that the concentration of dye in the discharge remained constant even though the flow rate varied. The release started on December 9, 1985 at 0500 EST (high slack water) and stopped at 1600 on December 14, 1985. The dye was tracked along six transect lines from the cove mouth to its interior using a Turner Design 111 flow-through fluorometer. Tracking commenced at the start of the dye release and ended on December 18, about 3 days after the end of the release. Measurements were collected the day before the release to provide background values for fluorescence. Fluorescence measurements were made approximately every 2 hours for the first day (daylight hours only) and then at every low tide (daylight hours) until the end of the experiment.

This dye release event was simulated with COASTMAP. Tidal currents for the period of the experiment were generated using the hydrodynamic model driven by tidal constituent data at the model open boundary for the time period of interest. The particle based pollutant transport model was then run to simulate the release. One hundred particles were used to represent the release. The horizontal dispersion coefficient was set at 2 m²/sec. The predicted particle locations were converted to concentration measurements by overlaying a grid and assigning the mass in each particle to a grid. Figure 7 shows a comparison of the model predictions to the observations at low tide on December 10, 1985, shortly after the release. The model predictions are broadly consistent with the observations both showing high concentrations near the discharge point and the same spatial extent of transport (i.e. compare the location of the 0.5 mg/l contour line).

To evaluate the model's ability to predict the flushing time in the area the observed and model predicted mean cove concentrations versus time are shown in Figure 8. The cove mean concentration is observed and predicted to increase with time as dye released from the discharge mixes with the cove waters. After about 150 hours the cove is observed and predicted to be in approximate equilibrium with the rate of dye input balanced by the amount flushed from the cove. When the dye release stops the dye continues to be removed by flushing and the concentration declines rapidly. The model predicted exponential rate of decrease is -0.035 per hour or a flushing time of about 1.2 days, where the flushing time is defined as the time for the concentration to decrease by e. This is in very good agreement with the observed value of -0.039 per hour or a flushing time of 1.07 days. The coefficients of determination of both the model and observed flushing times were very good (R^2 greater than 0.92).

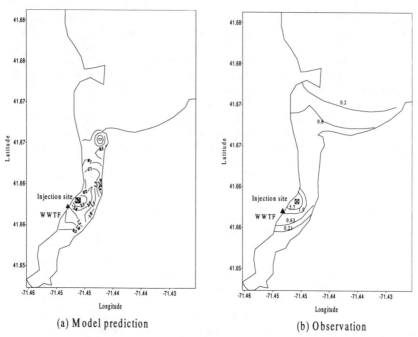

(a) Model prediction   (b) Observation

Figure 7   Comparison of model predicted and observed distribution of dye on December 10, 1985 for Greenwich Cove

The model was compared to a second dye release performed in Apponaug Cove in September 1996. Dye was continuously released for a period of 72 hours from a point in the upper cove, just south of the very narrow constriction, starting on September 20, 1996. The dye was premixed with receiving water to adjust its density. The freshwater flows for the time period were 0.25 m^3/sec in Apponaug Cove and 1.1 m^3/sec in Greenwich Cove. The water column concentrations were tracked using a towed fluorometer with GPS positioning for a period of five days after the release started. Measurements were generally made at high and low tide during daylight hours. The cove mean concentrations were computed and plotted versus time, as in the Greenwich Cove experiment, using the high and low tide data. Analysis of the cove mean dye concentration data gave an exponential decrease rate of -0.0806 per hour. This gives a flushing time of 12.4 hours for Apponaug Cove. Model predictions were performed for the dye release following the same strategy as used in Greenwich Cove and resulted in a flushing time of 17.2 hours. The predicted mean flushing time was substantially longer than the observed value. It is speculated that the observed flushing time was faster than predicted because of the vertical stratification that was present during the early stages of the experiment caused by a substantial rainfall event just prior to the dye experiment.

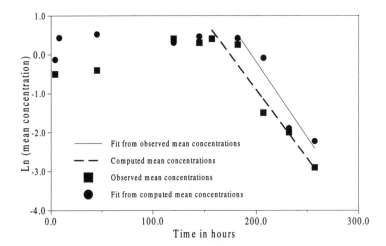

Figure 8   Observed and predicted mean cove dye concentrations versus time for the Greenwich Cove dye release study.

The particle based model was next used to estimate the flushing time of the central portion of Greenwich Bay. In this simulation the model was run for several weeks using tidal constituent data forcing at the model open boundary. The tidal current data were then stored for this period and used as input to the particle based model. The particle model was run assuming an instantaneous point release in the center of the bay. The horizontal dispersion coefficient was assumed at 2 m^2/sec. The simulation was then run for one week. The number of particles in a box that constituted the central portion of the bay versus time was plotted and the slope of this line used to determine the bay flushing time. This analysis gave a value of 3.5 days. If horizontal dispersion coefficients of 1 and 5 m^2/sec were used, flushing times of 3.9 and 2.6 days, respectively were obtained. These simulations showed that the flushing time decreased as the dispersion coefficient increased, consistent with theory.

Erikson (1998) performed a detailed flushing study for Greenwich bay using a standard box modeling methodology (Officer, 1980; Jayko and Swanson, 1988; Swanson and Mendelsohn, 1995). She divided the bay into seven, two layer boxes; with one box for each of the coves and three boxes for the main basin (western, center, and eastern end) (see Figure 10). She calculated the volume weighted salinities for the boxes based on an analysis of twenty seven (27) Graduate School of Oceanography surveys which collected salinity, temperature, and dissolved oxygen data every few weeks at 14 stations in Greenwich Bay and its coves from August 1995 to May 1997. The data were sampled at 0.5 to 1 m depth intervals. Freshwater inputs to each of the boxes was estimated from river discharge, ground water input, and discharge from the East Greenwich municipal treatment facility. She corrected the freshwater input estimates for rainfall and evaporative losses. Figure 9 shows a plot of the flushing time versus freshwater input for the central portion of Greenwich Bay. Each data point was calculated with the box model. A least squares regression line has been fit through the data and shows that the flushing time increases substantially as the freshwater input rate decreases. The regression is quite good with a coefficient of determination of 0.80. A prediction of the flushing time for a freshwater

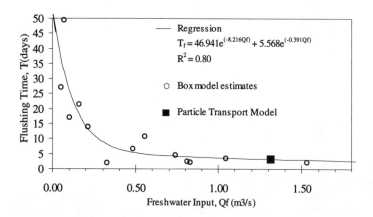

Figure 9   Flushing time versus freshwater input for Greenwich Bay based on application of a box model. Results from COASTMAP's particle model simulations are also shown for the mean freshwater input case.

input rate of 1.3 m³/sec (approximately the mean annual freshwater flow rate) was made using COASTMAP and is shown by the solid square. COASTMAP's prediction is in excellent agreement with Erikson's box model based estimate. Attempts to predict the flushing time for the very low freshwater flow events (0.2 m³/sec) showed results similar to the mean freshwater flow case. The lack of success in reproducing the box model results at very low flows appears to be due to an inadequate grid resolution in the vicinity of the principal freshwater sources and perhaps changes in horizontal mixing at low flow conditions.

In an Food and Drug Administration study (FDA, 1993) the fecal coliform levels and loading rates from the rivers and storm sewer overflows discharging into Greenwich Bay were monitored at twenty eight stations in the bay and from twenty six sources from April 5-14, 1993. There were two storm events between April 10 and 12, 1993 yielding a total rainfall of about 3.25 cm. The rainfall resulted in dramatically increased loads and levels of fecal coliform in the bay. Fecal levels increased from 7-10 MPN/100 ml before the storm throughout the bay to peak values of over 500 MPN/100 ml in Apponaug Cove and 30 MPN/100 ml in the central bay on April 13. Based on their analysis of the fecal concentration time history after the storm, FDA (1993) estimated a return to background levels within 3 to 4 days after the rainfall event ceased.

COASTMAP was applied to predict the behavior of fecal coliforms during this event. The two dimensional, vertically averaged circulation was simulated using the observed tidal elevations during the study to force the model. Using hydrodynamic model predicted currents, the particle based transport model was employed to predict the fecal coliform concentrations over time. The five largest loads (Apponaug Cove - 6.12E11MPN/day, WWTF-9.3E10 MPN/day, overflow 306 (Greenwich Cove) -1.64E10 MPN/day, and overflows 301 and 302 (Brush Neck Cove) - 1.38E10 MPN/day) were used as input. The loads reported by FDA were based on flows and concentrations that were measured once during each day and assumed that this value represented the load for the entire day. This procedure does not give an accurate sense of the real load from the source with time, particularly during storm events when loads vary rapidly. To address this problem the loads were scaled by a factor of 5 based on analysis of fecal coliform pollutographs from Apponaug Cove during a wet weather event in December 1994. This scaling is very approximate. The model assumed that the fecal coliforms were a conservative substance. The dispersion coefficient was set at 2 m²/sec. Figure 10 shows a plot of the model predicted and observed fecal coliform concentrations versus time from April 12 to 14, 1993 for Apponaug Cove (box 3), Greenwich Cove (box 5), and the western end of Greenwich Bay (box 6). The locations of the boxes are also shown in the figure. The model predictions are consistent with the observations in Apponaug Cove, substantially higher in Greenwich Cove and lower in the western end of Greenwich Bay. The overall agreement is poor. After the rainfall induced loading ceases the model predicts an exponential decrease of the fecal

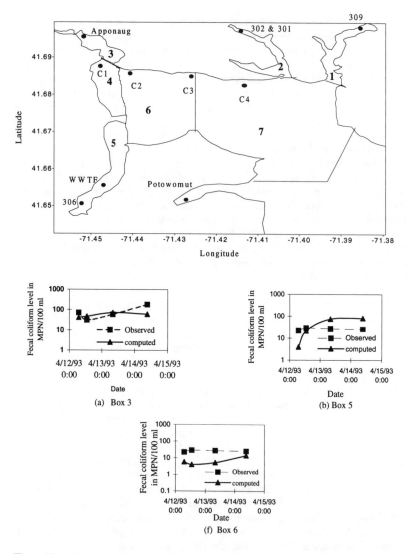

Figure 10   Box segmentation scheme used to represent Greenwich Bay and its coves. Comparison of model predicted and observed (FDA, 1993) box mean fecal coliform concentrations for boxes 3, 5 and 6 for April 12-14, 1993.

coliform concentrations at the rate of -0.01237 or a flushing time of 3.35 days. This value is in excellent agreement with that estimated by FDA (1993) of 3-4 days. Improvements in the model's ability to predict the fecal concentration levels is limited by the lack of adequate data to define the fecal coliform loading rate.

Conclusions

Application of COASTMAP to Greenwich Bay has allowed an improved understanding of the bay's circulation and pollutant transport dynamics. Analysis of the field data and results from the hydrodynamic model show that the circulation in the bay is primarily tidally forced. Maximum current speeds vary considerably throughout the bay but typically have peak values on the order of 15 cm/sec. Maximum current speeds increase in the narrow passages in the small coves. Currents generally decrease from east to west and show a generally low current area near the western end of the cove. The tidal currents are approximately 90 degrees out of phase with the surface elevations. The surface elevations are in phase, within a few minutes, across the entire system. Salinity variations in the bay are generally quite small (1 to 2 ppt) and well mixed vertically. Gradients in the horizontal and vertical are much stronger in the coves, particularly during high freshwater flow events.

Application of the COASTMAP's hydrodynamic and pollutant transport model has generally shown a good ability to estimate the flushing time in the two major coves (Apponaug and Greenwich) and the central basin in comparison to dye release experiments and a box model study of the bay. The model predicted flushing times are 1.2 days for Greenwich Cove, 0.7 days for Apponaug Cove, and 3.35 days for the central portion of the bay. The model has not performed as well at low freshwater input rates.

Comparison of model predictions to observations of fecal coliform concentrations following an April 1993 storm event were made. The model correctly showed the increase in concentrations immediately following the storm event but did poorly in estimating the correct levels. The model overpredicted the concentrations in some areas and underpredicted them in others. Improvement in the model's predictive performance was limited by the lack of an adequate characterization of the fecal coliform loading rates from the principal discharges during and following the storm. The model however predicted a flushing time of 3.4 days in very good agreement with the 3 to 4 day estimate of the flushing time made by FDA.

This study has shown the ability of the COASTMAP system to help understand the circulation and pollutant transport dynamics in a small embayment. Extensions of the system to operate as real time forecasting system and to incorporate other models (runoff, ecosystem) are straightforward and will be implemented.

Acknowledgments

The development of the COASTMAP system was sponsored by the University of Rhode Island's Ocean Technology Center. Support for the application of the system to Greenwich Bay was provided by the URI Sea Grant Program and performed under the Greenwich Bay collaborative.

References

Erikson, L., 1998. Flushing times of Greenwich Bay, RI: Estimates from freshwater inputs, Department of Ocean Engineering, University of Rhode Island, Narragansett, RI 02882.

Food and Drug Administration (FDA), 1993. Greenwich Bay, R.I. Shellfish growing area survey and classification considerations. U.S. Public Health Service. Food and Drug Administration, Davisville, Rhode Island

Muin, M. and M. L. Spaulding, 1996. Two dimensional boundary fitted circulation model in spherical coordinates, Journal of Hydraulic Engineering, Vol. 122, No 9, September, 1996, p. 512-521.

Muin, M. and M. L. Spaulding, 1997a. Three dimensional boundary fitted circulation model, Journal of Hydraulic Engineering, Vol. 123, No. 1, January, 1997, p. 2-12.

Muin, M. and M. L. Spaulding, 1997b. Application of three dimensional boundary fitted circulation model to the Providence River, Journal of Hydraulic Engineering, Vol. 123, No. 1, January 1997, p. 13-20.

Officer, C. B., 1980. Box model revisited, in *Estuaries and Wetland Processes*, editors P. Hamilton and R. B. McDonald, Marine Science Series, Vol. 11, Plenum Press, New York, p. 65-114.

Opishinski, T. and M. L. Spaulding, 1995. COASTMAP: An integrated system for environmental monitoring, modeling, and management, 7th Annual New England Environmental Exposition, Boston, Massachusetts, May 9-11, 1995.

Opishinski, T. and M. L. Spaulding, 1996. COASTMAP An integrated system for environmental monitoring and modeling: selected applications, University of Rhode Island, Ocean Technology Center, Narragansett, RI 02882.

Spaulding, M. L. and T. Opishinski, 1995. COASTMAP: Environmental monitoring and modeling, Maritimes, Vol. 38, No. 1, Spring 1995, p. 17-21.

Spaulding, M. L. and E. Howlett, 1995. A shell based approach to marine environmental modeling, Journal of Marine Environmental Engineering, Vol. 1, p. 175-198.

Spaulding, M. L., T. Opishinski, and S. Haynes, 1996. COASTMAP: An integrated monitoring and modeling system to support oil spill response, Spill Science and Technology Bulletin, Vol. 3, No. 3, p. 149-169.

Swanson, J. C. And D. Mendelsohn, 1995. BAYMAP: A simplified embayment flushing and transport model system, Proceedings of the 4 th International Conference on Estuarine and Coastal Modeling, San Diego, CA, October 26-28, 1995.

Swanson, J. C. and K. Jayko, 1988. A simplified estuarine box model of Narragansett Bay, prepared for the Narragansett Bay Project and Environmental Protection Agency Region I, ASA Report 85-11, Prepared by Applied Science Associates Inc, Narragansett, RI.

Turner, C. A., 1986. Dye study at Greenwich Cove, Narragansett Bay, RI, prepared by Applied Science Associates, Inc., Wakefield, RI, prepared for Battelle, Washington Environmental Program Office, Washington, DC.

Evaluating the Uncertainty in Water Quality Predictions - A Case Study

James D. Bowen[1], A.M. ASCE

Abstract

A method for assessing model result uncertainty is presented and applied to a case where a paper mill wastewater is discharged into an estuary in the Southeastern U.S.. Model result uncertainty was quantified by incorporating the uncertainty analysis into model calibration. The two-dimensional, laterally averaged model CE-QUAL-W2 was used to predict water quality conditions. The water quality model was calibrated against field measurements of longitudinal and vertical variations in salinity, dissolved oxygen, and biochemical oxygen demand (BOD) concentrations. A quantitative, multi-constituent criteria for acceptable calibration was used to identify plausible parameter sets. A collection of plausible parameter sets was then identified, and used to assess the uncertainty in dissolved oxygen prediction, and the uncertainty in predicted system response to a reduction in organic matter loading. A search procedure was also developed to minimize the calibration criteria statistic and to assess the range of model predictions. Plausible parameter sets differed widely in their parameter values, and they produced widely different dissolved oxygen concentration predictions. The system response to reduced loading, however, was found to be very similar between the plausible parameter sets.

Introduction

Water quality models are commonly used to determine suitable limits in wastewater discharge permits (e.g. Gong et al. 1997). To establish a permit limit, the model is used to provide quantitative impact assessments for various discharge scenarios. In substantiating the conclusions of these studies, modelers typically per-form an "uncertainty analysis," where the uncertainty of model predictions is quantified. The most common uncertainty analysis is the "sensitivity analysis", where variability in model results is measured as systematic changes are made to a single set of input

---

[1]Assistant Professor, Department of Engineering Technology, University of North Carolina at Charlotte, Charlotte NC 28223 (704) 547-3130

parameters established through model calibration (e.g., Bowen et al. 1992, Chapelle et al. 1994). The calibrated parameter set may be determined either through the traditional "trial-and-error" approach, through an optimization procedure (e.g. Neuman 1973, Rinaldi et. al 1979), or by solving an inverse parameter identification problem (e.g., Yeh 1986, Panchang and Richardson 1993, Shen and Kuo 1996).

Although a sensitivity analysis is often insightful, it may not give an accurate estimate of model uncertainty, for the following reasons:

- parameter ranges may be unrealistically large or small, since they are typically chosen without considering whether they produce a calibrated model,
- inter-parameter covariances are typically ignored, and
- synergistic effects among parameters are typically not considered.

The underlying problem with the standard sensitivity analysis procedure is that it is done independently of model calibration. In this article we present the results of a modeling study where the uncertainty of model results have been explicitly quantified by incorporating the uncertainty analysis into model calibration. The procedure and the analysis is developed for application to water quality prediction from an earlier model uncertainty analysis for groundwater flow (Brooks et al. 1994).

In the method developed here, many "plausible" parameters sets are selected from a larger collection of candidate sets. Each plausible parameter set produces model results that meet a quantitative calibration criteria. Model prediction uncertainty is assessed by determining the degree to which model results vary between these plausible parameter sets. This variability is examined both by running the model repeatedly with different plausible parameter sets, and by using an automated search technique that looks through the multi-dimensional parameter space to determine the range of model predictions. The automated search method is pursued here in lieu of Monte-Carlo based methods (e.g. Beven and Binley 1992) that would involve much more computational effort. The uncertainty analysis method is applied to a case study where an estimate is made of the degree to which a reduction in BOD loading would increase the dissolved oxygen concentrations in a tidally influenced river.

Study Area and Model Description

The uncertainty analysis technique was applied to a water quality study performed to predict dissolved concentrations in the Sampit River, South Carolina for various wastewater discharge scenarios. The water quality study included both an extensive hydrodynamic and water quality field study (Hickey et al. 1997) and a modeling study (Gong et al. 1997) that used the 2-d, laterally averaged water quality model CE-QUAL-W2 (Cole and Buchak 1995). The Sampit is a tidally influenced river that drains approximately 500 km^2 of coastal South Carolina and is surrounded by an extensive marsh system (Fig. 1). A paper mill wastewater enters the river on the ebb tide

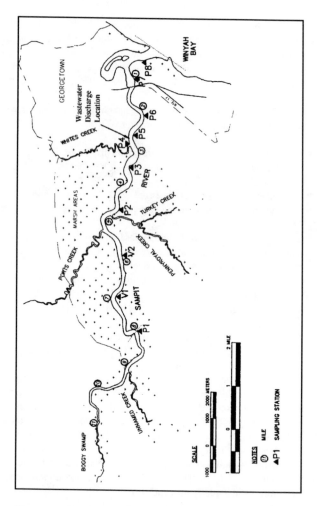

**Fig. 1. Sampit River Study Area, Wastewater Discharge
Location and Sampling Station Locations.**

only, from a surface discharge located approximately 5 km upstream of the river mouth
(Fig. 1).

The model analysis used a "pseudo steady-state" approach. Field studies and
model simulations focused on summer conditions when dissolved oxygen (DO)
concentrations are at a minimum. The state variables in the water quality model were
temperature, salinity, ammonia, nitrate-nitrite, dissolved oxygen, and carbonaceous

BOD. Ensemble averages of multiple sampling rounds were calculated, and were considered to represent a summer, steady state condition. Upstream boundary conditions for water quality state variables were set to steady values equal to the ensemble averages from the monitoring program. Downstream boundary elevations included the M2 tidal component only. Wastewater discharges, which occur for one hour, once per day, just after high tide, were based on plant discharge records. Downstream boundary conditions and wastewater conditions for water quality state variables were varied in time, based on results of the water quality monitoring study and plant discharge records. A steady and homogeneous sediment oxygen demand was used to represent benthic organic matter decomposition.

Uncertainty Analysis Procedure

The uncertainty analysis procedure can be summarized as follows:

1. calibrate the hydrodynamic model,
2. establish a "base case" water quality parameter set through model calibration,
3. establish fitness measures and criteria (i.e. determine what makes a "plausible" parameter set),
4. select parameters to vary for uncertainty analysis,
5. set up a "grid" of candidate parameter sets,
6. evaluate the degree-of-fit at each grid point,
7. examine prediction uncertainty for plausible sets, and
8. perform automated search of parameter space to determine:
   a. range of model fitness values, and
   b. range of system responses.

Steps 3 - 8 of the procedure are described in this article. Steps 1 and 2 of the procedure for this application are described by Gong et al. (1997). From this earlier work, the base case set of parameters was determined. The uncertainty analysis then examined how well this base case, and many other candidate parameter sets, performed against two quantitative fitness criteria.

Quantitative Fitness Criteria

Two separate measures were used to determine the degree to which model predictions agreed with observations from a monitoring program. The observations used for comparison were averages of four sampling rounds taken at different tide stages (high and low water slack, maximum ebb and flood current) during a survey in late August, 1996. Ten stations were sampled within the study area (Fig. 1). Comparisons of three measured water quality constituents (salinity, DO concentration, BOD concentration) were compared against model predictions. Vertical and longitudinal variations in each constituent were considered with this analysis (Table 1).

Table 1. Observed Data

Parameter	Vertical Location	Station									
		P1	V1	V2	P2	P3	P4	P5	P6	P7	P8
Salinity (ppth)	Top	2.5	3.0	3.0	3.3	3.2	-	3.5	3.5	3.5	3.5
	Bottom	2.7	3.0	3.0	3.3	3.2	-	3.9	4.5	6.4	8.0
DO (mg/l)	Top	3.1	2.8	2.8	2.7	3.1	-	2.5	3.6	4.0	3.8
	Bottom	3.6	3.6	3.4	3.4	3.5	-	3.6	3.5	3.8	3.8
BOD (mg/l)	Top	2.8	-	-	2.1	3.0	2.0	2.2	1.9	2.0	2.3
	Bottom	2.1	-	-	2.6	2.6	1.9	2.1	2.0	1.5	2.1
Channel depth @ low-water slack tide (m)		4.0	5.5	5.5	7.0	6.0	4.5	5.5	8.5	8.0	7.5

A multi-constituent, normalized fitness measure was used as the first of the two fitness criteria. This average "lack-of-fit" measure is defined as follows:

$$LOF = \frac{1}{n} \times \sum \frac{rms(observed - predicted)}{avg\ predicted} \qquad (1)$$

where LOF is the average normalized lack-of-fit, and n is the number of constituents (n=3 in this case). Because the model prediction of most interest was dissolved oxygen, a second measure, the normalized DO fit was also used. This measure, DO_{LOF}, is defined as follows:

$$DO_{LOF} = \frac{rms(observed\ DO - predicted\ DO)}{avg\ predicted\ DO} \qquad (2)$$

Selection of Variable Parameters

Approximately 20 adjustable parameters might be included in the uncertainty analysis. Because of computational limitations, and based on the work of Brooks et al (1994), it was decided to consider only 3 - 7 parameters as variables for the uncertainty analysis. The following considerations were used to select the variable parameters:

- select parameters that seem relatively uncertain, based on consideration of the field program and hydrologic conditions, and

- select those parameters that most likely affect the state variables that are used as part of the quantitative analysis of degree-of-fit (salinity, DO, and BOD in this case).

Five variable parameters were selected based upon this analysis (Table 2). A parameter specific range was selected for each of the five parameters. The ranges varied from a factor of 10 to 25 around the base case values. Two of the five parameters, BOD loading and freshwater runoff, were varied by factors relative to the base case. The five dimensional parameter space was then "gridded" by specifying three or four levels for each parameter. Every combination of parameter levels was then considered as a candidate parameter set, giving a total of 324 candidate sets (Table 2). The two fitness measures were then determined for each of the candidate parameter sets.

Table 2. Specification of Variable Parameters

Parameter	No. of Values	Grid Levels	Units
Longitudinal Dispersion	3	**1**, 5, 25	m²/s
Sediment Oxygen Demand	4	0.04, 0.10, **0.40**, 1.0	g/m²/d
BOD decay rate	3	0.032, **0.10**, 0.32	day⁻¹
Relative BOD Load	3	0.32, **1.0**, 3.2	* base
Relative Freshwater Runoff	3	0.10, 0.32, **1.0**	* base
Note: base case levels of each parameter are in **bold**			

Automated Search Procedure

An automated search procedure was developed to search through the entire parameter space for "optimal" parameter sets. To accomplish this, a utility program was written to create input files for arbitrary combinations of the five variable parameters. Two automated searches were conducted, having the following objectives:

1. minimize the multi-constituent lack-of-fit, (LOF), defined in Eq. 1, and
2. maximize the increase in DO concentrations for a 30% reduction in BOD load while maintaining an acceptable lack-of-fit.

Searches for minima and maxima were conducted with the Nelder and Mead simplex method (Dennis and Woods 1987). Since the functions to be maximized or minimized are not linear, the search procedure found only local extrema. Multiple searches were therefore performed to investigate variability in these local values. The search for the maximum DO increase was performed by maximizing the function, I_{max}, given as:

$$I_{max} = \max[\frac{DO_{70\%} - DO_{100\%}}{A} - \exp(LOF - B)] \qquad (3)$$

where $DO_{70\%}$ and $DO_{100\%}$ are the 5th percentile DO concentrations from the model predictions for 70% and 100% BOD loading respectively, A = DO normalization concentration = 0.02 mg/l, LOF is the multi-constituent lack-of-fit for the particular parameter set (see Eq. 1), and B is a LOF criteria value that was set equal to the LOF determined for the "base" set of parameters. The exponential term serves as a cost function that gets exponentially large as the lack-of-fit increases. This procedure is a computationally efficient approximation to a constrained optimization problem (Brooks et al. 1994).

Results

Grid Point Runs

Water quality modeling results were gathered from the five-dimensional parameter grid having 324 points, which represented variations in five independent model parameters. For the entire collection of 324 candidate parameter sets, the multi-constituent lack-of-fit (LOF) varied from 36% to 280% (Fig.2), with a median LOF of 60%. The DO specific fit measure (DO_{LOF}) varied from 32% to 77%, with a median value of 54%. The "base" case, which was grid point 1,3,2,2,3 (see Table 2), produced an LOF of 48.3% and a DO_{LOF} of 55.1%.

The lack-of-fit levels for the base case were chosen as the criteria for an acceptable lack-of-fit between model predictions and observations. A "plausible" parameter set was therefore one that produced an LOF of no more than 48.3% and a DO_{LOF} of no more than 55.1%. Looking separately at the criteria, 28% met the multi-constituent criteria, and 41% met the DO criteria, while 19% met both criteria (Fig. 2). Thus 62 of the 324 candidate parameter sets were considered to be plausible.

The plausible parameter sets were well distributed over the parameter space. No plausible parameter sets came from the highest SOD value of 1.0 g/m²/d or the highest relative runoff of 3.2 times the base case (Table 3). Only 4 of the 62 cases came from the highest BOD decay rate of 0.32 day⁻¹. Each of the remaining grid levels had at least 25% of the plausible parameter sets (Table 3).

A broad range of dissolved oxygen values were predicted for model runs using the various candidate parameter sets. The variability was due to several factors. For a particular parameter set, the model produced a range of DO concentrations, as expected, depending on the time and location of the prediction. In addition, there was variability

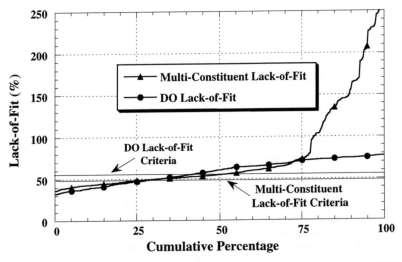

Fig. 2. Distribution of the Multi-Constituent and DO Lack-of-Fit Measures LOF and DO$_{LOF}$ for all 324 Candidate Parameter Sets.

Table 3. Number of Plausible Parameter Sets for the Various Grid Levels of the Five Variable Parameters

Parameter	Number of Plausible Parameter Sets at Grid Level				Total
	1	2	3	4	
Longitudinal Dispersion	18	17	27	-	62
Sediment Oxygen Demand	23	23	16	0	62
BOD Decay Rate	27	31	4	-	62
Relative BOD Load	36	26	0	-	62
Relative Freshwater Runoff	23	21	18	-	62

in the model predictions between runs with different parameter sets. For instance, the spatially and temporally averaged mean DO concentration varied from 0.0 to 5.2 mg/l for all 324 candidate parameter sets, and from 2.2 to 3.8 mg/l for the 62 plausible parameter sets (Fig. 3). The variability in maximum and minimum DO concentrations for the 324 and 62 run tests showed a similar relationship, with decreased, although

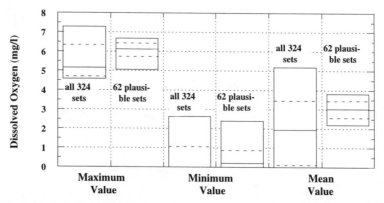

Fig. 3. Predicted Dissolved Oxygen Concentrations for all 324 Parameter Sets and for the 62 Plausible Parameter Sets. The bars indicate the range of values. The solid line through the interior of the bar represents the median value, while the dashed lines represent the 25th and 75th percentile values.

considerable, variability for the 62 run test. For the set of 62 plausible parameter sets, the maximum DO concentration varied from 5.0 mg/l to 6.7 mg/l, while the minimum DO concentration varied from 0.0 to 2.4 mg/l (Fig. 3).

As expected, runs with a BOD loading from the paper mill wastewater that was reduced by 30% from the base value had higher dissolved oxygen concentrations. The magnitude of the difference, however, was quite small. For the 62 plausible parameter sets, the median mean concentration increased from 3.01 mg/l to 3.06 mg/l (Fig. 4) when the wastewater BOD loading was reduced by 30%. Decreasing the loading resulted in a median minimum DO concentration that increased from 0.19 mg/l to 0.23 mg/l, while the median maximum concentration remained unchanged at 6.11 mg/l (Fig. 4). As a control, an additional set of runs simulated the river with no wastewater BOD loading. Higher DO concentrations resulted; the magnitude of the increase was consistent with the findings from the 30% BOD reduction cases (data not shown).

Decreasing the BOD loading by 30% increased the DO concentration; the magnitude of the increase varied between the plausible parameter set runs. For this comparison, the 5th percentile DO concentration was taken as the statistic of interest. This statistic was chosen because it depends both on the mean and variance of the DO distribution. It is also a useful statistic for consideration of wastewater discharge limits, where only infrequent violations of a water quality standard would be allowed. The increase in the 5th percentile DO concentration varied from 0.005 mg/l to 0.078 mg/l for the 62 plausible parameter sets (Fig. 5). The median increase in dissolved oxygen concentration was 0.035 mg/l.

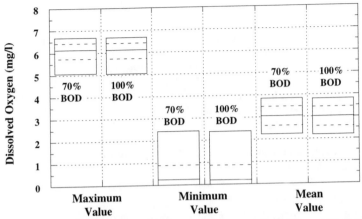

Fig. 4 Predicted dissolved oxygen concentrations for 100% and 70% BOD loading for the 62 plausible parameter sets. The bars indicate the range of values. The solid line through the interior of the bar represents the median value, while the dashed lines represent the 25th and 75th percentile values.

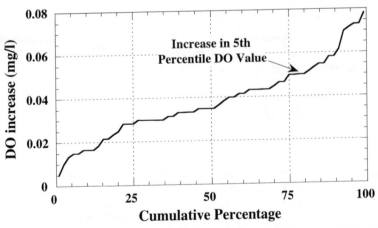

Fig. 5. Effect of a 30% load reduction of the 5th percentile DO concentration for the 62 plausible parameter sets.

Automated Searches

Two searches were conducted to find the set of parameters producing the minimum value for the multi constituent lack-of-fit, LOF. The first search started from the base case, grid point 1,3,2,2,3. The search routine finished after 150 iterations, with the LOF value decreasing from 48.3% to 30.6% (Table 4). A second search started from the grid point producing the lowest multi-constituent lack-of-fit (2,1,1,1,1). This search gave a different parameter set having a similar LOF value. In this case the LOF decreased from 36% to 34.5% and terminated after 137 iterations (Table 4).

Table 4. Results of automated searches to find the parameter set that produces the minimum lack-of-fit.

Search 1: starting from "base" case					
	Dispersion (m²/s)	SOD (g/m²/d)	BOD decay (day⁻¹)	BOD loading (* base)	Runoff (* base)
Initial Values	1.0	0.4	0.1	1.0	1.0
Final Values	1.5	-0.98	0.05	0.87	0.06
Lack-of-Fit			Number of Iterations		
30.6%			150		

Search 2: starting from grid point with best fit					
	Dispersion (m²/s)	SOD (g/m²/d)	BOD decay (day⁻¹)	BOD loading (* base)	Runoff (* base)
Initial Values	5.0	0.04	0.032	0.32	0.10
Final Values	5.03	0.01	0.025	0.36	0.06
Lack-of-Fit			Number of Iterations		
34.5%			137		

Automated searches were also conducted to find the set of parameters that gave the largest increase in DO concentration for a 30% decrease in wastewater BOD loading, while maintaining an acceptable level of fit. The search starting from the base case found a maximum DO increase of 0.17 mg/l after 191 iterations, more than twice the largest DO increase found in the grid point runs. The LOF value for this parameter set was 64.4 %. A second search was started from the grid point having the highest difference in the 5th percentile DO concentration (3,1,1,1,1). This search found a maximum difference in the 5th percentile DO concentration of 0.15 mg/l (Table 5). The LOF value for this case was 84.8%.

Table 5. Results of automated searches to find the parameter set that produces the maximum increase in DO concentration that results from a 30% load reduction.

Search 1: starting from "base" case					
	Dispersion (m^2/s)	SOD $(g/m^2/d)$	BOD decay (day^{-1})	BOD loading (* base)	Runoff (* base)
Initial Values	1.0	0.4	0.1	1.0	1.0
Final Values	1.53	-0.28	1.22	2.52	-5.34
Lack-of-Fit		DO Increase (mg/l)		Number of Iterations	
64.4%		0.17		191	

Search 2: starting from grid point with maximum DO increase					
	Dispersion (m^2/s)	SOD $(g/m^2/d)$	BOD decay (day^{-1})	BOD loading (* base)	Runoff (* base)
Initial Values	25.0	0.04	0.032	0.32	0.10
Final Values	20.2	-2.35	8.44	4.3	-9.78
Lack-of-Fit		DO Increase (mg/l)		Number of Iterations	
84.8%		0.15		142	

Discussion and Conclusions.

This study demonstrated the feasibility of using multiple parameter sets to perform a surface water quality impact assessment that includes a prediction uncertainty analysis. A previous modeling study (Gong et al. 1997) had identified one set of kinetic parameters (referred to as the base case) that produced an acceptable fit between water quality observations and model predictions. In our study, 324 candidate parameters sets were tested against two quantitative calibration criteria; 61 additional parameter sets were found to produce results that fit the observations as well as the base case. Any one of these 62 plausible parameter sets represents a reasonable description of the system under study, thus differences in the model results between the parameter sets truly represent one portion of model prediction uncertainty. Identification of this collection of plausible parameter sets therefore allowed for a prediction of model prediction uncertainty.

Model prediction uncertainty was found to be relatively large for predictions of water quality conditions, yet quite small for the prediction of the system response to an environmental manipulation. For instance, the range in predicted mean dissolved oxygen concentrations (100% BOD load) for the 62 plausible parameter sets was 1.6 mg/l (2.2 to 3.8 mg/l, Fig. 4), yet the range of increases in DO concentration for a 30% wastewater load reduction was only 0.073 mg/l (0.005 to 0.078 mg/l, Fig. 5). In addition, this study modeled only a single season (Summer 1996). Had interannual variability been considered, then prediction of water quality conditions would probably have been even less certain. Will model predictions of system response always be as certain as was seen in this case? Probably not, yet it does seem likely that system response prediction will generally be more certain than prediction of water quality conditions. This point should be kept in mind during project planning, when modeling project objectives and expectations are being discussed.

Selection of the variables to vary was found to be a difficult task. Like the earlier application of this method to groundwater flow (Brooks et al. 1994), it was found that it is practical to vary only a few variables, perhaps 3 - 7, depending on the simulation run time. Clearly this limit will grow as more computational power becomes available. In our study, where modeling was performed on a UNIX workstation, it took approximately one week to simulate all 324 cases. Unfortunately, since the number of candidate parameter sets grows exponentially with the number of variable parameters, it seems unlikely that anytime soon will we be able to consider all the parameters in a water quality simulation as variable. For this reason, future work on water quality modeling uncertainty analysis should address how best to pick the subset of parameters that are considered as variable.

Selection of the appropriate criteria for identifying plausible parameter sets is another difficult issue. In this study, the results of previous modeling (Gong et al. 1997) were used to establish the lack-of-fit criteria. Did the selection of the fitness criteria affect model prediction uncertainty? Clearly, uncertainty in predictions of water quality conditions would have decreased if the criteria had been more restrictive, i.e., if the maximum lack-of-fit for plausible parameter sets had been smaller. Would the uncertainty in prediction of system response also have decreased with a more restrictive criteria? The answer to this question seems neither obvious nor generally true. In this study, the system response (i.e., the increase in DO concentration with reduced waste-water loading) decreased slightly as the multi-constituent lack-of-fit increased (Fig. 6). Although a linear regression of the variables was significant at the 99% level, the variation in lack-of-fit explained only 16% (regression $R^2 = 0.4^2$) of the variability in system response. It is unclear whether the system response will always be so insensitive to changes in the fitness criteria.

As expected, the automated searches found parameter sets that produced better fits and larger system responses as compared to the grid point runs. The magnitude of the differences, however, was relatively small. The best fit for the grid point runs was 36%; it was 31% for the automated search. The maximum DO increase for the grid point runs was 0.08 mg/l; for the automated search it was 0.17 mg/l. A weakness of the current automated search method was identified during this study; there were no constraints on the magnitudes of the variable parameters. This was a problem, as the search to find the maximum DO increase produced negative parameter values (Table 5), which are not physically reasonable. Further development of this procedure to eliminate this problem is needed.

Both automated searches were found to require a large computational effort, as it required 100 - 400 model runs to locate a local minima or maxima. The search to find the maximum system response required the largest effort, as each function evaluation passed to the maximization routine was the result of two model runs, one at 100% BOD loading, and a second at 70% BOD loading. An additional difficulty of this search was the need to tune the normalization DO concentration (see Eq. 3) so that both the degree of fit and the DO increase were equally important in affecting the magnitude of the function to be evaluated.

In summary, when models are used for environmental management, prediction uncertainty should be quantified. Until now, modelers have generally relied upon a sensitivity analysis to estimate prediction uncertainty, while recognizing the method's limitations. This study has demonstrated the feasibility of a method that could provide a more accurate estimate of the uncertainty in water quality predictions.

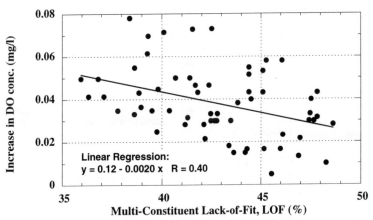

Fig 6. Effect of BOD loading reduction on DO concentrations for various levels of the multi-constituent lack-of-fit.

References

Beven, K., and Binley, A. (1992). The future of distributed models: Model calibration and uncertainty prediction. *Hydrol. Processes*, 6, 279-298.

Bowen, J.D., Galya, D. P., and Villars, M. T., (1993). Modeling the Impacts of Plankton Entrainment in a Tropical Bay. *In* <u>Hydraulic Engineering '93, Proceedings of the</u>

1993 Conference, Hsieh Wen Shen, S.T. Shu, and Fen Wen, eds., p. 1167 - 1171. American Society of Civil Engineers, New York, NY.

Brooks, R. J., Lerner, D. N., and Tobias, A. M. (1994). Determining the range of predictions of a groundwater model which arises from alternative calibrations. *Water Resour. Res.*, 30(11), 2993-3000.

Chapelle, A.,Lazure, P., and Menesguen, A. (1994). Modelling eutrophication events in a coastal ecosystem. Sensitivity analysis. *Estuarine, Coastal and Shelf Science*, 39, 529-548.

Cole, T. M., and Buchak, E. M. (1995). CE-QUAL-W2: A two-dimensional, laterally averaged hydrodynamic and water quality model, version 2.0, User Manual. Instruction Report E1-95-, Waterways Experiment Station, Vicksburg, MS.

Dennis, J. E. Jr., and Woods, D. J. (1987). Optimization on microcomputers: The Nelder-Mead simplex algorithm. *In New Computing Environments: Microcomputers in Large-Scale Computing*, A. Wouk, ed., pp. 116-122. Society for Industrial and Applied Mathematics, Philadelphia, PA.

Gong, G., Hickey, K., and Higgins, M. (1997). Hydrodynamic Flow and Water Quality Simulation of a Narrow River System Influenced by Wide Tidal Marshes. 5th International Conference on Estuarine and Coastal Modeling, Alexandria, VA.

Hickey, K., Gerath, M., Bowen, J., and Gong, G. (1997). Design of Field Program and Use of Field Measurements to Calibrate a Hydrodynamic and Water Quality Model of an Estuarine River System. 5th International Conference on Estuarine and Coastal Modeling, Alexandria, VA.

Neumann, S. P. (1973). Calibration of distributed parameter groundwater flow models viewed as a multiple-objective decision process under uncertainty. *Water Resour. Res.*, 9(4), 1006-1021.

Panchang, V. G., and Richardson, J. E. (1992). Inverse adjoint estimation of eddy viscosity for coastal flow models. *J. Hydr. Engrg.* 119(4), 506-524.

Rinaldi, S., Romano, P. and Sessa, R. S. (1979). Parameter estimation of Streeter-Phelps models. *J. Envir. Engrg.* 105(1), 75-88.

Shen, Y., and Kuo, A. Y. (1996). Inverse estimation of parameters for an estuarine eutrophication model. *J. Envir. Engrg.*, 122(11), 1031-1040.

Yeh, W. W. G. (1986). Review of parameter identification procedures in ground water hydrology: the inverse problem. *Water Resour. Res.*, 22(2), 95-108.

CONTROL OF MULTI-DIMENSIONAL WAVE MOTION
IN SHALLOW WATER

Brett F. Sanders[1], M. ASCE and Nikolaos D. Katopodes[2], M. ASCE

ABSTRACT

Extreme flooding events in recent years have renewed interest in flood mitigation techniques. Presented is a dynamic flood control strategy that is consistent with risk-based flood management policy. Control of flood waves is achieved using an adjoint sensitivity method (ASM) based on the shallow-water equations in both one and two spatial dimensions. The efficiency of the ASM allows active control of flood waves. A deliberate levee breach at an optimal time and location during a flooding event is considered as a flood mitigation measure. A second order finite-volume scheme is presented for the monotonic solution of the adjoint shallow-water equations.

INTRODUCTION

Recent flooding disasters in the U.S. have renewed interest in flood mitigation. The event that occurred along the Mississippi River and its tributaries in 1993 is most responsible for the current interest in flood control, but 1997 flooding in California's central valley, along the Ohio River, and along the Red River continue to demonstrate that a need exists for better flood mitigation measures.

Protection from flooding has historically been provided by structures designed for a specific event, often the 100 year flood. Recently though, flood management districts have embraced a risk based approach to flood mitigation. This approach has ended exclusive reliance on physical structures aimed to protect against floods, and encouraged the use of non-structural flood mitigation measures, such as flood insurance, early warning systems, and flood plain land management to mitigate the effects of floods.

A new flood control measure, the controlled levee breach, is introduced to miti-gate flooding events that exceed the design capacity of a flood control system. The measure is consistent with the risk-based approach because damage resulting from floods is minimized by discouraging flooding from occurring at locations where the

---

[1] Assistant Professor, Department of Civil and Environmental Engineering, University of California, Irvine, CA 92697
[2] Professor, Department of Civil and Environmental Engineering, University of Michigan, Ann Arbor, MI 48109

most financial or human loss would follow. This measure relies on shallow-water hydrodynamics to control a flood wave. A levee breach is designed to occur at an optimal location and time such that the resulting progressive and regressive depression waves lower the peak stage of a flood wave at a predetermined target location.

A fast and accurate method for controlling multi-dimensional wave motion in shallow water is developed to identify the optimal levee breach with arbitrary resolution. The method considers the hydrodynamic aspects of a specific incoming flood wave and chooses a levee breach time and location such that the depth at a target location, where flooding would lead to the greatest damage, is constrained. An adjoint sensitivity method (ASM) is used to characterize the hydrodynamic effect of every possible levee breach location and time on the incoming flood wave, i.e., the sensitivities.

The ASM is computationally efficient. It requires just a single solution of the shallow water equations followed by a single solution of the adjoint shallow water equations. The sensitivities are given following a sequential solution of both equations. Direct Sensitivity Methods (DSM), on the other hand, require thousands of sequential solutions of the governing equations to evaluate the same information, making these approaches too slow for active control. The computational efficiency of the ASM allows a levee breach location to be identified in real-time.

A finite-volume MUSCL (Monotone Upwind Scheme for Conservation Laws) approach is used to solve both the shallow-water equations and the adjoint shallow-water equations in either one or two spatial dimensions, depending on flood plain and channel geometry. The finite volume scheme uses Hancock's predictor corrector method for time integration and uses Roe's method to evaluate interface fluxes.

ADJOINT SENSITIVITY METHOD

The ASM has previously been applied in a variety of fields. In the atmospheric sciences, the ASM has been applied to assimilate data for general circulation models using optimal control methods (e.g. Hall and Cacuci, 1983; Hall et. al. 1982; Le Dimet and Talagrand, 1986; Marchuk, 1995). In the field of ground water hydrology, the ASM has also been extensively applied. Yeh (1986) provides a review of applications. In the field of surface water hydraulics and hydrology, however, the application of the ASM has been considerably less extensive. Piasecki and Katopodes (1997a, 1997b) and Katopodes and Piasecki (1996) applied the ASM to optimize hazardous waste discharges in rivers. Das and Lardner (1991) and Panchang and O'Brien (1989) estimated hydraulic parameters in one-dimensional tidal flow models. In addition, two-dimensional applications of the ASM have been made by Lardner (1993), Lardner, Al-Rabeh, and Gunay (1993), and Zou and Holloway (1995) in parameter identification and control applications. These surface water applications of the ASM, however, used truncated versions of the shallow-water equations which omit the nonlinear convective terms. Because these terms are appreciable in wave problems, such as that which results from a breached levee, Sanders (1997) recently derived equations which are adjoint to the full shallow-water equations in both one and two spatial dimensions, and provided a physical interpretation of the adjoint shallow-water variables.

The shallow-water equations describe long waves in open channels. The primary assumption in shallow water theory is that vertical accelerations are negligible, and hence the pressure distribution in the vertical remains hydrostatic. The shallow-water equations form the basis for the ASM used to control flood waves.

In one dimension, the hydrodynamic effect of a discharge from the channel to the flood plain is incorporated into the shallow-water equations by adding a lateral outflow term that not only removes mass from the channel, but also alters the momentum in the stream-wise direction. The one-dimensional continuity and momentum equations modified to include a lateral outflow are given as,

$$h_t + q_x + q^l = 0$$
$$q_t + \left(\frac{q^2}{h} + \frac{gh^2}{2}\right)_x - gh(S_o - S_f) - (q/h - u^l)q^l = 0 \tag{1}$$

where h is the depth of flow, q is the mass flux, or discharge, S_o is the bed slope, S_f is the friction slope, q^l is the lateral outflow per stream-wise length of channel, and u^l is the velocity in the stream-wise direction with which the lateral outflow leaves the channel.

To identify the optimal levee breach, the sensitivity of the objective function to changes in breach flow at all possible locations and times is identified. Using a DSM, each possible breach location and time requires an independent unsteady flow calculation. With a one dimensional model using N spatial nodes and N_T time levels, $N_T N$ independent unsteady flow calculations are needed to evaluate the necessary sensitivities. This is a prohibitive number of calculations for active control, so the ASM is instead applied. With the ASM, N_T sensitivities are computed with just a single solution of the shallow-water equations followed by a single solution to the adjoint shallow water equations. The following derivation indicates how the ASM is developed. A more thorough derivation is given by Sanders (1997).

Derivation

The objective function is introduced as an integral of a measuring function, r. Left in a general form, the objective function appears as,

$$J = \int_0^T \int_0^L r(h, q; x, t)dx\, dt \tag{2}$$

This objective function allows a specific aspect of the flow to be characterized once r is specified. Next, the continuity equation is multiplied by the adjoint variable $\phi(x, t)$, and the momentum equation is multiplied by the adjoint variable $\psi(x, t)$. The sum of these two products is then integrated over space and time and added to the objective function. The resulting expression appears as,

$$J = \int_0^T \int_0^L \left\{ r(h, q) + \phi \left[h_t + q_x + q^l \right] \right.$$
$$\left. + \psi \left[q_t + \left(\frac{q^2}{h} + \frac{gh^2}{2}\right)_x - gh(S_o - S_f) - (q/h - u^l)q^l \right] \right\} dx\, dt \tag{3}$$

The functional $r(h, q)$ given in Eq. 3 is chosen to reflect the threat of levee over-topping and is given as,

$$r(h, q) = \begin{cases} \frac{1}{2}(h - \bar{h})^2 \delta(x - x_T), & \text{if} \quad h(x_T, t) > \bar{h}(x_T) \\ 0, & \text{if} \quad h(x_T, t) \leq \bar{h}(x_T) \end{cases} \tag{4}$$

Hence, the objective function, J, is minimized when the depth never exceeds the levee height at the target location.

To arrive at the adjoint equations, the spatial and temporal operations in Eq. 3 are passed onto the adjoint variables using integration-by-parts. Taking the variation of the resulting expression, using non-reflecting adjoint boundary conditions, and using the initial conditions, $\delta h(x, 0) = \delta q(x, 0) = \phi(x, T) = \psi(x, T) = 0$, the adjoint equations follow as,

$$\phi_\tau + \left(\frac{q^2}{h^2}\right)\psi_x - gh\psi_x - g(S_o + 2S_f)\psi + \left(\frac{qq^l}{h^2}\right)\psi + \frac{\partial r}{\partial h} = 0$$
$$\psi_\tau - \phi_x - 2\left(\frac{q}{h}\right)\psi_x + 2\frac{ghS_f}{q}\psi - \left(\frac{q^l}{h}\right)\psi + \frac{\partial r}{\partial q} = 0 \tag{5}$$

and the sensitivity of the objective function to perturbations of lateral outflow is given by,

$$S_{i,j} = \left(\frac{\delta J}{\delta q^l(x_i, t_j)}\right) = \left[\phi - (q/h - u^l)\psi\right]\Big|_{(x_i, t_j)} \tag{6}$$

In two spatial dimensions, an appropriate boundary condition captures the hydrodynamic effect of a breached levee. The shallow-water equations in two spatial dimensions are written in terms of the depth, h, the discharge in the x-direction, p, and the discharge in the y-direction, q. The equations appear as,

$$h_t + p_x + q_y = 0$$
$$p_t + \left(\frac{p^2}{h} + \frac{gh^2}{2}\right)_x + \left(\frac{pq}{h}\right)_y - gh(S_o^x - S_f^x) = 0$$
$$q_t + \left(\frac{pq}{h}\right)_x + \left(\frac{q^2}{h} + \frac{gh^2}{2}\right)_y - gh(S_o^y - S_f^y) = 0 \tag{7}$$

where S_o^x and S_o^y are bottom slopes in the x and y directions, and S_f^x and S_f^y are friction slopes in the x and y directions, respectively. The objective function becomes,

$$J = \int_0^T \int_\Gamma r(h, p, q; x, t) \, d\Gamma \, dt \tag{8}$$

where Γ is the two-dimensional spatial domain. The measuring function, r, is now given as,

$$r(h, p, q; x, t) = \frac{1}{2}(h(x_T, y_T, t) - \bar{h}(x_T, y_T))^2 \delta(x - x_T) \, \delta(y - y_T) \tag{9}$$

where (x_T, y_T) is the target location that resides on $\partial \Gamma_1$, the boundary where levees are present.

The adjoint equations are obtained by procedures similar to the one-dimensional case. Sparing a complete derivation for the sake of brevity, the adjoint equations in two dimensions appear as,

$$\phi_\tau - \left(gh - \frac{p^2}{h^2}\right)\psi_x^x + \frac{pq}{h^2}(\psi_y^x + \psi_x^y) - \left(gh - \frac{q^2}{h^2}\right)\psi_y^y$$
$$-\psi^x g(S_o^x + 2S_f^x) - \psi^y g(S_o^y + 2S_f^y) + r_h = 0$$

$$\psi_\tau^x - \phi_x - \frac{2p\psi_x^x}{h} - \frac{q\psi_y^x}{h} - \frac{q\psi_x^x}{h}$$
$$-\psi^x gh\left(\frac{p}{q^2 + p^2} + \frac{1}{p}\right)S_f^x - \psi^y gh\left(\frac{p}{q^2 + p^2}\right)S_f^y + r_p = 0$$

$$\psi_\tau^y - \phi_y - \frac{p\psi_y^x}{h} - \frac{p\psi_x^y}{h} - \frac{2q\psi_y^y}{h}$$
$$-\psi^x gh\left(\frac{q}{q^2 + p^2}\right)S_f^x - \psi^y gh\left(\frac{q}{q^2 + p^2} + \frac{1}{q}\right)S_f^y + r_q = 0 \qquad (10)$$

where ϕ is the adjoint variable multiplying the continuity equation, ψ^x is the adjoint variable multiplying the momentum equation in the x-direction, and ψ^y is the adjoint variable multiplying the momentum equation in the y-direction.

The sensitivities, or more specifically, the sensitivity of the objective function to changes in boundary flux, are then given as a function of the adjoint variables,

$$S_{ijk} = \frac{\delta J}{\delta q(x_i, y_j, t_k)} = \phi(x_i, y_j, t_k) + u(x_i, y_j, t_k)\,\psi^x(x_i, y_j, t_k) \qquad (11)$$

where points (x_i, y_j) lie on $\partial \Gamma_b$ and $k = 1, \ldots, N_T$. The boundary $\partial \Gamma_b$ represents the levee breach boundary.

Limitations of the ASM

There are two important limitations of the ASM relevant to its application to identifying the optimal levee breach. First, the adjoint variables are assumed to be continuous and differentiable while the hyperbolic form of the adjoint shallow-water equations permits discontinuities. This paradox remains a topic of ongoing research; however, its ramifications are avoided because shock waves are expected only outside of the channel following a breached levee. Hence, the ASM is valid within the channel.

A second issue relevant to this application of the ASM is that sensitivities are valid for small perturbations of the flow field, and that a deliberately breached levee represents more than a small perturbation of the flow. This issue was investigated by Sanders (1997), who found that sensitivities provided by the ASM underestimated the relative effect of breaches upstream of the target location compared to breaches downstream of the target location. However, the ASM was found to accurately capture the optimal time for the breach, and due to its computational efficiency, the ASM remains an invaluable technique.

FINITE-VOLUME SOLUTION METHOD

In either one or two spatial dimensions, the procedure for identifying the optimal levee breach includes the numerical solution of the shallow-water equations, the adjoint shallow-water equations, and then the sensitivities.

Using the finite-volume method (FVM), the integral conservation form of the partial differential equations is directly discretized. This ensures that the scheme remains conservative even in the presence of shocks. When using the FVM on a structured grid, the spatial domain is discretized into N_x by N_y cells, and the conserved variables $U_{i,j}$ are defined at the center of each cell to be the algebraic average over the cell. The fluxes, F and G, are then evaluated at cell interfaces.

Second order spatial accuracy is obtained by considering the state variables to be piecewise linear in each cell. The variable extrapolation method of Van Leer, also known as the MUSCL (Monotone Upwind Scheme for Conservation Laws) approach, is used to obtain cell interface values for the state variables (1979). Second order time integration is then performed using a predictor-corrector approach known as Hancock's method (Van Albada, Van Leer, and Roberts, 1982). This approach solves the primitive form of the shallow-water equations to advance the solution to the half time level, evaluates interfaces fluxes by solving the Riemann problem approximately (Roe, 1981; Zhao et. al., 1996), and then advances the solution to the next full time level by solving the conservation form of the equations. Flux limiting is performed using the double minmod limiter to preserve monotonicity (Hirsch, 1990).

Boundary conditions are implemented by adding *ghost cells* immediately outside the computational space and by modifying the state variables and gradients accordingly in these cells. This treatment allows solid boundaries, inflow boundaries, and non-reflecting outflow boundaries to be implemented. Zhao et. al. (1996) present procedures for implementing various boundary conditions in shallow-water flow using the FVM.

The two-dimensional adjoint equations are placed into matrix form to present their numerical solution, as follows,

$$\Phi_\tau + \tilde{A}\Phi_x + \tilde{B}\Phi_y + \tilde{C}\Phi + R = 0 \tag{12}$$

where $\Phi = (\phi \; \psi^x \; \psi^y)^T$, $R = (r_h \; r_p \; r_q)^T$ and

$$\tilde{A} = \begin{pmatrix} 0 & \frac{p^2}{h^2} - gh & \frac{pq}{h^2} \\ -1 & -\frac{2p}{h} & -\frac{q}{h} \\ 0 & 0 & -\frac{p}{h} \end{pmatrix} \qquad \tilde{B} = \begin{pmatrix} 0 & \frac{pq}{h^2} & \frac{q^2}{h^2} - gh \\ 0 & -\frac{q}{h} & 0 \\ -1 & -\frac{p}{h} & -\frac{2q}{h} \end{pmatrix} \tag{13}$$

$$\tilde{C} = \begin{pmatrix} 0 & -g(S_o^x + 2S_f^x) & -g(S_o^y + 2S_f^y) \\ 0 & -gh\left(\frac{p}{q^2+p^2} + \frac{1}{p}\right)S_f^x & -gh\left(\frac{p}{q^2+p^2}\right)S_f^y \\ 0 & -gh\left(\frac{q}{q^2+p^2}\right)S_f^x & -gh\left(\frac{q}{q^2+p^2} + \frac{1}{q}\right)S_f^y \end{pmatrix} \tag{14}$$

As is indicated by the form of Eqs. 12, the adjoint equations naturally appear in non-conservation form. The shallow-water equations have both a non-conservative and a conservative form, so one might expect the adjoint shallow-water equations to also have a conservation form. However, this is not the case. No similarity transformation of the adjoint shallow-water equations is possible. Considering that the adjoint shallow-water equations are not derived from conservation laws per se, but rather are the result of manipulating conservation statements, the lack of a pure conservation form is justified. Nevertheless, by defining adjoint fluxes, the numerical schemes used for the solution of the shallow water equations can also be applied to solve the adjoint shallow water equations. For this reason, Eqs. 12 are rewritten in a conservative-like form as follows,

$$\boldsymbol{\Phi}_\tau + \mathbf{F}(\boldsymbol{\Phi})_x + \mathbf{G}(\boldsymbol{\Phi})_y + \mathbf{C}\boldsymbol{\Phi} + \mathbf{R} = 0 \tag{15}$$

where

$$\mathbf{F}(\boldsymbol{\Phi}) = \tilde{\mathbf{A}}\boldsymbol{\Phi}, \quad \mathbf{G}(\boldsymbol{\Phi}) = \tilde{\mathbf{B}}\boldsymbol{\Phi}, \quad \mathbf{C} = \tilde{\mathbf{C}} - \tilde{\mathbf{A}}_x - \tilde{\mathbf{B}}_y \tag{16}$$

A finite-volume approach using Hancock's predictor-corrector scheme is now applied to solve the adjoint shallow-water equations. The procedure is analogous to that used to solve the shallow-water equations. After limiting the derivatives of the state variables to preserve monotonicity, the non-conservative form of the adjoint equations, Eqs. 12, is solved to advance the solution to the half time level.

$$\boldsymbol{\Phi}_{i,j}^{n-1/2} = \boldsymbol{\Phi}_{i,j}^n - \frac{\Delta t}{2\Delta x}\tilde{\mathbf{A}}_{i,j}^n \delta_x \boldsymbol{\Phi}_{i,j}^n - \frac{\Delta t}{2\Delta y}\tilde{\mathbf{B}}_{i,j}^n \delta_y \boldsymbol{\Phi}_{i,j}^n - \frac{\Delta t}{2}\tilde{\mathbf{C}}_{i,j}^n \boldsymbol{\Phi}_{i,j}^n \tag{17}$$

Next, the interface fluxes are evaluated by solving a Riemann problem at each cell face. The fluxes are given as,

$$\mathbf{F}_{i+1/2,j}^{n+1/2} = \mathbf{F}\left(\boldsymbol{\Phi}_{i+\frac{1}{2}L,j}^{n+1/2}, \boldsymbol{\Phi}_{i+\frac{1}{2}R,j}^{n+1/2}\right), \qquad \mathbf{G}_{i,j+1/2}^{n+1/2} = \mathbf{G}\left(\boldsymbol{\Phi}_{i,j+\frac{1}{2}L}^{n+1/2}, \boldsymbol{\Phi}_{i,j+\frac{1}{2}R}^{n+1/2}\right) \tag{18}$$

where

$$\boldsymbol{\Phi}_{i-\frac{1}{2}R,j}^{n+1/2} = \boldsymbol{\Phi}_{i,j}^{n+1/2} - \frac{1}{2}\delta_x \boldsymbol{\Phi}_{i,j}^n, \qquad \boldsymbol{\Phi}_{i+\frac{1}{2}L,j}^{n+1/2} = \boldsymbol{\Phi}_{i,j}^{n+1/2} + \frac{1}{2}\delta_x \boldsymbol{\Phi}_{i,j}^n \tag{19}$$

$$\boldsymbol{\Phi}_{i,j-\frac{1}{2}R}^{n+1/2} = \boldsymbol{\Phi}_{i,j}^{n+1/2} - \frac{1}{2}\delta_y \boldsymbol{\Phi}_{i,j}^n, \qquad \boldsymbol{\Phi}_{i,j+\frac{1}{2}L}^{n+1/2} = \boldsymbol{\Phi}_{i,j}^{n+1/2} + \frac{1}{2}\delta_y \boldsymbol{\Phi}_{i,j}^n \tag{20}$$

Once the fluxes have been computed, the conservative-like form of the equations (Eqs. 15) is solved to advance the solution to the next time level.

$$\boldsymbol{\Phi}_{i,j}^{n-1} = \boldsymbol{\Phi}_{i,j}^n - \frac{\Delta t}{\Delta x}\left(\mathbf{F}_{i+1/2,j}^{n+1/2} - \mathbf{F}_{i-1/2,j}^{n+1/2}\right) - \frac{\Delta t}{\Delta y}\left(\mathbf{G}_{i,j+1/2}^{n+1/2} - \mathbf{G}_{i,j-1/2}^{n+1/2}\right)$$
$$-\Delta t\,\mathbf{C}_{i,j}^{n+1/2}\boldsymbol{\Phi}_{i,j}^{n+1/2} \tag{21}$$

It is not clear that Hancock's predictor-corrector approach is the ideal finite volume scheme for solving the adjoint equations. In the conservation-like form (Eqs. 15), derived so that interface fluxes could be defined, a source term, C, appears that includes derivatives of the solution to the direct problem. The direct problem is saved with each iteration prior to solving the adjoint problem, so the information needed to compute these derivatives is already in storage. However, computing spatial derivatives of the forward problem solution requires additional work per time step, making this finite volume approach computationally more expensive. Clues for a more efficient scheme that remains second order accurate and monotonic are provided in a recent paper by LeVeque (1997), and this remains a topic of current research.

Riemann Solver

An approximate Riemann solver is developed to evaluate the adjoint interface fluxes which follow from the conservative-like equations. The approach is identical to Roe's approximate Riemann solver used for the shallow water equations, only the eigenvectors and eigenvalues differ slightly. Writing the flux normal to the boundary as,

$$\mathbf{E} = \mathbf{F}\cos\theta + \mathbf{G}\sin\theta \qquad (22)$$

the interface flux is given as

$$\mathbf{E}_I = \frac{1}{2}\left(\mathbf{E}_L + \mathbf{E}_R\right) - \frac{1}{2}\sum_{k=1}^{3}|\bar{\lambda}|\Delta V_k\bar{\mathbf{r}}_k \qquad (23)$$

The quantities with an overbar imply that an average be taken. Unfortunately, no average is available which satisfies Roe's criteria. The algebraic average is thus used,

$$\bar{c} = \frac{c_L + c_R}{2}, \quad \bar{u} = \frac{u_L + u_R}{2}, \quad \bar{v} = \frac{v_L + v_R}{2} \qquad (24)$$

The matrix of right eigenvectors in this case is given by,

$$\mathbf{R} = \frac{1}{2c}\begin{pmatrix} -u_\perp + c & -2cu_\parallel & u_\perp + c \\ \cos\theta & 2c\sin\theta & -\cos\theta \\ \sin\theta & 2c\cos\theta & -\sin\theta \end{pmatrix} \qquad (25)$$

The eigenvalues are $\lambda_1 = -u_\perp - c$, $\lambda_2 = -u_\perp$, and $\lambda_3 = -u_\perp + c$. From $d\mathbf{V} = \mathbf{R}^{-1}d\mathbf{U} = \mathbf{L}d\mathbf{U}$, the wave strengths follow as,

$$\Delta\mathbf{V} = \begin{pmatrix} \Delta\phi + (\bar{u} + \bar{c}\cos\theta)\Delta\psi^x + (\bar{v} + \bar{c}\sin\theta)\Delta\psi^y \\ \sin\theta\Delta\psi^x + \cos\theta\Delta\psi^y \\ \Delta\phi + (\bar{u} - \bar{c}\cos\theta)\Delta\psi^x + (\bar{v} - \bar{c}\sin\theta)\Delta\psi^y \end{pmatrix} \qquad (26)$$

Boundary conditions for the adjoint problem are also implemented using ghost cells. State variables and gradients within ghost cells are specified such that the resulting Riemann problem solution yields the desired mass and momentum fluxes for reflecting and non-reflecting boundaries. The procedure is analogous to that used to implement boundary conditions for the shallow-water equations.

RESULTS

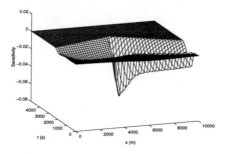

Figure 1: Sensitivities $\frac{\delta J}{\delta q^i}(x, t)$ with $N = 200$ and $N_T = 1000$.

Solution in One-Dimension

Sensitivities are now computed in one space dimension. A channel with length, $L = 10 \, km$, slope, $S_o = 0.00001$, and Chezy coefficient, $C = 80$, is used. Flow in the channel is initially steady and uniform with $q(x, 0) = 0.4648 \, m^s/s$ and $h(x, 0) = 1.5 \, m$. The upstream inflow is given as,

$$q(0, t) = q(0, 0) + 5sech^2(0.0018(t - 1500)) \tag{27}$$

This hydrograph generates a flood wave in which the discharge increases by an order of magnitude in a period on the order of an hour. The downstream boundary is non-reflective to simulate an infinite boundary.

The depth and discharge in the channel are computed following the solution of the shallow-water equations in one dimension. Given that $\bar{h} = 2.05 \, m$ at the target location, the adjoint system of equations (Eqs. 5) is now solved. The desired sensitivities are then computed using Eq. 6 and presented in Figure 1.

The solution to the sensitivities demonstrates the wave-like nature of the sensitivity information. That is, peak sensitivities both upstream and downstream of the target location occur at earlier and earlier times as distance from the target location increases. This is illustrated more clearly in Figure 2, where the maximum sensitivity (absolute value) found over time at each spatial location is plotted along with four points computed by a DSM. The four points computed using the DSM confirm that the ASM is accurate. Figure 3 presents the time when the maximum sensitivity occurs at each spatial location.

The physical implication of the sensitivity waves is that the location of a levee breach is not as critical as the timing of a breach. That is, a levee breach site can be chosen prior to an extreme flooding event, and so long as the breach is appropriately timed, the peak stage of the flood wave at the target location can be lowered.

Solution in Two-Dimensions

Sensitivities are now computed in two spatial dimensions. The Chézy coefficient is given as $C = 80$, $S_o^x = 0.00001$, and $S_o^y = 0$. Initially, a steady and uniform flow

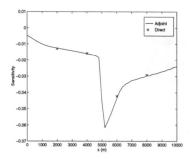

Figure 2: Maximum sensitivities at each   Figure 3: Time of maximum sensitivities
spatial location.   at each spatial location.

of $p(x, y, 0) = 0.3578 \, m^2/s$ and $q(x, y, 0) = 0 \, m^2/s$ with $h(x, t, 0) = 2 \, m$ is present
in a channel with $L_x = 1000 \, m$ and $L_y = 100 \, m$. Hard boundaries are present along
the sides of the channel ($y = 0$ and $y = 100 \, m$), and flow is prescribed on the inflow
boundary, $\partial\Gamma_3$, using the following inflow hydrograph,

$$p((x, y) \in \partial\Gamma_3, t) = q_o + sech^2(0.03(t - 120)) \tag{28}$$

with $q((x, y) \in \partial\Gamma_3, t) = 0$. The downstream boundary, $\partial\Gamma_2$, is non-reflective to allow
the flood wave to freely pass through. The target location is placed at $x_T = 500 \, m$
with $y_T = 0 \, m$. $\bar{h} = 2.05 \, m$, and $T = 300 \, s$.

The adjoint variables ϕ, ψ^x, and ψ^y are presented in Figure 4 at times $t = 250 \, s$,
$t = 200 \, s$, and $t = 150 \, s$. ϕ is plotted as contours while ψ^x and ψ^y are plotted as
vectors. The source in the adjoint problem is clear at $t = 250 \, s$, and consistent with
the one-dimensional result because the solution develops a distinct downstream and
upstream moving signal. The sensitivity of the objective function to perturbations
of flow through a levee is presented in Figure 5 at times $t = 250 \, s$, $t = 200 \, s$, and
$t = 150 \, s$.

The one dimensional ASM is appropriate for use when long reaches of rivers are
considered and when flow remains within a deep water channel. The two dimensional
ASM becomes necessary when flow leaves the channel and enters the floodplain and
when greater spatial resolution of the levee breach is required. When the ASM is
applied to control flow in real-time, the additional computational demand of the added
spatial dimension must be weighed against the need for greater resolution of the flow.

CONCLUSIONS

A dynamic flood control strategy is presented that is consistent with risk-based
flood management policy. Control of flood waves is achieved using an adjoint sensitivity method (ASM) based on the shallow-water equations in both one and two spatial
dimensions. The efficiency of the ASM allows active control of flood waves. A deliberate levee breach at optimal times and locations during a flooding event is shown

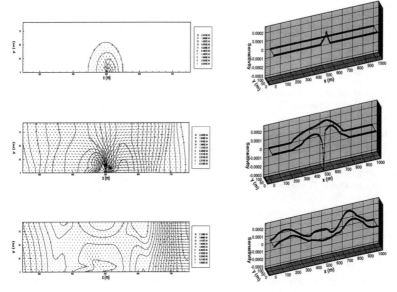

Figure 4: Adjoint equation solution at $t = 250\,s$ (top), $t = 200\,s$ (center), and $t = 110\,s$ (bottom). $S_o^x = 0.00001$, $S_o^y = 0$, and $C = 80$.

Figure 5: Sensitivities, S_{ijk}, at $t = 250\,s$ (top), $t = 200\,s$ (center), and $t = 150\,s$ (bottom). $S_o^x = 0.00001$, $S_o^y = 0$, and $C = 80$.

to reduce the peak stage of a flood wave at the target location. A second order finite-volume scheme is presented for the monotonic solution of the adjoint shallow-water equations.

REFERENCES

G.D. Van Albada, B. Van Leer, and W.W. Roberts, "A Comparative Study of Computational Methods in Cosmic Gas Dynamics," Astronomy and Astrophysics, Vol. 108, 1982, pp. 76-84.

S.K. Das and R.W. Lardner, "On the Estimation of Parameters of Hydraulic Models by Assimilation of Periodic Tidal Data," J. of Geophysical Research, Vol. 96, No. C8, 1991, pp. 15,187-15,196.

F.-X. Le Dimet and O. Talagrand, "Variational Algorithms for Analysis and Assimilation of Meteorological Observations: Theoretical Aspects," Tellus, Vol. 38A, 1986, pp. 97-110.

M.C.G. Hall and D.G. Cacuci, "Physical Interpretation of Adjoint Functions for Sensitivity Analysis of Atmospheric Models," J. of Atmospheric Sciences, Vol. 40, 1983, pp. 2537-2546.

M.C.G. Hall, D.G. Cacuci, and M.E. Schlesinger, "Sensitivity Analysis of a Radiative-Convective Model by the Adjoint Method," J. of Atmospheric Sciences, Vol. 39, 1982, pp. 2038-2050.

C. Hirsch, "Numerical Computation of Internal and External Flows", Vol. 2, John Wiley and Sons, 1990.

N.D. Katopodes and M. Piasecki, "Site and Size Optimization of Contaminant Sources in Surface Water Systems," ASCE J. of Environmental Engineering, Vol. 122, 1996, pp. 917-923, 1996.

R.W. Lardner, "Optimal Control of Open Boundary Conditions for a Numerical Tidal Model," Computer Methods in Applied Mechanics and Engineering, Vol. 102, 1993, pp. 367-387.

R.W. Lardner, A.H. Al-Rabeh, and N. Gunay, "Optimal Estimation of Parameters for a Two-Dimensional Hydrodynamic Model of the Arabian Gulf," J. of Geophysical Research, Vol. 98, No. C10, 1993, pp. 18,229-18,224.

B. Van Leer, "Towards the Ultimate Conservation Difference Scheme. V. A Second Order Sequel to Godunov's Method," J. of Computational Physics, Vol. 32, 1979, pp. 101-136.

R.J. LeVeque, "Wave Propagation Algorithms for Multi-Dimensional Hyperbolic Systems," J. of Computational Physics, Vol. 131, 1997, pp. 327-353.

G.I. Marchuk, "Adjoint Equations and Analysis of Complex Systems," Kluwer, 1995.

V.G. Panchang and J.J. O'Brien, "On the Determination of Hydraulic Model Parameters Using the Adjoint State Formulation," Modeling Marine Systems, Vol. I, 1989, pp. 5-18, A.M. Davies, ed., CRC Press.

M. Piasecki and N.D. Katopodes, "Control of Contaminant Releases in Rivers, I: Adjoint Sensitivity Analysis," ASCE J. of Hydraulic Engineering, Vol. 123, 1997a, pp. 486-492.

M. Piasecki and N.D. Katopodes, "Control of Contaminant Releases in Rivers, II: Optimal Design," ASCE J. of Hydraulic Engineering, Vol. 123, 1997b, pp. 493-503.

P.L. Roe, "Approximate Riemann Solvers, Parameter Vectors, and Difference Schemes," J. of Computational Physics, Vol. 43, 1981, pp. 357-372.

B.F. Sanders, "Control of Shallow-Water Flow using the Adjoint Sensitivity Method," Ph.D Dissertation, Department of Civil and Environmental Engineering, University of Michigan, 1997.

W. W.-G. Yeh, "Review of Parameter Identification Procedures in Groundwater Hydrology: The Inverse Problem," Water Resources Research, Vol. 22, No. 2, 1986, pp. 95-108.

D.H. Zhao, H.W. Shen, J.S. Lai, and G.Q. Tabios III, "Approximate Riemann Solvers in FVM for 2D Hydraulic Shock Wave Modeling," ASCE J. of Hydraulic Engineering, Vol. 122, 1996, pp. 692-702.

J. Zou and G. Holloway, "Improving Steady-State Fit of Dynamics to Data Using Adjoint Equation for Gradient Preconditioning," Monthly Weather Review, Vol. 123, 1995, pp. 199-211.

COASTAL FLOW MODELLING USING AN INVERSE METHOD
WITH DIRECT MINIMISATION

by
Dr. Graham Copeland
Department of Civil Engineering, University of Strathclyde,
107 Rottenrow, Glasgow, Scotland. G4 0NG
Tel 0141 548 3252 Fax 0141 553 2066
email g.m.copeland@strath.ac.uk

Abstract
The paper presents a flow model based on an inverse method which uses direct minimisation as a solution procedure. The theoretical background is described and an application to a full scale coastal engineering study of tidal flows past a headland is discussed. The operational flow model is packaged with a MS Windows graphical interface in an application called WOLF/$_{FLOW}$. The model is based on the non-linear, depth integrated (2-D), unsteady equations of motion (the shallow water equations [SWEs]) which are applied as weak constraints in the inverse problem. Either current or elevation data or combinations of both can be assimilated. The solution is found by minimising a global function which is the sum of the squares of the weighted residuals of the continuity, x and y momentum equations and the differences between data values and solution values of current and elevation. The solution method uses an conjugate gradient algorithm, CONMIM.

Introduction

Inverse models provide a way to build flow fields which are solutions to the governing equations of fluid motion <u>and</u> which are based on measurements located anywhere in the domain. Inverse models of the type described here are not driven by boundary/initial values in the flow variables like conventional models. A formal distinction can be made as follows: an initial or boundary value problem becomes inverse if some of the boundary or initial data is missing (i.e. the problem is ill-posed) and is replaced by information (data) about the desired solution located within the model domain, Bennett (1992), Bellomo and Preziosi (1995). This situation is typical of coastal modelling problems where boundary data is often poorly defined yet there exists good data within the model area.

The paper describes an operational model based on an inverse method which can assimilate large quantities of measured data within the model domain into the solution and does not require initial or boundary data in the flow variables. The method uses a direct minimisation technique and the governing equations (the SWEs) are included as a weak constraint on the solution. This allows the relative influences of data and equations on the final solution to be defined.

The model has been validated using idealised topographies in terms of data assimilation, flow diversion by topography and the influence of individual terms in the governing equations, Bayne (1997), Copeland & Bayne (1996). The model has been applied to the calculation of tidal flows in an area on the east coast of Scotland which were required as part of an outfall design study.

Theoretical Background

Background to the development

The main principle of this method is the minimisation of differences between observed and computed values of the flow variables whilst satisfying imposed constraints. The differences between the observed and computed values are described in a 'cost function' and the constraints

Another advantage of the direct minimisation method in particular is that there is no Courant-Friedrichs-Lewey (CFL) stability requirement. The direct minimisation method used here is a substitution method which does not involve any time integration and therefore the CFL condition does not affect the stability of the solution procedure (Bennett and Chua, 1994). However, for very large CFL numbers the effect of the time dependent terms in the governing equations on the solution diminishes so that each flow field in a series is effectively solved in isolation from its temporal neighbours and so the accuracy decreases.

There is also a disadvantage in that the predictive ability of the model appears to be limited because the governing equations are not integrated. The model will not predict flow separation in the lee of a headland for steady main stream flows (of course, if data on the eddies is available then this can be included to produce a solution which includes the eddies, Copeland (1997a)). However, for unsteady tidal flows there is evidence that secondary circulations are formed in the solution. This is the subject of continuing investigation by the author.

Methodology
The cost function
The first step in model construction was the development of the function to be minimised. The constrained cost function, $\mathbf{F}$, was defined in terms of the depth mean unsteady discharges u, v and water elevations η and the residuals in the governing equations. This cost function describes the differences between the measured data values and their computed model counterparts:

$$\mathbf{F} = \sum_k \sum_i \sum_j \begin{array}{l} w1 \; s1(continuity)^2 + \\ w1 \; s2(x \; momentum)^2 + w1 \; s2(y \; momentum)^2 + \\ w2_{i,j}^k s3(\widetilde{\eta}_{i,j}^k - \eta_{i,j}^k)^2 + \\ w3_{i,j}^k s4(\widetilde{u}_{i,j}^k - u_{i,j}^k)^2 + w3_{i,j}^k s4(\widetilde{v}_{i,j}^k - v_{i,j}^k)^2 \end{array} \tag{1}$$

where
k is the temporal indicator, i is the spatial indicator in the x direction, j is the spatial indicator in the y direction,

$\eta_{i,j}^k$ = computed elevation at (k, i, j)

$u_{i,j}^k$ = computed discharge per unit width in x direction at (k, i, j)

$v_{i,j}^k$ = computed discharge per unit width in y direction at (k, i, j)

and $\widetilde{\eta}_{i,j}^k$, $\widetilde{u}_{i,j}^k$, $\widetilde{v}_{i,j}^k$ are the measurements corresponding to the above state variables.

w1, w2, w3 represent weights and s1, s2, s3, s4 represented scaling factors.

'Continuity', 'x-momentum' and 'y-momentum' were defined to be the residuals of the discretised versions of the equations of tidal motion.

Direct minimisation
Once the constrained cost function had been constructed, gradient information with respect to each of the three model variables was obtained by calculating derivatives at each position in the computational domain. An algebraic substitution was made for the residuals of the equations of motion in these calculations. An example of these calculations is given below and the full gradient equations are given in Bayne (1997).

flow variables in the solution. This is equivalent to assuming that the (unknowable) flow variables in the boundary cells immediately *outside* the model grids are a perfect solution to the governing equations (because there is no data out there to require otherwise) and so adjustments may then be made to first guess flow variables in cells *on* the boundary to make them consistent with that. This does not, however, require that the residuals in the constraint equations become zero in these boundary cells. This is because the equations are only weak constraints and so data on or within the model boundary can be an influence on the solution. This Dirichlet boundary condition may be used at either spatial or temporal boundaries (when the problem is not periodic in time, see below).

2. Setting the gradient in the residuals to zero. This is the Neumann boundary condition. This results in a zero adjustment to the solution at the boundaries and is used at a closed boundary when the zero normal velocity component set in the first guess should remain unchanged. This condition is not physically meaningful in the temporal domain and is therefore unsuitable there.

3. The third was the periodic boundary condition whereby residuals on the boundary domain and immediately outside the solution domain were set to the equivalent value one period previously or subsequently. Clearly in the study of tidal dynamics, this is appropriate because a tidal wave is periodic in time. In this case, both the first guess and the solution are periodic and so it is quite appropriate to use a periodic condition for the residuals. Note that for coastal areas, this is generally unsuitable as a spatial boundary condition due to the very long wavelength of a tidal wave. However, periodic boundary conditions were particularly suitable for the temporal boundary and the best results were yielded when these were employed.

In this work, the following boundary conditions were applied:

- periodic temporal boundary conditions
- open boundary conditions in the spatial domain
- closed boundary conditions at any dry area.

The magnitude of the solution is not, therefore, determined by the boundary conditions, instead it is the magnitude of the internal data and of the 'first guess' (see below) based on this data which determines the magnitude of the final solution.

The first guess

The 'first guess' is the set of values given to the state variables at the outset of the minimisation process. It can be thought of as a trial solution which is improved by the direct minimisation procedure. The minimum found by the inverse method is that which is closest to the first guess. Therefore, for functions which do not have a unique solution, the first guess determines the area of the solution domain in which the solution lies. It is clearly advantageous that a good first guess be available. In order to achieve this, a useful strategy is to set the first guess values equal to the data values at all points where available. Interpolation to estimate missing neighbouring values then follows.

Advantages of inverse models

The major advantage of inverse methods over conventional solution methods is their ability to include a full data set into the model computations (e.g. the results of a field survey) and so produce a solution derived from data rich inputs.

(the equations of continuity and momentum conservation with the shallow water approximation) are appended to this by means of an appropriate algebraic device.

Some of the recently developed inverse models use variational analysis, Monteiro and Copeland (1995), Copeland (1994, 1997b). This is the classical approach and involves the construction of Euler-Lagrange partial differential equations (PDEs) from the constrained cost function of a system. By solving these equations the required minimum is obtained. However, these PDEs are difficult both to construct and to solve when the full set of momentum and continuity equations is considered. Therefore, other variational methods termed 'direct minimisation methods', which are conceptually much simpler, are increasingly being used, Legler, Navon and O'Brien (1989), Hoffman (1984). These involve minimising the constrained cost function using an iterative numerical algorithm. A direct method of minimisation was used in this work.

Similar problems in unsteady shallow water flows can also be solved by the adjoint method which involves the repeated integration of the SWEs with iterative adjustment of the open boundary condition(s), see Bogden et al (1996), Bowen et al (1995) and Griffin and Thompson (1996).

Strong and weak constraints

Sasaki (1970) defined both strong and weak variational constraint formalisms. In the strong constraint formalism, the constraining equations were appended to the cost function by means of Lagrange multipliers. This constrained function was then solved using a variational analysis subject to satisfying the constraints exactly. In the weak constraint formalism, on the other hand, the constraining equations in quadratic form were appended to the cost function by means of weighting factors. Then the constraints were only satisfied approximately in a least squares error sense. The weak constraint method is the more general and flexible of the two because it does not assume nor require that the constraining equations are satisfied exactly and so allows for the approximations inherent in their formulation. Instead, the relationship between the constraints and data can be varied according to the quantity and quality of the available data and to the level of approximation in the equations. The weak constraint formulation was used in this work.

Boundary conditions

Conventional solution methods for hydraulic problems have the form of initial and/or boundary value problems. These methods require a full compliment of initial and/or boundary values in the state variables (elevation and velocity) in order for the problem to be well-posed and hence solvable. Inverse methods are less reliant on boundary data because they have some data internal to the computational domain.

In particular, the direct minimisation method described here required no data for state variables at the boundaries. Only at no-flow boundaries (coastlines) was it necessary to give boundary conditions in the state variables, i.e. that there is no normal flow component. Instead, the method was developed so that boundary conditions in the residuals of the constraining governing equations were all that was required. This is satisfactory formally because the cost function is minimised with respect to the differences between observed and computed data values and the *residuals* of the governing equations (see below). It is the flow variables which best minimise these differences and residuals which form the solution. In the spatial and temporal domains there are 3 possible boundary conditions:

1. Setting the residuals to zero. This is the Dirichlet boundary condition. In this case the *gradient* in the residuals will not in general be zero. This allows a non-zero adjustment to the

Once expressions for the gradients had been calculated and a first guess obtained, the solution proceeded as follows: the first guess was substituted into the model equations to obtain the residuals of each equation; these were in turn substituted into the simplified expressions for the gradients to produce the gradients at the first guess, then the iterative minimisation algorithm CONMIN (see below) was applied. This provided new estimates of the state variables by way of the first guess and the gradient information. These new estimates were used to re-calculate the gradient so that the minimisation algorithm could be re-applied. This whole procedure was repeated until the residual of the constrained cost function reached a given convergence criterion.

The iterative minimisation algorithm used to minimise the cost function was a conjugate gradient descent (CGD) method by Shanno-Phua (1980). This was a Beale-restarted, quasi-Newton, memoryless, conjugate gradient algorithm named CONMIN. This is freely available at http://www.hensa.ac.uk/

Weights and scaling factors
The scaling factors, s1, s2, s3 and s4 in eq. 1 were needed for two reasons:
      (a) to scale each of the terms of the cost function to be of a similar magnitude (hence improving convergence), and
      (b) to make the terms of the model dimensionally consistent so that their summation was physically valid.
The scaling factors used were: s1 = 1 used to scale $(continuity)^2$; s2 = $(g\ hmean)^{-1}\ s^2 m^{-2}$ used to scale $(momentum)^2$; s3 = $(g\ hmean\ \Delta x^{-2})\ s^{-2}$ used to scale (elevation data)^2 and s4 = $\Delta x^{-2}\ m^{-2}$ used to scale (discharge data)^2 , where g is the acceleration due to gravity, and *hmean* is the average depth over the spatial domain.

The weights w_1, w_2, w_3 were used to determine the relationship between the equations and the data and determine how 'weak' the equations should be (w_2, w_3 can be assigned different values at each data point). As indicated above, the ability of the user to determine the weights of a function to be minimised gives the procedure its flexibility and also allows an indication of data quality. The weights also affect the size of the final solution and the rate of convergence.

The numerical scheme
An explicit, time and space centred finite differencing scheme was used in the formulation of the WOLF inverse model to approximate the derivatives in the governing equations of tidal motion and so transpose the PDEs into a form suitable for numerical solution by the conjugate gradient method algorithm.

Discretisation
The space and time derivatives in the 2D depth integrated equations of motion were both fully centred and had second order accuracy in space and time.
For example, the discretised continuity equation was:-

$$\frac{\eta_{i,j}^{k} - \eta_{i,j}^{k-1}}{\Delta t} + \frac{u_{i+1,j}^{k} - u_{i,j}^{k}}{\Delta x} + \frac{v_{i,j+1}^{k} - v_{i,j}^{k}}{\Delta y} = \lambda_{i,j}^{k} = 0(ideally) \qquad (2)$$

where λ is the residual which equals zero when the equation is satisfied exactly.

Calculation of gradients

The gradients were calculated by differentiating the discretised cost function with respect to the three state variables η, u and v in turn. This process was greatly simplified by using substitutions for the residuals of the continuity and momentum equations. These residuals were the remainders of the governing equations calculated at the most recently obtained values for the state variables. The following substitutions were used:

$$\lambda_{i,j}^{k} = continuity\ residual\ at\ (k,i,j) \tag{3}$$

$$\gamma_{i,j}^{k} = x\ momentum\ residual\ at\ (k,i,j) \tag{4}$$

$$\kappa_{i,j}^{k} = y\ momentum\ residual\ at\ (k,i,j) \tag{5}$$

As a simplified example, consider a cost function constrained by the continuity equation only :

$$\mathbf{F} = \sum_{k}\sum_{i}\sum_{j} \left(\frac{\eta_{i,j}^{k} - \eta_{i,j}^{k-1}}{\Delta t} + \frac{u_{i+1,j}^{k} - u_{i,j}^{k}}{\Delta x} + \frac{v_{i,j+1}^{k} - v_{i,j}^{k}}{\Delta y} \right)^2 \\ + (\widetilde{\eta}_{i,j}^{k} - \eta_{i,j}^{k})^2 + (\widetilde{u}_{i,j}^{k} - u_{i,j}^{k})^2 + (\widetilde{v}_{i,j}^{k} - v_{i,j}^{k})^2 \tag{6}$$

then the gradient with respect to the state variables η, u and v can be represented by:

$$\frac{\partial \mathbf{F}}{\partial \eta_{i,j}^{k}} = 2\left(\frac{\lambda_{i,j}^{k} - \lambda_{i,j}^{k+1}}{\Delta t} \right) - 2(\widetilde{\eta}_{i,j}^{k} - \eta_{i,j}^{k}) \tag{7}$$

$$\frac{\partial \mathbf{F}}{\partial u_{i,j}^{k}} = 2\left(\frac{\lambda_{i-1,j}^{k} - \lambda_{i,j}^{k}}{\Delta x} \right) - 2(\widetilde{u}_{i,j}^{k} - u_{i,j}^{k}) \tag{8}$$

$$\frac{\partial \mathbf{F}}{\partial v_{i,j}^{k}} = 2\left(\frac{\lambda_{i,j-1}^{k} - \lambda_{i,j}^{k}}{\Delta y} \right) - 2(\widetilde{v}_{i,j}^{k} - v_{i,j}^{k}) \tag{9}$$

The gradients for the full constrained cost function, eq. (1), were calculated in a similar way and are given in Bayne (1997).

Applications

Validation

The model was validated using idealised domains and simplified equations , Bayne (1997). Three important aspects of the developed model were validated in detail. These were its ability to accommodate data, the proper functioning of individual terms in the governing equations and the effect of different topographies. The effect of eddy viscosity was not assessed by Bayne (1997) but is the subject of current research by the author particularly in connection with headland flow.

Site Specific Study

As part of a study for a proposed marine wastewater outfall, the inverse flow model WOLF was used to describe the currents in an area on the east coast of Scotland near the border with England. The model domain extended 22.5 km alongshore. The model grids were rotated such that the x and y orthogonal axes lay approximately offshore and alongshore (along bearings 55° and 325° respectively relative to grid north) see Figure 1. The model was built with variable resolution so as to provide best resolution in a small bay near the proposed outfall. Four model grids were built with cell dimensions of 8.33 m, 25 m, 75 m and 225 m. The locations of the origins of each grid ensured that the grid cells nested together exactly in a 3:1 ratio. The four grids are outlined in Figures 1. The unsteady tidal flow was represented in each grid by a series of 25 flow fields separated by a time step of $\Delta t = 1800$ s.

Data

The first guess for water levels at each time step used in this application of WOLF was found by sinusoidal interpolation of data from a local tide gauge. This was sufficient since, unlike conventional models, the solution can be controlled by current data within the domain.
WOLF also used current flow data which was obtained from two sources:

   a) drogue tracks taken during a marine survey which produced 1037 data points for flow velocity at different times during a tide and for different tidal ranges, see Figure 2.

   b) acoustic Doppler current meter profiler (ADCP) which produced 617 depth averaged data points along each of two transects (shown in Figure 11).

These data were processed to produce flow data every half hour through a tidal cycle and referred to the time of local high water. The data were taken during different tidal ranges and were taken at different points in the water column. Scaling factors based on recording current meter data at the site were applied in order to unify the data to mean neap range and factors based on an assumed power law definition of the vertical profile in current speed were used to adjust flow data to depth average values.

Calculation of Flow Fields

The data coverage was spatially and temporally variable and some areas of the model at some times through a cycle required additional information to allow a good first guess to be made. In order to overcome these problems, the available data were analysed to indicate the flood and ebb maximum (neap) current speeds, the times of these maxima and the times of the slack waters at 'high water' (end of flood tide) and 'low water' (end of ebb tide). These scalar quantities were amenable to contouring and extending by simple interpolation. Examples of these analyses are shown in Figures 3 and 4. The real survey data were supplemented at selected locations by 'data' obtained by interpolation of this contoured values through a tidal cycle. This provided the extra data needed to make a reasonable first guess in areas where actual data was sparse.

The WOLF inverse flow model was used to impose continuity and consistency with the momentum equations in each of the four grids separately. In order to ensure that the separate grids were consistent with each other, the first guess for each was based on the same data set. In addition, the grids were computed in order starting with the 225 m grid through the 75 m and 25 m grids to the 8.33 m grids, the first guess for the next in the series was influenced by the final solution of the larger grid.

Validation

The validation of the flow model was achieved using two methods:

   i) the comparison of time series of current velocities from the rcm data and output from the flow model at the same points (a Eulerian approach),

   ii) the comparison of the trajectory of a particle in the model with that of equivalent drogue (a Lagrangian approach)

It was expected that good agreement (5% to 10%) could be achieved with the time series. Comparisons of trajectories, however, were expected to be less likely to achieve such good results because the accumulation of small errors could lead to small shifts relative to the strongly sheared currents in the area which in turn could lead to quite large changes in overall excursion.

Time Series of Velocities
The results from the 25 flow fields generated by the WOLF flow model were compared with the data for a mean neap tide taken from recording current meters (rcm's) deployed at two locations, points G and A. Examples of these comparisons are shown in Figures 5 and 6 which show good agreement in a current system which contains many higher harmonics and therefore has a high tide-to-tide variability. These rcm data were not assimilated by the model and so this was an independent validation.

Comparisons were also made between the results taken from the three larger grids (locations G and A were outside the 8.33 m grid). Figure 7 shows an example of the comparison between the 225 m and 25 m grids, a very good level of consistency between the grids was shown.

Trajectories
Seven long excursion drogue tracks were used to validate the model flow fields (these and subsequent validations used the nested series of grids so that the highest spatial resolution available was always used). Examples of the results are given in Figures 8, 9 and 10. Each shows the computed and measured trajectory. Most of the tracks show a reasonable or good agreement .

The wind conditions were very important to the successful drogue track simulations by the model. However, the flow fields produced by WOLF were computed without the effect of winds stress. This was added later during the particle tracking calculations. Wind induced currents were applied as a sheared surface layer with an approximately logarithmic profile. This profile has been evaluated, Copeland (1997b), in two forms:
     i) as a longshore current with a net transport,
     ii) as an onshore-offshore current with zero net transport.
A longshore current profile with net transport was used in water deeper than 20m to describe the alongshore ($325°$ - $145°$) wind induced velocity component. A profile with zero net transport was used to describe the onshore-offshore ($235°$ - $55°$) component in deeper water and both the alongshore and onshore-offshore components in shallower water (<20 m) nearer the coastline.

Examples of a final computed flow field in the 25 m grid is shown in Figure 11 along with the data which was assimilated into the solution at that point in the tidal cycle (HW+4.5).

Conclusions
An inverse model of unsteady tidal flow has been built and validated which uses a direct minimisation solution method and applies the depth averaged equations of motion as weak constraints. The model is able to assimilate large quantities of data into the solution but does not need flow or elevation data at the boundaries. Instead, boundary values in the residual of the constraint equations are required and these form a simple and convenient system. The method does not integrate the equations of motion forward in time but employs a substitution method to find the solution. This frees the procedure from the CFL stability requirement although accuracy is still affected at very large CFL numbers. The predictive ability of the model appears to be limited because the equations are not integrated. The model will not predict flow separation in the lee of a headland for steady flows. However, for unsteady tidal flows there is evidence of eddy generation

This is the subject of continuing investigation by the author. Of course, if data on eddy circulation is available this can be included directly into the solution.

Finally, the model has been applied to a full scale engineering study in which flow fields in four overlapping grids were computed with the assimilation of over 1600 items of current data.

Acknowledgements
The inverse model was developed with funding from the Engineering and Physical Sciences Research Council, UK by Dr. Gwen Bayne. The results from the case study are presented with kind permission of Mssrs. R.H. Cuthbertson and Partners, Consulting Engineers, Edinburgh and East of Scotland Water. The assistance of graduate student Scott Couch is also gratefully acknowledged.

References
Bayne, G.L.S., 1997. A direct minimization model to determine tidal flows on coastal waters, PhD Thesis, Dept. of Civil Engineering, University of Strathclyde

Bellomo, N. and Preziosi, L., 1995. Modelling mathematical methods and scientific computation. *CRC Mathematical Modelling Series*, Boca Raton.

Bennett, A.F., 1992. Inverse methods in physical oceanography. *CUP*, Cambridge.

Bennett, A.F. and Chua, B.S., 1994. Open ocean modelling as an inverse problem: the primitive equations. *Monthly Weather Review*, 122: 1326 1336.

Bogden, P.S., Malanotte_Rizzoli, P. & Signell, R. 1996. 'Open-ocean boundary conditions from interior data: Local and remote forcing of Massachusetts Bay. *J. Geophys. Res. 101, C3, pp 6487-6500.*

Bowen, A.J., Griffin, Hazen, D.A., Matheson, S.A., & Thompson, K.R., 1995. 'Shipboard nowcasting of shelf circulation', *Continental Shelf Res. 15, 1, pp115-128.*

Copeland , G.J.M., 1997a. Flow modelling using an inverse method with direct minimisation, *Proc. Conf. Coastal Zone '97*, Plymouth June 1997, publ. ASCE.

Copeland, G.J.M., 1997b. 'Computer modelling of tidal flows and effluent dispersion in the Firth of Forth, Scotland using an inverse method and particle tracking techniques', *J. Mar. Env. Eng'g, Volume 4/2 pp. 147-173.*

Copeland, G.J.M. & Bayne, G. L. S., 1996. 'Data Rich Models of Tidal Flows using Inverse Methods', *Hydroinformatics 96*, ed. Muller, pub. Balkema 1996, ISBN 90 5410 8525, pp. 239-244.

Copeland, G.J.M., 1994. 'An inverse method of kinematic flow modelling based on measured currents', *Proc.Inst. Civ. Engrs Wat., Marit. & Energy*, 1994, **106**, Sept., 249-258

Griffin, D.A. & Thompson, K.R. 1996. 'The adjoint method of data assimilation used operationally for shelf circulation'. *J. Geophys. Res. 101, C2, pp3457-3477.*

Hoffman, R.N., 1984. SASS wind ambiguity removal by direct minimisation. Part II: use of smoothness and dynamical constraints. *Monthly Weather Review*, 112: 1829 1852.

Legler, D.M., Navon, I.M. and O'Brien, J.J., 1989. Objective analysis of pseudostress over the Indian Ocean using a direct-minimization approach. *Monthly Weather Review*, 117: 709 720.

Monteiro, T.C.N. and Copeland, G.J.M., 1995. A hydrodynamic model of unsteady tidal flow in coastal waters based on an inverse method. *Arch. Mech.*, 47 (6): 1057 1071.

Sasaki, Y., 1970. Some basic formalisms in numerical variational analysis. *Monthly Weather Review*, 98 (12): 875 883.

Shanno, D.F. and Phua, K.H., 1980. Remark on Algorithm 500 - a variable method for unconstrained nonlinear minimization. *ACM Transactions on Mathematical Software*, 6 (4): 618 622

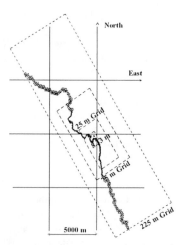

Figure 1 Location plan showing the four model grids.

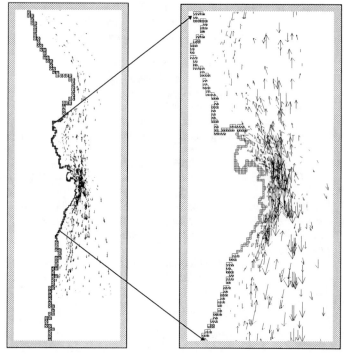

Figure 2   Vectorised drogue data, showing all data at all stages of the tide and at all tidal ranges which occurred during the survey.

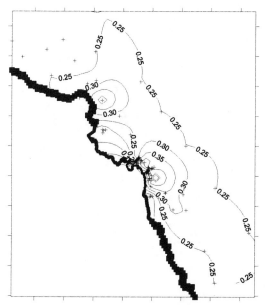

Figure 3   Contours of maximum flood speed for a mean neap tide (m/s) showing data points (+).

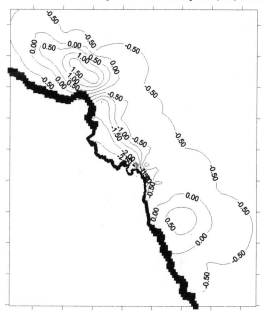

Figure 4   Contours of times of maximum flood current (hours relative to local high water).

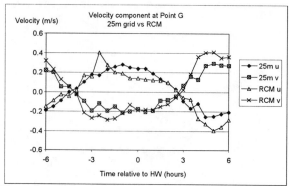

Figure 5 rcm data at Point G vs. results from 25m grid model.

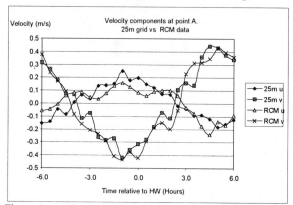

Figure 6. rcm data at Point A vs. results from 25m grid model.

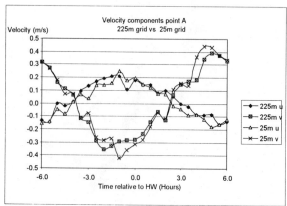

Figure 7 Results from the 225m grid model vs results from 75m grid model
at Point A.

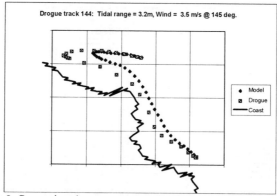

Figure 8 Comparison between measured and predicted drogue tracks on 27 April 1997 (showing 2 km grid squares).

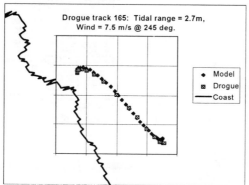

Figure 9 Comparison between measured and predicted drogue tracks on 1 May 1997 (showing 2 km grid squares).

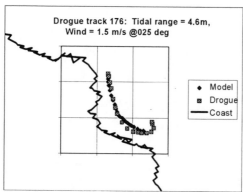

Figure 10 Comparison between measured and predicted drogue tracks on 7 May 1997 (showing 2 km grid squares).

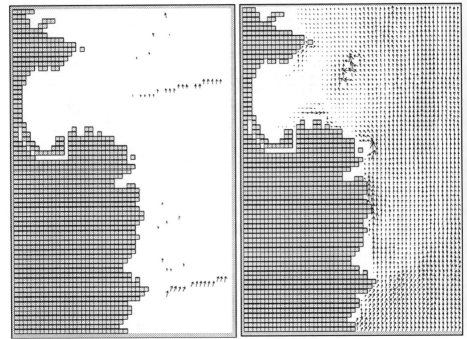

Figure 11 Comparison of input current data (drogue and ADCP) with the computed flow field (HW+4.5hrs)

Hydrodynamic Modeling of Johor Estuary – Johor, Malaysia

Johnny D. Martin, EIT[1] and Jeffrey G. Shelden, P.E., M. ASCE[2]

ABSTRACT

Hydrodynamic modeling of the Johor estuary was completed for the feasibility study of a future port located at Johor, Malaysia. The project consists of the construction of a new 100-hectare container port built in various phases (full buildout – 700+ hectares). The modeling was undertaken to investigate the effects of extensive channel dredging and land reclamation required for the port.

In order to quantify the port's environmental effects, tidal circulation and sedimentation modeling were performed for each construction phase of the project using the USCOE's RMA2 and STUDH numerical models. Dredging aspects of the project involve construction of a new 12 kilometer long, 13.5 meter deep channel requiring approximately 15 million cubic meters of dredging and associated disposal. Various channel configurations, widths and depths were evaluated to estimate annual maintenance dredging requirements for the channel as well as any changes to the natural erosional and depositional patterns at the site.

INTRODUCTION

The development of Tanjung Pelepas Port (TPP) required: (1) the construction of new port facilities at a previously undeveloped site and (2) dredging of a deep water navigation channel and turning basin. The development would also include extensive reclamation of existing low lying land and shallow water areas.

The proposed construction works would alter area bathymetry relative to existing conditions. Consequently, alignments for channels, berths and land reclamation were to be designed so as to produce favorable tidal current/sedimentation conditions and minimize adverse environmental impacts.

[1]Moffatt & Nichol Engineers, 2209 Century Drive Suite 500, Raleigh, NC 27612 (919) 781-4626
[2]Moffatt & Nichol Engineers, 2209 Century Drive Suite 500, Raleigh, NC 27612 (919) 781-4626

An initial study was carried out using limited published information during the preparation of the privatization proposal (October, 1993). While the study indicated that conditions appeared to be favorable, there was a strong recommendation that detailed studies, including mathematical modeling, be performed to assess hydrodynamic and sedimentation conditions at an early stage in the planning and design process [Ref. 1].

The primary objectives of the modeling efforts were to evaluate hydrodynamics (i.e. tidal elevations and currents) and sedimentation (bathymetric change) in and around the proposed port for: (1) existing conditions, and (2) project conditions (navigation channel and turning basin). Alternative channel configurations were to be studied to determine which would require the least annual maintenance dredging and provide the least risks to navigation.

PRE-DESIGN REVIEW AND DATA COLLECTION

Existing literature pertaining to the study area was collected and analyzed. The modeling work was based on the following data collection commissioned by others before Moffatt & Nichol's involvement with this project:

- Bathymetric survey covering the area surrounding the proposed port site and entrance channel

- Current meter readings carried out at various locations in the region

- Tidal level recordings taken at the proposed port site

- Wind data taken at various locations in the area

- Suspended solids and bed sample analysis taken at various locations in the region

These items were supplemented with bathymetric data taken from published admiralty charts where available.

Existing Literature and Information

The proposed port site is adjacent to Tanjung Pelepas at the mouth of the Sungai Pulai (river) (See Figure 1). The initial phase of the port development will be sited on reclaimed land to the south and east of Tanjung Pelepas. The land in this area is low lying and is protected by a dike constructed adjacent to the shoreline. Between the dike and the shore is a narrow strip of mangroves. A large area of shallow water exists to the south and east of the site. In fact, the 2m contour extends up to 5 km offshore in places (See Figure 1).

Admiralty Charts 2587 (Western Part of Johor Strait) and 2570 (Western Approaches to Singapore) indicate that the water depths in the Sungai Pulai adjacent to Tanjung Pelepas are up to 18m below Chart Datum. The influence of the Sungai Pulai is evident in the depth contours which initially extend to the south-southeast and eventually head south-southwest. As the Sungai Pulai thalweg extends south of the river mouth it gradually becomes shallower until it terminates at an offshore bar with a water depth of about 2m described on the charts as mud and sand. The bar is, however, a relatively short distance from the deep water of the Singapore Strait east of Tanjung Piai (See Figure 1).

Site visits and previous studies indicate that flow conditions in the region of Tanjung Pelepas are dictated by tidal current and river discharge which tend to maintain naturally deep water. The effect of this flow can still be observed 1.5 km south of Tanjung Piai, even though currents weaken as the Sungai Pulai widens into the embayment south of the site.

Bathymetry
Geometric information for the estuarial system was obtained from the bathymetric survey and supplemented with information from the admiralty charts. Depths with respect to chart datum were determined from Chart Nos. 2587 and 2570. Horizontal coordinates, x and y, for each corner node of an element and its corresponding water depth were digitized from these charts. The resulting mesh geometry was checked relative to the survey and admiralty charts. Alterations were then made to the mesh as deemed necessary to: (1) improve physical representation of the estuary and (2) improve model stability in areas of large depth gradients.

Current Meter Readings
Current velocity measurements were recorded (using ADCP's) over two 25 hour periods coinciding with spring and neap tides at various locations shown in Figure 1. Stations L1A, L1B and L1C are located along a line west of Tanjung Pelepas while stations L2A, L2B, and L2C are located along a line upstream of Tanjung Pelepas. Additional velocity measurements were taken at PT 1, 2, 3 & 4 at a later date over a 25 hour period coinciding with spring and neap tides.

Tidal Datums
Water level variations in the study area are generally dominated by astronomical tides although river discharge and storm tides may occasionally affect water levels. Tides in the region are semi-diurnal, and the nearest tidal reference station is located at

Tanjung Pelepas. Table 1 presents a summary of tidal datums for Tanjung Pelepas as determined by GEOMETRA Surveys [Ref. 2]. Chart Datum has been used in this study.

Table 1: Tidal Datums

Datum	Elevation (m - Chart Datum)
Highest Astronomical Tide (HAT)	3.896
Mean Higher High Water (MHHW)	3.189
Mean Lower High Water (MLHW)	2.690
Malaysian Land Survey Datum (MLSD)	1.668
Mean Sea Level (MSL)	1.668
Mean Higher Low Water (MHLW)	1.167
Mean Lower Low Water (MLLW)	0.706
Lowest Astronomical Tide (LAT)	0.000
Chart Datum (CD)	0.000

Wind and Waves

No wave data were available for the specific site, however, a review of wave conditions for the general region has been based on: (1) shipboard wave observations collected in the Malacca Strait, (2) directional wave gage measurements taken in the Malacca Strait, and (3) wave hindcasts. Overall, wave conditions within the Straits of Malacca are relatively mild with waves seldom exceeding significant heights of 1m and periods greater than 6 seconds (See Figure 1).

Wind records were obtained at the following locations: Senai Airport, Singapore Airport , Changi Meterological Station, Tengah Meterological Station, and JTC Flatted Factory, Jurong. The Jurong data were used for sedimentation modeling (See Figure 1).

The winds at Jurong are moderate with predominant winds from the north. Approximately 34% of the recorded winds are from the southeast and southwest quadrants which generate waves that can affect the port. Winds from the southeast and southwest quadrants seldom exceed 8 m/sec.

Although normal wind and wave conditions are moderate, storm events generate waves that suspend bottom sediments in the shallow areas of the region. As a result, relatively high suspended sediment concentrations can be experienced during storm events. Studies have shown that waves propagating over a muddy bottom (such as the offshore shoal in our case - see below) decay rapidly [Ref. 3]. This fact and the shallow water depths at the shoal are expected to limit heavy wave action to the offshore area.

Bottom Sediments

Seabed samples were collected at over 75 locations within the model area using a Van der Veen type grab sampler. The samples were analyzed for dry density, specific gravity and particle size distribution. The analyses showed that the bottom sediments range from fine to medium sands in the area between the Tanjung Pelepas shoreline and the offshore shoal while the offshore shoal is comprised of fine silts and clays.

2-D FLOW MODEL (TIDAL HYDRAULICS)

General Description

A 2-D hydraulic model developed by the U.S. Army Corps of Engineers, RMA-2, was utilized to determine flow velocities and discharges at the project site. RMA-2 is a finite element solution of the Navier-Stokes equations for turbulent flows. Friction is calculated with Manning's equation, and eddy viscosity coefficients are used to define the turbulent losses. A velocity form of the basis equation is used with side boundaries treated as either slip or static. The model automatically recognizes dry elements and corrects the mesh accordingly. Boundary conditions may be water-surface elevations, velocities, or discharges and may occur inside the mesh as well as along the edges. Wind stress can also be included to induce wind generated flows. The finite element grids of the Johor Estuary for the existing condition and two alternative channel alignments can be seen in Figure 2.

Calibration and Verification

Boundary Conditions

Ideal boundary conditions for the hydrodynamic simulation of the area would include tidal elevations along the offshore boundary of the finite element mesh and a river discharge along the Sungai Pulai at a location north of the reach of the tide. This was not possible in this case, however, as there was no bathymetric information for the Sungai Pulai far north of the site. In the absence of this data, the recorded flows within the Sungai Pulai (within the reach of the tide) were used as a second boundary condition for the model.

Based on the data described above, the boundary conditions used for calibration purposes of the existing condition mesh consisted of predicted time varying tidal elevations (from the British Admiralty Tide Tables) along the southern (offshore) boundary and time varying flow rates at the northwest boundary (upstream from Tanjung Pelpas along the Sungai Pulai). It was assumed that the flow entering at the northeast boundary was negligible since it is bordered by a causeway.

Roughness Parameters

The calibration parameters contained within RMA-2 are eddy viscosity and
Manning's 'n' coefficients which affect lateral mixing and bottom friction of the flow
system, respectively. Various combinations of these parameters were tested as part of
the calibration process. The values giving the best calibration results are listed in
Table 2.

Table 2: Roughness Parameters

Elevation (Chart Datum)	Eddy Viscosity (N*sec/m^2)	Manning's n
< -10m	12000	0.020
-5m to -10m	14350	0.025
0m to -5m	16750	0.030

Model Calibration

The hydrodynamic model was operated for the existing condition with a 15 minute
time step over a 25 hour period. Two calibration runs were conducted for the original
mesh; (1) neap (9/29/94-9/30/94), and (2) spring (10/7/94-10/8/94) tide events, when
velocity measurements were obtained along both L1 and L2 lines.

Comparisons of computed and predicted tidal elevations at Tanjung Pelepas, Tanjung
Bunga, and Johor Bahru are presented in Figure 3 for the neap tide period. Figure 4
shows comparisons of computed and measured flow velocities for the six locations
along the L1 and L2 lines, again for the neap tide period. Figure 5 shows the velocity
patterns of the region in the form of velocity vectors for the existing and project
conditions during the neap tide event.

In general, reasonably good agreement has been achieved in terms of both amplitude
and phase. Discrepancies, however, are observed. At Tanjung Pelepas and Tanjung
Bunga the model tends to slightly overpredict tidal amplitudes while at Johor Bahru
the model accurately simulates the tide range but the computed elevations are lower
than those predicted from tide tables. The use of predicted tides for boundary
conditions may account for some of the observed error.

The raw data points for the measured flow velocities are shown in Figure 4. In
addition, the figure for the L2 data points include a smoothed line that was fit through
the raw data points. The smoothed velocity values and measured depths were used to
calculate the flow rates (by continuity) used at the northwest boundary condition for
these calibration runs. It should be noted that positive velocities represent flood flow

while negative values represent ebb flow throughout this report. Generally speaking, for the neap tide event, the computed results fell within the range of the raw data points and showed good agreement between model simulation and measured data. For the spring tide event, the model results (not shown) are in good phase with the measured data while it is evident that ebb flow velocities are underpredicted. This discrepancy can be most likely attributed to using smoothed rather than actual velocity values to calculate the boundary condition flows. It is interesting, however, to note that the computed results are generally in good agreement with the smoothed values for the L2 data points with a slight tendency to overpredict the velocities.

Model Verification

Model verification runs were performed for the spring (10/24/94-10/25/94) and neap (11/12/94-11/13/94) tide events that occurred when velocity measurements were obtained at Points 3 and 4. Unfortunately, there were no velocity data taken at the L2 line during that period. Therefore, a linear regression analysis between flow rate and water surface elevation was performed to estimate a new boundary condition for the northwest boundary. The linear regression was performed on 9/29/94-9/30/94 and 10/7/94-10/8/94 velocities at L2 and tides at Tanjung Pelepas. Based on the results of this analysis, flow rates for the northwest boundary were calculated using tides predicted for Tanjung Pelepas.

Computed versus measured flow velocities are presented in Figure 4 for Points 3 and 4 during the spring tide event. Comparisons between computed and measured raw data indicate that the model simulated velocity amplitudes are less than the measured values, especially for Point 3, while phases are well predicted. With the extrapolation required to estimate boundary conditions, these simulated velocities were deemed to be in reasonable agreement with the measurements.

Based on the model calibration and verification results, it was concluded that the finite element model constitutes a reasonable representation of tidal hydrodynamics throughout the modeled estuarial system.

SEDIMENTATION MODELING

Sediment transport modeling results from STUDH provide an average sedimentation rate (erosion or deposition) approximation across each computational element. It should be noted, however, that the code does not compute a new flow field even though bottom bathymetry may change as a result of sediment transport. In this regard, the model simulation duration should not be too long so that changes in bottom bathymetry are small enough to justify not changing the flow field for the simulations. In general, long term sedimentation rates based on a shorter duration simulation can be overestimated because a new equilibrium between the flow field and sediment transport will be eventually reached.

Therefore, in the absence of measured shoaling rates at the site, it was assumed that the existing bathymetry is stable in terms of shoaling and erosion. This assumption is believed to be reasonable inasmuch as the naturally deep channels in the Sungai Pulai and the West Reach of the Johor Strait have been shown to be stable [Ref. 4]. Cohesive sediments with mean tides were used in sedimentation modeling, since bottom sediments offshore consist primarily of mud.

Concentration Data And Boundary Conditions

Water samples at the six current observation locations along the L1 and L2 lines were collected during high, low, flood, and ebb tides using a Jabsco pump water sampler. These samples were analyzed for total suspended solids (TSS), and results show that TSS varies from 5 mg/l to 125 mg/l with an average value of 20 mg/l.

Based on the full data set referenced above, cohesive sediment concentrations at the south and northwestern boundaries were estimated to be 40 and 20 mg/l, respectively, corresponding to the mean tides. The northeastern boundary concentration was determined internally within the model.

Model Parameters

As in the case of the hydrodynamic model, various combinations of parameters were tested in the calibration process. A final set of physical parameters was chosen for modeling based upon the sediment grain size distribution provided by GEOMETRA Surveys [Ref. 2].

Calibration Results

The calibrated hydrodynamic model was used as input to the numerical sediment transport code STUDH for the existing conditions. Long-term model simulations (of one month duration) were performed for the mean tide conditions. It is observed from the results that the area generally experiences negligible bathymetric change. Shoaling rates on the order of 1-2 cm/yr were calculated for the entire region. These model results are consistent with the expectations that the natural channels are self scouring and that the region is in a state of dynamic equilibrium.

Project Conditions

The calibrated hydrodynamic model was used as input to STUDH which was run for both Alternatives 1 & 2. Physical parameters determined in the sedimentation model calibration process were used to evaluate sedimentation rates for the alternative alignments.

Wind-induced waves play a significant role in sediment transport within the project area. In the absence of a rigorous analysis and the desired long-term wave and sediment data, results of steady wind/wave simulations were used to evaluate the factors which are expected to lead to channel shoaling. Specifically, the wind waves stir up sediments over the shallow areas of the shoal which can produce large suspended sediment concentrations. Once the winds and waves subside, sediments resettle to the bottom. Under existing conditions, bottom sediments are simply redistributed over the shoal while a dredged navigation channel would provide a sink for the suspended sediments.

To model this phenomenon numerically, sediment concentrations were varied from 20 mg/l at the northwest end of the model to 100 mg/l at the offshore end (range taken from data set). This was based on the fact that predominant winds were from the north causing waves to propagate southward and break on the offshore bar causing higher sediment concentrations there. The results for these model runs can be seen in Figure 6. The yearly sedimentation rates for the channels were estimated to be 850,000 m^3 for Alternative 1 and 880,000 m^3 for Alternative 2.

The above results should be viewed as preliminary subject to additional analyses. Specific additional efforts which are ongoing to refine the above estimates include: (1) refinement of wave climatology, (2) wave transformation studies to account for the reduction in wave heights which occur as they pass over the muddy shoal, and (3) a sensitivity study of sedimentation model parameters to account for the possible range of site conditions. Finally, it should be pointed out that the above and ongoing sedimentation modeling efforts are based on very limited field data. Additional field data that should be collected in order to improve sedimentation estimates are: (1) long-term suspended concentration measurements along the proposed channel alignments, (2) additional coincident measurements of currents, tides, and waves, (3) laboratory/field testing of bed sediments for in-situ density, and (4) laboratory testing to establish site-specific critical shear stresses for cohesive sediment erosion and deposition.

REFERENCES

1. Privatisation Proposal to Establish a Premier Integrated Seaport in Johor, Seaport Terminal (Johore) Sdn Bhd, October 1993.

2. Johor West Port - Hydrographic Survey and Data Collection, GEOMETRA Surveys, November 1994.

3. Bottom Mud Transport Due To Water Waves, Ph.D. Dissertation, Univ. of Florida, F. Jiang, 1993. Pg. 35-37.

4. Modeling of the Second Crossing to Singapore, HYDEC, 1993.

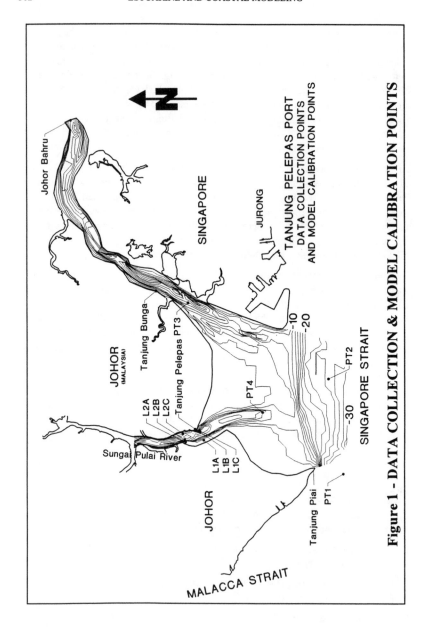

Figure 1 - DATA COLLECTION & MODEL CALIBRATION POINTS

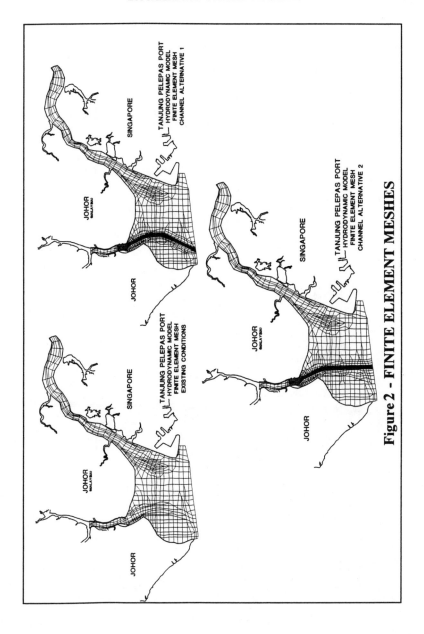

Figure 2 - FINITE ELEMENT MESHES

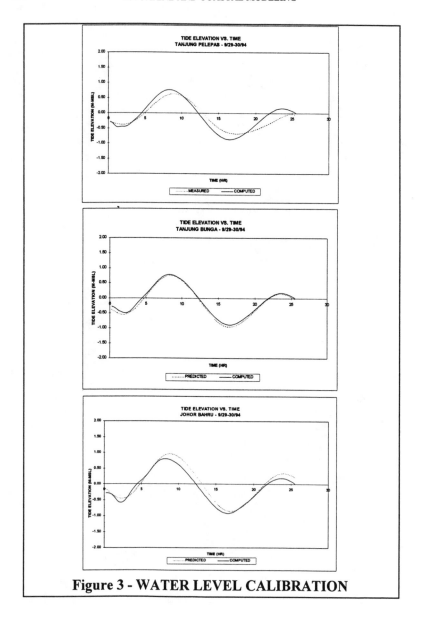

Figure 3 - WATER LEVEL CALIBRATION

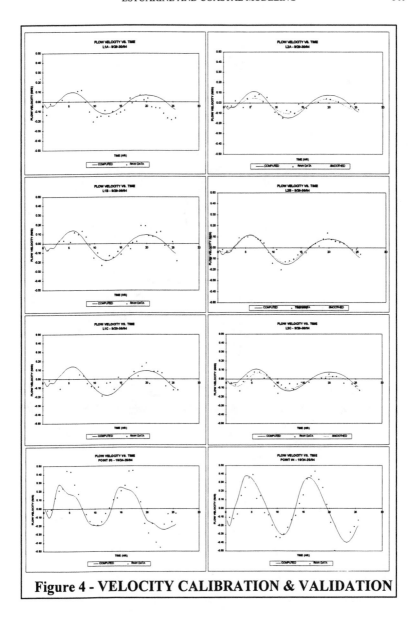

Figure 4 - VELOCITY CALIBRATION & VALIDATION

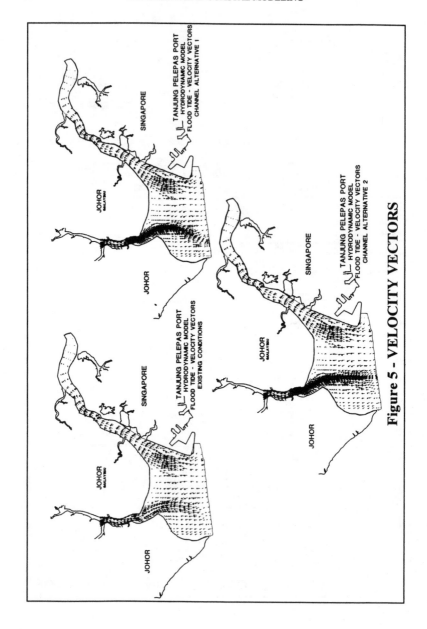

Figure 5 - VELOCITY VECTORS

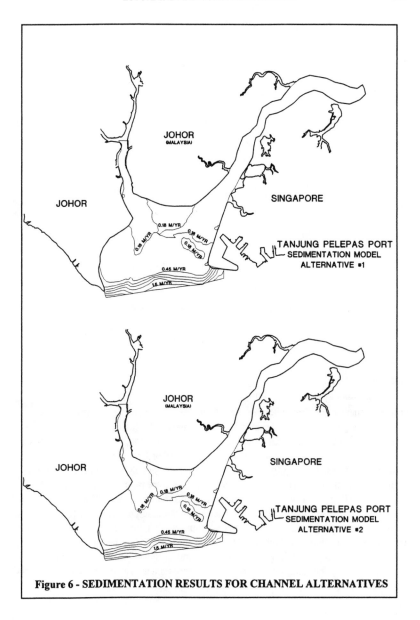

Figure 6 - SEDIMENTATION RESULTS FOR CHANNEL ALTERNATIVES

Hydrodynamic Modeling
Croatan / Pamlico Sound System, NC

Jeffrey G. Shelden, P.E.[1] , M. ASCE

Johnny D. Martin, EIT[2]

Abstract

A 2-d hydraulic model developed by the U.S. Army Corps of Engineers, RMA-2, was utilized to determine flow velocities and discharges at the proposed site of a new highway bridge crossing the Croatan Sound in eastern North Carolina. A finite element grid of the Albemarle, Croatan, Roanoke, Pamlico and Currituck Sounds was created using NOAA nautical charts of the region. In order to determine the flow velocities to be used during the design stage of the project and to determine potential scour and vessel impact velocities, hurricane events were modeled based on data presented by FEMA. These storms were placed on tracks 25 nautical miles apart for each model run resulting in a total of 42 simulations being performed. For nodes located along the bridge alignment, the maximum velocity and elevation was recorded and assigned a probability equivalent to that of the storm event being simulated. A statistical analysis was performed on this data, and these velocities were then incorporated into the scour and vessel impact analyses performed for the design of the bridge structure.

[1] Moffatt & Nichol Engineers, 2209 Century Drive Suite 500, Raleigh, NC 27612 (919) 781-4626
[2] Moffatt & Nichol Engineers, 2209 Century Drive Suite 500, Raleigh, NC 27612 (919) 781-4626

Introduction

The objective of this study was to determine the flow velocities for various return periods at the site of a new proposed bridge over the Croatan Sound (see Figure 1) in eastern North Carolina [Ref. 5]. These velocities were necessary to calculate the potential scour for different proposed foundation alternatives for the 5-year, 100-year, and 500-year storm events. Additionally, estimates of flow velocities were required for the vessel impact analysis performed for the bridge structure.

The Croatan Sound is part of a larger sound system which also includes the Pamlico, Albemarle, Currituke and Roanoke Sounds. Amein [Ref. 7] provides the following discussion of the hydrogaphy of the area.

"The Pamlico Sound is bordered by the mainland with its tributary rivers on the western side, and by the Outer Banks on the eastern side. At its northern end, it connects with the Albemarle Sound through the Croatan and Roanoke Sounds. In the southern direction, it is continuous with the Core Sound. The main tidal inlets that connect Pamlico Sound to the Atlantic Ocean are Oregon, Hatteras, and Ocracoke Inlets.

Pamlico Sound covers an area of approximately 1700 square miles. It is nearly 70 miles long in the southwest-northeast direction and 10 to 30 miles wide in the southeast-northwest direction, being narrowest at the northern end and widest opposite Hatteras Island.

Two large river systems, the Neuse-Trent and the Tar-Pamlico, discharge directly into the [Pamlico] Sound. Two other rivers, Chowan and Roanoke, empty into the Albemarle Sound which in turn discharges to Pamlico Sound through Croatan and Roanoke Sounds. In addition to the four river systems, there are many short, wide streams which contribute to the water supply of Pamlico Sound complex by draining the surrounding swampy areas.

The mouths of Oregon, Hatteras and Ocracoke Inlets are small as compared to the width and size of [Pamlico] Sound. Therefore, there are no perceptible lunar tides away from the inlets. The dominating factor in determining the flow pattern in the sound is the wind force. The currents in Pamlico Sound, which are relatively weak, depend mainly upon the direction and velocity of the wind and not upon tidal oscillations. The effect of mean annual freshwater inflow to the sound is also small. Therefore, except during the flooding season, the runoff current would easily be overpowered by the currents due to the wind friction."

Pre-Design Review And Data Collection

Previous studies of the North Carolina sound system were reviewed along with other pertinent sources of information about the project site. These are included in the Reference section.

Storm Surge

The FEMA Flood Insurance Study for Dare County [Ref. 3] included storm surges for the site which are shown in Table 1.

Table 1: Storm Surge

Return Period (yrs)	Stillwater Elevation (Feet-NGVD)	Stillwater Elevation (meters-NGVD)
5	4.3	1.31
10	4.8	1.46
50	5.8	1.77
100	6.1	1.86
500	6.7	2.04

Tidal Datums

There are no published tide gage records in the vicinity of the project site. However, the Tidal Analysis Branch of NOAA did have preliminary datum and benchmark elevations for Manteo, NC. Since this is the best available information at this time, it was adopted for this project and is shown in Table 2.

Table 2: Tidal Datums

Datum	Elevation (ft - NGVD)	Elevation (m - NGVD)
North American Vertical Datum 1988 (NAVD)	0.98	0.299
Mean Higher High Water (MHHW)	0.83	0.253
Mean High Water (MHW)	0.78	0.238
Mean Sea Level (MSL)	0.61	0.186
Mean Tide Level (MTL)	0.60	0.183
Mean Low Water (MLW)	0.42	0.128
Mean Lower Low Water (MLLW)	0.34	0.104
National Geodetic Vertical Datum (NGVD)	0.00	0.000

Field Reconnaissance and Supplementary Surveys

Velocity and Discharge Measurements

Water velocity measurements were collected by General Engineering along sections of Croatan Sound and nearby Roanoke Sound on February 18 and 19, 1997. Discharges for each cross-section were also calculated for use in calibrating the hydraulic model.

Profile Data

Profile data of Croatan Sound was collected by the Locations and Surveys Unit of NCDOT.

Water Surface Elevations

Although three tide gauges were installed by others, various problems occurred with them during their measurement periods resulting in no useful water level data being collected which was contemporaneous with the velocity and discharge measurements. Due to the time constraints on this project, a second mobilization to obtain water velocity measurements and water surface elevations concurrently could not be undertaken.

2-D Flow Model (Tidal Hydraulics)

General Description

A 2-d hydraulic model developed by the U.S. Army Corps of Engineers, RMA-2, was utilized to determine flow velocities and discharges at the project site. RMA-2 is a finite element solution of the Reynolds form of the Navier-Stokes equations for turbulent flows. Friction is calculated with Manning's equation and eddy viscosity coefficients are used to define the turbulent losses. A velocity form of the basis equation is used with side boundaries treated as either slip or static. The model automatically recognizes dry elements and corrects the mesh accordingly. Boundary conditions may be water-surface elevations, velocities, or discharges and may occur inside the mesh as well as along the edges. Wind stress can also be computed to induce wind generated flows. A finite element grid (shown in Figure 2) of the Albemarle, Croatan, Roanoke, Pamilico and Currituck Sounds was created using NOAA nautical charts of the region.

Calibration and Verification

The model was calibrated for the February 17, 18 and 19, 1997 time period. The recorded tide at the Army Corps of Engineers Field Research Facility (FRF) at Duck, NC was adjusted and used as a boundary condition at Oregon, Hatteras and Ocracoke

Inlets. Fresh water inflows calculated by Amein [Ref. 7] were used as boundary conditions at the Neuse, Pamilico, Chowan and Roanoke Rivers as well as at other minor tributaries.

Wind velocity measurements were obtained from National Weather Service stations at Cape Hatteras, Cherry Point and Elizabeth City, from the Manteo Airport and from FRF at Duck. These velocities were applied to the model as a boundary condition since the flows in the sound system are predominantly wind driven.

The model was calibrated by adjusting the Manning's "n" values and eddy viscosity coefficients in order to approximate the discharge measurements obtained in the field. Generally the magnitudes and trends of the discharges showed fairly close agreement with just a minor difference in phasing.

A comparison between the model calculated elevations and those recorded on the gages was attempted. However, as mentioned previously, the recorded values are highly suspect due to various problems incurred with the tide gages, and thus the model results with regard to water surface elevations could not be directly calibrated. However, the model was run to simulate Hurricane Donna and the Ash Wednesday Storm, and the storm surge results from these were compared to documented values throughout the sound system to assist in calibration of water surface elevations.

Due to the short time period available for this study, additional field data (flow velocities and discharges and water surface elevations) could not be collected at a second date, and thus verification runs could not be performed.

Results

In order to determine the final flow velocities to be used during the design stage of the project, hurricane events were modeled. Based on the data presented by FEMA [Ref. 3], hurricanes (or tropical storms) with maximum winds of 81 km/hr (50 mph), 137 km/hr (85 mph) and 177 km/hr (110 mph) were modeled moving in the Azimuth directions of 24°, 60°, and 348°. These storms were placed on tracks 25 nautical miles apart for each model run resulting in a total of 42 simulations being performed.

Figure 3 shows a typical wind field that was applied to the model for a hurricane with 177 km/hr (110 mph) maximum winds. Figure 4 presents a plot of the flow velocities and Figure 5 the storm surge elevations for a typical hurricane. The flow velocities typically range from 150-300 cm/sec (5-10 fps) through much of the sound system under this condition. The water surface elevations show a large buildup on the eastern side of Pamlico Sound and substantial head differences across many sections of the modeled area.

Figure 6 presents a time history of the flow velocities and water surface elevation at a point near the middle of the proposed bridge alignment for one of the 177 km/hr (110 mph) hurricane simulations.

The maximum water surface elevation along the proposed bridge alignment was recorded for each simulation. A probability was assigned to each elevation recorded that corresponded to the probability of the particular storm event that generated that elevation as presented by FEMA [Ref. 3]. A statistical analysis was then performed on this data to develop a probability distribution for the water surface elevations.

Similarly for nodes located along the bridge alignment, the maximum velocity was recorded and assigned a probability that was the same as that for the storm event being simulated. A statistical analysis was also performed on this data. For each of the FEMA storm surge elevations (shown in Table 1), the final probability from the statistical analysis described previously was noted, and the velocity for each node that had a corresponding probability was noted. These velocities are shown in Figure 7.

Based on these results, the proposed configuration of the foundation types for the bridge alternatives and discussions with the structural designers, two zones for flows were defined. These were the approach sections of the bridge where trestle type bent construction is proposed and the 6 LOA and Navigation spans (areas subject to vessel impact) where post and beam type bent construction will be used. Table 3 shows the design flows used.

Table 3: Design Flow Velocities (Metric)

Return Period (yrs)	Flow Velocity (cm per second)	
	Trestle and Approach Bridge Sections	6 LOA and Navigation Span Bridge Sections
5	137	168
100	183	213
500	198	229

The various model runs were reviewed and it was determined that the maximum flow angle to the proposed bents was about 40 degrees. This value was conservatively used for all bents since the structural designers recommended only calculating a single scour values for a given foundation. The only exception is the foundation for the navigation spans which is skewed in-line with the channel. A flow angle of 20 degrees was used for these foundation alternatives.

Bridge Scour Analysis

General Description

Scour analyses were conducted for the various foundation alternatives under consideration for the Bridge Type Study. The analysis was performed in accordance with the procedures outlined in FHWA HEC-18, "Evaluating Scour at Bridges - Third Edition" as modified by NCDOT and FHWA based on model tests for the previously designed Bonner Bridge project [Ref. 12]. These modified equations are also discussed by M. Salim and J.S. Jones of FHWA [Ref. 11].

For the post and beam foundation alternatives, HEC-18 recommends that scour be calculated for both the pile groups and the pile cap (if it is submerged) and the largest value computed used for design. However, the modified equations account for pile spacing and the location of the pile cap in the water column. Relative contributions to scour for both the pile group and the pile cap are calculated and added together to obtain the design scour depth for a given foundation type.

Long-Term Changes

Based on previous profile surveys performed by NCDOT at the existing US 64 bridge just north of the project site, it was determined that the site appears to be relatively stable, and thus, long term channel aggravation or degradation is assumed to be negligible.

Contraction Scour

Since the proposed bridge alignment resembles an "S" shape and, thus, is not straight nor directly perpendicular to the predicted flows, the reduction in flow area due to bridge construction is not readily apparent. For this analysis, it was assumed that the flow width of the upstream channel was equal to the bridge length. Then, for each foundation alternative, the perpendicular area of the bents was deleted to determine the contracted flow width. This resulted in contraction scour values less than 0.15 m (0.5 feet); relatively insignificant. While other more rigorous methods of determining the change in flow width might be applicable, it was felt that due to the large natural flow width compared to the structure, any procedure would likely result in fairly small contraction scours which would be outweighed by the local scour at the bents.

Local Scour

Local scour was calculated for the 11 foundation alternatives under consideration by URS / Greiner Engineering and the 10 foundation alternatives considered by Moffatt and Nichol Engineers / Wilbur Smith and Associates (WSA). It was assumed that the flow was at a 40 degree angle to the bridge to the foundations that were not skewed. Refinements to the angle of flow for the bents in the curved sections of the bridge were not made based on recommendations from the structural designers that such precision would not be utilized in the foundation design. For the foundations which are proposed to be skewed so they are parallel to the channel, a flow angle of 20 degrees was used. Scour depths ranged up to 10.5 meters for a 500-year event.

References

1. Eastern North Carolina Hurricane Evacuation Study, NC Division of Emergency Management, 1987.

2. Final Report on The Albemarle Pamlico Coupling Study, Leonard Pietrafesa and Gerald Janowitz, NCSU, May, 1991.

3. Flood Insurance Study for Dare County, FEMA, 1992.

4. HEC-18, Evaluating Scour at Bridges, Third Edition, Federal Highway Administration.

5. Hydraulic Modeling and Scour Analysis, NCDOT R-2551, Manteo Bypass Over Croatan Sound, Moffatt & Nichol Engineers, 12 September 1997.

6. Hydrology and Circulation Patterns in the Vicinity of Oregon Inlet and Roanoke Island, North Carolina, James Singer and C.E. Knowles, NCSU, June 1975.

7. Mathematical Modeling of Circulation and Hurricane Surge in Pamlico Sound, North Carolina, Michael Amein and Damodar Airan, November, 1976.

8. Meteorological Criteria for Standard Project Hurricane and Probable Maximum Hurricane Windfields, Gulf and East Coasts of the United States, NOAA Technical Report NWS 23, September 1979.

9. Numerical Simulation of Oregon Inlet Control Structures' Effect on Storm and Tide Elevations in Pamilico Sound, Coastal Engineering Research Center, April, 1984.

10. Replacement of the Herbert C. Bonner Bridge - Hydraulic and Scour Analyses, Analysis of Flume Modeling Experiments and Scour Estimates for Replacement Bridge Piers, Parsons Brinckerhoff Quade and Douglas, Inc., January, 1997.

11. Replacement of the Herbert C. Bonner Bridge - Hydraulic and Scour Analyses, Expert Panel Meeting Background Information, Parsons Brinckerhoff Quade and Douglas, Inc., April, 1996.

12. Replacement of the Herbert C. Bonner Bridge - Hydraulic and Scour Analyses, Parsons Brinckerhoff Quade and Douglas, Inc., Revised March, 1995. Revised April, 1997.

13. The Physical Oceanography of Pamlico Sound, Leonard Pietrafesa, et.al., NCSU, November 1986.

Figure 1: Vicinity Map

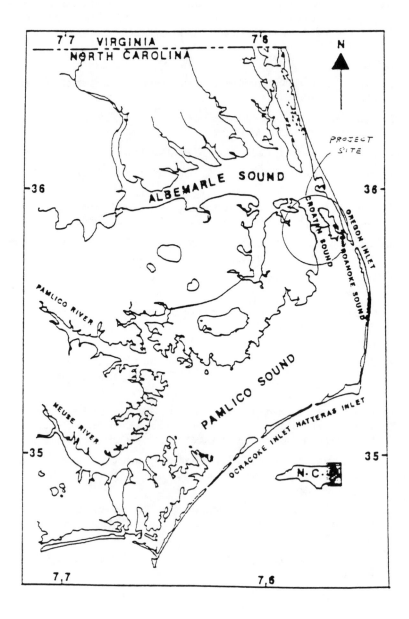

Figure 2: Finite Element Mesh

Figure 3: Typical Hurricane Wind Field

Figure 4: Typical Flow Velocities

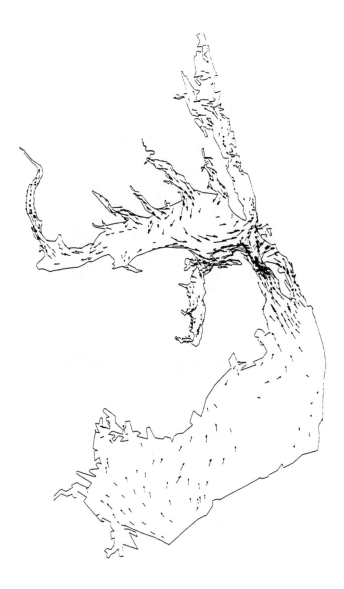

Figure 5: Typical Storm Surge Elevations (feet-MSL)

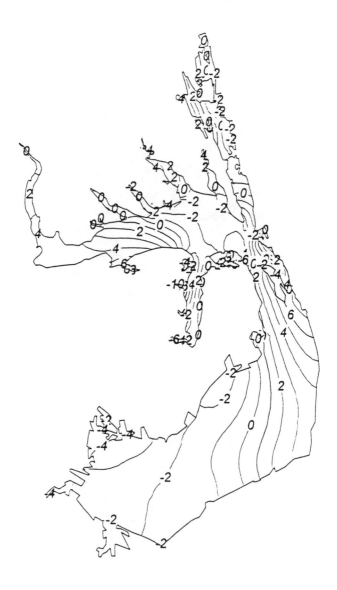

Figure 6: Flow Velocities and Water Surface Elevation at Middle of Proposed Bridge Alignment

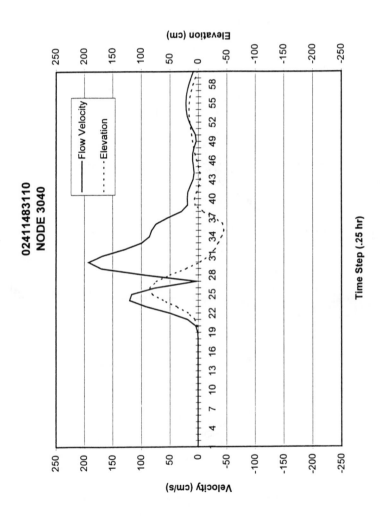

Figure 7: Design Flow Velocities along Proposed Bridge Alignment

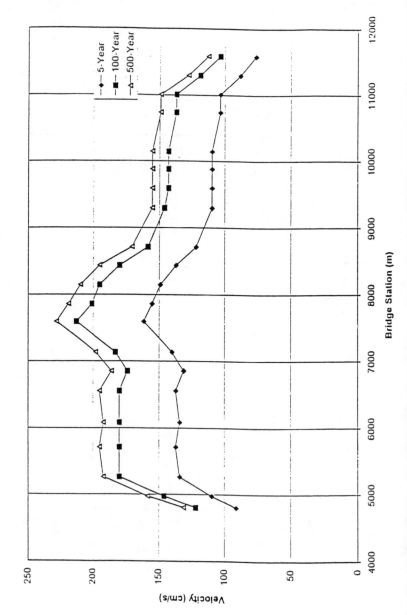

EMPIRICAL TECHNIQUES FOR A PRIORI PARAMETER ESTIMATION

By Thomas O. Herrington,[1] Associate Member ASCE, and Michael S. Bruno,[2] Member, ASCE

ABSTRACT: Inverse methods have allowed for the development of modeling strategies to solve for flow variables which satisfy the governing equations of complex hydrodynamic models while remaining as close as possible to observed data. Insufficiencies in observational data can be overcome through the use of *a priori* parameter estimates. The estimated parameters are systematically adjusted through fitting techniques until the model results approach the measured data. An alternative analysis method for determining the relative contribution of input data parameters for use in models employing inverse methods is empirical modeling. Empirical modeling does not make *a priori* assumptions about the data but instead allows the data itself to determine the form of the relationships within constraints determined by the structure of the selected empirical relations. A multivariable polynomial autoregressive analysis with exogenous variables (PARX model) is shown to produce effective empirical fits to measured nearshore current data. A nonlinear model developed to predict the observed current structure was able to explain 82% of the measured variability of the time series record and reproduce observed complex frequency distributions of tide and current energy densities.

INTRODUCTION

Estuarine and coastal circulation models require as input a number of spatially- and temporally- varying and competing physical forcings. Dominant forcing mechanisms which drive nearshore flows include, e.g. tidal motions, wind stress, wind waves, swell, freshwater inflow, density driven circulation, atmospheric pressure variation, bottom friction, eddy viscosity, etc. The interaction of the various forcing mechanisms generates complex three-dimensional current fields which do not lend themselves to simplified numerical solutions. In numerical model development the

---

[1] Res. Asst. Prof., Dept. of Civ., Envir. and Ocean Engrg., Stevens Inst. of Tech., Hoboken, NJ 07030

[2] Prof., Dept. of Civ., Envir. and Ocean Engrg., Stevens Inst. of Tech., Hoboken, NJ 07030

appropriate value, or weight applied, to each parameter is usually established with the help of available data by the process of model "tuning". This tuning process is often a laborious and sometimes arbitrary process, owing to the uncertainties in the relationship between parameters and the model variables corresponding to the data (Panchang and Richardson, 1993). An alternative, is to use inverse modeling strategies to solve for variables which satisfy the governing hydrodynamic equations while remaining as close as possible to the observed data.

Inverse models symmetrically adjust, in a least squares sense, the observational data, the model parameters, and other nonexactly known parameters through a set of equations which describe the physical model (Wunsch and Minster, 1982). Increased model efficiency can be obtained through the use of variational adjoint methods which strive to minimize the misfit between observed data and model output resulting from the improper selection of model parameters (Panchang and Richardson, 1993; Ullman, 1996). The inverse method, while constraining the model output to the range of observed data, is extremely sensitive to the magnitude and weight given to the initial (a priori) parameter estimates. Different initial estimates may yield different minima which may not be the global minimum or the desired solution (Panchang and Richardson, 1993).

The inclusion of prior information in parameter estimation serves to reduce the variance in the fitted parameters. However, placing excessive weight of the a priori information overwhelms the actual data, leading to final parameter values which are essentially equal to the prior estimates (Ullman, 1996). Additionally, in many cases, the fitting technique is computationally intensive and requires the modeler to make some a priori expectations about the magnitude and interactions of the initial estimated parameters. An alternative analysis method which does not make a priori assumptions about the measured data and can be used in conjunction with inverse modeling is to utilize empirical modeling techniques. Empirical modeling allows the data itself to determine the form of the relationships within constraints determined by the structure of the selected empirical relations (Vaccari and Christodoulatos, 1992). Here, a multivariable polynomial autoregressive analysis with exogenous variables (PARX model) is shown to produce effective empirical fits to measured nearshore current data.

EMPIRICAL MODELS

Empirical models predict the value of a single dependent variable using the current and previous values of a number of independent variables. The form of the predictive function can be linear or nonlinear depending on the modeling technique utilized. The linear empirical model is the most commonly utilized and best known technique for modeling time series data. These models, known as autoregressive (AR) models, predict future values of a time series based on linear combinations of previous values. If the time series is a function of multivariate data sets, a linear combination of other external variables is included in the model. Such models are known as autoregressive with exogenous variables (ARX) models. An example of an ARX model is;

$$N_j = a_0 + a_1 N_{j-1} + a_2 N_{j-2} + a_3 M_{j-1} + e_j \tag{1}$$

where e_j is the error in the prediction of N_j and M_j is an independent external variable. Another similar model would be developed for M_j. The coefficients a are selected according to some criterion such as the minimization of the sum of the squares of the errors e_j. The terms $j-1$ and $j-2$ are lag terms such that $j-1$ would be the value of the variable at the previous time step.

In spite of its simple form the ARX model is capable of describing complex interactions within a time series data set. However, the model ignores significant functions such as interactions between variables and nonlinearities which would improve the performance of the model. Relationships between variables can be included in the model through the use of cross-product terms. Nonlinear terms can be incorporated into the model through the use of higher powers of one or more variables. The inclusion of such terms in an empirical model produces solutions in the form of a multivariate polynomial. An advantage of multivariate polynomials is that they are relatively free of restrictions on the interaction between variables. For example, a complete multivariate polynomial model can contain, as a subset, a truncated multidimensional Taylor series approximation to any fundamental model. If lagged variables are considered, then the model becomes a polynomial autoregressive with exogenous variables (PARX) model (Chen and Billings, 1989). A PARX model has the form:

$$N_{i,j} = \sum_{p=1}^{n_p} a_p \cdot Z_p + e_{i,j}$$
$$Z_p = X_1^{b_{p,1}} \cdot X_2^{b_{p,2}} \dots X_m^{b_{p,m}}$$
$$X_n = I_{k,j-l} \tag{2}$$

such that $N_{i,j}$ is the predicted parameter i at time step j and is a linear combination of n_p terms, Z_p, where each term is the product of m independent variables X, raised to a power b, and multiplied by the coefficient a_p. In the application of the model, the powers are drawn from a list of candidate exponents, such as $b \in \{1,2,3\}$. The independent variable consists of any one of the variables $I_{k,j-l}$, where k may or may not be equal to i, and the lag l is drawn from a list of candidate lags. The list is usually a sequence of integers such as $l \in \{1,2,3\}$, but it may be a discontinuous list, such as $l \in \{1,2,12\}$, to describe longer period effects.

The basic formulation of a PARX model is determined by the structure of the fit terms. The appropriate number of time lags, exponents, and multiplicands must be selected to determine the physical structure of the model. For example, a model consisting of one multiplicand, no time lags and an exponent of 1 would be utilized to formulate a linear model. Once the structure of the model has been determined, the identification of the candidate terms and the magnitude of any coefficients are calculated through a fitting procedure. Since the PARX model is linear in the coefficients, a linear least-squares procedure is used for model fitting. The best fit is achieved through an iterative stepwise regression procedure, similar to that used with multilinear regression, which selects, from all of the possible terms, only those which contribute significantly to the predictive capability of the model. The significance of individual terms in the model are evaluated through the use of the t-statistic. The model only accepts those terms with t-statistics at the 95% confidence level or greater based on the actual time series data. Of all the possible combinations of terms, the model with the maximum R^2 value is selected, where R^2 represents the proportion of variability explained by the model. Since polynomial models sometimes exhibit explosive behavior, the output of the polynomial is constrained to the range of data used in fitting the empirical expression. The use of such a restraint is justified since no empirical correlation should be considered valid outside the range of the data used in the fitting process. This obviously demands that the data include values covering the expected range of the variables being examined.

Once the best possible fit to the time series is achieved, the predictive ability of the model is evaluated by utilizing it to estimate an independent time series. The goodness of the fit is calculated based on the sum of the squares of the error (SSE). The simplest criterion utilized is the proportion of the variance in the data not explained by the model, calculated as the ratio of the sum of the squares of the error and the total sum of squares (TSS):

$$\gamma^2 = SSE / TSS \tag{3}$$

The value of γ^2 can be used to compare the performance of various models. Values of γ^2 greater than 1.0 indicate that the model prediction is worse than using the mean of the fitting data.

MODEL APPLICATION

Field Measurements and Analysis

A seven day time series of wave and near-bottom current measurements were obtained 60 m inshore of the western edge of the main ebb channel of Townsends Inlet, Avalon, New Jersey, in September, 1993 (Figure 1). The wave and current meter was mounted 1 m above the sea bed, 180 m south of the southern inlet jetty, in a mean water depth of 3.0 m. In this configuration the meter was exposed to both nearshore wave and current action as well as strong tidal currents. The meter was initialized to provide measurements of the pressure and two orthogonal components of velocity at 2 Hz over 9 minute burst samples, at one hour intervals (1080 half second samples of P, Vn, and Ve).

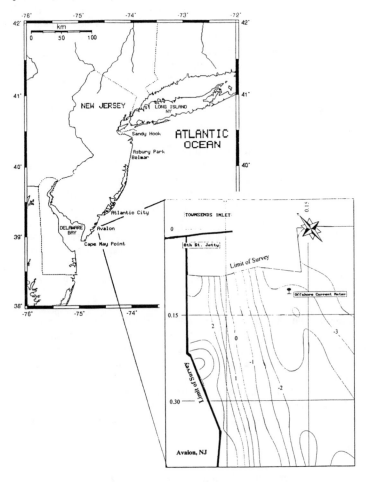

Figure 1. Location Map, Avalon, New Jersey and Vicinity
Distances in km, Elevations in m (NGVD 1929)

The measured time series is presented in Figure 2. The tidal elevation record is plotted in the upper panel, followed by the surface wind observations from Atlantic City, the measured significant wave height ($H_s = 4(m_0)^{1/2}$), the peak wave direction, and the 8.5 minute average current measured 1 m above the bottom. All of the recorded vectors were decomposed into shore-parallel and cross-shore components. Since the axis of the ebb channel of the inlet lies exactly on a bearing of 0° Magnetic North (MN), all of the recorded directions are reported relative to Magnetic North. Thus, shore-parallel components are parallel to the axis of the inlet channel and cross-shore components are perpendicular to the inlet channel. Positive values of the shore-parallel and cross-shore components are directed to magnetic north and east, respectively. Directions are given as the direction of propagation (i.e. a shore-parallel wind of -5 m/s is a wind directed from the north). The x-axis is scaled in tenths of days Eastern Daylight Savings Time (EDT) so that 20.5 would represent the 20th day of the month at 1200 hours.

The energy density spectra for the tidal elevation and shore-parallel component (U-Comp) of velocity is presented in Figure 3. Energy density spectra were calculated for the zero mean records of tidal elevation and the shore-parallel current component for the 7 day data set. The time series were divided into five 64 hour data segments and ensemble averaged with 50% overlap, providing spectral estimates with 10 degrees of freedom and a 95% confidence interval of 25 to 150 percent of the estimates. The cross-spectra between the tidal elevation and shore-parallel component of velocity was calculated to examine the coherence and phase difference information, and are shown in the lower two panels of Figure 3. The dashed line on the coherence plot indicates the point above which values are significantly different from zero at the 95% confidence level.

Examining Figure 3, two peaks are evident in the tidal elevation energy density, one centered at the semidiurnal frequency (0.08 cph) and one centered at a frequency of 0.02 cph. Three peaks in kinetic energy are present in the shore-parallel velocity component. Peaks of equal magnitude (0.175 (m/s)^{2}s) are centered at 0.02 cph and the semidiurnal tidal frequency and a lesser peak of 0.13 (m/s)^{2}s is centered at a frequency of 0.16 cph. The shore-parallel current is incoherent with the tidal elevation at the 95% confidence level, indicating the possibility of a nonlinear relationship between the tide and shore-parallel current component. The tidal energy and shore-parallel current components are in phase at the semidiurnal and 0.16 cph frequencies and slightly out of phase (-20°) at 0.02 cph.

In an effort to determine the forcing mechanisms responsible for the energy density distribution outside of the semidiurnal frequency, the data sets can be filtered to remove the frequency components associated with the tide. A harmonic analysis was performed on the measured water surface elevation and shore-parallel velocity data by utilizing a best least squares fit of each tidal constituent to the measured data.

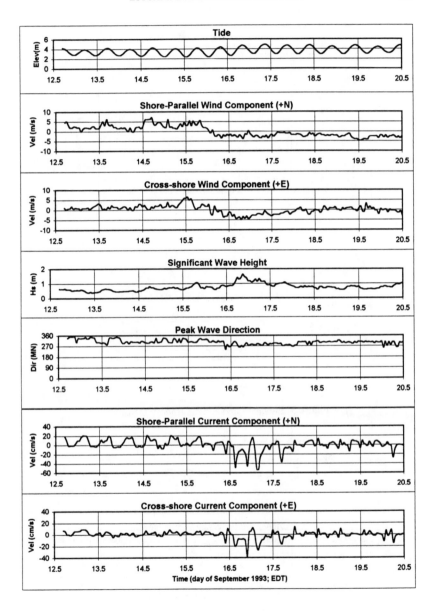

Figure 2. September, 1993 Observations

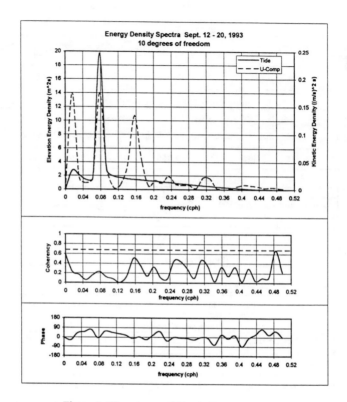

Figure 3. Elevation and Kinetic Energy Density

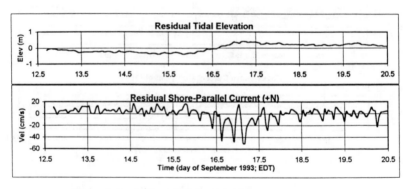

Figure 4. September, 1993 Residual Time Series Record

The residual surface elevation and shore-parallel current component are presented in Figure 4. As expected, the filtered surface elevation record is very close to zero. Note the linear rise in the residual elevation record between day 15.5 and 17.5 in response to the change in forcings during the period of increased wave height at the site.

Examination of the residual shore-parallel current indicates the presence of high frequency components throughout the data set. Comparing the wind and wave time series to the residual current record, the 6 hour periodicity within the shore-parallel current component occurs during the period in which an increase in the significant wave height occurred (day 15.5 to 18.5). The lack of periodicity within the time series of significant wave height and shore parallel wind component precludes the use of spectral analysis to determine the frequency distribution of wave energy density over the observation period. An alternative method is therefore needed to analyze the relationship between the observed forcings and the resulting shore-parallel current structure.

Empirical Time Series Modeling

Model Development

The initial step in the development of a PARX model is the definition of the structure of the candidate terms by creating a list of exponents, time lags, and multiplicands. Since the desired outcome of the empirical modeling effort is to develop some fundamental insights into the interactions of the wind, wave, and tidal forcing, the exponent list was initially constrained at 3 terms {-1, 1, 2} and the multiplicands at 2. A series of fits were performed with increasing time lags {0, 1 , 2, 3, 4, 5, 6} and it was found that the candidate terms using lags greater than 2 were never selected by the model. Finally, it was determined that removing the negative exponent always reduced the performance of the model.

The second step in the model formulation is to determine the form of the independent variables needed to develop the most parsimonious model to predict the dependent (shore-parallel current) variable. A sensitivity analysis performed using measured and derived variables from the data set determined that the incident wave characteristics were best described by the wave amplitude, a, for magnitude, the shore-parallel wave number component, k_x , for direction, and the wave frequency, ω. Directly defining dimensionless parameters such as kh and ka always resulted in fits which were less accurate than if the model selected the terms individually. The measured shore-parallel wind, W, filtered to remove spikes in the data, and the change in wind speed over time, dW/dt, were determined to best describe the wind forcing. The variation of water depth, h, due to the tide was incorporated into the model through the use of the recorded pressure measurements.

Utilizing the six basic parameters (a, k_x, ω, W, dW/dt, h) and the model structure described above, PARX models were developed to predict either the residual or measured (complete) shore-parallel current velocities. During the development of each model a linear fit to the data was initially attempted by setting the exponent and multiplicand values to 1 and the time lag to 0. The linear model was then used to seed the development of a nonlinear model with the predetermined structure described previously. Finally, a second "purely" nonlinear model was developed without the influence of an initial linear model in the system. Out of the three developed models the one with the highest R^2 value was chosen as the best predictive model.

The implementation began with an attempt to fit the initial 150 hours of the September 1993 data set. The developed model was then tested to see how well it predicted the shore-parallel current velocity over the entire time series. All but 40 hours of the data set was chosen to create the fit so as to provide a time series that contained well-defined periods of low and high wave energy. The resulting empirical model would therefore be expected to predict the shore-parallel current velocity over most expected combinations of wind, wave, and tide forcing.

Model Results

The initial model developed utilized the wave terms (a, k_x, ω) to predict the residual shore-parallel current velocity. A purely nonlinear model was determined to provide the best fit to the data. However, the maximum R^2 value obtained was only 0.35. The model only contained 2 terms both with t-statistics greater than 2. The model was able to predict the dominant trend in the measured current but failed to resolve the variability within the measured time series record. It can be concluded from the model results that the wave parameters alone are not solely responsible for the observed residual current structure.

In the development of the second model to predict the residual shore-parallel current velocity the wind forcing was incorporated into the fit through the use of the two wind terms (W and dW/dt). Incorporating the shore-parallel wind velocity terms into the model increased the R^2 value of the purely nonlinear model to 0.55. The developed model utilized 7 terms, all of which had t-statistics greater than 2. The effect of including the average shore-parallel wind component and the rate of change of wind velocity terms was an increase in the ability of the model to resolve some of the peaks in the time series data, however, it was unable to predict the smaller peaks in the data set.

The poor fits obtained by both models indicate that the wind and wave forcing terms are not the only mechanisms responsible for the observed variability in the residual shore-parallel current velocity. Since the only terms removed from the original data set were those with tidal frequencies, an interaction between the tidal elevation and measured shore-parallel currents must be present. Placing the tidal elevation variable into the PARX model dictates the use of the model to predict the measured (complete) shore-parallel current component. Various combinations of the wind, wave, and tide parameters were utilized to produce a model which best fit the observed current structure. In the end, the best fit to the measured September, 1993 data set was obtained utilizing all 6 parameters $(a, k_x, \omega, W, dW/dt, h)$. The final predictive model was a purely nonlinear model with an R^2 value of 0.82. The model contained 16 terms, all of which had t-statistics above 2.0. The model prediction for the entire September 1993 shore-parallel current component resulted in a γ^2 value of 0.20.

Figure 5 presents the measured (solid line), fit (dotted line) and predicted (dashed line) shore-parallel current component. The fitted nonlinear model, spanning the first 150 hours (day 12.75 to 19.0) of the time series, is able to resolve, to a reasonable degree, most of the peaks contained within the shore-parallel current record. The fit does quite well in resolving the strong southerly directed flow between day 16.5 and 17.5, only missing the magnitude of the initial strong spike in velocity. The predicted shore-parallel current over the final 36 hours (day 19.0 to 20.5) of the time series indicates that the model is able to predict the overall magnitude and direction of the current but loses some of the higher resolution peaks contained in the data.

A determination of the degree to which the developed model can simulate the structure of the measured shore-parallel current component can be obtained by analyzing the frequency distribution of energy within the predicted time series. Figure 6 presents the frequency distribution of the energy density of the measured (U-Meas) and fit (U-Fit) shore-parallel current component for the September, 1993 time series. The fitted nonlinear model is able to recreate the frequency distribution of the shore-parallel current component quite accurately. The kinetic energy density measured at 0.02 and 0.08 cph is slightly underpredicted but the predicted energy at the 0.16 cph peak is accurately represented.

The empirical equation developed by the PARX model to predict the shore-parallel component of velocity, U_{SP}, is

$$U_{SP} = K_W + K_{dW} + K_{AH} + K_K \qquad (4)$$

where

$K_W = [\ 0.015157(\omega^2 / W_{t\text{-}1}) - 0.0016022\ h^2_{t\text{-}2}W_{t\text{-}1}\ ] + [\ 0.13338\ a^2 + 0.011704\ \omega_{t\text{-}1}$
$\quad + 0.00058704\ /(k\ W^2)]\ W$

$K_{dW} = [0.0020512\ (\ dW/dt\ )/k - 0.0017450\ (\ dW/dt\ )_{t\text{-}1}/k_{t\text{-}1} - 0.0025645\ k_{t\text{-}1}/(dW/dt\ )_{t\text{-}2}\ ]$

$K_{AH} = [0.010630\ (\ h^2\ /a_{t\text{-}2}\ ) - 0.010994(\ h^2_{t\text{-}1}\ /a_{t\text{-}2}\ )] + [53.695/h - 92.342/h_{t\text{-}1}$
$\quad + 40.090/h_{t\text{-}2} - 0.087996\ h^2_{t\text{-}2}\ ]\ a^2_{t\text{-}1} + 0.10636\ a_{t\text{-}1}\ h_{t\text{-}2}$

$K_K = [\ -0.0000040479/\ (k_{t\text{-}1}\ k_{t\text{-}2}\ )\ ]$

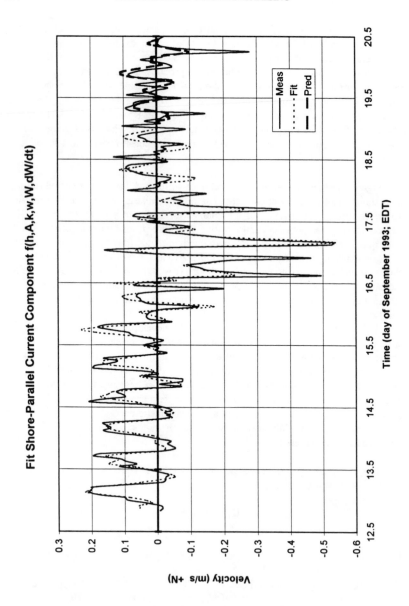

Figure 5. Empirical Fit of Measured Shore-Parallel Current

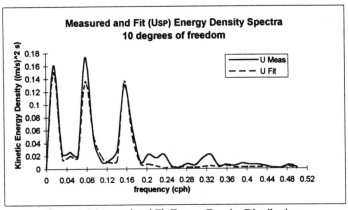

Figure 6. Measured and Fit Energy Density Distributions

A determination of the dominant terms in Equation 4 can be obtained by comparing the contribution of each forcing term utilized to fit the shore-parallel current component. Figure 7 is a comparison plot of the contribution of each predictive term in Equation 4 relative to one another. The four terms in the empirical equation are those terms which are a function of the wind velocity, K_W, rate of change of the wind velocity, K_{dW}, tidal elevation and wave amplitude, K_{AH}, and the wave number, K_K. Examining Figure 7a and 7c, it is apparent that the dominant tidal structure of the predicted current velocity during these periods is driven by K_{AH}, the terms containing both the change in water surface elevation, h, and the wave amplitude, a. Additionally, it is evident that the frequency of the forcing produced by the tidal elevation and wave amplitude terms increases during the time period of strong southerly-directed currents (Figure 7b). As Figure 8 indicates, the frequency distribution of the kinetic energy density associated only with the terms containing both the variation of water depth, h, and the wave amplitude, a, verifies that the nonlinear interaction between the two parameters is responsible for the measured current variability at 0.16 cph. A similar analysis indicated that the wind energy was centered at a frequency of 0.02 cph, indicating that the wind energy oscillates at a period of about 50 hours, coinciding with the shore-parallel current at 0.02 cph.

The relative contributions of the remaining terms in Equation 4 (Figure 7) indicate that the predicted current is strongly modified by the magnitude and direction of the shore-parallel wind velocity term, which can be of equal or greater magnitude than the water depth and wave amplitude terms. The shore-parallel wind velocity and the water depth - wave amplitude terms combine to produce the strong southerly-directed currents observed during the period of increased wave height (Figure 7b). These two predictive terms are slightly modified by the rate of change of the wind velocity and the wave number terms during the wave event.

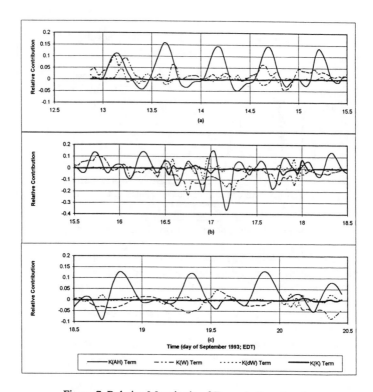

Figure 7. Relative Magnitude of Terms in Equation 4

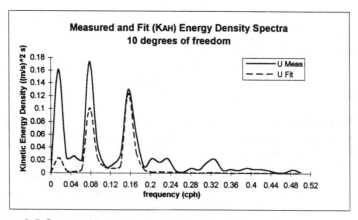

Figure 8. Influence of Water Depth and Wave Amplitude on Energy Distribution

CONCLUSIONS

An analysis of nearshore wave and current forcing adjacent to Townsends Inlet, New Jersey, revealed the presence of current energy at a super-harmonic of the semidiurnal tidal frequency. A filtering analysis applied to the data failed to derive a relationship between the current structure and measured physical forcings at the site. A multivariable polynomial autoregressive analysis with exogenous variables (PARX model) was able to produce effective empirical fits to measured nearshore current data. The model predicted the observed current structure and reproduced the frequency distributions of tide and current energy densities.

The developed empirical expression indicated that temporally-varying nonlinear interactions between the surface elevation and the incident wave amplitude generates the observed current structure adjacent to the inlet. In the development of accurate nearshore hydrodynamic models, ignoring such non-intuitive interactions among the physical forcing parameters in the nearshore region may produce erroneous model results or the inability to tune the model through *a priori* parameter estimates. By analyzing the observational data through empirical fitting techniques, the actual magnitude and interaction among the physical parameters can be defined by the data prior to modeling. Such *a priori* knowledge should improve the efficiency and accuracy of hydrodynamic models, especially those using inverse methods.

APPENDIX. REFERENCES

Chen, S., and Billings, S.A. (1989). "Representation of nonlinear systems: the NARMAX model." *Int. J. Control,* 49(3), 1,013-1,032.

Panchang, V.J. and Richardson, J.E. (1993). "Inverse adjoint estimation of eddy viscosity for coastal models." *J. Hydr. Engrg.,* ASCE 119(4), 506-524.

Ullman, D.S. (1996). "Tidal model parameter estimation using data assimilation." PhD dissertation, SUNY, Stony Brook, NY.

Vaccari, D.A. and Christodoulatos, C. (1992). "Generalized multiple regression with interaction and nonlinearity for system identification in biological treatment processes." *Trans. Instr. Soc. of Am.,* 31(1), 97-102.

Wunsch, C.M. and Fancois, J. (1982). "Methods for box models and ocean circulation tracers: mathematical programming and nonlinear inverse theory." *J. Geophys. Res.,* 87(C8), 5647-5662.

APPLICATION OF MULTIGRID TECHNIQUE TO HYDRODYNAMIC EQUATIONS

Prasada Rao and Scott A.Yost
Department of Civil Engineering, University of Kentucky,
Lexington, KY 40506-0281

ABSTRACT

In this paper we present a second-order accurate Finite Volume scheme for computing free surface flows. To overcome the limitation of the standard explicit schemes, a multigrid technique has been coupled to the numerical approach. The present effort has been motivated by the recognition that the time step in the explicit computation is very small, because of stability constraints. The effect of coupling a multigrid approach on the steady-state solution is numerically investigated. The present formulation is tested on a range of two-dimensional flows, paving the way for its potential application to larger magnitude problem.

INTRODUCTION

Numerical modeling of flows in channels and estuaries has been done largely using Finite Difference and Finite Volume schemes, mainly because of their simplicity in implementation and physical interpretation. Though this can be justified, since the time dependent flow equations are hyperbolic for both sub-critical and super-critical flows, the computational effort required to obtain a steady-state solution is very large. For time-dependent flow equations, the stationary solution is obtained by using a false transient approach. In this approach, time is taken as the iteration parameter. The computations are done over a period of time to ensure the time dependent flux terms vanish. Since the time step is limited by the CFL stability condition, the number of iterations required for obtaining the steady state is large, which limits the applicability of explicit schemes. To

338

this end multigrid methods are becoming valuable procedures in speeding up computations. Furthermore, multigrid methods are independent of the discretization approach. Though multigrid methods for elliptical system of equations are quite mature, the development of an efficient and robust method for hyperbolic systems is still a challenging problem. In this paper, we present a multigrid formulation with a finite volume discretization for solving the 2D depth averaged flow equations.

BACKGROUND

The pioneering work in the applicability of Finite Volume (FV) schemes has been done in parallel by Jameson et al. (1981), Ni (1982) and Denton (1981). Though Jameson et al. had used a Runge-Kutta procedure and Ni a Lax Wendroff approach for solving the algebraic equations, both met with success independently. A FV approach guarantees an optimal representation of physics as the formulation is based on the integral form of conservation equations for mass and momentum. An advantage reflected by this approach is the convenient application over inhomogeneous meshes with the simple use of Cartesian representation of coordinates.

The family of FV schemes can be classified as either cell centered or cell vertex, based on the location where flow variables are defined. In the cell centered category [Fig. 1-a], the flow variables are associated with the center of the control volume, which is surrounded by the four grid nodes. In the cell vertex schemes [Fig. 1-b], the variables are associated with the cell vertices. Visualizing the two approaches represented in Fig. 1 indicates that for irregular grids a cell center approach would involve further approximations (interpolations) in computing the cell center values of flow components. Morton and Suli (1991) had analytically studied the above two approaches, and concluded that the cell vertex family is more versatile. Hence, in this work a cell vertex FV scheme is used. For a uniformly computational domain, the cell vertex FV scheme reduces to a finite difference one. Ni (1982) had used a Lax-Wendroff type FV formulation for computing the flow variables. Based on his original approach, several methods have been derived (Hall 1984). The FV discretization used in the present study is explicit, second-order accurate in space and first order accurate in time. A reduction of accuracy in time is acceptable, since the interest is in steady-state conditions. A characteristic feature of all the second and higher order accurate schemes lies in producing dispersive errors in the numerical solution. The procedure outlined by Jameson (1981) has been used here to smoothen the numerical oscillations.

The transient solution has been accelerated to steady state by coupling a multigrid technique to the numerical approach. Ni (1982), Johnson (1985) and Jameson (1984) have shown that using a sequence of nested coarser grids, improvements can be made with no loss of accuracy. While the focus of the work is on steady state, Yost and Prasad (1998) have studied the applicability of multigrid for accelerating the convergence for transient simulations.

GOVERNING EQUATIONS

Starting from the Navier-Stokes equations, the two-dimensional flow equations are obtained by integrating them along the vertical direction. Integrating the equations along the depth (from bed level to water surface) results in some of loss of information, but this is minimal. The final equations that represent the continuity and the momentum equations in the x and y directions can be written as (Chaudhry 1993),

$$\frac{\partial h}{\partial t} + \frac{\partial}{\partial x}(hu) + \frac{\partial}{\partial y}(hv) = 0 \tag{1}$$

$$\frac{\partial}{\partial t}(hu) + \frac{\partial}{\partial x}(hu^2) + \frac{\partial}{\partial x}\left(\frac{gh^2}{2}\right) + \frac{\partial}{\partial y}(huv) = 0 \tag{2}$$

$$\frac{\partial}{\partial t}(hv) + \frac{\partial}{\partial x}(huv) + \frac{\partial}{\partial y}(hv^2) + \frac{\partial}{\partial y}\left(\frac{gh^2}{2}\right) = 0 \tag{3}$$

Where h is the flow depth, g is the acceleration due to gravity; u and v are the depth-averaged velocities, defined as

$$u = \frac{1}{h}\int_{z_b}^{z_b+h} \bar{u}\, dz \qquad\qquad v = \frac{1}{h}\int_{z_b}^{z_b+h} \bar{v}\, dz$$

With z_b representing the bottom elevation. While Eqs. 2 and 3 describe flow in horizontal frictionless channels; sloping and non-smooth channels can be simulated by adding the appropriate source terms to the right hand side of Eqs. 2 and 3.

NUMERICAL FORMULATION

The finite volume discretization approach for the above equations is described in detail in this section. To keep uniformity in the discussion, the above equations are recast in matrix notation as

$$\frac{\partial U}{\partial t} + \frac{\partial F}{\partial x} + \frac{\partial G}{\partial y} = 0 \tag{4}$$

Where the elements of matrices U, F and G are

$$U = \begin{bmatrix} h \\ hu \\ hv \end{bmatrix} \qquad F = \begin{bmatrix} hu \\ hu^2 + \dfrac{gh^2}{2} \\ huv \end{bmatrix} \qquad G = \begin{bmatrix} hv \\ huv \\ hv^2 + \dfrac{gh^2}{2} \end{bmatrix}$$

The goal of a steady-state simulation is to drive the net flux into any control volume to zero. Numerically, it is equivalent to integrating Eq. 4 over a cell Ω to give

$$\iint_{\Omega} \left[F(U)_x + G(U)_y \right] dxdy = 0 \qquad (5)$$

Applying the divergence theorem to Eq. 5, the resulting boundary integral can be written as

$$\int_{\partial\Omega} \left[F(U)dy - G(U)dx \right] = 0 \qquad (6)$$

In discrete form Eq. 6 reduces to the algebraic sum of all the terms across all the cell faces. With reference to Fig. 2, for any control volume C (Fig. 2a), this can be written as

$$\sum_{i=1}^{4} \left[F(U)_i \Delta y_i - G(U)_i \Delta x_i \right] = 0 \qquad (7)$$

As we are solving the unsteady equations, the residual term given by the left side of Eq. 7 is the change of flux that is occurring with in a control volume. With reference to Figure 2, for any control volume C, Eq.7 can be written as,

$$\Delta U_c = \frac{\Delta t}{\Delta A_c} \left\{ \begin{array}{l} \left[\frac{F_1 + F_2}{2}(y_2 - y_1) - \frac{G_1 + G_2}{2}(x_2 - x_1) \right] \\ -\left[\frac{F_3 + F_4}{2}(y_3 - y_4) - \frac{G_3 + G_4}{2}(x_3 - x_4) \right] \\ +\left[\frac{G_1 + G_4}{2}(x_4 - x_1) - \frac{F_1 + F_4}{2}(y_4 - y_1) \right] \\ -\left[\frac{G_2 + G_3}{2}(x_3 - x_2) - \frac{F_2 + F_3}{2}(y_3 - y_1) \right] \end{array} \right\} \qquad (8)$$

In Eq. 8, the subscripts 1-4 denote the values of the flow variables at the grid nodes surrounding the control volume C, ΔA_c is the area of control volume, Δt is the time step and ΔU_c is the flux change occurring in the control volume. As the flux change occurring in control volume C needs to be transferred to its surrounding grid nodes 1-4, Ni (1982) had developed the "distribution formulae", which can be written as

$$\left(\delta U_1\right)_c = 0.25\left[\Delta U_c - \frac{\Delta t}{\Delta A_c}\Delta F_c - \frac{\Delta t}{\Delta A_c}\Delta G_c\right]$$

$$\left(\delta U_2\right)_c = 0.25\left[\Delta U_c - \frac{\Delta t}{\Delta A_c}\Delta F_c + \frac{\Delta t}{\Delta A_c}\Delta G_c\right]$$

$$\left(\delta U_3\right)_c = 0.25\left[\Delta U_c + \frac{\Delta t}{\Delta A_c}\Delta F_c + \frac{\Delta t}{\Delta A_c}\Delta G_c\right] \tag{9}$$

$$\left(\delta U_4\right)_c = 0.25\left[\Delta U_c + \frac{\Delta t}{\Delta A_c}\Delta F_c - \frac{\Delta t}{\Delta A_c}\Delta G_c\right]$$

where ΔU_c given by Eq. 8 , and

$$\Delta F_c = \Delta f_c \Delta y^l - \Delta g_c \Delta x^l$$
$$\Delta G_c = \Delta g_c \Delta x^m - \Delta f_c \Delta y^m \tag{10}$$

in which $\Delta f_c = \dfrac{\partial F_c}{\partial U_c}$, $\Delta g_c = \dfrac{\partial G_c}{\partial U_c}$ and

$$\Delta x^l = 0.5\left(x_2 + x_3 - x_1 - x_4\right) \qquad \Delta x^m = 0.5\left(x_3 + x_4 - x_1 - x_2\right)$$
$$\Delta y^l = 0.5\left(y_2 + y_3 - y_1 - y_4\right) \qquad \Delta y^m = 0.5\left(y_3 + y_4 - y_1 - y_2\right) \tag{11}$$

The flow properties at node 1, at the new time level are then calculated as

$$U_1^{n+1} = U_1^n + \delta U_1$$
$$\delta U_1 = \left(\delta U_1\right)_A + \left(\delta U_1\right)_B + \left(\delta U_1\right)_C + \left(\delta U_1\right)_D \tag{12}$$

Equation 12 can be used for calculating the flow variables at all the interior nodes. Specifying the proper boundary conditions completes the computation at all the grid nodes, after which the solution can be marched along the time axis.

The solution obtained by Eq. 12 is second-order accurate and hence contains numerical oscillations. A satisfactory simulation requires smoothening of these oscillations. In the present work, the procedure outlined by Jameson(1981) has been used. As Jameson's viscosity has found wide application in computational hydraulics, its not dealt here. We refer the readers to Chaudhry (1993) for a detailed explanation.

MULTIGRID ALGORITHM

Multigrid methods use a hierarchy of grids. Its philosophy is to carry out the iterations on progressively coarser grids. As the number of grid nodes on coarser levels is less, the number of unknowns are reduced. The residual values of the solution on the coarse grid node are then prolongated back to the fine-grid nodes. For structured grids, as in the present one, the coarse grids are obtained by skipping the alternate nodes in each flow direction. The same procedure can be repeated to generate coarser and coarser grids. The maximum coarse level that can be reached depends more on the nature of the numerical approach than on the problem (equations) under study. For a hyperbolic system of equations, which are sensitive to the grid dimensions, a very large coarse grid may result in a loss of accuracy (Wesseling, 1992). Our experience shows that larger coarse-grids, comprised of 8 or more fine-grids did not yield improved results.

A possible reason for this could be the damping mechanism. As the damping mechanism in the present code acts only once for every time step, an increase of grid dimension at coarser levels may deteriorate the numerical solution. This argument motivates further study into the influence time stepping has on coarser level computations. Since a specific grid spacing is accompanied by an optimum time step, from the stability point of view one can anticipate that increasing the time step on coarse grid levels results in a faster convergence rate. This feature is incorporated in the present code. As the coarser levels are generated (each level larger by a factor of 2 over the previous level, Fig. 2-b) the optimum time step on the coarser levels is calculated from the stability condition. This increase in the time step on coarser levels results in a rapid wave propagation across the boundaries, speeding convergence.

The final aspect in the multigrid formulation lies in prolongating the solution from the coarse grids on to its surrounding fine grid nodes (Wesseling 1992). The objective of prolongation, which is also addressed in the multigrid literature as post smoothing, lies in smoothening the coarse-grid solution on to the fine-grid domain. As this smoothening is analogous to the effects achieved by artificial viscosity, prolongation of the coarse-grid solution was not performed. Hence as shown in Fig.2, a one to one mapping of coarse-grid node to its corresponding fine-grid node was carried out. Formally the whole procedure can be written as (Fig. 3),

i)   With the given initial conditions at the initial/known time level (n), and the adopted computational domain, calculate the time step from the CFL stability criteria

ii) At this new time level (n+1), compute the values of the flow variables at all the interior nodes, using Eq. (12). Making use of appropriate boundary conditions complete the computation at all the grids (fine grid nodes).

iii) Generate the coarse grids by skipping the alternate fine grids. With the modified grid spacing and time step, recompute the value of flow variables at all the coarse grid nodes using equations 8-12

iv) Repeat step(iii) until the maximum coarse grid is reached

v)   Apply Jamesons artificial viscosity at all the nodes to dampen the oscillatory nature of the obtained solution

vi) Reassign the flow variables at all the nodes as initial values, check the convergence criteria and go to step (i)

APPLICATION

The FV scheme along with the multigrid approach is applied to a wide variety of flows in this section. A courant number of *0.8* was used in the evaluation of time step from the CFL stability criteria. A dissipation constant of *0.2* was used in the computation of Jameson's artificial viscosity. Though the final goal is to apply the above techniques on the Green Bay, we present the preliminary results to test the efficiency of above algorithms on various flow patterns that are commonly encountered.

Case I. The objective of this test problem is to act as a code validating measure, both for the FV scheme and for the multigrid technique. A channel of length of *5m* x *0.5m* was considered for analysis. The cross section of the channel is rectangular. The initial flow conditions in the channel are a depth of *0.01m* and zero discharges throughout the computational domain. At the upstream end, a flow depth of *0.03m*, a longitudinal discharge of *0.06m²/s* and zero transverse discharge are specified and are held constant for all time periods. At the downstream end, all the flow variables were interpolated from the interior nodes; this represents a uniform outflow. The computational domain was divided to *40x9* control volumes. Flow is said to be steady, when the water depth at all the grid nodes is uniform at *0.03m*. Figure 4 is a plot of the number of iterations required for the solution to converge to steady versus the residual error. The results indicate that by generating a coarse grid of *4x4* the number of iterations can be reduced by half. For a system of hyperbolic equations, the saving in terms of CPU is within the range of published results (Ni, 1982). The corresponding plot with the same initial and boundary conditions for a trapezoidal cross-section is shown in Fig. 5. The cross-sectional slope of the channel was *1* in *100*. The results indicate that for coarser grids, some rounding off errors are evident. However, it should be noted that these errors become manifest only for results of higher accuracy, a factor that can be used as a trade-off in many engineering problems.

Case II. This problem deals with the capacity of Finite volume scheme to absorb irregular grids in its computational domain and the effect of these irregular domains have on the multigrid approach. A critical problem of simulating a shock wave is considered, as simulating shocks with multigrids is considered to be a critical test (Wesseling 1992). The channel geometry is shown in Fig.6. A *0.305m* width channel was gradually reduced to *0.1525m* over a length of *1.45m*. The initial conditions and the boundary conditions are the same as in Case I. At the sidewalls the surface tangency condition, representing a solid free-slip wall boundary, is imposed while calculating the flow components. The computational domain has been illustrated in Fig. 7. Because of the symmetry, the simulation was done for half-width of the channel. At the centerline, a reflection boundary condition has been used. Figure 8 shows the evolution of the shock front in the transition along the sidewall and along the centerline. The result obtained with a coarse grid, equal to doubling the fine grid in each direction, is also shown. The plot indicates that in spite of having an incoming flow with a Froude number of 4, not much information is lost, on the coarser levels, indicating the reliability of the present formulation.

Case III. The effect of cross flow in the channel on the present formulation is studied in this section. The definition sketch along with the flow variables is shown in Fig. 9. A cross flow was introduced in the channel between the grid nodes *8* and *12*. The initial conditions were zero discharges and a flow depth of *0.1m*. As the governing equations are devoid of any effective stress terms, the capacity of the present model to simulate the circulation is not studied. From the computational hydraulics side, the major consideration in any algorithm is its capacity to treat flows over irregular domains, proper representation of discontinuities, computational efficiency and treatment of viscous effects. In our investigation until now, we have confined to the first three. Figure 10 presents the convergence histories and the corresponding CPU for different grids. A convergence factor of 2-3 was observed in either case.

CONCLUSIONS

This paper presents the advantages of coupling a multigrid approach to a Finite Volume discretization of the two-dimensional transient flow equations. The focus was on steady-state simulation. The resulting algorithm was applied to a wide variety of free surface flows. The factors that affect the performance of a multigrid approach and the possible alternatives to accelerate the convergence of the solution are discussed. Though multigrid methods promise to be encouraging tools, their application to hydrodynamic problems is in infant stage. As prospective users start applying the multigrid algorithms to simulate various flow problems, their efficacy can be better understood.

REFERENCES

1) Chaudhry, M.H. (1993) Open Channel Flows, Prentice Hall, NY.

2) Denton, J.D (1981) An improved time marching method for turbomachinery flow calculation, Proceedings of the IMA Conf. held at the Univ. of Reading.

3) Hall, M.G (1984) Fast Multigrid solution of the Euler equations using a Finite-Volume scheme of Lax-Wendroff type, RAE Technical Report 84013.

4) Jameson (1983) Solution of the Euler equations by a Multigrid method, Applied Mathematics and Computation 13, 327

5) Jameson, A., Schmidt, W. & Turkel (1981) Numerical solutions of the Euler equations by finite volume methods using Runga-Kutta time stepping schemes, AIAA 14th Fluid and Plasma Dynamics Conference, Palo Alto, CA,AIAA,81-1259

6) Jesperson, D.C (1994) Recent developments in multigrid methods for the steady Euler equations, Von Karman institute for fluid dynamics, Lecture series 1994-04.

7) Johnson, G.M and Chima, R.V (1985) Efficient solution of the Euler equations and Navier-Stokes equations with a vectorized multigrid algorithm, AIAA 23(1) 23-32

8) Li C.P (1985) A Multigrid factorization technique for the flux-split Euler equations, Lecture notes in Physics, Springer Verlag (Vol. 218)

9) Morton, K.W. and Suli, E. (1991) Finite Volume methods and their analysis, IMA Jl. of Numerical Analysis, 11(2) 241-260

10) Ni, R.H (1982) A Multigrid scheme for solving the Euler Equations, AIAA, 20 (11)

11) Yost S.A and Prasad Rao (1998) Coupling Multigrid methods with Finite Elements for Time Dependent Solutions of Hydrodynamic Systems, Proceedings of 10th Int. Conf. on Finite Elements in fluids, Univ.of Arizona, Tucson.

12)Wesseling, P. (1992) An Introduction to Multigrid Methods, John Wiley & Sons, NY.

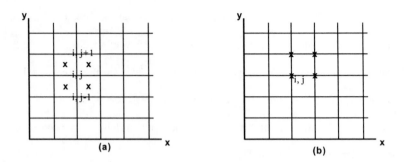

Figure 1. Family of Finite Volume Schemes (*x*- computational location of flow variables). (a) Cell centered (b) Cell vertex

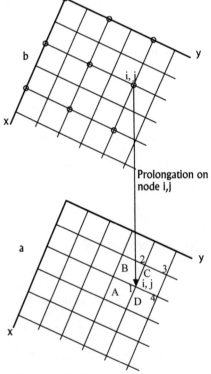

Figure 2. Illustration of prolongation between coarse and fine grids. (a) fine grid (b) coarse grid.

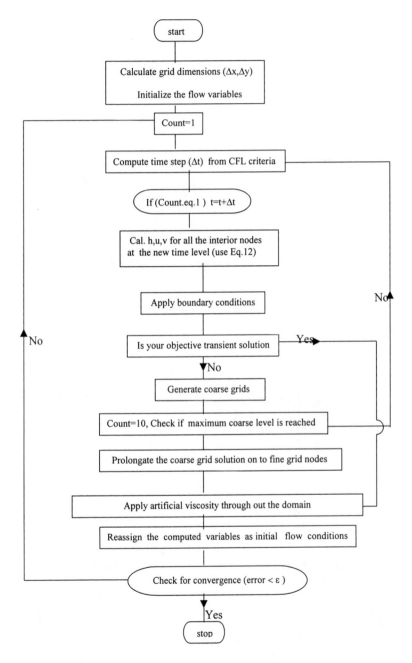

Figure 3.Computational sequence using multigrid approach

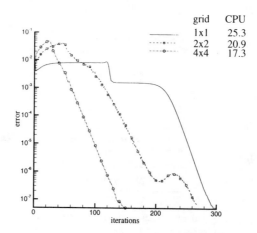

Figure 4. Convergence Histories for Case I: Rectangular Channel.

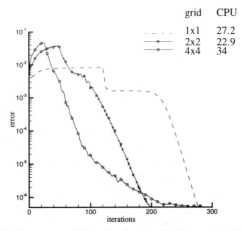

Figure 5. Convergence Histories for Case I: Trapezoidal Channel.

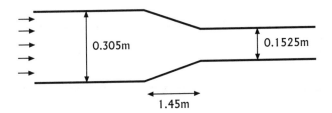

Figure 6. Definition Sketch for flow in a Contraction.

Figure 7. Computational domain for flow in transition

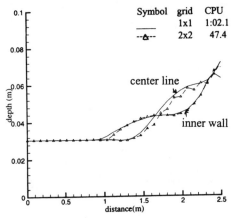

Figure 8. Flow profiles with and without coarse grid computations.

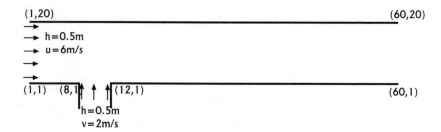

Figure 9. Definition Sketch for Cross Flow.

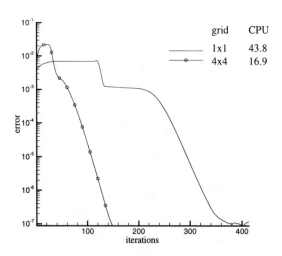

Figure 10. Convergence Histories for Cross Flow into Channel.

Development of Model-Based Regional Nowcasting/Forecasting Systems

Bruce B. Parker [1]

Abstract

The National Ocean Service is developing nowcast/forecast systems for bays and harbors using numerical oceanographic models forced with real-time data and with forecast wind fields from atmospheric models. This effort presently involves regional high-resolution hydrodynamic numerical models of Chesapeake Bay, Galveston Bay, and the Port of New York and New Jersey, with forecast boundary conditions provided by either an East Coast ocean model, a Gulf of Mexico ocean model, or an individual shelf model, each driven by forecasts from the NWS ETA weather model. Nowcasts and 24-hour forecasts of water level and/or currents are presently produced twice daily and presented in graphical form on a WWW Home Page for evaluation purposes.

Forecast water levels and currents from these systems will make navigation in these waterways safer and more efficient by moving the state of the art beyond astronomical tide and tidal current predictions (provided for decades in the NOAA Tide and Tidal Current Tables) and beyond recently developed real-time oceanographic systems. Forecast currents will also improve oil spill clean-up operations by better predicting the movement of the spill.

These model systems must not provide misinformation to the mariner, so emphasis is being placed on solving those scientific problems that will improve the accuracy of the nowcasts and forecasts.

The majority of this paper deals with the special problems associated with developing and implementing a regional operational nowcast/forecast model system and the various approaches for solving such problems.

1 Coast Survey Development Laboratory, National Ocean Service,
  NOAA, N/CS1, Silver Spring, MD 20910

INTRODUCTION

The Coast Survey Development Laboratory (CSDL) in the National Ocean Service (NOS) in NOAA is developing nowcast/forecast systems for bays and harbors using numerical oceanographic models forced with real-time data and with forecast meteorological fields from weather forecast models. The dissemination of the resulting predictions is via the World Wide Web and other mechanisms. In this paper we briefly describe the motivation for developing these systems and an overview of the regional and coastal nowcast/forecast projects that CSDL is working on. The majority of the paper is devoted to a discussion of the special problems that must be dealt with in developing and implementing a regional operational nowcast/forecast model system.

THE NEED FOR NOWCAST/FORECAST MODEL SYSTEMS

The primary motivation for CSDL's development of regional nowcast/forecast model systems is to support safe and efficient navigation in U.S. harbors and waterways. Accurate knowledge of water levels (in conjunction with accurate charts) is necessary for the mariner to avoid ship groundings. Accurate knowledge of currents is important for ship maneuvering and the avoidance of collisions with other ships or with bridges and piers. Such maritime accidents can lead to hazardous spills and the loss of life and property. If a hazardous spill should occur, accurate current speeds and directions are needed to help predict the movement of the spill so that it can be efficiently cleaned up before it can hit a beach or mangrove swamp or some other habitat.

For decades mariners have depended on the NOAA Tide and Tidal Current Tables for the best prediction of water levels and currents. Such Tables provide very accurate predictions of the *astronomical* tide and tidal current, but they cannot predict the important effects of wind and river flow. This deficiency has become more important in recent decades, when cargo ships and oil tankers became so large that they could not enter U.S. ports except near times of high water. Since every additional inch of ship's draft can means hundreds of thousands of dollars of additional cargo, it is important to know water levels as accurately as possible. Another limitation of these Tables is that they provide predictions only at selected locations, and so may not provide enough spatial coverage, especially for currents, which can vary dramatically over short distances.

The first step in providing the mariner with more accurate water level and current information was the implementation of real-time oceanographic systems. Such information, however, is useful only for that immediate moment in time when the measurement is made, and perhaps up to a couple of hours into the future, if conditions do not change too quickly. Since it can take hours to load a ship and hours to transit a bay, mariners need to know what the water level and currents will be as much as 24 hours into the future. Ships coming into port also need this information that far ahead to plan their arrival for the most favorable conditions. For the most effective clean up of oil spills, one needs to know well ahead of time if the currents that transport the spill are going to change direction. Since only a few locations can be instrumented, the spatial coverage of real-time data can be inadequate for showing dangerous eddies or current shears, or for identifying oil convergence zones after a spill. For any predictions more than a couple hours into the future the mariner must still go back to the Tide and Tidal Current Tables, which will only provide astronomical tidal predictions at a few selected locations.

The solution to providing accurate water level and current predictions up to 24 hours into the future is the development of regional oceanographic nowcast/forecast model systems. These model systems also provide necessary spatial coverage, since they provide predictions at hundreds of locations. In many harbors there are requirements for real-time information at more locations than there are sensors, so nowcasts produced by running a model in real time also provide useful "real-time" information at these other locations.

Although the primary motivation has been to support safe and efficient navigation and oil spill drift prediction, there appear to be other areas where nowcasts and forecasts of various oceanographic parameters should be useful (Parker, 1996). Forecast currents could have a number of useful applications in the environmental, ecosystem, and sediment transport fields, since currents play an important role in most biological, chemical, and geological processes in the coastal zone. The currents in a bay flush out pollutants, stir up and transport sediments (and attached pollutants), and move larvae and juveniles out of and into estuaries. Forecast currents could warn of particular natural events, or could be used to predict the near-term consequences of a man-produced event. Water temperature and salinity directly affect habitats and the species of marine plants and animals that live there. Forecast salinity and temperature fields could warm of stratified situations that may lead to anoxic conditions in bottom waters, or to red tides or other toxic blooms in the upper waters. (The high-resolution forecast wind fields being developed [see below] to drive a bay model, will also have other uses. These fields are critical to forecasting wave conditions, which is important not only to the commercial maritime community, but also to the recreational boating community. Forecast winds and other parameters are critical to air quality prediction. High resolution nowcast winds are needed to assess the transport and diffusion of atmospheric pollutants, and their deposition in the Bay.)

SPECIAL ASPECTS OF A NOWCAST/FORECAST MODEL SYSTEM

There are many special considerations when one is implementing a numerical hydrodynamic model for use in a nowcasting/forecasting system, as compared with, for example, such a model's use in a research process study.

The need for greater accuracy. Predictions from a nowcast/forecast model system will be used to make operational decisions that can have important safety and economic consequences. Required accuracies are determined by these operational needs. One must have enough water under the keel of the ship to keep it from running aground, but if one chooses too large a safety margin then shipping companies will lose money by not having loaded enough cargo. For this case, accuracies on the order of 15 cm (six inches) may be needed.

To prove that a model system is working well it is not enough just to show two similar curves representing model predictions and actual data. Even root-mean-square (RMS) differences between observations and predictions over some test period can be misleading. What counts is the accuracy of what the user (e.g. the mariner) sees, i.e. each individual prediction on which an operational decision is based. What counts is not how well a model system does on average, but how well it does for each wind event.

If the model is going to be wrong, it is safer to be off on the low side for water levels, i.e. telling the mariners they have less water under keel than they really have (since the other way can lead to a grounding). It may, however, be safer to be off on the high side for currents when a ship is turning and docking, because a large ship needs low current speeds to maneuver safely. Current direction is a critical factor in predicting oil spill movement, or determining right-of-way between two ships approaching each other. In most bays current direction can be fairly accurately predicted except near times of slack water. Thus, accurately predicting the time of slack water can be the most critical item for some ports.

Of equal importance is the ability to tell the user *what the probability is that a particular forecast will be accurate*. This cannot be some average probability based on the model system's overall past performance. For it to be useful to the mariner the probability estimate must be calculated for each nowcast and forecast and must vary according to the type of weather event. For a mature fully operational forecast model system, such probability estimates could come from an ensemble averaging system, based on how different the various forecasts are. Prior to the implementation of an ensemble averaging system, however, errors in past forecasts will need to be correlated with details of each weather event (such as predominant wind direction, atmospheric stability, type of front passage, etc.) in order to ascertain a pattern in such errors that might be used to estimate the accuracy of a future forecast based on the type of weather event. (For nowcasts, the actual real-time data compared with the nowcast will provide the user with information on how well the model system is doing. If data assimilation is used in producing the nowcasts, data from a couple of locations should not be assimilated, but instead used to determine quality indicators of the nowcasts.)

The dependency on real-time oceanographic and meteorological data and data fields. One may have a calibrated, verified, accurate model, but if the real-time data input into this model are inadequate in coverage or quality, then the nowcasts produced by the system will not meet user accuracy requirements. Nor will the forecasts that were initialized with these nowcasts. Thus, an important part of testing a model is determining the minimum data inputs needed to produce required accuracies, and then assuring that there will be real-time access to such data.

Real-time data are required for open boundary forcing, the most critical being real-time water level data at (or outside) the bay entrance, which must accurately represent the water level signal propagating in from the continental shelf. Real-time wind stress on the surface of the bay is a critical input, if the bay is large enough for the wind over the bay to produce a significant portion of the water level variation. In this case real-time wind data *fields* are required, of high enough resolution to accurately represent the wind direction. Such fields must be created from wind measurements, including some taken from instruments over the water. Real-time river discharges may also be required depending on the particular bay, as well as real-time temperature and salinity at the entrance, and temperature of river water.

In addition to real-time data at the model boundaries for direct forcing, real-time data from the interior of the model area are needed to provide a real-time indication of the quality of the predictions by showing how close the nowcast is to reality at particular locations. Real-time data from the interior of the model can also be used for data assimilation to improve nowcasts, especially where some of the forcing fields may not be as complete or as accurate as one would like [see discussion below]. The multiple model runs with varying forcing fields carried out in the assimilation scheme also provide estimated error bars for the nowcasts.

The dependency on forecasts from other meteorological and oceanographic model systems. When running the bay model system in the forecast mode, the entrance-, surface-, and river-forcing conditions must come from other forecast model systems. Depending on the particular bay this can include one or all of the following: (1) a coastal/shelf oceanographic forecast model; (2) an over-the-bay high-resolution weather forecast model; and (3) a larger-scale weather forecast model; and (4) a river forecast model. (See Figure 1.)

The **coastal/shelf oceanographic forecast model** provides the forecast entrance water levels (as well as forecast salinity and temperature if needed). For small bays the subtidal variations in water level will be dominated by the water level signal produced by winds over the continental shelf and propagated into and up the bay. An **over-the-bay high-resolution weather forecast model** is needed to provide forecast wind fields which will drive the bay model. This is most critical for large shallow bays where direct wind forcing significantly affects water levels and currents. The high resolution is often necessary to produce accurate forecast wind directions (especially for a bay with complex geometries and/or varying topography around it). A **larger-scale weather forecast model** serves two critical functions. First, it is needed to drive the coastal/shelf oceanographic forecast model with forecast winds, heat flux, moisture flux, etc. (and it may be one reason why the coastal model is not providing accurate water level forecasts at the entrance to the bay model). Second, it is needed to provide boundary conditions for the over-bay high-resolution weather forecast model. In the U.S. the two primary candidates are either NWS's AVIATION model, which covers the entire northern hemisphere at one-degree resolution, or NWS's ETA model, which is presently operating at a 29-km resolution. The **river model** provides forecast discharges. Since this usually changes fairly slowly and the delays from precipitation over the river drainage basin to the discharge into the bay are usually long and well known, this should not be a difficult part of the problem. It may even be possible to use real-time data as a reasonable substitute for this. For large rivers, discharge can have a direct and significant influence on the currents and on the water levels (via the frictional affect on the river currents and the frictionally caused nonlinear interaction between the river current and the tidal current). River discharge will also affect the salinity and density of the bay, which then affects the currents, and even indirectly the water levels (through possible stratification effects on vertical momentum transfer).

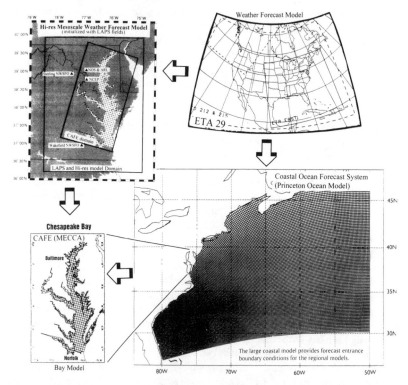

Figure 1. The Chesapeake Bay forecast model requires forecast entrance boundary conditions from the East Coast forecast model, which is driven by NWS's ETA 29-km weather forecast model. Forecast wind fields over the bay are provided by a high-resolution mesoscale weather forecast model, whose boundary conditions are provided by ETA 29, and whose initials conditions are provided by real-time data fields from the LAPS system.

The need for ongoing quality control of the predictions in an operational environment. Maritime decision making will typically be on time scales of minutes to 24 hours, so some type of 24-hour-7-days-a-week monitoring of the quality of the nowcasts and forecasts will be needed, as well as of the quality of the data going into the model. Procedures must also exist to handle missing data, which may involve using backup information (e.g. real-time data instead of forecasts at the model open boundaries). Error estimates and forecast probabilities are an important part of how the predictions are provided to the user. There are also issues concerning data standards and prediction standards, as well as model system standardization, that must be dealt with.

The need for timely delivery of information to users in an understandable form. Model nowcasts and forecasts must be delivered to users in a timely manner and in a form that they easily make use of in their decision making. An important part of this is providing some indicator of the likely accuracy of the predictions (e.g. probability envelopes or error bars). Mariners such as pilots are very educated users who have years of experience in and knowledge of the bays they work in. For years they have relied on tide and tidal current predictions (and sometimes corrected them according to their own knowledge of local effects or of changing conditions not reflected in the Tide or Tidal Current Tables). Thus, it is informative to put nowcasts and forecasts on the same time series plots with predicted tide or tidal current curves, in order to provide useful points of reference. Similarly the predicted and measured wind information (that went into producing the water level and current forecasts) presented on time series plots (as well as synoptic presentations over a map) can also be useful. The wind fields themselves are something the mariner cares about (as well as the waves generated by those wind fields). Modern computer and communications technology provide a number of mechanisms for delivering this information to the user (see below).

Providing predictions in a timely manner puts pressures on the modeling system. A user certainly cannot wait hours for a nowcast or short-term forecast. This means that the mechanisms for delivering the input data to the model must be fast (i.e. updated in minutes) and reliable (i.e. operating all the time). It also means that the models must compute these predictions quickly. This puts certain requirements on computer power so that the models can run fast enough. It also may determine what model is used and whether it is a 2-D or 3-D model. If, for example, only water level forecasts are needed, a 2-D model may be accurate enough, and such a 2-D model will run much faster than a comparable 3-D model. For water level one may also get away with a model grid with less resolution.

Demands for timeliness are less of a computational problem for nowcasts since only a few run cycles are needed to update the previous nowcast (unless some type of data assimilation scheme is used). The timeliness of a nowcast may be affected, however, by late "real-time" data inputs or missing data that must be compensated for in some fashion. Timely forecasts are more of a problem because 24-hour runs will be carried out, and these can not be done until forecast information has been received from other forecast models, which themselves need time to produce their forecasts. Additional time may be taken if some type of ensemble averaging procedure is used. Data input and computing strategies must be developed to maximize timeliness and deal properly with missing or late inputs from these other sources (over which one may have little or no control).

THE FLOW OF DATA AND INFORMATION THROUGH A NOWCAST/FORECAST MODEL SYSTEM

The components of a nowcast/forecast model system are shown in Figures 2 and 3, along with the flow of data and information through them. The **bay model** is the center of both the nowcast and the forecast parts of the system. The initial conditions for the first nowcast are provided by a hindcast typically covering the previous 48 hours. Each succeeding nowcast is initialized by the previous nowcast, with forcing at the boundaries provided by real-time data. Each forecast is initialized by the most recent nowcast, and then run into the future with forcing at the boundaries provided by forecasts from other models (a coastal ocean/shelf forecast model, a weather forecast model, and a river forecast model).

An **automated data server** receives, organizes, and inputs the real-time oceanographic and meteorological data into the bay model. Although these data should be quality controlled (QC) before reaching this server, some final QC must still be done along with automated judgements made with respect to their usage in the model, in particular if data are totally missing or of unacceptable quality. Of critical importance for larger bays (especially if they are shallow) is having a **system to produce the best possible real-time wind fields** from all available real-time wind observations. Some type of a **data assimilation system** (and multiple runs of the model) may also be needed to improve the nowcasts, by making use of available real-time data from within the model regime (see below).

When the model system is used to produce a 24-hour forecast, it will start with initial conditions provided by the most recent nowcast. A **coastal ocean/shelf model** will provide forecast water levels at the entrance to the bay. If this coastal model does not provide accurate tidal forcing (which would require good open-ocean tidal constituents), it can be used to provide only the wind-induced water level change, with the tidal forcing at the entrance gotten from tidal harmonic constants derived from data (and acknowledging that some nonlinear interactions between subtidal surge and the tide will not be included). The winds to drive this coastal model (as well as the heat fluxes, etc.) come from a large-scale weather model. This weather model also provides boundary conditions for a **regional higher-resolution mesoscale weather forecast model over the bay** to provide winds to drive the bay model. Forecast river discharges will come from a **river model** (or real-time discharge data will be used instead).

Such forecast models may not always be available or may not provide good forecasts under some conditions. In such cases some use may be made of real-time data to provide some estimate of what the forecast conditions might be. For example, in some cases *persistence* of real-time water levels may provide a substitute for a forecast entrance boundary condition because subtidal water level changes due to winds over the shelf may change slowly and there is a lag in the response of the waters over the shelf to the wind. For some areas real-time water level data from gauges up and down the coast from the bay entrance could be used, especially if there is a progression of the coastal water level signal along the coast. The use of real-time data, however, is not likely to provide good results much beyond 6 hours. Some type of an **ensemble averaging system** could also be tried in order to improve model forecasts, running the model numerous times with varying forcing conditions. As with the use of data assimilation in producing nowcasts, the inherent sensitivity analyses would also produce estimates of possible errors.

The nowcasts and forecasts are then handled by an **information system** to provide user-friendly output. As already mentioned, these must be in a form that can be easily understood by the user, and easily made use of in their decision making.

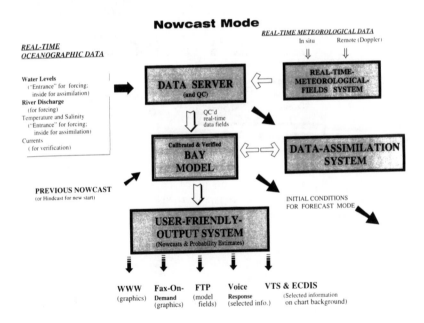

Figure 2. The flow of data and information through the nowcast part of a nowcast/forecast model system.

This will therefore include graphical (often time series) outputs as well as indicators of the likely accuracy of the predictions (e.g. probability envelopes or error bars). The system will send this information to the user via a variety of convenient mechanisms such as the World Wide Web and other Internet services, fax on demand, cellular phone calls to voice response systems, Vessel Traffic Service (VTS) systems, and electronic chart and display systems (ECDIS). Entire predictions files can be sent to other systems via FTP.

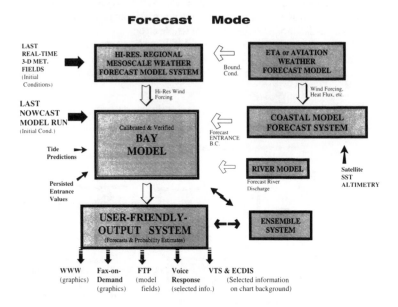

Figure 3. The flow of data and information through the forecast part of a nowcast/forecast model system.

METHODOLOGY FOR IMPROVING THE ACCURACY OF NOWCASTS AND FORECASTS

As mentioned earlier, the accuracy demands for a nowcast/forecast model system are greater than for a model used for research purposes because of the nature of the operational decisions that will be based on its predictions. One must be especially careful that such a system does not provide "misinformation" to the mariner, since there could be serious consequences. Once a model system has been implemented, with most or all of the components mentioned in the previous section, and the resulting nowcasts and forecasts have been compared with actual data, the detective work begins to track down the sources of errors in the nowcasts and forecasts. One must carefully look at all components of the system in finding ways to improve the skill of such model predictions. Underlying this is the need for a systematic approach and a standardized consistent skill assessment methodology.

Improving the Bay/Harbor Model. The first consideration is the model for the bay or harbor. The steps described below for improving this model predominantly involve hindcasts and sensitivity studies.

The first consideration should always be to get the very best bathymetry and shoreline available. If the bathymetry is taken from a nautical chart several things should be remembered. First, a nautical chart shows depths relative to mean lower low water (MLLW), not relative to mean sea level, which is the reference level for a model. Thus, half the tide range (at each location) must be added to each depicted depth sounding. Since the spatial variation in the tide range over the region is best determined by the model, this may involve an iterative process, whereby the latest and best model predictions of average tide range at each grid point will be used to improve the half-tide-range adjustment to the depths taken from the chart. Second, each nautical chart sounding is the shallowest of the measured depths taken in that immediate vicinity. Nautical charts are used by the navigation community and are important for helping the mariner avoid ship groundings, so it is always safer to tell the mariners the shallowest depth they can expect in a particular area. Thus, the average depth may be greater than depicted on the chart. For more exact determination of bathymetry, one can use the original denser survey data on which the chart was based, which are usually still available from NOS or from the National Geophysical Data Center (also in NOAA). Third, the bathymetry shown on the chart may be based on old surveys and may not accurately represent present conditions. Funding has not been available to resurvey all the waterways of the U.S., but the largest ports and their approaches have usually been given the highest priority. Navigation channels in major ports typically have the most recent data, since they are dredged by the Army Corps of Engineers to maintain them to a specified depth.

All these items related to bathymetric data are a more serious problem for shallow-water areas, since any errors will be a larger percentage of the actual depth, and will thus more significantly affect frictional energy loss and the generation of nonlinear effects. When attempting to calibrate or improve the results of a 2-D depth-integrated model, choices are sometimes made between adjusting the depth or adjusting a friction coefficient. Although adjusting the friction coefficient would usually seem to be the wiser choice (since depths have both a frictional and a continuity effect), it is still important to start with the most accurate depths possible.

Improving a model's predictions may in some cases require increasing the grid resolution. This solution may depend on the particular needs for the nowcasts and forecasts, for example, whether accurate current predictions are needed in narrow navigation channels, or whether current shears or eddies are an important problem. However, if transport to particular parts of a bay or harbor is through one or more narrow waterways, then more parameters than just currents may be affected by too coarse grid resolution. There are a number of well known approaches to achieving higher resolution in particular areas, such as using a finite-element model with its very adaptable triangular grid cells, or, using some type of two-way nested grid in a finite-difference model.

Improving a model's predictions may also require going from a 2-D model to a 3-D model. The latter is certainly required if accurate predictions of currents, temperature, or salinity are needed. However, water level predictions may even be affected, for example, by stratification affects on vertical momentum transfer, changing the effect of bottom friction or wind.

In trying to calibrate and verify a model, the usual starting place is to accurately reproduce the tides. This is certainly the right place to start, but one must be sure to take advantage of the well-known periodicities in the tide to determine whether the model has the correct frequency response. The various nonlinear terms in the equations of motion have different frequency effects, and the generation and amplification of the various overtides and compound tides (in the model predictions versus in the data) can provide valuable clues concerning (for example) frictional effects versus transport effects. Simulations with tidal-only forcing at the entrance must be long enough to calculate all the most important tidal constants (including these overtides and compound tides) using harmonic analysis.

A critical aspect of verifying a model is its reproduction of wind effects, since these are usually the most important nontidal effects and probably the reason why a nowcast/forecast model system is needed in the first place. The problem is that this part of the model verification is very data dependent, and there may not be enough locations with good wind data. If one does a model simulation using available wind data to drive the model, and the water level and current predictions have errors, one can not be sure whether the errors are due to the model or due to the wind data. And since wind data with adequate spatial coverage and resolution (including measurements from over the water) are rare, this is an especially serious problem. One approach is to try to find time periods where the wind data from the available sensors is spatially uniform (at least in direction) and then run model simulations for such periods. One must also carry out sensitivity studies in order to determine the magnitude of the changes in water level and currents that result from variations in wind speed and direction, not just varying the wind over the entire bay, but also simulating fronts passing over the bay and other typical weather situations. If one is wind data poor for a particular area, it is important to understand the magnitude of the possible errors that may result from such a data deficiency. There is also the possibility that atmospheric stability and waves will have enough of an effect on how effectively wind stress transfers momentum to the bay waters, that these effects may also have to be included.

Sensitivity studies varying other parameters (besides wind) are also important to model verification. When varying depths or friction coefficients or other parameters, intuition (even based on theoretical knowledge and experience with analytical models) doesn't always work. Only through sensitivity studies can one see how a model will respond to different situations and, in terms of providing predictions to a user, how large the errors may be.

Another important consideration, again related to data and most important for large bays, is whether the water level data being used for the ocean entrance open boundary condition truly represents the signal coming in from the continental shelf. Water level data from a gauge at the entrance to a large bay will include some effect of the bay's dynamics. If a bay could be driven by a water level signal from a truly semi-infinite ocean (as is typically specified for analytical models) then the entrance water level gauge would not be affected by the bay's dynamics. But as soon as one adds some reality including shallower depths out onto the shelf, one must worry about the entrance forcing condition. Setting up an boundary condition some distance offshore from the bay entrance will work better, but then one must be careful how one determines the tidal and wind-driven forcing at those offshore points (offshore water level measurement being the safest approach, if that is operationally feasible). For smaller bays, a gauge right at the entrance should work well in most situations.

Improving the Real-Time Input Oceanographic Data and Meteorological Fields.
One may have an excellent model, carefully calibrated for all situations, and completely verified for all available historical data, but if the data available in real time to be input into the model is inadequate in coverage or quality, then one will still produce poor nowcasts. Forecasts that use these nowcasts as their initial conditions will then also be poor.

For example, as mentioned at the end of the last section, if the real-time water level data to be used to force the model of a large bay at its entrance is from a location not sufficiently outside of the entrance and away from the effects of the bay dynamics, then nowcasts produced by the model system will be significantly in error (especially near the head of the bay) during particular wind situations. If one cannot install new gauges to better represent the real-time open boundary condition, one will have to determine through careful trial and error how best to determine the real-time entrance boundary condition from those gauges that are available. This will likely mean treating the tidal and nontidal signals separately (and paying a small price in losing some accuracy due to the lost nonlinear interactions). Using other real-time gauges up and down the coast from the bay entrance one should be able to get some idea of the subtidal signal induced by winds over the shelf. The tidal signal to force the model at offshore locations can be produced with harmonic constants derived from the entrance gauge, but with suitable phase shifts and changes in amplitude so that the tidal signal produced by the model at the gauge will match that from data analysis.

For large bays the real-time wind data fields to drive the model are critical to accurate nowcasts. In some bays with complex geometry and/or surrounding topography with high elevations, these wind fields must have a high enough resolution or the wind directions will not be accurate enough. And these fields must be based on at least some over-water wind observations or the wind speeds will be too low. The bottom line is to obtain as much real-time wind data as possible, with some over-water locations. At that point, the next critical question is how to carry out the spatial interpolation to create an accurate wind field. This may be some straight forward statistical technique, or a kinematic approach, or even a dynamic approach.

Improving the Input Forecast Fields. Analogous to the statement above made for nowcasts, no matter how good a bay model is, if the meteorological and open boundary oceanographic forecasts input into the model are bad then the forecasts produced by the bay model will be bad.

For all bays models, no matter how large or small the bay, the forecast water level condition at or just outside the entrance will be critical. For small bays the subtidal water level signal from wind blowing over the continental shelf will be the dominant nontidal signal in the bay. This means that the coastal/shelf forecast model must do a good job, and this opens up a whole series of similar accuracy considerations for that model.

To produce a good forecast the coastal model must start from good initial conditions, which means that real-time data fields over the coastal ocean become important. For good coastal water level nowcasts, this means producing good analyzed wind fields from available in situ and remote wind data and probably the assimilation of coastal water level data. For good offshore currents, temperature, and salinity, the model will require the assimilation of satellite altimetric sea surface heights and sea surface temperatures. Another issue is whether the coastal model must extend into a major bay (however coarsely), or, if it does not extend into the bay, how far the influence of the bay extends out into the coastal model regime and the size of the resulting errors near the entrance. This then affects how far the bay model's open boundary condition must extend out over the shelf.

To produce a good forecast the coastal model must be driven by good wind and heat flux forecasts from a large-scale atmospheric forecast model. For large bays, forecast wind fields over the bay will be critical. This may require higher-resolution fields from a regional mesoscale weather forecast model for the bay. This model will get its boundary conditions from the same large-scale weather forecast model that drives the coastal ocean forecast model, and its 3-D initial conditions from the real-time meteorological field system used in the nowcast part of the system.

Obtaining improved forecasts from other forecast models is the most difficult part of developing a regional nowcast/forecast model system, both because of the complex problems involved (esp. in forecasting accurate wind fields using the atmospheric models), and because one may have little control over those other forecast models. Sometimes the ocean forecast models, using output from the weather forecast models, can provide useful feedback that may help improve the weather models. If the forecasts from these other models are not accurate for all situations, one must find ways to determine when those less accurate situations occur so it can be reflected in the regional model's probability estimates or error bars.

Developing Data Assimilation Techniques. Given the problems in obtaining real-time data to drive a bay model (in particular, real-time over-bay wind fields and real-time "entrance" forcing), the assimilation of real-time data from the model's interior becomes a potentially important way to improve model nowcasts (as well as the forecasts that start from those nowcasts). Most data assimilation studies have so far been with coastal models (or global models), in particular the assimilation of sea surface heights (SSH) from satellite altimetry data and the assimilation of sea surface temperatures (SST) from satellite AVHRR data. However, the *time scales of the mesoscale circulation* that these data help a coastal model maintain are much longer than the *weather and tidal time scales* of concern in a bay. (It must also not be forgotten that a coastal model also has this second shorter time scale, and that real-time wind field information over the continental shelf is certainly inadequate, so that some time of assimilation of the water level data from coastal gauges should help improve the coastal model's predictions of water level and currents.)

There appears to be virtually no experience with data assimilation on these shorter times scales in a bay model (or in a coastal model). Not all the lessons learned so far in assimilating SSH and SST data into those coastal models used to predict mesoscale phenomena may be applicable, but certainly many of the same mathematical techniques will be tried. Whatever technique we try, we are essentially looking for ways to use the model as a dynamic spatial interpolator amongst the locations with real-time data.

One normally starts with the simplest method, which is also usually the "cheapest" one computationally, and this an especially important consideration in an operational system where timeliness is critical. Thus, some type of "nudging" technique will be tried, whereby the model is constrained to be correct at those grid points nearest the locations where real-time data are available (and is nudged toward those values over a certain number of time steps). This is essentially *forcing* the model at each of these locations with a signal that is the difference between the model prediction and the real-time data. Even if this is done without numerical instability problems, there is still the question of how these forced-difference signals will propagate to the rest of the model grid points. In the case of water levels, for example, where these differences may be due to some combination of inadequate wind field forcing and poor entrance forcing, how likely is it that these difference signals will propagate in way that will emulate improved wind fields or improved entrance forcing? Of course, the more locations there are with real-time data the more likely it is that nudging could work, since those difference-signals won't have to propagate very far.

More likely to work perhaps is a technique that focuses on the suspected inadequacies of particular forcing data and data fields, and which uses the real-time data at all the locations to provide the means to modify these forcing fields to get a closer match between model predictions and data. For example, to improve model-predicted nontidal water levels, one approach might be to vary the wind fields over the bay and do multiple model runs, and to use some type of technique to minimize the least-squares difference between the nowcast nontidal water levels and the real-time nontidal water levels (i.e. real-time water level data minus tide predictions) at the gauges inside the bay. Likewise, the entrance forcing could be varied, or both the entrance forcing and the wind field forcing could be varied together..

There are a variety of related mathematical techniques to try (see Malanotte-Rizzoli and Tziperman, 1996), including least-squares inverse methodology, optimal interpolation, Kalman filtering, as well as other less rigorous/optimal (but less costly) techniques. The creativity may come in modifying one of these techniques in a way that can take advantage of our knowledge of the likely inadequacy of the forcing fields, so that a reasonably efficient and successful methodology can be devised.

If we are trying to nowcast salinity and temperatures, we are much more likely to have better real-time measurements inside the bay than outside the bay entrance. But poor real-time entrance temperature and salinity forcing data is only a small part of this problem, since circulation and mixing are so critical and these go back to problem of inadequate wind and entrance water level forcing (assuming we have the tidal part right).

The one real-time parameter that probably will not be useful for data assimilation is current data, since it varies significantly, often dramatically, over short spatial distances horizontally. (Even for model verification purposes, currents must be used with caution in areas with changing bathymetry and shoreline, since they can vary significantly over the space of one model cell, even for high resolution models.)

At the very least, multiple model runs involved in any data assimilation methodology will demonstrate the sensitivity of the model to the varied forcing fields and provide a better estimate of the possible errors associated with a each nowcast, thus providing the user with an indicator of the likely accuracy of that particular nowcast. (Real-time data at a few locations can also be left out of the assimilation for comparison with the nowcasts as another indicator of their accuracy. There will be a trade off here, however, whether such data is better served as part of the assimilation or as part of the quality indication.)

Developing Ensemble Averaging Techniques. As already mentioned, forecasts from a bay model depend on forecasts from other oceanographic, meteorological, and hydrologic models, which can vary considerably in quality, and which, like the bay model, also depend on initial conditions based on data that may have been inadequate. An important questions is whether there is anything that can be done to improve this situation.

Ensemble averaging is routinely used with weather forecast models to try to improve weather forecasts. Ensembles of 30-50 forecasts are typically used, with each forecast based on slightly different initial conditions, selected according to some chosen technique (for example, based on errors that would grow the fastest over one or two days, or based on errors from earlier forecasts). Of course, this is done at a greater cost and demands the fastest computers, and even then requires particular compromises, such as a decrease in the complexity of the model to save computer time.

To use an ensemble averaging technique to try to improve forecasts from a bay model, one could take an approach similar to the weather forecast models, i.e. run the model numerous times with differing initial conditions. One way to obtain these different initial conditions would be to use the different nowcasts created during the data assimilation procedure mentioned above. This would, however, still ignore the forecast forcing conditions coming from other models, which may cause more errors than the initial conditions from the bay model. If a regional weather forecast model is being used to provide over-bay forecast winds, and that model is run many times to produce an ensemble average, selected individual forecast runs might be chosen to drive the bay model. Similarly, one could vary the entrance open boundary water levels by running the coastal ocean model many times using the outputs from selected individual runs of the large-scale atmospheric model carried out during its ensemble scheme. In this case one might also include persisted or statistically forecast water levels.

Of course, all these different varying forecast parameters could lead to an overwhelming number of combinations and too many computer runs to accomplish in the allotted time. Sensitivities associated with these varying parameters must be understood in order to limit the computer runs to those that have the most important effect on the bay model forecasts. As with the use of data assimilation in producing nowcasts, the inherent sensitivity analyses associated with ensemble averaging would also provide estimates of error bars associated with the final forecast, which as said many times, is important for the user of these forecasts.

NOWCAST/FORECAST MODEL SYSTEMS UNDER DEVELOPMENT AT CSDL/NOS

The nowcast/forecast modeling efforts in CSDL presently include three regional projects in Chesapeake Bay, Galveston Bay, and the Port of New York and New Jersey (NY/NJ), as well as two larger-scale coastal projects (the entire U.S. East Coast, and the Gulf of Mexico, including individual shelf models) to provide forecast open boundary conditions for the regional model systems. Other papers in this volume provide more details about the model systems being developed in these projects (see Bosley and Hess; Schmalz; O'Connor, et al; Schultz and Aikman; and Kelley, et al). Two of the regional projects (NY/NJ and Galveston Bay) are at locations where NOS has installed Physical Oceanographic Real-Time Systems (PORTS). In Chesapeake Bay real-time water level and meteorological data are available from several stations in NOS's National Water Level Observation Network (NWLON). Two other PORTS locations, Tampa Bay and San Francisco Bay, also have nowcast modeling efforts (at the University of South Florida, and at the U.S. Geological Survey, respectively) discussed in papers in this volume (see Vincent, et al; and Cheng).

With respect to nontidal forcing, what the three CSDL regional project areas have in common is a connection to a shelf that is wide and fairly shallow (versus the U.S. West Coast where the shelf is narrower and deeper), so that wind effects are important. These three areas cover a range of dynamic situations. NY/NJ is fairly small in area and has a relatively wide entrance. Its subtidal water level variations are dominated by the wind-induced signal propagating in from the shelf. Chesapeake Bay also has a wide entrance, but it is so long that winds blowing over the Bay account for roughly half the nontidal variation in water level. Galveston Bay is larger in area and shallower than NY/NJ, has a much smaller entrance, and it has a length-to-depth ratio that is almost as large as for Chesapeake Bay. With regard to river discharge into the bay, Galveston Bay has a larger "flow ratio" (proportion of fresh water entering the bay during a tidal cycle divided by the tidal prism) during its average *low* flow period than Chesapeake Bay or NY/NJ have during their average *high* flow period (but it is vertically homogeneous all the time because of its very shallow depths).

In all these projects many of the special problems of a nowcast/forecast model system, mentioned in this paper, are being worked on. Automated model systems have been implemented and tested, with nowcasts and 24-hour forecasts presently produced (at least) twice daily and graphically portrayed on a restricted-access WWW Home Page for evaluation purposes.

The Chesapeake Bay and NY/NJ models use forecast open boundary conditions from an East Coast model (See Figure 1), presently either: (1) the Coastal Ocean Forecast System (COFS) developed jointly by CSDL, NWS's National Centers for Environmental Prediction (NCEP), and Princeton University, and driven by forecast wind fields from NWS's 29-km ETA weather forecast model (Kelley et al, 1997), or (2) the extra-tropical surge model developed by NWS's Techniques Development Lab and driven by forecast wind fields from NWS's Aviation weather forecast model. The Galveston model system will initially use an in-house model of the Louisiana-Texas (LATEX) shelf, which will be driven by ETA-29 wind fields.

For real-time wind fields over Galveston Bay, Texas A&M is testing a basic statistical spatial interpolation technique (used previously for winds over the LATEX shelf) using all available wind observations around the Bay. CSDL is also testing a two-step Barnes technique. For real-time wind fields over Chesapeake Bay a more ambitious approach is being implemented with a number of partners from inside NOAA, i.e. the Forecast Systems Laboratory (FSL) and the Air Resources Laboratory (ARL) in the Office of Oceanic and Atmospheric Research (OAR), NCEP (in NWS), and two NWS Forecast Offices (at Sterling and Wakefield, Virginia). The Local Analysis and Prediction System (LAPS), developed by FSL, is being implemented for the Chesapeake Bay region on a workstation at the Sterling WFO. LAPS is a all-purpose meteorological data ingest system which accepts all types of in-situ and remotely sensed data and produces 3-D fields in real time. LAPS also serves as a platform for a regional weather forecast model, which for this project will be a high-resolution workstation version of NWS's ETA model implemented by NCEP and ARL. The standard operational 29-km ETA model will provide the boundary conditions, and the initial conditions will come from the LAPS nowcast fields. (Another approach for producing nowcast wind fields is being tried in San Francisco by USGS; see Cheng in this volume.)

A number of other universities are involved in various aspects of these projects, including the University of Maryland (looking at in-bay data assimilation techniques, and the lead partner for the real-time Chesapeake Bay Observing System [CBOS]), Old Dominion University (also a partner in CBOS, and also modeling Chesapeake Bay), and the Virginia Institute of Marine Sciences (the third CBOS partner, and interested in using the forecast Chesapeake winds for wave forecasting).

With the proof-of-concept phase of these projects completed, the major effort is now going toward improving the skill of the nowcasts and forecasts. The initial improvements to the a particular model system depend on the particular dynamic situation. As mentioned above, for Chesapeake Bay, where over-bay winds are critical, and where ARL has shown that high-resolution in a mesoscale model is necessary in order to get the wind directions right, the LAPS system initializing a high-resolution ETA model is being implemented. Also, to improve the nowcasts, additional water level sensors are being installed outside the entrance to the Bay, since the gauge at the Chesapeake Bay Tunnel near the entrance is probably not providing a good open boundary condition for the nontidal signal coming in off the shelf. New York Harbor has a complex geometry with important navigation channels (important from both a navigation point of view, and from a hydrodynamic point of view). That project is running hourly nowcasts with a 3-D model (the Princeton Ocean Model), and although run in a barotropic mode, it still has showed significant improvements over the 2-D version. Techniques for improving resolution in the navigation channels are being worked on with Princeton. For Galveston Bay, along with the real-time wind fields work mentioned above, real-time and forecast river discharge boundary conditions will be obtained from the local NWS River Forecast Office. Data assimilation and forecast ensemble averaging have not yet been tried, but will be eventually.

REFERENCES

Kelley, John G.W., Frank Aikman III, Laurence C. Breaker, and George L. Mellor, 1997. Coastal ocean forecast. *Sea Technology*, **38**(5), pages 10-17.

Malanotte-Rizzoli, P. and E. Tziperman, 1996. The oceanographic data assimilation problem: overview, motivation and purposes. In: Modern Approaches To Data Assimilation In Ocean Modeling, P. Malanotte-Rizzoli (Ed.), Elsevier Science, New York, pp 3-17.

Parker, Bruce B., 1996. Monitoring and modeling of coastal water in support of environmental preservation. *J. of Marine Science and Technology*, vol. **1**, pages 75-84.

Real-Time Predictions of Surge Propagation

Norman W. Scheffner[1], Member ASCE
and Patrick J. Fitzpatrick[2]

Abstract

This paper describes the development and implementation of a procedure for forecasting the propagation of a hurricane surge. The approach requires the successive application of three components. The first is an interactive input interface in which the user obtains an updated marine advisory forecast from web site locations such as that of the National Hurricane Center. The known and projected locations of the storm are then used to define input to the second component of the procedure, a Planetary Boundary Layer (PBL) hurricane model.

The PBL model employs the vertically-averaged primitive equations of motion for predicting hurricane-generated wind and atmospheric pressure fields which are computed at each node in a translating uniform grid and interpolates them onto the computational grid of the third component of the procedure, the depth-integrated finite element long-wave hydrodynamic ADvanced CIRCulation (ADCIRC) model. The PBL-based wind and pressure fields are input to the ADCIRC model to compute the propagating storm surge over the documented and projected track of the storm. Surface elevation and currents corresponding to the last known location of the hurricane eye are archived in "hot start" files. Projected track information are then used to estimate surge propagation from the last known location of the eye. New simulations can then made for each updated forecast.

Introduction

Real-time predictions of the hurricane track are currently made by the National Hurricane Center (NHC). However, predictions of the storm surge accompanying the

[1]Senior Research Hydraulic Engineer, U.S.Army Engineer Waterways Experiment Station, Vicksburg, MS 38180-6119
[2]Assistant Professor of Meteorology, Jackson State University, Jackson, MS 39217-0460

event of interest are based on results of the finite difference long-wave model SLOSH (Jelesnianski et al. 1992). These prediction are made from precomputed surge elevation data bases for specified regions based on approximations of storm characteristics such as: 1) landfall basin, 2) direction of propagation, 3) forward speed, and 4) storm severity. Although these results may be ideal for long-range (more than 24 hours from landfall) forecasts of surge elevation potential, they may not be realistic for estimating the spatial distribution of storm surge for a specific event as it approaches land or makes landfall. This is especially true if the track of the event rapidly changes direction and intensity prior to impacting the coast.

This paper presents a forecast application of the long-wave hydrodynamic model ADCIRC for predicting the spatial and temporal distribution of tropical events based on the actual prediction of storm path and intensity. An example application of this capability is given for Hurricane Danny as it made landfall in the vicinity of Mobile Bay, Alabama in July 1997.

The simulation procedures reported in this paper require only the known and estimated location of the eye and central pressure (or maximum wind) of the hurricane as input to the PBL hurricane model. The PBL model then generates wind and pressure fields which are input to the long wave model ADCIRC. The modeling procedure is to simulate the propagation of the storm surge over the documented storm track and archive elevation and current data into "hot start" files at that known hurricane eye location. Projected track forecasts are then used to estimate surge propagation from the last documented location. As the updated position of the hurricane eye is documented, a new hot start file is archived. New simulations can then be made for every updated hurricane forecast in a "leapfrog" fashion.

Because the projection of the hurricane surge ultimately depends on the hydrodynamic model which requires input from the hurricane model, the components of the prediction methodology will be presented in reverse order. Therefore, the following sections briefly describe: 1) the ADCIRC model, 2) the PBL model, and 3) the interactive interface between the NHC track predictions and the PBL/ADCIRC model. Finally, an example simulation for Hurricane Danny is presented.

It should be noted that the model is not tuned to optimize the reproduction of storm surges for specific locations. The values given by the NHC are directly used as input to the PBL model from which wind and pressure are used to compute a surge distribution. Although some simplifications are introduced in this procedure which may introduce error (i.e., assumed symmetrical storm and estimated radius to maximum winds), it is felt that the approach of modeling the actual track and minimum pressure corresponding to the event is more accurate than estimating surge based on idealized segments of the actual event.

The ADCIRC Model

The ADCIRC model was initially developed under the Dredging Research Program (DRP), a 6-year program funded by the Office of the Chief of Engineers (Griffis et al. 1995). The model was developed as a family of 2- and 3-Dimensional finite element based codes (Luettich et al., 1992; Westerink et al., 1992) with the capability of:

a. Simulating tidal circulation and storm surge propagation over very large computational domains while simultaneously providing high resolution in areas of complex shoreline and bathymetry. The targeted areas of interest include continental shelves, nearshore areas, and estuaries.

b. Properly representing all pertinent physics of the 3-dimensional equations of motion. These include tidal potential, Coriolis, and all nonlinear terms of the governing equations.

c. Provide accurate and efficient computations over time periods ranging from months to years.

In two dimensions, model formulation begins the depth-averaged shallow water equations for conservation of mass and momentum subject to incompressibility, Boussinesq, and hydrostatic pressure approximations. Using the standard quadratic parameterization for bottom stress and neglecting baroclinic terms and lateral diffusion/dispersion effects, the following set of conservation statements in primitive, non-conservative form, expressed in a spherical coordinate system, are incorporated in the model (Flather, 1988; Kolar et al., 1993):

$$\frac{\partial \zeta}{\partial t} + \frac{1}{R \cos\phi} \left[\frac{\partial UH}{\partial \lambda} + \frac{\partial (UV\cos\phi)}{\partial \phi} \right] = 0 \qquad (1)$$

$$\frac{\partial U}{\partial t} + \frac{1}{r \cos\phi} U \frac{\partial U}{\partial \lambda} + \frac{1}{R} V \frac{\partial U}{\partial \phi} - \left[\frac{\tan\phi}{R} U + f \right] V =$$

$$-\frac{1}{R \cos\phi} \frac{\partial}{\partial \lambda} \left[\frac{p_s}{\rho_0} + g(\zeta - \eta) \right] + \frac{\tau_{s\lambda}}{\rho_0 H} - \tau_* U \qquad (2)$$

$$\frac{\partial V}{\partial t} + \frac{1}{r\cos\phi}U\frac{\partial V}{\partial \lambda} + \frac{1}{R}V\frac{\partial V}{\partial \phi} - \left[\frac{\tan\phi}{R}U + f\right]U =$$

$$-\frac{1}{R}\frac{\partial}{\partial \phi}\left[\frac{p_s}{\rho_0} + g(\zeta-\eta)\right] + \frac{\tau_{s\phi}}{\rho_0 H} - \tau_* V \qquad (3)$$

where t represents time, λ,ϕ are degrees longitude (east of Greenwich is taken positive) and degrees latitude (north of the equator is taken positive), ζ is the free surface elevation relative to the geoid, U,V are the depth averaged horizontal velocities, R is the radius of the Earth, $H = \zeta + h$ is the total water column depth, h is the bathymetric depth relative to the geoid, $f = 2\Omega \sin \phi$ is the Coriolis parameter, Ω is the angular speed of the Earth, p_s is the atmospheric pressure at the free surface, g is the acceleration due to gravity, η is the effective Newtonian equilibrium tide potential, ρ_0 is the reference density of water, $\tau_{s\lambda},\tau_{s\phi}$ are the applied free surface stress, and τ_* is given by the expression $C_f(U^2 + V^2)^{1/2}/H$ where c_f equals the bottom friction coefficient.

The momentum equations (Equations 2 and 3) are differentiated with respect to λ and τ and substituted into the time differentiated continuity equation (Equation 1) to develop the following Generalized Wave Continuity Equation (GWCE):

$$\frac{\partial^2 \zeta}{\partial t^2} + \tau_0\frac{\partial \zeta}{\partial t} - \frac{1}{R\cos\phi}\frac{\partial}{\partial \lambda}\left[\frac{1}{R\cos\phi}\left(\frac{\partial HUU}{\partial \lambda} + \frac{\partial(HUV\cos\phi)}{\partial \phi}\right) - UVH\frac{\tan\phi}{R}\right]$$

$$\left[-2\omega\sin\phi HV + \frac{H}{R\cos\phi}\frac{\partial}{\partial \lambda}\left(g(\zeta-\alpha\eta) + \frac{p_s}{p_0}\right) + \tau_* HU - \tau_0 HU - \frac{\tau_{s\lambda}}{\rho_0}\right]$$

$$-\frac{1}{R}\frac{\partial}{\partial \phi}\left[\frac{1}{R\cos\phi}\left(\frac{\partial HVV}{\partial \lambda} + \frac{\partial HVV\cos\phi}{\partial \phi}\right) + UUH\frac{\tan\phi}{R} + 2\omega\sin\phi HU\right]$$

$$-\frac{1}{R}\frac{\partial}{\partial \phi}\left[\frac{H}{R}\frac{\partial}{\partial \phi}\left(g(\zeta-\alpha\eta) + \frac{p_s}{\rho_0}\right) + (\tau_*-\tau_0)HV - \frac{\tau_{s\phi}}{\rho_0}\right]$$

$$-\frac{\partial}{\partial t}\left[\frac{VH}{R}\tan\phi\right] - \tau_0\left[\frac{VH}{R}\tan\phi\right] = 0 \qquad (4)$$

The ADCIRC-2DDI model solves the GWCE in conjunction with the primitive momentum equations given in Equations 2 and 3.

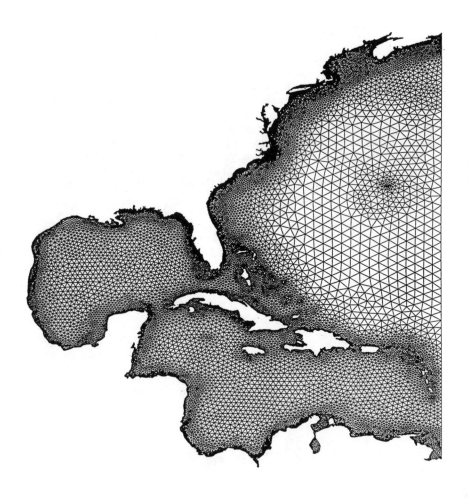

Figure 1 East Coast, Gulf of Mexico, and
Caribbean Sea Computational Grid

The ADCIRC model was rigorously tested to demonstrate proficiency in accomplishing the three goals specified above for model development. These goals were accomplished by first demonstrating accuracy in reproducing analytical solutions of a quarter annulus flow regime and a steady-state wind setup and release slosh test. Next, the model was applied to the North Sea Benchmark data set (Werner and Lynch 1988). Results showed the model to be capable of acceptably reproducing elevations and currents at 11 elevation stations and 8 velocity locations (Luettich et al. 1992) for an 11 tidal constituent based surface elevation boundary condition.

Finally, a proof of concept application was made to the entire Gulf of Mexico in which the model was used to successfully replicate surface elevations and currents for both tides and storm surges. Success of the Gulf of Mexico application led to the development of the East Coast, Gulf of Mexico, and Caribbean Sea computational grid shown in Figure 1. The grid has a minimum node-to-node resolution of approximately 800 m resulting in a Courant criteria-based time step of 30 sec. Flexibility of the unstructured grid can be quantified by the ratio of maximum element area to minimum element area of over 27,000. This 30,000 node grid represents the basis for the storm surge prediction capability described in this paper. In addition to the grid shown in Figure 1, computational grids for the eastern North Pacific Ocean from the Aleutian Islands to Peru, Samoa, the Hawaiian Islands, the Yellow Sea and Sea of Japan, and the South China Sea and Indonesia have been prepared and linked to the hurricane model.

The PBL Model

The wind field model used in conjunction with the ADCIRC model is that developed by Cardone (Cardone, Greenwood, and Greenwood 1992). This model simulates hurricane-generated wind and atmospheric pressure fields by solving the equations of horizontal motion which have been vertically averaged through the depth of the planetary boundary layer. Additionally, a moving coordinate system is defined such that its origin always coincides with the moving low pressure center of the eye of the storm p_c. Therefore, the standard equations of motion are transformed into the following relationships in Cartesian coordinates:

$$\frac{\partial u}{\partial t} + u\frac{\partial u}{\partial x} + v\frac{\partial u}{\partial y} - fv = \frac{1}{\rho}\frac{\partial p_c}{\partial x}\frac{\partial}{\partial x}\left[K_H\left(\frac{\partial u}{\partial x} + \frac{\partial v}{\partial y}\right)\right] + \frac{C_D}{h}|V|u$$

$$\frac{\partial v}{\partial t} + u\frac{\partial v}{\partial x} + v\frac{\partial v}{\partial y} + fu = \frac{1}{\rho}\frac{\partial p_c}{\partial y}\frac{\partial}{\partial y}\left[K_H\left(\frac{\partial u}{\partial x} + \frac{\partial v}{\partial y}\right)\right] + \frac{C_D}{h}|V|v$$

where (u,v) are the wind speeds in (x,y) directions, ρ is the mean air density, p_c is the pressure field representing the tropical cyclone, K_H is the horizontal eddy viscosity coefficient, C_D is the drag coefficient, h is the depth of the planetary boundary layer, and V is the magnitude of the wind velocity. The model includes parameterizations of the momentum, heat, and moisture fluxes together with surface drag and roughness formulations.

An exponential pressure law is used to generate a circularly symmetric pressure field situated at the low pressure center of the storm:

$$p_c(r) = p_0 + \Delta p e^{-(R/r)}$$

where p_0 is the pressure at the center or eye of the storm, $\Delta p = p - p_0$ is the pressure anomaly with p taken as an average background or far field pressure, R is the scale radius, often assumed equivalent to the radius to maximum wind, and r is the radial distance outward from the eye of the storm.

Wind speeds generated by the model are converted to surface wind stresses using the following relationship proposed by Garratt (1977):

$$\frac{\tau_x}{\rho_0} = C_D \frac{\rho_{air}}{\rho_0} |V| u$$

and

$$\frac{\tau_y}{\rho_0} = C_D \frac{\rho_{air}}{\rho_0} |V| v$$

where τ_x, τ_y are the wind stresses in the x and y directions, respectively, $\rho_{air}/\rho_0 = 0.001293$ is the ratio of the air density to the average density of seawater, and C_D is the frictional drag coefficient.

The PBL model requires a series of input "snapshots" consisting of a set of meteorological parameters defining the storm at various stages in its development or at particular times during its life. These parameters include: latitude and longitude of the storm's eye; track direction and forward speed measured at the eye; radius to maximum winds; central and peripheral atmospheric pressures; and an estimate of the geostrophic wind speed and direction. The radius to maximum winds is approximated using a nomograph that incorporates the maximum wind speed and the atmospheric pressure anomaly (Jelesnianski and Taylor, 1973). Peripheral atmospheric pressures were assumed equal to 1013 millibars (mb), and the geostrophic wind speeds were specified as 6 knots and have the same direction as the storm track. The PBL model also requires a "histogram" file containing the hourly location of the eye of the storm.

The PBL model computes a stationary wind and pressure field distribution corresponding to each of the specified snapshots on a nested grid, composed of five subgrids. Each subgrid measures 21 by 21 nodes in the x- and y-directions, respectively, and the centers of all subgrids are defined at the eye of the hurricane. Although the number of nodes composing each subgrid is the same, the spatial resolution is doubled for each successive grid. For this study, the center grid with the finest resolution had an Δx and Δy grid spacing of 5 km. Incremental distances for the remaining subgrids were 10, 20, 40, and 80 km. These fixed grids translate with the propagating storm.

The hurricane translational or forward motion is incorporated into PBL model calculations by adding the forward and rotational velocity vector components. A non-linear blending algorithm is then incorporated to generate a nested grid field of wind and pressure for each hour during the life of the storm event. The location of each grid field is defined with respect to the location of the storm eye contained in the histogram file described above. These hourly wind and pressure fields are then interpolated from the PBL nested grid onto the hydrodynamic model grid and subsequently stored for use by the ADCIRC model. Although the PBL model is idealized and modifications in stress are not made for changing sea state and/or landfall, surge results have been shown to be acceptable (Mark and Scheffner 1997 and Scheffner et al. 1994). The following sections describes the interface between the NHC prediction and the PBL/ADCIRC models which leads to a predicted storm surge distribution due to the projected path.

The NHC Track Prediction Interface

Primary input to the PBL model are the snapshots of the eye of the hurricane at some specified time interval. The first level interfacing of storm location and intensity for the PBL model was developed in support of a data base of historical storm surges corresponding to 134 historical events over the east coast and Gulf of Mexico. Selection criteria, specified in Scheffner et al. (1994), included a 300 mile minimum distance from landfall to the U.S. coastline and minimum pressure less than or equal to 994 mb. The updated HUrricane DATabase (HURDAT) developed by the NHC (Jarvinen, Neumann and Davis 1988) was used to select events of interest and provide input to an interface program to generate input to the PBL model. The HURDAT summarizes all hurricane and severe tropical storm events that occurred in the North Atlantic Ocean, Gulf of Mexico, and Caribbean Sea over the period of 1886 through the present.

Information contained in the HURDAT data base includes latitude and longitude of the eye of the storm, central pressure in mb, and maximum wind speed in nautical miles per hour (knots) at 6-hr time intervals during the entire event. An example HURDAT format for Hurricane Danny is shown in Figure 2 in which the starting date is shown to be 16 July 1997 and M=12 indicates a storm duration of 12 days. The following 12 lines indicate the 6-hour value for the longitude and latitude of the eye, maximum wind speed, and central pressure of the eye beginning at 0000 Zulu.

```
48310 07/16/1997 M=12  5 SNBR= 956 DANNY        XING=1 SSS=1
48315 07/16*0000000    0    0*0000000    0    0*2740926  25 1013*2750925  30 1013*
48320 07/17*2770923   30 1011*2790920   30 1007*2830914  40 1003*2860910  50 1002*
48325 07/18*2890902   55  997*2920899   65  992*2950894  70  990*2970890  70  988*
48330 07/19*2980884   70  984*3010881   65  987*3030880  70  984*3040879  65  986*
48335 07/20*3030876   60  991*3040875   45  998*3060874  35 1001*3080874  30 1004*
48340 07/21*3100875   25 1006*3130876   20 1009*3170876  20 1010*3210872  20 1011*
48345 07/22*3290871   20 1011*3320868   20 1012*3340866  20 1013*3370863  20 1013*
48350 07/23*3400860   20 1012*3410852   20 1012*3420845  20 1012*3430837  20 1012*
48355 07/24*3440824   20 1012*3460807   20 1010*3520792  30 1004*3640767  40 1000*
48360 07/25*3750735   50  996*3860716   50  995*4000704  50  995*4070699  50  994*
48365 07/26*4070696   45  995E4040680   45  998E4060656  40 1003E4100630  40 1004*
48370 07/27E4170604   40 1004E4280560   40 1004E4400480  30 1005*0000000   0    0*
48375 HR LA1 AL1
```

Figure 2 HURDAT Format for Hurricane Danny

Development of a tropical storm surge prediction capability therefore requires only the development of a systematic procedure for accessing information on the projected storm path and generating an exact HURDAT formatted file for input to the PBL model. The first step in performing this real-time storm surge forecast is retrieving the NHC marine advisory forecast from their web site:

http://www.nhc.noaa.gov

or through a private weather vendor who provides the National Weather Service "Family of Service" products. The marine advisory forecast, which is issued every 6 hours at 0300, 0900, 1500, and 2100 Zulu time, contains the NHC intensity and track forecasts out to 72 hours. The track forecasts contain projected latitudes and longitudes, as well as intensity (defined as the maximum sustained wind speed in knots). An example advisory forecast for Tropical Storm Danny is shown below:

WTNT24 KNHC 171443
TCMAT4
TROPICAL STORM DANNY FORECAST/ADVISORY NUMBER   4
NATIONAL WEATHER SERVICE MIAMI FL   AL0497
1500Z THU JUL 17 1997

AT 10 AM...CDT...1500 UTC...A TROPICAL STORM WARNING AND A HURRICANE WATCH ARE IN EFFECT FROM CAMERON LOUISIANA TO ORANGE BEACH IN EASTERN ALABAMA.

TROPICAL STORM CENTER LOCATED NEAR 28.3N 91.9W AT 17/1500Z POSITION ACCURATE WITHIN 60 NM

PRESENT MOVEMENT TOWARD THE NORTHEAST OR 35 DEGREES AT

4 KT

ESTIMATED MINIMUM CENTRAL PRESSURE 1007 MB
MAX SUSTAINED WINDS 45 KT WITH GUSTS TO 55 KT
34 KT....... 30NE 60SE 30SW 0NW
12 FT SEAS.. 30NE 60SE 30SW 0NW
ALL QUADRANT RADII IN NAUTICAL MILES

REPEAT...CENTER LOCATED NEAR 28.3N 91.9W AT 17/1500Z
AT 17/1200Z CENTER WAS LOCATED NEAR 28.1N 92.2W

FORECAST VALID 18/0000Z 28.7N 91.7W
MAX WIND 50 KT...GUSTS 60 KT
50 KT... 25NE 25SE 0SW 0NW
34 KT... 50NE 100SE 30SW 0NW

FORECAST VALID 18/1200Z 29.3N 91.0W...INLAND
MAX WIND 55 KT...GUSTS 65 KT
50 KT... 25NE 25SE 0SW 0NW
34 KT... 50NE 100SE 30SW 0NW

FORECAST VALID 19/0000Z 30.0N 90.5W...INLAND
MAX WIND 30 KT...GUSTS 40 KT

REQUEST FOR 3 HOURLY SHIP REPORTS WITHIN 300 MILES OF 28.3N
91.9W

EXTENDED OUTLOOK...USE FOR GUIDANCE ONLY...ERRORS MAY BE
LARGE

OUTLOOK VALID 19/1200Z 31.0N 90.0W...INLAND
MAX WIND 25 KT...GUSTS 35 KT

OUTLOOK VALID 20/1200Z 33.5N 89.0W...INLAND
MAX WIND 25 KT...GUSTS 35 KT

NEXT ADVISORY AT 17/2100Z

AVILA

Currently each forecast is downloaded off the NHC homepage using an automated C
program. This retrieval process produces a directory of files, each representing a
separate forecast issued every 6 hours.

The next step is "parsing" out the necessary track and intensity forecast locations from the family of advisory files. This is accomplished by using a Practical Extraction and Report Language (PERL) script. PERL is a programming language optimized for scanning arbitrary text files and extracting information from those text files. The NHC advisories contain consistent phrasing, for example, current intensity always follows the phrase "MAX SUSTAINED WINDS," and predicted intensity always follows the phrase "MAX WIND." Therefore, latitude, longitude, and intensity, can be easily extracted, reformatted, and used to generate HURDAT formatted input files.

The PERL script parses out the track and intensity data from the total set of advisory files searching for current observations ("PRESENT LOCATION") and forecast "FORECAST VALID" data. The result is a single file containing known location and maximum speed data followed by forecast location data. If an estimate of central pressure is not given in the marine advisory, central pressure is estimated from the NHC prediction for maximum velocity according to the following relationship:

$$p_0 = 1013.0 - (0.1478\, V_{max})^{1.55}$$

where 1013.0 is the standard atmospheric pressure in millibars and V_{max} is the maximum wind speed from the NHC advisory.

This file of known/forecast location, speed, and pressure data are then interpolated to full days corresponding to 0000, 0600, 1800, and 2100 hours using a an Akima spline routine. Because this time increment corresponds exactly with the HURDAT/PBL interfacing software, a separate program inputs the known/forecast file to generate a HURDAT format file (as in Figure 2) which is used to generate the snapshot and histogram input files for the PBL model. The PBL model automatically interpolates hourly wind and pressure data onto the ADCIRC grid and the ADCIRC model simulates the spatial and temporal distribution of surge over the entire computational domain shown in Figure 1.

Because the computational grid covers the entire east coast of the United States, Gulf of Mexico, and Caribbean Sea, forecasts may be done for any hurricane in the domain without having to modify grids or adjust boundary conditions. An example simulation can be shown for Hurricane Danny using the final path archived in the HURDAT and shown in Figure 2. Figure 3 shows the 6-hourly locations of the eye of the Hurricane and Figure 4 presents the spatial distribution of maximum surge.

Verification of computed surges were made to four prototype gage hydrographs reported by the U.S. Army Corps of Engineers, Mobile District (Scheffner 1998). Data were in the form of 5-day hydrographs of water level at four locations. Although the ADCIRC model has the capability of simulating tide in conjunction with hurricanes, the results presented here are for surge only. Therefore,

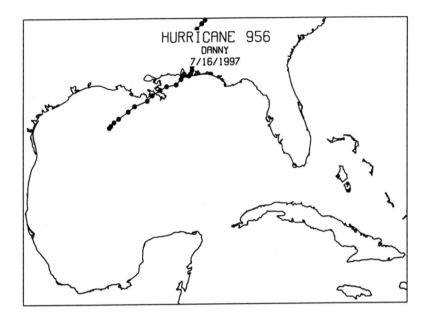

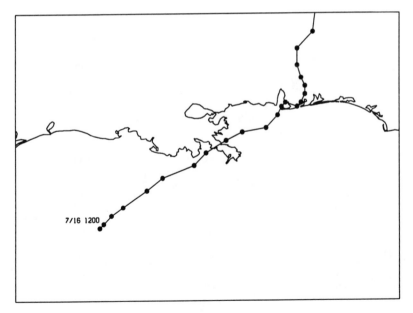

Figure 3 Track information for Hurricane Danny

the Prototype hydrographs were analyzed and the surge extracted. Table 1 shows the comparison of ADCIRC results and the comparable measured surge. As noted above, no adjustments in bathymetry, shoreline orientation, wave setup, or windfields were made for the ADCIRC simulations, therefore the results shown in Table 1 and Figure 4 result directly from the initial simulations. Although Hurricane Danny was not intense, the simulated surges are reasonably accurate and could be used as a realistic basis for predicting storm surge.

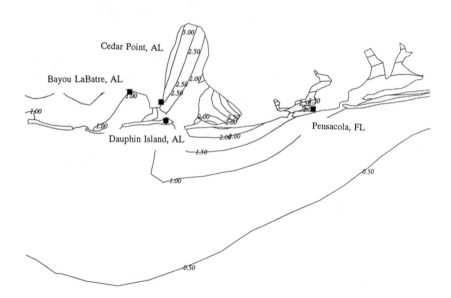

Figure 4 Spatial distribution of maximum
storm surge for Hurricane Danny

Table 1 Comparison for ADCIRC results with measurements for Hurricane Danny

Station	ADCIRC - ft, msl	Measured - ft, msl
Bayou LaBatre, AL	1.0	1.2
Cedar Point, AL	2.5	1.5
Dauphin Island, AL	1.5	1.0
Pensacola, FL	2.0	1.8

Conclusions

A systematic procedure for performing real-time estimates of tropical storm surge is presented in this paper. The approach is based on surge forecast information from the NHC marine advisory forecast or through a private weather vendor who provides the National Weather Service "Family of Service" products. These data provide input to the PBL hurricane model which provides wind and pressure input to the ADCIRC hydrodynamic model. The computational grid used by ADCIRC covers the entire east coast, Gulf of Mexico, and Caribbean Sea; therefore hurricane simulations can routinely be performed for any Hurricane reported within the North Atlantic Ocean domain of the National Hurricane Center. Hydrodynamic simulations then provide a spatial distribution of surge which is based on known and projected track and intensity information. The procedures described in this paper will be made operational for the 1998 Hurricane season.

Because the procedure is based on real-time predictions of hurricane behavior, the authors believe the predicted surge elevations represents a more accurate approach to surge predictions for near landfall conditions than the use of stored data bases of hurricanes characterized by generic combinations of parameters. The conclusion of the paper is that, although the procedures presently in use by the National Weather Service are appropriate for long range forecasts, the ADCIRC technology should be used in conjunction with the existing technology for near landfall predictions.

Acknowledgments

The research described in this paper was performed at the US Army Engineer Waterways Experiment Station and at Jackson State University. Permission to publish the material was granted by the Chief of Engineers.

References

Cardone, V.J., C.V. Greenwood, and J.A. Greenwood, Unified Program for the Specification of Hurricane Boundary Layer Winds Over Surfaces of Specified Roughness, Coastal Engineering Research Center, U.S. Army Engineers, C.R.-CERC-92-1, September, 1992.

Flather, R.A., A Numerical Model Investigation of Tides and Diurnal-Period Continental Shelf Waves along Vancouver Island, J. of Physical Oceanography, 18, pp. 115-139, 1988.

Garratt, J.R., Review of Drag Coefficients Over Oceans and Continents, Monthly Weather Review, 105, pp. 915-929, 1977.

Griffis, F.H., Jettmar, C.E., and Pagdadis, S, "Dredging Research Program Benefit

Analysis," Technal Report DRP-95-8, U.S. Army Engineer Waterways Experiment Station, Vicksburg, MS, September 1995.

Jarvinen, Brian R., Neumann, Charles J., and Davis, Mary A. S. "A Tropical Cyclone Data Tape for the North Atlantic Basin, 1886-1983: Contents, Limitations, and Uses," NOAA Technical Memorandum NWS NHC 22, March 1984.

Jelesnianski, Chester P., Chen, Jye, and Shaffer, Wilson A., "SLOSH: Sea, Lake, and Overland Surges From Hurricanes," NOAA Technical Report NWS 48, National Weather Service, Silver Spring, MD, April 1972.

Jelesnianski, C. P. and Taylor, A. D. "A Preliminary View of Storm Surges Before and After Storm Modifications," NOAA Technical Memorandum ERL WMPO-3, Weather Modification Program Office, Boulder, CO, 1973.

Kolar, R.L., W.G. Gray, J.J. Westerink, and R.A. Luettich, Shallow Water Modeling in Spherical Coordinates: Equation Formulation, Numerical Implementation, and Application, J. Hydraul. Res., in press, 1993.

Luettich, R.A., J.J. Westerink, and N.W. Scheffner, ADCIRC: An Advanced Three-Dimensional Circulation Model for Shelves, Coasts and Estuaries, Report 1: Theory and Methodology of ADCIRC-2DDI and ADCIRC-3DL, Technical Report DRP-92-6, Department of the Army, 1992.

Mark, David J. and Scheffner, Norman W., "Coast of Delaware Hurricane Stage-Frequency Analysis," Miscellaneous Paper CHL-97-1, U.S. Army Engineer Waterways Experiment Station, Vicksburg, MS, January 1997.

Scheffner, Norman W., Personnel communication with Mobile Districe personnel of surge elevations for Hurricane Danny, 1998.

Scheffner, N.W., Mark, D.J., Blain, C.A., Westerink, J.J., and Luettich, R.A., "ADCIRC: An Advanced Three-Dimensional Circulation Model for Shelves, Coasts and Estuaries Report 5: A Tropical Storm Data Base for the East and Gulf of Mexico Coasts of the United States," Technical Report DRP-92-6, U.S. Army Engineer Waterways Experiment Station, Vicksburg, MS, August 1994.

Werner, F.E., and Lynch, D.R., "Tides in the Southern North Sea/English Channel: Data Files and Procedures for Reference Computations," Dartmouth College Numerical Laboratory Report, Dartmouth, NH, 1988.

Westerink, J.J., R.A. Luettich, C.A. Blain, N.W. Scheffner, ADCIRC: An Advanced Three-dimensional Circulation Model for Shelves, Coasts, and Estuaries, Report 2: Users Manual for ADCIRC-2DDI, Coastal Engineering Research Center, U.S. Army Engineers, 1992.

A PREDICTIVE MODEL OF SEDIMENT TRANSPORT

Wilbert Lick[1] (member), Zenitha Chroneer[1], Craig Jones[1], and Rich Jepsen[1]

Abstract

Erosion rates of relatively undisturbed sediments can now be measured as a function of depth in the sediments and as a function of shear stress using a recently developed flume called Sedflume. Because of this, a sediment transport model has been developed which can utilize this information and take into account erosion rates which are highly variable in both the horizontal and vertical directions. For purposes of verifying the model, resuspension experiments were done in an annular flume at shear stresses from 1 to 8 dynes/cm^2. On the basis of erosion rate data from Sedflume, calculations were then made to predict the suspended sediment concentrations in the annular flume. Good agreement between the calculated and observed sediment concentrations was obtained and serves to partially verify the model.

Introduction

In order to predict the transport and fate of sediments in aquatic systems, the processes of erosion, settling, and deposition must be understood and quantified. For real sediments, which generally consist of particles with varying sizes and composition, these processes are generally not well understood nor accurately quantified. Because of this, numerical models of sediment transport describe these processes on the basis of empirical or semi-empirical functions; the parameters that appear in these functions are then determined by comparing the results of numerical calculations with field measurements when available, i.e., the model is

---

[1]Department of Mechanical and Environmental Engineering, University of California, Santa Barbara, CA 93106.

calibrated or fine-tuned. However, the model needs to be re-calibrated for different sites and also for different conditions at the calibrated site. In particular, it is doubtful that the model is valid for predicting sediment transport and fate during big events, e.g., large floods on rivers and large storms on lakes and oceans, simply because data for model calibration is generally not obtained during these events.

The goal of our present research is to develop a predictive model of sediment transport, i.e., a model in which the governing parameters can be determined a priori on the basis of laboratory experiments or field tests without extensive calibration or fine-tuning of these parameters. Of the three processes mentioned above (erosion, settling, and deposition), the emphasis here is on the description of erosion. Settling through the water column depends on the flocculation of particles into aggregates and the settling speeds of these aggregates. This has been described extensively (e.g., see Lick and Lick, 1988; Burban et al, 1989, 1990; Lick et al, 1993) and is reasonably well understood in principle but not yet well quantified. Deposition, especially the deposition of flocculated particles during flows, is not well understood and needs significant work for an adequate understanding.

An accurate description of erosion is especially important because of the large variations (quite often by orders of magnitude) in sediment properties and hence erosion rates that are possible in sediment beds. Erosion rates vary significantly not only in the horizontal direction (e.g., from soft, cohesive, easily erodible sediments in shallow, near-shore, slow-flowing areas of a river to coarse, non-cohesive, difficult-to-erode sediments in deeper, rapidly-flowing parts of a river) but also in the vertical direction (from unconsolidated, easily erodible sediments at the sediment-water interface to very compact, difficult-to-erode sediments at depth).

A unique flume, called Sedflume (McNeil et al, 1996), has been developed and tested which can measure sediment erosion rates as a function of depth in the sediments and as a function of shear stress for relatively undisturbed cores of real sediments as well as for reconstructed sediments. Because these measurements can now be done and this type of information on erosion rates is potentially available, a numerical model of sediment transport has been developed which can utilize this information. This model and preliminary verification of this model is discussed here.

In the following section, previous results on the measurement of erosion rates by means of Sedflume are briefly described. Examples of the large variations in erosion rates for different sites and with depth are given. In the preliminary verification of the transport model which is described later, results of resuspension experiments by means of an annular flume are used. For this reason, a brief description of the annular flume is also given in the next section. In the third section, the basic sediment transport model is described. Results on erosion rates obtained from Sedflume are then used in the sediment transport model to predict suspended sediment concentrations in the annular flume. These results are also presented and discussed in the third section. A summary and concluding remarks are given in the final section.

Sediment Erosion and Resuspension

Two devices that are used to measure erosion and resuspension properties of sediments are Sedflume and an annular flume. In Sedflume experiments, the rates at which sediment is eroded are measured directly. In contrast, in annular flume experiments, the concentration of the suspended sediments is measured; from this measurement, the net amount of sediment that is resuspended can then be determined. In the steady state, this suspended sediment concentration is a dynamic equilibrium between the rate of sediment erosion and the rate of deposition. Although the net resuspension can be determined from measurements in the annular flume, erosion rates can not be determined directly. These devices are described briefly below.

Sedflume

Sedflume is shown in Figure 1 and is essentially a straight flume which has a test section with an open bottom through which a rectangular cross-section coring tube containing sediment can be inserted. This coring tube is 1 m long and has a cross-section which is 10 cm by 15 cm. Water is pumped through the flume at varying rates and produces a turbulent shear stress at the sediment-water interface in the test section. This shear stress is known as a function of flow rate from standard pipe flow theory. As the shear produced by the flow causes the sediments in the core to erode, the sediments are continually moved upwards by the operator so that the sediment-water interface remains level with the bottom of the test and inlet sections. The erosion rate is then recorded as the upward movement of sediments in the coring tube. For more details of the apparatus and procedure, see McNeil et al (1996).

By means of Sedflume, erosion rates of relatively undisturbed sediments at high shear stresses and with depth have been measured in the Trenton Channel of the Detroit River in Michigan, the Lower Fox River in Wisconsin, Lake Michigan, Long Beach Harbor, and a dump site offshore of New York Harbor (McNeil et al, 1996; Taylor and Lick, 1996; Jepsen et al, 1997b). These tests have illustrated the large differences in sediment erosion rates (by as much as several orders of magnitude) at different sites, with depth in the sediments, and as a function of shear stress. Tests have also been done using reconstructed sediments in order to determine the effects of bulk density and particle size on erosion rates (Jepsen et al, 1997a; Roberts et al, 1997).

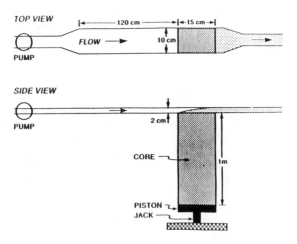

Figure 1. A schematic diagram of Sedflume.

Examples of typical measurements of erosion rates of real sediments are shown in Figures 2 and 3 (McNeil et al, 1996). The first core consisted of dark gray silt in the upper half and dark silt mixed with gray clay in the lower half. The sediments had an average particle size of 10 to 15 μm which was fairly constant with depth. The sediments were also permeated with small gas bubbles on the order of 1 mm in diameter. It can be seen that the erosion rates increase as the shear stress increases and are a strong function of depth, i.e., they are higher near the surface and decrease rapidly with depth, especially for the lower shears. This behavior is characteristic of a fine-grained cohesive sediment and is in contrast to the erosive behavior of a coarse-grained, non-cohesive sediment which is more uniform with depth.

Figure 3 illustrates the erosion rates for a stratified sediment where the properties and associated erosion rates can vary by an order of magnitude or more in distances of a few centimeters. Sharply stratified layers of clay, sand, and peat and the associated rapid changes in erosion rates are not unusual for the sediments in the areas that we have investigated.

The field and laboratory tests that we have conducted have demonstrated that erosion rates depend on at least the following parameters: bulk density (or water content), particle size distribution as well as average particle size, mineralogy, organic content, and amount and sizes of gas bubbles. Unfortunately, at the present time, it is not possible to unambiguously relate erosion rates to bulk properties, and therefore it is not possible to predict erosion rates based on a knowledge of bulk properties.

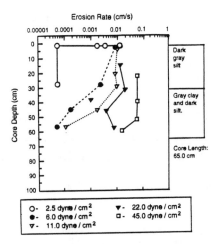

Figure 2. Erosion rate as a function of depth with shear stress as a parameter. Site in the Trenton Channel of the Detroit River in Michigan. Sediments are fine-grained and cohesive.

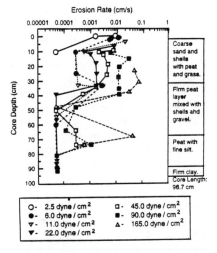

Figure 3. Erosion rate as a function of depth with shear stress as a parameter. Site in the Trenton Channel of the Detroit River in Michigan. Sediments are highly stratified.

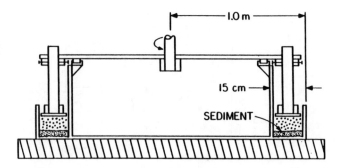

Figure 4. A schematic diagram of an annular flume.

<u>Annular Flume</u>

A schematic diagram of the annular flume used in our laboratory is shown in Figure 4. The flume is 2 m in diameter while the annulus, which contains a sediment bed and overlying water, is 15 cm wide. The water depth in the present experiments was 7.6 cm. The rotating lid produces a turbulent flow which in turn exerts a shear stress on the sediment-water interface. This shear stress causes the sediment to erode. A steady state is reached when the suspended sediment concentration is such that the rate of erosion equals the rate of deposition. The shear stresses at the sediment-water interface were determined by measuring the boundary layer velocity profiles and then deducing the shear stress from standard turbulent flow theory. For more details of the flume and procedures, see Fukuda and Lick (1980) and MacIntyre et al (1990).

Although the main flow in the flume is in the azimuthal direction, a secondary flow is also present; it is inwards near the sediment-water interface, upwards at the inner wall of the flume, outwards near the lid-water interface, and down at the outer wall. This flow is generally small by comparison with the main flow, does not significantly modify the shear stress at the sediment-water interface, but can have significant consequences as far as sediment transport is concerned as will be shown below.

Modeling Sediment Dynamics

In order to use the data from Sedflume, a layered sediment bed model was developed. In each layer of the bed, an erosion rate as a function of shear stress is specified; this erosion rate can also be specified so as to vary in the horizontal direction. In the present calculations and for the sediments used in our experiments, the erosion rates were determined by means of Sedflume and were assumed to vary only in the vertical direction. As sediment is eroded, the erosion rate at the sediment surface changes as deeper sediment layers (which generally have different erosion rates) are exposed and subsequently erode. Conversely, as sediments are deposited, new layers of sediment are formed. Because of a lack of information on the erosion rates of newly deposited sediments, these latter sediments are presently assigned an erosion rate equal to the erosion rates of the sediments that were initially at the surface. Different size classes of sediments and associated settling speeds and critical stresses for deposition can also be specified. For the present calculations, these are given below.

For purposes of verification of the model, experiments were done with identical sediments and compaction times in both Sedflume (to determine erosion rates) and in the annular flume (to determine suspended sediment concentrations). These experiments were done at shear stresses of 1, 2, 4, and 8 dynes/cm². Results for erosion rates as a function of shear stress and as a function of depth in the sediments as measured in Sedflume and as used in the calculations are shown in Figure 5. It can be seen that (a) at a particular depth, the erosion rate increases as the shear stress increases and (b) at a particular shear stress, the erosion rate decreases rapidly with depth, i.e., at a constant shear stress, the erosion rate of the surficial sediments will decrease rapidly with time as the uppermost layers erode and lower layers are exposed and subsequently eroded.

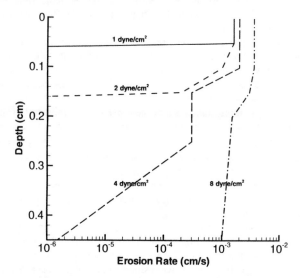

Figure 5. Erosion rates as a function of depth with shear stress as a parameter for present experiments. Shear stresses of 1, 2, 4, and 8 dyne/cm².

In the calculations presented here, five size classes of sediments were assumed, each initially containing 20% of the total mass of the sediments. On the basis of measurements, these size classes had average sizes of 2, 4, 10, 25, and 130 µm and settling speeds of 4, 16, 100, 600, and 15,000 µm/s. A critical stress for deposition, τ_{cd}, is defined such that particles of a certain size deposit only for shear stresses equal to or less than this critical value. For the above five size classes, the critical stresses for deposition were assumed to be 0, 1, 2, 5, and 10 dynes/cm^2; these values were based on qualitative observations of sediment deposition in the annular flume.

With this data, the numerical model was then used to predict the suspended sediment concentrations for the experiments in the annular flume. Much to our surprise, the calculations disagreed with the observations by one to three orders of magnitude. These differences could not be reduced significantly by fine-tuning the parameters in the model.

In order to investigate the reasons for these differences, the experiments were redone and were more closely observed. The results for the erosion rates from the Sedflume experiments and for the suspended sediment concentrations from the annular flume experiments were almost identical to those obtained previously. All features of the flow and transport in Sedflume were as expected. However, close observation of the flow and sediment transport in the annular flume demonstrated the sources of the problem.

From these observations, one reason for the differences between the calculations and observations was determined to be the following. During an experiment in the annular flume, it was observed that eroded sediments in the form of chunks or flocs tended to collect and subsequently deposit and consolidate in the flow stagnation areas where the sediment-water interface meets the side walls (see Figure 4). This was more significant at the inner wall; here, many of the flocs which were moved inwards by the secondary flow could not be convected upwards by the weak flow in the corner and tended to settle there in a volume which had a triangular cross-section. At 1 dyne/cm^2, this volume was approximately 5 mm in height, 1.5 cm in width, and had a length equal to the circumference of the inner wall. Sediments also collected in the stagnation region near the outer wall, but the amount collected there was generally much smaller than that near the inner wall.

As the shear stress increased, this volume decreased somewhat but the concentration of sediments in this volume increased so that the total amount of sediment in this volume remained approximately constant. The total mass of sediments in the stagnation regions was estimated to be equivalent to a sediment resuspension of 40 mg/cm^2.

A second reason for the discrepancy between calculations and observations is what can loosely be described as bed load, i.e., the consolidated sediments (which were relatively fine and cohesive) tended to erode in chunks which were then transported horizontally near the sediment-water interface. These chunks eventually disintegrated with time but generally caused a suspended sediment concentration near the sediment-water interface which was greater than the sediment concentration away from this interface. The sediment concentration in the middle of the water column is what is normally measured and, in the present experiments, does not give an accurate measurement of the total amount of sediment in resuspension. This effect was negligible at a shear stress of 1 dyne/cm^2, but increased with shear stress. At 8 dyne/cm^2, the mass of sediment in this bed load was estimated to be 1 to 3 times the amount of sediment collected in the stagnation regions.

A comparison of the amount of sediment eroded in the annular flume (a) as determined from measured suspended sediment concentrations, (b) corrected as described above including eroded sediment depositing in stagnation regions and bed load, and (c) as determined from numerical calculations based on Sedflume data is shown in Figure 6. Also shown is the possible range ($\pm$50%) in our estimate of the observed net resuspension. The large discrepancies between (a) and both (b) and (c) are evident. However, it can also be seen that the agreement between the corrected observations and the calculations is reasonably good.

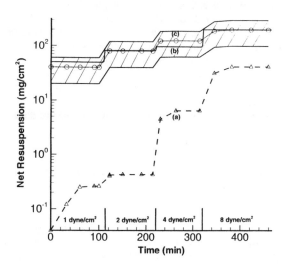

Figure 6. Net sediment resuspension as a function of time for experiments at shear stresses of 1, 2, 4, and 8 dyne/cm^2. Comparison of (a) observed, (b) corrected, and (c) calculated results. Also shown is the possible range ($\pm$50%) in our estimates of net resuspension.

Summary and Concluding Remarks

A sediment transport model has been developed which can take into account sediment erosion rates which are highly variable in both the horizontal and vertical directions. Experiments were conducted in Sedflume (in order to determine erosion rates as a function of depth) and in an annular flume (to determine suspended sediment concentrations). The experiments were done at shear stresses of 1, 2, 4, and 8 dyne/cm^2. Calculations were then done using the Sedflume data so as to predict the suspended sediment concentrations in the annular flume. The initial comparison of the calculated and observed results was poor. After correcting for sediment in the annular flume that was eroded and subsequently either deposited in stagnation regions of the flume or was transported as quasi-bed load near the sediment-water interface, the agreement between calculations and observations was quite good.

Because of the high variability of sediment erosion rates, especially with depth, the correct determination of erosion rates is essential in predicting the effects of big events (e.g., large floods on rivers and large storms in lakes and oceans) where deep layers of sediment may be exposed and subsequently eroded. A quantitative knowledge of erosion rates is also essential for the long-term prediction of sediment and contaminant transport and fate in surface waters since it is the large events which generally dominate this long-term fate.

In order to have a truly predictive model without the necessity of fine-tuning or calibrating parameters, it is necessary to not only have a knowledge of erosion rates of in-place sediments, but also to know erosion rates of depositing and newly deposited sediments. Work on this has been initiated.

The other major area in which knowledge is inadequate is the deposition of sediments during flows, especially for flows of different magnitudes and for different types of sediments. Once this knowledge is obtained, modeling of the transport and fate of sediments and associated contaminants can be done in a truly predictive manner.

Acknowledgements

This research was funded by the U.S. Environmental Protection Agency and by the University of California Campus Laboratory Collaboration Program.

References

Burban, P.Y., J. Lick, and W. Lick. 1989. The flocculation of fine-grained sediments in estuarine waters, J. Geophysical Research, Vol. 94, pp. 8323-8330.

Burban, P.Y., Y. Xu, J. McNeil, and W. Lick. 1990. Settling speeds of flocs in fresh and sea waters, J. Geophysical Research, Vol. 95, pp. 18213-18220.

Fukuda, M. and W. Lick. 1980. The entrainment of cohesive sediments in fresh water, J. Geophysical Research, Vol. 85(C5), pp. 2813-2824.

Jepsen, R., J. Roberts, and W. Lick. 1997a. Effects of bulk density on sediment erosion rates, Water, Air and Soil Pollution, Vol. 99, pp. 21-31.

Jepsen, R., J. Roberts, and W. Lick. 1997b. Long Beach Harbor sediment study, Report, Department of Mechanical and Environmental Engineering, University of California, Santa Barbara, CA.

Lick, W., H. Huang, and R. Jepsen. 1993. The flocculation of fine-grained sediments due to differential settling, J. Geophysical Research, Vol. 98, pp. 10279-10288.

Lick, W. and J. Lick. 1988. On the aggregation and disaggregation of fine-grained sediments, J. Great Lakes Research, Vol. 14, pp. 514-523.

MacIntyre, S., W. Lick, and C.H. Tsai. 1990. Variability of entrainment of cohesive sediments in freshwater, Biogeochemistry, Vol. 9, pp. 187-209.

McNeil, J., C. Taylor, and W. Lick. 1996. Measurements of erosion of undisturbed bottom sediments with depth, J. Hydraulic Engineering, Vol. 122, pp. 316-324.

Roberts, J., R. Jepsen, and W. Lick. 1997. Effects of particle size and bulk density on the erosion of quartz particles, Report, Department of Mechanical and Environmental Engineering, University of California, Santa Barbara, CA.

Taylor, C. and W. Lick. 1996. Erosion properties of great lakes sediments, Report, Department of Mechanical and Environmental Engineering, University of California, Santa Barbara, CA.

Physical Processes Affecting the
Sedimentary Environments of Long Island Sound

Richard P. Signell[1], Harley J. Knebel[1], Jeffrey H. List[1] and Amy S. Farris [1]

Abstract

A modeling study was undertaken to simulate the bottom tidal-, wave-, and wind-driven currents in Long Island Sound in order to provide a general physical oceanographic framework for understanding the characteristics and distribution of seafloor sedimentary environments. Tidal currents are important in the funnel-shaped eastern part of the Sound, where a strong gradient of tidal-current speed was found. This current gradient parallels the general westward progression of sedimentary environments from erosion or nondeposition, through bedload transport and sediment sorting, to fine-grained deposition. Wave-driven currents, meanwhile, appear to be important along the shallow margins of the basin, explaining the occurrence of relatively coarse sediments in regions where tidal currents alone are not strong enough to move sediment. Finally, westerly wind events are shown to locally enhance bottom currents along the axial depression of the Sound, providing a possible explanation for the relatively coarse sediments found in the depression despite tide- and wave-induced currents below the threshold of sediment movement. The strong correlation between the near-bottom current intensity based on the model results and the sediment response as indicated by the distribution of sedimentary environments provides a framework for predicting the long-term effects of anthropogenic activities.

Introduction

Long Island Sound is a major east-coast estuary located adjacent to the most densely populated region of the United States. Because of the enormous surrounding population, the Sound has received anthropogenic wastes and contaminants from various sources (Wolfe et al., 1991). As part of its Coastal and Marine Geology Program, the U.S. Geological Survey is conducting a regional study program designed to understand the processes that distribute sediments and related

---

[1]Woods Hole Field Center, U. S. Geological Survey, Woods Hole, MA 02543-1598. E-mail: rsignell@usgs.gov, hknebel@usgs.gov, jlist@usgs.gov, afarris@usgs.gov

contaminants in the Sound. Knowledge of the bottom-current regime is crucial both in understanding the distribution of bottom sedimentary environments in the Sound and in predicting the long-term fate of wastes and contaminants which have been introduced there.

There have been numerous observational, theoretical and modeling studies concerning the currents in Long Island Sound. Many of the observational and theoretical studies pertaining to the interaction of bottom currents with the sea floor characteristics are summarized in a series of review papers by Gordon and Bokuniewicz (Gordon, 1980; Bokuniewicz and Gordon, 1980a; Bokuniewicz and Gordon, 1980b; Bokuniewicz, 1980). In these papers, they determine that the character of the seabed is controlled primarily by tidal currents, with a lesser role played by estuarine circulation and storms. Previous modeling studies have explored the M_2 tidal response (Kenefick, 1985) and the interaction of estuarine, wind-driven and tidally-driven circulation during a realistic simulation including forcing by observed wind, heat flux, tide and river input (Schmalz, 1993; Schmalz et al, 1994). These studies indicate that the bottom currents in the eastern Sound are dominated by density-driven circulation and tidal residuals, whereas in the central and western Sound other currents are more important than the tidal residuals. In this paper, we use numerical simulations to further define the contribution of three processes that potentially move sediments: tides, wind waves and storm-induced currents.

The Study Region

Long Island Sound is located between Connecticut and Long Island, New York, on the east coast of the United States (Figure 1). It is approximately 150 km long, 30 km wide, and the average water depth is 24 m. A 30-40 m deep axial depression runs east-west through the western half of the Sound, and water depths reach more than 100 m at the eastern entrance to the Sound. The Sound opens into the offshore waters of Block Island Sound on its eastern end and is connected to New York Harbor through the East River on its western end. The system is approximately in quarter-wave resonance with the semi-diurnal tide, resulting in a threefold increase in tidal range from about 0.8 m on the eastern end to more than 2.2 m on the western end. Strong tidal currents in excess of 120 cm/s are found in the constricted eastern entrance.

Sedimentary Environments

Knebel et al. (1997) recently outlined the general distribution of modern seafloor environments in Long Island Sound, identifying four categories of environments based on an extensive regional collection of sidescan sonar data. These categories included: (1) erosion or nondeposition; (2) coarse-grained sediment sorting; (3) sediment sorting and reworking; and (4) fine-grained deposition. In the funnel-shaped eastern end of the Sound, they found a westward progression of bottom environments ranging from erosion or nondeposition at the narrow

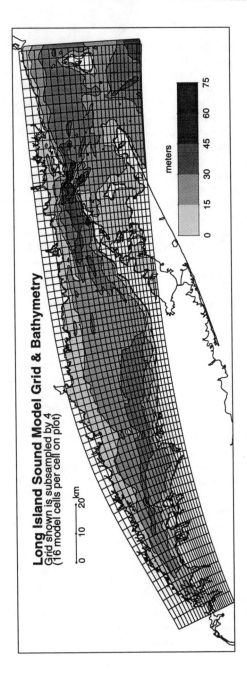

Figure 1: Long Island Sound model grid and bathymetry. The average depth is 24 m. There is a axial depression 30–40 m deep in the central/western Sound, and the maximum depth exceeds 100 m in the contricted eastern entrance to the Sound. The curvilinear model grid is subsampled by a factor of 4 for clarity. The actual grid cell sizes are between 200 and 400 m.

eastern entrance to the Sound, through an extensive area of bedload transport and sediment sorting, to a region of fine-grained deposition. The broader western Sound, on the other hand, is comprised largely of depositional environments except in local areas of topographic relief where there is a patchy distribution of various other environments. An extensive treatment of the bottom sedimentary environments in Long Island Sound is currently being completed as part of the U.S. Geological Survey regional study program (Knebel et al, 1998, submitted manuscript). Preliminary analysis indicate that winnowing of sediments occurs along the shallow margins and along some segments of the axial depression of the Sound.

Circulation Modeling

To address the bottom currents associated with tides and strong wind events, we configured a high-resolution model of Long Island Sound capable of representing topography at the 1–2 km scale. We used the Estuary, Coastal, and Ocean Model (Blumberg and Mellor, 1987) with 10 evenly spaced sigma levels and 300 x 100 grid cells in a curvilinear domain (Figure 1). This resulted in a typical grid spacing of 200–400 m over most of the Sound. The model was run with uniform density because modeling of estuarine circulation was beyond the scope of this study. For open boundary conditions at the eastern, open-ocean end, we specified elevation by M_2 tidal constituent data interpolated from a detailed finite-element tidal model of the East and Gulf Coast (Luettich and Westerink, 1995). For the western end, we specified the M_2 amplitude and phase from the NOS tidal data. We used a uniform wind stress for the simulation of wind-driven currents, and since only tidal heights were specified along the open boundary, the sub-tidal elevation was effectively set to zero. Thus, only the local wind effect was simulated. At the bottom boundary the roughness length z_0 was set to 0.67 cm, equivalent to a drag coefficient of $C_d = 0.003$ at 10 m above the bed. This value of C_d (applied to depth-averaged currents) was found to produce good results in the tidal modeling study of Kenefick (1985). The model was run for 5 tidal cycles, with results saved every 10 lunar minutes over the last cycle. An internal time step of 186.3 seconds was used, with an external time step of 9.31 seconds. The coefficient in the Smagorinsky horizontal viscosity parameterization was set to 0.05.

Tidally-Driven Bottom Currents

As an indicator of the intensity of the bottom currents driven by typical tides, the maximum bottom velocity over the course of the tidal cycle was calculated at 1 m above bottom. The results show strong bottom currents in excess of 50 cm/s in the constricted eastern end of the Sound, but the peak speed decreases westward as the width of the Sound increases (Figure 2). In general, the eastern third of the Sound has bottom tidal speeds between 30 and 60 cm/s, the central third of the Sound has speeds between 20 and 30 cm/s and the western third of the

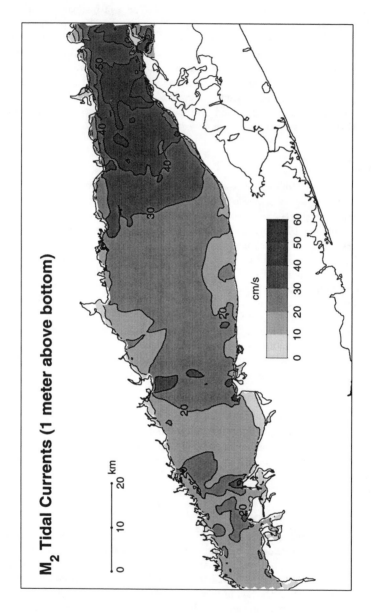

Figure 2: Maximum tidal currents 1 meter above bottom, driven by M_2 tidal forcing on the open boundaries.

Sound has speeds less than 20 cm/s. Local enhancements of bottom tidal currents exist near headlands and atop cross-Sound shoal complexes in the western Sound; in places the currents exceed 30 cm/s. There is a clear correspondence between the tidal-current distribution and the western progression of sedimentary environments in the eastern part of the Sound as outlined by Knebel et al. (1997). Here, areas of erosion or nondeposition, bedload transport, and sediment sorting occur where the bottom tidal currents exceed about 30 cm/s. These speeds are consistent with the theoretical calculations of strengths of near-bottom currents needed to move the sediments that typically compose these environments (fine sand and coarser) (Knebel et al., 1997). For a bottom roughness of 0.5 cm, the currents required at 1 m above the bottom are 27 cm/s for fine sand (0.125 mm diameter) and 37 cm/s for coarse sand (0.5 mm diameter) (e.g. Butman, 1987).

While the spatial gradients of the strength of the tidal currents explain the general distribution of sedimentary environments in the eastern Sound, the asymmetry in the ebb-flood tidal currents can give rise to small-scale residual circulation and divergence-convergence of bedload transport that can help explain the local maintenance of selected features in the Sound (Figure 3). Over the Long Sand Shoal, for example, the tide-induced bottom residual currents indicate clockwise sediment transport and convergence. This suggests a continuous mechanism for supplying sand to sustain the shoal.

Wave-Driven Bottom Currents

In addition to tidal currents, the orbital currents associated with waves generated by local winds could be a significant mechanism of bottom sediment resuspension. To better understand the resuspension potential throughout the Sound, we simulated the patterns of bottom orbital currents in the basin with the numerical wave-prediction model, HISWA (HIndcasting Shallow water WAves, Holthuijsen et al., 1989). HISWA computes steady-state wave heights on a rectangular grid over complex topography by solving the wave action balance equation. It includes the simultaneous effects of wave generation by wind, wave propagation including shoaling and refraction, and wave dissipation through bottom friction and breaking. An incoming wave may be specified as a boundary condition, although this was not used in Long Island Sound because of the nearly fully enclosed nature of the Sound.

A square computational grid was constructed with dimensions 220 x 220 km and grid spacing of 300 m in the wind direction, 600 m perpendicular to the wind direction. This grid was centered on Long Island Sound, allowing prediction of waves generated by wind from all points of the compass. We computed 144 HISWA simulations of the bottom wave orbital velocity maximum, U_b, for winds of 2.5, 5.0, 7.5, 10.0, 12.5, 15.0, 17.5, 20.0 and 22.5 m/s, each at 16 directions equally spaced around the compass.

An example of predicted U_b for winds of 15 m/s from the east-northeast

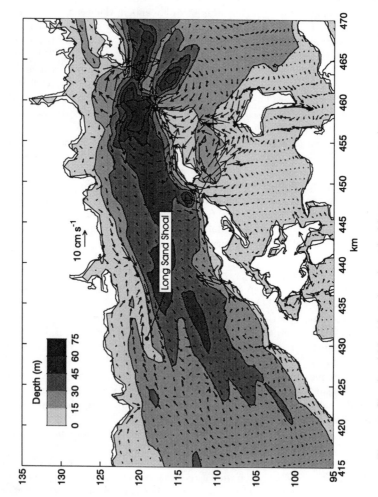

Figure 3: Simulated M_2 tide-induced residual bottom currents (1 m above bottom).

(typical of a strong winter northeaster) is shown in Figure 4. Under these strong storm conditions, the significant wave height ranges from 1.5 to 2 m, with typical periods of 4–6 seconds. The bottom velocity ranges from less than 5 cm/s in water deeper than about 20 m to more than 20 cm/s in water shallower than about 10 m, generally found within a few kilometers of the coast. The wave velocity necessary to resuspend fine-grained muds is approximately 15 cm/s (Komar and Miller, 1975). Thus wave-induced bottom velocities during strong wind events could explain the winnowing of sediments observed along the shallow margins of the Sound.

In order to calculate a long-term estimate of U_b throughout the region, the set of 144 model simulations of bottom orbital velocity were weighted with the wind distribution over a 12 year period (Nov 1984 – Dec 1996) from the NOAA Ambrose Light meteorological station. These results are presented as the percentage of time that U_b is expected to exceed values thought to represent the threshold for the initiation of sediment resuspension. An example is given in Figure 5 for a threshold of 15 cm/s. Similar to the northeasterly storm example, the percentage of time that U_b exceeds threshold values is greatest in a thin strip around the periphery of the Sound, and the threshold value is exceeded less than 0.001 percent of the time in water depths greater than about 20 m. This is consistent with estimates of wave influence by Bokuniewicz and Gordon (1980b).

Wind-Driven Bottom Currents

In addition to driving surface waves, strong wind events in the Sound generate bottom currents which may influence the distribution of sedimentary environments. Observations of low-passed (33 hour) bottom currents and winds show strong correlation at zero lag (Blumberg, 1997, personal communication); thus it is appropriate to examine the steady response of the Sound to wind. Similar to the steady wind response in a long lake (Csanady, 1973), the currents in Long Island Sound respond most efficiently to the along-axis wind component, and circulation is generally downwind in the shallows and against the wind in the deeper reaches (Figure 6). A west-southwesterly wind of 10 m/s blowing along the axis of the sound generates the strongest bottom currents along the coast in the downwind direction. In the axial depression, however, there is a local maximum of bottom current intensity directed in the upwind direction. Winds from the west drive a westward current which adds to the westward near-bottom estuarine inflow along the depression which has a magnitude of about 5 cm/s (Schmalz et al., 1994). The westward wind-driven flow also reinforces the ambient flood tidal currents of 15–20 cm/s. Thus, westward-directed currents along the axial depression can at times reach speeds of more than 30 cm/s. In contrast, storm winds from the east drive an eastward-directed bottom current that opposes the estuarine flow and, therefore, decreases the magnitudes of the currents in the depression. From analysis of the Ambrose Light wind data, westerly low-frequency wind events having wind speeds of at least 10 m/s occur about 10–20 times a

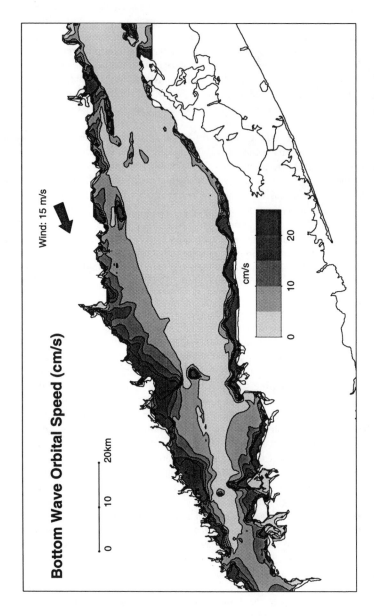

Figure 4: Simulated RMS bottom wave orbital velocities resulting from a 15 m/s east-northeasterly storm.

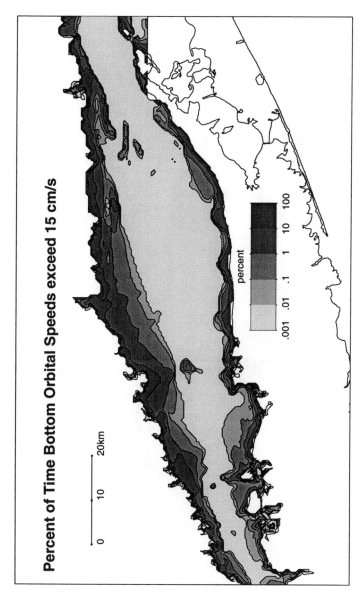

Figure 5: Percentage of time that the RMS bottom wave orbital velocities exceed 15 cm/s, based of 12 years of wind data from the NOAA Ambrose Light meteorological station.

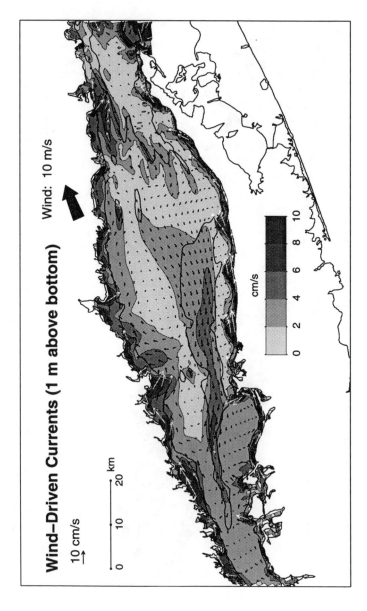

Figure 6: Simulated near-bottom currents (1 m above bottom) during a moderate west-northwesterly wind event (10 m/s).

year chiefly during the winter months. Thus, during westerly winds events and during the incoming tide, the combination of flood tidal currents, the estuarine flow, and the westward wind-driven currents may explain the observed sediment winnowing in the axial depression.

Conclusions

The results of this study provide a general framework of bottom currents in Long Island Sound. In the funnel-shaped eastern part of the Sound, the gradient of tidal-current speeds parallels a westward progression of sedimentary environments (Knebel et al., 1997). Currents here are sufficient to move sediments of fine sand and coarser and to produce coarse lag deposits in areas of erosion or nondeposition as well as winnowed finer sands in areas of bedload transport and sediment sorting. Although the tidal-current regime can explain most general aspects of the distribution of bottom environments, our modeling indicates that the tidal currents are too weak to move sediments along the nearshore margins of the Sound, and sediment transport by waves may be more important. In these shallow regions, the bottom orbital speeds associated with surface waves are strong and are sufficient to resuspend fine-grained sediments (muds) about 1–10% of the time. The frequency of sediment movement drops dramatically with water depth, and waves have essentially no effect in water depths greater than about 20 m. Westerly wind events are shown to locally enhance estuarine and tidal bottom currents along the axial depression of the Sound, providing a possible explanation for the relatively coarse sediments found in the depression. Work is continuing on the development of high-resolution models of bedload and suspended-load transport to further increase our understanding of these processes.

Acknowledgements

John Evans developed analysis and graphical tools that greatly facilitated this study. Ralph Lewis and Muriel Grim supplied us with bathymetry data that made construction of a high-resolution digital bathymetric grid possible.

References

Blumberg, A.F. and Mellor, G.L., 1987, A description of a three-dimensional coastal model, in Three-Dimensional Coastal Ocean Models, N. Heaps [ed], Coastal and Estuarine Sciences, v. 4, p. 1-16.

Bokuniewicz, H.J., 1980, Sand transport at the floor of Long Island Sound, *Advances in Geophysics*, v. 22, p. 107-128.

Bokuniewicz, H.J., and Gordon, R.B., 1980a, Storm and tidal energy in Long Island Sound, *Advances in Geophysics*, v. 22, p. 41-67.

Bokuniewicz, H.J., and Gordon, R.B., 1980b, Sediment transport and deposition in Long Island Sound, *Advances in Geophysics*, v. 22, p. 69-106.

Butman, B., 1987, Physical processes causing surficial-sediment movement, in Georges Bank, R.H. Backus [ed.], MIT Press, Chapter 13, p. 147-162.

Csanady, G.T., 1973, Wind-induced barotropic motions in long lakes, *Journal of Physical Oceanography*, v. 3, 429-438.

Gordon, R.B., 1980, The sedimentary system of Long Island Sound, *Advances in Geophysics*, v. 22, p. 1-40.

Holthuijsen, L.H., Booij, N., and Herbers, T.H.C., 1989, A prediction model for stationary, short-crested waves in shallow water with ambient currents, *Coastal Engineering*, v. 13, p. 23-54.

Kenefik, A.M., 1985, Barotropic M_2 tides and tidal currents in Long Island Sound: a numerical model, *Journal of Coastal Research*, v. 1, no. 2, p 117-128.

Knebel, H.J., Rendigs, R.R., Signell, R.P., Poppe, L.T., List, J.H. and Buchholtz ten Brink, M.R., 1997, Seafloor environments in Long Island Sound: implications for contaminant dispersal in a large urbanized estuary (abs): Geological Society of America, Abstracts with Programs, v. 29, no. 6, p. A-90.

Knebel, H.J., Signell, R.P., Rendigs, R.R., Poppe, L.J., and List, J.H., 1998, Seafloor environments in Long Island Sound, submitted to Marine Geology.

Komar, P.D. and Miller, M.C., 1975, Sediment threshold under oscillatory waves, in Proceedings 14th Conference on Coastal Engineering, American Society of Civil Engineers, New York, p. 756-775.

Luettich, R.A, Jr. and Westerink, J.J., 1995, Continental shelf scale convergence studies with a barotropic tidal model, in *Quantitative Skill Assessment for Coastal Ocean Models*, D. Lynch and A. Davies [eds.], Coastal and Estuarine Studies Series, v. 48, p. 349-371, American Geophysical Union Press, Washington, D.C.

Schmalz, R.A., 1993, Numerical decomposition of Eulerian residual circulation in Long Island Sound, *Proceedings of the Third International Conference on Estuarine and Coastal Modeling*, ASCE Press, p. 294-308.

Schmalz, R.A., Devine, M.F., and Richardson, P.H., 1994, Residual circulation and thermohaline structure, Long Island Sound Oceanography Project Summary Report, Volume 2, NOAA Technical Report NOS-OES-003, National Oceanic and Atmospheric Administration, Rockville, MD, 199 pages.

Wolfe, D.A., Monahan, R., Stacey, P.E., Farrow, D.R.G., and Robertson, A., 1991, Environmental quality of Long Island Sound: assessment and management issues, *Estuaries*, v.14, p. 224-236.

Development of an Experimental Nowcast/Forecast System for Chesapeake Bay Water Levels

Kathryn Thompson Bosley[1] and Kurt W. Hess[1]

ABSTRACT

The National Oceanic and Atmospheric Administration/National Ocean Service (NOAA/NOS) has undertaken the Chesapeake Area Forecasting Experiment (CAFE) with the goal of improving the accuracy of water levels predictions for Chesapeake Bay. Although tidal predictions based solely on astronomical forcing have been traditionally used by mariners, nontidal forcing is significant in the Bay and at times dominates. CAFE is designed to incorporate tidal and nontidal (wind stress and coastal water level) effects to produce numerical model-based nowcasts and forecasts of total water level in the Bay twice daily for the commercial and recreational maritime communities.

INTRODUCTION

Mariners operating in the Chesapeake Bay presently use traditional NOAA/NOS astronomically based tidal products as aides to navigation. However, nontidal forcing is significant in the Bay and at times completely overwhelms the tidal signal, causing safety problems for mariners who have made decisions based on the tidal predictions (Figure 1). Over 1,000 groundings were reported in just five of the Nation's estuaries (New York/New Jersey, Boston, San Francisco, Tampa, and Houston/Galveston) from 1981-1995 (Kite-Powell, 1996). In addition to the safety concern, accurate knowledge of water level provides an economic advantage as well. Each additional foot of draft available for bulk carriers and container ships results in an estimated revenue increase of between $36,000 and $288,000 per ship transit.

---

[1]NOAA/National Ocean Service, N/CS13 1315 East-West Highway, Silver Spring, MD 20910

NOS has undertaken a project named the Chesapeake Area Forecasting Experiment (CAFE) which is aimed at enhancing predictions of water level within the Chesapeake Bay (Bosley, 1996). The objective of CAFE is to produce twice daily nowcasts and forecasts of total water level in the Bay to be used by the commercial and recreational maritime community. CAFE is now running in experimental mode on NOS computers and the output is being evaluated. In the existing experimental system the nowcast mode is forced with water level observed at the Bay mouth and wind observations from two Bay locations. The forecast mode is forced with a forecast of entrance water level (Chen et al., 1993) and forecast windfields from the National Weather Service (NWS) atmospheric meso-Eta model (Black, 1994).

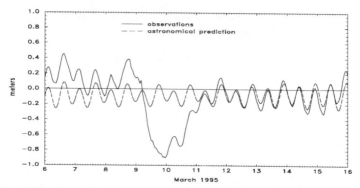

Figure 1. The observed (solid) and astronomically predicted (dashed) water level at Baltimore during a winter storm in March 1995. The rapid draining of the Harbor beginning March 9th was caused by strong northwesterly winds.

Many local users have expressed interest in CAFE. Both the Maryland and Virginia Pilot Associations would like to have accurate forecasts of water level to be used during transit through the Bay. Likewise, the local port authorities plan to use CAFE to make sound decisions regarding vessel loading and draft. The US Navy, with huge facilities in the Hampton Roads area, has asked to be included in the first test release of CAFE products. Storm surge forecast responsibility rests with the NWS, and thus NWS is extremely interested not only in obtaining CAFE nowcasts and forecasts, but also in contributing to the development and improvement of CAFE (Cobb, 1997). Private and general public users have also been identified. The CAFE homepage (chartmaker.ncd.noaa.gov/ocs/csdl/cafesim.html) has attracted interest from a variety of small marina operators and recreational boaters. The academic community, primarily at the University of Maryland and Old Dominion University, is interested both in using CAFE in the classroom environment and in helping to solve some of the remaining scientific issues related to CAFE nowcasts and forecasts.

Chesapeake Bay forecasting is presently the most advanced component of NOS's national program for water level forecasting (Parker, these proceedings). On the national scale, a predictive model for the whole East Coast (Schultz and Aikman, these proceedings) and for the Gulf of Mexico (O'Connor, et. al., these proceedings) are under development to provide ocean boundary conditions to regional models. The regional models for Chesapeake Bay, as described herein, as well as for New York Harbor and Galveston Bay (Schmalz, these proceedings) can use these ocean boundary conditions as forcing to predict water levels and currents at specific locations within bays and estuaries.

This paper documents the present experimental CAFE system, presents the results of skill assessment of recent nowcast and forecast runs, as well as outlines the continued research and development which is planned.

PHYSICAL OCEANOGRAPHY of the CHESAPEAKE BAY

Water level fluctuations in Chesapeake Bay are strongly determined by the interaction of the astronomic tide with the Bay's bathymetry and by local and remote wind systems. The astronomic tide is the most regular forcing and is the easiest to simulate. The tide range at the entrance is approximately 1 m, and the length of the estuary (320 km from the entrance to the mouth of the Susquehanna River), along with the shallow water (the average depth is 8 m and maximum is 53 m) mean that the tide, as a progressive wave, requires about 12 hours to move from the entrance to the upper Bay. Thus the entrance and upper Bay can be experiencing a high tide, while at the same time the central Bay can be at or near low tide. Because of the long time delays in tide wave progression, any numerical model will be quite sensitive to depth and bottom friction parameters.

Local winds are an important force determining water level variations, with subtidal variations at Baltimore on the order of 1-2 m. The most common wind systems are associated with the frequent passage of mid-latitude atmospheric low pressure systems across the Bay. Because of the Bay's extremely long north-south extent, winds in the lower Bay can be quite different in speed and direction from those prevailing over the upper Bay. This diversity makes nowcasting and forecasting meteorological effects especially difficult, and is compounded by the fact that there are relatively few over-water wind observations available. In addition, short-duration storms may have time scales near the Bay's resonant period of approximately 2 days (Hess, 1987; Chuang and Boicourt ,1989).

In addition to local wind forcing, remote forcing, which is manifest in the non-tidal water level variation on the continental shelf just outside the Bay entrance, is equally important. Low pressure systems moving up the Eastern Seaboard generate coastal waves that propagate both northward (forced) and southward (trapped) along the shelf. These non-tidal waves enter the Bay and travel up the Bay with the same free gravity wave speed as the tidal wave. Depending on arrival time, the non-tidal wave can either reinforce or counteract the astronomical tide and local wind effects. Bathymetric effects of the entrance itself may alter the amplitude of the incoming wave in ways not yet fully understood.

Observations of the variables necessary for nowcasts and forecasts are made by NOAA and are available in real time. Water level gauges at 11 locations around the Bay provide data at 6-minute intervals (Figure 2). Wind observations at two locations, one in the upper Bay at Thomas Point Light and the other in the lower Bay on the Chesapeake Bay Bridge Tunnel (CBBT), provide hourly wind speed, direction, gust information, and atmospheric pressure. Measurement of water levels at the Bay entrance are not yet available, so at present the non-tidal forcing is approximated using the water level observed at CBBT.

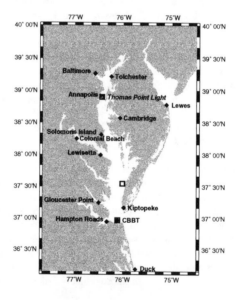

Figure 2. Location of the water level (♦) and wind stations (□) used in CAFE. The unlabeled box is the position of the Rappahannock front range light where meteorological instrumentation will soon be deployed.

THE EXPERIMENTAL SYSTEM

The existing experimental CAFE system was designed and implemented as a proof of concept for regional nowcast/forecast systems. The intention was to initiate an experimental forecast system that over time would provide a system performance record and an archive of model results which will both be used in the transition to an operational system. The present system runs automatically twice a day on NOS computers, displays the results both in hard copy and on a restricted access WWW site, and archives the results. The twice daily procedure is:

- obtain wind data from the NOAA station at Thomas Point Light and the CBBT and water level from CBBT,
- run a nowcast numerical simulation using the above inputs as forcing and the restart file produced by the previous 12-h nowcast (see schematic-Figure 3),
- obtain forecast windfields from the NWS Eta model (Black, 1994) and forecast coastal subtidal water level from the NWS extra-tropical storm surge model (Chen, et. al., 1993),
- run a 24-hour forecast using the above inputs as forcing and the restart file produced by the present 12-h nowcast (see schematic-Figure 4).

Forcing fields for the nowcast are generated from observations of water level and wind. A wind field is produced every model time step from the two wind observations. Northward and eastward components of the wind in the lower Bay are set equal to those at CBBT, and in the upper Bay they are set equal to those at Thomas Point. In the middle Bay, each component is linearly interpolated in space between the lower and upper values. If either of the sensors does not report, data from the other are used to fill in. Wind stress is then calculated from the wind components. The open boundary condition is produced by taking the observations of water level at CBBT and applying a phase and amplitude factor. These factors account for the fact that CBBT is located three grid cells into the model domain rather than at the true open boundary. Astronomical predictions are substituted for gaps of greater than three hours.

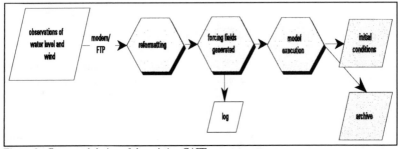

Figure 3. Conceptual design of the existing CAFE nowcast system.

A nowcast model run spanning the 12 h elapsed since the previous nowcast run is executed using the initial conditions saved from the previous nowcast run. The results are archived. The nowcast runs occur at approximately 0700 and 1900 and are valid from 1800 the previous day until 0600, and for 0600 until 1800 respectively (all times are in UT). The hour lag allows for collection and preprocessing of the necessary input data.

Forcing fields for the forecast are generated from forecasts of subtidal water level at the Bay mouth and forecast winds over the Bay. Windfields are generated from the NWS Eta 29-km 3-h forecast wind (Black, 1994). The forecast of subtidal water level is produced by the NWS Techniques Development Laboratory (TDL) (Chen et al., 1993). The open boundary condition is produced by adding the TDL subtidal forecast to the astronomically predicted water level and then applying the same factors which are used in the nowcast mode. As in the nowcast scheme, astronomically predicted water level is substituted if TDL forecasts are missing for greater than three hours. Gaps of three hours or less are filled by linear interpolation. An additional feature of the forecast mode is that the difference between the last observation of water level at CBBT and the concurrent forecast for CBBT is calculated and used to create a ramp function which is applied over the first three hours of the forecast. This feature was added because the forecast and observations are often different and thus produced a large discontinuity in the early hours of the forecast.

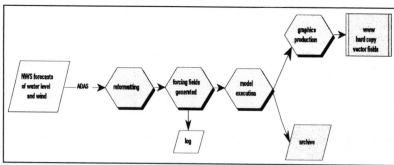

Figure 4. Conceptual design of the existing CAFE forecast system.

A model forecast for the next 24 h is made using the initial conditions from the current nowcast. The results are archived as well as input into a suite of graphics production software. The forecast runs are initiated at 0740 and 1940, but often delays in the arrival of NWS forecast data cause the CAFE forecast to be delayed.

The numerical code which is used in both the nowcast and forecast mode of CAFE is the Model for Estuarine and Coastal Circulation Assessment (MECCA) (Hess,

1987; 1997). The MECCA code solves the hydrodynamic equations of momentum, mass, salt, and heat conservation. It is three-dimensional in space, uses a vertical sigma coordinate, has a time-varying free surface, and incorporates non-linear horizontal momentum advection. It includes a three-dimensional time variable horizontal diffusion based on the Smagorinsky diffusivity (see Tag et al., 1979) and vertical turbulent diffusion based on a mixing length and Richardson number-dependent reduction (Munk and Anderson, 1948). For the horizontal momentum equations, the external gravity wave mode is split from the internal mode. Variables are placed on an Arakawa C-grid with square cells in the horizontal and at uniform intervals along a sigma-stretched vertical coordinate. External-mode momentum is solved with an alternating-direction, semi-implicit method in the horizontal. The salinities, temperatures, and internal-mode velocities are solved with a semi-implicit method in the vertical. In recent years, some modelers have encountered certain problems with sigma coordinate systems (Haney, 1991). Many of these problems can be ameliorated by using uniformly- spaced sigma levels and by subtracting the spatially-averaged density before computing the horizontal gradient. MECCA has both these features.

Chesapeake Bay runs are made on a 55 x 34 cell grid with bathymetry that was developed at the US Naval Academy (Hoff, 1990). The grid cell size is 5.6 km, the minimum cell depth is 1m and maximum is 19 m (Figure 5). The model is run in barotropic mode. Upper reaches of rivers are modeled as narrow channels.

Graphical output was designed to be easily made available via fax-on-demand (Figures 6). This output along with animation of hourly 2-dimensional water level and hourly wind forcing vectors are placed on a restricted access website.

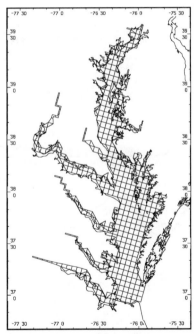

Figure 5. The MECCA model grid. Grid cells are nominally five kilometers on a side.

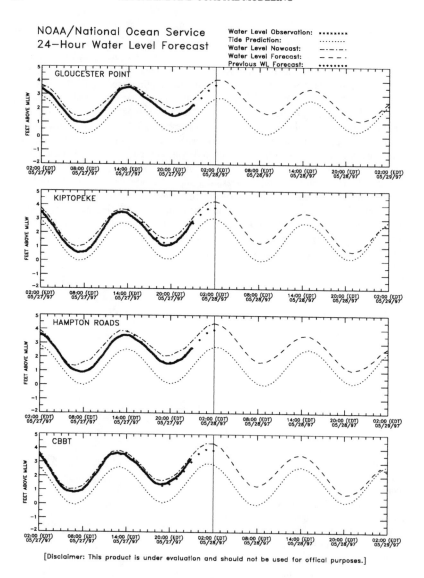

Figure 6. Sample of water level time series output for the lower Bay. A similar plot is produced for the stations in both the middle and upper Bay.

RECENT RESULTS

The nowcast portion of CAFE was begun in March 1996, with an initial hindcast beginning January 1, 1996, and forecasting was begun in July 1996. Thus we have selected the one year common period beginning July 1, 1996 to perform skill assessment of both the nowcast and the forecast systems. The results for the nowcast runs reveal a RMS difference which is largest near the head of the Bay at Tolchester (12.0 cm) and decreases to a minimum near the mouth at CBBT (4.5 cm) (Figure 7). This geographic distribution of the error is not surprising given that observed water level is used to force the nowcast at the entrance to the Bay.

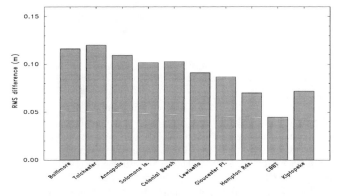

Figure 7. RMS difference between the observed and nowcast hourly water levels during the one year comparison period. Baltimore, Tolchester, Colonial Beach and Lewisetta have less than a complete year (8760 hours) of observed data (8684, 8702, 5170, and 8615 hours respectively).

Another measure of model skill is the number of hours that the absolute value of the difference between the observed and the nowcast water level exceeds a given threshold. During the one year comparison period, the number of hours (out of a possible 8760) that the difference was larger than 30 cm was 124 (1.4 %) at Baltimore and 7 hours (0.1%) at CBBT (Figure 8). These seemingly small differences must be tempered by the fact that the maximum difference at Baltimore during the comparison period was -64 cm indicating that the model nowcast more water than was present. Keeping in mind navigational safety, this type of overestimation is extremely dangerous. Another important consideration is that the accuracy of the forecast is very dependent on the accuracy of the nowcast, because the nowcast provides the initial conditions for the forecast. Thus even small errors in the nowcast can translate into larger errors in the forecast.

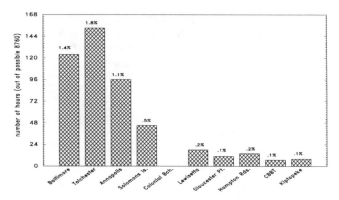

Figure 8. Number of hours (percentage of hours is labelled above each bar) during the study period that the absolute value of the difference between the observed and nowcast water level exceeded 30 cm. Colonial Beach is not included due to incomplete observational series.

Due to the occasional unavailability of forcing data, the forecasting system was somewhat unreliable throughout the initial 6 months of operation. Thus only 5765 hours out of a possible 8760 are available for assessment. Nonetheless some interesting trends exist. The RMS difference between the observed and forecast water level was largest at CBBT (20.9 cm) and considerably less in the upper reaches of the Bay, including 14.8 cm at Baltimore (Figure 9). The first 12 of each 24 h forecast were included in the analysis. The number of hours (out of a possible 5765) that the absolute value of the difference between the observed water level and the forecast water level exceeded 30 cm was 336 (5.9 %) at Baltimore and 844 (14.6%) at CBBT (Figure 10). These results present an interesting contrast to the skill of the nowcasts. Unlike in the nowcast mode, the skill of the forecast is better in the upper Bay than in the lower. This result suggests the importance of the open boundary forcing. The large errors which are present only three grid cells into the domain must result from errors in the forecast of the open boundary condition.

The initial skill of both the CAFE nowcasts and forecasts has been deemed acceptable for a beta release to a limited number of users. Feedback from this test group will be used to develop CAFE products which will best meet the needs of the maritime community. Research aimed at nowcast and forecast improvement will continue during the beta test in order to increase the accuracy of CAFE products.

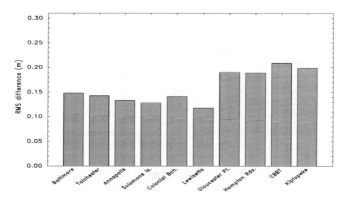

Figure 9. RMS difference between the observed and forecast (first 12 h of each twice daily run) hourly water levels during the one year comparison period. At Tolchester, Colonial Beach, Lewisetta, Gloucester Point, and Hampton Roads fewer than 5765 hours were analyzed (5511, 3360, 5611, 5688, and 5460 respectively).

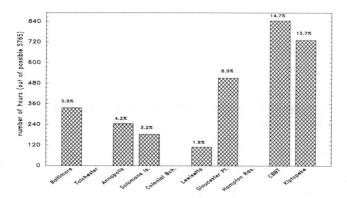

Figure 10. Number of hours (percentage of the hours compared is labelled above each bar) during the study period that the absolute value of the difference between the observed and forecast water level exceeded 30 cm. Tolchester, Colonial Beach, and Hampton Roads are not included due to incomplete observational series.

ON-GOING RESEARCH TOWARD ENHANCEMENT

Two areas of research will be addressed to insure that CAFE continues to improve the quality of its water level nowcasts and forecasts. These scientific issues focus, not surprisingly, around the two major forcing functions for the numerical model: winds over the Bay and coastal ocean water level (Bosley, 1996). These research needs were identified by MECCA hindcast runs and by preliminary skill assessment of the experimental system output.

Analysis of simulations of water level in the Bay have revealed that the quality of the simulation is highly dependent upon the quality of the windfield which is used to force the model run. Many storms which transit the Bay have a high degree of spatial variability, and so the two point measurements of wind used in the nowcast cannot adequately resolve the structure of the windfield. Likewise, there are events which are not well represented in the atmospheric forecast model which is currently available. New advances in atmospheric mesoscale analysis and forecasting are actively being investigated for use in CAFE (Bosley, 1997). Because very few over-water measurements of wind exist in the Bay, we are deploying a meteorological package on the front range light of the Rappahannock Shoal Channel (see Figure 2). Data from this site will be compared with data from land-based stations in order to develop correction factors for over-water winds.

The open boundary conditions at the Bay's entrance are critical to accurate nowcasts and forecasts produced by CAFE. In the nowcast mode real-time water level observations are used to force the model. We have seen, however, situations where the measurements taken at the Bay mouth are "contaminated" with a water level signal from the Bay itself, and do not accurately represent the true water level propagating into the Bay. To address this concern, we are deploying a tide gauge at Chesapeake Light tower which is approximately 2.6 km offshore from the mouth of the Bay. Data from this station will be used to determine the relationship between water level on the coast and water level measured at CBBT. Research is needed to determine how best to combine water level observations near the entrance in order to produce the most accurate open boundary condition. The quality of the forecast open boundary condition is dependent on the skill of the large coastal ocean model from which the boundary conditions are taken and on the skill of the atmospheric model which drives these large scale coastal ocean models. The optimum method for using the output of large scale coastal ocean models as input forcing for the Bay model is yet be determined.

CONCLUDING REMARKS

The nowcast/forecast has been running every day, with few exceptions, since March 1996. After 19 months of operations, we believe that most of the possible system failures have been experienced, and steps have been taken to correct them. Weaknesses in the observational system have been identified, and we have begun to address them by deploying a tide gauge at Chesapeake Light and a wind sensor at Rappahannock Light. The forecast output is still under evaluation, especially since extreme storm forcing events are relatively infrequent and the system's performance during these critical events needs to be thoroughly analyzed and understood. Also, given the potential legal liability NOS faces, additional thought is being given as to when the forecast will not be issued. Finally, plans are being made to assemble and educate a representative user group to help us assess the forecast products's utility and accuracy.

Acknowledgements- The authors gratefully acknowledge the efforts of the following people who have contributed to the development of the Chesapeake Area Forecasting Experiment. John Cassidy wrote much of the display software for the CAFE website; Michael Evans, Tom Bethem, and Kenneth Gormally implemented the data gathering routines; John Kelley developed the Barnes interpolation scheme which will be used to produce analyzed windfields, Charles Sun developed and Eugene Wei maintains the automatic data retrieval and archiving system; Bradford Wynn, John Stepnowski, and Oliver Jones are deploying additional sensor packages within the Bay; and the US Coast Guard continues to provide ship support.

REFERENCES

Black, T. L. (1994). The new NMC mesoscale Eta Model: Description and Forecast Examples. **Weather and Forecasting**, 9, pp.265-278.

Bosley, K. T. (1996). Toward a Nowcast/Forecast System for Water Levels in the Chesapeake Bay. Proceedings of the Oceans 96/Marine Technology Society Meeting September 23-26, 1996 , Ft. Lauderdale, FL.

Bosley, K. T. (1997). The Application of the Local Analysis and Prediction System (LAPS) to the Chesapeake Bay Area in Support of Oceanographic and Atmospheric Forecasting. A NOAA internal proposal.

Chen, J., W. Shaffer, and S. Kim (1993). A Forecast Model for Extratropical Storm Surge. **Advances in HydroScience and Engineering**, 1, 1437-1444.

Chuang, W. S. and W. C. Boicourt, (1989). Resonant seiche motions in the Chesapeake Bay. **Journal of Geophysical Research**, 94, 2105-2110.

Cobb, H. (1997). Short-fused coastal flooding event of April 23,1997. NWS internal memorandum.

Haney, R. L. (1991). On the pressure gradient force over steep topography in sigma coordinate ocean models. **Journal of Physical Oceanography**, 21, 610-619.

Hess, K. W. (1987). Constituent tides and tidal currents in the Chesapeake Bay (abstract). EOS, 68(16), 336.

Hess, K. W. (1989). MECCA Program Documentation. US Department of Commerce NOAA Technical Report NESDIS 46 258 pp.

Hess, K. W. (1997). MECCA2 Program Documentation. unpublished manuscript.

Hoff, M. (1990). A Chesapeake Bay Circulation Model. US Naval Academy, Annapolis, 26 pp.

Kite-Powell, Hauke L. (1996). Formulation of a Model for Ship Transit Risk. An MIT Sea Grant College Program Report No. 96-19.

Munk, W. H. and E. R. Anderson (1948). Notes on the theory of the thermocline. **Journal of Marine Research**, 7, 276-295.

O'Conner, W. P., F. Aikman III, E. J. Wei, and P. H. Richardson. Comparison of observed and forecasted sea levels along the West Florida Coast. These proceedings.

Parker, B. P., Development of model-based regional nowcasting/forecasting systems. These proceedings.

Schmalz, R. A., Development of a nowcast/forecast system for Galveston Bay. These proceedings.

Schultz, J. R., and F. Aikman III, Evaluation of subtidal water level in NOAA's Coastal Ocean Forecast System for the US East Coast. These proceedings.

Tag, P. M., F. W. Murray, and L. R. Koenig (1979). A comparison of several forms of eddy viscosity parameterization in a two-dimensional cloud model. **Journal of Applied Meteorology**, 18, 1429-1441.

Real-Time Data Acquisition and Modeling in Tampa Bay

Mark Vincent[1], David Burwell[1], Mark Luther[1], Boris Galperin[1]

Abstract

A real-time data acquisition system has been successfully coupled to the Blumberg-Mellor ECOM-3D model of Tampa Bay, Florida. The data acquisition system consists of 2 stations of the new Coastal Monitoring Network (CMN) and 8 stations of the Physical Oceanographic Real-time System (PORTS). Since March 1997 a protocol has been operational to conduct continuous nowcast model simulations with boundary conditions updated every twelve minutes. Recently, a new protocol has been implemented which conducts 48-hour forecast simulations twice a day. The forecast mode is still undergoing active development and testing. Preliminary skill assessment using data sets from the nowcast mode indicates good agreement between the model and observational data. One of the future objectives of this research is to implement an on-call oil-spill response program capable of providing circulation data and predicted trajectories of real oil-spill plumes. Both Lagrangian particle and Eulerian modeling approaches are currently being tested. Online information, including real-time data and simulations is currently available at http://ompl.marine.usf.edu.

Introduction

The central focus of this research is the integration of a real-time oceanographic data acquisition system with the Blumberg-Mellor ECOM-3D numerical circulation model of Tampa Bay, Florida. This technology conducts model simulations in hindcast, nowcast and 48 hour forecast modes, and is expected to become a valuable bay management tool. One target objective of this research is to ultimately implement an on-call oil-spill response program capable of providing circulation data and predicted trajectories of real oil-spill plumes. Included in this paper are overviews of the data acquisition and dissemination

---

[1]Department of Marine Science, University of South Florida, 140 7th Avenue South, St. Petersburg, FL 33701

system, the numerical circulation model and several model applications and simulations.

Data Acquisition System

The oceanographic and atmospheric data acquisition systems used in this study, consist of eight stations from the Physical Oceanographic Real-Time System (PORTS), and from two new stations constructed by the University of South Florida as part of the Coastal Monitoring Network (CMN) water-level network. A product of this instrumentation is a spatially and temporally resolved data set of the bay, that provides a test bed for model integration and skill assessment.

Ports

PORTS is an operational real-time environmental data acquisition and dissemination system in Tampa Bay, established in June 1992.

The PORTS system was established by the National Ocean Service, (NOS) for the shipping interest in the bay, and is funded through state and county agencies for that purpose. Since 1993 the PORTS has been housed at the University of South Florida, St. Petersburg Campus in St. Petersburg Florida, where the data is collected via line of sight radio, and disseminated in near real-time by telephone modem, automated phone reports, and on the world wide web.

This system of sensors is comprised of four Next Generation Water Level Stations (NGWLS), three stations with Acoustic Doppler Current Profilers (ADCP), and one meteorological station that provides wind speed and direction at 20 meters together with barometric pressure and air temperature (Table 1). The data is telemetered to USF every six minutes.

The NGWLS, which utilize Bartex acoustic transducers, are located in St. Petersburg (sp) at the USCG base, in Old Port Tampa (opt) in Picnic Island Park, also at the CSX Rock Port Terminal (pt) in Tampa, and in Port Manatee (pm). The four NGWLS are calibrated and surveyed by NOS to insure that national standards for water levels are maintained. The tide stations each have a Young anemometer that report wind speed and direction, at approximately 10 meters.

There are three RDI ADCPs in the bay, that report the current speed and direction in approximately 1 meter bins from 2 meters above the bed to approximately 2 meters below the surface. These instruments also collect bottom water temperatures at their respective sites. One is located near the opt tide station (vtam), one under the Sunshine Skyway Bridge center span (vsky), and one in the bay north-west of the pm tide station (vman) that also provides wind speed and direction.

	Water Level	Winds	Currents	Salinity	Other Meteoro-logical	Water Temp
sp	yes	yes	no	no	no	no
opt	yes	yes	no	no	no	no
pt	yes	yes	no	no	no	no
pm	yes	yes	no	no	no	no
vtam	no	no	yes	no	no	bottom
vman	no	yes	yes	no	no	bottom
vsky	no	no	yes	no	no	bottom
ccut	no	yes	no	no	yes	no
ami	yes	yes	no	yes	yes	surface
ekey	yes	yes	no	yes	no	surface

Table 1: Sensor arrays on the Tampa Bay PORTS and CMN Stations

Finally there is a Coastal Climate meteorological package on top of the C-Cut range tower (ccut) in the center of the bay that provides wind speed and direction, and other meteorological parameters including, barometric pressure, and air temperature (Figure 1).

Coastal Monitoring Network

The CMN is a network of tide and current meters that are presently being developed to study the West Florida Shelf. USF has installed two water level stations at the mouth of Tampa Bay as part of this network, as well as to provide real-time boundary conditions for the ECOM-3D numerical model. The stations are located on the north-east end of Egmont Key and on the north end of Anna Maria Island, and thus bracket the mouth of the bay at the open boundary of the model domain (Figure 1). The Egmont Station is a NGWLS together with an anemometer at 10 meters, and near surface water temperature and salinity. The Anna Maria station has, in addition to the instrumentation at Egmont, other meteorological instruments that measure, air temperature, relative humidity, barometric pressure, and precipitation (Table 1). These sites report via radio to USF every 12 minutes.

All sensors are maintained to conform to NOS standards, and are compared daily by remote monitor to insure accuracy.

Circulation Modeling System

The numerical circulation model used for this research is a version of the Blumberg-Mellor (1987) Estuarine Coastal and Ocean Model (ECOM-3D). The original application of this model to Tampa Bay was by Galperin et al.

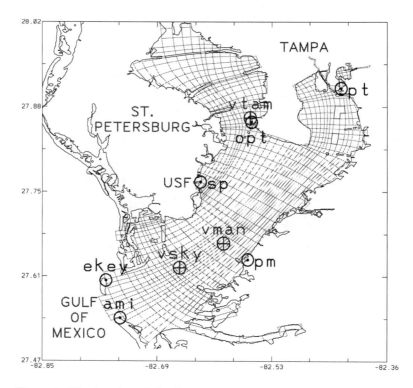

Figure 1: The location of the Tampa Bay PORTS and CMN tide, met., and current meter stations and the model grid.

(1992a, 1992b). This class of model, also known as the Princeton Ocean Model (POM), has been extensively deployed and described in the literature. Briefly, the ECOM-3D solves the three-dimensional time dependent equations for conservation of mass, momentum, heat and salinity. Vertical turbulent mixing is parameterized by an embedded second moment Mellor-Yamada turbulence closure model. Horizontal diffusion is provided by Smagorinsky diffusivity. The model equations are transformed to curvilinear orthogonal coordinates in the horizontal axes and bottom and free surface following sigma coordinate in the vertical axis.

For this study of Tampa Bay, a 40 x 57 grid (Figure 1) with 11 sigma layers is used. The internal three-dimensional mode and external two-dimensional mode use a 240 sec. and 24 sec. split time step respectively.

Open boundary conditions are provided by the measured sea surface elevation, temperature and salinity at the mouth of Tampa Bay. Presently temperature and salinity boundary conditions are applied uniformly over the depth of the boundary, which studies have shown to be a reasonable approximation (Zervas, 1993). The free surface boundary conditions include wind stress, mass flux and heat flux. The flux of mass, heat and salinity is included for rivers and other large discharges located around the perimeter of the bay.

Simulation Protocols

Currently, nowcast, forecast and hindcast simulations of the circulation model are performed in parallel, on a network of three Silicon Graphics workstations.

Hindcast

The nowcast protocol automatically archives the components of the model input data sets on an end of month basis. This allows for the rapid generation of hindcast data sets, with which to perform model skill assessment tests or circulation investigations.

Nowcast

Since March 1997, a protocol has been operational to capture the data stream from the Egmont Key CMN Station, error check the data, prepare and archive model data sets, compute the number of time steps and perform the nowcast ECOM-3D simulations. During normal operating conditions (case 1), a complete set of boundary conditions is received every 12 minutes. Using the output from the last nowcast run as initial conditions, the model is run forward to the most recent boundary condition received. With this approach the model lags real-time by 12 minutes or less.

One important feature of the nowcast protocol is the capability to automatically continue model simulations during periods of communication loss with the remote stations (case 2). The open boundary water level in this case is computed from 37 tidal constituents superimposed on the average sub-tidal setup observed over the last 2 hours. Salinity and temperature profiles are held constant at the last observed values, while the wind stress is ramped to zero. Another similar feature allows for automatic generation of model data sets after any shut down of the USF computer system (case 3), such as in the event of a severe lightning strike. Obviously the filled data sets may degrade over an extended period, but seem more than adequate to bridge data gaps on the order of minutes to several hours, while maintaining model simulations and avoiding the need for additional model spin-up. Although the case 2 and case 3 contingencies are fully automated, they are seldom invoked.

Forecast

In late September 1997 a test version of a 48 hour forecast protocol was implemented. This automated protocol uses the results of the most recent nowcast simulation as initial conditions. Similar to case 2 discussed above, the predicted open boundary water level is computed from 37 tidal constituents superimposed on the recent average sub-tidal setup (i.e. persistence). Eventually several other approaches will be evaluated for providing the predicted non-tidal component of the water level. The forecast wind speeds and directions are provided from output of the Nested Grid Model (NGM), which is obtained from the local National Weather Service office. This information is obtained twice daily at approximately noon and midnight (local time), and provides wind data on three hour increments. The forecast protocol performs 48 hour forecast simulations twice a day at 0200 and 1400 hours local time.

Presently the forecast protocol only uses the most recent nowcast output as initial conditions. An improvement planned for implementation in the near future is the ability to conduct rapid simulations with archived initialization files (i.e. on the order of hours to days old). This feature will be beneficial in responding to emergency search and rescue and oil-spill catastrophes.

Model Performance

Preliminary skill assessment has been conducted for the model in hindcast mode, using the identical boundary condition data sets prepared and used by the nowcast model mode. Since the nowcast model became fully operational in March 1997, data sets after this period were considered for calibration and validation. Calibration was conducted for April 1997, using March as the spin-up period. Validation was performed for May, using March and April as the spin-up period.

A systematic varying of model parameters was performed, resulting in 96 calibration simulations. Parameters varied included the Smagorinsky horizontal mixing coefficient (HORCON), the bottom friction coefficient (BFRIC), bottom roughness coefficient (ZOB), and a global scaling of the bathymetry. Optimal calibration results were found when these parameters were set to: HORCON=0.02 ; BFRIC=0.0025; ZOB=0.004; and depth scaling of 0.925. This configuration was held fixed for the subsequent validation simulation.

The performance of the model measured against the PORTS observational data was determined using the model evaluation statistics developed by Hess and Bosley (1992) and used previously in Tampa Bay for the Tampa Oceanography Project (TOP)(Hess, 1994). In addition, cross correlation analysis between the model output and observational data were performed.

Archived six minute interval data from four water level stations St. Petersburg (sp), Old Port Tampa (opt), Port Tampa (pt), and Port Manatee (pm) and two ADCP current stations Skyway Bridge (vsky) and Old Port Tampa (vtam) were available for comparison. The ADCP bin selected from each site was

station	lag (hr)	cross corr	R	Drms	Dp	Gw	Arms (m)	Lm (hr)	Lrms (hr)
sp	-0.200	0.997	0.487	0.022	0.045	1.038	0.023	-0.142	0.297
opt	-0.267	0.997	0.542	0.030	0.055	0.995	0.017	-0.157	0.270
pt	-0.200	0.995	0.587	0.032	0.054	1.025	0.026	-0.096	0.262
pm	-0.067	0.997	0.472	0.019	0.041	1.048	0.022	-0.055	0.300
Global Values				0.026	0.049	1.027	0.022	-0.112	0.282
Skill Parameters				SD=95.1%		SA=95.8%		SL=95.5%	

Table 2: Results of the May 1997 Validation for Water Levels

approximately 3 meters below mean sea level. Both of the ADCP stations are located at the bottom of the main shipping channels.

Results from the validation are provided in Tables 2 and 3. All tables share a common format. The first three columns are the station name, the lag of highest cross correlation, and the cross correlation respectively. The remaining columns (4-10) are the statistics computed by the method of Hess and Bosley (1992). For each parameter (water level or currents), R is the mean extrema range, Drms is the rms difference (between the model and data), Dp is the Drms non-dimensionalized via R, Gw is the gain ratio of model extrema to data, Arms is the rms difference of extrema values, Lm is the mean time lag of extrema, and Lrms is the rms extrema lag. Here extrema are defined as maxima or minima that are separated by more than two hours.

The primary parameters which are used by this method to evaluate the model are SD (6 minute difference skill), SA (extrema amplitude skill), SL (extrema lag skill), Gw (the weighted gain), and Lm (the mean lag). The parameters SD, SA, and SL are non-dimentionalized measures of the rms differences.

For water levels, agreement between the model and observational data for the validation period was very good, with all three skill parameters (SD, SA, SL) exceeding 95%. The station averaged model gain (Gw) was very close to unity at 1.027. A slight model lead of extrema (Lm) was observed at all stations, ranging from -3.3 minutes to -9.42 minutes. The cross-correlation coefficient for all stations exceeded .995 while the lag of highest cross correlation (based on 4 minute steps) ranged from a model lead of -4.0 minutes to a model lead of -16.0 minutes.

Validation scores for the current comparisons were less than those for water level. The skill assessment parameters ranged from 87.1% to 92.0%. The model currents lagged the observational extrema (Lm) by a station averaged value of 6.96 minutes. The model gain (Gw) was moderately less than unity, averaging

station	lag (hr)	cross corr	R	Drms	Dp	Gw	Arms (m)	Lm (hr)	Lrms (hr)
vsky	0.000	0.985	1.275	0.113	0.088	0.860	0.136	0.040	0.639
vtam	0.067	0.980	1.176	0.084	0.071	0.858	0.101	0.186	0.960
Global	Values			0.098	0.080	0.859	0.118	0.116	0.807
Skill	Parameters			SD=92.0%		SA=90.4%		SL=87.1%	

Table 3: Results of the May 1997 Validation for Currents

86%. Several factors may explain this slight under prediction including: slight differences between the principal axis of model and observational currents, and the location of the ADCPs in deeper sub-grid areas of the shipping channels, that are not resolved by the model grid.

Trajectory Simulations

Two approaches were applied to test the possibility of using the model to track either a surface drifting object (i.e. oil, drifting ship, person in the water, etc.) or a neutrally buoyant contaminant (i.e. emulsified oil, biological organisms, etc.).

The first approach used ECOM-3D output velocity fields to drive an external Lagrangian drifter model, and the second used the passive tracer subroutines of ECOM-3D.

External Lagrangian Model

A hindcast simulation was used to drive the external Lagrangian model, for a period when surface drifters were released and tracked over a portion of the bay. The external model used the surface velocity field product from ECOM-3D to calculate the position in time of 100 simulated drifters initialized identically with the release of the real drifters. The resulting drift tracks are shown in Figure 2.

This first test established that the coefficients on the Markovian dispersion in the drifter model were reasonable, and provided evidence that the external Lagrangian model could be used as proposed. After half a tidal cycle the mean distance between the model and the real drifters was less than one kilometer.

Passive Tracer Trajectories

Simultaneously with the external drifter model simulation an internal conservative passive tracer simulation was conducted. The output, indicated the

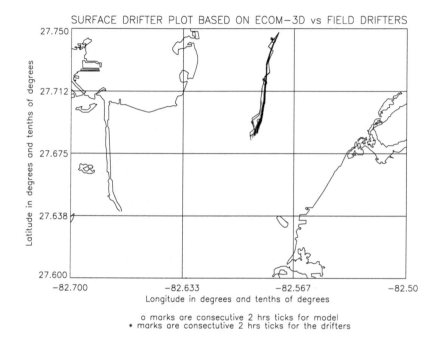

SURFACE DRIFTER PLOT BASED ON ECOM—3D vs FIELD DRIFTERS

o marks are consecutive 2 hrs ticks for model
* marks are consectutive 2 hrs ticks for the drifters

Figure 2: The drifter tracks from the external Lagrangian drifter model vs. the drift buoy tracks.

surface plume trajectory of the internal passive tracer had a larger spread than the external Lagrangian model, possibly due to numerical dispersion.

To date both the passive tracer and surface drifter model show promise in the ability to track both surface trapped species (external Lagrangian drifter) and plumes that are neutrally buoyant (internal passive tracer).

Data Dissemination and Products

Presently the real-time data, together with nowcast and forecast model products are disseminated via the world wide web at http://ompl.marine.usf.edu/, a few examples are shown in Figures 3, 4, and 5. Figure 3 contains the plan view surface velocity field for the nowcast product as well as a cross section near the Sunshine Skyway Bridge of the horizontal velocity field with depth, from the nowcast model. Figure 4 contains the plan view of surface salinity and a cross section of salinity from the mouth of the bay to the head of the bay. Figure 5 is the previous 48 hour time series of real-time data versus model nowcast tides for pt and pm.

Future plans include, direct transmission of pertinent nowcast/forecast products to the United States Coast Guard, and field response teams, as well as bay management personnel.

Summary and Future Work

To date, protocols are operational that provide continuous nowcast model simulations, and twice-daily 48 hour forecast test simulations. Preliminary skill assessment indicates good agreement between the nowcast model and real-time data. Additional development and testing of the forecast mode is planned. Both the use of the internal passive tracer and the external Lagrangian trajectory model show promise for future applications and research involving search and rescue efforts, oil-spill trajectory predictions, and biological dispersion studies.

An initial test indicated the external Lagrangian model was accurate to less than 1 kilometer over half a tidal cycle, in hindcast mode. Longer duration tests using remotely tracked drifters are needed to further advance this application.

There is still the need for additional testing and evaluation, and also the need to disseminate the model products in near real-time to USCG, Harbor Pilots, bay management personnel, etc. for their use.

The ultimate goal is to produce a nowcast/forecast product that is readily available to agencies for use in many aspects of shipping, search and rescue, and hazardous spill containment in Tampa Bay.

Acknowledgments

The principle funding for this ongoing research is provided by the Environmental Protection Agency (EPA). Additional funding and resources are provided by the Greater Tampa Bay Marine Advisory Council (GTBMAC), the National Weather Service (NWS) and the Office of Naval Research (ONR). The authors thank the contributions of Dr. Meredith Haines, Linae Boehme, and Haihua Liu at USF for assistance in data acquisition and programming, and Andrew Nash of the NWS for providing the forecast wind data. The use of ECOM-3D is made possible by Alan Blumberg of HydroQual Inc.

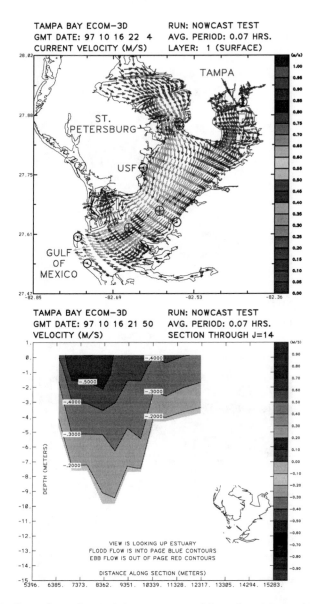

Figure 3: Example products of the nowcast model: surface velocity (top), cross section velocity (bottom)

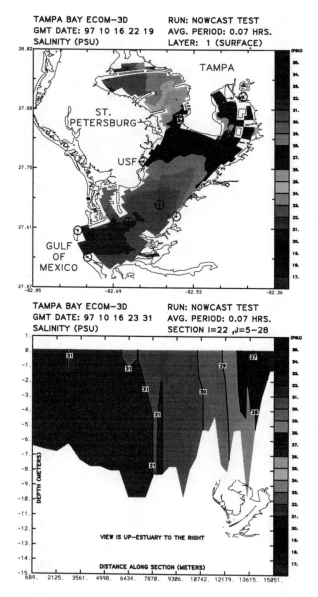

Figure 4: Example products of the nowcast model: surface salinity (top), cross section salinity (bottom)

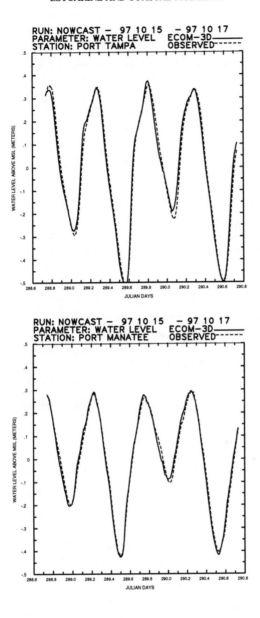

Figure 5: Example comparison of the nowcast model and the real-time data: pt (top), pm (bottom)

References

Blumberg, A. F. and Mellor, G. L. 1987. A description of a three-dimensional coastal ocean circulation model, in Three-Dimensional Coastal Ocean Models, Heaps, N. S. (Ed.), American Geophysical Union, Washington, DC 1- 16.

Galperin, B., Blumberg A. F., and Weisberg, R. H., 1992a. A time-dependent three-dimensional model of circulation in Tampa Bay, in Proceedings, Tampa Bay Area Scientific Information Symposium 2, Tampa, FL, February 27 - March 1, 1991, 77 -98.

Galperin, B., Blumberg A. F., and Weisberg, R. H., 1992b. The importance of density driven circulation in well mixed estuaries: the Tampa experience, in Proceedings of the 2nd International Conference on Estuarine and Coastal Modeling, Tampa , FL, November 13-15, 1991, 332 -343.

Hess, K., 1994. Tampa Bay Oceanography Project: Development and Application of the Numerical Circulation Model, NOAA Technical Report NOS OES 005. NOAA, National Ocean Service, Office of Ocean and Earth Science, Silver Spring, MD. 90 pp.

Hess K. and Bosley K., 1992. Techniques for validation of a model for Tampa Bay. in Proceedings of the 2nd International Conference on Estuarine and Coastal Modeling, Tampa , FL, November 13-15, 1991, 83 - 94.

Zervas C. E. (Editor), 1993. Tampa Bay Oceanography Project: Physical Oceanographic Synthesis. NOAA, National Ocean Service, Office of Ocean and Earth Sciences, Silver Spring, MD. 175 pp.

Development of a Nowcast/Forecast System
for Galveston Bay

Richard A. Schmalz, Jr.[1], M. ASCE

Abstract

The National Oceanic and Atmospheric Administration installed a Physical Oceanographic Real Time System (PORTS) in June 1996 in Galveston Bay. Water surface elevation, currents at prediction depth (4.6m) as well as near-surface and near-bottom temperature and salinity, and meteorological information are available at six-minute intervals. To complement the PORTS a nowcast/forecast system is being developed using the Blumberg-Mellor (1987) three-dimensional hydrodynamic model. The model has been applied to Galveston Bay in support of a Differential Global Positioning System hydrographic survey (Schmalz, 1996). Herein, the grid system, boundary and initial conditions and results of the astronomical tide calibration for May 1995 will be initially reviewed. To enable a robust nowcast/forecast capability the model has been extended to include drying/wetting, and a flux corrected transport scheme to treat the large horizontal salinity gradients. The numerical algorithms for both of these extensions will be presented along with results obtained with the extended model for three one month hindcasts. Simulated water level results for the October 1994 flood and January 1995 "Northers" will first be presented, followed by simulated water level and current results for the April 1996 PORTS test period. For April 1996, results will also be presented for a one-way coupled fine resolution Houston Ship Channel model. In concert with the hindcast results, an initial nowcast/forecast system has been designed and a preliminary evaluation of system forecast inputs conducted. In conclusion, the hindcast results, initial nowcast/forecast system design, and evaluation of system inputs are reviewed and plans for additional experiments are discussed.

Introduction

The National Oceanic and Atmospheric Administration installed a Physical Oceanographic Real Time System (PORTS) patterned after Bethem and Frey (1991)in June 1996 to monitor

---

1 Oceanographer, National Oceanic and Atmospheric Administration, National Ocean Service, Coast Survey Development Laboratory, Oceanographic Programs, 1315 East-West Highway, Rm 7824, Silver Spring, Maryland 20910. (301)-713-2809 x104; e-mail: dick@ceob-sg380.nos.noaa.gov.

Galveston Bay. Water surface elevation, currents at prediction depth (4.6m) as well as near-surface and near-bottom temperature and salinity, and meteorological information are available at six-minute intervals for five, three, and four stations, respectively, as shown in Figure 1 in the format illustrated in Figure 2. To complement the PORTS a nowcast/forecast system has been designed based on the National Ocean Service (NOS) Galveston Bay three-dimensional hydrodynamic model and the National Weather Service (NWS) Aviation atmospheric model. To simulate currents within the Houston Ship Channel (HSC), a finer resolution three-dimensional HSC model has been developed. The Galveston Bay model is used to provide Bay wide water level and near entrance current forecasts as well as to directly provide water levels, density, and turbulence quantities to the HSC model for use in a one-way coupling.

Herein, we will first describe the Galveston Bay model (GBM) and its calibration to May 1995 astronomical tide. Model extensions to include drying/wetting and flux corrected salinity transport are next presented. Hindcast results for the extended GBM are presented for the October 1994 flood and the January 1995 "Northers". The development of the one-way coupled high resolution HSC model is then presented in conjunction with the April 1996 hindcast. Results of an initial evaluation of system forecast inputs (wind/pressure fields, subtidal water level, and streamflow) are also presented. To conclude, plans for additional system experiments and refinements are discussed.

Galveston Bay Model Astronomical Tide Calibration

A three-dimensional sigma coordinate Galveston Bay and near shelf (GBM) model has been developed (Schmalz, 1996) based on a version of the Blumberg and Mellor (1987) model extended to orthogonal curvilinear coordinates (see Mellor, 1996). The GBM computational grid in Figure 3 consists of 181x101 horizontal cells (dx = 254-2482m, dy = 580-3502m) with 5 levels in the vertical. The model was spun up from rest over the first day, 1 May. The initial salinity and temperature fields were constructed from climatological considerations using a grid patch method developed for Long Island Sound (Schmalz, 1994). A sea surface temperature (SST) specification was used in lieu of heat flux. Climatological river inflows (Orlando et al., 1993) were included for the Trinity and San Jacinto Rivers and Buffalo Bayou. The salinity and temperature boundary conditions were determined based on National Marine Fisheries Service cruise data (Temple et al., 1977). Water surface elevations were specified at grid cells (3,2), (60,2), (120,2), (180,2), and (180,32) by modifying one year least square analysis (Schureman, 1958) harmonically derived tidal constituents at Freeport, Galveston Pleasure Pier, Galveston Pleasure Pier, Port Bolivar, and High Island, Texas, respectively. In general, at the offshore stations, phases were adjusted to account for up to a one hour advance in phase and 1.02 amplitude adjustment. A linear spatial interpolation along the open boundary was used between the appropriate pair of boundary cells. Results of demeaned simulated water surface elevations versus harmonically reconstructed tidal signals are given in terms of an rms difference and a dimensionless (0 no shape error to 1 total shape error) relative error (Willmott et al., 1985) and ranged from 4 cm and 0.01-0.02, respectively, at Galveston Pleasure Pier and Pier 21 to 5 cm and 0.04, respectively, at Eagle and Morgans Points. To study the impact of long period constituents contained in the one-year least squares analysis (Schureman, 1958) a 29-day harmonic analysis of simulated water levels was performed. Based on this analysis, a constituent amplitude weighted gain (model/observation) and phase difference in hours (model - observation) was computed and an rms error estimated (see Hess, 1994) as given in line 1 of Table 1. Rms errors were reduced to 3 cm, indicating that some error (order 2 cm) is contained in the longer period tidal response. Weighted gains were order 0.9 over the lower bay, indicating that the tidal response is 10 percent damped. Weighted phase errors are within one hour. Simulated level 3 current principal flood directions were computed and compared with March and April 1996 PORTS station observations given in Table 2. In general, May 1995 astronomic principal flood directions are within 30 degrees of PORTS station observations, which include meteorological effects. Results of a 29-day harmonic analysis of principal direction level 3 currents are compared in line 1 of Table 3 with a 29-day harmonic analysis of March and April 1996 PORTS prediction depth currents. The weighted gain decreased significantly as one proceeds from Bolivar Roads in the entrance up to Morgans Point at the head of the Bay with rms

Table 1. Galveston Bay Tide Station 29-Day Harmonic Water Level Analysis
Line (1,2,3) = = = (May 1995 GBM, April 1996 GBM, April 1996 HSC)

Station	Gain (-)	Phase (hr)	RMS (cm)
Galveston Pier 21	0.87	-0.04	3
677-1450	0.85	0.13	3
	0.88	0.04	3
Eagle Point	0.97	0.32	3
677-1013	0.91	0.98	4
	0.90	0.79	4
Morgans Point	0.89	-0.68	3
677-0613	0.87	0.01	3
	0.87	0.08	3

Table 2. Galveston Bay PORTS Station Principal Flood Direction Analysis

Station	Model (deg True)			Observation (deg True)
	GBM (5/95)	GBM (4/96)	HSC (4/96)	(3-4/96)
Bolivar Roads	350	342	321	322
Redfish Bar	336	336	331	322
Morgans Point	317	313	318	341

Table 3. Galveston Bay PORTS Current Station 29-day Harmonic Principal Direction Analysis at prediction depth (4.6m).
Line (1,2,3) = = = (May 1995 GBM, April 1996 GBM, April 1996 HSC)

Station	Gain (-)	Phase (hr)	RMS (cm/s)
Bolivar Roads	0.86	-0.50	17.8
	0.78	0.36	17.0
	0.63	-0.27	26.8
Redfish Bar	0.72	0.06	19.7
	0.70	-0.17	14.4
	0.70	-0.66	16.6
Morgans Point	0.32	-1.45	16.0
	0.42	-0.77	13.1
	0.68	-0.49	8.6

errors order 20 cm/s. Based on these results, the model tidal current as well as tide response is damped.

Galveston Bay Model Extensions

To further improve the Bay model, heat flux, drying/wetting, and flux-corrected salinity transport schemes were incorporated. The latter two are described herein, as it was found that a SST specification was more favorable than heat flux as discussed below.

Drying/Wetting Scheme

The scheme developed by Hess (1994) in Tampa Bay is modified for application in Galveston Bay to simulate winter time "Northers", during which northerly winds of up to 40 knots persist over the Bay associated with cold front passages. An x-direction flow width reduction factor, $wx_{i,j}$ based on upstream vertically integrated velocity, $u_{i,j}^n$, and cell depth, $d_{i,j}^n$, is computed at the beginning of each external mode time step using the relation:

$$wx_{i,j} = \min(\ 1, \max(\ 0, (d_{i-1,j}^n - d_u)/d_T)\ )\ ) \qquad u_{i,j}^n \geq 0$$
$$wx_{i,j} = \min(\ 1, \max(\ 0, (d_{i,j}^n - d_u)/d_T)\ )\ ) \qquad u_{i,j}^n < 0 \qquad (1)$$
$$\text{Note } wx_{i,j} \in (0,1) \text{ for } d_{*,j} \in (d_u, d_T + d_u), \quad \text{where } * = (i, i-1).$$

An analogous relationship is used to specify the y-direction flow width reduction. In the Galveston Bay model application, $d_u = 0.25$m and $d_T = 0.5$m. Since the model is written in horizontal area format, one multiplies the x-direction flow width, $dx = 0.5(dx_{i,j} + dx_{i-1,j})$ by $wx_{i,j}$ and the analogous expression for dy by $wy_{i,j}$ to reduce the horizontal fluxes at each sigma level. The above linear cell depth relationship is used to reduce cell face flow widths when cell water depths drop below 0.75m and fully eliminates flow paths when water depths drop below 0.25m. For water depths greater than 0.75m no reduction in flow width is made. A five day test wind loading case, in which winds were ramped from 0 to 40 kts out of the north during day one, held constant at 40 kts out of the north for the next two days, then ramped to zero over the fourth day, and held at zero over the final day was used. May 1-5, 1995 astronomical tide conditions were specified with a -50 cm subtidal water level along the open boundaries ramped analogously to the wind. A large section of Trinity Bay dried and then reflooded after the wind and subtidal water level signal went to zero. In addition to the flood width reduction, it is necessary to reduce the wind stress over cells with small water depths. The following approach was utilized for the above test case and in subsequent simulations:

$$r_t = \min(\ 1, \max(\ 0, (\bar{d}_* - d_0)/(d_1 - d_0))\ )\ ) \qquad * \equiv (x,y) \qquad (2)$$
$$\text{where } \bar{d}_x = 0.5(d_{i,j}^n + d_{i-1,j}^n) \text{ and } \bar{d}_y = 0.5(d_{i,j}^n + d_{i,j-1}^n).$$

The factor r_t was applied to the surface wind stress terms with $d_0 = 0.5$m and $d_1 = 1.5$m, respectively. In general, the constants d_u and d_T in the cell width reduction formulas and d_0

and d_l in the wind stress reduction relationship must be determined for each application and are a function of the wind event strength and of the estuarine tidal range, bathymetry, and morphology. The scheme fails if a negative water depth is computed in the external mode.

Flux Corrected Transport Scheme

A second order van Leer-type upstream-biased transport scheme (Lin et al., 1994) has been implemented to treat the very sharp horizontal salinity gradients in Galveston Bay. The scheme was obtained from Dr. Sirpa Hakkinen, NASA-GSFC, and modified for application to the shallow water Galveston Bay region. Consider the following parameter, $f_{i,j,k}^m$, to represent grid cell salinity at internal mode time level m. The scheme corrects the flux based on grid cell upstream velocity, $u_{i,j,k}^m$, x-direction cell width, $dx_{i,j}$, and internal mode time step length, ΔT, in the following manner:

$$
\begin{aligned}
\Delta \bar{F}_{i,j,k}^m &= 0.5(f_{i+1,j,k}^m - f_{i-1,j,k}^m) \\
fmin &= \min(f_{i-1,j,k}^m, f_{i,j,k}^m, f_{i+1,j,k}^m) \\
fmax &= \max(f_{i-1,j,k}^m, f_{i,j,k}^m, f_{i+1,j,k}^m) \\
co &= 0.5(1 - u_{i,j,k}^m)\Delta T/dx_{i-1,j} \qquad u_{i,j,k}^m > 0 \\
co &= 0.5(1 + u_{i,j,k}^m)\Delta T/dx_{i,j} \qquad u_{i,j,k}^m \leq 0 \\
XFLUX &= u_{i,j,k}^m(f_{i-1,j,k}^m + co\Delta \bar{F}_{i-1,j,k}^m) \qquad u_{i,j,k}^m > 0 \\
XFLUX &= u_{i,j,k}^m(f_{i,j,k}^m + co\Delta \bar{F}_{i,j,k}^m) \qquad u_{i,j,k}^m \leq 0.
\end{aligned}
\tag{3}
$$

Analogous relationships hold for the y-direction , YFLUX, and sigma direction, ZFLUX. The XFLUX and YFLUX terms are multiplied by $\bar{d}_x = 0.5(d_{i,j}^m + d_{i+1,j}^m)wx_{i,j}$ and $\bar{d}_y = 0.5(d_{i,j}^m + d_{i,j+1}^m)wy_{i,j}$. The XFLUX, YFLUX, and ZFLUX terms replace the original quantities used in Subroutine ADVT (refer to Mellor, 1996). Note the scheme employs a single increment from time level m to level $m+1$ and hence, the diffusion terms are evaluated at time level m in the standard manner.

Hindcast Experiments

River inflows, water level residual forcings, wind and atmospheric pressure fields were included in all hindcasts. Average daily flows were obtained from the USGS Houston Office, for Buffalo Bayou at Piney Point, Texas, Trinity River at Romayor, Texas and Lake Houston near Sheldon, Texas via a stage vs discharge relation. NDBC buoy 42020 (3m Discus) and 42035 (3m Discus) and C-MAN station S-2 Sabine and S-4 Port Aransas, Texas observations were obtained along with NWS surface weather observations at Houston IAH, Port Arthur, and WSO Galveston, Texas. Wind and pressure fields were developed over the model domain via two-step Barnes (1973) interpolation. Subtidal water levels at Galveston Pleasure Pier were used along the entire GBM open boundary. Texas Water Development Board (TWDB) salinity and temperature data were melded with climatological salinity and temperature data to form the initial density fields, surface temperature fields, and offshore boundary salinity and temperature conditions.

During the October 1994 and January 1995 hindcast, the heat flux formulation developed by Martin (1985) was employed. In general, the heat flux formulation predicted temperatures

3-5 degrees cooler than the observations, and additional work is needed to further calibrate the flux parameters. For the April 1996 hindcast, a SST specification was employed. The mid-depth temperature comparisons were within 1 degree, and since SST information is available via PORTS, this approach has been adopted.

To test the flux-corrected transport scheme, a one-month simulation of October 1994 was performed. During 17-18 October, the flood of record occurred on both the San Jacinto and Trinity Rivers, with average daily flows on each river above 100,000 cfs. For the next week, the water in Galveston Bay remained fresh. Demeaned water level comparisons are given in columns 1 and 3 in Table 4. Rms errors are order 10 cm with the relative errors order 0.05 indicating excellent agreement in shape characteristics. The water level response shown in Figure 4 at Morgans Point shows the success of the model in capturing the major storm of Julian day 292. The model overprediction of water level may be due to the lack of a overland/marsch flooding algorithm. The present scheme will only allow an orginally wet cell to dry and then subsequently wet. The flux-corrected salinity response at Port Bolivar is compared with TWDB observations in Figure 5. Note the excellent agreement between the simulated and observed abrupt decrease of order 28 psu in salinity on Julian day 292. The lack of agreement order 3-4 psu prior to Julian day 292 is due to the difference in the climatological and observed initial salinity. The excellent replication of the advection of the large horizontal salinity gradient over Julian days 294-305 is also to be noted. The flux corrected salinity scheme exhibits no under or overshooting and positivity.

To test the drying/wetting algorithm, the January 1995 period was considered, during which several "Northers" occurred and portions of Trinity Bay near Round Point dried. Simulated water level responses at Round Point are compared with observations in Figure 6. Note the absence of low water on days 14,15,19,23,24,29, and 30 indicated by the clipping of the observed water level signals on low tide. The simulated water levels are in general agreement with the observations except that no clipping occurs and hence there is no loss of low water cell width flow reduction factors remain greater than zero. Additional knowledge on the spatial extent of the wetting/drying region is needed to further verify the drying/wetting scheme. Demeaned simulated water levels throughout the Bay are compared with observations in columns 2 and 4 in Table 4 and are less than 10 cm in rms and 0.05 in relative error.

Table 4. Demeaned Water Level GBM vs Data Intercomparisions

Station	RMS (cm)		Relative Error (-)	
	(10/94)	(1/95)	(10/94)	(1/95)
Galveston Pleasure Pier	8	8	0.04	0.02
Galveston Pier 21	5	6	0.03	0.02
Morgans Point	9	7	0.04	0.02
Clear Lake	8	7	0.04	0.02
Eagle Point	8	8	0.07	0.03
Port Bolivar	11	10	0.11	0.06
Round Point	-	9	-	0.03

During April 1996, PORTS current meters were in test operation and this period was used to evaluate both water level and current response. To simulate currents within the Houston Ship Channel (HSC), a fine resolution channel model (HSCM) was developed. The refined channel grid was developed in three sections based on the Wilken (1988) elliptic grid generation program patterned after Ives and Zacharais (1987). Each grid section was linked in order to develop the final composite channel grid (see inset Figure 3) consisting of 71 x 211 horizontal cells (dx=63-1007m, dy=133-1268m) with the same 5 sigma levels as in the

GBM. In both models, bathymetry is based on historical hydrographic surveys (Pullen et al., 1971). However, the HSC bathymetry was incorporated into the HSCM grid based on Corps of Engineers channel survey data given on nautical charts.

The two models were then nested in a one-way coupling scheme, wherein GBM water surface elevation, salinity, temperature, turbulent kinetic energy, and turbulent length scale time histories were saved at 6-minute intervals to provide boundary conditions to drive the HSCM. For salinity, temperature, turbulent kinetic energy, and turbulent length scale a one-dimensional (normal to the boundary) advection equation is used. On inflow GBM values are advected into the HSCM domain, while on outflow HSCM internal values are advected through the boundary. Open lateral boundary coupling is accomplished by: 1) specifying open boundary cells on HSC, 2)locating the corresponding (nearest neighbor) open boundary cells on Bay grid, 3) determining HSCM initial conditions via nearest neighbor on Bay grid and vertical sigma (GBM) - depth - sigma (HSCM) interpolation. Lateral flow boundary coupling is achieved by specifying river inflow cells on HSC grid and by using the corresponding flow and salinity and temperature boundary signals. Inflows and salinity and temperature boundary conditions are the same for Buffalo Bayou and San Jacinto Rivers, while the Trinity River is not included in HSCM. Surface boundary coupling is accomplished by placing the SST field on HSCM grid via nearest neighbor from the GBM grid. Wind and atmospheric pressure fields are directly determined on HSCM grid via 2-step Barnes (1973) interpolation.

April 1996 demeaned water level comparisons are presented in columns 1 and 3 and in columns 2 and 4 in Table 5 for the GBM and HSCM, respectively. Simulated water levels are nearly identical in each model(see HSCM response at Morgans Point in Figure 7), indicating that the coupling mechanics and grid topologies (grid structure and bathymetry) are compatible. Rms errors are order 5 cm and relative errors less than 0.05. Level 3 simulated principal direction currents are directly compared with observed currents in columns 1 and 3 and in 2 and 4 of Table 6 for the GBM and HSCM, respectively, and are shown in Figure 8 at Morgans Point for the HSCM during a ten day mid-month period. Both models produce excellent agreement in shape with GBM currents closer in agreement with observations near the entrance and HSCM results improved over GBM results at Morgans Point.

Table 5. Demeaned Water Level GBM and HSCM Model vs April 1996 Intercomparisons

	RMS (cm)		Relative Error (-)	
Station	(GBM)	(HSCM)	(GBM)	(HSCM)
Galveston Pleasure Pier	9	-	0.06	-
Galveston Pier 21	7	8	0.06	0.06
Morgans Point	5	5	0.02	0.02
Clear Lake	7	-	0.04	-
Eagle Point	5	5	0.02	0.02
Port Bolivar	7	7	0.06	0.06
Lynchburg Landing	6	6	0.02	0.02
Manchester Dock	7	8	0.02	0.03

Table 6. Principal Direction Current at Prediction Depth GBM and HSCM Model vs April 1996 Intercomparisions

	RMS (cm/s)		Relative Error (-)	
Station	(GBM)	(HSCM)	(GBM)	(HSCM)
Bolivar Roads	21	26	0.04	0.07
Morgans Point	17	13	0.34	0.11

Due to data quality issues, Redfish Bar currents were not considered. 29-day harmonic and principal flood direction analyses analogous to those previously discussed for the May 1995 simulation were performed and are given in lines 2 and 3 in Tables 1-3 for the GBM and HSCM, respectively. Salinity and temperature rms errors were comparable in both models and ranged from 1-4 PSU and were order 1.5 deg C, respectively.

Nowcasting/Forecasting System Design

The data delivery system consists of an Automated Data Archival System (ADAS) maintained by the NOS Coast Survey Development Laboratory (CSDL) in which NWS Aviation Model wind/pressure fields are automatically downloaded to CSDL machines. Additional scripts decode NWS Techniques Development Laboratory (TDL) storm surge water levels at Galveston Pleasure Pier. The NWS Western Gulf River Forecast Center (WGRFC) uploads to CDSL anonymous ftp 3 day 6 hour interval forcasted river flow and stage for the Trinity River at Liberty , Texas and Lake Houston Dam near Sheldon, Texas, respectively. In addition, the previous day's hourly discharges at Liberty, Texas on the Trinity River and at Piney Point, Texas on Buffalo Bayou and stage for Lake Houston Dam near Sheldon, Texas are uploaded. A decode script accesses and decodes the Houston/Galveston PORTS screen (see Figure 2) every 6 minutes and stores daily station files.

An initial design of a nowcasting/forecasting system compatible with the above ADAS has been completed. The design concept is modular such that refined hydrodynamic models can be readily substituted for the initial models. To this end, a separate nowcast/forecast program has been developed to establish hydrodynamic model forecast inputs. The program utilizes the following ten step procedure:

1) Setup 24 hour nowcast and 36 hour forecast time periods,
2) Predict astronomical tide,
3) Predict astronomical currents,
4) Read PORTS screeen and develop station time series,
5) Develop GBM subtidal water level signal,
6) Assimilate PORTS salinity and temperature data into GBM and HSCM initial conditions,
7) Establish GBM and HSCM salinity and temperature boundary conditions,
8) Establish GBM and HSCM SST forcings,
9) Establish USGS observed and NWS/WGRFC forecast freshwater inflows, and
10) Establish PORTS based and Aviation Model wind and pressure fields.

Time series files for predicted water surface elevation, and principal direction prediction depth currents are generated as well as PORTS time series data files for water levels, currents, salinity, temperature, wind, and atmospheric pressure. Time series analysis programs to plot nowcast and forecast results in conjunction with the above time series files for both models have also been incorporated within the system.

System Forecast Input Analysis

The system forecast inputs consist of NWS Aviation model winds, NWS/TDL forecasted subtidal water levels at Galveston Pleasure Pier, and NWS/WGRFC forecasted freshwater inflows. Analysis results are presented for each input in turn below.

NWS Aviation model winds at 995 mb were adjusted to 10m by using a factor of 0.846. The first twelve hours of the 00UTC and 12UTC forecast winds (3 hour interval) were composited into daily files over April 1996 and interpolated to observation station locations.

RMS difference, relative error (Willmott et al., 1985), bias, gain, correlation coefficient and standard error were computed versus observations and are given in Table 7. RMS differences in wind speed ranged from 3.1 to 4.0 m/s. Relative errors were below 0.5. Correlation coefficients were below 0.3. Additional work is in progress to extend the time period of the analysis to assess the stability of the bias and gain on a monthly basis. Based on this initial assessment, low resolution continental atmospheric model results may need to modified locally to provide regional estuarine forcings. This has been accomplished partially in this work through the use of the model land/water cell masks in the overwater/overland adjustment of windspeed. NWS/TDL forecast subtidal water levels using the above composition of 00UTC and 12UTC forecasts at Galveston Pleasure Pier were compared with observed residuals obtained by subtracting a one- year least squares harmonic constant prediction including the long period constituents from the observed water levels. Both series were demeaned and the rms difference computed during January 1997. The rms difference was approximately 29 cm with the ratio of the NWS/TDL standard deviation to the standard deviation of the observed residual order 2. The NWS/TDL series was then adjusted by dividing by the above ratio and the rms was recomputed at 13 cm. Efforts are in progress to extend the time period of the analysis to investigate the stability of the variability ratio on a monthly basis.

Table 7. Aviation Model 10 m Forecast Windspeed (m/s) Analysis Results

Parameter	Houston	Galveston	42035
RMS	3.05	3.14	4.03
Relative Error	0.43	0.46	0.47
Bias	3.57	4.92	5.77
Gain	0.23	0.17	0.17
Correlation Coefficient	0.28	0.28	0.19
Standard Error	2.14	1.77	2.91

Streamflow data were obtained from the United States Geological Survey from April to September 1996 for the three major inflows. During this period below normal stream flows occurred and the use of a one-day persistence forecast resulted in monthly rms differences of the same order as the mean flow. During low flow conditions, the one-day persistence forecast on the Buffalo Bayou was appropriate. During August and September significant rainfall/runoff events occurred and the one day persistence forecast was significantly in error. However, the flows on the Buffalo Bayou constitute in most instances less than 20 percent of the total freshwater inflow.

Review and Future Plans

From Table 2, principal flood directions appear to be slightly improved in the HSCM relative to the GBM. Based upon comparison of Table 1 with Tables 4-5 and of Table 3 with Table 6, a major portion of the water level and current errors are in the astronomical tidal component in both models. As a result, additional experiments will focus on improving tidal response. Refined tidal boundary conditions and further adjustment of bottom friction will be considered. Since the GBM appears to be damped an increase of tidal amplitude of order 10 percent will be investigated. Within the HSCM, a velocity/transport boundary condition will be explored in addition to further adjustments of the present internal mode radiation scheme. Additional work on specifying the subtidal signal along the GBM open boundary will also be undertaken.

In concert with this effort, additional tests based on individual forecast experiments will be conducted to refine the nowcast/forecast system. After further refinement, the system will be tested over a 1-3 month period in quasi-operational mode during which 00UTC nowcast/forecast cycles will be executed.

In the longer term, we will pursue the development of an ongoing nowcast system, which will directly provide the initial conditions for twice (00UTC and 12UTC) daily forecasts. These forecasts will be more efficiently conducted with reduced processing time and be more robust with automated quality control of a PORTS expanded to include current, salinity, and temperature measurements off the entrance onto the near shelf.

Acknowledgements

Dr. Bruce B. Parker, Chief of the Coast Survey Development Laboratory, conceived of the this project and provided leadership and critical resources. Dr. Kurt W. Hess supplied a wealth of information on modeling shallow bay systems. Special acknowledgement is made to Mr. Philip H. Richardson, who developed the hindcast experiment wind and atmospheric pressure field interpolation program and performed the forecast system input analysis.

References

Barnes, S. L., 1973: Mesoscale objective map analysis using weighted time-series observations, *NOAA Technical Memorandum ERL NSSL-62*, National Severe Storms Laboratory, Norman, OK.

Bethem, T. D., and H. R. Frey, 1991: Operational physical oceanographic real-time data dissemination. *Proceedings, IEEE Oceans 91*, 865 - 867.

Blumberg, A. F., and G. L. Mellor, 1987: A description of a three-dimensional coastal ocean circulation model. *Three-Dimensional Coastal Ocean Models*, (ed. Heaps), American Geophysical Union, Washington, DC., 1 - 16.

Hess, K. W., 1994: Tampa Bay Oceanography Project: Development and application of the numerical circulation model. NOAA, National Ocean Service, Office of Ocean and Earth Sciences, *NOAA Technical Report NOS OES 005*, Silver Spring, MD.

Ives, D. C. and R. M. Zacharias, 1987: Conformal mapping and orthogonal grid generation, Paper No. 87-2057, *AIAA/SAE/ASME/ASEE 23rd Joint Propulsion Conference*, San Diego, CA.

Lin, S-J., W. C. Chao, Y. C. Sud, and G. K. Walker, 1994: A class of the van-Leer transport schemes and its application to the moisture transport in a general circulation model, *Monthly Weather Review*, 122, 1575-1593.

Martin, P.J, 1985: Simulation of the mixed layer at OWS November and Papa with several models, *Journal of Geophysical Research*, 90, C1, 903-916.

Mellor, G.L., 1996: Princeton Ocean Model Users Guide, Program in Oceanic and Atmospheric Sciences, Princeton University.

Orlando, S. P., L. P. Rozas, G. H. Ward, and C. J. Klein, 1993: Salinity characterization of Gulf of Mexico estuaries, *NOAA Tech Memorandum, Office of Ocean Resources Conservation and Assessment*, Silver Spring, MD.

Pullen, E. J., W. L. Trent, G. B. Adams, 1971: A hydrographic survey of the Galveston Bay system, Texas, 1963-1966, *NOAA Technical Report NMFS SSRF-639*, Seattle, WA.

Schmalz, R. A., 1994: Long Island Sound Oceanography Project Summary Report, Volume 1: application and documentation of the Long Island Sound three-dimensional circulation model. NOAA, National Ocean Service, Office of Ocean and Earth Sciences, *NOAA Technical Report NOS OES 03*, Silver Spring, MD.

Schmalz, R. A., 1996: National Ocean Service Partnership: DGPS-supported hydrosurvey, water level measurement, and modeling of Galveston Bay :development and application of the numerical circulation model. NOAA, National Ocean Service, Office of Ocean and Earth Sciences, *NOAA Technical Report NOS OES 012*, Silver Spring, MD.

Schureman, P., 1958: Manual of harmonic analysis and prediction of tides. *U.S. Department of Commerce, Coast and Geodetic Survey, Special Publication No. 98 [revised 1940 edition, reprinted 1988]*, Rockville, MD.

Temple, R. F., D. L. Harrington, J. A. Martin, 1977: Monthly temperature and salinity measurements of continental shelf waters of the northwestern Gulf of Mexico, 1963-1965., *NOAA Technical Report NMFS SSRF-707*, Rockville, MD.

Wilken, J. L., 1988: A computer program for generating two-dimensional orthogonal curvilinear coordinate grids, *Woods Hole Oceanographic Institution (unpublished manuscript)*, Woods Hole, MA.

Willmott C. J., S.G. Ackleson, R.E. Davis, J.J. Feddema, K. M. Klink, D.R. Legates, J. O'Donnell, and C. M. Rowe, 1985: Statistics for the evaluation and comparison of model, *Journal of Geophysical Research*, 90, 8995-9005.

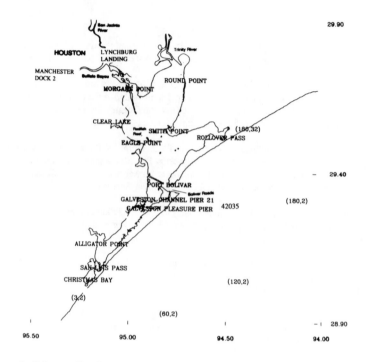

Figure 1. Galveston Bay Base Map with PORTS station locations indicated. Ordered pairs denote Galveston Bay Model grid cell coordinates at water level boundary signal locations.

```
                        Houston/Galveston PORTS
                        National Ocean Service/NOAA
                      at  1:00 pm CST  March 31, 1997
...............................................................
            TIDES              :             CURRENTS
Pleasure Pier        3.1 ft.   : Bolivar Roads    1.7 kts.(F), 302xT
Pier 21              2.4 ft.   : Redfish Bar      1.5 kts.(F), 320xT
Bolivar Roads        2.0 ft.   : Morgans Point    0.5 kts.(F), 337xT
Eagle Point          1.8 ft.   : (F)lood,(S)lack,(E)bb,towards xTrue
Morgans Point        1.7 ft.,Rising :..............................
...........................................:  Salinity  S.G. Srfc Temp
   Bottom Water Temp. (ADCP) : Pleasure Pier                    69xF
Bolivar Roads        69øF      : Bolivar Roads    21.0 psu 1.015 66xF
Redfish Bar          72øF      : Eagle Point       3.0 psu 1.001 68xF
Morgans Point        70øF      : Morgans Point     5.0 psu 1.003 71xF
...............................................................
   METEOROLOGICAL        Wind Speed/Dir        Air Pressure    Air Temp
Pleasure Pier    17 knots from ENE, gusts to  22  1021 mb,Fallin  64xF
Bolivar Roads    12 knots from ENE, gusts to  22  1023 mb,Fallin  65xF
Eagle Point      17 knots from E  , gusts to  24  1024 mb,Fallin  65xF
Morgans Point    11 knots from E  , gusts to  15  1023 mb,Rising  65xF
...............................................................
For a description of PORTS, please contact Capt. Stephen Ford of TAMUG at
(409) 740-4471 or NOAA PORTS Office at (301) 713-2806
```

Figure 2. Houston/Galveston PORTS Screen Data Format.

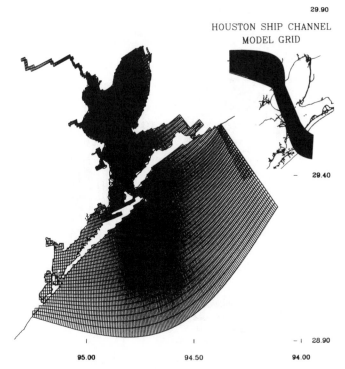

29.90

HOUSTON SHIP CHANNEL
MODEL GRID

29.40

28.90

95.00   94.50   94.00

Figure 3. Galveston Bay Model grid with fine resolution Houston Ship Channel Model grid inset.

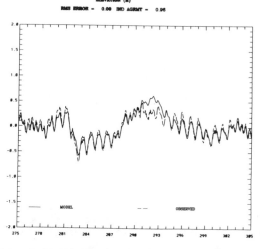

ELEVATION (M)
RMS ERROR = 0.00 IND AGRMT = 0.96

MODEL   OBSERVED

Figure 4. Galveston Bay Model simulated versus observed water level at Morgans Point during October 1994. (Note IND AGMT equals one minus dimensionless relative error.)

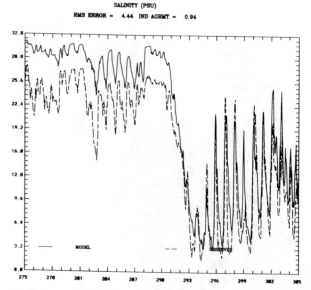

Figure 5. Galveston Bay Model simulated versus observed salinity at Port Bolivar during October 1994. (Note IND AGMT equals one minus dimensionless relative error.)

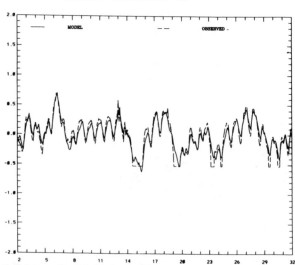

Figure 6. Galveston Bay Model simulated versus observed water level at Round Point during January 1995. (Note IND AGMT equals one minus dimensionless relative error.)

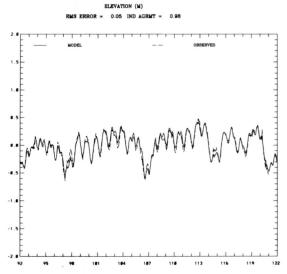

Figure 7. Houston Ship Channel Model simulated versus observed water level at Morgans Point during April 1996. (Note IND AGMT equals one minus dimensionless relative error.)

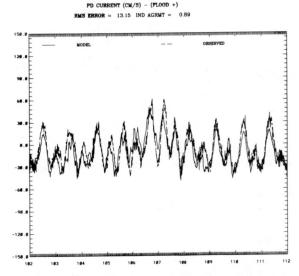

Figure 8. Houston Ship Channel Model simulated versus observed principal direction current at Morgans Point during April 1996. (Note IND AGMT equals one minus dimensionless relative error.)

Morphodynamic Responses to Extratropical Meteorological Forcing on the Inner
Shelf of the Middle Atlantic Bight: Wind Waves, Currents, and Suspended Sediment
Transport

Sung-Chan Kim[1], L. Donelson Wright[2], Jerome P.-Y. Maa[3], and Jian Shen[4]

Abstract

A simulation of the nearshore morphodynamics on the inner shelf of the Middle
Atlantic Bight, for a one-month period during October 1994 with prevailing
extratropical systems, was attempted through a series of models—a wind wave
model, a circulation model, and a wave-current combined boundary layer model. The
STWAVE model was used to calculate spectral transformations of waves in shallow
water through nonlinear action density fluxes. A three-dimensional circulation model
(HEM-3D) provided the near bed mean current structures responding to the
meteorological forcing. The waves and currents calculated were input to a wave-
current boundary layer model to calculate the bottom boundary layer
characteristics—skin friction, bottom stress, and bottom roughness. Suspended
sediment fluxes were obtained from the concentration and current profiles. Four
phases around a storm passage—the pre-storm calm period, the onset of a
northeaster, the peak of a storm, and the post-storm period—were investigated.
Changes in peak frequencies and wave heights in shoaling waves were realized by
the STWAVE model. Both the directions and the magnitudes of bottom currents
from the HEM-3D model agree with measurements. The HEM-3D model
underestimates bottom stresses under high wave conditions, which can be improved
by including a wave-current boundary layer model.

[1] Visiting Assist. Prof., Virginia Institute of Marine Science, School of Marine Science, College of
William and Mary, Gloucester Point, VA 23062
[2] Prof., Virginia Institute of Marine Science, School of Marine Science, College of William and Mary,
Gloucester Point, VA 23062
[3] Member ASCE, Assoc. Prof., Virginia Institute of Marine Science, School of Marine Science,
College of William and Mary, Gloucester Point, VA 23062
[4] Postdoctoral Res. Assoc., Virginia Institute of Marine Science, School of Marine Science, College of
William and Mary, Gloucester Point, VA 23062

Introduction

The coastal morphodynamic system is defined as the complete assemblage of boundary constraints, driving forces, fluid dynamic processes, morphologies, and sequences of evolutionary changes (e.g. Wright, 1995; Fig 1). To properly simulate the processes in the natural world, we need to assess not only the coastal depositional forms and their temporal and spatial variability, but also the patterns of fluid dynamic behavior which are linked to the forms through bi-directional causality. The inner shelf couples the wave dominated surf zone and the geostrophic and tidal current dominated outer continental shelf. The increased friction effect caused by decreased water depths toward the coast over the inner shelf thus induces a wave-current boundary layer problem (e.g., Grant and Madsen, 1986).

Over the Middle Atlantic Bight (MAB), extratropical systems have posed significant impacts on hydrodynamics and subsequent sediment transport. The dominant role of such systems as "northeasters" have been acknowledged in many studies (e.g., Dolan and Davis, 1994). Recently, Kim et al (1997) identified the morphodynamic responses with varying extratropical conditions through the combined effects of synoptic-scale and local-scale meteorological events on boundary layer dynamics. They examined the bottom boundary layer measurement data over an inner shelf off Duck, North Carolina (Fig 2) during the 25-day period starting October 5, 1994 by Virginia Institute of Marine Science (VIMS).

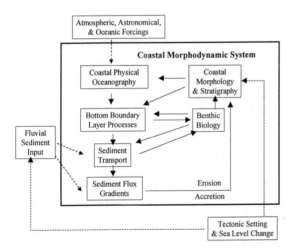

Fig 1. Morphodynamic system (Wright, 1995). In this study, we focused on the one way processes from physical oceanography to sediment transport.

In this study, we attempt to model the morphodynamic behavior in order to complement such valuable yet sparse observation data. Atmospheric forcing was provided by observation at the US Army Corps of Engineers Field Research Facility (FRF) at Duck, North Carolina. Offshore wave data, as well as supplemental wind data, were given by National Oceanic and Atmospheric Administration (NOAA) Buoy 44014 (36°34'59"N, 74°50'01"W) about 180 km offshore from Duck, North Carolina. The VIMS tripod measurements are used to verify the model results. Wind-waves are simulated and current and turbulence fields are modeled concurrently. A bottom boundary layer model then calculates bottom stress and suspended sediment concentration, which determines suspended sediment flux.

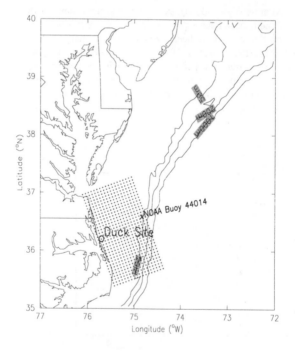

Fig 2. Study area. Circle represents the VIMS tripod measurement site location. X is the NOAA Buoy 44014 location. The 4.5-km grids for the HEM-3D model are shown. For STWAVE, the 2.25-km grids are used.

Meteorology

The weather pattern over the MAB October 1994 was typical of an extratropical origin. Two significant high wind events occurred during this period (Fig 3). Relatively strong winds (over 10 m/s) were sustained over a one-week period between 10 and 17 October. This event started with a northeaster characterized by a strong anticyclone and was highlighted with an intensifying cyclone approaching and retracting from this area. The second event around October 27 was of short duration.

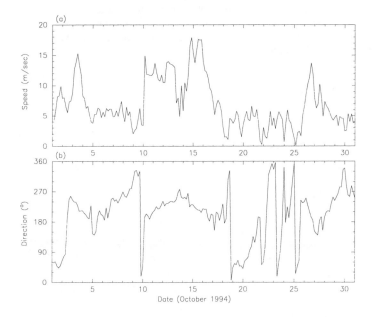

Fig 3. Surface winds: (a) speed (m/s); (b) direction (degrees clockwise from north) from FRF anemometer data at 19.5 m above MSL.

Waves

The STWAVE model is based on a series of studies (Resio, 1987, 1988; Resio and Perrie, 1989) in which spectral action density instead of spectral energy density represents the existence of a depth independent equilibrium range of a wind wave spectrum in shallow water. Another important result of these works is the existence of a natural limit to the evolution of the frequency of the spectral peak into lower

frequencies. A steady-state condition is assumed, which is reasonable for the
extratropical condition. Offshore directional wave spectra serve as input. As the
waves propagate toward the coast with refraction by topography (kinematics), local
wind effect, wave-wave interaction, and energy dissipation by wave breaking
(dynamics) interact to give energy spectra at each grid point.

Square grids with a 2.25-km cell spacing (a 37 by 75 matrix) were generated.
Directional wave data from NOAA Buoy 44014 provided offshore wave condition.
FRF wind data were used as local forcing input on each grid point. Fig 4 shows the
wave transformation toward the coastline. Refraction caused the wave rays to
become perpendicular to the isobaths. Wave period increase as well as wave height
decrease toward the coastline during the 10/10 – 10/17 wind event visibly shows the
spectral growth with source terms from atmospheric forcing, wave-wave
interactions, and bottom friction effects.

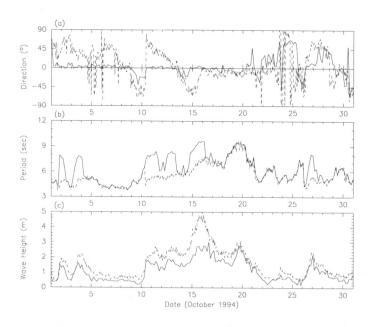

Fig 4. Modeled (solid line) wave (a) direction, (b) period, and (c) height at Duck site
on 20-m isobath. Dashed line represents offshore wave conditions given at NOAA
Buoy 44014.

Currents

By doubling the grid length of the wave model grid, we generated an 18 by 38 matrix of 4.5-km grids for the hydrodynamic calculation. Vertically, we chose 20 layers. The model (HEM-3D) is described elsewhere in this volume (Shen et al.). It is able to reproduce the bottom currents measured by the VIMS tripod deployment (Fig 5). Control of local wind over the hydrodynamics is apparent. During the prevailing northeaster between 10/10 and 10/13, bottom currents rapidly accelerated to about 30 cm/sec and the direction became predominantly alongshore. With the approaching storm system around 10/16, the maximum speed reached 40 cm/sec and the directions were alongshore but with cross-shore variation.

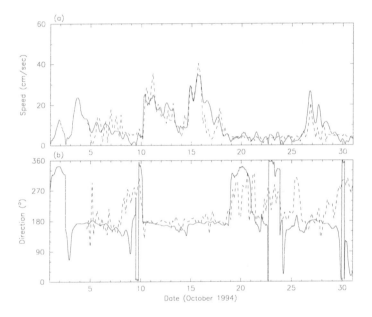

Fig 5. Bottom current (a) speed (cm/sec) and (b) direction (degrees clockwise from the positive y-axis) evaluated at 50 cm above bottom. Solid lines represent the model whereas dashed line represent observation from VIMS tripod.

Boundary layer dynamics and suspended sediment transport

To examine the role of wave-current interaction, we applied Grant and Madsen's (1986) 2-layer linear eddy viscosity model (GM model) in which thin wave-current combined boundary layer ($\sim \delta_{cw}$) is nested beneath current boundary layer ($\sim \delta_c$). Suspended sediment concentration induced stratification is accounted by the Monin-Obukhov length scale, L_c. Glenn and Grant (1987) gives the stability parameter at height, z, from the bed as

$$\frac{z}{L_c} = \frac{\kappa\,z}{u_{*c}^3} \sum_n g(s_n - 1) w_{fn} C_{mn}(z) \tag{1}$$

where g is gravitational acceleration and κ is von Karman constant. For each sediment class n, s_n is relative density and w_{fn} is settling velocity.

Mean concentration is given by

$$C_{mn}(z) = \begin{cases} C_{mn}(z_{0r})\left(\dfrac{z}{z_{0r}}\right)^{\gamma w_{fn}/\kappa u_{*cw}} & \text{for } z < \delta_{cw} \\[3mm] C_{mn}(\delta_{cw})\left(\dfrac{z}{\delta_{cw}}\right)^{\gamma w_{fn}/\kappa u_{*c}} \exp\left(-\dfrac{\beta w_{fn}}{\kappa u_{*c}} \int_{\delta_{cw}}^{z} \dfrac{dz}{L_c}\right) & \text{for } z > \delta_{cw} \end{cases} \tag{2}$$

Here β (~ 4.7) and γ (~ 0.74) are constants. z_{0r} is reference height for concentration.

Mean velocity is given by

$$u(z) = \begin{cases} \dfrac{u_{*c}}{u_{*cw}} \dfrac{u_{*c}}{\kappa} \ln\left(\dfrac{z}{z_0}\right) & \text{for } z < \delta_{cw} \\[3mm] \dfrac{u_{*c}}{\kappa}\left(\ln\left(\dfrac{z}{z_0}\right) + \beta \int_{\delta_{cw}}^{z} \dfrac{dz}{L_c}\right) & \text{for } z > \delta_{cw} \end{cases} \tag{3}$$

Here, z_0 is roughness length scale. Bottom current from the HEM-3D model and wave data from the STWAVE model are used for input. Bottom sediments are described in Kim et al. (1997).

Compared to the VIMS tripod data, the HEM-3D model underestimates the bottom friction velocity, u_{*c}, during the storm peak (Fig 6). In the HEM-3D model, we assume constant apparent roughness, z_{0c}. Over an inner shelf, z_{0c} is increased due to wave-current interaction (e.g., Grant and Madsen, 1979), bedforms (e.g., Wiberg and Harris, 1994), sediment transport (e.g., Xu and Wright, 1995), and suspended sediment concentration induced stratification (e.g., Glenn and Grant, 1987), which in turn changes u_{*c}. The GM model improves the underestimation of u_{*c} from HEM-3D (Fig 6). This indicates the importance of the role of the wave-current boundary layer in sediment transport over an inner shelf. One has to note that the error of z_{0c} estimation from log-fit is big compared to u_{*c}, thus we have to content with the order of magnitude agreements between model and observations.

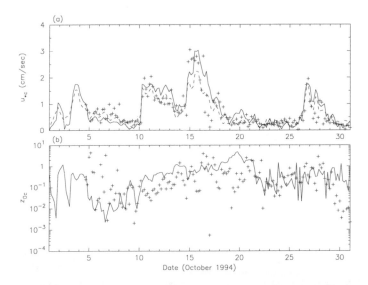

Fig 6. Bottom boundary layer characteristics; (a) bottom friction velocity, u_{*c}, in cm/sec and (b) apparent bottom roughness, z_{0c}, in cm. Cross represents log-fit estimation from VIMS tripod measurement data. Solid line represents the GM model results whereas dashed line represents the HEM-3D model result.

Suspended sediment flux, Q, was calculated as the depth integration of the product of mean velocity, u(z) and mean concentration, $C_m(z)=\Sigma_n C_{mn}(z)$:

$$Q = \int_{z_{0r}}^{\zeta} uC_m dz \qquad (4)$$

The calculated bottom 2 m suspended sediment fluxes are shown in Fig 7. Alongshore flux is an order of magnitude higher than cross-shore flux. The flux events clearly respond to meteorological events. Prevailing N to NE to E winds cause primarily southerly bottom currents thus strong southerly sediment flux. A secondary, but more important, down-welling component causes offshore flux of suspended sediment under these systems.

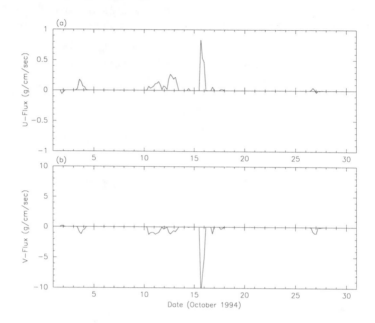

Fig 7. Suspended sediment flux (g/cm/s) for bottom 2 m calculated from the GM model: (a) cross-shore flux (positive offshore) and (b) alongshore flux (positive north).

Concluding remarks

This study complements the VIMS tripod measurements of bottom boundary layer dynamics and suspended sediment transport, demonstrating the control of morphodynamics by extratropical systems over the MAB inner shelf. The two layer wave-current bottom boundary layer model with suspended sediment concentration induced stratification, driven by the results from the concurrent modeling of waves and bottom currents, appropriately reproduce the bottom friction velocity, u_{*c}, and apparent roughness, z_{0c}, estimated from the measurement.

The STWAVE model's steady state assumption may work with long time scales of extratropical systems. The kinematics and dynamics of the STWAVE model appears to be reasonable qualitatively. The HEM-3D model also seems to give satisfactory bottom current. The use of GM model using the results from the concurrent wave and circulation models proves the necessity of including more realistic bottom boundary model. Mastenbroek et al. (1993) showed the dependency of surface drag on wave ages, which may be substantial for circulation models. This may be another reason to link a circulation model to a wave model. However, before developing a feedback mechanism or a hybrid model for a circulation model with a wave model, more careful quantitative examination is desirable.

Acknowledgements

This study was supported by the National Science foundation (OCE-9123513). Dr. Donald T. Resio of US Army Corps of Engineers generously provided the STWAVE model. Dr. John M. Hamrick developed the base code of the HEM-3D model. We owe thanks to Dr. John D. Boon, Dr. Albert Kuo, and Mr. G. McAllister Sisson for their support and advice. We also thank reviewers for the constructive criticism. This is Contribution Number 2092 from the Virginia Institute of Marine Science.

References

Dolan, R., and Davis, R. E. (1994) Coastal storm hazards (ed. by Finkl, Jr. C. W. Coastal hazards: Perception, susceptibility and mitigation) *Journal of Coastal Research*, Spec. Issue, **12**, 103-114

Glenn, S. M., and Grant, W. D. (1987) A suspended sediment stratification correction for combined wave and current flows. *Journal of Geophysical Research*, **92**, 8244-8264

Grant, W. D., and Madsen, O. S. (1979) Combined wave and current interaction with a rough bottom. *Journal of Geophysical Research*, **84**, 1797-1808

Grant, W. D., and Madsen, O. S. (1986) The continental shelf bottom boundary layer. Annual Review of Fluid Mechanics, **18**, 265-305

Kim, S.-C., Wright, L.D., and Kim, B.-O. (1997) The combined effects of synoptic-scale and local scale meteorological events on bed stress and sediment transport on the inner shelf of the Middle Atlantic Bight. *Continental Shelf Research*, **17**, 407 – 433

Mastenbroek, C., Burgers,G., and Janssen, P.A.E.M. (1993) The dynamical coupling of a wave model and a storm surge model through the atmospheric boundary layer. *Journal of Physical Oceanography*, **23**, 1856-1866

Resio, D.T. (1987) Shallow-water waves. I: Theory. *Journal of Waterway, Port, Coastal, and Ocean Engineering*, **113**, 264–281

Resio, D. T. (1988) Shallow-water waves. II: Data comparison. *Journal of Waterway, Port, Coastal, and Ocean Engineering*, **114**, 50–65

Resio, D.T. and Perrie, W. (1989) Implication of an f^{-4} equilibrium range for wind-generated waves. *Journal of Physical Oceanography*, **19**, 193–204

Shen, J., Sisson, M., Kuo, A., Boon, J., and Kim, S. (*this volume*) Three dimensional numerical modeling of the tidal York River system, Virginia.

Wiberg, P. L. and Harris, C. K. (1994) Ripple geometry in wave-dominated environments. *Journal of Geophysical Research*, **99**, 775-789

Wright, L. D. (1995) Morphodynamics of inner continental shelves. *CRC Press*, 241 pp

Xu, J. P. and Wright, L. D. (1995) Tests of bed roughness models using field data from the Middle Atlantic Bight. *Continental Shelf Research*, **15**, 1409-1434

A PARTICLE METHOD FOR SEDIMENT TRANSPORT
MODELING IN THE JADE ESTUARY

Frank Bergemann[1], Günther Lang[1], and Gerd Flügge[1]

ABSTRACT

A method for the off-line computation of particle tracks in a time-dependent 2D-depth-averaged flowfield is presented. The hydro-numerical method TRIM-2D is applied to compute the underlying flowfield. Within the particle method, the Reynolds-averaged movement of fluid particles as well as the movement of sediment particles can be described. Deposition and erosion of sediment particles on the sea bed depends on the balance between the settling velocity and the turbulent velocity component in the vertical direction. For the outer Jade Estuary, transport paths of sediments obtained from the particle method, from the bedload transport model of van Rijn and from a Sediment Trend Analysis[TM] being based on samples of bottom sediment are compared. The results from *PARTRACE* and the van Rijn model agree well, but the differences to the Sediment Trend Analysis are severe. The conclusion is that the quality of the computed flow field needs to be improved.

INTRODUCTION

In the Jade Estuary located within the German Bight (see Fig. 1), a complicated tidal flow modified by tidal flats evolves. Along with it a delicate transport of majorly non-cohesive sediments is observed. Millions of German Mark are spent every year for dredging to keep the shipping channel to the Wilhelmshaven harbor at the required water depth of 20 meters. For unknown reasons the necessary amount of dredging varies strongly along the channel, the most demanding sections lying at its southern end near Wilhelmshaven harbor and along a stretch to the north of the island Wangerooge. Our intention is to investigate the effects of regulation measures proposed to reduce the accretion of sediment in the channel.

The hydro-numerical (HN) model *TRIM-2D* developed by Casulli and Cheng (1993) is used to compute the flow in the Jade Estuary for several tidal cycles. In *TRIM-2D*, the 2D depth-averaged conservation equations of mass and momentum of the fluid are solved on a rectangular grid by a finite difference method. *TRIM-2D* also allows to compute the transport capacity of bedload sediments from one of two offered semi-empirical models. In addition, we developed a particle tracking method called *PARTRACE* (*PAR*ticle *TRACE* program) for the off-line computation of the movement of different types of particles within a 2D depth-averaged flowfield.

[1]all authors: Federal Waterways Engineering and Research Institute (BAW), Coastal Department, Wedeler Landstraße 157, D-22559 Hamburg, Germany, fax:+49-40-81908-373, email:<last name>@hamburg.baw.de

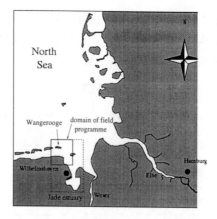

Fig. 1: Overview map of the German Bight.

In the first part of this study, the physical background and some implementation aspects of the particle method are described. Details that cannot be given here can be found in a program manual available from the WWW homepage of the BAW-AK: http://www.hamburg.baw.de/hnm/partrace/partrace-de.htm (written in german). The second part of this study deals with the sediment transport regime in the outer Jade Estuary around the island Wangerooge. Sediment transport paths from a *PARTRACE* computation are compared to results of two other methodologies, the first being the bedload transport model of van Rijn (1989) which is available within *TRIM-2D*. The other distinct methodology is the Sediment Trend Analysis (STA[TM])[1] (McLaren and Bowles 1985) which comprises of a statistical analysis of the grain-size distribution of bottom sediment. In order to utilize this method, an extensive field programme was undertaken in the area of interest to collect samples of bottom sediment. From the comparison of these results interesting conclusions about the quality of the numerical models can be drawn.

PHYSICAL MODELING

Input quantities and grid structure

PARTRACE requires the following quantities as input from the HN model: the grid in the horizontal plane, a definition of a triangulation of that grid, the bathymetry H, the water-level elevation η, and the 2D depth- and Reynolds-averaged horizontal velocity components U and V. The triangular grid is vertically projected to the sea bed and to

---

[1] Sediment Trend Analysis is a trademark by GeoSea Consulting (UK) Ltd.

the water level surface. This forms a 3D mesh consisting of a single layer of prismatic cells in which the particle movement is simulated. The cell structure in the horizontal plane and in 3D is shown in Fig. 2.

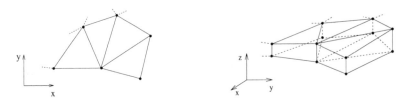

Fig. 2: Grid structure in the horizontal plane and in 3D.

Equations of motion

In our model, the inertia of a particle is neglected. The equations of motion of a particle in the three spatial directions are written as a system of ordinary first-order differential equations

$$
\begin{aligned}
\dot{x}_p &= U + u' \\
\dot{y}_p &= V + v' \\
\dot{z}_p &= w' + w_k - w_s
\end{aligned}
\tag{1}
$$

Here, $(\dot{x}_p, \dot{y}_p, \dot{z}_p)$ denote the time-derivative of the position of the particle, i.e. its velocity components. The first column of the set of relations defined in Eq. (1) contains the components of the mean flow velocity. The entries in the second column (u', v', w') denote stochastic velocity components that describe turbulent mixing. Next, w_k denotes the vertical velocity that makes a particle horizontally advected with velocities U and V follow the vertical convergence or divergence of streamlines caused by the inclination of the water-level surface and of the sea-bed against the horizontal plane. This effect must be computed in retrospect because a 2D depth-averaged method does not solve for a vertical velocity component. Finally, w_s indicates the settling velocity of a particle. The path of a particle in the flowfield is determined by solving Eq. (1) using a Runge-Kutta scheme due to Cash and Karp, see Press et al. (1992).

Different types of particles can be modeled by Eq. (1). For example, if (u', v', w') and w_s are set to zero a particle merely follows the main motion of the water body. Such a particle is called here a fluid particle. A substance dissolved in the fluid, e.g. a salt particle is modeled by leaving w_s at zero but switching on (u', v', w'). To describe sediment particles, all four contributions on the right hand side (r.h.s.) of Eq. (1) must be active.

Interpolation

The time-dependent quantities η, U and V at some grid node i are linearly interpolated in time. In case of η this reads

$$\eta_i(t) = a_i^\eta t + b_i^\eta, \tag{2}$$

where

$$a_i^\eta = \frac{\eta_{i,j+1} - \eta_{i,j}}{t_{j+1} - t_j}, \qquad b_i^\eta = \frac{\eta_{i,j} t_{j+1} - \eta_{i,j+1} t_j}{t_{j+1} - t_j}, \qquad t_j \le t \le t_{j+1}. \tag{3}$$

Here, t_j, t_{j+1} are the time levels adjacent to time t and η_{ij} denotes the η-value at node i and time level j.

The spatially dependent variables H, η, U and V are bilinearly interpolated from the nodes of the triangle 1,2,3, within which the particle lies in to the location of the particle (x, y). For example, for H this can be written as

$$H(x, y) = F_0^H + F_x^H x + F_y^H y. \tag{4}$$

The interpolation coefficients F_0^H, F_x^H, F_y^H follow from the condition, that at the nodes of the triangle, Eq. (4) must reproduce the node values H_1, H_2, and H_3. This yields

$$\underline{F}^H = B^{-1} \underline{H}, \tag{5}$$

where

$$B = \begin{pmatrix} 1 & x_1 & y_1 \\ 1 & x_2 & y_2 \\ 1 & x_3 & y_3 \end{pmatrix}, \qquad \underline{F}^H = \begin{pmatrix} F_0^H \\ F_x^H \\ F_y^H \end{pmatrix}, \qquad \underline{H} = \begin{pmatrix} H_1 \\ H_2 \\ H_3 \end{pmatrix}, \tag{6}$$

and B^{-1} is the inverse matrix to B. Eq. (5) and (6) likewise hold for η, U and V.

Combination of the above equations gives the following space-time interpolation formula for η

$$\eta(x, y, t) = C_1^\eta t + D_1^\eta + (C_2^\eta t + D_2^\eta) x + (C_3^\eta t + D_3^\eta) y, \tag{7}$$

where

$$C_i^\eta = (B^{-1})_{ij} a_j^\eta, \qquad D_i^\eta = (B^{-1})_{ij} b_j^\eta \qquad i = 1, 2, 3. \tag{8}$$

In Eqs. (8), the summation convention is used. Similar relations hold for U and V.

Inclination-induced velocity

The inclination-induced velocity w_k can be found from the continuity equation

$$\frac{\partial w}{\partial z} = -\left(\frac{\partial u}{\partial x} + \frac{\partial v}{\partial y}\right), \tag{9}$$

where (u, v, w) are the 3D flow velocity components. Since in our model particles are advected by the depth-averaged velocities U and V, u and v must be replaced by U and V in Eq. (9). Then the r.h.s. of Eq. (9) becomes independent from z, so that Eq. (9) can be integrated to yield $w = w_k = a z + b$. Similar to Eq. (3), the straight-line coefficients a and b can be expressed in terms of the boundary values of w_k at the sea bed and at the water-level surface. The boundary values of w_k can be determined by evaluating the condition that there is no mass mass flux through the sea bed and the water-level surface. One finally obtains

$$w_k = \left(\frac{n_b^x}{n_b^z} U + \frac{n_b^y}{n_b^z} V\right)\frac{\eta - z}{H + \eta} + \left(\frac{n_\eta^x}{n_\eta^z} U + \frac{n_\eta^y}{n_\eta^z} V + C_1^\eta + C_2^\eta x + C_3^\eta y\right)\frac{H + z}{H + \eta}. \tag{10}$$

Here, (n_b^x, n_b^y, n_b^z) and $(n_\eta^x, n_\eta^y, n_\eta^z)$ are the normal-vector components of the sea bed and the water-level surface, both pointing into the flow, respectively. Furthermore, $(C_1^\eta, C_2^\eta, C_3^\eta)$ are given by Eq. (8) and H, η, U, and V are evaluated from Eqs. (4) and (7).

Turbulent velocity components

To determine the turbulent velocity components in the horizontal plane u' and v', the mixing-coefficient model in connection with a Monte-Carlo method is used, e.g., Dick and Schönefeld (1996) . Since the mixing coefficients can depend on direction, the co-ordinate system is rotated such that, at the location of the particle, the x-axis is aligned with the flow and the y-axis points in the transverse-flow direction. For example, for u' we have

$$u' = \sqrt{\frac{3D_x}{2\Delta t}}(2R_f - 1), \tag{11}$$

where D_x is the turbulent mixing coefficient (unit m^2/s), Δt is the time step used in the Runge-Kutta scheme and R_f is a uniform random number within $[0, 1]$. According to Flügge (1982), D_x is assumed to be

$$D_x = D_x^0 + \alpha_x(H + \eta)\sqrt{U^2 + V^2}. \tag{12}$$

D_x^0 and α_x denote the neutral mixing coefficient and the mixing factor, respectively which have been measured in various estuaries, see Flügge (1982) . In the transverse-flow direction D^0 and α can have different values. In this simplified model any secondary effects caused by strong spatial variations in the mixing coefficients are neglected.

The mixing coefficient model suffers the drawback that in Eq. (11) the time step from the Runge-Kutta method Δt appears. One can conclude that when using this model, only the average distance traveled by the particle due to turbulent mixing, but not the turbulent velocities themselves, are in general physically realistic. Within our set-up, this has no negative consequences in the horizontal directions. In the vertical direction, however, a better turbulence model is needed as will be seen later. A statistical turbulence model for w' can be derived as follows: Assuming that w' follows a Gaussian distribution, random values of w' can be generated by, see Press et al. (1992)

$$w' = \sigma\sqrt{-2\ln R_{f_1}}\,\sin\left(2\pi R_{f_2}\right)\,, \tag{13}$$

where σ is the standard deviation of w', and R_{f_1} and R_{f_2} are uniform random numbers within $[0, 1]$. To determine the unknown value of σ, we assume that the turbulence is isotropic. Then $\sigma = \sqrt{2/3k}$ holds, where k is the mean turbulent kinetic energy. According to an algebraic turbulence model described by Uittenbogaard et al. (1992), k can be related to the mean flow by

$$k = 2c_\mu^{-\frac{1}{2}}L_t^2 D_{ij}D_{ij}\,, \qquad c_\mu = 0.09, \tag{14}$$

where D_{ij} is the velocity gradient

$$D_{ij} = \frac{1}{2}\left(\frac{\partial u_i}{\partial x_j} + \frac{\partial u_j}{\partial x_i}\right)\,, \qquad i,j = 1,2,3\,. \tag{15}$$

In many estuaries the characteristic length scale in the vertical direction is much smaller than the one in the horizontal plane so that the partial derivatives of the horizontal velocity components with respect to z give the largest contribution to k in Eq. (14). The 2D depth-averaged velocity field, however, does not depend on z. Thus, in order to achieve the right order of magnitude for k from Eq. (14), we introduce the approximations

$$\frac{\partial U}{\partial z} \approx \frac{U}{H+\eta}, \qquad \frac{\partial V}{\partial z} \approx \frac{V}{H+\eta}\,. \tag{16}$$

Neglecting the contribution of the vertical velocity, it follows that

$$k = c_\mu^{-\frac{1}{2}}L_t^2\left\{\frac{U^2+V^2}{(H+\eta)^2} + 2\left[\left(\frac{\partial U}{\partial x}\right)^2 + \left(\frac{\partial V}{\partial y}\right)^2\right] + \right. \tag{17}$$
$$\left. + \left(\frac{\partial U}{\partial y}\right)^2 + \left(\frac{\partial V}{\partial x}\right)^2 + \frac{\partial U}{\partial y}\frac{\partial V}{\partial x}\right\}\,,$$

where L_t is the turbulent length scale and the partial derivatives of U and V are determined by differentiation of Eq. (7) applied U and V. The turbulent length scale is approximated to be proportional to the water depth

$$L_t = c(H + \eta). \tag{18}$$

The constant c has to be found by calibration and is currently set to 0.02. Again, any secondary velocity terms due to strong variations in the mixing coefficient σ are neglected in this model.

Settling velocity

We assume that sediment particles are spherical and noncohesive. The settling velocity is determined from a balance of forces acting on the particle due to gravity, buoyancy and drag. Using an empirical c_D-law of a sphere given by Raudkivi (1976), the following conditional relation can be derived

$$
\begin{aligned}
24Re_p(1 + 0.15Re_p^{0.687}) - F^\star &= 0 && F^\star \leq F^\star_{crit}, \\
Re_p &= \sqrt{F^\star/c_{D,\infty}} && F^\star > F^\star_{crit}.
\end{aligned}
\tag{19}
$$

Here, $Re_p = 2w_s r_p/\nu$ is the particle Reynolds number, where r_p and ν denote the particle radius and the kinematic viscosity of the fluid, respectively. The dimensionless parameter $F^\star$ is defined as

$$F^\star = \frac{32}{3} \frac{g r_p^3}{\nu^2} \frac{\rho_s - \rho}{\rho}, \tag{20}$$

where ρ, ρ_s and g indicate the density of the fluid and the particle and the gravity constant, respectively. The critical value of $F^\star$ is related to other constants by $F^\star_{crit} = c_{D,\infty} Re_{crit}^2$, where according to Raudkivi (1976) $c_{D,\infty} = 0.424$ and $Re_{crit} = 1000$. Once ρ, ρ_s, ν and r_p are prescribed, the particle Reynolds number is determined either from Eq. $(19)_1$ or Eq. $(19)_2$, depending on the fixed value of $F^\star$. The settling velocity can then be calculated from the definition of Re_p. The implicit relation defined in Eq. $(19)_1$ is solved by the Newton Iteration method, see Press et al. (1992).

Interaction at cell boundaries

Whenever a particle attempts to intersect one of the five boundaries of the cell, in which it is located, the particle is moved exactly to this intersection point. The Newton method is again applied to compute the time-of-flight for the particle to reach it. Eventually, the particle movement is continued in the neighboring cell, whereas if a particle hits the water-level surface it is specularly reflected. If a particle tries to intersect the sea bed or an impermeable side boundary it stays at the intersected boundary. In this case the solution procedure of Eq. (1) is nevertheless continued, for if the direction of the particle velocity vector $\dot{\underline{x}}_p$ changes accordingly, the particle is bound to become free again. Changes of $\dot{\underline{x}}_p$ can be due to tidal variations of U and V as well as to the random changes of the turbulent velocity components.

Deposition and erosion

The solution procedure described above already accounts for deposition and erosion of sediments so that no critical parameters for the onset of these processes are needed. Let us consider a particle lying at the sea bed that for the moment is assumed to be horizontal. In this case a particle located at the bed erodes, i.e., it gets lifted-up into the water column, whenever its vertical velocity component $\dot{z}_p = w_p$ is positive, whereas it remains deposited at the bed when w_p is negative. Experience shows that the velocity w_k is generally much smaller in magnitude than w_s and w'. Therefore, since the contribution of w_s in Eq. (1) is always negative, w' must be positive and must exceed w_s in magnitude to erode the particle from the bed. The standard deviation σ of the turbulent component w' is seen from Eqs. (13) and (17) to be proportional to the turbulent length scale $L_t = c(H + \eta)$. Thus, the time fraction where the sediment particle is eroded, i.e., its transport rate depends on the value of the constant c which is the only à priori unknown parameter in the model.

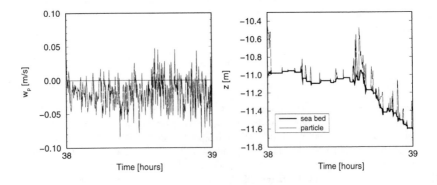

Fig. 3: Left: vertical velocity of a sediment particle; right: z-location of the particle and of the sea bed at the position of the particle.

An example for erosion and deposition of a sediment particle can be seen in Fig. 3 obtained from an application of *PARTRACE* to the flow in the Jade Estuary. In the left diagram, the vertical velocity of the particle w_p, and in the right diagram its z-coordinate together with the z-coordinate of the bed at the position of the particle are plotted for the same time interval. By comparing both diagrams, it is seen that during time phases where the vertical particle velocity w_p is mainly negative, the particle stays deposited at the bed. Conversely, significant erosion of the particle takes place whenever w_p is often positive.

Grid mode

In order to facilitate the analysis of sediment transport, a special mode of operation has been incorporated into *PARTRACE* which is called grid mode. It works as follows: first, a rectangular grid of point sources is generated inside the computational domain, whose resolution in the in the x- and y-direction is set by the user. From each source, one or several particles are launched all at the same time but tracked only for a short time, say 5 minutes. In this way short particle paths originating from the nodes of the source grid are obtained. This launching and tracking of particles is repeated in short intervals, e.g., every 10 minutes, the whole process continuing in time for one or several complete tidal cycles. To the end of the program run, all particle paths originating from the same source node are averaged. This results in an area-wide plot of tidal-averaged or residual particle transport paths.

SEDIMENT TRANSPORT IN THE OUTER JADE ESTUARY

PARTRACE computation

PARTRACE is applied to the flow in the western portion of the outer Jade Estuary indicated by the solidly drawn rectangle in Fig. 1. The flow was calculated previously with *TRIM-2D*, whereby the computational domain encompassed the whole Jade (dashed box in Fig. 1), and the Weser river including its entrance delta to the North Sea. In the *TRIM*-computation the boundary condition for η on the seaward outer boundary of the domain was determined by interpolation and extrapolation of meter data measured at certain locations on the boundary in June 1990. In the *TRIM* computation, no wind forcing was applied at the free surface. A good calibration of the HN model to meter data recorded at certain locations within the domain during the simulation period was achieved by chosing the the bottom friction coefficient and the turbulent momentum exchange coefficient of the HN model appropriately. A good calibration of the HN model to meter data obtained within the flowfield The *PARTRACE* computation was performed in the grid mode, the time frame of tracking extended over 4 complete tidal cycles, i.e. 49 hours and 40 minutes. The number of nodes of the source grid was (40×45) in the x- and y-direction, respectively. The tracking duration for individual particles and the temporal separation of the particle launching instances were the same as stated above, 5 minutes and 10 minutes, respectively. The diameter of the sediment particles was set to $2r_p = 0.2 \ mm$ which is a rough average of the grain sizes encountered in this region. Non-cohesive sediments are highly dominant in the domain of interest so that disregarding the effects of cohesive sediments is well justified in this case. The settling velocity of a model sediment particle was found from Eq. (19) to be $w_s = 1.94\ cm/s$. The constant in Eq. (18) was assumed to be $c = 0.02$.

Van Rijn model

As another means to determine the sediment transport in the flow, the transport of bed-load material was computed from the semi-empirical model by van Rijn (1989) which is available with *TRIM-2D*. According to the van Rijn model, the bedload transport capacity within unit time (unit: kg/m) in the x- and y-direction is given by

$$q_x^{tot} = q \left(\frac{U}{\sqrt{U^2 + V^2}} - \beta \frac{\partial H}{\partial x} \right), \qquad q_y^{tot} = q \left(\frac{V}{\sqrt{U^2 + V^2}} - \beta \frac{\partial H}{\partial y} \right), \qquad (21)$$

where

$$q = 0.053 \left[\left(\frac{\rho_s}{\rho} - 1 \right) g \right]^{\frac{1}{2}} d_{50}^{1.5} (D^*)^{-0.3} T^{2.1}. \qquad (22)$$

In Eq. (21) the parameter β controls the contribution of the bottom gradient, in other words, that of gravity on the bedload transport and is assumed to have the value 0.3. In Eq. (22) d_{50} denotes the grain diameter at which the integral over the grain-size distribution reaches 50 percent of its total value and is assumed to be $d_{50} = 0.25$ mm. Furthermore, D^* indicates sedimentological grain diameter and T denotes the dimensionless bottom shear stress. For a the detailed elaboration of the model and its parameters the reader is referred to van Rijn (1989). The van Rijn model is merely algebraic so that no boundary conditions for (q_x^{tot}, q_y^{tot}) are needed. Moreover, the coupling between bedload transport and the flow field is neglected in the model. The bedload transport capacity can be interpreted as the rate with which sediment is transported under the provision that an abundance of erodable material is available at the specified time and location.

Statistical grain size analysis

The third methodology that we can compare to is the so-called Sediment Trend Analysis (STA[TM]) (McLaren and Bowles 1985). In this method a statistical analysis of samples of bottom sediment taken in the domain is performed. In the present case about 920 samples were collected in an extensive field programme. The grid of sample locations that was employed had a linear spacing between 500 m and 1 km. The method proceeds by determining the grain-size distribution and its first three statistical moments called mean, sorting, and skewness for every sample. According to the theory of McLaren and Bowles (1985), any transport trend, for example accretion or erosion of sediment, is associated with certain known combinations of either positive or negative spatial gradients of these moments. A transport trend path which in the method always connects sample locations is found, when the statistical score-value of that trend reaches a maximum. The statistical score-value in turn is determined by considering all pairs of sample locations on that path and testing if the changes of the statistical moments within a pair of samples is in agreement with characteristics of the trend in question. From the Sediment Trend Analysis, only the directions of transport but not the magnitude of the transport rate can be determined.

Results

In Fig. 4, the transport paths of sediments in the outer Jade Estuary around the island Wangerooge as obtained from the methods described above are shown. From left to right, the plots indicate, the tidal-averaged values of (q_x^{tot}, q_y^{tot}) according to the van Rijn model (Eq. (21), the tidal-averaged sediment paths from the *PARTRACE* computation, and the transport paths derived from the Sediment Trend Analysis. In blank regions in the right plot the magnitude of bedload transport capacity is below a certain threshold value and can be neglected. In case of the result from the particle method the scale of the vectors is arbitrary so that the vector magnitudes in the plots of the numerical results cannot be directly compared. In all three plots the shipping channel is also drawn.

The right plot is a summary of the drawing originally obtained from the Sediment Trend Analysis which is too fine-featured to be clearly visible at the size of the plot. Here, the water depth is indicated by different tones of grey shading whereby the darkness of the shading increases with water depth. High areas are white with the exception of the shipping channel which, though the deepest part of the topography, is drawn in white as well to improve the visibility of the transport paths. The region between Wangerooge and the main land is a wadden area. In the section of the shipping channel marked by the two vertical lines sediment accretion is very pronounced and much maintenance by dredging is necessary here.

Discussion

In the bedload transport capacity from the van Rijn model a large peak is observed where the shipping channel bends south which in our opinion should not be taken too seriously. This is because the magnitude of the bedload transport capacity not due to gravity q from Eq. (22) is roughly proportional to u^4, where u is taken to denote the magnitude of the velocity . Therefore disturbances in u are strongly amplified and may lead to unrealistically high differences in the magnitudes of q in the flowfield.

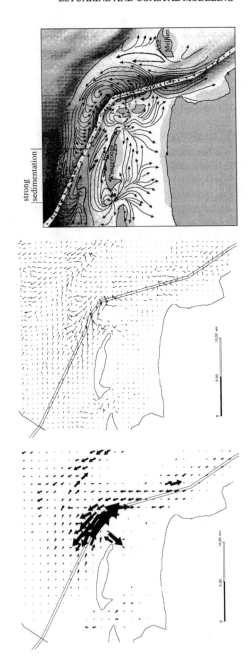

Fig. 4: Sediment transport paths in the outer Jade Estuary: left: bedload transport model of van Rijn (1989) center: particle method, right: Sediment Trend Analysis (STA) (McLaren and Bowles 1985) based on samples of bottom sediment.

Some differences between the results from van Rijn model and the particle method are apparent. First, the *PARTRACE* result does not show strong peaks as the van Rijn model, a fact for which an explanation has not been found yet. Secondly, near the shipping channel the transport vectors from the particle method are more perpendicularly oriented to the fairway than according to the van Rijn model. Better correspondence may be achieved here by increasing the parameter β that controls the influence of the bottom topography on the transport capacity. In this respect the particle method shows an advantage over the bedload transport model, because the influence of gravity enters into the method accurately through the well-determined value of the settling velocity of the grain particle. Finally, in certain areas some weak scatter can be observed in the result from the particle method which is missing in the van Rijn model. This is due to the Monte Carlo-type schemes of Eqs. (11) and (13) used to generate turbulent velocity components in the particle method.

Despite these differences, the overall agreement between the van Rijn model and the particle method is remarkably good. This must be seen in view of the fact that, though the velocity field entering into both methods as input is the same, both methods work very differently. Evidently, the velocity field strongly dictates the result of the sediment transport models investigated so that other details of these methods seem to be of minor importance.

Since the transport paths according to the particle and van Rijn models are quite similar, only the results from *PARTRACE* and from the Sedment Trend Analysis are compared further. Good agreement is observed in the regions between the islands Wangerooge and Spiekeroog at the eastern end of Wangerooge. The circulation pattern to the north and the parting of the paths to the south is well captured by the particle method. Furthermore, near the northern shore of Wangerooge, a west-to-east transport trend is seen in both results. However, in the region around the north-eastern exit of the shipping channel, the sediment trend analysis shows remarkable S-shapes and turnings that are not present in the *PARTRACE* result. Beyond the eastern and northern side of the channel and also at its southern end, the agreement between Sediment Trend Analysis and the particle method is poor. Most transport patterns disagree and any local agreement must be viewed as coincidental. Only the most easterly path from the Sediment Trend Analysis which is likely to come from the Weser river seem to correspond to a similar trend in the *PARTRACE* result. In the latter result, however, this trend is shifted to the north-east.

CONCLUSION

In conclusion, the comparison between the numerical methods and the Sediment Trend Analysis is not satisfactory and an improvement is highly desirable in view of the maintenance problems to be remedied in this region. Since the influence of the flow field seems to be highly dominant within the sediment transport models investigated, the computed flow field does not seem to approximate the real flow during the time interval captured by the Sediment Trend Analysis well enough. This may be due to 3D effects, to the effects of waves, wind or of singular events such as storms which all are not taken into account in the HN model. On the other hand, the outcome from the Sediment Trend Analysis may as well reveal the short term effects of flow conditions that are different from those assumed in the HN model.

Not all these problem areas can be attacked at once. Future work is aimed to apply 3D HN models to the flow in this estuary and to extend *PARTRACE* to 3D. It will be interesting to see if the numerical transport models give better results if a 3D flow is taken as their basis.

REFERENCES

Cheng, R.T., Casulli, V. and Gartner, J.W., (1993). "Tidal, residual, intertidal mudflat (TRIM) model and its applications to San Fransisco Bay, California." *Estuarine, Coastal and Shelf Science*, 36, 235-280.

Dick, S. and Schönefeld, W., (1996). "Water transport and mixing in the North Frisian Wadden Sea – Results of numerical investigations." *German Journal of Hydrography*, 48, 27-48.

Flügge, Gerd, (1982). "Transport- und Ausbreitungsmechanismen in Flüssen und Tideästuarien unter besonderer Berücksichtigung der Strömungsturbulenz." PhD dissertation, Institut für Bauingenieur- und Vermessungswesen of the University of Hannover.

http://www.hamburg.baw.de/ (1997). *Internet homepage of the Federal Waterways Engineering and Research Institute, Germany.*

McLaren, Patrick, Bowles Donald (1985). "The effects of sediment transport on grainsize distributions." *Journal of Sedimentary Petrology*, 55, 457-470.

Press, W.H., Teukolsky, S.A., Vetterling, W.T. and Flannery, B.P., (1992). *Numerical recipes in Fortran.* Cambridge University Press.

Raudkivi, A.J., (1976). *Loose boundary hydraulics.* Pergamon Press.

Uittenbogaard, R.E., van Kester, J.A.Th.M. and Stelling, G.S., (1992). "Implementation of three turbulence models in TRISULA for rectangular horizontal grids." *Technical report*, Delft Hydraulics, Delft, NL.

van Rijn, L.C., (1989). Handbook – Sediment transport by currents and waves. *Report H 461*, Delft Hydraulics, Delft, NL.

Resuspension and Advection of Sediment During Hurricane Andrew on the Louisiana Continental Shelf

Timothy R. Keen[1] and Scott M. Glenn[2]

Abstract

Validated hindcast storm currents and waves are used to drive a coupled bottom boundary layer-sediment transport model for Hurricane Andrew. Shear stresses and suspended sediment profiles are computed using a combined wave-current bottom boundary layer model with variable bottom roughness. The sedimentation model calculates suspended sediment profiles, and suspended and bed load transport rates. Sediment erosion and deposition is found using mass conservation equations with advection and diffusion terms. The average model-predicted combined resuspension-advection sediment thickness during a 1-hour interval at the storm peak is less than 0.04 m, with a maximum of more than 2.6 m.

1. Introduction

This paper discusses a coupled bottom boundary layer-sedimentation model for predicting sediment resuspension and transport on continental shelves. This work is motivated by the need to better understand shallow marine sequences and to predict sediment fluxes in the modern marine environment. The coupled model is introduced in section 2, and its application to Hurricane Andrew is described in section 3. Hindcast winds, waves, and currents are discussed in section 4. The bottom boundary layer is examined in section 5, and sedimentation results are examined at three times during the storm in section 6.

---

[1] Naval Research Laboratory, Oceanography Division, Stennis Space Center, MS 39529
[2] Institute of Marine and Coastal Sciences, Rutgers University, New Brunswick, NJ 08903

481

2. The coupled bottom boundary layer-sedimentation model

2.1. Definitions

Sediment resuspension is mainly a result of wave action and, therefore, it operates at the time scale of the wave period. To first order, if no currents are present, the same volume of sediment is redeposited repeatedly (reworked) with no net erosion or deposition. When a mean flow exists, this suspended sediment is transported. If, however, the net suspended sediment flux is zero at a grid point, any sediment which is removed is replaced from upstream. Again, there is a depth of resuspension by the combined wave and current flow equivalent to the amount of sediment held in suspension by the combined flow.

When a net loss of mass occurs at a grid point as a result of advection, the resulting decrease in bed elevation is herein termed erosion H_E (Fig. 1). An increase in bed elevation associated with a net gain in mass is termed deposition H_D. The resuspension depth H_R is the equivalent thickness of sediment suspended in the water column when averaged over a wave period.

Mass (and elevation) changes in the bed are associated with steady currents, which vary at long time scales (1 hour in this study) compared to storm wave periods, which are on the order of 10 seconds. If the wave field changes slowly, the average resuspension depth can be assumed constant over the time interval of the steady flow. Thus, as the bed elevation changes in response to erosion and deposition, wave reworking (Fig. 1) extends below the bed to a depth which is in equilibrium with the wave-current field. As a result, the resuspension depth is superimposed on erosion and deposition.

Several reference levels are also defined in figure 1. The apparent erosion depth Z_A is the maximum reworking depth during a time interval. The reference elevation Z_R is the erosion depth during a time interval, and the bed elevation Z_B is the height of the bottom above the initial state ($Z_B = 0$ initially). These reference levels may be defined over time intervals ranging from one time step to an entire model simulation. Thus, for longer time intervals they function as cumulative reference levels. This paper discusses resuspension and advection during one model time step (1 hour) and Z_A, Z_R, and Z_B are thus *instantaneous* reference levels.

The instantaneous apparent bed thickness, resulting from advection and resuspension, is given by $B_C = Z_B - Z_A$. This is the thickness of the sediment between the resuspension depth and the sea floor. This bed will be referred to as the **storm bed** hereinafter. Note, however, that for a time interval less than the event duration, this bed represents both transported sediment within the bed and sediment which remains in suspension above the bed. The instantaneous **transport bed** thickness, defined as $B_T = Z_B - Z_R$, represents sediment which has been transported by steady currents to its final deposition site–i.e., it originated elsewhere. The thickness of sediment which has been suspended and redeposited in the same location comprises the **resuspended bed**, defined as $B_R = Z_R - Z_A = B_C - B_T$. The resuspended bed is a

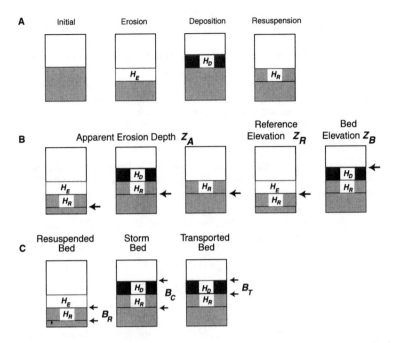

Figure 1. Definitions: **(A)** bed elevation changes; **(B)** reference levels; and **(C)** bed thickness. See text for explanation.

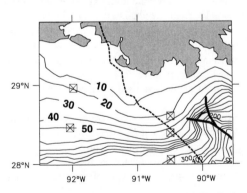

Figure 2. Bathymetry map of study area. The dashed line is the storm track, the heavy black line is the axis of Mississippi Canyon, and the squares are the locations of moorings used in validating the wind, wave, and current hindcasts.

convenient unit for keeping track of the sediment reworking depth. Whether or not these "beds" can ultimately be identified as discrete layers depends largely on the wave history at a point; for example, it is possible for a transport bed B_T to be discrete because deposition from a storm flow often occurs as the flow passes from a region of strong wave action to a region of weak wave action. In this case, resuspension by oscillatory waves will not uniformly mix the newly deposited sediment into a pre-existing bed. If, however, wave action is moderate and deposition is slow enough, it is expected that sediment will be mixed into the bed as it is deposited.

2.2. The bottom boundary layer model (BBLM)

Glenn and Grant (1987) present a suspended sediment stratified BBLM for wave-dominated flows on continental shelves (NBM87). Keen and Glenn (1994) describe an unstratified version of this BBLM with a fixed bottom roughness k_b that is applicable for an extended range of wave and current conditions (WCM93). The WCM93 model has been modified as in Glenn and Grant (1987) to calculate the bottom roughness and suspended sediment stratification parameter from given sediment, wave, and current values. The extended-range, stratified model is called NBM96. It has been used to examine the temporal and spatial variability of the bottom boundary layer during Hurricane Andrew (Keen and Glenn, 1997b).

2.3. The sedimentation model

The sediment concentrations computed by the BBLM do not reflect the inhibiting influence of bed armoring, as finer material is preferentially removed. This process is incorporated by using an active layer h_A which represents that part of the bed which interacts with the flow during one time step (Raheul et al., 1989). We expect the active layer during low flow conditions exceeding the initiation of motion criteria to be proportional to the ripple height. During high flow conditions, we expect that it is proportional to the height of the near-bed transport layer. Following Grant and Madsen's (1982) approach for bottom roughness, we assume that the active layer is the sum of these components: $h_A = \eta + h_{TM}$, where η is the ripple height and h_{TM} is the height of the near-bed transport layer. This active layer relates the sediment profile to the physical bottom parameters used to calculate the wave-current shear stress.

The bed concentration C_{bn} for a specific size class is given initially by the assumed grain size distribution; however if H_R exceeds h_A, then C_{bn} is reduced by the ratio, h_A / H_R, to simulate armoring. New profiles of currents, sediment concentrations, and the stability parameter are then calculated at the grid point. This reduction procedure is repeated until the resuspension depth of the size class does not exceed h_A.

After the sediment and current profiles have been computed at each grid point by NBM96, a bed conservation equation is solved for each size class. The size-fractionated mass fluxes are integrated over the bottom boundary layer to form the advection terms for the erosion/deposition equation. In addition to advection by the steady current, a turbulent diffusion term uses the Smagorinsky formulation (Mellor, 1993).

Mass gains or losses for each sediment size class at a grid point are used to compute changes in bed elevation following van Niekerk et al. (1992). An increase in elevation is deposition H_D. A decrease in elevation is the erosion depth H_E. If the erosion depth due to a size class exceeds h_A, the reference concentration for that size class is reduced in the same way as for resuspension. Using the new reference concentration, the current profiles, size-fractionated sediment profiles, and stability parameter profiles are recomputed for the entire grid. This iterative procedure is applied at each grid point for each sediment size class in suspension.

As a result of this numerical procedure, the thickness of erosion or deposition for each size class at each grid point is found at the new time step. Thus, changes in bed elevation, bottom roughness (i.e., ripples), and the sediment distribution at each grid point are known at the new time.

3. Hindcast method

Hurricane Andrew (Fig. 2) entered the Gulf of Mexico at 1200 GMT August 24, 1992, with a central pressure of approximately 950 mb. The eye made landfall southwest of New Orleans 48 hours later. Computing sediment resuspension, advection, and deposition requires that bottom currents and wave parameters (i.e., bottom orbital speed and amplitude, and wave propagation direction) be available at sufficient temporal and spatial resolution to drive the BBLM-sedimentation model. Bottom currents at a horizontal resolution of approximately 5 km were computed by the Princeton Ocean Model (hereinafter POM; Mellor, 1993), and wave orbital parameters are calculated from significant wave heights and mean periods, as well as local water depths, using linear wave theory. The POM uses 20 sigma levels, with surface resolutions of 0.06 m and 26 m for the smallest and largest water depths, respectively.

The circulation model has been run and compared with and without wave-current interaction (Keen and Glenn, 1997a, 1998), and bottom currents from a previous shallow water current hindcast which used WCM93 to compute bottom friction (Keen and Glenn, 1997a) were available. Feedback from the bottom boundary layer model to the ocean circulation model, in the form of enhanced bottom friction, can be important (Spaulding and Isaji, 1987), but wave-current enhanced bottom friction does not significantly alter regional flow patterns (Keen and Glenn, 1995). The stratification correction could reduce bottom friction somewhat but not as much as turning off the wave-current interaction. Since the regional patterns are reasonable and the hindcast bottom currents are from a validated model, the details of the bottom flow are calculated by the stratified model (NBM96). Using a coupled BBLM to compute bottom stresses and near-bed current profiles makes very high near-bottom resolution unnecessary in POM.

Hindcast wind and wave fields between 1000 GMT Aug. 24 and 0000 GMT Aug. 27 were available (Cardone et al., 1994). The mean and standard deviation of the wind speed differences are 0.22 m s^{-1} and 2.3 m s^{-1}, respectively. The mean difference between hindcast and measured significant wave heights is 0.05 m, and the standard deviation of the differences is 0.69 m. The hindcast setup for the circulation model is discussed by Keen and Glenn (1998) and the storm flow has been described by Keen and Glenn (1997a); only the bottom currents will be examined below. The average model-predicted peak surface current speed error is 0.25 m s^{-1}, and the corresponding timing error is 4.4 hours early prediction by the model. The speed error is biased by poor performance at a mooring in a water depth of 47 m where winds were near the maximum.

The NBM96 model is used to compute the combined wave-current shear stresses as well as sediment profiles in the present study. The bottom currents used by the BBLM and the sedimentation model are interpolated to a reference height of 1 m from currents calculated at the σ levels in POM, using a law-of-the-wall boundary layer profile. The bottom boundary layer and sedimentation computations are completed using a time step of 1 hour. The sediment grain size distribution of the bed (Table 1) consists of 5 classes with a mean diameter d_m of 88.39 μm (very fine sand). This sediment distribution was chosen to reasonably match observations (Snedden, 1985) while minimizing clay-size particles, which are cohesive and not properly treated by the model.

Table 1. Sediment size classes available in bed.

Size Class	Available Weight %	Mean (μm)	Bottom (μm)	Top (μm)
1	4.77	4.19	2.00	9.00
2	24.29	19.24	9.00	41.20
3	41.89	88.39	41.20	189.50
4	24.29	406.13	189.50	870.60
5	4.77	1866.07	870.60	4000.00

4. Hindcast storm winds, waves, and currents

The hindcast storm conditions and model-predicted sedimentation are examined at three times: (1) before the storm peak is represented by output at 1324 GMT Aug. 25; (2) the storm peak within the study area occurred at approximately

0135 GMT Aug. 26; and (3) a waning storm flow was present at 1345 GMT, several hours after the eye made landfall.

As the hurricane approached the Louisiana study area, the winds were from the northeast with magnitudes less than 20 m s^{-1}, and significant wave heights (Fig. 3a) were generally less than 2.5 m on the shelf. The strongest winds and highest waves were located east of the storm track. Model-predicted bottom currents (Fig. 4a) are westward on the inner shelf, and velocities are greater than 0.3 m s^{-1} in water depths less than 20 m.

As the eye passed over the moorings, the hindcast winds at the eastern moorings are greater than 30 m s^{-1}. Significant wave heights (Fig. 3b) exceed 7 m in water depths less than 30 m east of the storm path, whereas such large waves are only predicted in water deeper than 50 m on the west side of the track. Hindcast bottom currents (Fig. 4b) reveal a convergent coastal current system during the storm peak, with broad shelf-wide westward flow between 90°W and 91°W, where bottom currents are greater than 0.5 m s^{-1}, and a weaker eastward flow near 92°W. These flows meet near 91.3°W and turn offshore west of the storm track. Bottom flow on the outer shelf east of the storm track is directed into Mississippi Canyon (see Fig. 2).

After the hurricane eye made landfall, the wind speed fell below 20 m s^{-1} and hindcast significant wave heights (Fig. 3c) are about 2.5 m along the Louisiana coast, with waves larger than 3 m still predicted east of the track. The circulation model predicts eastward flow (Fig. 4c) at the bottom with speeds of about 0.3 m s^{-1} in water depths less than 25 m. Bottom currents are much weaker in deeper water, however, and display a more complex pattern.

5. The bottom boundary layer

As the eye approaches the study area, the bottom boundary layer (Fig. 5a) west of the storm track is current dominated, as indicated by u_r/u_b (the ratio of the reference current to the wave velocity at the top of the wave boundary layer), while waves and currents are of similar magnitude east of the track. The maximum shear velocity u_{*cw} (Fig. 6a) exceeds 0.08 m s^{-1} near the storm track in a water depth of 35 m and near the coast. Ripples dominate the physical roughness at this time.

The ratio u_r/u_b (Fig. 5b) is about 0.5 over most of the shelf during the storm peak, indicating that combined wave-current shear stresses should dominate. The maximum shear velocity u_{*cw} (Fig. 6b) is highest near the storm track where the largest waves are predicted, with a peak of more than 0.2 m s^{-1}. Model-predicted values greater than 0.06 m s^{-1} are found in water as deep as 80 m east of the storm track, and at progressively shallower depths towards the west. Ripples dominate bottom roughness wherever u_r/u_b is greater than 1, and thus are predicted on the outer shelf near the storm track and to the east, and in progressively shallower water to the west. Near-bed sediment transport dominates bottom roughness where u_r/u_b is less than 1, because the high maximum shear stresses within the wave boundary layer wash out ripples and produce a flat bed dominated by sheet sediment transport.

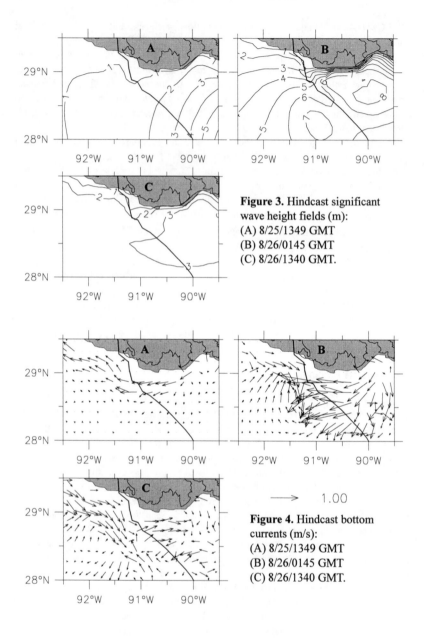

Figure 3. Hindcast significant
wave height fields (m):
(A) 8/25/1349 GMT
(B) 8/26/0145 GMT
(C) 8/26/1340 GMT.

$\longrightarrow$ 1.00

Figure 4. Hindcast bottom
currents (m/s):
(A) 8/25/1349 GMT
(B) 8/26/0145 GMT
(C) 8/26/1340 GMT.

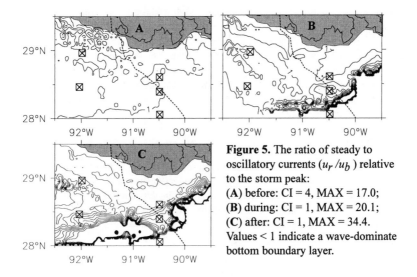

Figure 5. The ratio of steady to oscillatory currents (u_r / u_b) relative to the storm peak:
(**A**) before: CI = 4, MAX = 17.0;
(**B**) during: CI = 1, MAX = 20.1;
(**C**) after: CI = 1, MAX = 34.4.
Values < 1 indicate a wave-dominate bottom boundary layer.

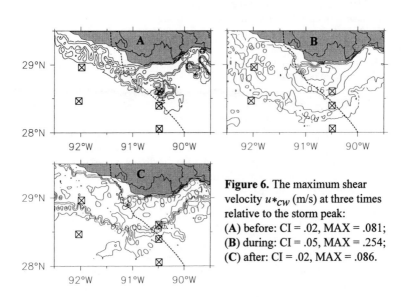

Figure 6. The maximum shear velocity $u*_{cw}$ (m/s) at three times relative to the storm peak:
(**A**) before: CI = .02, MAX = .081;
(**B**) during: CI = .05, MAX = .254;
(**C**) after: CI = .02, MAX = .086.

Between the nearshore transport-dominated and offshore ripple-dominated bottoms, ripples and sediment transport contribute equally to physical bottom roughness.

Several hours after the eye made landfall, u_r/u_b (Fig. 5c) is less than 1 at the coast and near the storm track to a depth of 20 m. The maximum shear velocity u_{*cw} (Fig. 6c) exceeds 0.07 m s^{-1} along a swath landward of the 20 m isobath and east of the storm track where waves are slightly larger. Ripples dominate for all water depths after the storm peak. The near-bed transport layer exceeds 0.01 m at only a few locations in shallow water.

6. Sedimentation

The storm sedimentation is discussed using the instantaneous storm bed thickness B_C, transport bed thickness B_T, and the resuspension bed thickness B_R. These variables represent material either in suspension or deposited during a single model time step of 1 hour. They should thus be considered as representing the relative contribution of resuspension and advection to bed generation and not as discrete layers.

6.1. Storm approach

Maximum model-predicted shear velocities exceed 0.03 m s^{-1} near the coast and immediately east of the storm track, decreasing to the west in the direction of bottom flow. A maximum storm bed thickness (Fig. 7a) of 3.5 cm is predicted where these large gradients occur. Otherwise, B_C is less than 1 cm. The bulk of this sediment is in suspension (Fig. 8a), as indicated by the similarity of the isopleths of B_R and B_C (compare Figs. 7a and 8a).

A transported bed (Fig. 9a) is limited to areas where bottom currents and shear velocities decrease together in the direction of flow. Resuspension is also reduced at these locations because u_{*cw} is less than 0.01 m s^{-1}. This sediment has been transported to these depositional sites from upstream. The landward boundary condition used in the model supplies the sediment fluxes into a grid point calculated by the sediment transport model, thus simulating erosion in water depths not represented by the model grid. This eroded sediment tends to be deposited not far from the coast. Such coastal erosion and deposition is evident where B_T is greater than 0.4 cm thick at that coast.

6.2. The storm peak

As the hurricane eye passed over the study area, model-predicted bottom currents (Fig. 4b) greater than 0.5 m s^{-1} are widespread and flow converges west of the storm track. The maximum shear velocities at this time are greater than 0.06 m s^{-1} over most of the inner shelf and peak at more than 0.25 m s^{-1} near the coast. The storm bed (Fig. 7b) exceeds 260 cm near the track, and averages less than 6 cm along the storm track as well as to the west. The storm bed is less than 1 cm east of the track on the middle shelf.

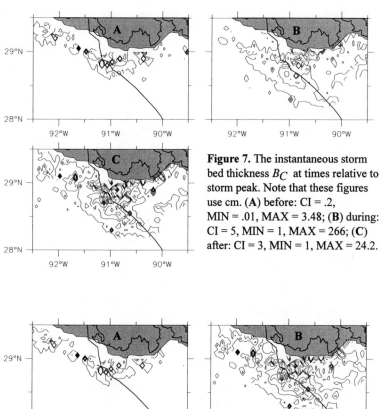

Figure 7. The instantaneous storm bed thickness B_C at times relative to storm peak. Note that these figures use cm. (**A**) before: CI = .2, MIN = .01, MAX = 3.48; (**B**) during: CI = 5, MIN = 1, MAX = 266; (**C**) after: CI = 3, MIN = 1, MAX = 24.2.

Figure 8. The instantaneous re-suspension bed thickness B_R at times relative to storm peak. Note that these figures use cm. (**A**) before: CI = .2, MIN = .01, MAX = 3.25; (**B**) during: CI = 2, MIN = 1, MAX = 20.1; (**C**) after: CI = 3, MIN = 1, MAX = 24.2.

Sediment resuspension depths greater than 1 cm (Fig. 8b) are broadly distributed west of the storm track and a maximum B_R of more than 24 cm is predicted near the track in shallow water. Resuspension also occurs at the head of Mississippi Canyon (see Fig. 2) in a water depth of 100 m, where B_R exceeds 5 cm.

Sediment was deposited in the west (Fig. 9b) in linear beds which parallel the isopleths of shear velocities in figure 6b, as well as in several small beds near the coast east of the storm track, with B_T less than 5 cm. A maximum transport bed thickness of 265 cm is predicted at 91.5°W on the inner shelf where bottom flow is convergent. Resuspension and transported beds are both predicted where u_s/u_b (Fig. 5b) is near 1.

6.3. After landfall

Storm deposition decreases significantly after the eye makes landfall. The hindcast storm flow immediately east of the storm track is eastward on the inner shelf and shear velocities (Fig. 6c) exceed 0.05 m s^{-1} over most of the area. As a result, the storm bed (Fig. 7c) is greater than 4 cm thick on the inner shelf near the storm track. Local maxima exceed 20 cm. Along the coast to the east, the storm bed is more than 1 cm thick. A storm bed thicker than 1 cm is also predicted west of the track and B_C exceeds 4 cm near the head of Mississippi canyon.

Despite the reduction in wave heights after landfall, resuspension dominates sedimentation (Fig. 8c), while advection (Fig. 9c) produces an irregular bed with a maximum thickness of 4.1 cm. The broad offshore flow west of the track (Fig. 4c) has deposited a series of beds on the midshelf. Local deposition is also distributed near the coast throughout the study area.

7. Conclusions

Sedimentation during a strong flow event such as a hurricane can be viewed as resulting from resuspension by oscillatory waves and currents (a resuspended bed), and deposition from a steady flow (a transported bed). The observed apparent bed at a location is commonly referred to as a "storm bed." This storm bed consists of both types of bed, but mixing of sediment by resuspension will act to homogenize the sediment as it is deposited. Observations of storm deposition during Hurricane Andrew do not exist, but previous work has shown good qualitative comparison between model-predicted sedimentation and cores from the Texas shelf (Keen and Slingerland, 1993). Furthermore, the estimated error in bed shear stresses is about 10% (Keen and Glenn, 1997b).

Because of the complex spatial and temporal relationships between the storm currents and bottom stresses, the evolving storm bed changes with time. The early storm bed during Hurricane Andrew is restricted to the coast and is less than 1 cm thick except for local areas. During the storm peak, resuspension is greatest near the storm track whereas a storm bed about 1 cm thick is predicted west of the track. Local deposition exceeds 200 cm where strong flow convergence occurred. After the eye makes landfall, the storm bed exceeds 1 cm to water depths of 60 m near the track, and attains a maximum thickness of 242 cm immediately east of the track at 28.9°N. Most of the instantaneous storm bed is associated with sediment in suspension above the bed.

Acknowledgments. The first author was funded by Program Element 62435N sponsored by the Office of Naval Research. The second author was funded by the Middle Atlantic Bight National Undersea Research Center (MAB96-10) and the Office of Naval Research (N00014-95-1-0457). LATEX mooring data was supplied by the Minerals Management Service (MMS), although the contents of this paper do not necessarily reflect the policies of the MMS.

References

Cardone, V. J., Cox, A. T., Greenwood, J. A., Evans, D. J., Feldman, H., Glenn, S. M. and Keen, T. R. (1994) Hindcast Study of Wind, Wave and Current Fields in Hurricane Andrew–Gulf of Mexico. Final Report to Minerals Management Service, Herndon, Virginia, 150 pp.

Glenn, S. M. and Grant, W. D. (1987) A suspended sediment stratification correction for combined wave and current flows. Journal of Geophysical Research, 92, 8244-8264.

Grant, W. D. and Madsen, O. S. (1979) Combined wave and current interaction with a rough bottom. Journal of Geophysical Research, 84, 1797-1808.

Grant, W. D. and Madsen, O. S. (1982) Moveable bed roughness in unsteady oscillatory flow. Journal of Geophysical Research, 87, 469-481.

Keen, T. R. and Glenn, S. M. (1994) A coupled hydrodynamic-bottom boundary layer model of Ekman flow on stratified continental shelves. Journal of Physical Oceanography, 24, 1732-1749.

Keen, T. R., and S. M. Glenn (1995) A coupled hydrodynamic-bottom boundary layer model of storm and tidal flow in the Middle Atlantic Bight of North America, Journal of Physical Oceanography, 25, 391-406.

Keen, T. R. and Glenn, S. M. (1997a) Shallow water currents during Hurricane Andrew. Journal of Geophysical Research, in revision.

Keen, T. R. and Glenn, S. M. (1997b) A suspended sediment stratified model of the bottom boundary layer on the Louisiana continental shelf during Hurricane Andrew. Continental Shelf Research, in review.

Keen, T. R. and Glenn, S. M. (1998) Factors influencing hindcast skill for modeling shallow water currents during Hurricane Andrew. Journal of Atmospheric and Oceanic Technology, 15, 221-236.

Keen, T. R. and Slingerland, R. L. (1993) Four storm-event beds and the tropical cyclones that produce them: a numerical hindcast. Journal of Sedimentary Petrology, 63, 218-232.

Mellor, G. L. (1993) Users Guide for a Three-Dimensional, Primitive Equation, Numerical Ocean Model. Report to the Institute of Naval Oceanography, Atmospheric and Oceanic Sciences Program, Princeton University, Princeton, New Jersey, 35 pp.

Raheul, J. L., Holly, F. M., Belleudy, P. J., and Yang, G. (1989) Modeling of river-bed evolution for bedload sediment mixtures. Journal of Hydraulic Engineering, 115, 1521-1542.

Snedden, J. W. (1985) Origin and sedimentary characteristics of discrete sand beds in modern sediments of the central Texas shelf. Ph.D. dissertation, Louisiana State University, Baton Rouge, La., 247 pp.

Spaulding, M. L., and T. Isaji (1987) Three dimensional continental shelf hydrodynamics model including wave current interaction, in Three-Dimensional Models of Marine and Estuarine Dynamics, edited by J. C. J. Nihoul and B. M. Jamart, 405-426, Elsevier, New York.

van Niekerk, A., Vogel, K. R., Slingerland, R. L. and Bridge, J. S. (1992) Routing of heterogeneous size-density sediments over a movable stream bed: Model development. Journal of Hydraulic Engineering, 118, 246-262.

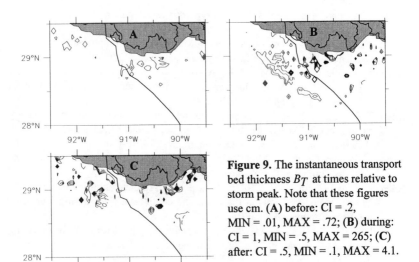

Figure 9. The instantaneous transport bed thickness B_T at times relative to storm peak. Note that these figures use cm. (**A**) before: CI = .2, MIN = .01, MAX = .72; (**B**) during: CI = 1, MIN = .5, MAX = 265; (**C**) after: CI = .5, MIN = .1, MAX = 4.1.

Three-Dimensional Numerical Modeling of the Tidal York River System, Virginia

J. Shen[1], M. Sisson[2], A. Kuo[3], J. Boon[3], and S. Kim[4]

ABSTRACT

The York River system is one of the western shore tributaries of the Chesapeake Bay. It is presently the site of several major ongoing research projects at the Virginia Institute of Marine Science. Since the water circulation and salinity distribution are of importance to these research projects, the application of a three-dimensional high resolution hydrodynamic model to simulate the circulation and salinity transport in the York River system becomes essential. A three-dimensional Hydrodynamic-Eutrophication Model (HEM-3D) developed by the Virginia Institute of Marine Science (VIMS) was applied to the system, including the York River and the tidal portions of the Pamunkey and the Mattaponi Rivers. To obtain high resolution of the hydrodynamic field and salinity distribution of the system, a grid size with 250 m × 250 m was applied to the York River. This paper documents the procedures of calibration and verification of the hydrodynamic portion of the model. Comparisons of model results and field observations show that the model accurately reproduces the tide, current, and salinity in the York River system. The calibrated model is ready to be applied to a variety of environmental problems.

1. Postdoctoral and Research Associate, School of Marine Sci./Virginia Inst. of Marine Sci., The Coll. of William & Mary, Gloucester Point, VA 23062.

2. Marine Scientist Sr., School of Marine Sci./Virginia Inst. of Marine Sci., The Coll. of William & Mary, Gloucester Point, VA.

3. Prof., School of Marine Sci./Virginia Inst. of Marine Sci., The Coll. of William & Mary, Gloucester Point, VA.

4. Visiting Asst. Prof., School of Marine Sci./Virginia Inst. of Marine Sci., The Coll. Of William & Mary, Gloucester Point, VA.

495

INTRODUCTION

The tidal York River system is one of the western shore tributaries of the Chesapeake Bay. The York River system is formed by the confluence of the Pamunkey and Mattaponi Rivers about 53 km from the point of the York River entry into the Chesapeake Bay. The mean tidal range at the mouth is 0.70 m and the long term mean freshwater discharges near the heads of the tide in the Pamunkey and Mattaponi Rivers are about 32.2 m^3s^{-1} and 16.7 m^3s^{-1}, respectively. The surface salinities at the mouth range from 15 - 24‰ and the 1 ‰ isohaline is normally found 65 - 90 km from the mouth. The York River is presently the site of several major ongoing research projects at the Virginia Institute of Marine Science (VIMS), such as the York River regional ecosystem modeling project, the contaminant and sediment transport project, the York River blue crab study, and the water quality modeling study. The water circulation and salinity distribution are of importance to these research projects. This led to the application of a three-dimensional high resolution hydrodynamic and water quality model to simulate the circulation, salinity transport, and water quality in the York River system. This paper documents the procedures of calibration and verification of the hydrodynamic portion of the model. Comparisons of tidal elevation, current, and salinity between model results and field observations are presented in this paper. The results show that the model accurately reproduces the tides, currents, and salinities of the York River.

BRIEF MODEL DESCRIPTION

VIMS has been developing a general purpose three-dimensional Hydrodynamic-Eutrophication Model, hereinafter referred to as HEM-3D. The Environmental Fluid Dynamics Code (EFDC; Hamrick, 1992a, 1996) constitutes the hydrodynamic portion of HEM-3D. The EFDC model developed by Hamrick (1992a) resembles the widely used Blumberg-Mellor model (Blumberg and Mellor 1987) in both the physics and the computational scheme utilized. The model solves the three-dimensional continuity and free surface equations of motion. It responds to surface wind stress, heat and salinity fluxes, freshwater discharge, and specification of tidal forcing. The Mellor and Yamada level 2.5 turbulence closure scheme was implemented in the model (Mellor and Yamada 1982; Galperin et al. 1988). The model uses stretched (or sigma) vertical coordinates and Cartesian or curvilinear, orthogonal horizontal coordinates; therefore, it is easy to handle irregular shoreline and bottom topography. The model uses a second-order accurate, three-time-level finite difference scheme with an internal-external mode splitting procedure to separate the internal shear or baroclinic mode from the external free surface gravity wave. The external mode solution uses a semi-implicit scheme to allow large time steps which are constrained only by the stability criteria of the explicit central difference or upwind advection scheme used for the nonlinear accelerations. The model simulates density and topographically-induced circulation as well as tidal and wind-driven flows, and spatial and temporal distributions of salinity, temperature and suspended sediment concentration. It is uniquely suited to estuaries and sub-estuary components (tributaries, marshes, wet and dry littoral margins). The model has been applied to a wide range of environmental studies in the Chesapeake Bay system and other systems including simulations of power plant cooling water discharges (Hamrick et al. 1995), mixing and dilution of discharges into the York River (Hamrick 1992b), simulation of oyster and crab larvae transport, tidal circulation of the James River (Hamrick 1992c), salinity transport in the Indian River Lagoon (Moustafa and Hamrick 1994), and suspended sediment transport in the James River.

A water quality model with twenty-one state variables and a sediment process model with twenty-seven state variables have been linked to EFDC to form HEM-3D (Park et al. 1995). The model simulates the spatial and temporal distributions of water quality state variables including dissolved oxygen, suspended algae, various components of carbon, nitrogen, phosphorus and silica, and fecal coliform bacteria. The sediment process model, upon receiving the particulate organic matter deposited from the overlying water column, simulates their diagenesis and resulting fluxes of organic substances and sediment oxygen demand back to the water column. The reader is referred to the above referenced articles for a detailed description of the model aspects.

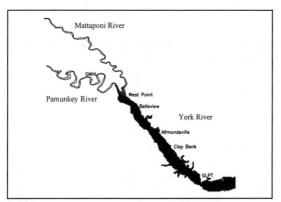

Figure 1. Computation Grid for the York River System.

MODEL CALIBRATION

The model domain is shown in Fig. 1. The model grid combines both curvilinear and Cartesian grids. The grid size is 250 m × 250 m in the Cartesian portion of the domain and 250 m to 1000 m in the curvilinear portion. Eight layers were used vertically. Because there are many shallow embayments and marshes adjacent to the mainstem of the river, especially in the two tributaries, correct modeling of their contribution is very important for an accurate simulation of tidal propagation (Kuo and Park 1995). The original EFDC model has been further modified to handle these marshes in the York River system. The shallow embayments and marshes were treated as storage areas, which allow exchange of water mass between the mainstem and the shoals. The momentum lost only occurs during rising tide at the surface layer and it is treated explicitly. Since the system is shallow, water temperature is almost homogeneous. Water temperature is treated as a constant during model simulation. 25C° and 10C° are specified during summer and winter simulations, respectively.

The bottom roughness height was calibrated by simulating an equilibrium-state condition of the mean tide characteristics of the system. Freshwater inflows equal to the long-term means at the heads of the tide in the Pamunkey and Mattaponi Rivers were used as up-river boundary conditions. A specified characteristic of an incoming wave, (i.e., the M_2 constituent), with a free radiation of an outgoing wave was used as the open boundary condition. Since the long-term tidal record is at Gloucester Point, 11 km from the river mouth, the amplitude of a simple sinusoidal incoming wave (M_2) was adjusted so that the range of the model prediction matched the mean range measured at Gloucester Point. The bottom roughness height was calibrated such that the ranges and phases of the tide predicted by the model matched the mean ranges and phases tabulated in the NOAA Tidal Tables at various stations along the River. Fig. 2 compares the model results with the tabulated mean tide characteristics in the Tidal Tables. The results show that the model well reproduces
the mean tide characteristics of the system.

The model simulation of tidal propagation was further confirmed by simulation of seven constituents and bottom roughness heights were refined if necessary. The model was forced by a characteristic incoming wave of seven constituents (Table 1) and the outgoing wave was allowed to radiate freely at the mouth. The amplitude and phase of each constituent obtained from the measurements near the mouth (Goodwin Islands) were initially used as forcing characteristics. The amplitude and phase of each constituent of the incoming wave were adjusted to match the results of model predictions versus observations, both amplitudes and phases, at Gloucester Point. The model simulation period was three months. The model results of the last 29 days were used for harmonic analysis. The harmonic analysis results of the surface elevation at various stations along the River are listed in Table 1. These results show that the propagation of the tide along the River can be satisfactorily predicted by the model.

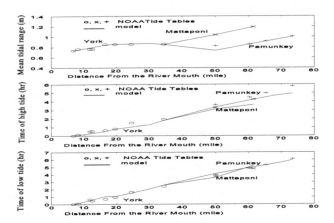

Figure 2. Mean Tidal Range Calibration.

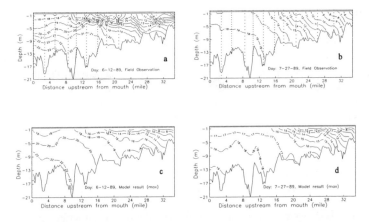

Figure 3. Comparison of Salinity Distribution for VIMS Slackwater Surveys and HEM-3D Predictions for Both June 12 and July 27, 1989 (a, b are observations at slack before ebb and c, d are HEM-3D predictions of maximum salinity during a tidal cycle).

To simulate the salinity distribution in the system, the open boundary condition for salinity is based on the specification of inflowing salinity during flood tide. Outflowing salinities are calculated using upwind salinities immediately inside the open boundary. When the flow at the open boundary changes from outflow to inflow, the model provides a linear interpolation of inflowing salinity based on the last outflowing salinities and the specified incoming salinity. The slackwater surveys conducted during the period of May 30, 1989 to August 29, 1989 were used for the model calibration. During the period of field survey, a tide gauge was also installed at the mouth to provide a time series record of the tide. Two parameters are calibrated by comparing contours of the salinity regime for the York mainstem for both VIMS slackwater surveys and HEM-3D predictions on 7 separate dates in the summer of 1989. The first parameter is the coefficient in the Smogorinsky (1963) subgrid scale horizontal diffusion formulation, which was found to be 0.05 from the mouth to Gloucester Point and 0.01 upriver from Gloucester Point. The second parameter is a lag period from the beginning of flood tide to the point in time at which the specified salinity is attained at the mouth, which is about 3 hours. An example of comparisons of salinity distributions for VIMS slackwater surveys and HEM-3D predictions for both June 12, 1989 and July 27, 1989 are shown in Fig. 3. The results show that the model satisfactorily predicts the spatial pattern of salinity distributions.

MODEL VERIFICATION

The model capability of reproducing surface elevation, current fields, and salinity was verified by direct comparison of model simulation with the prototype time series data. Several model runs were conducted during periods of abnormal as well as normal freshwater discharges. The model was forced by the hourly surface elevation measured at Gloucester Point and the daily mean freshwater discharges at the heads of the tide in the Pamunkey and Mattaponi. A comparison of predicted and observed surface elevations at three stations (Sweet Hall, Belleview, and West Point) during different time periods is shown in Fig. 4. It can be seen that the model can accurately predict surface elevation during periods of different freshwater discharges, especially, during the period of May 2 to May 27, 1986 (Fig. 4a), which was affected by a pulse of high freshwater discharge. To verify the salinity and current, a model simulation was conducted from November 15, 1989 to January 9, 1990. A time series of hourly surface elevation and daily maximum salinity were prescribed at Gloucester Point because these two parameters were available at Gloucester Point during that period of the field survey. The daily mean freshwater discharges at the heads of the tide in both the Pamunkey and Mattaponi were used as up-river boundary conditions. During the period of model simulation, the daily average wind speed and resultant wind direction from the VIMS Gloucester Point station were used to specify the

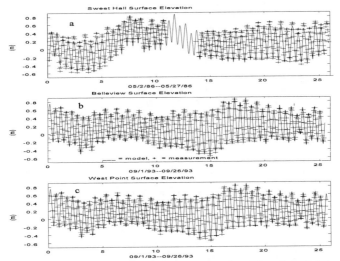

Figure 4. Time Series Comparison of Surface Elevation (Solid Lines are HEM-3D Predictions and +'s are Observations).

meteorological forcing. The model simulated tide, current, and salinity from November 15, 1989 to January 9, 1990. A comparison of current near the surface at Allmondsville from December 7, 1989 to January 9, 1990 is presented in Fig. 6. Comparison of observed salinities against HEM-3D predictions at both 7.7 and 3.7 meters depth at both Clay Bank and Allmondsville is shown in Fig. 5. The results of model verification show that the calibrated model can accurately represent the prototype of the system.

DISCUSSIONS

During the processes of salinity calibration and verification, it was found that the surface wind is very important for the accurate simulation of salinity distribution of the River. Wind affects the hydrodynamic field at the York River by transferring momentum through the surface layer which consequently changes mixing and residual circulation of the system. Fig. 7 displays a salinity time series comparison of observed salinity against model predictions without imposing wind stress at Allmondsville. It can be seen that salinity was predicted to be consistently too low

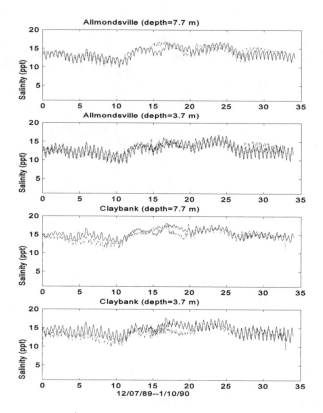

**Figure 5. Salinity Time Series Comparison Between Model
Prediction and Field Measurements (Solid Lines are Model
Results and Dotted Lines are Observations).**

near the surface while too high at the bottom, indicating not enough salt was
transported upward. Without the surface wind stress needed to enhance vertical
mixing, the model predictions become too low at the surface. The model results
using daily averaged wind speed and resultant wind direction measured at the VIMS
station at Gloucester Point show that inclusion of wind stress improves the agreement
between modeled and measured results considerably (Fig. 5). The model predictions
of the salinity are still not very satisfactory around December 22 - 25. The bottom

salinities at Clay Bank are over-predicted, whereas those at Allmondsville are under-predicted. Lower salinities are predicted in January 1990 at Allmondsville. These discrepancies are probably due to the bay conditions that were changed by the wind events, such as surface set-up and tilting of the pycnocline (Kuo and Park 1992; Park and Kuo 1994). These influences are not incorporated into the model calibration. To simulate more accurately both estuary and sub-estuary influences, a larger model domain to include the bay is required for the future model development.

One of the phenomena of the York River is that it undergoes a periodic stratification-destratification resulting from tidal oscillation over the spring-neap cycle. The fortnightly cycle of stratification - destratification has long been recognized and it appears to play a significant role in phytoplankton dynamics and dissolved oxygen dynamics in the York River (Haas 1977; Haas et al. 1981). The model capability of simulation of this pattern is verified by examining the results of model simulation from May 30, 1989 to August 29, 1989. An example of model prediction of periodic stratification-destratification during the spring-neap cycle is shown in Fig. 8, in which the longitudinal-vertical representation of salinity distribution at slack before ebb along the deep channel of the York is presented. Note that the figure contains time-slices aligned with the indicated tidal history at the mouth of the York River. It can be seen that model can correctly simulate the fortnightly cycle of stratification-destratification of the York River.

The typical bathymetry of the York River consists of a deep channel along the axis of the estuary flanked transversely with large areas of shoals. This particular bathymetry has a tendency for front formation along the channel. Observations conducted in the York River show that shear front aligned to the axis of the estuary can be observed near the slack tide (Huzzey 1988; Huzzey and Brubaker 1988). Density differences were observed between the water over the shoal regions and that of the main channel. Flow convergence developing near the slack before ebb results in aggregation of material along the channel. The transverse variability in density and flow play an important role in variation of transverse circulation. The model capability of simulating shear fronts of the York River is examined by conducting a model simulation forced with long term mean freshwater discharge and seven tidal constituents. A sequence of salinity and velocity distributions near the slack tide at various transverses along the axis is examined. The fronts are developing at both slack before ebb and slack before flood every lunar cycle. Strong transverse circulation and flow convergence can be observed at slack before ebb at Clay Bank. Fig. 9 shows an example of instantaneous salinity and velocity distributions at Clay Bank. It can be seen that a flow convergent region can be identified near the north flank of the channel. The front lasts about 1.5 hours with transverse velocity about 10 cm. The results are consistent with the observations and other model experiments

(Nunes and Simpson 1985; Levinson and O'Donnell 1996). The model results also show that the front formed at slack before flood is relatively weak compared to that developed at slack before ebb. Since the shear front is a small scale feature, the resolution of this model is still not fine enough to show the fine velocity structure. For the present model configuration, the shoals at the north of the channel are represented by two to three cells. To simulate the front accurately, higher lateral resolution is required.

CONCLUSION

A high resolution three dimensional model was applied to study the tidal circulation and salinity transport of the York River system, a western tributary of the Chesapeake Bay. The model was calibrated using the predicted mean tide characteristics from the NOAA Tide Tables and the field data collected by VIMS. The model accurately predicts the characteristics of the tidal propagation, in terms of harmonic constituents as well as mean tide, along the river system. The model predictions of the spatial salinity distribution agree very well with the field measurement.

The capability of HEM-3D to reproduce the hydrodynamic field of the prototype estuary was further verified by several model simulation runs. The agreement between model prediction and measured results are demonstrated by comparison of various time series data, including surface elevation, current, and salinity. It is found that inclusion of wind stress affects model simulation of residual circulation of the estuary and results in an improvement of model-field agreement in salinity considerably. The agreement between model predictions and field measurements indicate that the calibrated model properly represents the prototype estuary and it is ready to be applied to a variety of environmental problems of the tidal York River estuary.

ACKNOWLEDGMENTS

We owe special thanks to Dr. John M. Hamrick, developer of the base code of HEM-3D. We thank Drs. Carl T. Friedrichs, David A. Evans, Jerome P.-T. Maa, Zhaoqing Yang for providing critical comments and many helpful suggestions during the processes of model application. Messrs. Steven Snyder and Sam Wilson provided invaluable assistance through their participation in the field program which collected the observational data used in this project and Ms. Nancy Wilson provided assistance in computer facility. We are especially grateful to Dr. Robert J. Byrne for providing guidance and defining the goals of the modeling initiative through which the application of HEM-3D model to the York River system has been made possible.

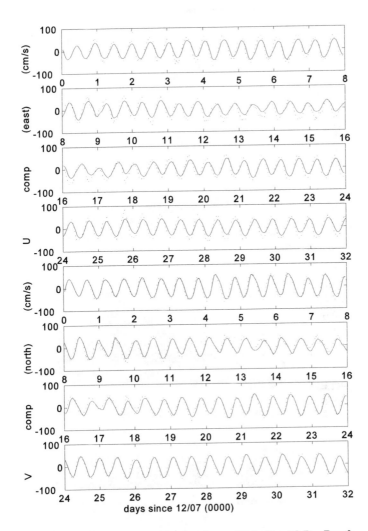

Figure 6. Time Series Comparison of Velocity at 3.7 m Depth at Allmondsville (Solid Lines are HEM-3D Results and Dotted Lines are Observations).

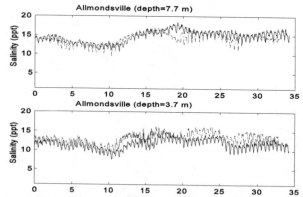

Figure 7. Salinity Time Series Comparison Between Model
Prediction Without Wind Stress and Field Measurements
at Allmondsville (Solid Lines are Model Results and Dotted
Lines are Observations).

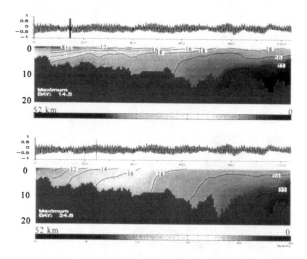

Figure 8. An Example of Longitudinal Salinity Distribution in
the York River (a is spring-tide and b is neap-tide).

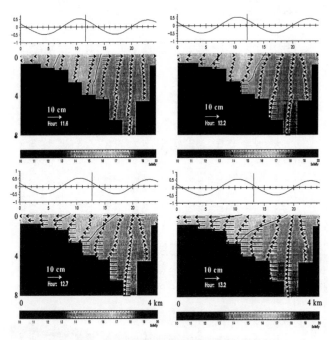

Figure 9. An Example of Front Simulation at Claybank (Looking Upriver, The Vectors Represent Transverse Flow).

Table 1. Comparison of Harmonic Analysis Results for Surface Elevation.

		M2		S2		N2		K1		M4		O1		M6	
		A	P	A	P	A	P	A	P	A	P	A	P	A	P
Goodwin Isl.:	measured	0.317	3.5	0.056	-2.3	0.079	-2.0	0.052	7.1	0.011	2.7	0.036	0.9	0.003	1.0
	modeled	0.321	3.7	0.061	-2.8	0.070	-2.4	0.052	7.4	0.008	2.2	0.039	0.7	0.004	0.5
Glou. Pt.:	measured	0.340	3.8	0.064	-2.3	0.074	-2.1	0.053	7.2	0.004	-2.9	0.040	0.6	0.004	-1.4
	modeled	0.340	3.8	0.065	-2.7	0.074	-2.3	0.053	7.4	0.005	-3.0	0.039	0.7	0.004	-3.5
Claybank	measured	0.366	4.7	0.066	-1.9	0.084	-1.7	0.062	8.0	0.006	-2.5	0.040	0.7	0.004	-0.6
	modeled	0.383	4.4	0.067	-2.0	0.076	-1.6	0.055	8.1	0.010	-2.7	0042	1.3	0.005	-3.2
Allmondsville:	measured	0.385	4.3	0.063	-1.7	0.053	-3.0	0.055	6.9	0.006	-3.1	0.035	1.9	0.004	-0.8
	modeled	0.389	4.6	0.067	-1.8	0.076	-1.4	0.055	8.2	0.010	-2.3	0.042	1.5	0.005	-3.0
Belleview:	measured	0.384	5.3	0.081	-0.8	0.097	-0.4	0.046	9.1	0.021	-2.4	0.031	2.9	0.004	-1.9
	modeled	0.386	5.3	0.061	-1.1	0.069	-0.6	0.054	9.0	0.028	-2.1	0.044	2.3	0.013	2.3
West Point:	measured	0.398	5.7	0.070	-0.5	0.087	9.5	0.057	9.5	0.036	-2.2	0.036	3.4	0.008	-1.7
	modeled	0.369	5.7	0.055	-0.6	0.064	9.5	0.053	9.5	0.044	-2.0	0.044	2.7	0.018	-2.0
Sweet Hall:	measured	0.303	-4.3	0.057	1.2	0.056	3.0	0.050	-11.8	0.020	-0.3	0.042	6.1	0.012	0.6
	modeled	0.315	-4.9	0.040	1.2	0.056	3.0	0.049	-12.5	0.020	-0.3	0.043	4.7	0.020	3.3
Elsing Green:	measured	0.396	-3.2	0.052	3.6	0.052	3.4	0.056	-11.0	0.037	1.6	0.047	6.4	0.015	1.3
	modeled	0.401	-3.8	0.051	2.5	0.062	3.1	0.054	-11.5	0.028	1.1	0.046	5.7	0.021	0.5
Indian Res.:	measured	0.429	-5.7	0.075	1.0	0.067	0.6	0.052	11.4	0.038	-2.6	0.026	1.5	0.018	-1.1
	modeled	0.463	-5.7	0.067	0.6	0.079	1.1	0.057	10.4	0.016	-0.2	0.047	3.6	0.023	-0.8
Walkerton:	measured	0.467	-4.1	0.067	2.5	0.102	2.3	0.043	-11.2	0.048	0.9	0.034	5.5	0.026	0.6
	modeled	0.494	-4.9	0.080	1.6	0.105	2.0	0.072	-12.9	0.059	0.5	0.059	4.5	0.015	0.6

Note: A = amplitude (m) and P = phase (hour) relative to 00:00 1/1/1989

This paper is a contribution of the Virginia Institute of Marine Science, No. 2089.

APPENDIX I. REFERENCE

Blumberg, A. F., and Mellor, G. M. (1987). A description of a three-dimensional coastal ocean circulation model. In: *three-Dimensional Coastal Ocean Models, Coastal and Estuarine Science*, 4, Heaps, N. S., ed., American Geophysical Union, pp. 1-19.

Galperin, B., Kantha, L. H., Hassis, S., and Rosati, A. (1988). A quasi-equilibrium turbulent energy model for geophysical flows. *J. Atmos. Sci.*, 45: 55-62.

Haas, L. W. (1977). The effect of the spring-neap tidal cycle on the vertical salinity structure of the James, York and Rappahannock Rivers, Virginia, U.S.A. *Estuarine and Coast. Marine Sci.*, 5: 485-496.

Haas, L. W., Hastings, S. J., and Webb, K. L. (1981). Phytoplankton response to a stratification-mixing cycle in the York River estuary during late summer. In: *Estuarines and Nutrients*, B. J. Neilson and L. E. Cronin, eds., The Humana Press Inc., Clifton, New Jersey, 619-636.

Hamrick, J. M. (1992a). A three-Dimensional Environmental Fluid Dynamics Computer Code: Theoretical and computational aspects. *Special Report in Applied Marine Science and Ocean Engineering, No. 317.* The Coll. of William and Mary, Virginia Inst. of Marine Sci., 63 pp.

Hamrick, J. M. (1992b). Preliminary analysis of mixing and dilution of discharges into the York River. *A Report to the Amoco Oil Co.* The Coll. of William and Mary, Virginia Inst. of Marine Sci., 40 pp.

Hamrick, J. M. (1992c) Estuarine environmental impact assessment using a three-dimensional circulation and transport model. *Estuarine and Coastal Modeling*, Proceedings of the 2nd International Conference, M. L. Spaulding et al., eds., ASCE, New York, 292-303.

Hamrick, J. M., Kuo, A. Y., and Shen, J. (1995). Mixing and Dilution of the Surry Nuclear Power Plant Cooling Water Discharge into the James River. *A report to Virginia Power Company, Richmond.* The Coll. of William and Mary, Virginia Inst. of Marine Sci., 76 pp.

Hamrick, J. M. (1996). User's manual for the environmental fluid dynamics computer code. *Special Report in Applied Marine Science and Ocean Engineering, No. 331.* The Coll. of William and Mary, Virginia Inst. of Marine Sci., 223 pp.

Huzzey, L. M. (1988). The lateral density distribution in a partially mixed estuary. *Estuarine, Coast. and Shelf Sci.* 9:351-358.

Huzzey, L. M., and Brubaker, J. M. (1988). The formation of longitudinal fronts in a coastal plain estuary. *Journal of Geophysical Research*, 93(C2),1329-1334.

Kuo, A. Y., and Park, K. (1992). Transport of hypoxic water: an estuary-subestuary exchange. In: D. Prandle, ed., *Dynamics and Exchanges in Estuaries and Coastal Zone*, AGU Coast. and Estuarine Sci., Washington, D.C., 599-615.

Levinson, A. V., and O'Donnell, J. (1996). Tidal interaction with buoyancy-driven flow in a coastal plain estuary. In *Buoyancy Effects on Coastal and Estuarine Dynamics*. D. G. Aubrey and C. T. Friedrichs, eds, AGU, Washington DC, 265-281.

Moustafa, M. Z., and Hamrick, J. M. (1994). Modeling circulation and salinity transport in the Indian River Lagoon. *Estuarine and Coastal Modeling*, Proceeding of the 3rd International Conference, M. L. Spaulding et al., eds., ASCE, New York, 381-395.

Nunes, R. A., and Simpson, J. H. (1985). Axial convergence in a well-mixed estuary. *Estuarine, Coast. and Shelf Sci.* 20: 637-649.

Park, K., and Kuo, A.Y. (1994). Numerical modeling of advective and diffusive transport in the Rappahannock Estuary, Virginia. In: *Estuarine and Coastal Modeling*, Proceedings of the 3rd International Conference, M. L. Spaulding et al., Eds., ASCE, New York, 461-474.

Park, K., Kuo, A.Y., Shen, J., and Hamrick J. M. (1995). A three-dimensional hydrodynamic-eutrophication model (HEM-3D): Description of water quality and sediment process submodels. *Special Report in Applied Marine Science and Ocean Engineering* No. 327. The Coll. of William and Mary, Virginia Inst. of Marine Sci., Gloucester Point, 102 pp.

Smogorinsky, J. (1963) General circulation experiments with the primitive equations, Part I: the basic experiment. *Mon. Wea. Rev.*, 91, 99-152.

Modeling thermal structure and circulation in Lake Michigan

Dmitry Beletsky[1] and David J. Schwab[2]

Abstract

A three-dimensional primitive equation numerical ocean model, the Princeton model of Blumberg and Mellor (1987), was applied to Lake Michigan for the 1982-1983 study period. The model has a terrain following (sigma) vertical coordinate and the Mellor-Yamada turbulence closure scheme. This two-year period was chosen because of an extensive set of observational data including surface temperature observations at permanent buoys and current and temperature observations from subsurface moorings. The emphasis of this paper is on the large-scale seasonal variations of thermal structure and circulation in Lake Michigan.

The hydrodynamic model of Lake Michigan has 20 vertical levels and a uniform horizontal grid size of 5 km. The model is driven with surface fluxes of heat and momentum derived from observed meteorological conditions at eight land stations and two buoys from April 1982 to November 1983. The model was able to reproduce all of the basic features of the thermal structure in Lake Michigan: spring thermal front, full stratification, deepening of the thermocline during the fall cooling, and finally an overturn in the late fall. The largest currents occur in the fall and winter when temperature gradients are lowest and winds strongest. Large-scale circulation patterns tend to be cyclonic (counterclockwise), with cyclonic circulation within each subbasin. All these facts are in agreement with observations.

Introduction

There has been significant progress in hydrodynamic modeling of short-term processes such as water level fluctuations due to seiches or storm surges in the Great Lakes (Schwab, 1992). On the other hand, long-term circulation modeling efforts have been rare. For example, since the pioneering works of Simons (1974, 1980) created the basis for numerical studies of circulation and thermal structure in the

---

[1] Cooperative Institute for Limnology and Ecosystems Research, University of Michigan, Ann Arbor, MI 48105-2945

[2] NOAA Great lakes Environmental Research Laboratory, Ann Arbor, MI 48105-2945

Great Lakes, Lake Michigan experienced only two long-term modeling exercises: Allender and Saylor (1979) simulated three-dimensional circulation and thermal structure for an 8-month period, and Schwab (1983) studied circulation with a two-dimensional barotropic model also for an 8-month period.

Currently, with increases in computer power, seasonal variations in thermal structure and circulation can be more easily studied using three-dimensional hydrodynamic models. Many oceanographers and limnologists have used the Princeton Ocean Model (POM) of Blumberg and Mellor (1987) in the ocean, coastal areas, and lakes. In particular, over the past 5 years, POM has been adapted for use in the Great Lakes and has been successfully applied to Lake Erie as part of the Great Lakes Forecasting System (Schwab and Bedford, 1994), and to Lake Michigan (Beletsky et al., 1997).

In this paper, the Princeton model was applied to Lake Michigan in support of the EPA Lake Michigan Mass Balance Project (LMMBP). Model output is being used as an input for sediment transport and water quality models to study the transport and fate of toxic chemicals in Lake Michigan for the 1982-1983 study period. This 2-year period was chosen for the model calibration because of an extensive set of observational data (Fig. 1) including surface temperature observations at two permanent buoys, and current and temperature observations during June 1982 - July 1983 at several depths from 15 subsurface moorings (Gottlieb et al., 1989). The emphasis of this paper is on the large-scale seasonal variations of thermal structure and circulation in Lake Michigan.

There is no ice modeling component in the present version of the model, which can be a problem for the annual cycle modeling in general, because ice cover can cause significant changes in winter circulation patterns in a large lake. The Great Lakes are usually at least partially ice-covered from December to April. Maximum ice extent is normally observed in late February, when ice typically covers 45% of Lake Michigan (Assel et al., 1983). The 1982-83 winter was very mild and therefore, Lake Michigan was practically ice-free during the whole period of our study.

Model description

The Princeton model (Blumberg and Mellor, 1987) is a nonlinear, fully three-dimensional, primitive equation, finite difference model that solves the equations of fluid dynamics. The model is hydrostatic and Boussinesq so that density variations are neglected except where they are multiplied by gravity in the buoyancy term. The model uses wind stress and heat flux forcing at the surface, zero heat flux at the bottom, free-slip lateral boundary conditions, and quadratic bottom friction. Horizontal diffusion is calculated with a Smagorinsky eddy parameterization (with a multiplier of 0.1) to give greater mixing coefficient near strong horizontal gradients. Horizontal momentum and thermal diffusion are assumed to be equal. The equation of state (Mellor, 1991) calculates the density as a function of temperature, salinity, and pressure. For applications to the Great Lakes, the salinity is set to a constant

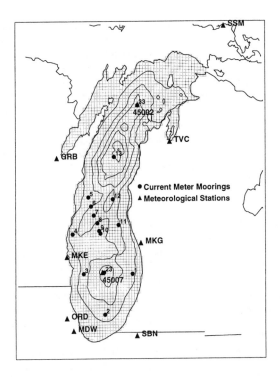

Figure 1: Observation network and computational grid. Isobaths every 50 m.

value of 0.2 ppt. The equations are written in terrain following sigma coordinates ($\sigma = z/h$, where h is depth) in the vertical, and in tensor form for generalized orthogonal curvilinear coordinates in the horizontal. The equations are written in flux form, and the finite differencing is done on an Arakawa-C grid using a control volume formalism. The finite differencing scheme is second order and centered in space and time (leapfrog).

A body of water such as a lake has two separate types of motions, the barotropic (density independent) mode and the baroclinic (density dependent) mode. The Princeton model uses a mode splitting technique that solves the barotropic mode for the free surface and vertically averaged horizontal currents, and the baroclinic mode for the fully three-dimensional temperature, turbulence, and current structure. This necessitates specifying separate barotropic and baroclinic mode time steps in accordance with the Courant-Friedrich-Lewy computational stability criterion.

The model includes the Mellor and Yamada (1982) level 2.5 turbulence closure parameterization. The vertical mixing coefficients for momentum and heat are calculated from the variables describing the flow regime. The turbulence field is described by prognostic equations for the turbulent kinetic energy and turbulent length scale.

The hydrodynamic model of Lake Michigan has 20 vertical levels and a uniform horizontal grid size of 5 km (Fig. 1). To initialize the model, we used surface temperature observations at two buoys (45007 and 45002) located in the southern and northern parts of the lake, respectively. The model run starts on March 31, 1982. Vertical temperature gradients are very small because of convection during that time of the year when the water temperature is less than the temperature of maximum density (4°C). Therefore, we set vertical temperature gradients to zero, but retained horizontal gradients. The initial velocity field in the lake is set to zero. The model run ends on November 20, 1983, after 600 days of integration.

Forcing functions

In order to calculate heat and momentum flux fields over the water surface for the lake circulation model, it is necessary to estimate wind, temperature, dew point, and cloud cover fields at model grid points. Meteorological data were obtained from the eight NWS weather stations and two NDBC buoys. The marine observation network is shown in Figure 1. These observations form the basis for generating gridded overwater wind, temperature, dew point, and cloud cover fields.

Three main steps are required to develop overwater fields from the marine observation data base: 1) height adjustments, 2) overland/overlake adjustment, and 3) interpolation. First, measurements must be adjusted to a common anemometer height. Ship observations are usually obtained at considerably higher distances above the water surface than buoy measurements. Measurements are adjusted to a common 10 m height above the water surface using profile methods developed by Schwab (1978). This formulation adjusts the wind profile for atmospheric stability and surface roughness conditions. Liu and Schwab (1987) show that this formulation for the overwater wind profile is effective in representing typical conditions in the Great Lakes.

The second problem in dealing with the combination of overland and overwater measurements is that overland wind speeds generally underestimate overwater values because of the marked transition from higher aerodynamic roughness over land to much lower aerodynamic roughness over water. This transition can be very abrupt so that wind speeds reported at coastal stations are often not representative of conditions only a few kilometers offshore. Schwab and Morton (1984) found that wind speeds from overland stations could be adjusted by empirical methods to obtain fair agreement with overlake wind speeds measured from an array of meteorological buoys in Lake Erie. For the eight NWS weather stations in Fig. 1 we apply the empirical overland-overlake wind speed adjustment from Resio and Vincent (1977). Air temperature and dew point reports from the overland stations are adjusted with similar empirical formulas.

Finally, in order to interpolate meteorological data observed at irregular points in time and space to a regular grid so that it can be used for input into a circulation model, some type of objective analysis technique must be used. The complexity of the analysis technique should be compatible with the complexity of the observed data, i.e., if observations from only a few stations are available, a best-fit linear variation of wind components in space might be an appropriate method. If more observations can be incorporated into the analysis, spatial weighting techniques can be used. For LMMBP we used the nearest-neighbor technique, with the addition of a spatial smoothing step (with a specified smoothing radius). We found that the nearest-neighbor technique provided results comparable to results from the inverse power law or negative exponential interpolation functions.

After we have produced hourly gridded overwater fields for wind, dew point, air temperature, and cloud cover, the momentum flux and heat flux can be calculated at each grid square in the three-dimensional lake circulation model at each model time step. To calculate momentum flux, the profile theory described above for anemometer height adjustment is used at each grid square at each time step to estimate surface stress, using the surface water temperature from the circulation model. This procedure provides estimates of bulk aerodynamic transfer coefficients for momentum and heat. Surface heat flux, H, is calculated as

$$H = H_{sr} + H_s + H_l + H_{lr}$$

where H_{sr} is short-wave radiation from the sun, H_s is sensible heat transfer, H_l is latent heat transfer, and H_{lr} is long wave radiation. The heat flux procedure follows the methods described by McCormick and Meadows (1988) for mixed-layer modeling in the Great Lakes. H_{sr} is calculated based on latitude and longitude of the grid square, time of day, day of year, and cloud cover. H_s and H_l are calculated using the bulk aerodynamic transfer coefficients described above. H_{lr} is calculated as a function of T_a, T_w, and cloud cover according to Wyrtki (1965). McCormick and Meadows (1988) showed that this procedure works quite well for modeling mixed-layer depth in the Great Lakes. The gross thermal structure generated in the three dimensional model using these heat flux fields is similar to the profile that would be obtained from a one dimensional model. However, there is considerable horizontal variability in the three dimensional temperature field due mainly to wind forcing.

Model results and comparison with observations

Temperature

The most distinctive feature of the physical limnology of the Great Lakes is a pronounced annual thermal cycle (Boyce et al., 1989). By the end of fall, the lakes usually become vertically well-mixed from top to bottom at temperatures near or below the temperature of maximum density for freshwater, about 4°C. Further cooling during winter can lead to inverse stratification and ice cover. Springtime warming tends to heat and stratify shallower areas first, leaving a pool of cold water (less than 4°C and vertically well-mixed because of convection) in the deeper parts of the lake. In spring, stratified and homogeneous areas of the lake are separated by a sharp thermal front, commonly known as the thermal bar. Depending on meteorological conditions and depth of the lake, the thermal bar may last for a period of from 1 to 3 months. Stratification eventually covers the entire lake, and a well-developed thermocline generally persists throughout the summer. In the fall, decreased heating and stronger vertical mixing tend to deepen the thermocline until the water column is again mixed from top to bottom. When the nearshore surface temperature falls below the temperature of maximum density, the fall thermal bar starts its propagation from the shoreline toward the deeper parts of the lake. Thermal gradients are much smaller during this period than during the springtime thermal bar.

The model was able to reproduce all of the basic features of thermal structure of Lake Michigan during the 600 day period of study: spring thermal bar, full stratification, deepening of the thermocline during the fall cooling, and finally an overturn in the late fall (Fig. 2). Observed temperatures from surface buoys and subsurface moorings were compared to model output (Fig. 3). The comparison is quite good for the horizontal distribution and time evolution of the surface and bottom temperature, but it is worse in the thermocline area. In addition, the model predicted internal waves are much less pronounced than in observations. We think that because the model tends to generate excessive vertical diffusion, the modeled thermocline is too diffuse and hence temperature fluctuations are decreased. On the other hand, the simulation of the surface temperature is much more accurate, which shows correct calculation of heat fluxes near the surface. We should also note that the model performs better in the second year (at least near the surface - we do not have subsurface observations for the second year summer) which we attribute to the gradual adjustment of the temperature field to the boundary conditions as the model solution drifts away from the rather crude initial conditions.

Currents

Wind-driven transport is a dominant feature of circulation in the lakes. As shown by Bennett (1974), Csanady (1982), and others, the response of an enclosed basin with a sloping bottom to a uniform wind stress consists of longshore, downwind currents in shallow water, and a net upwind return flow in deeper water. The streamlines of the flow field form two counter-rotating closed gyres, a cyclonic gyre to the right of the wind and an anticyclonic gyre to the left (in the northern hemisphere). As the wind relaxes, the two-cell streamline pattern rotates cyclonically within the basin, with a characteristic period corresponding to the lowest mode topographic wave of the basin (Saylor et al., 1980). Numerical models approximating actual lake geometry have proven to be effective in explaining observed short-term circulation patterns in lakes

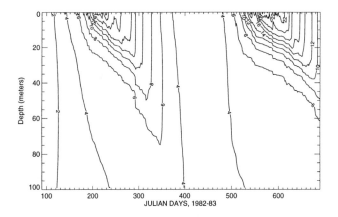

Figure 2: Simulated mean temperature profile, March 31, 1982- November 20, 1983 (600 days).

(Simons, 1980; Schwab, 1992). The results of these modeling exercises show that the actual bathymetry of each of the Great Lakes tends to act as a combination of bowl-shaped sub-basins, each of which tends to support its own two-gyre circulation pattern.

Besides bathymetry and geometry, two other important factors tend to modify the simple two-gyre lake circulation model described above, namely non-uniform wind forcing and stratification. Thus, during the stratified period, longshore currents frequently form a single cyclonic gyre circulation pattern driven by onshore-offshore density gradients. The effect of horizontal variability in the wind field enters through the curl of the wind stress field (Rao and Murty, 1970). Any vorticity in the forcing field is manifest as a tendency of the resulting circulation pattern toward a single gyre streamline pattern, with the sense of rotation corresponding to the sense of rotation of the wind stress curl. Because of the size of the lakes, and their considerable heat capacity, it is not uncommon to see lake-induced mesoscale circulation systems superimposed on the regional meteorological flow, a meso-high in the summer (Lyons, 1971) and a meso-low in the winter (Petterssen and Calabrese, 1959). There are also indications that nonlinear interactions of topographic waves can contribute to the mean single gyre cyclonic circulation (Simons, 1985).

Recent long-term current observations in Lake Michigan suggested a cyclonic large-scale circulation pattern, with cyclonic circulation within each subbasin, and anticyclonic circulations in ridge areas (Gottlieb et al., 1989). Our model results coincide with their conclusions (Fig. 4). To study seasonal changes in circulation patterns, we averaged model results over two 6-month periods: from May to October (summer period), and from November to April (winter period), approximately

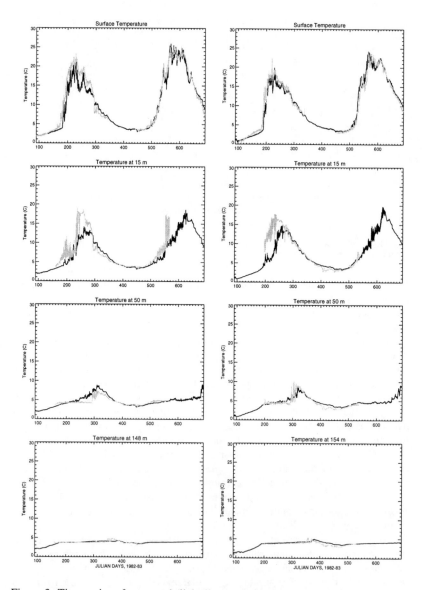

Figure 3: Time series of measured (light line) and simulated (dark line) temperature. Left: buoy 45007 and mooring 23. Right: buoy 45002 and mooring 33.

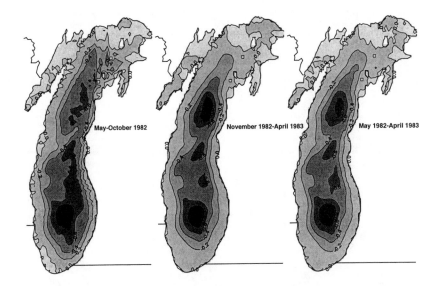

Figure 4: Normalized streamfunctions for summer, winter, and annual circulation. Negative values (dark color) indicate cyclonic vorticity, positive values (light color) anticyclonic.

reflecting stratified and non-stratified periods. Circulation is more organized and more cyclonic in winter than in summer, which is in agreement with Gottlieb et al. (1989) and earlier findings of Saylor et al. (1980). Because winter circulation is stronger than summer circulation, annual circulation looks very similar to the winter circulation (Fig. 4).

We also compared progressive vector diagrams of simulated and observed currents (Fig. 5). The largest currents occur in the fall and winter, when temperature gradients are lowest, but winds are strongest. Nearshore currents appear to be much stronger than offshore currents, in agreement with existing conceptual models and observations (Csanady, 1982). Yet, the mean current is relatively weak. For example, according to Fig. 5 data, it will take a passive particle about a year to complete a round trip of Lake Michigan, which would be about 1000 km. The point to point comparison of currents was successful mostly in the southern basin, which is characterized by a relatively smooth bathymetry. It was more successful in fall-winter months than in summer, most probably because the horizontal resolution of the model is not adequate for proper simulation of baroclinic processes with horizontal length scales comparable to the Rossby deformation radius (which is around 5 km for summer months). In addition, model resolution was too coarse to describe precisely

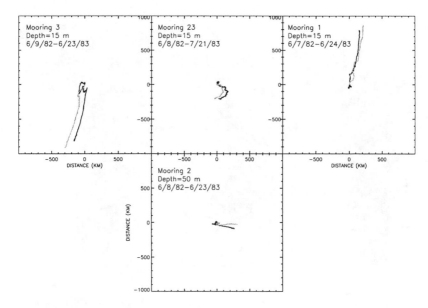

Figure 5: Progressive vector diagrams of measured (dark line) and simulated (light line) currents in 1982-1983.

the dynamics in the areas of strong depth gradients, even in the fall and winter when lake dynamics are essentially barotropic.

Discussion and conclusions

Our general conclusion is that the model realistically simulated the large scale thermal structure and circulation on the 5 km grid. We want to mention some problems, though. First, the model did not predict temperature as well in the thermocline area as it predicted near the surface. To study the effect of vertical resolution on the vertical temperature gradients, we carried out a model run with 39 sigma levels, e.g. double the vertical resolution. In this run we noticed only moderate improvement in the thermocline area. We also ran the model with a zero horizontal diffusion to test for artificial diffusion along sigma surfaces. Again, we did not notice a significant improvement in model results. Therefore, we think that in order to determine whether the problem lies in the calculation of vertical mixing or elsewhere, the next logical step will be a conduction of numerical experiments with a one-dimensional lake model.

Second, the horizontal resolution of the model was only sufficient for description of the large-scale circulation patterns. Point to point comparison of observed and simulated currents was successful only in fall-winter months in the southern basin characterized by a smooth bathymetry. Obviously, 5 km horizontal grid resolution is too coarse to adequately resolve baroclinic motions in summer. The point to point comparison was worst in the areas of strong depth gradients.

The circulation was stronger in winter than in summer, and also more organized and cyclonic, which is in agreement with observations. Since the lake is essentially homogeneous in winter, there are two most probable explanations: existence of stronger cyclonic wind vorticity in winter, or existence of residual mean cyclonic circulation driven by nonlinear interactions of topographic waves. More research is needed to clarify the role of each factor.

Acknowledgments

This research was partially supported by the USEPA Lake Michigan Mass Balance Project. This is GLERL Contribution Number 1066.

References

Allender, J.H. and J.H. Saylor, 1979. Model and observed circulation throughout the annual temperature cycle of Lake Michigan. J. Phys. Oceanogr. 9: 573-579.

Assel, R.A., F.H. Quinn, G.A. Leshkevich, and S.J. Bolsenga, 1983. Great Lakes Ice Atlas. National Oceanic and Atmospheric Administration.

Beletsky, D., W.P. O'Connor, D.J. Schwab, and D.E. Dietrich. 1997. Numerical simulation of internal Kelvin waves and coastal upwelling fronts. J.Phys.Oceanogr. 27(7), 1197-1215.

Bennett, J.R., 1974. On the dynamics of wind-driven lake currents. J. Phys. Oceanogr. 4(3), 400-414.

Blumberg, A.F. and G.L. Mellor, 1987. A description of a three-dimensional coastal ocean circulation model. Three dimensional Coastal Ocean Models, Coastal and Estuarine Sciences, 5, N.S. Heaps [ed.] Amer. Geophys. Union, Washington, D.C., pp 1-16.

Boyce, F.M., M.A. Donelan, P.F. Hamblin, C.R. Murthy, and T.J. Simons, 1989. Thermal structure and circulation in the Great Lakes. Atmos.-Ocean, 27(4), 607-642.

Csanady, G.T., 1982. Circulation in the Coastal Ocean. D.Reidel Co., Dordrecht, Holland, 279 pp.

Gottlieb, E.S., J.H. Saylor, and G.S. Miller, 1989. Currents and temperatures observed in Lake Michigan from June 1982 to July 1983. NOAA Tech. Memo. ERL GLERL-71, 45 pp.

Liu, P.C., and D.J. Schwab, 1987. A comparison of methods for estimating u* from given uz and air-sea temperature differences. J. Geophys. Res. 92(C6), 6488-6494.

Lyons, W.A., 1971. Low level divergence and subsidence over the Great Lakes in summer. Proc. 14th Conf. Great Lakes Res., Int. Assoc. Great Lakes Res., 467-487.

McCormick, M.J., and G.A. Meadows, 1988. An intercomparison of four mixed layer models in a shallow inland sea. J. Geophys. Res., 93(C6), 6774-6788.

Mellor, G.L., 1991. An equation of state for numerical models of oceans and estauries. J. Atmos. Oceanic Technol., 8, 609-611.

Mellor, G.L., and T. Yamada, 1982. Development of a turbulence closure model for geophysical fluid problems. Rev. Geophys. Space Phys., 20(4), 851-875.

Pettersen, S. and P.A. Calabrese, 1959. On some weather influences due to warming of the air by the Great Lakes in winter. J. Meteor.,16, 646-652.

Rao, D.B., and T.S. Murty, 1970. Calculation of the steady-state wind-driven circulation in Lake Ontario. Arch. Meteor. Geophys. Bioklim., A19, 195-210.

Resio, D.T. and C.L. Vincent, 1977. Estimation of winds over the Great Lakes. J. Waterway Port Coast. Ocean Div., ASCE, 102, 265-283.

Saylor, J.H., J.C.K. Huang, and R.O. Reid, 1980. Vortex modes in Lake Michigan. J. Phys. Oceanogr. 10(11), 1814-1823.

Schwab, D.J., 1978. Simulation and forecasting of Lake Erie storm surges. Mon. Weather Rev. 106, 1476-1487.

Schwab, D.J., 1983. Numerical simulation of low-frequency current fluctuations in Lake Michigan. J. Phys. Oceanogr. 13(12), 2213-2224.

Schwab, D.J., 1992. A review of hydrodynamic modeling in the Great Lakes from 1950-1990 and prospects for the 1990's. In: A.P.C. Gobas and A. McCorquodale (Eds), Chemical dynamics in fresh water ecosystems. Lewis Publ., Ann Arbor, MI, 41-62.

Schwab, D.J. and K.W. Bedford, 1994. Initial implementation of the Great Lakes Forecasting System: a real-time system for predicting lake circulation and thermal structure. Water Poll. Res, J. Canada, 29(2/3), 203-220.

Schwab, D.J. and J. A. Morton , 1984. Estimation of overlake wind speed from overland wind speed: a comparison of three methods. J. Great Lakes Res. 10(1), 68-72.

Simons, T.J.,1974. Verification of numerical models of Lake Ontario: I. Circulation in spring and early summer. J. Phys. Oceanogr. 4: 507-523.

Simons, T.J., 1980. Circulation models of lakes and inland seas. Can. Bull. if Fisheries and Aquat. Sci., Bull. 203, 146pp.

Simons, T.J.,1985. Reliability of circulation models. J. Phys. Oceanogr.15, 1191-1204.

Wyrtki, K., 1965. The average annual heat balance of the North Pacific Ocean and its relation to ocean circulation. J. Geophys. Res. 70, 4547-4599.

Hydrographic Data Assimilation on Georges Bank

Daniel R. Lynch and Christopher E. Naimie[1]

Abstract
    The problem of initial condition estimation from hydrographic data is explored in the context of a tidally energetic system. Observational System Simulation Experiments reveal potentially serious errors arising from the space-time aliasing of hydrographic features which are transported in tidal time. These are shown to lead to structured errors in hindcast circulation fields initialized from such observations. A limiting "agnostic" case describes the minimum standard which an initial condition estimation procedure must meet. Suggestions for improving the observability of the hydrography are made and involve modifications of both observational and simulation protocol.

Introduction
    As part of the USGLOBEC program, we are producing hindcasts of the Georges Bank circulation. (See figure 1.) The observations consist of velocity (ADCP) and hydrographic (CTD) surveys over a limited area, roughly 150x300 km. Each survey requires approximately 10 days to complete. These data must be assimilated into a circulation model which will be useful for ecological purposes. A candidate assimilation procedure is as follows:

- Obtain a "best prior estimate" of the circulation by initializing with the observed (T,S) and forcing with observed wind; boundary conditions are taken from a climatological archive;

- Compare the model velocity with the ADCP observations;

- Invert the velocity errors to improve the boundary conditions.

[1]Dartmouth College, 8000 Cummings Hall, Hanover, NH 03755

Experiences with the inversion portion of the process are described elsewhere (Lynch et al 1998). Herein we discuss the estimation of the temperature and salinity fields.

Implied in the above procedure is that the CTD data, obtained in tidal time and over a 10-day period, may be used as an initial condition following suitable interpolation. Of course these data are not synoptic – containing tidal as well as subtidal variability. Our goal here is to evaluate the adequacy of this procedure.

The Georges Bank regime is strongly influenced by tides. Tidal velocities dominate the velocity record as well as the the mixing regime; and tidal rectification in the presence of stratification contributes substantially to the subtidal circulation. On the steep sides of the Bank, the tidal excursion is of order 10 km, with resultant vertical isopycnal motion of order 30 m. The mixing front therefore moves routinely of order 10 km across the Bank, and the pycnocline migrates of order 30 m vertically, all on a 12-hour basis (Naimie, 1996). These motions will inevitably affect the CTD sampling and are potential sources of serious error unless the modeled tidal regime is capable of assimilating the error and reestablishing the underlying signal.

Our approach is to construct "Observational System Simulation Experiments" or OSSE's. We take the output of a simulation model as "Truth"; sample it in a realistic fashion; and study the errors created by the sampling. Key questions concern the persistence of these errors after they are injected into a simulation: how long do they remain in the simulation? and what are their effects on the circulation?

Hydrography
Truth

We concentrate on the climatological mean conditions during May-June. This is a period of significant stratification on the Bank and is of critical ecological interest within USGLOBEC. Our focus on climatology implies that tidally-driven processes will be the only source of temporal variability.

The wide-area climatological solution of Lynch et al (1997) is the most realistic depiction of the Georges Bank circulation available. This is a nonlinear, 3-D simulation incorporating tidal-time baroclinicity and level 2.5 turbulence closure. We use this simulation to drive our limited-area simulation, using the same model and wind stress but simulating only the Bank within the 150m isobath. Dirichlet boundary conditions are enforced on sea surface elevation. Surface temperature is nudged toward climatology with a time constant of one-half day. The surface salinity flux is set to zero.

Like previous simulations of Georges Bank, this simulation reveals significant hydrographic variability both in space and time. There is a mixing front between the 60 and 80m isobaths surrounding the Bank, separating well-mixed and stratified regions. This is evident in figure 2, which depicts the True temperature sampled instantaneously at $t = 0$ (the initial conditions herein). Figure 4 shows the Eulerian horizontal tidal excursions on the Bank; spatial structures in figure 2 can be expected to advect accordingly. The interaction of these effects is pronounced on the steep northern flank of the bank. Figure 3 shows the temperature structure at two points in the tide, as well as the mean and tidal variation of the temperature in an Eulerian sense.

Clearly, sampling within this complex space-time variability presents serious challenges. Should the sampling and estimation procedure reproduce figure 2, then the simulation would be perfect.

Sampling and Estimation

We sample this solution based on an actual USGLOBEC cruise (EN265, 1995). The CTD stations are indicated in figure 5. Sampling began in the southwest corner of the Bank and proceeded counterclockwise around the Bank. The complete sampling required 9.8 days to complete, which is typical within this program. The CTD data are mapped onto the finite element grid by 2-D Objective Analysis (Bretherton et al, 1976) on level surfaces, ignoring temporal variations. The assumed spatial correlation was locally anisotropic with principal axes aligned with the topography. Typical length scales were of order 30 km, with roughly a factor of 2 distortion over steep topography. The software employed was the OAX package developed at Bedford Institute (He et al., 1997).

The resulting temperature field appears in figure 6; and the associated Objective Analysis (hereinafter OA) error map is shown in figure 7. The OA error is small, indicating that within the assumptions given, we have a good estimate. However the actual discrepancy from Truth is a different story, as illustrated in figure 8. In the central, well-mixed area, the differences are minor, consistent with the hydrographic and mixing regimes. But significant discrepancies are obvious in the stratified portion of the Bank.

Two effects contribute to this error – the sparsity of the spatial sampling, and its nonsynopticity. The spatial sampling error may be isolated by sampling the grid instantaneously at time=0. The resulting error (Truth - OA) is displayed in figure 9. As in figure 8, the discrepancies are less than 1 °C within the mixing regime near the center of the Bank and exceed 3 °C in some locations for the stratified periphery of the Bank. It is also noteworthy that the discrepancy exceeds 1 °C for isolated regions near the western extent of the domain; where the hydrographic structure is not fully captured by the sampling pattern.

Circulation

Here we describe the outcome of several simulations. Each was run for 10 days, and differed only in the (T,S) initial conditions. In each case the (T,S) boundary conditions consisted of (nudging toward the climatology,zero flux) at the surface, with vertical diffusion active but all other transports ignored. Effectively, we nudged toward the time-averaged Truth at the surface, but leave initial conditions largely unchanged below the pycnocline. We focus below on the time-averaged circulation, a critical indicator of retention and loss in the Bank system.

Truth

The True circulation, initialized as in figure 2, is illustrated in figure 10. The familiar features of the Georges Bank circulation are apparent: a general clockwise circulation with a strong, concentrated jet on the northern flank; a broader jet over the Northeast Peak and southern flank, which bifurcates into northward (recirculating) and westward flows at Great South Channel (near the southwest corner of the mesh); a tight recirculation cell around the central cap of the Bank. These features are well-established in previous circulation studies (e.g. Naimie 1996).

Simulation #1: CTD-based hydrography

The time-mean circulation initialized from the CTD data as estimated above appears in figure 11. Figure 12 shows the discrepancy from Truth. There are systematic but small errors throughout the mixed zone; significant, organized errors in the northern flank and Great South Channel regions (labeled NF and GSC in figure 1, respectively); and significant errors, with smaller spatial scales, elsewhere around the periphery of the Bank. The errors in the NF and GSC regions amount to an apparent slowdown of the Georges Bank gyre on the northern flank, and to a reduction of its retentive tendency at the crucial Great South Channel branch point, manifest as a decrease in Northward flow at the Western edge of the Bank relative to Truth.

Figure 14 compares tidal-time snapshots of temperature on a roughly North-South cross-bank transect. The True spatial variation on this transect is approximately $2°C$. Errors in the range .5 to 1.5 $°C$ are present in the stratified regions; the central well-mixed area is essentially error-free. Figure 15 shows the same comparison for along-bank (normal to the transect) current speed at this point in time.

Simulation #2: Agnostic hydrography

In this simulation we examine the extreme case: no reliable CTD data and no prior estimate of (T, S). The "agnostic" simulation was therefore conducted in constant-density mode, and otherwise forced identically. The errors relative to truth are illustrated in figure 13. The error pattern is qualitatively similar to the CTD-based simulation, with a slower and less retentive gyre than Truth. The size of this error, however, is significantly larger than from Simulation #1. It is apparent that the CTD-based estimation, while imperfect, is successful relative to this very poor result.

Discussion

These experiments are obviously idealistic; success here does not imply success in real estimation. In particular, several features favor estimation success here:

- The only source of variability is the tidal forcing

- Pressure boundary conditions and wind stress are known exactly

- There is no observational error.

- The nudging surface condition seeks to restore Truth irrespective of the initialization

Nevertheless, we find these OSSE's reveal important aspects of the Georges Bank system, and pose some challenges to the successful estimation of its hydrographic structure.

The agnostic case, Simulation #2, sets out a null or entry-level standard of performance: a procedure for incorporating data must do better than one which ignored the related physics. Clearly the agnostic simulation shows that the baroclinic processes are crucial and failure to consider them would be unacceptable.[2] Reducing the agnostic error is a measure of successful estimation.

Simulation #1, initialized from the CTD data, shows promise in this context but still retains important errors. It is successful in reducing the size of the agnostic error. But the error structure remains intact and there is a systematic underestimation of speed and retention. This procedure therefore clearly needs improvement. Directions for improvement include the following:

- Move the simulation mesh seaward. Getting the boundaries away from the principal tide-topography interaction areas will undoubtedly make the estimation problem for boundary conditions easier. This recommendation was also reached in the related study of the ADCP inversion on this mesh (Lynch et al 1998).

---

[2]In previous studies we have found that the baroclinicity affects the circulation not only directly, through the pressure gradient, but indirectly, through the stratification-dependent reduction of vertical mixing and viscosity. Both are key dynamical players in the Bank regime.

- Revise the (T,S) boundary conditions. In the present simulations these create the potential for persistence of initial condition estimation error below the pycnocline.

- Change the sampling. The present uniform coverage of the Bank is not necessary. It is possible to use fewer CTD stations in the mixed zone, since the stratification is nil there, and perhaps rely on alternate measures of surface temperature. This would allow greater concentration on the stratified portions of the Bank. Spending more effort to get T(t) at a smaller number of stations might well prove cost-effective.

- Develop the space-time covariance structure in the (T,S) fields from the seasonal tidal simulation. Since the tidal motion dominates the short-term advective and mixing regime, it should be possible to describe a useful covariance structure based on climatology. The OA mathematics is rooted in the assumption that such can be developed; in the present case simulation is the obvious way to proceed for tidal variability.

- Develop a "feature model" for the Bank hydrography. This would reduce drastically the number of degrees of freedom used to fit the CTD data, and simultaneously exert physics-based constraints on the outcome.

Most of these ideas share a common thread: that the Bank's tidal regime, along with seasonal heating, maintains complex space-time hydrographic structures fixed to the topography. To the extent that this is true, one can rely on the internal physics to develop these structures. The key themes are to avoid overconstraining these internal mechanisms via the estimation of IC's and BC's; and to avoid dictating the persistence of estimation errors.

The problems described here remain unsolved in any general way. They limit the proper investigation of circulation and ecology in tidally energetic areas. We hope that this short discussion will help to frame the research agenda for both observational, theoretical, and computational scientists.

Acknowledgements

This work was supported by the National Science Foundation through the USGLOBEC program (USGLOBEC contribution #97); and by the Office of Naval Research.

References

Bretherton, F.P., R.E. Davis and C.B. Fandry, "A technique for objective analysis and design of oceanographic experiments applied to MODE-73", *Deep Sea Research* **23**, 559-582 (1976).

He, I., R. Hendry and G. Boudreau, "OAX Demonstration and Test Case." http://dfomr.dfo.ca/science/ocean/coastal_hydrodynamics/oax.html (1997).

Lynch, D.R., M.J. Holboke and C.E. Naimie, "The Maine coastal current: spring climatological circulation", *Continental Shelf Research*, **17**:6, 605-634 (1997).

Lynch, D.R., C.E. Naimie and C.G. Hannah, "Hindcasting the Georges Bank Circulation, Part I: Detiding", *Continental Shelf Research* (in press, 1998).

Naimie, C.E., "Georges Bank residual circulation during weak and strong stratification periods - Prognostic numerical model results", *J. Geophysical Research* **101**, C3, 6469-6486 (1996).

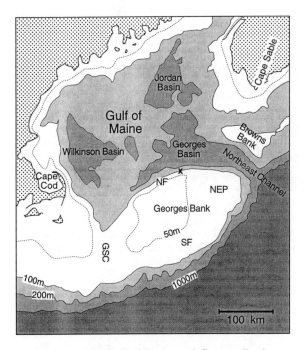

Figure 1: The Gulf of Maine and Georges Bank.

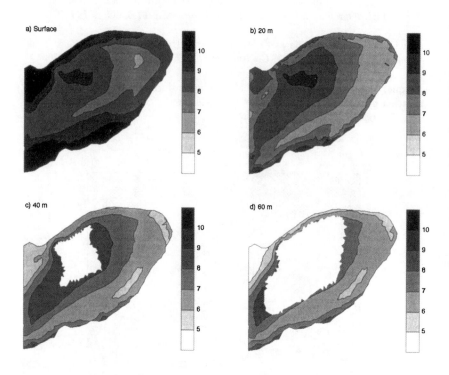

Figure 2: Truth: horizontal sections of temperature at time=0 (Units: $°C$). a) Surface; b) Depth of 20m; c) Depth of 40m; d) Depth of 60m.

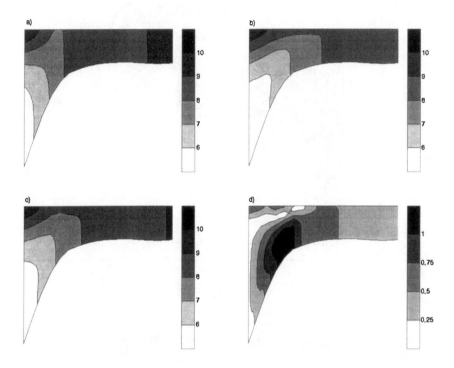

Figure 3: Truth: northern flank tidal dynamics as evidenced by vertical sections of temperature ($°C$). a) Temperature snapshot within tidal time; b) Temperature snapshot 6.2 hours later; c) Tidally-averaged temperature; d) Tidal-time variation in temperature.

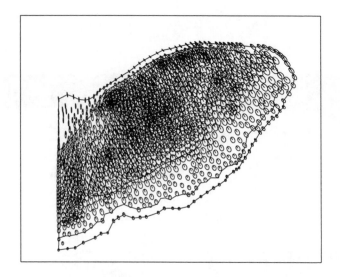

Figure 4: Truth: tidal displacement ellipses.

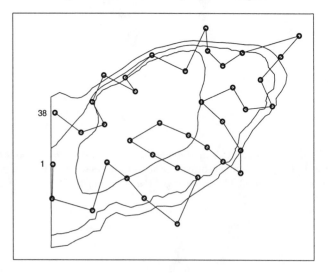

Figure 5: Sampling pattern for USGLOBEC Cruise EN265, April 1995.

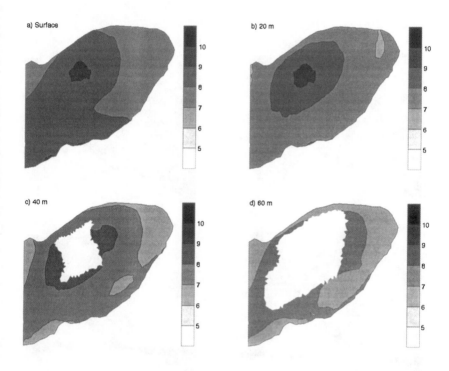

Figure 6: CTD-based sampling and estimation: Horizontal sections of OA estimate of temperature (Units: $°C$).

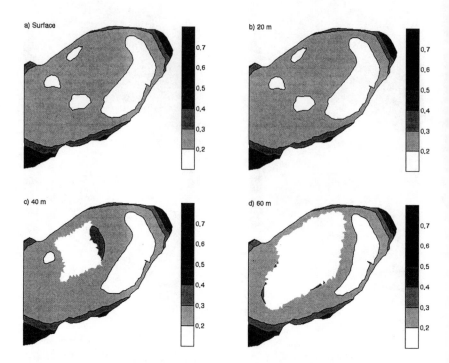

Figure 7: Cruise-based sampling and estimation: Horizontal sections of OA estimate of temperature error. This reflects the statistical assumptions specified during the estimation.

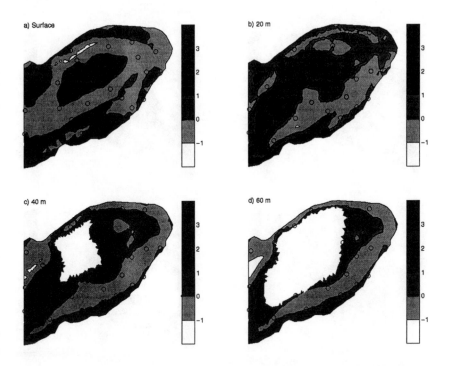

Figure 9: Horizontal sections of temperature error associated with an observational strategy which samples all locations in figure 5 at time=0 (Truth minus OA of sampled data; Units: $\Delta°C$). Locations of CTD samples utilized for the estimation at each horizontal level are indicated by the circles in each panel.

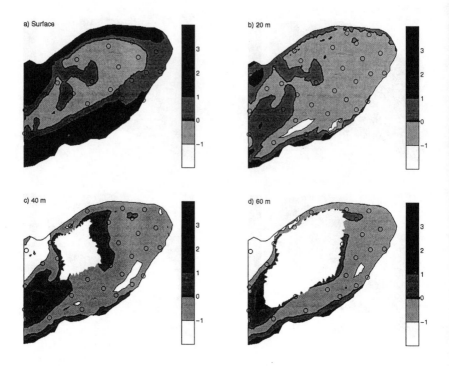

Figure 8: CTD-based sampling and estimation: Horizontal sections of actual temperature error (i.e. truth minus OA of CTD-based temperature sample set; Units: $\Delta°C$). Locations of CTD samples utilized for the estimation at each horizontal level are indicated by the circles in each panel.

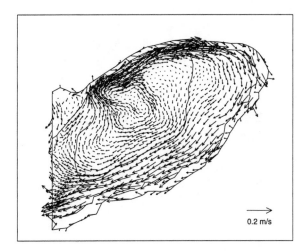

Figure 10: Truth: vertically averaged velocity for the simulation initialized with actual initial conditions.

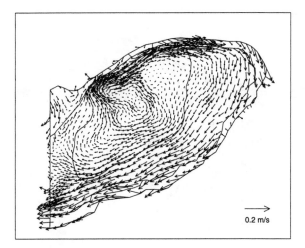

Figure 11: Simulation #1: vertically averaged velocity for the simulation initialized with OA of CTD-based hydrography.

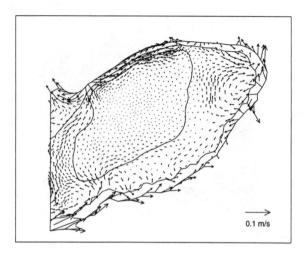

Figure 12: Truth minus Simulation #1: discrepancy in vertically averaged velocity for the CTD-based hydrography case.

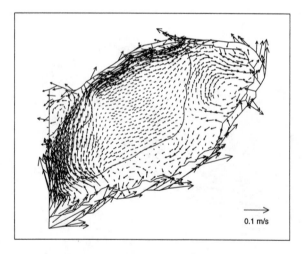

Figure 13: Truth minus Simulation #2: discrepancy in vertically averaged velocity for the agnostic hydrography case.

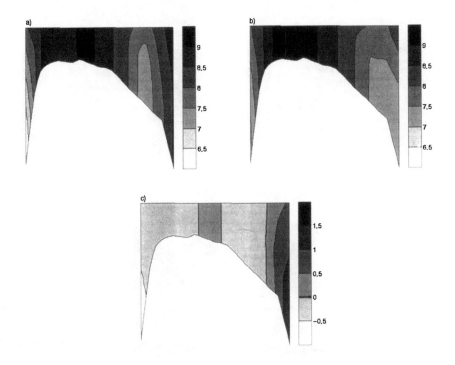

Figure 14: Tidal-time snapshots of temperature on a transect perpendicular to the long axis of the Bank (Units: $°C$). a) Truth; b) Simulation #1; c) Discrepancy between simulation #1 and truth.

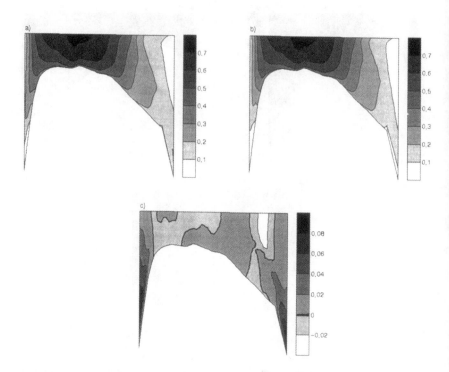

Figure 15: Same as figure 14, but showing current speed normal to the section (i.e. along-bank speed; Units: ms^{-1}).

Open Boundary Conditions for use in Coastal Models

R. P. Matano[1], E. D. Palma[2], J. M. Mesias[1], P. T. Strub[1]

Abstract: This article describes the performance of several open boundary conditions when applied to the Princeton Ocean Model, a 3-D primitive equation model designed for coastal studies. The presentation is divided in two parts. The first part evaluates the performance of radiation conditions, characteristic methods, and relaxation schemes applied to the barotropic mode. According to these experiments the open boundary conditions with the best overall performance are the relaxation schemes proposed by Martinsen and Engedahl (1987), and the radiation type condition suggested by Flather (1976), supplemented with a local solution suggested by Røed and Smedstad (1984) for wind-driven problems. The second part of this study evaluates the performance of a selected set of open boundary conditions for the fully 3-D model. The particular problem to be discussed is the wind driven upwelling of a coastal region. The first set of experiments are conducted in an idealized domain to allow comparisons of the numerical results with theory. Following the idealized experiments the model is used to investigate the coastal upwelling of the Chilean coast.

1. Introduction

Most problems in coastal oceanography are defined within spatial domains that are so large that it is not feasible to compute numerical solutions over the entire region. A common approach to tackle this problem is to calculate the numerical solution in the sub-domain of interest, and to impose open boundary conditions (OBCs) on the artificial

---

[1]College of Oceanic & Atmos. Sc., Oregon State University, Corvallis OR 97331-5503
[2]Dept. de Física, Univ. Nacional del Sur, Bahia Blanca, Argentina.

boundaries. Experience shows that the numerical behavior of any particular OBC strongly depends on its numerical implementation, the general characteristics of the model, and the nature of the problem to be investigated. The objective of this study is to evaluate the performance of several OBCs proposed in the literature when applied to the Princeton Ocean Model, a three-dimensional, primitive equation model of widespread use in coastal oceanography. The evaluation is divided in two parts. The focus of the first part is on passive OBCs applied to the barotropic mode, while the second part address the baroclinic mode. The term "passive OBCs" implies that the dynamics of the flow at the open boundaries are determined by the interior circulation. This is in contrast to "active OBCs" where flow conditions are specified a-priori. Passive OBCs are particularly useful if the circulation at the open boundaries is unknown or an integral part of the problem. The approach followed in this study was to select a suite of OBCs and test their performance in a series of standardized experiments. Three types of OBCs were chosen: a) radiation conditions, b) characteristic methods, c) relaxation schemes. The numerical experiments were designed to emphasize flow conditions dominated by wind forcing and/or wave radiation. This strategy follows the steps of Chapman (1985), Hayashi et al. (1986), Røed and Cooper (1987), and Tang and Grimshaw (1996), the most important differences being the model used, experimental set-up, and the type of OBCs tested. Because of these differences however, the recommendations that stem from this study differ from those offered previously. In particular we do not recommend the use of Orlanski's type conditions as a zero-order approach for the barotropic mode. Our conclusion is that the relaxation method proposed by Martinsen and Engedhal or the radiation type scheme proposed by Flather (1976), complemented by the local solution proposed by Røed and Smedstad (1984), might offer better alternatives. This holds, in particular, for flows with strong non-linear components or variable wind forcing. It has been shown that the performance of a given OBC can be boosted through inverse methods (Shulman and Lewis, 1995). Such possibility has not been explored here, instead the focus has been to determine the best OBCs for different dynamical situations.

This article has been organized as follows: Section 2 contains a brief summary on past studies and a description of the numerical model and the OBCs schemes. Sections 3 and 4 describes the numerical experiments and, finally, section 4 summarizes and discusses the results.

2. Numerical Model

The numerical model to be used is the Princeton Ocean Model (POM). The model equations and the numerical algorithms used to solve them have been described in detail by Blumberg and Mellor (1987). The numerical experiments were conducted in a zonal channel on a f-plane. The channel was 1000 km long and 500 km wide (e. g. Palma and Matano, 1990). The model had a constant resolution of 20 km in the horizontal and 15 vertical sigma levels with high resolution at the top and bottom. Two bathymetric profiles used in our experiments. The first is a bottom with constant slope whose depth increases from 50 to 110 m. The second is a hyperbolic tangent profile with a minimum depth of 20 m at the coast and a maximum of 2000 m offshore. These profiles were chosen to allow comparisons between our results and previous experiments (e.g., Chapman, 1985; Tang and Grimshaw, 1996), and to resolve physical phenomena of particular interest in the future use of the model (e. g., upwelling processes or the propagation of continental shelf waves).

3. Barotropic experiments

The OBCs for the barotropic mode tested in this study can be divided into three categories: 1) radiation conditions, 2) characteristic methods, 3) relaxation methods. Other conditions used in the past, such as clamped (e.g. zero surface elevation), and gradient (e.g. zero normal gradient of the sea surface elevation), will not be considered here since there is enough evidence of their reflective properties (Chapman, 1985, Hayashi et al, 1986, Røed and Cooper, 1986).

Radiation Conditions	Analytic Form
GWI (Gravity Wave)	$\phi_t \pm C_o \phi_x = 0$ $\;with\; C_o = \sqrt{gh}$
BKI (Blumberg & Kantha)	$\phi_t \pm C_o \phi_x = -\dfrac{\phi}{T_f}$
FLA (Flather radiation)	$U = \pm \left(\frac{C_o}{h}\right)\eta$
FRO (Flather + local solution	$U - U_o(t) = \pm \frac{C_o}{h}\left[\eta - \eta_o(t)\right]$
ORE (Orlanski Explicit) ORI (Orlanski Implicit) SOE (Somerfeld Explicit) SRE (Idem but 2nd order) MOI (Camerlengo & O'Brian)	$\phi_t \pm C\phi_x = 0$ $\quad$ with $C = \pm \dfrac{\phi_t}{\phi_x}$;
Characteristic Methods	
HOC (Characteristic)	$\dfrac{\partial(UD)}{\partial t} = 0.5 C_o \dfrac{\partial(UD \pm C_o\eta)}{\partial x} + F_x$
Relaxation Methods	
SPO (Sponge layer + ORI)	In sponge layer $C_d = \varepsilon(x_B \pm x) + C_{di}$
MAR (Flow relaxation scheme)	$\phi = (1-\alpha)\phi; \;\; \alpha = 1 - tanh\left[m0.5(x - x_B)\right]$
MRO (MAR + local solution)	$\phi = \alpha \phi_o(t) + (1-\alpha)\phi$

Table I. Description of the OBCs tested with the numerical experiments.

Table I summarizes the main characteristics of the selected OBCs schemes and the names by which they will be referred during the discussion. A complete description of these conditions, as well as their numerical implementation, can be found in Palma and Matano (1996).

Three numerical experiments, each of which focuses on a dynamic phenomenon of particular interest, were used for the evaluation of the different OBCs schemes: 1) oceanic adjustment to the sudden switch on of a constant wind, 2) relaxation of an initial perturbation in sea surface elevation, 3) the oceanic response to the passage of a traveling storm. The wind driven experiment permits the study of the OBCs behavior in forced flows with a non-zero steady state solution. The barotropic adjustment to an initial perturbation in sea surface elevation is an example of flows dominated by wave radiation. Finally, the passage of a traveling storm combines the characteristics of the previous two experiments i.e., direct wind forcing and wave radiation.

3. 1 Wind driven spin-up: The first experiment investigates the adjustment of a channel flow to an impulsive alongshelf wind-stress forcing. A zonally uniform wind-stress, $\tau=0.1$ Pa, is switched on at $t=0$ and kept constant during the 10 days of integration time. The bathymetry used in this experiment resembles a shallow shelf of constant slope . To assess the OBCs performance we compared the steady state, analytical solutions with those obtained from a benchmark, non linear, numerical run with the open boundaries replaced by cyclic conditions.

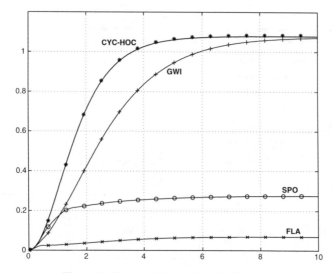

Figure 1: Time evolution of the kinetic energy.

During the spin-up of the model Ekman transports along the solid boundaries generate cross-shore pressure gradients that lead to an alongshore geostrophic flow. In the steady state the energy input by the wind is dissipated by bottom friction in the alongshore flow. Fig. 1 shows the time evolution of the average mechanical energy of most representative experiments; the solid line represents the benchmark case while the other lines corresponds to the cases using OBCs. HOC, MOI, and SOE (not shown) produce results similar to those of the benchmark experiment, the difference between curves is negligible. GWI, and ORE (not shown) show a slightly slower temporal evolution, but they reach the same steady state as the benchmark experiment. Among the Orlanski type OBCs, SRE (not shown) has the poorest performance, still approaching the steady state at day 10. The solution using FLA and MAR reaches an incorrect steady state value after a short period; its numerical scheme prevents the normal evolution of the cross-shore surface slope at the open boundaries which reduces the alongshore transport. For these two schemes it was assumed that no information on the boundary was available so that the computed values were relaxed to zero. These OBCs performed correctly when a local correction was added (FRO and MRO). The experiment using SPO converges towards smaller steady state values than those of the benchmark experiment. This is a consequence of the increased bottom friction in the sponge region, which reduces the alongshore flow and the cross-shelf Ekman flow in the outside layers. Similar results (not shown) were obtained using BKI with a relaxation time of 10 hours (approximately equal to the time it takes a perturbation to travel along the domain), in this case the result is a partial clamping of the boundary. Experiments decreasing the relaxation time to 4 and 1 hours resulted in a stronger clamping of the boundary and therefore in a poorer performance of BKI.

3.2 Barotropic relaxation: This experiment was chosen to test the reflection properties of the various OBCs schemes in flows dominated by wave radiation. It consists of the free adjustment of a sea surface displacement in a basin which includes a shelf and a slope. The initial perturbation is a symmetrical mound centered at the middle of the channel, with maximum elevation of 1 m at the southern boundary and linearly decreasing offshore (e. g. Palma and Matano, 1990). The benchmark experiment for this test was conducted in a closed basin with the same meridional extension, and bottom topography, as the cases with OBCs but with a zonal extension of 10,000 km. The analysis was focused on the central 1,000 km of the basin, and the time integration was stopped before reflections from the closed boundaries could reach the domain of interest. The results of the benchmark experiment are summarized in the left top panel of Fig.2, which shows the time evolution of the sea-surface elevation at an alongshore line 20 km from the southern boundary. The oceanic adjustment to the initial perturbation is accomplished by the propagation of surface gravity and edge waves, with the continental shelf acting as a wave guide. The initial disturbance splits into two mounds of almost equal amplitude traveling alongshore, a Kelvin wave moving to the right and a

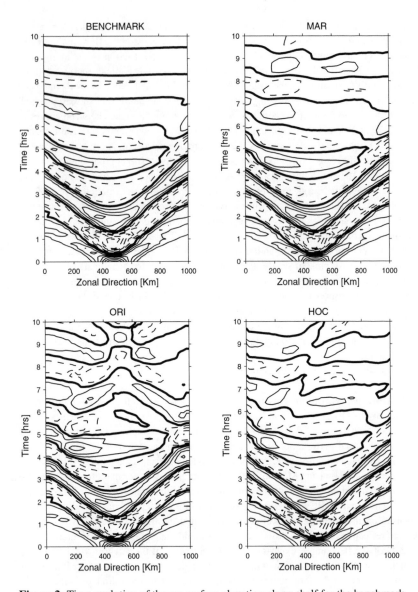

Figure 2: Time evolution of the sea surface elevation along shelf for the benchmark and several OBCs cases.

fundamental mode edge wave moving to the left. The phase speed of both wave packets is almost the same and equal to the non-dispersive long wave speed, rotation playing a minor role. As the depression left behind oscillates, new edge waves are generated until approximately hour 6, where most of the initial energy is lost by radiation. Subsequently, the basin adjusts to equilibrium with small oscillations of nearly 2.3 hours period. The time evolution of the average mechanical energy in this experiment shows that during the spin-down period, energy losses are mostly related to wave radiation, with bottom and lateral friction playing only negligible roles.

The behavior of the most representative OBCs schemes is shown in the remaining panels of Fig. 2. The tilting of sea surface contours is associated with the phase speed of surface gravity waves. The relaxation schemes MAR and SPO and the characteristic method HOC behave similarly to the benchmark experiment, allowing a perfect propagation of incoming waves. Among the radiation type OBCs the best performances were those of FLA and GWI, not surprisingly since they use the phase speed of the dominant signal, and SRE which outperforms all the other Orlanski type OBCs. Of the remaining OBCs the worst performance was that of MOI (not shown) which starts to show reflections as soon as the first wave packets reaches the open boundaries. The remaining Orlanski type OBCs, namely SOE, ORI, and ORE perform well until the end of the spin-down process and then start to produce spurious reflections. This behavior seems to be associated to the fact that, by the end of the spin-down, the little energy that remains makes difficult the calculation of the appropriate phase speeds. This hypothesis seems to be strengthened by the fact that SPO performs well for the entire integration period.

3.3 The oceanic response to the passage of a storm. The objective of this experiment is to test the behavior of the various OBCs schemes in flows dominated by direct wind forcing and wave radiation. We attempted to model the transient forcing associated with the passage of a tropical, or extra-tropical, cyclone over a coastal region. Following Røed and Cooper (1987) the wind stress is generated by a cyclone of Gaussian shape, and maximum wind stress of 3 Pascal, translating at 8 m.s^{-1} from the northwest to the southeast. The storm has a radius of 100 km and enters the domain at x= 250 km, y= 500 km, leaving 24 hours later at x= 750 km, y= 0.

Since there is no known analytical solution to this problem, the cases using OBCs were compared with an experiment conducted in the extended domain described in the previous section. The oceanic adjustment to the storm forcing is achieved through the generation of surface gravity waves and continental shelf waves. For a positive Coriolis parameter there are high frequency edge waves propagating alongshore in both directions (East-West), and low frequency coastal shelf modes propagating in the positive x-direction. Because of the closed offshore boundary there are also two Kelvin modes propagating with the coast to their right. After 36 hours the center of the storm is near the middle of basin, as indicated by the depression following the wake of the storm. Wind forcing over the shelf area generates continental shelf waves that propagate eastward, the first wave packet reaches the eastern boundary after approximately 50 hours. Alongshore wavelengths, calculated from the model output, vary from approximately 500 km to 600 km, with a phase speed of about 6 m.s^{-1}. There is good agreement between the observed characteristics of the first wave-packet and the lowest barotropic shelf modes. Using the computer code developed by Brink and Chapman (1987), we evaluated the dispersion relation and spatial structure of the two lowest barotropic shelf modes corresponding to the bathymetry used. For a mode-1 wave 550 km long the estimated speed is approximately 5.8 m.s^{-1}. The dominance of the low-mode shelf response to cyclones can be related to the similarity in space and time scales of the forcing and these waves (Tang and Grimshaw, 1996). In our experiment the translational speed of the storm in the direction of propagation of the shelf waves is 5.6 m.s^{-1} which is comparable with the phase speed of the mode 1 shelf wave and much less than the phase speed of Kelvin edge waves.

The results of the experiments using different OBCs are summarized in Fig. 3 which shows the time evolution of the eastward energy flux (extended domain results are dashed).

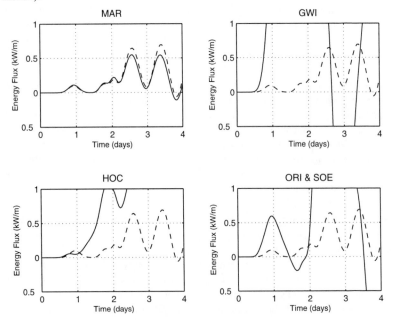

Figure 3: Time evolution of energy fluxes at the eastern boundary.

Of all the conditions tested only MAR, FLA, and SPO were able to handle the passage of the different wave packets without much reflection; MAR having the best performance. Although not shown in the figure, all these schemes were tested during an extended period of 10 days without significant reflections. HOC also shows a satisfactory response when supplemented by the integral constraint. The reflective behavior of the remaining OBCs appears to be related to the highly dispersive nature of the problem. A numerical calculation of the dispersion relation for this type of topographic waves indicates that waves with wavelengths smaller than 1000 km will have large differences in phase speeds. The numerous peaks of the energy flux diagram seems to confirm this.

To test the OBCs in a less dispersive case we conducted a new set of experiments but using the storm with a radius of 200 km described previously. In this case the dominant mode has a wavelength of 850 km. Although at such wavelengths dispersion effects are still present they do not play an important role in the relatively short trajectory that the perturbation has to travel to reach the open boundary. In contrast to the previous test, which showed the wind energy dispersed by a train of wave packets, this experiment was characterized by a single bundle of energy, with a peak centered at around day 2. For this particular case MAR, FLA, and SPO performed as well as in the previous case. The most noticeable improvements were those of GWI and SOE. These improvements were boosted by the addition of an integral constraint, on mean sea level, which forced volume conservation. The better performance of GWI seems also to be related to the fact that in this case the surface gravity wave speed is more representative of the whole wave packet. HOC still shows reflections that are probably caused by the strong forcing at the open boundary. It should be noted that the use of integral constraints for GWI and SOE, whether on sea level or transport, did not change the performance of these OBCs in the highly dispersive case.

Røed and Cooper (1987) investigated the behavior of several OBCs to the passage of a storm in a basin of constant depth. Since for this particular model set-up the transient response is dominated by the propagation of a single Kelvin mode, their results are better compared with our weakly dispersive simulation. Røed and Cooper reported anomalous values of mean sea level in experiments using an explicit version of MOI and an explicit version of Orlanski's condition, which also considered oblique radiation. The results that they obtained using SPO and HOC were similar to those described here, including HOC's reflections at the upstream (western) boundary. Tang and Grimshaw (1996) tested the Orlanski's type radiation conditions using a non linear, inviscid model and a hyperbolic tangent profile bathymetry. In agreement with our results they also observed high reflections during the passage of dispersive waves through the open boundaries, although in their case, SRE worked better than in our experiments. They did not report any problems related to mass conservation.

4. Baroclinic Experiments.

The previous suite of experiments allowed us to identify relative advantages and disadvantages of several OBCs schemes applied to the barotropic mode. For the 3-D simulations it is necessary to specify conditions for the barotropic component of the flow, the baroclinic velocities, and temperature and salinity. In our tests we used the following schemes: a) MOA: MRO for the barotropic model, ORI for the baroclinic mode and the temperature equation, b) BBA: BKI for the barotropic and baroclinic modes and an advection scheme for temperature, c) HOA: HOC for the barotropic mode, ORI for the baroclinic mode and an advection scheme for temperature.

The selected test is the baroclinic response of a coastal ocean to upwelling favorable winds. The first experiment was conducted in the channel domain, initialized with a vertical density profile that depended only on temperature, and forced with a constant wind stress. In the second experiment the channel geometry was replaced by the Chilean coast, the analytical density profile with the annual mean climatological values of density and salinity, and the constant winds with values derived from the ECMWF climatology and coastal observations. Since there is no analytical solution to the non-linear problem the test cases were compared to an experiment run in a cyclic domain.

4.1 Channel experiments: As the wind starts to blow its surface stress drives an offshore Ekman drift and a coastal upwelling jet. The offshore scale of the jet is approximately equal to the internal Rossby radius (~ 10 km). The time evolution of the model is shown in Fig. 4 which displays the time vs. sea surface elevation at a point close to shore and in

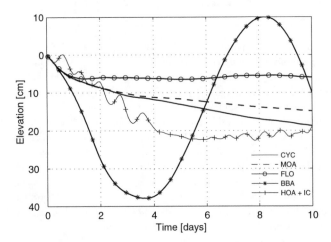

Figure 4: Time evolution of the sea surface elevation.

the middle of the basin for representatives experiments. Of all the schemes only MOA produces a similar, although slower, response than the cyclic case. FLO and BBA shows an earlier tendency towards a wrong steady state with energy levels much lower than expected. Exploratory experiments with HOA presented a steady emptying of the basin after a few days of integration. To improve its response we imposed an integral constraint (IC in the figures). The IC prevented volume changes but increased the amount of reflections at the open boundary. Experiments replacing HOC by GWI and ORI which performed rather well in the along-shelf wind barotropic experiment showed the same tendencies.

The most clear measure of the success of the different OBCs schemes is a comparison of cross-shelf sections at the middle of the basin. Fig. 5 shows a composite of temperature and velocity fields for the most representative schemes. The upper left panel represent the benchmark case. To compensate the offshore Ekman drift the model draws deeper waters. In Fig. 5 the upwelling process is represented by upward tongues of deep waters over the continental shelf. Since the upwelling event moves cold over warmer water this result in an increase in vertical mixing that leads to the quasi vertical temperature structure observed over the shelf break. The schemes with the best performance were MOA and HOA (+IC).

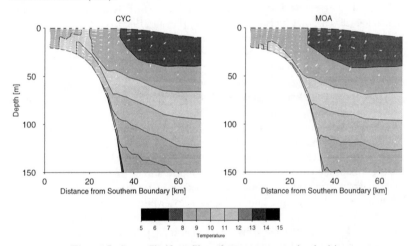

Figure 5: Cross-Shelf profiles of temperature and velocities

Fig. 5 indicates that the amount of upwelling produced by the best performing scheme (MOA) is not as strong as the cyclic case. In this regard it should be remembered that although the cyclic case is the best benchmark we could think of, particular care should be taken when making individual comparisons. The experiments using OBCs schemes distinguish between inflow and outflow conditions i.e., whether the fluid is entering or leaving the domain. Temperature and salinity are restored to their "observed" values at the inflowing points (i.e. the eastern open boundary). This is not the case in the experiment using cyclic conditions which allow cold water to circulate continuously along the model domain. In this regard it should be expected that the overall temperature patterns in the cases using OBCs were warmer than in the case using cyclic conditions.

4.2 Simulation of the circulation in the Gulf of Arauco: The final set of experiments used one of the OBC schemes with the best result in the idealized case (MOA) to simulate the upwelling conditions in the Gulf of Arauco (Chile) (Fig. 6). According to observations typical upwelling events in this region consist of cold plumes that extend 50-150 km offshore, and the generation of energetic cyclonic and anticyclonic eddies. The annual mean wind distribution of this region is characterized by equatorward patterns (upwelling favorable) that are enhanced during the spring-summer period and evolve into poleward winds (downwelling favorable) during the spring.

The horizontal resolution of the model used in this experiment varies from a maximum of 3.5 km over the shelf area to a minimum of 13 km in offshore. It was forced with wind stress values corresponding to the 1993 year derived from the ECMWF climatology and ground observations collected along the Chilean coast, and initialized with annual climatological values of temperature and salinity (Levitus, 1983). Fig. 6 shows a snapshot of velocities and temperature field during the fall season. The velocity fields shows a meandering alongshore jet, and cyclonic eddies which propagates northward. Below the surface layer, an onshore flow is formed that upwells colder and denser waters near the coast. This process generates a strong front that propagates westward and northward with speeds of ~ 0.5-1 km/day. At this early stage a bottom boundary layer is generated, with a relatively strong onshore and upward flow (~0.2 m/s) crossing most of the basin. Confined to the bottom deep shelf, there is also an offshore southward but weaker flow which interacts with the equatorward current. The flow in the bottom boundary current shows intensity fluctuations that are correlated with the variability of the wind stress forcing. Temperature fields show minimum temperatures of 10° C along the coast with some cold plumes extending offshore. At the southern part of the domain, the equatorward jet meanders following the bottom topography, while it generates cyclonic and anticyclonic eddies. These eddies increase in size and intensity while propagating northward. The large, cyclonic, eddy observed near 35° C has a northward propagation speed of approximately 3-4 km/day. Further south, and west of Punta Lavapie, a similar but less intense cyclonic eddy is observed offshore of the coastal jet. The formation of these eddies appear to be related to shear instabilities occurring between the coastal jet and the poleward undercurrent. The instability process is corelated to strong fluctuations of the local wind forcing. Cross-shelf sections of velocity and energy fluxes along the northern boundary show the outward displacement of the eddies like the one just described seemed not to be perturbed by the OBCs schemes.

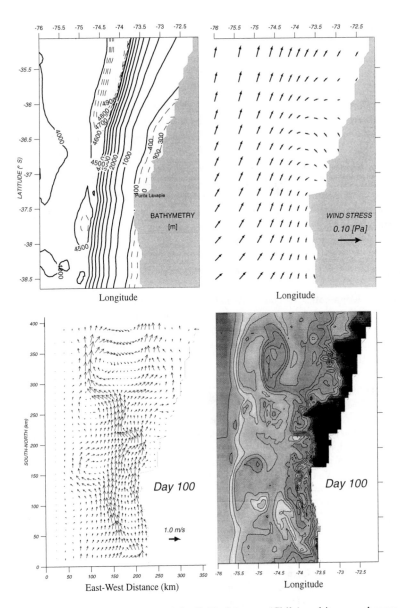

Figure 6: Top panels: bathymetry of the Gulf of Arauco (Chile) and its annual mean wind stress. Bottom: surface velocity and temperature from a model simulation

5. Conclusions.

Although POM has been extensively used in coastal studies most of those studies have been conducted with active OBCs i.e., imposed inflows and there has been relatively little research on POM's performance using passive OBCs. The objective of this article has been to show what OBCs had the best performance for POM in a selected set of experiments, and how those results compare with previous studies. Unlike previous comparative studies, which recommended the use of Orlanski's type radiation schemes (Chapman, 1985; Tang and Grimshaw, 1996), or the characteristic method (Røed and Cooper, 1987), the OBCs with the best overall performance in our numerical experiments were the relaxation scheme proposed by Martinsen and Engedahl (1987) and the radiation type condition suggested by Flather (1976), supplemented with a local solution for the wind driven cases (MRO and FRO). None of these conditions have been tested in previous comparative studies nor, to the best of our knowledge, have FLA been used in conjunction with the local solution proposed by Røed and Smedstad (1984). The different judgments on OBCs performances can be, in part, rationalized in terms of inter-model differences; in particular on the inclusion (or not) of non-linear terms.

Acknowledgments: The comments and suggestions of two anonymous reviewers are gratefully acknowledged. This study was supported by the National Aeronautics and Space Administration grants NAGW 2475 and JPL 958128 and National Science Foundation contract OCE 9402856. E. D. Palma was supported by Universidad Nacional del Sur (Bahia Blanca, Argentina).

References.

Blumberg, A. F. and G. L. Mellor, A description of a three-dimensional coastal ocean circulation model, in *Three-Dimensional Coastal Ocean Models, Coastal Estuarine Sci.,* vol., 4, edited by N. Heaps, pp. 1-16, AGU, Whashington, D. C., 1987.

Brink, K. H., and D. C. Chapman, Programs for Computing Properties of Coastal-Trapped Waves and Wind-Driven Motions Over the Continental Shelf and Slope (second edition), *Technical Report WHOI-87-24,* 119 pp. Woods Hole Oceanographic Institution, 1987.

Chapman, D. C., Numerical treatment of cross-shelf open boundaries in a barotropic coastal ocean model. *J. Phys. Oceanogr.,* 15: 1060-1075, 1985.

Flather, R. A., A tidal model of the northwest European continental shelf, *Mem. Soc. R. Sci. Liege, Ser. 6, 10,* 141-164, 1976.

Hayashi, T., D. A. Greenberg, and C. J. R. Garrett, Open Boundary conditions for numerical models of shelf sea circulation, *Cont. Shelf Res., 5(4),* 487-497, 1986.

Martinsen, E. A., and H. Engedahl, Implementation and testing of a Lateral Boundary Scheme as an Open Boundary Condition in a Barotropic Ocean Model, *Coastal Eng., 11,* 603-627, 1987.

Orlanski, I., A simple boundary condition for unbounded hyperbolic flows, *J. Comput. Phys., 21*, 251-269, 1976.

Palma, E. D., and R. P. Matano, 1997, On the implementation of open boundary conditions to a General Circulation Model: The barotropic mode. *J. Geophys. Res., 103*, 1319-1341.

Røed, L. P., and O. M. Smedstad, Open boundary conditions for forced waves in a rotating fluid, *SIAM J. Sci. Stat. Comput., 5*: 414-426, 1984.

Røed, L. P., and C. Cooper, C., Open Boundary Conditions in Numerical Ocean Models, in *Advanced Physical Oceanographic Numerical modelling, NATO ASI Series, (C)*, vol., 186, edited by J.J. O'Brien, pp. 411-436, D. Reidel Publishing Company, 1986.

Røed, L. P., and C. Cooper, A study of various open boundary conditions for wind-forced barotropic numerical ocean models, in *Three-dimensional Models of Marine and Estuarine Dynamics*, edited by J.C.J. Nihoul, and B. N. Jamart, pp. 305-335, Elsevier, Amsterdam, 1987.

Shulman, I., and J. Lewis, 1995. Optimization approach to the treatment of open boundary conditions. *J. Phys. Oceanogr., 25*, 1006-1011.

Tang, Y., and R. Grimshaw, Radiation Boundary Conditions in Barotropic Coastal Ocean Numerical Models, *J. Comput. Phys., 123*, 96-110, 1996.

Impact of Satellite Derived Cloud Data on Model Predictions of Surface Heat Flux and Temperature: A Lake Erie Example

Yi-Fei P. Chu[1], Keith W. Bedford[2]

Abstract

Accurate information on cloud cover and surface heat flux is critical to the success of surface water model predictions. To predict temperature, a numerical model requires surface heat flux values which are calculated based on cloud cover, wind speed and other meteorological variables. Existing surface heat flux formulations overestimate the values due to the underestimation of cloud cover and predicted surface water temperatures are correspondingly higher than that observed from measured data. Lack of adequate spatial and temporal resolution of cloud data and improper representation of cloud cover in the interpolation scheme are the two major sources hypothesized to cause the heat flux over-prediction. To correct this problem, satellite-derived cloud cover data are used in this article to compute the surface heat flux. Since the Geo-stationary Operational Environmental Satellite (GOES) provides extremely high spatial and temporal resolution data, accurate cloud cover information can be obtained to provide a better surface heat flux estimation. The purpose of this study is to determine the effect of satellite-derived cloud cover data on surface heat flux estimates and evaluate the impact of resulting heat flux on Lake Erie surface water temperature. In this study, more than 3600 GOES-8 satellite images have been analyzed to derive cloud cover information. We found that in general, adequately robust cloud cover results in a 50% reduction in the magnitude of solar radiation. Therefore the new heat flux formulation predicted Lake Erie surface water temperatures will drop about 2 degrees C in average and compare much more favorably to observed values.

---

[1]Graduate Research Associate, Dept. of Civil and Environmental Eng. and Geodetic Sci. The Ohio State University, Columbus, Ohio 43210

[2]Professor and Chair, Dept. of Civil and Environmental Eng. and Geodetic Sci., The Ohio State University

Introduction and Objective

Surface heat flux is the key component in hydrologic models and ocean circulation models. Of interest to this article, it is also the required input data for the Great Lakes Forecasting System (GLFS) in order to predict Lake Erie's thermal structure (Bedford and Schwab, 1990; Bedford and Schwab, 1991). Since the in-situ surface heat flux measurements are unavailable in the Great Lakes, we currently use a heat flux model based on an energy balance concept to estimate surface heat flux. This heat flux model requires cloud cover and other meteorological information in order to calculate radiation transfer terms and net surface heat flux. Recent studies show the predicted Lake Erie surface temperatures are about 3 degrees C higher than that observed from measured data at the end of cooling season. Lack of adequate spatial and temporal resolution of cloud data and improper representation of cloud cover in the interpolation scheme are the two major sources hypothesized in this article to cause the surface heat flux over-prediction. In this study, more than 3600 high resolution satellite-derived cloud cover data are used to compute the surface heat flux.

Current Cloud Data Collection Problems

Currently, cloud data are collected at eight land-based stations around Lake Erie, then these data are interpolated by using a nearest neighbor technique. Even though the observation data are collected on an hourly basis, there are still several factors which prevent obtaining accurate cloud cover, surface heat flux estimates and satisfactory Lake Erie temperature predictions.

First, stations report cloud condition in oktas where one okta is defined as one eighth of the sky. Normally, the surface observer's "sky dome" is roughly a 50 km diameter circle above the geographic location of the surface observation site. The reported cloud cover information is the percentage of cloud amount in that 50 km diameter circle. For example, an observer might report 4 oktas indicating that there is 50% sky cover by the cloud, but the 50% cloud cover could be either above the lake or land because all the stations are land-based so the correct cloud cover over the lake is unknown unless under clear sky (0 octa) or fully cloudy(8 octas) conditions. Second, the surface heat flux model and radiation transfer terms were developed during a time when the effect of clouds was incorporated without spatial and temporal variability. At this time, surface heat flux is computed in a spatially robust way wherein the heat flux algorithm and cloud cover are applied to every numerical grid cell in the model, a single average value cloud cover over the surface of the lake is not a proper representation for the cloud cover. For example, a 50% average cloud cover over Lake Erie means that half of Lake Erie is cloudy and half is clear sky. Third, the nearest neighbor technique is used to interpolate all the meteorological variables including air temperature, dew point temperature, wind speed and cloud cover. This procedure make sense for all the meteorological variables, except cloud cover where a sharp cutoff between the cloud and no cloud zone is required. Cloud cover therefore does not obey presently available interpolation function procedures which typically depend upon the distance between the analyzed grid and observed value. In addition, a "ring effect" happens in many of the interpolated occasions when one station reports 8 octas cloud and another station reports clear sky condition; this in turn causes a series of alternating cloud cover bands to occur ranging from 8 to 0 octa. All the above factors contribute to the underestimation of cloud cover.

To remedy the above problems, a proper method that can provide the spatial and temporal variabilities of cloud cover on the scale of the model grid and correct cloud representation must be developed. GOES-derived cloud data appear to be an excellent source to solve these problems. The cloudiness of each grid point at each hour can be determined by the actual GOES pixel value.

Surface Heat Flux Model

Currently, GLFS uses a heat flux model based on the heat balance concept to estimate hourly surface heat flux. The net surface heat flux is the algebraic sum of the following four components: incoming shortwave radiation from the sun, longwave radiation transfer, sensible heat flux transfer and latent heat flux transfer. The computer code is based on a subroutine written by Dr. M. J. McCormick of GLERL (McCormick and Meadows, 1988), using various authors' algorithms for the individual components.

The equations for hourly global solar radiation are based on the solar zenith angle and the amount of cloudiness (Cotton, 1979; Guttman and Mathews, 1979). The procedure first calculates solar radiation for clear sky conditions and then modifies it for cloudy conditions by including the cloud effects. The clear sky rate derives from a cubic regression of observed rates as a function of sun location, latitude, zenith angle and day of the year. Longwave radiation heat transfer is the Stephan-Boltzmann black-body radiation modified by surface emissivity effects and is a function of cloud cover, air vapor pressure, water-air temperature difference and absolute water temperature based on Wyrtki's (1965) procedure. The latent heat transfer rate is due to evaporation or condensation and is calculated from the wind speed, momentum drag coefficient, air density, and specific humidity difference between air and water surface. The sensible heat transfer rate represents heat dissipation across the air-water interface due to any temperature differential. It's calculated using the wind speed, air density, air-water temperature differences and a heat transfer coefficient. Wind speed is corrected based on Schwab's algorithm (Schwab and Morton,1984).

The surface heat flux model and it's individual equations are described as follows:

$$H = H_{SR} + H_{LR} + H_L + H_S$$

where: H_{SR} = shortwave radiation from sun; H_{LR} = longwave radiation heat transfer; H_L = latent heat transfer; and H_S = sensible heat transfer.

H_{SR}: Shortwave radiation from sun

$$SRCS = a_0 + a_1 \cos ZA + a_2 \cos^2 ZA + a_3 \cos^3 ZA$$
$$SR = SRCS \cdot CLD$$
$$CLD = c_0 + c_1 OPQ + c_2 OPQ^2 + c_3 OPQ^3$$

where: SRCS = clear sky global radiation; SR = the hourly global radiation; $\cos ZA$ = solar zenith angle; a_0, a_1, a_2, a_3 = regression coefficient; c_0, c_1, c_2, c_3 = regression coefficients for cloud cover; and OPQ = the number of tenths of opaque cloudiness.

H_{LR}: Longwave radiation heat transfer

$$H_{LR} = sc \cdot \theta w^4 (0.39 - 0.05 ea^{1/2})(1 - kC^2) + 4sc \cdot \theta w^3 (Tw - Ta)$$

where: s = the ratio of the radiation of the sea surface to that of a blackbody; c = Stefan-Boltzmann constant, θw = absolute temperature of the water; ea = vapor pressure of the air in millibars; C = cloud cover; k = 0.67 depends on latitude; Tw = water temperature; and Ta = air temperature.

H_S: Sensible heat transfer

$$H_S = C_h \cdot C_P \cdot \rho_A \cdot u_W \cdot \Delta T$$

where: C_h = bulk heat coefficient; C_P = specific heat of air at constant pressure; ρ_A = density of the air; u_W = wind velocity; and ΔT = air water temperature difference.

H_L: Latent heat transfer

$$H_L = C_d \cdot qL \cdot \rho_A \cdot u_W \cdot (h_A - h_W)$$
where: C_d = drag coefficient; qL = latent heat of vaporization; h_A = specific humidity of air; and h_W = specific humidity at water surface.

GOES Data Acquisition and Image Processing

NASA launched the first Geostationary Operational Environmental Satellite GOES-1 in 1975. In the spring of 1995, the first of NOAA's next generation of geostationary satellites, GOES-8, was launched. GOES satellites orbit in the earth's equatorial plane at an attitude of 35,000 km and their west-to-east motion is matched to the earth's rotation, so it permits extremely high frequency observations over a very large area. The visible/infrared spin scan radiometer (VISSR) on board GOES was designed to provide frequent visible and IR images of cloud patterns for meteorological applications. The GOES-8 imager instrument has a five-band multispectral channels that provide data with 1 km spatial resolution and a temporal resolution of 1 hour with 10-bit precision (1024 gray scale). An example of a GOES image is provided in Figure 1.

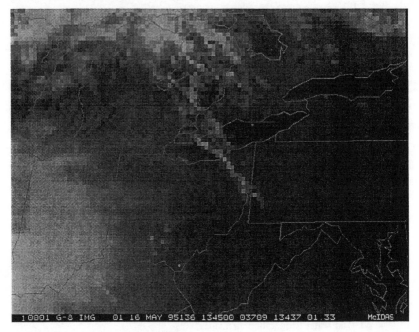

Figure 1. Example of GOES image

Since the incoming GOES images include the whole United States and only the Lake Erie data are needed in this study, the first step is to select a subregion containing only the Lake Erie portion and save them for further image processing. The total number of GOES images used in this study is 3600.

Several image processing routines are applied to the GOES image to generate a gridded cloud input file for the interpolation program and lake circulation model. First, each image was displayed on a workstation screen, then the image was rotated 27 degree clockwise so the Lake Erie image can align with the numerical model grid correctly. Then, the rotated Lake Erie image was resampled to a 209 by 57 array to match the model grid system. Once images are rotated and resampled, a cloud threshold testing algorithm is applied and a threshold value is determined. Any grid point has a digital count value higher than the cloud threshold will be assigned fully cloudy (8 octas) and any value lower than the threshold will be assigned as clear sky (0 octa), and finally a new file containing Lake Erie cloud information with value either 0 or 8 is created. At the same time, coordinates of ground control points, threshold cloud value and cloud cover percentage over Lake Erie for every single images are computed and recorded.

Sensitivity Test and Season Long Simulation

As an initial hypothesis test it is useful to study the sensitivity of the cloud cover on the surface heat flux calculation before incorporating GOES data into the existing surface heat flux model. A preliminary short-term sensivity test has been performed to evaluate the cloud cover effect quantitatively. A two-week period (Julian days 136 to 150) has been selected to calculate surface heat flux and its components under various cloud conditions. There is a two-week "spin up" period before the evaluation, so the effect of initial temperature can be minimized. Three cases with different cloud cover (0 octa, 4 octas, 8 octas) were tested with all the cloud fields kept spatially homogeneous or constant so the effect of cloud cover can be easily determined without the interference from the spatial variation.

The season long simulation was performed from May 16 (Julian day 136) to October 31 (Julian day 304), 1995 which covers the heating, transitional and cooling phases of Lake Erie. Simulation results based on three different cloud options: 1) no cloud condition (0 octa), 2) with station-reported octa cloud data plus nearest neighbor interpolation, and 3) with GOES derived cloud cover. Surface temperatures, at grid points coincident with NOAA buoys 45005 and 45132 are stored so a direct evaluation between the computed values and observation can be made. The location of two NOAA buoys is shown in Figure 2.

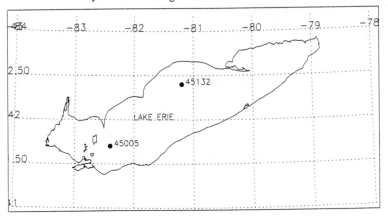

Figure 2. Location of NOAA buoy 45005 and 45132

Data and Tools Used in Evaluation

Standard statistical evaluation tools will be used in this research: statistical measurements such as mean, variance, standard deviation, root mean square error (RMSE) and absolute difference are calculated. All the statistic measures are computed on a two week interval. Figure 3. is a time series of surface water temperature measured from NOAA buoy 45005 during the heating season.

Model Results

The effects of cloud on surface heat flux under different cloud conditions are tested and the results are shown in Figure 4. First, cloud cover does have a significant impact on the solar radiation term calculations. The maximum value for solar radiation is about 950 watts/m^2 under clear sky conditions, compared to 800 watts/m^2 under 4 octas cloud cover and only 350 watts/m^2 under cloudy conditions. For the no cloud and 4 octa cloud cases, the longwave radiation magnitude ranges from 0 to -50 watts/m^2 and only differs by about 10watts/m^2 , while for the 8 octa case, the longwave radiation terms are considerably higher and stay on the positive side all the time. The presence of clouds yields slightly higher sensible heat flux at the end of testing period. The values for sensible heat flux range from 0 to 250 watts/m^2, depending on the magnitude of wind stress. The latent heat flux does not fluctuate much and contributes only a small portion to the net surface heat flux. The net surface heat flux under 4 octa clouds is only slightly lower than the value under clear sky condition, but the reductions are significant under 8 octa cloud cover. The net surface heat flux values under cloudy conditions are about half of the values under clear sky conditions. This suggests that surface heat flux will be heavily overestimated if cloud cover is underestimated between 4 octa to 8 octa zone.

Figures 5 and 6 show the time series plot of measured and computed Lake Erie surface temperature at location 45005. Temperatures under three different cloud conditions (No cloud, nearest neighbor cloud and GOES derived cloud covers) were computed and plotted. The root mean square error and difference between measured and computed temperature are summarized in Table 1 and 2. Under no cloud and nearest neighbor cloud conditions, temperatures were over predicted the entire heating and cooling season except the first two weeks. Smaller root mean square error and difference indicate that GOES derived cloud cover do improve the temperature predictions and the computed temperature values are much closer to the observed data. However, even with GOES derived cloud, the temperature is still about 1.5 degree C higher than the measured data during the last two weeks (Julian days 290-304).The large temperature difference and RMSE at buoy 45132 in the first six weeks (Julian day 136-178) are due to instrument malfunction.

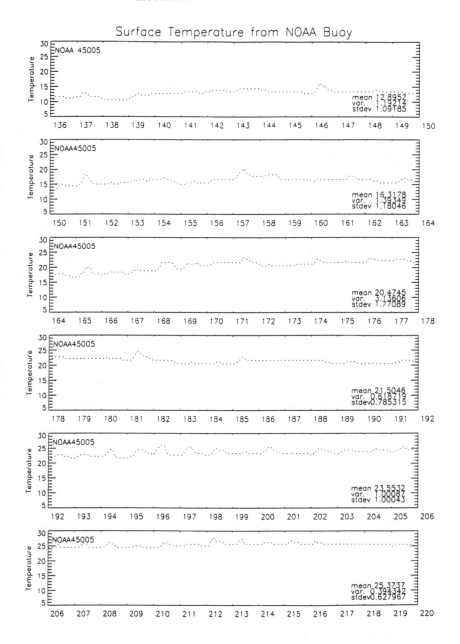

Figure 3. Surface temperature measured from NOAA buoy 45005

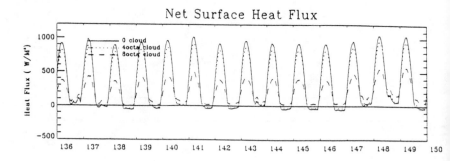

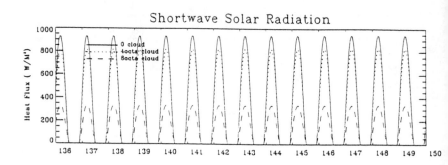

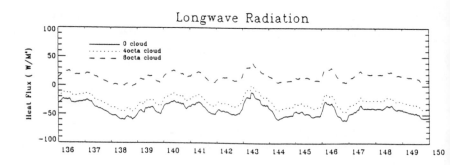

Figure 4. Heat flux, short and longwave radiation values during JD136-150

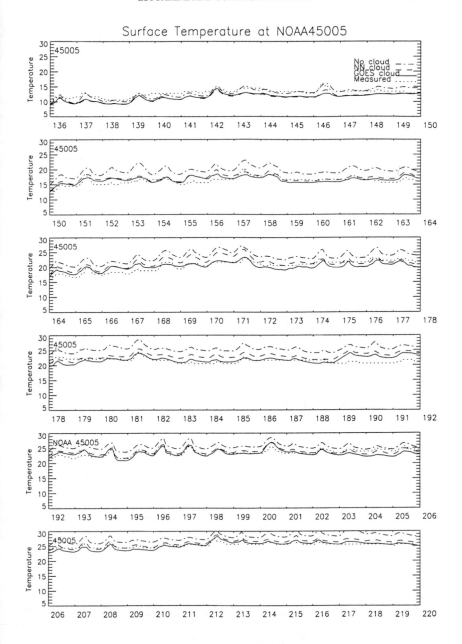

Figure 5. Comparison between measured and computed temperature at 45005

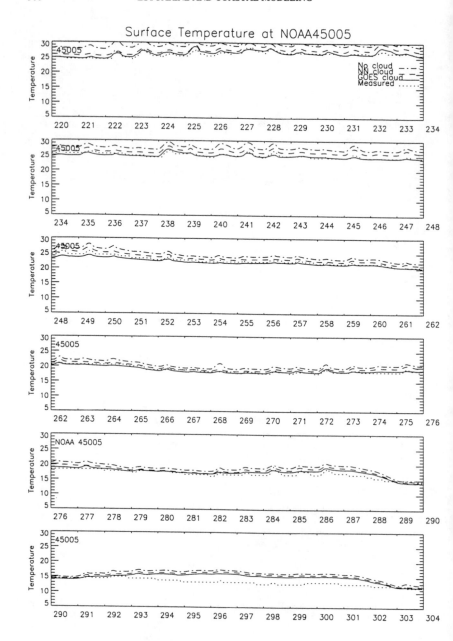

Figure 6. Comparison between measured and computed temperature at 45005

Table 1. Statistical summary of results for buoy 45005

45005	No Cloud		Nearest Neighbor		GOES Cloud	
	DIFF.	RMSE	DIFF.	RMSE	DIFF	RMSE
136-150	-45.55	0.858	-487.08	1.557	-466.18	1.499
150-164	1112.08	3.412	286.61	1.217	97.51	0.978
164-178	1047.45	3.254	389.56	1.678	-48.17	1.137
178-192	1349.11	4.135	540.54	2.003	141.47	1.281
192-206	688.47	2.197	113.92	0.859	-89.72	0.952
206-220	895.53	2.798	252.13	0.956	-32.52	0.663
220-234	1075.44	3.230	421.64	1.335	83.95	0.517
234-248	872.75	2.624	425.28	1.324	-45.87	0.420
248-262	593.34	1.793	264.30	0.819	-171.04	0.600
262-276	509.60	1.581	229.87	0.795	-21.53	0.367
276-290	591.00	1.919	321.91	1.164	167.51	0.750
290-304	854.32	2.786	644.72	2.220	498.83	1.822

Table 2. Statistical summary of results for buoy 45132

45132	No Cloud		Nearest Neighbor		GOES Cloud	
	DIFF.	RMSE	DIFF.	RMSE	DIFF.	RMSE
136-150	2083.18	8.378	1672.98	7.922	1521.62	7.914
150-164	881.91	4.001	-224.42	3.151	-656.59	3.722
164-178	3679.23	14.439	3010.27	13.259	2645.24	11.935
178-192	1308.37	3.946	495.44	1.578	26.49	0.451
192-206	1108.96	3.342	452.81	1.452	67.26	0.713
206-220	1185.10	3.563	560.11	1.781	151.78	0.906
220-234	1047.04	3.132	369.37	1.167	-31.11	0.413
234-248	949.64	2.874	364.39	1.224	-137.58	0.642
248-262	1104.27	3.317	693.89	2.137	229.41	0.964
262-276	1055.35	3.188	750.37	2.283	463.08	1.426
276-290	790.50	2.374	496.65	1.501	329.05	1.018
290-304	652.62	1.970	458.89	1.423	340.16	1.099

Summary and conclusions

A surface heat flux model and temperature field of a three-dimensional lake circulation model have been evaluated from mid May to the end October, 1995 using traditional interpolated cloud cover and satellite derived hourly cloud fields. Based on the model results and evaluation, several conclusions can be drawn from this preliminary study:

1) Cloud sensitivity tests prove that cloud cover does play an important role in surface heat flux models. The presence of clouds can result in a 50% reduction in the magnitude of solar radiation.

2) GOES-8 derived cloud cover does solve the cloud cover underestimation problem in the surface heat flux model and hence improves model performance and temperature predictions. Therefore hourly GOES-8 derived cloud cover data could and should be used in the surface heat flux model and daily Lake Erie nowcast operations.

3) The surface heat flux formulation with the cloud information derived from the GOES satellite yields predicted Lake Erie surface temperature drops of 2 degree C on average which compares much more favorably with the observed values.

4) The Lake circulation model can simulate Lake Erie surface temperature well when the correct surface heat flux and cloud cover are specified.

5) Based on the lessons learned from processing these images manually, we found that the GOES-8 data stream and transmission are very reliable and the images are extremely well registered. A procedure to download and process real-time GOES-8 images has been developed and can be fully automated.

Future Research

The following two areas of future research are suggested.

1) Use GOES data to estimate surface heat flux directly.

The solar radiation component in the previous heat flux model is still a statistical regression model based on a climatological record. Further improvements on the heat flux calculations need a new approach for this component since it contributes more than 80% of the total heat flux. It is possible to estimate the radiation directly from the satellite imagery. However, direct measurement of solar radiation on Lake Erie is needed to calibrate and verify the values derived from satellites.

2) Compare the GOES derived surface heat flux and ETA heat flux.

There is no doubt that in the hindcast or nowcast mode the GOES derived cloud cover will generate better surface heat flux estimates and more accurate lake surface temperature. However, an accurate present time meteorological conditions and calculated heat flux does not necessarily guarantee the prediction of accurate future conditions in the forecast mode. The ETA model does provide forecasted heat flux but at a much coarser resolution. The question of which approach to use for more accurate lake temperature predictions can not be answered until a full evaluation has been performed.

References

Bedford, K. and D. Schwab, 1990, Preparation of Real-Time Great Lakes Forecasts, Cray Channels, Cray Res. Inc., MN, Summer issue, pp. 14-17.

Bedford K. and D. Schwab, 1991, The Great Lakes Forecasting System - Lake Erie Nowcasts/Forecasts, Proc., Marine Technology Society Annual Conference (MTS '91), Marine Technology Society, Washington, DC, pp. 260-264.

Chu, Y.P., K.W. Bedford, C.J.Merry and J.S. Hobgood, 1994, "Impact of GOES Data on Surface Heat Predictions". National Conference on Hydraulic Engineering, ASCE, Buffalo, NY, pp207-211.

Cotton, G.F., 1979, "ARL models of global solar radiation," in: Hourly solar radiation - surface meteorological observations. Solmet, vol.2, final report, National Climate Center, Dept. of Energy.

Guttman, N.B. and J.D. Mathews, 1979, Computation of extraterrestrial solar radiation, solar elevation angle, and true solar time of sunrise and sunset, in: Hourly solar radiation - surface meteorological observations. Solmet Vol.2, Final report, Dept. of Energy, pp. 41-54.

McCormick, M.J. and G.A. Meadows, 1988, An Intercomparison of Four Mixed Layer Models in a Shallow Inland Sea, J. Geophysical Res., Vol. 93, No. C6, pp. 6774-6788.

Schwab, D.J., and J.A. Morton, 1984, Estimation of overlake wind speed from overland wind speed: a comparison of three methods, J. Great Lakes Res. 10(1):68-72

Wyrtki, K., 1965, The average annual heat balance of the north Pacific Ocean and its relation to ocean circulation, J. Geophysical Res. 70, p 4547-4559

Three-Dimensional Model Simulations off the West Coast of Vancouver Island

M.G.G. Foreman[1] and P.F. Cummins[1]

Abstract

The three-dimensional, prognostic, finite element model, QUODDY4, is used to calculate baroclinic tidal and residual flows off the west coast of Vancouver Island. Although strong summer upwelling in the region is known to support high biological productivity and a lucrative commercial fishery, recent fluctuations in the biomass of several species (salmon, herring, hake) and their correlation with variable oceanic conditions has illustrated the need to better understand these biophysical linkages. In this numerical study it is demonstrated that upwelling, arising solely from bottom mixing in combination with the bottom boundary condition, produces many of the sub-tidal circulation features observed in the region. Tidal currents are also shown to have more accurate representations than in earlier barotropic simulations. These results are milestones toward our eventual goal of building a biophysical model for the region.

Introduction

The successful application of the three-dimensional, prognostic, finite element model, QUODDY4, to U.S. GLOBEC studies around Georges Bank (Lynch et al., 1996; Naimie, 1996; Werner et al. 1993) was the main reason that it was also chosen for GLOBEC Canada studies off the western continental margin of Vancouver Island. Although prognostic finite difference models have been used to study the tides (Flather, 1987; Flather, 1988) and buoyancy-driven Vancouver Island Coastal Current (Masson and Cummins, 1998) in this region, previous finite element model applications (Foreman and Walters, 1990; Foreman et al., 1992, Foreman and Thomson, 1997) have been restricted to either barotropic or diagnostic simulations. Although these calculations were performed with triangular grids that permitted high horizontal resolution in regions of interest (such as along the shelf break), the available numerical techniques did not permit a similar

---

[1]Institute of Ocean Sciences, P.O. Box 6000, Sidney, B.C., V8L 4B2.

degree of temporal resolution or the inclusion of baroclinic effects. The emergence of QUODDY4 has provided a tool for including these effects while maintaining the flexibility of variable spatial resolution.

The western continental margin of Vancouver Island is characterized by relatively intense summer upwelling that has traditionally supported high biological productivity and a lucrative commercial fishery. However, recent large fluctuations in the biomass of various species (salmon, herring, hake) and their correlation with variable oceanic conditions (especially temperature) has highlighted the need to better understand the biophysical linkages and processes that cause the physical change. It is hoped that such an understanding will allow better management of the fishery and permit an advanced warning of the potential impact of (imminent) events such as the 1997 El Niño.

Summer currents along the Vancouver Island shelf are spatially complex. Tidal streams are predominantly diurnal with considerable variation in their magnitudes due to the presence of diurnal period shelf waves (Crawford and Thomson, 1984; Foreman and Thomson, 1997). Whereas typical alongshore winds from the northwest force a southeastward shelf-break current (Freeland et al., 1984), a buoyancy flux due to river runoff generates the Vancouver Island Coastal Current (Thomson et al., 1989; Hickey et al., 1991; Masson and Cummins, 1998) that flows counter to the prevailing wind.

In this presentation, we describe the application of QUODDY4 to the western continental margin of Vancouver Island. We find that the tidal currents have more accurate representations than in earlier barotropic simulations, and that coastal upwelling and some observed sub-tidal circulation features can be produced simply from bottom mixing in combination with the bottom boundary condition. This phenomenon, as described in Cummins and Foreman (1998), has not been represented in any of the previous diagnostic models for the region, yet we believe that it could prove to be relevant to the local ecosystem.

The western continental margin of Vancouver Island is characterized by an extensive shelf, steep slope, and broad continental rise (Figure 1). As delineated by the 200 m depth contour, the continental shelf attains a width of about 65 km seaward of Juan de Fuca Strait, narrows to about 5 km off Brooks Peninsula, then widens again to as much as 150 km in Queen Charlotte Sound. Off southern Vancouver Island the bathymetry has a particularly convoluted geometry consisting of numerous shallow banks separated by a series of roughly 100-m deep basins and the 400-m deep and approximately 10-km wide Juan de Fuca Canyon. Lesser canyons such as Tully, Nitinat, Barkley, Clayquot, and Nootka lie north of Juan de Fuca and, although they do not bisect the shelf, they are suspected to play an important role in upwelling.

The Model and Grid

The finite element model, QUODDY4, used in these calculations is identical to that described in Lynch et al. (1996) and Lynch et al. (1998). It combines solutions of the salinity and/or temperature transport equation with solutions of the nonlinear, three-dimensional, shallow-water equations. The vertical mixing of momentum, heat, and mass is represented by a level 2.5 turbulence closure scheme (Mellor and Yamada, 1982; Galperin et al., 1988) while in the horizontal, a Smagorinsky-type closure provides a shear and mesh-scale dependent eddy viscosity (Smagorinsky, 1963). In short, QUODDY4 may be thought of as a finite element counterpart of the Princeton Ocean Model (Blumberg and Mellor, 1987). See Lynch et al. (1996) for further details on the governing equations and the numerical techniques used to solve them.

The triangular grid for the model calculations is a refinement of the one used in Foreman and Thomson (1997) (henceforth FT97). Generated using the TRIGRID software developed by Henry and Walters (1992), this grid has 30872 nodes, 60147 elements, and a horizontal resolution that varies from approximately 12 km in deep

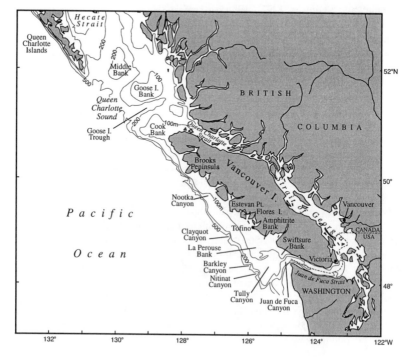

Figure 1: Bathymetry of the west Vancouver Island shelf showing the 100, 200, and 500 m depth contours.

waters to less than 0.2 km along the shelf break. In accordance with the studies of Hannah and Wright (1995), Luettich and Westerink (1995), and Heaps et al. (1988), resolution was increased so that within each triangle $\Delta h/h$ is less than 0.4 for all depths greater than 200 m. Linear basis functions are used to approximate all variables and under each horizontal node there are 21 vertical nodes (i.e. sigma coordinates) whose sinusoidal spacing is identical to that used in Lynch et al. (1996). In order to avoid the spurious flows that can be generated through sigma-coordinate discretizations of the baroclinic pressure gradient (Haney, 1991), this piecewise linear approximation of density is used to interpolate vertically prior to calculating the horizontal gradients.

Topographic Upwelling

In an idealized process study, Cummins and Foreman (1998) demonstrated that vertical mixing coupled with the condition of zero diffusive heat flux through the sea floor, i.e.,

$$K_v \frac{\partial T}{\partial n}\Big|_{z=-h} = 0 \tag{1}$$

(where T is temperature, K_v is the vertical turbulent mixing coefficient, and n is normal to the sea floor) leads to a distortion of isothermal surfaces near the bottom (Figure 2). This distortion was shown to produce baroclinic pressure gradients that drive upslope and cyclonic flows in the bottom boundary layer adjacent to a bank, and anticyclonic downwelling flows outside the bottom boundary layer.

This physical process has been established by Phillips et al. (1986) in a series of laboratory tank experiments with intensified turbulent mixing near the bottom.

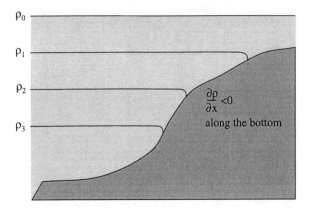

Figure 2: Schematic of the pressure gradients arising from the condition of zero diffusive density flux through the sea floor, and uniform vertical mixing coefficients.

They demonstrated that mixing in combination with this boundary condition gives rise to a secondary circulation comprised of upslope flow near the bottom and downslope motion outside the boundary layer. Further consequences of this boundary condition have been studied analytically by numerous fluid dynamicists such as Wunsch (1970), Phillips (1970), Thorpe (1987), and Garrett et al. (1993). A summary of their results can be found in Cummins and Foreman (1998).

It should be pointed out that the derivative in equation 1 is approximated by $\partial T/\partial z$ in both QUODDY and the Princeton Ocean Model (Blumberg and Mellor, 1987). Although this introduces a small error in the pressure gradient term (proportional to the slope), it is a good approximation except over very steep topography.

In this regional modeling study for the west coast of Vancouver Island, we demonstrate that vertical mixing coupled with equation 1 produces both upwelling along the continental shelf and residual surface currents with many of the summer circulation features that have been observed by current meters. Although we do not suggest that this mechanism is the sole cause of these flow patterns, we do feel that it may play a contributing role.

The first simple run was performed with the horizontally uniform sigma-t profile shown in Figure 3 as an initial condition. This profile was calculated from the same set of summer CTD observations that were used in FT97. No direct forcing was imposed, and closed $(\mathbf{u} \cdot \mathbf{n} = 0)$ conditions were assumed not only along the

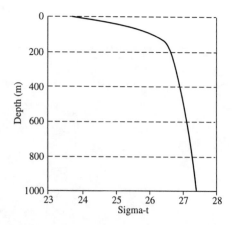

Figure 3: Upper portion of the sigma-t profile that was used to initialize the first simulation. Values below 1000 m were simply a linear extrapolation of those between 900 and 1000 m.

coast, but also along the sea boundaries. In the absence of a radiation condition (that is still under development for QUODDY4), these closed boundaries, in combination with a horizontal no-flux condition on sigma-t, were thought to provide minimal boundary contamination to solutions in the model interior. The bottom friction coefficient was assumed to be 0.01 and the vertical eddy viscosity calculated by the Mellor-Yamada (1982) turbulence closure was not allowed to drop below 0.005 m^2/s. The Smagorinsky coefficient was set at 2.0 in the model interior and, to compensate for the lack of radiation boundary conditions, was increased to 10.0 in a sponge layer adjacent to the sea boundaries. (We acknowledge that both the minimum vertical eddy viscosity and the Smagorinsky coefficient are too large, and are presently conducting runs with more realistic values.) The time step was 20 seconds.

Figures 4 and 5 show the surface and bottom flows after five days. Although these flows have not yet reached steady state in deep regions (Cummins and Foreman (1998) suggest that hundreds of days may be required), they do clearly show upwelling along the shelf break, and in particular, at the canyons. The surface flows along the shelf are generally toward the northwest (opposing those at the bottom) and display many characteristics of the buoyancy-driven Vancouver Island Coastal Current (Thomson et al., 1989; Hickey et al., 1991). A southwesterly current is also seen to be flowing along the continental slope and is consistent with the wind-driven Shelf Break Current that is present in summer (Freeland et al., 1984; Thomson et al., 1989). Numerous cyclonic eddies are seen over depressions and canyons. The eddy southwest of the entrance to Juan de Fuca Strait is consistent with the Tully Eddy that has been shown to be associated with upwelled California Undercurrent Water (Freeland and Denman, 1982), while the clockwise eddies over the heads of Barkley and Nitinat Canyons are also of biological interest. Barkley Canyon is the site of ongoing GLOBEC Canada field work, while extensive acoustic backscatter surveys conducted around Nitinat Canyon in 1988 (Mackas et al., 1997) showed high concentrations of euphausiids and hake positioned above the 150–300 m depth contours.

Though still noisy because they have not yet reached a steady-state, the bottom currents in Figure 5 clearly show flow up Nitinat, Juan de Fuca, and Tully Canyons. Off northern Vancouver Island, analogous flows are seen to be moving up Goose Island Trough, into Queen Charlotte Strait, and onto Cook Bank.

These model results demonstrate that the process described in detail by Cummins and Foreman (1998) applies in more general settings. In particular, we have demonstrated that upwelling along the continental margin off Vancouver Island can arise simply from vertical mixing and the bottom boundary condition (1), and need not be wind-driven.

The Inclusion of Tides

In order to extend the realism of the previous simulation, a second simple run was performed with M_2 and K_1 tidal forcing. With elevation specified boundary conditions taken from the FT97 model, these two constituents were turned on with linear ramping over 96 hours. The background stratification and baroclinic residual flows were initialized with the final values from the first simulation, and the associated boundary elevations were saved and specified along with the tidal forcing. This second simulation was run for six days with an harmonic analysis of the tidal and residual signals over the last day.

Contour plots of the M_2 and K_1 elevation co-amplitudes and co-phases calculated by this analysis were very similar to those displayed in Figure 4 of FT97, and had comparable accuracy. RMS differences between modeled and observed M_2 amplitudes and phases at 42 sites within the model domain were 2.7 cm and 1.1° respectively, while the analogous differences for K_1 were 1.4 cm and 1.7° respectively. Plots of the K_1 clockwise component of co-amplitude and co-phase,

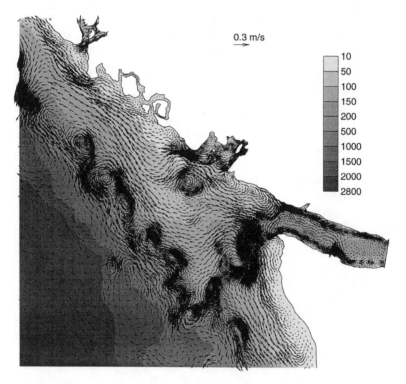

Figure 4: Surface currents after five days for the run initialized with the horizontally uniform sigma-t profile shown in Figure 3. Bathymetry is in meters.

illustrating the nature of the diurnal shelf wave originating at the entrance of Juan de Fuca Strait, were also similar to those shown in Figure 5 of FT97. However, the presence of a realistic density field did provide better agreement with the phase (and hence wavelength) values computed from historical time series.

Vertical profiles of the M_2 and K_1 current ellipses at the same locations as those shown in Figure 6 of FT97 generally displayed better agreement between the modeled and observed values. For example, at sites VT4 and E01 off Tofino and Estevan Point respectively, the K_1 model phases were now only 8° and 10° larger than the observed values, as opposed to the 24° and 40° respective differences shown in FT97. These systematic improvements mean that the wavelength of the K_1 shelf wave is longer, and more accurate, than in FT97.

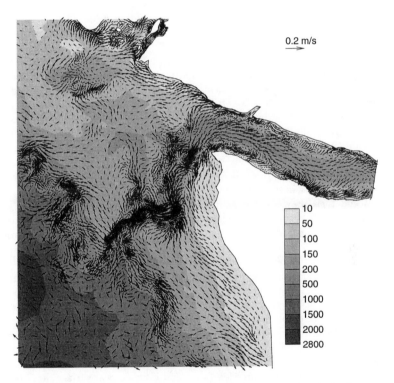

Figure 5: Bottom currents after five days for the run initialized with the horizontally uniform sigma-t profile shown in Figure 3. Bathymetry is in meters.

The newly simulated M_2 ellipses at the same four locations also displayed better agreement with the historical observations. However, they do not yet seem to have represented the internal tide behaviour accurately. (This is probably due to the horizontally uniform stratification, and vertical mixing and Smagorinsky coefficients that are too large.) Nevertheless, internal tides are present in the simulation. Figure 6 shows amplitudes of the M_2 vertical velocity (w) at 100 m depth. Large values are evident in Juan de Fuca Strait and along the continental slopes west of Cook Bank, Brooks Peninsula, and Queen Charlotte Sound. (If w is assumed to be vertically constant, each mm/s of amplitude corresponds to a vertical excursion of about 7 m.) Figure 12 of Cummins and Oey (1997) shows the continental slope off Cook Bank to be one of the primary generation regions of M_2 baroclinic energy flux along the northern coast of British Columbia, and the radiating pattern of decreasing w amplitudes seen in our Figure 6 certainly suggests propagation to the southwest. A secondary peak offshore from Cook Bank may indicate the wavelength of these internal tides, though its proximity to the western boundary could mean that it is simply an artifact of the boundary condition and/or the sponge layer.

A similar plot for K_1 shows w amplitudes as large as 3.0 mm/s in Juan de Fuca Strait and along the continental slope off La Perouse Bank. (In this case, each mm/s corresponds to a vertical excursion of about 14 m.) However, as K_1 is subinertial at these latitudes, free internal wave propagation is not possible and the vertical motions are trapped over the slope, enhancing the shelf waves. (Figure 10 in Cummins and Oey (1997) shows this behaviour off the north coast of the Queen Charlotte Islands.)

Regions with large M_2 and K_1 vertical motions will also have relatively large turbulence. This, in turn, may increase the encounter rates between various predators and their prey. In their simulations around Georges Bank, Werner et al. (1996) found that turbulence-enhanced contact rates were necessary to reproduce observed cod and haddock larvae growth rates. (Without turbulence, the growth was too small.) Future work is planned to determine whether or not turbulence plays a similar role off the continental margin of Vancouver Island.

Unfortunately, the surface residual flows arising from this second run are strongly affected by the specified boundary conditions. Away from these boundaries, the density, flow, and elevation fields are able to adjust due to their interaction with the tides but, near the specified boundary this adjustment is constrained by the specified values. Consequently, these boundaries tend to act as barriers for the background flows. In this second run, sea surface elevations over the deep regions of the model domain decrease from their initial values. As the boundary values remain fixed, a depression forms and cyclonic near-surface flows develop to provide a geostrophic balance. While the onshelf flows are similar to those shown in Figure 4, along the shelf break the previously southeastward flows have reversed. However, Figure 7 shows that the bottom flows are not affected to the same extent. Compared to Figure 5, stronger upwelling is evident in Juan de

Fuca Canyon and similar onshelf flows emanate from Tully, Nitinat, and Barkley Canyons. A clockwise eddy is now visible around western Swiftsure Bank due to tidal rectification.

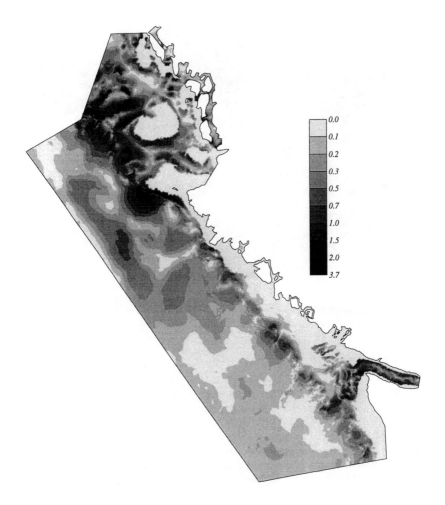

Figure 6: M_2 vertical velocity amplitudes (mm/s) at 100 m depth.

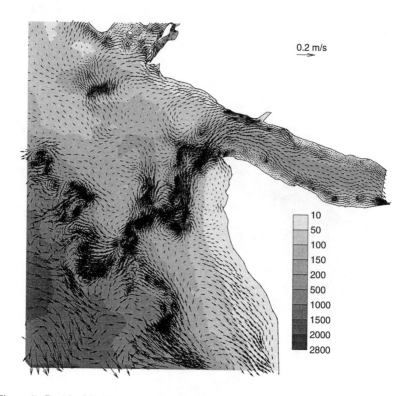

Figure 7: Residual bottom currents calculated from an harmonic analysis over the last day of a six-day run that combined the buoyancy flows illustrated in Figures 4 and 5 with M_2 and K_1 forcing. Bathymetry is in meters.

Summary and Discussion

The preceding presentation has demonstrated that vertical mixing coupled with a condition of zero diffusive density flux through the sea floor leads to topographic upwelling and flows that display many of the summer circulation patterns observed along the continental margin of Vancouver Island. Though we do not suggest that this process is the sole causal mechanism for these flows, our numerical simulations have demonstrated that it may be a contributing factor.

Residual bottom flows were seen to move up Juan de Fuca and Tully Canyons while the flows around Nitinat Canyon were shown to display upwelling and eddy patterns that are consistent with the Mackas et al. (1997) ADCP surveys. This reinforces their premise that these physical features are largely responsible, either directly or indirectly, for the high aggregations of euphausiids and hake observed around the head of the canyon.

The M_2 and K_1 tidal currents were also seen to be represented more accurately than with previous barotropic simulations. The wavelength of the K_1 shelf wave displayed closer agreement with historical current meter observations, and the strong generation of M_2 internal tides along the continental margin west of Cook Bank was consistent with the findings of Cummins and Oey (1997). The presence of significant vertical motions west of Cook Bank and Brooks Peninsula, in Juan de Fuca Strait, and along the continental slope off La Perouse Bank suggests that these regions may be biologically productive due to turbulence-enhanced, predator-prey encounter rates similar to those found off Georges Bank by Werner et al. (1996).

A more complete evaluation of this application of QUODDY4 to the western continental margin of Vancouver Island is underway and will be reported in future manuscripts.

Acknowledgements

We thank Luc Cuypers for processing the CTD data; Sok Kuh Kang, Claire Smith, and Peter Chandler for helpful comments; Dan Lynch, Chris Naimie, Charles Hannah, and Monica Holboke for guidance on the use of QUODDY4; Antonio Baptista and Paul Turner for their visualization software; and Trish Kimber for assistance with the figures. We also acknowledge support under the GLOBEC Canada initiative jointly funded by the Natural Sciences and Engineering Research Council and the Department of Fisheries and Oceans.

Appendix I. References

Blumberg, A.F., and Mellor, G.L. (1987). "A description of a three-dimensional coastal ocean circulation model." In *Three-Dimensional Coastal Ocean Models*, C.N.K. Mooers, ed., Coastal and Estuarine Sciences, Vol. 4, Amer. Geophys. Union, Washington.

Crawford, W.R., and Thomson, R.E. (1984). "Diurnal period shelf waves along Vancouver Island: A comparison of observations with theoretical models." *J. Phys. Oceanogr.*, 14(10), 1629–1646.

Cummins, P.F., and Oey, L.-Y. (1997). "Simulation of barotropic and baroclinic tides off northern British Columbia." *J. Phys. Oceanogr.*, 17(5): 762–781.

Cummins, P.F., and Foreman, M.G.G. (1998). "A numerical study of circulation driven by mixing over an isolated topographic bank." *Deep-Sea Res.* In press.

Flather, R.A. (1987). "A tidal model of the northeast Pacific." *Atmos.-Ocean*, 25(1): 22–45.

Flather, R.A. (1988). "A numerical investigation of tides and diurnal-period continental shelf waves along Vancouver Island." *J. Phys. Oceanogr.*, 18(1): 115–139.

Foreman, M.G.G., and Walters, R.A. (1990). "A finite element tidal model for the southwest coast of Vancouver Island." *Atmos.-Ocean*, 28(3), 261–287.

Foreman, M.G.G., Thomson, R.E., Lynch, D.R., and Walters, R.A. (1992). "A finite element model for three-dimensional flows along the west coast of Vancouver Island." *Estuarine and Coastal Modeling*, 2nd Int. Conf./WW Div. ASCE, Tampa, Florida, Nov. 13–15, 1991. pp. 574–585.

Foreman, M.G.G., and Thomson, R.E. (1997) "Three-dimensional model simulations of tides and buoyancy currents along the west coast of Vancouver Island." *J. Phys. Oceanogr.*, 27, 1300–1325.

Freeland, H.J., and Denman, K.L. (1982). "A topographically controlled upwelling center off southern Vancouver Island." *J. Mar. Res.*, 40, 1069–1093.

Freeland, H.J., Crawford, W.R., and Thomson, R.E. (1984). "Currents along the Pacific coast of Canada." *Atmos.-Ocean*, 22(2), 151–172.

Galperin, B., Kantha, L.H., Hassid, S., and Rosati, A. (1988). "A Quasi-equilibrium turbulent energy model for geophysical flows." *J. Atmos. Sci.*, 45, 55–62.

Garrett, C., MacCready, P., and Rhines, P. (1993). "Boundary mixing and arrested Ekman layers: Rotating stratified flow near a sloping boundary." *Ann. Rev. Fluid Mech.*, 25, 291–323.

Haney, R.L. (1991). "On the pressure gradient force over steep topography in sigma coordinate ocean models." *J. Phys. Oceanogr.*, 21, 610–619.

Hannah, C.G., and Wright, D.G. (1995). "Depth dependent analytical and numerical solutions for wind driven flow in the coastal ocean." In *Quantitative Skill Assessment for Coastal Ocean Models*, D.R. Lynch and A.M. Davies, eds., Coastal and Estuarine Studies, Vol. 47, Amer. Geophys. Union, 125–152.

Heaps, N.S., Huthnance, J.M., Jones, J.E., and Wolf, J. (1988). "Modelling of storm-driven shelf waves north of Scotland – I. Idealized models." *Cont. Shelf Res.*, 8, 1187–1210.

Henry, R.F., and Walters, R.A. (1992). "A geometrically-based, automatic generator for irregular triangular networks." *Comm. Appl. Num. Meth.*, 9, 555–566.

Hickey, B.M., Thomson, R.E., Yih, H., and LeBlond, P.H. (1991). "Velocity and temperature fluctuations in a buoyancy-driven current off Vancouver Island." *J. Geophys. Res.*, 96(C6), 10507–10538.

Luettich, R.A., and Westerink, J.J. (1995). "Continental shelf scale convergence studies with a barotropic tidal model." *Quantitative Skill Assessment for Coastal Ocean Models*, D.R. Lynch and A.M. Davies, eds., Coastal and Estuarine Studies, Vol. 47, Amer. Geophys. Union, 349–372.

Lynch, D.R., Ip, J.T.C., Naimie, C.E., and Werner, F.E. (1996). "Comprehensive Coastal Circulation Model with Application to the Gulf of Maine." *Cont. Shelf Res.*, 16(7), 875–906.

Lynch, D.R., Holboke, M.J., and Naimie, C.E. (1998). The Maine coastal current: Spring climatological circulation. *Cont. Shelf Res.* In press.

Mackas, D.L., Kieser, R., Saunders, M., Yelland, D.R., Brown, R.M., and Moore,

D.F. (1997). "Aggregation of euphausiids and Pacific hake (*Merluccius productus*) along the outer continental shelf off Vancouver Island. *Can. J. Fish. Aquat. Sci.*, 54, 2080–2096.

Masson, D., and Cummins, P. (1998). "Numerical simulations of a buoyancy-driven coastal countercurrent off Vancouver Island." Submitted to *J. Phys. Oceanogr.*

Mellor, G.L., and Yamada, T. (1982). "Development of a turbulence closure model for geophysical fluid problems." *Rev. Geophys. Space Phys.*, 20, 851–875.

Naimie, C.E. (1996). "Georges Bank residula circulation during wesk and strong stratification periods: Prognostic numerical results." *J. Geophys. Res.*, 101(C3), 6469–6486.

Phillips, O.M. (1970). "On flows induced by diffusion in a stable stratified fluid." *Deep-Sea Res.*, 17, 435–443.

Phillips, O.M., Shyu, J., and Salmun, H. (1986). "An experiment on boundary mixing: mean circulation and transport rates." *J. Fluid Mech.*, 173, 473–499.

Smagorinsky, J. (1963). "General circulation experiments with primitive equations I. The basic experiment." *Mon. Weather Rev.*, 91, 99–164.

Thomson, R.E., Hickey, B.M., and LeBlond, P.E. (1989). "The Vancouver Island coastal current: Fisheries Barrier and conduit." In *Effects of ocean variability on recruitment and evaluation of parameters used in stock assessment models.* R.J. Beamish and G.A. McFarlane, eds., *Can. Spec. Publ. Fish. Aquatic Sci.*, 108, Ottawa, Ont.

Thorpe, S.A. (1987). "Current and temperature variability on the continental slope", *Trans. Roy. Soc. London, Series A*, 331, 1516–1520.

Werner, F.E., Page, F.H., Lynch, D.R., Loder, J.W., Lough, R.G., Perry, R.I., Greenberg, D.A., and Sinclair, M.M. (1993). "Influences of mean advection and simple behaviour on the distribution of cod and haddock early life stages on Georges Bank." *Fish. Oceanogr.*, 2(2), 43–64.

Werner, F.E., Perry, R.I., Lough, R.G., and Naimie, C.E. (1996). "Trophodynamic and advective influences on Georges Bank larval cod and haddock." *Deep-Sea Res. II*, 43(7-8), 1793–1822.

Wunsch, C. (1970). "On oceanic boundary mixing." *Deep-Sea Res.*, 17, 293–301.

A THREE-DIMENSIONAL OCEAN CIRCULATION
MODEL WITH WAVE EFFECTS

James K. Lewis[1]

ABSTRACT

   A three-dimensional circulation model is presented which uses information on surface waves to produce wave-induced circulation effects. Forms of the momentum and
continuity equations are used which produce a flow field that reflects wave-induced
flow as well as flow due to winds, tides, density structure, etc. A critical aspect is
investigated for including wave dissipation by specifying viscous effects reflective of
ocean waves. Results of test cases are presented for near-shore regions in which wave
effects might be considerable in terms of radiation stresses, wave-enhanced bottom
friction, Coriolis wave stress, wave dissipation, and wave setup. It is shown that the
impact of Stokes drift can substantially increase the flushing of shallow embayments
even in the presence of the wave-enhanced bottom friction. Also, simulations show
that there is a considerable impact of three-dimensional calculations and wave-enhanced bottom friction in determining vertically-averaged longshore currents resulting from radiation stresses due to breaking waves.

1. INTRODUCTION

   Wave effects can have a significant impact on the circulation of nearshore and
coastal regions. The realm of ocean modeling provides one of the best means of assessing and predicting the impact of wave-related phenomena since simulations can
be conducted for tidal-, wind-, or density-induced flows with and without wave effects. Wave-enhanced bottom friction or water column stresses induced by wave
breaking are the more common wave effects included in hydrodynamic models.
There is also Stokes drift, forcing at the surface of the water column (virtual tangential surface stress) and bottom of the water column (bottom streaming) as well as a
body force within the water column (Coriolis wave stress). Because of the importance of wave effects, a three-dimensional, hydrodynamic circulation model was developed which incorporates all the major wave-induced circulation effects. The
model is a hydrostatic, wave-averaged model, so the actual simulation of surface
waves is not considered here. The model uses as input information from a model of
surface waves to determine the total impact of a wave field on circulation produced
by winds, tides, and density variations. Although the incorporation of specific wave
effects into ocean models has been accomplished previously, this model incorporates
all the major wave effects so that we can investigate the interaction of different wave-
related phenomena. While the inclusion of some wave-effects is well established, the
inclusions of others are not. For these effects, we consider different formulations and
provide simulation test cases that indicate differences between various formulations.

---

[1]Ocean Physics Res. & Dev., 207 S. Seashore Ave., Long Beach, MS 39560

The virtual tangential surface stress and wave dissipation effects necessitate the specification of a vertical eddy viscosity related to ocean surface waves. But the turbulence closure schemes employed in most ocean circulation models are crafted to reflect turbulent viscous dissipation effects relative to the typically larger spatial and temporal scales of geophysical fluid dynamics. Vertical viscous effects when dealing with waves can be expected to be smaller than those that would be calculated by such schemes because of the smaller spatial and temporal scales of surface waves. As will be shown, the great difference in the scales of waves and of those of tide-, wind-, and density-induced flows poses some problems in formulating the momentum equations of the model.

2. WAVE EFFECTS

Here we assume that surface wave parameters are provided from some source for each grid cell in the circulation model. These include wave height, wave period, wavelength, and direction of wave propagation. This information is used to determine all the wave effects in the model governing equations. We will utilize a version of the circulation model of Blumberg and Mellor (1987). In that model, the basic momentum and continuity equations are of the form

$$DU/Dt - f\,V = -\rho^{-1}\,\partial P/\partial x + \partial(K_H\,\partial U/\partial x)/\partial x + \partial(K_H\,\partial U/\partial y)/\partial y + \partial(K_z\,\partial U/\partial z)/\partial z,$$

$$DV/Dt + f\,U = -\rho^{-1}\,\partial P/\partial y + \partial(K_H\,\partial V/\partial x)/\partial x + \partial(K_H\,\partial V/\partial y)/\partial y + \partial(K_z\,\partial V/\partial z)/\partial z,$$

$$\partial W/\partial z = -\partial U/\partial x - \partial V/\partial y \quad \text{or} \quad \partial\eta/\partial t = -(\partial(H\,\bar{U})/\partial x + \partial(H\,\bar{V})/\partial y),$$

where $U = (U,V,W)$ is the non-wave current field, U is the x-directed speed component, V is the y-directed speed component, W is the vertical velocity, η is the sea surface height anomaly, t is time, and the overbars indicate vertical averages. Also, D/Dt is the total derivative, f is the Coriolis parameter, ρ is the water density, P is pressure, K_H is the coefficient of horizontal eddy viscosity, and K_z is the coefficient of vertical eddy viscosity.

In the following, we present modifications of these basic equations to illustrate how the Blumberg-Mellor model expressions were modified to include various wave effects. Except for the Stokes drift, the wave effects were added as source terms in the momentum equations. The manner in which various wave effects are represented and incorporated into the model comes from the existing literature. No distinction is made between how the wave effects were developed (empirical parameterizations, first-order wave theory, second-order wave theory, etc.).

STOKES DRIFT

Stokes drift is the wave-induced, wave-averaged velocity $u_{St} = (u_{St},v_{St})$ such that

$$|u_{St}| = \sigma\,a^2\,k\,\cosh\,(2k\,(H + z))/(2\,\sinh^2\,(kH)) \tag{1}$$

where σ is the wave frequency, a is the wave amplitude, k is the wave number, H is the water depth, and z is positive upward from the surface of the water. The model uses (1) and the direction of wave propagation to calculate (u_{St},v_{St}) for each grid cell. When model outputs are required, the model-predicted velocities in the horizontal always have (u_{St},v_{St}) added to them in order to reflect the total Eulerian flow field. It is important to note that, following the conventions of the wave research community, the spatial variations of (u_{St},v_{St}) are not considered when determining the non-linear advective and horizontal shearing stress terms in the momentum equations. Exclusion of the Stokes drift from these terms will be revisited in the Discussion.

Another impact of Stokes drift is wave setup when there exist spatial variations of depth and/or u_{St}. In terms of our model equations, this impact is included in the continuity equation in the form of

$$\partial W/\partial z = -\partial(U+u_{St})/\partial x - \partial(V+v_{St})/\partial y \quad \text{or} \quad \partial\eta/\partial t = -(\partial H(\ \bar{U} + \bar{u}_{St})/\partial x + \partial H(\ \bar{V} + \bar{v}_{St})/\partial y).$$

RADIATION STRESSES

Longuet-Higgens and Stewart (1964) detail the physics related to breaking waves in nearshore regions of coastal areas. The breaking waves result in forces on the water column, referred to as radiation stresses. The form and inclusion of these radiation stresses in hydrostatic, time-dependent, numerical models has been considered in a number of studies (e.g., Yamaguchi, 1986; O'Connor et al., 1992). Radiation stresses can be included in the momentum equations as forces which are constant with depth but change with the horizontal variations of wave parameters. The momentum equations that include radiation stresses are

$$DU/Dt - f\,V = - \rho^{-1}\,\partial P/\partial x + \partial(K_H\,\partial U/\partial x)/\partial x + \partial(K_H\,\partial U/\partial y)/\partial y +$$
$$\partial(K_z\,\partial U/\partial z)/\partial z - [\partial S_{xx}/\partial x + \partial S_{xy}/\partial y]/\,(\rho H),$$

$$DV/Dt + f\,U = - \rho^{-1}\,\partial P/\partial y + \partial(K_H\,\partial V/\partial x)/\partial x + \partial(K_H\,\partial V/\partial y)/\partial y +$$
$$\partial(K_z\,\partial V/\partial z)/\partial z - [\partial S_{xy}/\partial x + \partial S_{yy}/\partial y]/\,(\rho H),$$

where S_{xx}, S_{yy}, and S_{xy} are the radiation stresses given by

$$S_{xx} = S_{11}\cos^2\theta + S_{22}\sin^2\theta$$
$$S_{yy} = S_{11}\sin\theta\cos\theta - S_{22}\cos\theta\sin\theta$$
$$S_{xy} = S_{11}\sin^2\theta + S_{22}\cos^2\theta$$
$$S_{11} = E\,(2\,C_g/C - 0.5), \quad S_{22} = E\,(C_g/C - 0.5)$$

with E being the wave energy per unit area of sea surface ($\rho\,g\,a^2/2$), C_g the group velocity, C the phase speed, and θ the direction of wave propagation relative to the x axis.

WAVE-ENHANCED BOTTOM FRICTION

The most common wave-related modification to hydrodynamic models has been the enhancement of the bottom friction coefficient C_d as a result of the combined motion of waves and currents in the bottom boundary layer (Signell et al., 1990). The circulation model uses a logarithmic boundary layer formulation to calculate the bottom friction coefficient. Without wave effects, a value of $z_0 = 0.003$ m is used in the boundary layer formulation as the distance above the bottom where the logarithmic velocity profile goes to zero. The bottom drag coefficient is then specified in the model as:

$$C_d = \text{MAX}\,(2.5\text{x}10^{-3},\,[\,0.4\,/\ln\,(\,z_{bottom}\,/\,z_0\,)\,]^2\,) \tag{2}$$

where z_{bottom} is equal to half the thickness of the bottom level of the model grid cell.

In this study, the enhancement of C_d due to waves follows Glenn and Grant (1987) and is included in the model equations in the typical manner of other numerical models (e.g., Signell et al., 1990) for determining the wave-modified C_d. The Glenn and Grant scheme is based on a coupling of monochromatic, small amplitude (linear) waves with a steady current, and iteration is required to compute C_d. In such formulations, the presence of the turbulent boundary layer induced by surface waves results in the need for using a bed roughness scale k_b for calculating the bottom drag coefficient as opposed to z_0 in equation (2) (Fredsoe and Diegaard, 1992). Following Nikuradse (1932), it has been found that the bed roughness k_b is approximately z_0 x 30, and this is used in the model.

WAVE-INDUCED SURFACE BOUNDARY LAYER

Longuet-Higgens (1953) showed that ocean waves produce a thin boundary layer at the surface of the water column. This boundary layer can impact the flow within the interior of the water column (especially in shallower water) through the stress it exerts at the surface of the water column. The virtual tangential surface stress acts on the top few millimeters of the water column in the direction of wave propagation (Longuet-Higgens, 1953; Longuet-Higgens and Stewart, 1960) and is given by

$$2 \mu \sigma a^2 k^2 \coth k H \quad \text{at } z \cong 0 - \beta, \text{ with } \beta \text{ small}$$

where μ is an eddy viscosity.

As in previous studies, the virtual tangential surface stress was included in the model as an addition to any wind stress on the surface layer of the model. However, before utilizing the above expression, we must specify a value for μ. Longuet-Higgens (1969a,b) used μ equal to the molecular viscosity of water ($\sim 10^{-6}$ m^2/s) while Madsen (1978) used a much larger, more wind velocity-related, eddy viscosity. Researchers have yet to ascertain the exact value of μ. The results of test simulations using both smaller and larger values of μ are presented.

CORIOLIS WAVE STRESS

The Coriolis wave stress is a peculiar effect of waves in a rotating system, so we shall provide some background as to this phenomena. It was Ursell (1950) who first pointed out the problem of associating the mean flow of Stokes drift with a wave field on a rotating earth. He showed that a steady Lagrangian mean current induced by a wave field was not consistent with circulation conservation in an invisid, rotating system. Nearly 20 years later, Pollard (1970) and Hasselmann (1970) showed analytically that a wave-averaged Reynolds stress arises as a result of the wave velocity component parallel to the wave crests being in phase with the vertical wave velocity component in a rotating frame. For a progressive wave moving in the x direction, the resulting wave velocity field $\mathbf{u} = (u,v,w)$ is such that, in an Eulerian frame work,

$$\partial <v,w>/\partial z = f u_{St} \tag{3}$$

where $< >$ denotes an average over the wave period. Thus, the invisid, linear, wave-averaged governing equations assuming no external pressure gradients and no horizontal variations of the mean Eulerian flow $\mathbf{U} = (U,V,W)$ are

$$\partial U/\partial t - f V = 0, \quad \partial V/\partial t + f U = - \partial <v,w>/\partial z \tag{4}$$

with the Coriolis wave stress, $\partial <v,w>/\partial z$, in the crest-parallel direction. It is easily seen that the solutions to (4) using (3) are $U = A \cos ft - u_{St}$, $V = -A \sin ft$. Thus, the Coriolis wave stress drives a current which oscillates at the inertial period but has a mean flow that exactly mirrors the Stokes drift. Therefore, the total, wave-averaged, current regime (after including the Stokes drift) is one that oscillates at the inertial period but with no net transport.

However, the fact remained that wave fields do indeed result in mean drift currents in a rotating system. Madsen (1978) provided a formulation that resulted in wave-averaged net transports by including vertical viscous stresses. In his formulations, wave-average viscous effects are formulated as $K_z \partial^2 U/\partial z^2$ where K_z is a constant vertical viscosity coefficient and $\mathbf{U}$ is the wave-average response to the wave forcing. Again considering a progressive wave moving in the x direction, the wave-average governing equations with no external pressure gradients and no horizontal variations of the mean Eulerian flow $\mathbf{U}$ become

$$\partial U/\partial t - f V = K_z \partial^2 U/\partial z^2 \quad \partial V/\partial t + f (U + u_{St}) = K_z \partial^2 V/\partial z^2. \tag{5}$$

Solutions to the above equations result in an oscillatory components of U and V as well as mean drift components. As K_z becomes larger, the mean flow becomes larger, resulting in a larger net drift of a water parcel (after adding in the Stokes drift component in the x direction). Thus, the vertical viscous stresses modify the mirror image of the Stokes drift in the -x direction, slightly decreasing the speeds nearer the surface and slightly increasing the speeds below the surface. Because of the modification of the mirror image of the Stokes drift by viscous stresses, not all of the Stokes drift at the surface of the water column is canceled out when we look at the total velocity $U+u_{St}$.

Madsen's research provided a physical basis for a wave field to result in mean transports in a rotating system. However, if the vertical variations of the mirror image of the Stokes drift can result in a shearing stress on the water column, it is reasonable to assume that u_{St} itself could result in a shearing stress. The implications are considerable. Analytical solutions to the wave-average force balance equations which include vertical shearing stresses of the Stokes drift itself show that the Coriolis wave stress and vertical shearing stresses would generate inertial oscillations but *no net flow*.

Jenkins (1987) addressed the problem of viscous effects on a wave field for the realistic case of K_z varying with depth. Jenkins analytical expressions for the wave-averaged flow are of the form (again, for a progressive wave traveling in the x direction and ignoring the surface vorticity layer due to the virtual tangential surface stress)

$$\partial U/\partial t - f\,V = \partial(K_z\,\partial(U + u_{St})/\partial z)\,\partial z + D_w \qquad \partial V/\partial t + f\,(U + u_{St}) = \partial(K_z\,\partial V/\partial z)\partial z \quad (6)$$

where D_w is a momentum source due to wave dissipation. In Jenkins' formulation, the wave dissipation term is given by

$$D_w = -\,\partial(K_z\,\partial u_{St}/\partial z)\,\partial z \,-\, \partial K_z/\partial z\;\partial u_{St}/\partial z.$$

A component of D_w exactly cancels the shearing stress of the Stokes drift in $\partial(K_z\,\partial(U + u_{St})/\partial z)\,\partial z$. This leaves a net source of momentum due to wave dissipation of $D_w = -\partial K_z/\partial z\;\partial u_{St}/\partial z$. In typical situations, $-\partial K_z/\partial z\;\partial u_{St}/\partial z$ would accelerate the water in the direction of wave propagation.

Thus, Jenkins' research provided a physical justification for neglecting the vertical shearing effect of Stokes drift. But his work had much greater implications in our understanding of the net drift associated with surface waves. Previously, the conceptual model was that net drift was a result of water particles traveling faster near the wave crest and slower near the wave trough: i.e. simply Stokes drift. But without the wave dissipation source term D_w (and still neglecting the flow that would be induced by the virtual tangential surface stress), the momentum equations in (6) imply that we will have no net flow due to the wave field after reaching steady-state. Jenkins' results indicate that it is actually wave dissipation effects that result in a net drift after reaching steady-state conditions, and the net drift is not a residual of the Stokes drift but a function of the vertical variations of the Stokes drift as well as K_z.

We based the formulations of the momentum equations used in the model on the work by Jenkins (1987). These are

$$DU/Dt - f\,(V + v_{St}) = -\,\rho^{-1}\,\partial P/\partial x + \partial\,(K_H\,\partial U/\partial x)/\partial x + \partial\,(K_H\,\partial U/\partial y)/\partial y$$
$$+ \partial\,(K_z\,\partial U/\partial z)/\partial z - [\partial\,S_{xx}/\partial x + \partial\,S_{xy}/\partial y]/\,(\rho\,H) - \partial K_z/\partial z\;\partial u_{St}/\partial z,$$

$$DV/Dt + f\,(U + u_{St}) = -\,\rho^{-1}\,\partial P/\partial y + \partial\,(K_H\,\partial V/\partial x)/\partial x + \partial\,(K_H\,\partial V/\partial y)/\partial y \qquad (7)$$
$$+ \partial\,(K_z\,\partial V/\partial z)/\partial z - [\partial\,S_{xy}/\partial x + \partial\,S_{yy}/\partial y]/\,(\rho\,H) - \partial K_z/\partial z\;\partial v_{St}/\partial z$$

where $\boldsymbol{D}_w = (-\partial K_z/\partial z\;\partial u_{St}/\partial z, -\partial K_z/\partial z\;\partial v_{St}/\partial z)$ are the components of the net wave dissipation.

3. THE HYDRODYNAMIC MODEL

The hydrodynamic model used in this work is an adaptation of the model of Blumberg and Mellor (1987) as modified by Lewis et al. (1994). The model is based on the hydrostatic, shallow water, primitive equations for momentum, salt, and heat. Sub-grid scale turbulence is specified using the schemes of Mellor and Yamada (1982) in the vertical and Smagorinsky (1963) in the horizontal. For additional information on the model, the reader is referred to Blumberg and Mellor (1987).

Three model domains were used in test simulations for this work. Domain 1 was a deep water region (2600 m depth) that covered 250 km x 250 km with a 10 km horizontal grid dimension. A constant temperature and salinity was used in the model, and the latitude was set to 27.7°N. The distribution of vertical levels is shown in Table 1. Domain 2 was a domain of a coastal embayment, a channel that was closed on three sides and had an average depth of 10 m. The channel was 20 km long (500 m grid resolution) and 2.5 km wide (250 m grid resolution). The vertical levels were at 1 m spacing, with the temperature and salinity constant. Domain 3 was for a shoreline region with a water depth which varied linearly from 15 cm along the shoreline grid cells to 9 m at the offshore boundary. The model domain extended 1800 m in the longshore direction (at 60 m resolution) and 480 m in the cross-shelf direction (at 30 m resolution). The temperature and salinity were also constant for these simulations, and the vertical levels were at a 15 cm spacing.

Table 1. Layer characteristics (m) for the deep water model domain.

Layer	Top of Layer	Bottom of Layer	Layer Thickness
1	0	0.3	0.3
2	0.3	0.7	0.4
3	0.7	1.1	0.4
4	1.1	1.5	0.4
5	1.5	2.1	0.6
6	2.1	2.7	0.6
7	2.7	3.5	0.8
8	3.5	4.4	0.9
9	4.4	5.4	1.0
10	5.4	6.6	1.2
11	6.6	7.8	1.2
12	7.8	9.1	1.3
13	9.1	11.8	2.7
14	11.8	14.1	2.3
15	14.1	16.9	2.8
16	16.9	20.1	3.2
17	20.1	24.0	3.9
18	24.0	28.5	4.5
19	28.5	33.8	5.3
20	33.8	40.0	6.2
21	40.0	60.0	20.
22	60.0	80.0	20.
23	80.0	100.0	20.
24	100.0	150.0	50.
25	150.0	200.0	50.
26	200.0	400.0	200.
27	400.0	1000.	600.
28	1000.	1500.	500.
29	1500.	2000.	500.
30	2000.	2600.	600.

To limit pressure gradients due to horizontal sea level variations in Domains 1 and 2 and the offshore boundary of Domain 3, a longwave radiation condition was used for the vertically-averaged velocities at the open boundaries. The condition was of the form of

$$U = \pm \, \eta \, (g/H)^{\frac{1}{2}}$$

where U is the barotropic velocity perpendicular to the model boundary, η is the surface height anomaly of the boundary grid cell, and g is the acceleration due to gravity. The vertical structure at the open boundaries was specified using a Sommerfeld (1949) radiation condition:

$$\partial U/\partial t \pm C \, \partial U/\partial n = 0$$

where U is the velocity at a given vertical level, n is the coordinate perpendicular to the open boundary, and C is the propagation speed at the given depth as determined from vertical density variations and normal mode theory (Veronis and Stommel, 1956). A zero-gradient condition was used for the two cross-shelf open boundaries for Domain 3 to allow setup along the shoreline.

4. TEST SIMULATIONS

DEEP WATER SIMULATIONS

The deep water model domain was used to perform simulations dealing with the impacts of the Coriolis wave stress, the vertical shearing stress of the Stokes drift, the wave dissipation factor $\mathbf{D}_w$, and the virtual tangential surface stress. The simulations provide a means of quantifying the impacts of the above wave effects and how the wave dissipation factor modifies these impacts. The numerical simulations performed in the deep water domain were forced with a wave traveling in the +x direction with the same characteristics used by Jenkins (1987). The virtual tangential surface stress was initially neglected by setting the value of μ to zero, and there was no wind stress. In addition, the non-linear advective terms were neglected, and K_H was set to zero for these particular simulations. The vertical eddy viscosity was determined directly from the second-order closure scheme of Mellor and Yamada (1982) but was modified for some of the simulations.

The first simulation was with the Coriolis wave stress, $\mathbf{D}_w = 0$, and the Stokes drift excluded from the vertical shearing stress term (along the lines of the momentum balance put forward by Madsen, 1978). At time zero, the total current is that of the Stokes drift. With time, the current develops into the mean currents shown in Fig. 1a along with inertial oscillations. In a second simulation, the Stokes drift was included in the vertical shearing stress term but with $\mathbf{D}_w = 0$. The results are shown in Fig. 1b. As predicted analytically, there is no net drift.

A third simulation was performed following Jenkins' formulations in equations (6). The results are shown n Fig. 1c. With the larger values of K_z determined by the Mellor-Yamada scheme, the net currents are unrealistically large, with surface currents in the direction of wave propagation being 106 cm/s. But a fourth simulation used $\mathbf{D}_w = 0.1 \times (-\partial K_z/\partial z \, \partial u_{St}/\partial z, -\partial K_z/\partial z \, \partial v_{St}/\partial z)$, reducing the Mellor-Yamada K_z's by 90% but only in the wave dissipation source term. In this case, the net flow (shown in Fig. 1d) is more realistic in magnitude, with the x-directed component being almost identical to the Stokes drift, especially in the surface layers.

These test simulations point out the problem of directly utilizing the Mellor-Yamada vertical eddy viscosities when dealing with wave effects. It appears that, when dealing with the wave source term $\mathbf{D}_w$, we need to modify the Mellor-Yamada K_z's downward to obtain realistic wave-induced flow fields. But for the vertical shearing stress terms in (6), we will utilize the Mellor-Yamada viscosities directly so that the model will produce realistic fields of currents for tidally, wind, and density induced flows. Thus, the rest of the simulations in this paper use

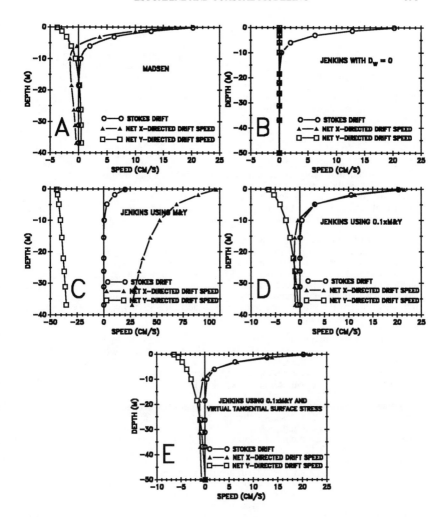

Fig. 1. The vertical distribution of the speed of the Stokes drift and net drift rates when the deep water basin is forced with a 0.88 m amplitude, 4.5 s wave: a) using Madsen (1978) type dynamics, b) using Jenkins (1987) dynamics but with $D_w = 0$, c) using Jenkins (1987) dynamics with D_w non-zero but using the K_z's determined by the Mellor-Yamada turbulence closure scheme, d) using Jenkins (1987) dynamics with D_w non-zero but using one-tenth of the K_z's determined by the Mellor-Yamada turbulence closure scheme, and e) using Jenkins (1987) dynamics with D_w non-zero but using one-tenth of the K_z's determined by the Mellor-Yamada turbulence closure scheme plus $\mu = 0.01 \times K_z(\text{top-most})$ for each grid cell in a model domain.

$$\boldsymbol{D}_w = 0.1 \times (-\partial K_z/\partial z \; \partial u_{St}/\partial z, \; -\partial K_z/\partial z \; \partial v_{St}/\partial z).$$

Another simulation was made based on the expressions in (6) but with $\mu = 10^{-6}$ m^2/s (molecular viscosity). The results were essentially unchanged from those shown in Fig. 1d. We then made simulations with the μ specified to be a function of the viscosity nearest the top of the water column as determined by the Mellor-Yamada turbulent closure scheme. Since the virtual tangential surface stress is another wave source term, the above test results would imply that μ might be at least an order of magnitude smaller than the K_z(top-most) calculated using the Mellor-Yamada scheme. Moreover, if the virtual tangential surface stress is acting over only the top millimeter or so of the water column, mixing length theory would lead us to expect that μ might be even two orders of magnitude smaller than K_z(top-most) from Mellor-Yamada. Test simulations used μ equal to 0.1 and 0.01 of the top-most K_z for each grid cell calculated using the Mellor-Yamada scheme. The results of both tests are very similar to those shown in Fig. 1d. There are small additional increases in the speeds of the currents. Based on the results of these simulations, when the virtual tangential surface stress is turned on for the following tests, we use $\mu = 0.01 \times K_z$(top-most) from the Mellor-Yamada scheme at each grid cell.

WIND AND WAVES IN A SHALLOW EMBAYMENT

Signell et al. (1990) had simulated the impact of waves on a steady wind-driven circulation in narrow, shallow bays with different cross-sectional shapes. The study was to consider the sensitivity of wind-induced flushing to both the cross-sectional shape of the bay as well as the effects of waves. However, the only wave effect considered was the wave-enhanced bottom friction. The test simulations discussed below included other wave effects and non-linear advection and diffusion. As mentioned above, the viscosities used by the wave source terms were reduced Mellor-Yamada viscosities.

The first bay simulated is the 10 m constant depth channel which is opened at one end (Domain 2). As in Signell et al. (1990), the bay was forced with a constant wind stress of 1 dyn/cm^2 and the Coriolis parameter was set to zero. The resulting current profile at a location toward the closed end of the channel is shown in Fig. 2, a reproduction of the results in Signell et al. (1990). The flux out of the bay (referred to as the exchange flux) for this situation is 311.9 m^3/s. Fig. 2 also shows the results when the enhanced friction of a 50 cm amplitude, 5 s period wave, constant throughout the domain, is included. The results indicate a reduction of the bottom current by 28%, resulting in a reduced exchange flux of 292.9 m^3/s. The corresponding simulation by Signell et al. (who used a cubic to approximate the structure of the vertical eddy viscosity), produced a 35% reduction in the bottom current and, accordingly, a similar reduction in the flushing of the channel.

Also shown in Fig. 2 are the current profiles when other wave effects are included. When using the Madsen physics at 0° latitude, the only factors to come into play is the additional transport associated with the Stokes drift and the virtual tangential surface stress. The Stokes drift had a significant impact, resulting in an increased exchange flux of 402.7 m^3/s. At steady state, the additional sea level pressure gradient resulting from Stokes drift induces a upwind flow that balances the downwind Stokes transport, and this causes the dramatic increase in the overall exchange flux. The results of the Jenkins dynamics (which, at 0° latitude, means the addition of Stokes drift, virtual tangential surface stress, and wave dissipation) further increases the exchange flux to 411.5 m^3/s. The wave dissipation increases the flow downwind, and this adds to the exchange flux.

The impacts of other factors on the exchange flux in the 10 m flat bottom channel are presented in Table 2. Being at 45°N with enhanced bottom friction reduces the exchange flux from 292.9 m^3/s to 285.0 m^3/s, a result of some of the flow being diverted to the cross-channel direction by the Coriolis effect. With all wave effects turned on and using either the Madsen or Jenkins dynamics, the additional factors

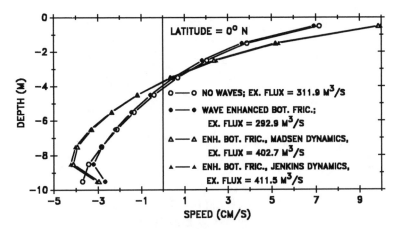

Fig. 2. Current profiles of flow in a 10 m channel induced by a 1 dyn/cm^2 wind plus the enhanced bottom friction of a 50 cm amplitude, 5 s period wave. The currents are 14.5 km from the open end of the 20 km channel. Positive values are toward the closed end of the channel.

compared to the 0° latitude simulations is the impact of the Coriolis effect and the Coriolis wave stress. The exchange flux for the Madsen dynamics is 390.5 m^3/s while that for the Jenkins dynamics is 399.7 m^3/s. The Coriolis wave stress had a negligible impact on the exchange flux in this situation. The Coriolis wave stress attempts to generate a cross-channel flow, but the sides of the channel result in sea level pressure gradients that tend to counter the cross-channel flow. Thus, the cross-channel flow never reaches the magnitude to generate the mirror image of the Stokes drift in the upwind direction.

We also reproduced the simulations of Signell et al. (1990) for a channel which had a slanted bottom going from 2 m on one side of the channel to 18 m on the other side (still a cross-sectional average depth of 10 m). The mechanics of the flow in the slanted channel is considerably different than that of the flat channel. The shallow half of the slanted channel is dominated by the downwind flow while the return

Table 2. Model-predicted exchange fluxes for a flat bottom, 10 m deep channel using various elements of the coupled hydrodynamic-wave equations.

Driving	Exchange Flux (m^3/s)
Wind, No Waves, Latitude = 0°	311.9
Wind, Wave-Enhanced Friction, Latitude = 0°	292.9
Wind, Latitude = 0°, Enhanced Friction, Stokes Drift, Wave Effects After Madsen (1978)	402.7
Wind, Latitude = 0°, Enhanced Friction, Stokes Drift, Wave Effects After Jenkins (1987)	411.5
Wind, No Waves, Latitude = 45°N	301.4
Wind, Wave-Enhanced Friction, Latitude = 45°N	285.0
Wind, Latitude = 45°N, Enhanced Friction, Stokes Drift, Wave Effects After Madsen (1978)	390.5
Wind, Latitude = 45°N, Enhanced Friction, Stokes Drift, Wave Effects After Jenkins (1987)	399.7

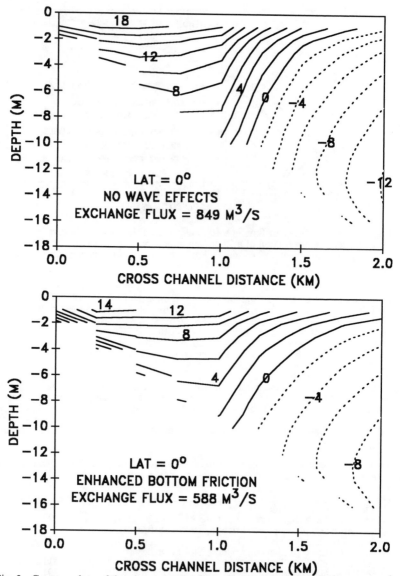

Fig. 3. Contour plots of the along-channel flow (cm/s) induced by (top) a 1 dyn/cm^2 wind and (bottom) a 1 dyn/cm^2 wind plus the enhanced bottom friction of a 50 cm amplitude, 5 s period wave. The currents are 14.5 km from the open end of the 20 km channel. Positive currents are toward the closed end of the channel.

Table 3. Model-predicted exchange fluxes for a slanted bottom channel with a 10 m average depth using various elements of the coupled hydrodynamic-wave equations.

Driving	Exchange Flux (m^3/s)
Wind, No Waves, Latitude = $0°$	848.9
Wind, Wave-Enhanced Friction, Latitude = $0°$	588.1
Wind, Enhanced Friction, Latitude = $0°$, Stokes Drift, Wave Dynamics After Madsen (1978)	654.5
Wind, Enhanced Friction, Latitude = $0°$, Stokes Drift, Wave Dynamics After Jenkins (1987)	668.4
Wind, No Waves, Latitude = $45°$N, Shallow Water on RHS of Channel	844.0
Wind, No Waves, Latitude = $45°$N, Shallow Water on LHS of Channel	845.7
Wind, Wave-Enhanced Friction, Latitude = $45°$N, Shallow Water on RHS of Channel	580.0
Wind, Wave-Enhanced Friction, Latitude = $45°$N, Shallow Water on LHS of Channel	607.8
Wind, Enhanced Friction, Latitude = $45°$N, Wave Effects After Madsen (1978), Shallow Water on RHS of Channel	638.3
Wind, Enhanced Friction, Latitude = $45°$N, Wave Effects After Madsen (1978), Shallow Water on LHS of Channel	644.4
Wind, Enhanced Friction, Latitude = $45°$N, Wave Effects After Jenkins (1987), Shallow Water on RHS of Channel	653.2
Wind, Enhanced Friction, Latitude = $45°$N, Wave Effects After Jenkins (1987), Shallow Water on LHS of Channel	659.3

upwind flow dominates the lower layers of the deep half of the channel. On approaching the closed end of the channel, there is a cross-channel flow that takes water from the shallow side of the channel to the deep side.

The wind-induced flow through the channel at a latitude of $0°$ is shown Fig. 3. The change in bathymetry results in a nearly three-fold increase in exchange flux, an increase two times smaller than that predicted by Signell et al. (1990). Also shown in Fig. 3 is the flow with the enhanced bottom friction added. The exchange flux was reduced by 31%, approximately the same as that predicted by the simulations by Signell et al. (1990).

Simulations at a latitude of $0°$ were also made with the Madsen dynamics and the Jenkins dynamics (both with the virtual tangential surface stress). As in the flat bottom test cases, a significant increase in exchange flux occurs, a result of the inclusion of the downwind Stokes drift. The exchange flux with the Madsen dynamics is 654.5 m^3/s while that with the Jenkins dynamics is slightly greater (because of the wave source term) at 668.4 m^3/s (Table 3).

The impacts of other factors on the exchange flux of the slanted channel are presented in Table 3. As before, the Coriolis effect (no wave effects) reduces the exchange flux by a small amount at $45°$N. With only wave-enhanced bottom friction at $45°$N, the exchange flux is reduced as before. But in this case the amount of the reduction is dependent on whether the shallow water is on the right-hand-side (RHS) or left-hand-side (LHS) of the channel, being greater with the shallow water on the LHS. With the shallow water on the RHS of the channel, the induced cross-channel flow at the end of the channel is hindered by the Coriolis effect. However, when the shallow water is on the LHS of the channel, the Coriolis effect gives an additional acceleration in the cross-channel direction, and this results in a slightly larger exchange flux.

Similar to the flat-bottom case, the inclusion of Stokes drift and other wave effects increases the exchange flux (Table 3). Once again, the exchange flux is slightly larger when the shallow water is on the LHS of the channel.

BREAKING WAVES IN SHALLOW WATER

A radiation stress simulation was made using the shallow water model domain (Domain 3) with the wave parameters shown in Fig. 4 (courtesy of Dr. J. Smith, Coastal Engineering Research Laboratory, Waterways Experiment Station). These show a 4 second wave with an offshore amplitude of 67.5 cm and angle of incidence of ~30 degrees moving shoreward. The drop in amplitude at ~300 m offshore represents energy loss due to wave breaking.

Simulations were performed in a vertically-averaged mode (constant drag coefficient C_d of 2.5×10^{-3}) as well as a full three-dimensional mode. The latitude of these simulations was $0°$, and the time required to reach a steady-state flow was ~20 minutes of simulated time with a 5 second time step. The vertically-integrated (2-D) simulation had radiation stresses only and no other wave effect. In addition, the non-linear advective and horizontal shearing stress terms were neglected. The predicted setup and vertically-averaged longshore currents resulting from the 2-D simulation are shown in Fig. 5. The maximum setup was 7.9 cm, and the maximum longshore current speed of 127 cm/s occurred at the region of drop in wave amplitude.

The results of a fully three-dimensional (3-D) simulation are also shown in Fig. 5. The drag coefficient calculated for the bottom boundary layer for the three-dimensional simulation ($C_d = 1.54 \times 10^{-2}$) results in a considerably smaller average current, with a maximum of ~75 cm/s. Three-dimensional simulations were then made with enhanced bottom friction, then with Stokes drift added, then by adding the Coriolis wave stress by setting the latitude to $45°$, and then finally adding wave dissipation and the virtual tangential surface stress. The predicted maximum setups and vertically-averaged longshore current speeds for each of these three-dimensional simulations are presented in Table 4. Without wave dissipation, the maximum setups are practically identical to the simulations, 7.9-8.1 cm. However, the inclusion of wave dissipation results in a significant increase in sea level setup, over 106 cm. This increase is a result of the greater shoreward momentum resulting from the wave dissipation terms.

The wave-enhanced bottom friction reduced the maximum vertically-average current by an additional 14% (to about 64 cm/s), but the other simulations indicate that the Stokes drift, Coriolis wave stress, wave dissipation, and the virtual tangential surface stress had little impact. Thus, the overall reduction of the predicted maximum current speed when compared to the results of the two-dimensional simulation is ~50%.

One final simulation was made with the non-linear advective terms included in the momentum equations. The results are presented in Table 4 and show a maximum setup of about 107 cm with a significant reduction in the predicted vertically-averaged longshore current, only 44.8 cm/s. Tests show that the non-linear advective term responsible for the dramatic reduction of the maximum longshore current is $\partial(UV)/\partial x$

Table 4. Maximum sea level setup (cm) and maximum vertically-averaged longshore current (cm/s) induced by the radiation stresses determined from the wave field depicted in Fig. 4. Wave effects other than radiation stresses used in each three-dimensional simulation are listed on the left of the table.

Wave Effects	Max Setup	Max Longshore Current
Enhanced Bottom Friction	7.9	64.3
Enh. Bottom Fric., Stokes Drift	8.1	65.7
Enh. Bot. Fric., Stokes Drift, Coriolis Wave Stress @ 45oN	7.9	65.7
Enh. Bot. Fric., Stokes Drift, Cor. Wave Stress, Wave Dissipation, Virtual Tang. Surface Stress	107.2	60.5
Enh. Bot. Fric., Stokes Drift, Cor. Wave Stress, Wave Diss., Virtual Tang. Surf. Stress, Non-Linear Advective Terms	107.5	44.8

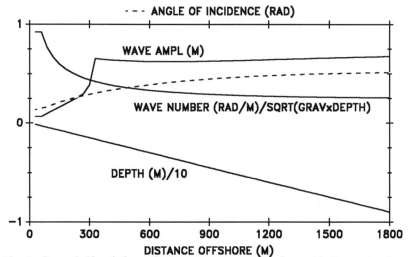

Fig. 4. Cross-shelf variations of wave parameters used to force a shallow water domain with a linearly increasing depth. Variations in the longshore direction were zero. The period of the wave was 4 seconds.

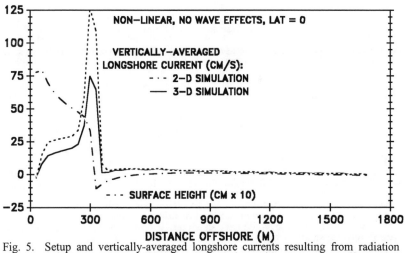

Fig. 5. Setup and vertically-averaged longshore currents resulting from radiation stresses: a) for two- and three-dimensional simulations without any wave effects other than radiation stresses, and b) maximums for three-dimensional simulations with various wave effects.

in the longshore-directed equation of motion (x being cross-shelf). The net cross-shelf U flow is offshore (positive) all across the model domain, primarily to balance the onshore Stokes drift. The spatial gradients of the longshore flow V are positive from the shoreline out to the maximum longshore current (~300 m offshore) and then negative offshore of the location of maximum current. The longshore gradients of U also increase out to the approximate location of the maximum longshore current and then decrease further offshore. Altogether, there is a strong advective flux of longshore momentum across the domain toward deeper water, and this leads to the reduction of the speed of the maximum longshore current induced by radiation stresses.

5. DISCUSSION AND CONCLUSIONS

A critical factor in the formulation of the momentum equations for a model that is to include wave effects appears to be the specification of viscous effects associated with surface waves. As pointed out previously, for viscous effects for wave source terms (i.e., D_w and μ), we would expect the magnitude of the viscosities to be smaller than those determined by the Mellor-Yamada scheme. When we pick viscosities for D_w which are 10% of the values determined by the Mellor-Yamada scheme (which makes the viscosities similar to those used by Jenkins, 1987), the currents generated by the wave field in the deep water basin are very close to the Stokes drift itself except for the crest-parallel current component. Choosing $\mu = 0.01$ x K_z(top-most) from the Mellor-Yamada scheme appears to be reasonable considering the small scales over which the virtual tangential shearing stress is working (order of 10^{-3} m). However, it is pointed out that such specifications for μ and the K_z's used in determining D_w may be reasonable but must be verified by observations.

It is possible that wave-related viscosities can be determined by an appropriate scaling of the Mellor-Yamada viscosities. But for breaking waves, a totally different turbulent closure scheme might need to be developed for assessing wave-related viscosities. The Mellor-Yamada viscosities are parabolic in structure, tending toward zero at the surface and bottom of the water column. But data presented by Battjes and Sakai (1981) show that the Reynold's stress $\langle U'W'\rangle$ in water under breaking waves has relatively large values at the very surface of the water column. With $\langle U'W'\rangle = K_z \partial U/\partial x$, this suggests a non-parabolic-like structure of eddy viscosity under breaking waves (i.e., K_z not tending to zero at the surface). Thus, the vertical variability of K_z near the top of the water column could be considerably less than those indicated by a scaled Mellor-Yamada scheme, and this would significantly impact the determination of Jenkins' wave dissipation term. This is likely the cause of the unrealistically large setup of 106 cm predicted by the model with wave dissipation effects in the simulation of radiation stresses.

Another factor of consideration for circulation models with wave effects is the inclusion of the Stokes drift in the appropriate terms in the momentum equations. The Jenkins dynamical formulation incorporates the vertical shearing effects of the Stokes drift. And the Coriolis wave stress can be thought of as the impact of the Stokes drift acting as a background flow field. If the Stokes drift is considered as a background flow, then its components would be included also in the non-linear advective terms and the horizontal turbulent Reynolds stress terms. For example, consider the significant and unexpected reduction of the maximum longshore current induced by the radiation stresses when the non-linear advective terms were included in the momentum equations. With the inclusion of Stokes drift in the non-linear advective terms, the radiation stress simulation would have an advective term of the form $\partial((U + u_{St})V)/\partial x$ (v_{St} is relative small in this case). Tests show that the resulting momentum flux would not have resulted in the reduction of the maximum longshore current since $U \approx -u_{St}$. Thus, the reduction of the longshore current speed is strictly a result of not considering the flux of momentum due to Stokes drift in the non-linear advective terms.

It is not uncommon to include the advective effects of Stokes drift on the transport of scalar fields (e.g., Skyllingstad and Denbo, 1995), so inclusion in the momentum equations is not unreasonable. And the impact of the spatial variability of Stokes drift on momentum flux has been used for decades in the study of Langmuir circulation (Leibovich, 1977). In most situations, we would not expect the lateral variations of the Stokes drift to have much of an impact on the horizontal flux of momentum. But the exceptions are those situations in which the Stokes drift induces a balancing flow, and in those cases it may be very important to include Stokes drift in the non-linear advective terms (and likely the horizontal turbulent stress terms) to obtain reliable predictions of the wave-averaged flow field.

The shallow embayment test case shows wave-enhanced bottom friction can greatly reduce the flushing of coastal bays. However, other wave effects, mostly Stokes drift, can significantly increase the exchange fluxes of such bays. Wave dissipation also increases the flushing, but the Coriolis wave stress and the virtual tangential surface stress have little impact.

The radiation stress simulations show that we can obtain very different predictions of the vertically-averaged longshore currents depending on the formulation of shearing stress effects. A fully three-dimensional simulation results in a significantly larger C_d than that used in the two-dimensional simulation. Understandably, this leads to smaller maximum longshore currents. Moreover, the inclusion of wave-enhanced bottom friction further reduces the maximum longshore current. Overall, the three-dimensional simulation with enhanced bottom friction result in about a 50% reduction of the predicted maximum, vertically-averaged, longshore current when compared with that predicted by the two-dimensional simulation.

The results of the radiation stress simulations imply that the application of the model physics, at least in the three-dimensional cases, is not appropriate. Within breaking waves, turbulence just below the water surface dominates. As suggested above and in the results of Battjes and Sakai (1981), the Mellor-Yamada scheme cannot be utilized with radiation stress formulation of Longuet-Higgens and Stewart (1964) to properly represent the turbulent kinetic energy distribution. Although recent work indicates that a Mellor-Yamada type scheme will work for breaking waves if the turbulent kinetic energy dissipation under the waves is specified (Terray et al. 1996; Burgers, 1997), such TKE formulations are generally based on deep-water wave data and include a parameterization of the surface friction velocity u_*, a function of wind speed. Clearly, additional research is required for the three-dimensional simulation of energy dissipation in shallow water for waves that break from shoaling effects only.

ACKNOWLEDGMENTS

This work was supported through a contract between HydroQual, Inc., and the Office of Naval Research, Naval Ocean Modeling and Prediction program. Thanks are forwarded to Drs. A. D. Kirwan, A. Blumberg, and J. Kiahatu for their helpful comments and Dr. R. Signell for his help concerning some of the subtle details of the wave-enhanced bottom friction formulation.

REFERENCES

Battjes, J. A., and T. Sakai, 1981: Velocity field in a steady breaker. *J. Fluid Mech.*, 111, 421-437.

Blumberg, A. F., and G. L. Mellor, 1987: A description of a three-dimensional coastal ocean circulation model. *Three Dimensional Coastal Models*, Coastal and Estuarine Sciences, 4, N. S. Heaps (ed.), Amer. Geophys. Union Geophysical Monograph Board, 1-16.

Burgers, G., 1997: Comments on "Estimates of kinetic energy dissipation under breaking waves". *J. Phys. Oceangr.*, 27, 2306-2307.

Fredsoe, J., and Diegaard, R., 1992: *Mechanics of Coastal Sediment Transport.*

Advanced Series on Ocean Engineering, Vol. 3. World Scientific, New Jersey. 369 pp.

Glenn, S. M., and W. D. Grant, 1987: A suspended sediment stratification correction for combined wave and current flows. *J. Geophys. Res.*, 92 (C8), 8244-8264.

Hasselmann, K., 1970: Wave-driven inertial oscillations. *Geophys. Fluid Dyn.*, 1, 463-502.

Jenkins, A. D., 1987: Wind and wave induced currents in a rotating sea with depth-varying eddy viscosity. *J. Phys. Oceanogr.*, 17, 938-951.

Leibovich, S., 1977: On the evolution of the system of wind drift currents and Langmuir circulation in the ocean, Part 1, Theory and averaged current. *J. Fluid Mech.*, 79, 715-743.

Lewis, J. K., Y. L. Hsu, and A. F. Blumberg, 1994: Boundary forcing and a dual-mode calculation scheme for coastal tidal models using step-wise bathymetry. In *Estuarine and Coastal Modeling III: Proceedings of the 3rd International Conference.* (M. Spaulding et al., eds.) Am. Soc. Civil Eng., 422-431.

Longuet-Higgins, M. S., 1953: Mass transport in water waves. *Philos. Trans. Roy. Soc.*, A254, 535-581.

Longuet-Higgins, M. S., 1969a: On the transport of mass by time-varying ocean currents. *Deep-Sea Res.*, 16, 431-447.

Longuet-Higgins, M. S., 1969b: A nonlinear mechanism for the generation of sea waves. *Proc. Roy. Soc. London*, A311, 371-389.

Longuet-Higgins, M. S., and R. W. Stewart, 1960: Changes in the form of short gravity waves on long waves and tidal currents. *J. Fluid Mech.*, 8, 565-583.

Longuet-Higgins, M. S., and R. W. Stewart, 1964: Radiation stresses in water waves: a physical discussion, with applications. *Deep Sea Res.*, 11, 529-562.

Madsen, O. S., 1978: Mass transport in deep water waves. *J. Phys. Oceanogr.*, 8, 1009-1015.

Mellor, G. L., and T. Yamada, 1982: A hierarchy of turbulence closure models for planetary boundary layers. *J. Atmos. Sci.*, 31, 1791-1896.

Nikuradse, J., 1932: Gesetzmassigkeiten der turbulenten Stromung in glatten Rohren. VDI-Forschungsheft 356., Berlin.

O'Connor, B. A., H. Kim, and K.-D. Yum, 1992: Modeling siltation at Chukpyon Harbour, Korea. In *Coastal Modeling of Seas and Coastal Regions* (P. W. Partridge, ed.). Elsevier Applied Science, New York. 397-410.

Pollard, R. T., 1970: Surface waves with rotation: an exact solution. *J. Geophys. Res.*, 75, 5895-5898.

Signell, R. P., R. C. Beardsley, H. C. Graber, and A. Capotondi, 1990: Effect of wave-current interaction on wind-driven circulation in narrow, shallow embayments. *J. Geophys. Res.*, 95(C6), 9671-9678.

Skyllingstad, E. D., and D. W. Denbo, 1995: An ocean large-eddy simulation of Langmuir circulations and convection in the surface mixed layer. *J. Geophys. Res.*, 100(C5), 8501-8522.

Smagorinsky, J., 1963: General circulation experiments with the primitive equations, I. The basic experiment. *Mon. Wea. Rev.*, 91, 99-164.

Sommerfeld, A., 1949: *Partial Differential Equations: Lectures in Theoretical Physics, Vol. 6.* Academic Press, New York.

Terray, E. A., M. A. Donelan, Y. C. Agrawal, W. M. Drennan, K. K. Kahma, A. J. Williams III, P. A. Hwang, and S. A. Kitaigorodskii, 1996: Estimates of kinetic energy dissipation under breaking waves. *J Phys. Oceanogr.*, 26, 792-807.

Ursell, F., 1950: On the theoretical form of ocean swell on a rotating earth. *Mon. Not. Roy. Astron. Soc., Geophys. Suppl.*, 6, 1-8.

Veronis, G., and H. Stommel, 1956: The action of variable wind stresses on a stratified ocean. *J. Mar. Res.*, 15, 43-75.

Yamaguchi, M., 1986: A numerical model of nearshore currents based on a finite amplitude wave theory. In *Proc. 20th Coastal Eng. Conf.* (B. L. Edge, ed.), ASCE, New York. Vol. 1, 849-863.

COMPARISON OF OBSERVED AND FORECASTED SEA LEVELS ALONG THE WEST FLORIDA COAST

William P. O'Connor[1,2], Frank Aikman III[1], Eugene J. Wei[1], and Philip H. Richardson[1]

ABSTRACT

A barotropic version of the Princeton Ocean Model has been configured for the Florida region as part of a NOAA National Ocean Service coastal ocean prediction system. The model domain extends from Cuba to the Florida Panhandle and includes both the Florida Gulf and Atlantic coasts. The orthogonal curvilinear grid spacing is 5' (approximately 9 km) in both longitude and latitude. The model is forced at the open boundaries by the tide consisting of four semi-diurnal and four diurnal tidal constituents. The model is also forced by wind stress and atmospheric pressure over the domain, which were obtained from the NOAA National Weather Service step-coordinate Eta Model output at 29 km resolution. The study covers a 14-month period from September 1995 through October 1996. For comparison purposes, the observed hourly water levels at Clearwater Beach, St. Petersburg, Naples, and Key West were obtained from water level gauge data. The observed and modeled hourly water levels were compared for each month and have root mean square differences of about 0.10 m. A 30-h low pass filter was applied to the 14-month time series of both the observed and modeled data in order to compare the subtidal components. This analysis yielded root mean square differences of about 0.08 m in subtidal water levels.

[1]Coast Survey Development Laboratory, NOAA/NOS, N/CS13, 1315 East-West Highway, Silver Spring, MD 20910-3282.
[2]UCAR Visiting Scientist

INTRODUCTION

The importance of timely water level and current information for ports in support of marine navigation and environmental management is noted by Parker (1994, 1996), who discusses the coastal forecasting systems now being developed for a number of bays and coastal regions along the U.S. coast. The NOAA National Ocean Service (NOS) is developing forecast systems for the Port of New York and New Jersey (Wei and Sun 1998), the Chesapeake Bay (Bosley 1996, Bosley and Hess 1998), and for Galveston Bay (Schmalz 1996, 1998). Parker (1996) points out the need to supply water level boundary conditions for the regional forecasting systems around these ports. This will be accomplished for East Coast ports by the Coastal Ocean Forecasting System that is being operationally developed at the NOAA National Centers for Environmental Prediction (Aikman *et al.* 1996, 1998; Kelley *et al.* 1997). This paper reports on a model of the Florida Shelf (FS) intended to supply water level information for regional models of Tampa Bay and Florida Bay. The shelf model will be used to provide open boundary conditions to a Corps of Engineers barotropic finite element model (RMA-10) of Florida Bay, as well as to possibly provide open ocean forcing for a Tampa Bay forecast system under development by the University of South Florida (Vincent *et al.* 1998) and the NOS.

The two major components of any forecasting system are the ocean model for prediction of the sea conditions, and the atmospheric forcing input to that model. In this report we first describe the ocean model for the FS domain. The tidal calibration and wind forcing used to drive the model are then described. A 14-month simulation was made and the model output water levels were compared with observed water level gauge data. Correlation statistics and plots are shown for these water levels.

OCEAN MODEL

The numerical ocean model used is the Princeton Ocean Model (POM) written by Blumberg and Mellor (1987). Since we are primarily interested in water levels, we made use of only the barotropic mode of the POM, that solves for the water level and the vertically averaged horizontal velocities. This is adequate to model the barotropic dynamics of tides and wind forced water levels. The fully nonlinear system is written in flux form and includes two horizontal momentum equations and the continuity equation. The momentum equations include terms for the Eulerian time derivative, horizontal advection, Coriolis acceleration, quadratic bottom friction, Smagorinsky eddy diffusion, horizontal pressure gradient due to sea level slope, and surface wind stress forcing. The coefficient of bottom friction is 0.0025 and the constant that multiplies the Smagorinsky eddy diffusivity is 0.10. The finite differencing scheme for these equations is centered in space and time on the Arakawa-C grid. A further description of this barotropic model and its application to estuarine and coastal water level problems is given by Blumberg (1977) and Blumberg and Kantha (1985).

The model FS model domain extends from 22.6 ° N to 29.9 ° N in latitude, and from 85.9 ° W to 79.0 ° W in longitude.   This includes the region from Cuba to the Florida Panhandle and both the Florida Atlantic and Gulf coasts (Fig. 1), and is considerably larger than the shelf proper. Orthogonal curvilinear coordinates are used with 5' grid spacing in both latitude and longitude.   The grid domain then consists of 89 grid points in the north-south direction, and 84 grid points in the east-west direction. The grid spacing in latitude is 9.27 km, and the average grid spacing in longitude is 8.31 km. The digitized bottom topography was obtained from the DBDB5 (National Geophysical Data Center 1985) 5' bathymetry, supplemented by the NOS15 (National Geophysical Data Center 1988) 15" bathymetry. The regions of Tampa and Florida Bays were updated from bathymetric charts. The minimum depth was taken to be 1.5 m. The maximum depth is 3500 m in the Gulf basin. The model was run with a time step of 20 s without violating the Courant stability criterion.

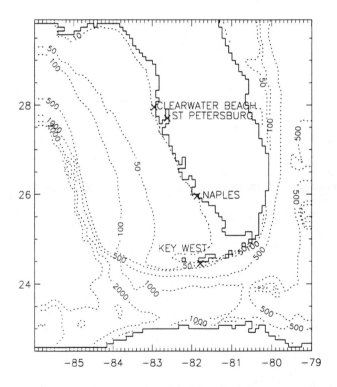

Fig. 1 Model Florida Shelf domain with the locations (×) of four NOS water lev gauges. Depths in meters.

TIDAL FORCING

The tides are enhanced across the West Florida Shelf, compared to their deep water values. Along the West Florida coast the amplitudes generally increase northward with the maxima occurring between Tampa Bay and the Florida Panhandle, where the dominant tides and their maximum amplitudes are the M_2 (0.4 m), K_1 and O_1 (0.2 m), and S_2 (0.15 m), with the others of lesser magnitude (see, for example, Westerink *et al.* 1993).

The FS model domain is significantly smaller than the ocean basin scale of tidal forcing, so that we do not need to include the tidal forcing directly in the equations of motion, but must specify tidal boundary conditions on the open boundaries. This is not always a trivial proceedure, and therefore our domain (Fig. 1) has the open boundaries at a considerable distance from our regions of primary interest, the West Florida Shelf near Tampa Bay and Florida Bay. The open boundaries are mostly in deep water where the tides vary more gradually in space than across the shelves, and are better represented with numerical models. The POM is forced with four semi-diurnal (M_2, N_2, S_2, K_2) and four diurnal (O_1, K_1, P_1, Q_1) tidal constituents. The amplitudes and phases of these constituents are specified on the open boundaries, based on the data of Schwiderski (1980), who used a global ocean tidal model of $1° \times 1°$ resolution, and assimilated observed tidal data along the coasts until the model converged to the appropriate tide cycle. This resolution does not adequately resolve the continental shelves where tidal dissipation is large. However, the POM model we used has $5' \times 5'$ resolution, which should provide adequate representation of the shelf dissipation processes. Because of the difference in model grids, it was necessary to adjust the values of the Schwiderski (1980) data on the open boundaries to yield the observed tidal data at the coasts. The amplitudes and phases of the tidal constituents used to calibrate our model were the corresponding tidal harmonics determined from the NOS tide gauge stations at Clearwater Beach, Naples, and Key West (Fig. 1). The first two stations are along the open coast while the third is on an island. These three stations are representative of the Gulf water levels. For each tidal constituent, if necessary, the Schwiderski (1980) data were modified slightly in amplitude and phase on the open boundaries until the POM model data near these station locations were within ten percent of the amplitude and within one hour in time of the corresponding tidal harmonic determined from the observed NOS gauge data at these coastal stations.

During the tidal simulations the tide height ζ_T, which is the sum of the eight tidal constituents given by the Schwiderski (1980) data, was specified on the boundaries. At the open boundaries the normal component of velocity was specified as

$$\upsilon_n = (g/h)^{1/2} (\zeta - \zeta_T) \tag{1}$$

where $g = 9.8$ m s^{-2}, h is the depth, and ζ is the sea level at the first interior point. At the boundary the tangential velocity component was taken to be the same value as that of the first interior point.

ATMOSPHERIC FORCING

The atmospheric forcing was the output from the numerical weather prediction Eta Model (Black 1994) from the NOAA National Centers for Environmental Prediction. The Eta Model is the operational model for North America and the surrounding oceans at 29 km resolution. The daily 0000 UTC forecast outputs at 3-h intervals out to 24 h were saved, for the region covering the Gulf of Mexico. These data included the 10 m surface winds and surface pressures. These data fields were then intepolated to the 9 km grid for the FS model domain. During the model simulations, the data were temporally interpolated to 10 minute intervals. The 10 m Eta Model winds were used to calculate the surface wind stress components with the equations

$$\tau_x = \rho \, C_D \, u \, (u^2 + v^2)^{1/2} \tag{2}$$

$$\tau_y = \rho \, C_D \, v \, (u^2 + v^2)^{1/2}$$

where the constant air density is $\rho = 1.2$ kg m^{-3} and the Large and Pond (1981) drag coefficient

$$C_D = 0.0012 \, , \qquad\qquad U < 11 \text{ m s}^{-1} \tag{3}$$

$$C_D = (0.49 + 0.065 \, U) \times 10^{-3} \, , \, U \geq 11 \text{ m s}^{-1}$$

varies linearly with the windspeed $U = (u^2 + v^2)^{1/2}$. The same boundary condition (equation 1) was used for the simulation with tides and winds together.

MODEL SIMULATION

A simulation was made with tides and the Eta Model winds forcing the POM for the 14-month period September 1995 through October 1996. The POM hourly water level outputs were saved for the grid points closest to Clearwater Beach, St. Petersburg, Naples, and Key West. The observed NOS hourly water level gauge data were also obtained for these stations. The observed water levels contain strong annual and seasonal components which are partially steric temperature effects, and were not included in the model forcing at the boundaries. Therefore, the annual and seasonal signals were removed prior to the comparisons, as discussed below.

Two comparisons were made to determine how well the model forecasting system can predict water levels. First, we investigated how well the model forecasts the total water level due to winds and tides. Monthly comparisons were made between the observed and modeled water levels for each of the stations. Both the observed and modeled water levels were demeaned for each month, which removes the annual and seasonal signals. The correlation coefficient (cc) and root mean square difference (rmsd) statistical comparisons were made for these 14 demeaned monthly time series. These statistics were similar for each of the months for a given station, and between stations.

The respective statistics for each station were averaged over the 14 months and the results are shown in the first two columns of Table 1. A representative value for the cc is 0.90, and for the rmsd it is 0.10 m. The statistical results for St. Petersburg in Tampa Bay were comparable to those at the open ocean stations. The water level curves for Clearwater Beach and St. Petersburg for the month of February 1996, and the Eta Model wind vectors near Tampa Bay, are shown in Fig. 2. The water level time series for Naples and Key West for February 1996, and the Eta Model wind vectors over Florida Bay are shown in Fig. 3. Both these figures clearly show that the north-south winds which are parallel to the West Florida coast have the greatest effect on sea level at Clearwater Beach, St. Petersburg, and Naples due to Ekman transport. Winds from the north cause flow away from this coast and lower water levels, as occurred on February 4-7, 12-13, and 17. Winds from the south cause flow toward this coast and higher water levels, as occurred on February 2-3, 14-15, and 19-20.

Table 1. Correlation coefficient (cc) and root mean square difference (rmsd) in meters between model output and observations.

	Avg of 14 monthly statistics tides and winds		14-month detided series	
	cc	rmsd	cc	rmsd
Clearwater Beach	0.89	0.12	0.66	0.08
St. Petersburg	0.91	0.09	0.71	0.08
Naples	0.95	0.08	0.55	0.07
Key West	0.86	0.09	0.13	0.08
Average	**0.90**	**0.095**	**0.51**	**0.078**

Second, we investigated the effect of the wind forcing on the model water levels. The model output water levels were processed with a 30-h low pass filter to remove the semi-diurnal and diurnal tidal signals. For the observed water levels, the annual and semi-annual components were determined by a least squares algorithm, were subtracted from the data, and a 30-h low pass filter was applied to the resulting residual. The cc and rmsd statistics are given in the last two columns of Table 1. A representative value of the rmsd is 0.08 m. A representative value of the cc for the three stations of Clearwater Beach, St. Petersburg, and Naples is 0.64. The very low value 0.13 for Key West lowers the average cc to 0.51. This low cc is most likely due to the fact that the Key West station is on an island, with no coastal set-up or set-down. The 14-month long, 30-h low pass filtered time series for the observed and modeled water levels for Clearwater Beach, Naples, and Key West are shown in Figs. 4-6, respectively.

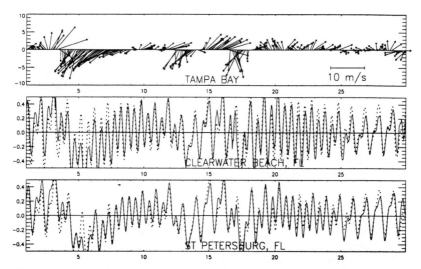

Fig. 2. February 1996. Top: Eta Model winds for the Tampa Bay region. North is toward the top of page. Lower two: the observed (solid) and model output (dashed) demeaned hourly water levels (m) for Clearwater Beach and St. Petersburg.

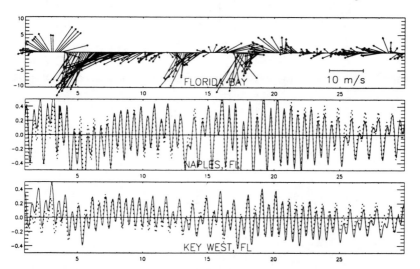

Fig. 3. February 1996. Top: Eta Model winds for the Florida Bay region. North is toward the top of page. Lower two: observed (solid) and model output (dashed) demeaned hourly water levels (m) for Naples and Key West.

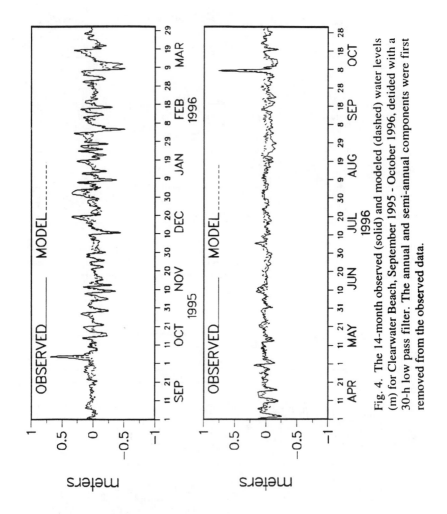

Fig. 4. The 14-month observed (solid) and modeled (dashed) water levels (m) for Clearwater Beach, September 1995 - October 1996, detided with a 30-h low pass filter. The annual and semi-annual components were first removed from the observed data.

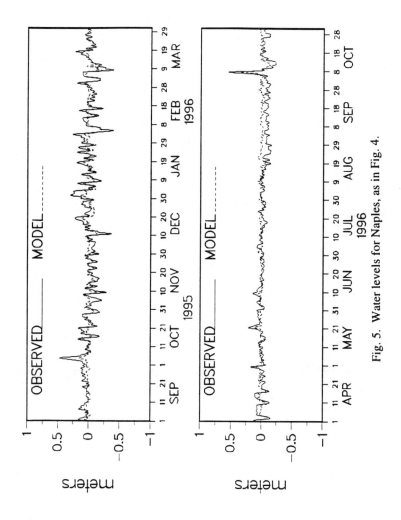

Fig. 5. Water levels for Naples, as in Fig. 4.

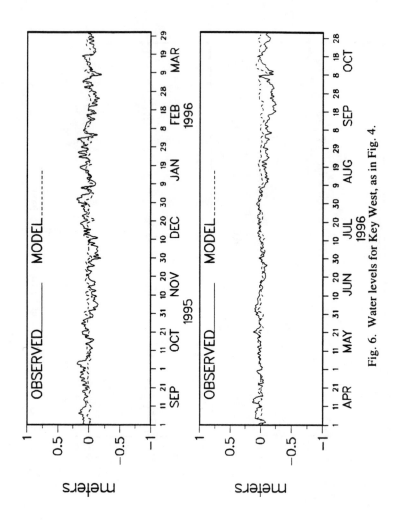

Fig. 6. Water levels for Key West, as in Fig. 4.

SUMMARY AND DISCUSSION

The objective of this NOS work is to develop coastal forecast models with sufficient skill to provide reliable water level boundary conditions to regional ports, including New York and New Jersey, as well as Chesapeake, Tampa, Galveston, and Florida Bays. A barotropic version of the Princeton Ocean Model has been set up for the West Florida Shelf that can provide water level boundary conditions to regional models of Tampa Bay and Florida Bay. The shelf model is forced by tides and the forecast winds from the NOAA numerical weather prediction Eta Model. The quality of the forecast water levels is primarily determined by the quality of the Eta Model forecast winds. The wind driven set-up and set-down along the West Florida coast are well represented in these model simulations, as are the tidal fluctuations. The results indicate that with accurate wind forecasts, the shelf model should improve water level forecasts for regional models of Tampa Bay and Florida Bay. It is expected that the addition of a 24-h nowcast, driven by analysed Eta wind fields, and preceeding the forecast cycle (Aikman et al. 1998) will provide a better initial model state and improve the water level forecasts.

ACKNOWLEDGMENTS

The authors wich to thank Lloyd C. Huff, Kurt W. Hess, Christopher N. K. Mooers, and Richard A. Schmalz, Jr. for reviewing the manuscript. One of the authors (WPO) was funded by the NOAA Coastal Ocean Program Office South Florida Ecosystem Restoration Prediction and Modeling Program through a University Corporation for Atmospheric Research (UCAR) Visiting Scientist appointment.

REFERENCES

Aikman, F. III, G.L. Mellor, T. Ezer, D. Sheinin, P. Chen, L. Breaker, K. Bosley, and D.B. Rao, 1996: Towards an operational nowcast/forecast system for the U.S. East Coast. *Modern Approaches to Data Assimilation in Ocean Modeling,* P. Malanotte-Rizzoli, Ed., Elsevier Oceanographic Series 61, Amsterdam, The Netherlands, 347-376.

Aikman, F. III, E.J. Wei, and J.R. Schultz, 1998: Water level evaluation for the coastal ocean forecast system. *Second Conference on Coastal Atmospheric and Oceanic Prediction and Processes,* Jan. 11-16, 1998, Phoenix, AZ, Amer. Meteor. Soc., 1-6.

Black, T.L., 1994: The new NMC Mesoscale Eta Model: Description and forecast examples. *Wea. Forecasting,* **9,** 265-278.

Blumberg, A.F., 1977: Numerical tidal model of Chesapeake Bay. *Journal of the Hydraulics Division,* ASCE, **103,** 1-10.

Blumberg, A.F., and L.H. Kantha, 1985: Open boundary conditions for circulation models. *Journal of Hydraulic Engineering,* ASCE, **111,** 237-255.

Blumberg, A.F., and G.L. Mellor, 1987: A description of a three-dimensional coastal ocean circulation model. *Three-Dimensional Coastal Ocean Models, Coastal and Estuarine Sciences,* Vol. 4, N.S. Heaps, Ed., Amer. Geophys. Union, Washington, DC, 1-16.

Bosley, K.T., 1996: Research aimed at prediction of water levels in the Chesapeake Bay.

Conference on Coastal Oceanic and Atmospheric Prediction, Jan. 28-Feb. 2, 1996, Atlanta, GA, Amer. Meteor. Soc., 268-271.

Bosley, K.T., and K.W. Hess, 1998: Development of an experimental nowcast/forecast system for Chesapeake Bay water levels. *Estuarine and Coastal Modeling, Proceedings of the 5th International Conference,* Oct. 22-24, 1997, Alexandria, VA, M.L. Spaulding, Ed., (this proceedings).

Kelley, J.G.W., F. Aikman III, L.C. Breaker, and G.L. Mellor, 1997: Coastal ocean forecasts. Real-time forecasts of physical state of water level, 3-D currents, temperature, salinity for U.S. East Coast. *Sea Technology,* **38**(5), 10-17.

Large, W.G., and S. Pond, 1981: Open ocean momentum flux measurements in moderate to strong winds. *J. Phys. Oceanogr.,* **11**, 324-336.

National Geophysical Data Center, 1985: Worldwide gridded bathymetry - DBDB5 5 minute latitude/longitude grid, data announcement 85-MGG-01, NOAA/NGDC, Boulder, CO.

National Geophysical Data Center, 1988: NOS Hydrographic Data Base - Expanded, Digital Bathymetric Data for U.S. Coastal Waters, data announcement 87-MGG-12, NOAA/NGDC, Boulder, CO.

Parker, B.B., 1994: Real-time oceanographic data for ports. *Mariners Weather Log,* **38**(3), 12-16.

Parker, B.B., 1996: Monitoring and modeling of coastal waters in support of environmental preservation. *J. Mar. Sci. Technol.,* **1**, 75-84.

Schmalz, R.A., Jr., 1996: *National Ocean Service Partnership: DGPS-Supported Hydrosurvey, Water Level Measurement, and Modeling of Galveston Bay: Development and Application of the Numerical Circulation Model,* NOAA Tech. Rep. NOS OES 012, 136 p.

Schmalz, R.A., Jr., 1998: Design of a nowcast/forecast system for Galveston Bay. *Second Conference on Coastal Atmospheric and Oceanic Prediction and Processes,* Jan. 11-16, 1998, Phoenix, AZ, Amer. Meteor. Soc., 15-22.

Schwiderski, E.W., 1980: On charting global ocean tides. *Rev. Geophys. Space Phys.,***18**, 243-268.

Vincent, M., D. Burwell, B. Galperin, and M. Luther, 1998: An operational real-time data acquisition and circulation modeling system in Tampa Bay, Florida. *Estuarine and Coastal Modeling, Proceedings of the 5th International Conference,* Oct. 22-24, 1997, Alexandria, VA, M.L. Spaulding, Ed., (this proceedings).

Wei, E.J., and C.L. Sun, 1998: Development of a New York/New Jersey Harbor water level and current nowcast/forecast system. *Second Conference on Coastal Atmospheric and Oceanic Prediction and Processes,* Jan. 11-16, 1998, Phoenix, AZ, Amer. Meteor. Soc., 11-14.

Westerink, J.J., R.A. Luettich, Jr., and N. Scheffner, 1993: *ADCIRC: An Advanced Three-Dimensional Circulation Model for Shelves, Coasts, and Estuaries; Report 3, Development of a Tidal Constituent Database for the Western North Atlantic and Gulf of Mexico, Technical Report DRP-92-6,* US Army Corps of Engineers, Washington, DC, 154 p.

Finite Element Model of Wassaw Sound
with Synthetic Marsh Flooding Boundary Conditions

Thomas F. Gross[1] and Francisco E. Werner[2]

Abstract-

The circulation of Wassaw Sound, a Georgia estuary forced by three meter tidal range, was modeled to provide precise prediction of surface currents to aid the 1996 Olympic Yachting events. The tidal circulation was computed with the free-surface, 3-D, nonlinear, finite element QUODDY model developed by Lynch and Werner (1991) and Lynch *et al.* (1996). Using the finite element model which includes tidal and wind forcing to generate the three dimensional structure of the circulation of Wassaw Sound, we demonstrate the special boundary conditions needed to correctly account for drainage from extensive marsh systems typical of the US South Atlantic Bight. Tidal forcing from outside the sound is used to specify the interior response of tidal creeks and channels. By synthesizing the back water areas of the marsh flats, the volume of the tidal prism can be correctly modeled without the artificial specification of interior tidal heights.

Model results were distributed daily to all the Olympic competitors along with the daily meteorology reports. The daily predicted tides could be precomputed months in advance, however the wind forcing could not predicted. We attempted to calculate tidal and surface currents with wind forcing based on the previous day's winds and predictions of the sea-breeze made for the next day. Moving the model from a research tool to this nearly real time predictive mode proved to be much more difficult than anticipated.

1 Introduction

The circulation of coastal Georgia is dominated by strong M2 tides which flood extensive salt marshes behind the barrier islands. These Spartina marshes are drained by steep banked creeks which branch repeatedly. Ultimately the volume of flow through the mouth of the sounds is determined by the tidal prism which is comprised by the volume of the main tidal channels and the area of the marshes. A numerical model of the tidal circulation of the sounds should either resolve all of the marsh creeks with a wetting and drying model of the marsh, or account for the volume flux of the marshes through a mass flux boundary condition applied at the mouth of the secondary marsh creeks. The latter technique was developed to simulate the currents of Wassaw Sound for use by the 1996 Olympic yachting contestants.

Typically the open water boundary condition is specified by the time dependence of the water height (tidal constituents on outer edge of domain), Dirichlet

---

[1]Assoc. Professor, Skidaway Institute of Oceanography, (Presently: NOAA/CSDL rm 7881,1315 East-West Hwy, Silver Spring, MD 20910, tom.gross@noaa.gov)

[2]Professor, Marine Sciences Curriculum, University of North Carolina, Chapel Hill, NC 27599-3300, cisco@marine.unc.edu

condition, or by specifying the transport through the boundary (solid walls, or river flow above head of tide), Neumann condition. Lynch and Holboke (1997) demonstrated the equivalence of both specifications and how a strong enforcement of the transport constraint may be utilized. However intuitively we recognize that the flow and tidal height at the mouth of a marsh creek is determined by external tidal forcing of the sound and the geometry of the marshes which the creek drains. The problem would be over specified if either height or transport were given *a priori*. A boundary condition for the rivers was used which solves for the surface height at the mouth of each river. The volume flux and salt flux up the river out of the domain is continuously calculated and used to specify a salt concentration on reversal when the flow is back into the domain. The inflowing salt concentration is rather arbitrary, but can be controlled so that salt that passes back and forth through the boundary is conserved.

2 Methods

2.1 The Model

The hydrodynamic model used is "Quoddy" a free-surface, 3-D, nonlinear finite element model developed by Lynch and Werner (1991) and Lynch *et al.* (1996). It determines the tidal gravity wave solution by solving the advective Navier Stokes equations. The sea level height is calculated by the convergence of the advected water masses. The three dimensional (QUODDY) model resolves $\mathcal{O}(1$ km) length scales, *i.e.*, $\mathcal{O}(100$ m) grid spacing, and has been used by Kapolnai *et al.* (1996) to trace and delineate sub-tidal time scale mixing and dispersion processes in an idealized domain. It incorporates Mellor Yamada (1982) 2.5 level turbulence closure, operates in tidal time, and computes the evolution of tracer and density (baroclinic) fields prognostically. Variable horizontal and vertical resolution are facilitated by the use of unstructured meshes. The model has the ability to resolve the details of tidal shear dispersion by directly including the shear length scales and a turbulent bottom boundary layer closure.

Bathymetric surveys of the interior of Wassaw Sound and the adjacent inner shelf region were completed by NOAA during the Fall 1994 and Spring 1995. This data gave bathymetry with resolution of better than 100 m spacing throughout the region of interest. Since Wassaw Sound, like most South Atlantic Bight (SAB) sounds, has slowly varying bed bathymetry due to motions of sand bars and accretion to barrier islands this bathymetric data set represents the only up to date mapping of an SAB sound. (The previous NOAA survey was completed in 1974.)

The semi-diurnal tides on the SAB shelf propagate as Poincaré waves, with phase lines parallel to shore and resulting tidal ellipses' principal axes perpendicular to the isobaths (Wang *et al.* 1984; Pietrafesa *et al.* 1985, Werner *et al.* 1993). Over 80% of the tidal signal is contained in the semi-diurnal band and over 80% of that signal is due to the M2 tide. Initial model runs used a single M2 tidal constituent, forcing an off-shore Poincaré wave. The outer forcing was then specified by the North-South phase shift of the M2 Poincaré wave with amplitude modulated by a six component tide (M2 S2 N2 K2 K1 O1) based on the NOAA tidal tables for the period spanning the Olympic events. The gain factor based on the inshore response to the M2 tide adjusted the outer boundary forcing to give proper height at the shore location.

2.2 Riverine Boundary Conditions

The boundary conditions for the inland estuarine rivers were adapted for the peculiarities of the Georgia salt marsh systems. Little data is available for the inland sea surface heights so direct driving of the interior circulation by specifying sea surface at the head of the tidal channels is not a possibility. Because the different rivers and creeks can be phase shifted by only ten or fifteen minutes these heights would have to be prescribed with an accuracy (better than 1 min.) not attainable from field measurements. An alternative method is used wherein the tidal prism flooding the back-river areas is specified and allowed to determine the sea surface by continuity.

Within the sounds there is an almost infinite hierarchy of drainage channels which communicate with the marsh system. These were modeled by specifying the area or volume of water drained by the primary channels. The currents and total tidal excursion are expected to depend not only on the changes in sea level height, but on the tidal prism as well. Although the main body of the sound and tidal rivers can contain a large part of the tidal prism, a significant volume is drained out of the distributary channels which finger through the large marsh areas between the barrier islands and land.

The parameterization of the flooding and draining of the backwater marshes was related to the sea level height along the distributary channel open boundary, the flow across that boundary and the area behind the channel (*i.e.*, into which, or from which, the flooding and draining takes place). The boundary condition states that flow through the open boundary is resisted by the rise of sea level beyond the open boundary and is related to the backwater tidal prism. Sea level rises in proportion to the area flooded and the flow through the open boundary's cross-section. For shelf circulation, only the major sounds and rivers need be specified. For nearshore flows the tidal creeks must be added, with mud flats and the distributary creeks possibly becoming important to circulation within the sound.

Along the open boundary of the primary channel of each distributary channel system, the area of the backwater tidal prism, A, is specified. The total volume moving through the river, V, is spread across this surface area to derive the changing sea level:

$$V(t) = \int U(t)[H(t) + H_0]W\,dt \quad \text{where} \quad H(t) = \frac{V(t) - V_0}{A}.$$

Here V is backwater volume, U is flow through the head of the river or distributary channel which is $[H(t) + H_0]$ deep and W wide, and the mean backwater volume and depths are denoted by V_0 and H_0. The boundary condition is implemented on the sea surface height through

$$\frac{dH}{dt} = \frac{U(t)[H(t) + H_o]W}{A}.$$

The extreme case of A becoming very small is equivalent to an impermeable boundary and a condition on the normal velocity is a better way to specify this limit. The other extreme is for large A. For instance when the tide is flooding and A is very large, the height at the boundary will grow too slowly compared to the height within the model domain and a large pressure gradient will become established which forces a comparably large velocity. As the velocity increases the hydraulic jump condition is approached. This is the tidal bore found in places like the Bay of Fundy and cannot be accurately modeled by imposing this boundary condition. Thus the limiting velocity is for a Froude number condition ($F_r = U/\sqrt{gH}$) which must be less than unity and in practice is $\mathcal{O}(0.1)$. Therefore, independent of the actual area behind the head of the channel, the area will be defined by the tidal prism to be the equivalent area which can possibly be drained. If the backwater area is much larger, then one may expect that phase shifts may become important. The magnitude of U would still be limited as would the magnitude of A and H. At the primary channels' open boundary, nonlinear tides associated with the upstream propagation of the tide in the shallow channels are also expected as described by Aubrey and Speer (1985) and Aubrey and Friedrichs (1988).

The backwater areas were roughly estimated for this study by making measurements on charts. Another estimate of the backwater area and volume could be empirically provided by current and tidal height measurements at the mouths of the distributary channels. The backwater area drained by a channel could be empirically determined by measuring the volume flux (velocity) and the time-varying sea level change at the mouth of the root creek. These simple measurements would be made with a current meter and pressure gauge placed at the mouth of the creek for one or two tidal cycles, preferably during extreme spring tides. The relation of the rate of change of tidal height to velocity provides a hypsometric curve for the area behind the channel mouth. Kjerfve *et al.* 1991 showed that errors of 50 - 100% may occur using this method because of cross channel variability of currents. There are about ten primary creeks which open into Wassaw Sound, the Wilmington and Bull Rivers, which drain a total area of about 40 km^2. It is important to have the correct total backwater area drainage modeled to provide the full tidal prism as observed at the mouth of the sound.

2.3   Finite Element Implementation of Tidal Creek B.C.

The implementation of this boundary condition in the FEM is slightly different from the familiar Neumann and Dirichlet conditions. To demonstrate this, the methodology and nomenclature featured in Lynch and Holboke 1997,(LH97), is used. The 3-D continuity equation is vertically integrated to give the equation for time rate of change of sea surface:

$$\frac{\partial H}{\partial t} + \nabla_{xy} \cdot \int_{-h}^{\zeta} \mathbf{v} dz = r$$

where $\mathbf{v}$ is velocity, H is total depth from seabed, z=-h, to sea surface, $z = \zeta$, $\nabla_{xy}\cdot$ is horizontal divergence and r is a vertically integrated source term (e.g. "waterfall" river source or rainfall). The wave equation solution uses the time derivative of the above equation and the Navier Stokes rate of change of velocity equation to provide the velocity terms. The finite element solution involves writing the weighted residual form with the Galerkin test functions, ϕ_i, the inner product notation $\langle , \rangle$, a numerical constant τ_o and an integration by parts, yielding equation 4 of LH97 :

$$\langle \frac{\partial^2 H}{\partial t^2}\phi_i \rangle + \langle \tau_o \frac{\partial H}{\partial t}\phi_i \rangle - \langle \nabla_{xy}\phi_i \cdot \left((\frac{\partial}{\partial t} + \tau_o) \int_{-h}^{\zeta} \mathbf{v} dz \right) \rangle + \oint (\frac{\partial}{\partial t}+\tau_o)q\phi_i ds = \langle (\frac{\partial}{\partial t}+\tau_o)r\phi_i \rangle$$

The boundary transport integral:

$$(\frac{\partial}{\partial t} + \tau_o)Q = (\frac{\partial}{\partial t} + \tau_o) \oint q\phi_i ds$$

is composed of typical boundaries, where the Neumann or Dirichlet conditions are applied as described in LH97, and the open boundaries of the rivers draining the marshes. For each river draining an area of A_b, the change in volume of the flooded marsh drains through the boundary:

$$Q_b = A_b \frac{\partial H}{\partial t}$$

Thus for:

$$(\frac{\partial}{\partial t} + \tau_o)Q_b = A_b \frac{\partial^2 H}{\partial t^2} + \tau_o A_b \frac{\partial H}{\partial t}$$

The full equation for test function ϕ_i becomes:

$$\langle \frac{\partial^2 H}{\partial t^2}\phi_i \rangle + A_b \frac{\partial^2 H}{\partial t^2}\delta_{ib} + \langle \tau_o \frac{\partial H}{\partial t}\phi_i \rangle + \tau_o A_b \frac{\partial H}{\partial t}\delta_{ib}$$

$$-\langle \nabla_{xy}\phi_i \cdot \left((\frac{\partial}{\partial t} + \tau_o) \int_{-h}^{\zeta} \mathbf{v} dz \right) \rangle + (\frac{\partial}{\partial t} + \tau_o) \oint q\phi_i ds(1 - \delta_{ib}) = \langle (\frac{\partial}{\partial t} + \tau_o)r\phi_i \rangle$$

where the Krondeker delta δ_{bi} specifies the river boundary for the ϕ_i which adjoins the river boundary.

2.4   Marsh storage of salt

The salinity boundary condition for a reversing river is not well posed. During flood tide, salt moves out of the domain and a Neumann, gradient condition should be used. During ebbing tide, salt previously stored in the marsh is discharged as a decreasing salinity and a Dirichlet condition is specified. The model must continuously test the direction of flow through the riverine boundaries and switch between the boundary condition types. Three problems afflict this method: 1) An estuarine flow up a river will have a vertical shear which can produce simultaneous in and out flows near slack water. 2) Large open boundaries may have horizontal gradients with flow reversals. 3) During the in-flow the salinity value is required but is not an intrinsic of the model. A solution to the first two problems is the imposition of a "sponge" layer inside the boundaries where artificially enhanced vertical and horizontal mass diffusion mix away the gradients. The "sponge" layer may also provide a solution to the third problem by allowing a specified salinity to be maintained on the boundary. Without the enhanced mixing a strong and unstable gradient may develop next to the boundary during out flow. However the solution to the three problems may be found in the geometry of the Georgian tidal rivers. The tidal rivers and creeks are all quite shallow with a few isolated "holes" of up to 20 meters, but most also shoal between river bends to less than three or four meters. In the shallow areas they are well mixed vertically and are too shallow to allow much baroclinic pressure gradient. Although the secondary residual circulation of a channelized estuary may have cross channel reversing flow the effect is small compared to the daily tidal velocity. It is, however, an important factor to tidally averaged physics. Thus the tidal rivers have been found to have a well defined reversal and the switching between boundary condition types appears stable. However on the open ocean boundary the horizontal shear is always significant and reversal almost always occurs somewhere along the boundary. The present model runs avoid this problem by specifying the salinity on the oceanic boundary and not running the model through time to the point where fresh water from the estuaries is allowed to reach the boundary (less than 10 days).

Upon reversal to the marsh emptying ebb flow direction, the salinity of the marsh water must be specified at the mouth of the creek. The salt balance up the rivers is maintained by calculating the salt flux out of the sound up into the river-marsh during flood tide. The salt which flooded the marshes is then "replayed out" on the ebbing tide guaranteeing salt conservation. There is an integral constraint on the total salt to exit the marshes, but the temporal variation must be defined by a model of mixing within the marsh. The salt volume of the high water, flooded marsh of a river is

$$V_S = \int_{T_l}^{T_h} \oint \int_{-h}^{\zeta} \mathbf{v} S dz ds dt$$

and the water volume is

$$V_f = \int_{T_l}^{T_h} \oint \int_{-h}^{\zeta} \mathbf{v} dz ds dt$$

where T_l and T_h are the times of the last low and high water, respectively. The average salinity of this marsh's tidal prism is

$$\overline{S} = V_S / V_f$$

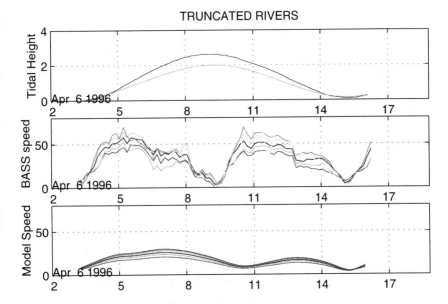

Figure 1: Truncated rivers model run. Time series of tidal height and speed at station "B" (Fig. 2). Dark tidal height curve is BASS data, grey is modeled. The five BASS current meters are plotted indicating the shear of the velocity profile (43, 78, 125, 257, 447 cm.). The modeled heights span the water column, (variable depths, constant proportion of water depth).

The salinity of the waters leaving the marsh during ebb is given as a simple function of the ebbing water volume, such that the flooded salt content will be returned to the sound/ocean. We also require that the given initial high salinity of maximum flood, S_f decreases towards the source salinity, S_r of the river. The source salinity is required as the rivers may have a fresh water source or the next tide may drain more water than entered. A simple quadratic,

$$S(V) = A + BV + CV^2$$

has the minimal degrees of freedom required to satisfy the constraints:

$$S(V = 0) = S_f$$

$$S(V = V_f) = S_r$$

$$\overline{S} = \frac{1}{V_f} \int_0^{V_f} A + BV + CV^2 dV$$

where the ebbing volume $V(t)$ is

$$V(t) = \int_{T_h}^{t} \oint \int_{-h}^{\zeta} \mathbf{v}\, dz\, ds\, dt$$

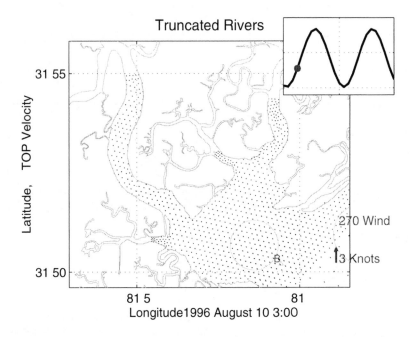

Figure 2: Surface vector velocity with truncated rivers. The velocity vectors are drawn flowing away from the dot at their base. The "B" marks the location of the BASS tripod deployment.

The coefficients A, B and C are solved and used in the replay function to specify the ebb flow salinity as a function of time through the function of ebbing volume, $V(t)$. Flow in excess of V_f is kept at the fixed salinity specified for the river, S_r.

$$S(t) = \begin{cases} A + BV(t) + CV(t)^2 & V(t) < V_f \\ S_{river} & V(t) \geq V_f \end{cases}$$

3   Results

To demonstrate the necessity of the additional "realistic" attributes of the river area boundary condition, wind forcing and baroclinic pressure gradients a series of model runs will be presented. Comparison between model runs is useful to evaluate the effects of the additional attributes, but some real field data would also be useful. A short deployment of the BASS current meter system, (Williams *et al.* 1987), was made in Wassaw Sound in April 1996. The BASS tripod was placed on the north flank of the Wilmington River channel near the center of the sound (Fig. 2). The BASS measured speed at five heights above the sea bed (

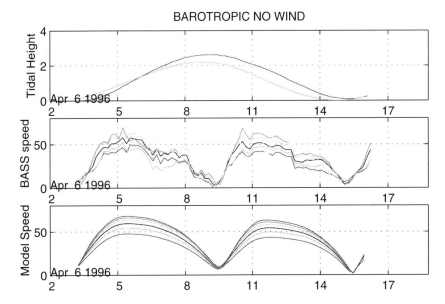

Figure 3: Open rivers model run. Time series of tidal height and speed at station "B" (Fig. 2). Dark tidal height curve is BASS data, grey is modeled. The five BASS current meters are plotted indicating the shear of the velocity profile (43, 78, 125, 257, 447 cm.). The modeled heights span the water column, (variable depths, constant proportion of water depth).

43, 78, 125, 257, 447 cm above bed). Figure 1, a typical tidal cycle demonstrates the rapid speed increase during onset of ebb tide (hours 10-15) and the more symmetrical speed curve of the flood tide. Low water slack is very short while high water slack may last almost an hour. The vertical shear is indicated by the spread of the lines representing current at different levels above the sea bed, Fig. 1. The shear is of comparable magnitude on flood and ebb indicating little baroclinic pressure gradient effects.

3.1 Truncated Rivers

When the rivers are simply truncated with a no normal flow boundary condition the tidal prism exchanged through the mouth of the sound is reduced. This in turn decreases the speeds throughout the sound. Fig. 1 compares the modeled speeds at the same location as the BASS tripod data. Although the sea surface heights are similar (they were specified to be similar) the speeds are less than those measured. The vector plot, Fig. 2, shows how the constraints at the head of the rivers has led to low speeds throughout the sound.

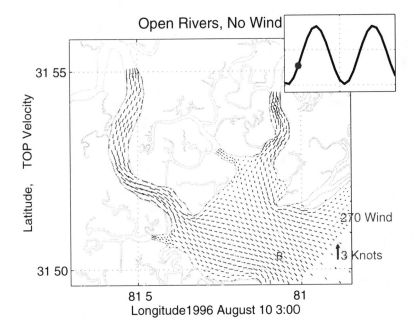

Figure 4: Surface vector velocity with open rivers. The velocity vectors are drawn flowing away from the dot at their base. The "B" marks the location of the BASS tripod deployment.

3.2   Flow Through Rivers

The river area boundary condition allows a larger tidal prism than that obtained from only the main channels. The nearly doubled volume of water results in stronger flow throughout the sound. Fig. 3 demonstrates that the speeds are now very close to those measured with the BASS current meters. The vector velocity plot (Fig. 4) shows that maximum currents are flowing through the Wilmington River. The smaller Bull River also controls current in the eastern part of Wassaw Sound. Fig. 5 is the Olympic race overview plot showing the slow veering of currents in the area of the outer race circles.

3.3   Wind Driven Flow

The Olympic sailors are more concerned with the effects of the wind on their sails than the effects of the wind on the currents, but there will be some modification of the tidal currents. The National Weather Service was providing detailed weather predictions for the area and had several buoys and met stations set up for the race area. Superimposed on the large scale wind was a daily sea breeze compo-

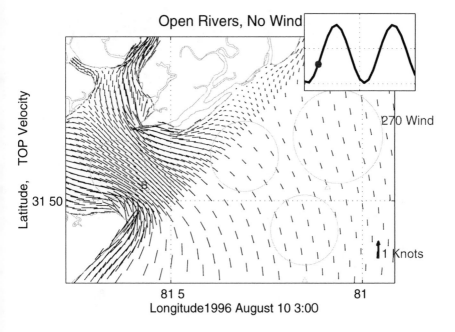

Figure 5: Surface vector velocity with open rivers. This wider view of Fig. 4 includes the Olympic race circles marking the areas used for different events. Within Wassaw sound where the currents were most variable and strong, only the windsurfing events were held.

nent blowing lightly seaward in the morning, and rising with a landward breeze maximum in late afternoon. A common wind during the period of the races was a mean 10 knot wind from the South with a 5 knot sea breeze superimposed. This wind was simulated with the model. Fig. 6 shows a slight shift in current vectors toward the North when compared with the wind free run of Fig. 5. A mean current is set up of about 2-10 cm/s throughout the outer race area.

3.4   Salinity Flux and Baroclinic Forcing

Little data were available to construct an accurate salinity field. A gradient of 4 ppt from the head of the Wilmington River to the oceanic salinity at the mouth of Wassaw sound was specified as an initial condition. After ten days of spin up the fresh water became confined to the channels of the two rivers, Fig. 7. The effects of inclusion on the baroclinic pressure gradient on the vector velocities of Fig. 5 and 6 were negligible and the comparable plots no different. The subtle effect of baroclinic forcing was not of use to the yacht racers. However the tidally averaged estuarine circulation caused by the baroclinic field is necessary

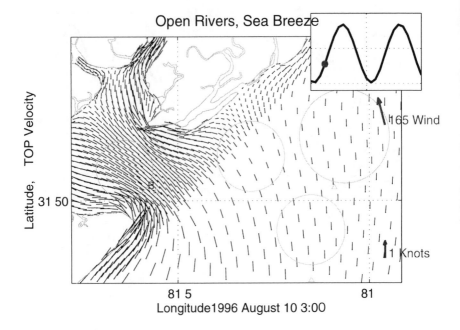

Figure 6: Surface vector velocity with open rivers and with wind. Careful comparison with Fig. 5 will reveal a veering toward the North and speed increase of a fraction of a knot due to the inclusion of wind forcing.

to transport calculations.

4 Conclusions

The finite element model proved capable of modeling the intricate shoreline of Wassaw Sound and produced vector velocity fields found to be of use to the Olympic sailors. Of the "realism" enhancements, marsh draining, wind forcing and baroclinic pressure fields, the marsh area model was the most essential. Presumably a suite of measurements to establish the hypsometric curve of the marshes would have the most value in improving the accuracy of the model. The wind forcing was not important inside the sound, but outside the 2-10 cm/s currents would be significant to the keel boat racers. The baroclinic forcing is probably not necessary when modeling currents to be used by yacht racers. As a research tool the model works quite well. But in its role providing daily updated current predictions based on meteorological wind forecasts it proved cumbersome to use on a daily basis as turnaround time on a model run was in excess of ten hours (with a SUN SPARCstation 10).

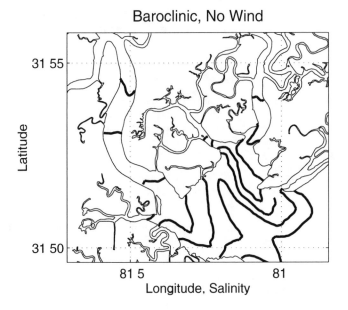

Figure 7: Distribution of salinity after 10 day spin up. Contours are in 0.5 ppt increments, ranging from 28.5 - 31.5 ppt.

BIBLIOGRAPHY

Aubrey, D.G. and C.T. Friedrichs (1988) Seasonal climatology of tidal nonlinearities in a shallow estuary. In: *Hydrodynamics and Sediment Dynamics of Tidal Inlets*, Eds. Aubrey, D.G. and L. Weishar, Springer-Verlag, New York, 103-124.

Aubrey, D.G. and P.E. Speer (1985) A study of nonlinear tidal propagation in shallow inlet/estuarine systems. Part I: Observations. *Est. Coastal Shelf Science*, **21**, 185-205.

Kapolnai, A., F.E. Werner and J.O. Blanton (1996) Circulation, mixing and exchange processes in the vicinity of tidal inlet: A numerical study. *J. Geophys. Res.*, **101(C6)**, 14253-14268.

Kjerfve, B., L.B. Miranda, E. Wolanski (1991) Modelling water circulation in an estuary and intertidal salt marsh system. *Neth. J. Sea Res.* **28(3)**, 141-147.

Lynch, D.R. and M.J. Holboke (1997) Normal flow boundary conditions in 3D circulation models. *Int. J. Numerical Methods in Fluids*, **25**, 1185-1205.

Lynch, D.R., J.T.C. Ip, C.E. Naimie, F.E. Werner (1996) Comprehensive Coastal Circulation Model with Application to the Gulf of Maine. *Continental Shelf Research*, **16**, 875-906.

Lynch, D.R. and F.E. Werner (1991) Three-dimensional hydrodynamics on finite elements. Part II: nonlinear time-stepping model. *International Journal for Numerical Methods in Fluids*, **12**, 507-533.

Mellor, G.L. and T. Yamada (1982) Development of a turbulence closure model for geophysical fluid problems, *Reviews of Geophys. Space Phys.*, **20**, 851-875.

Pietrafesa, L.J., J.O. Blanton, J.D. Wang, V. Kourafalou, T.N. Lee and K.A. Bush (1985) The tidal regime in the South Atlantic Bight. In: *Oceanography of the southeastern U.S. continental shelf*, Atkinson, L.P., D.W. Menzel, K.A. Bush, Eds., Coastal and Estuarine Series 2, American Geophysical Union, Washington, D.C., 63-76.

Wang, J.D., V. Kourafalou and T.N. Lee (1984) Circulation on the continental shelf of the southeastern United States. Part II: Model development and application to tidal flow. *J. Phys. Oceanogr.*, **14**, 1013-1021.

Werner, F.E., J.O. Blanton, D.R. Lynch and D.K. Savidge (1993) A numerical study of the continental shelf circulation of the U.S. South Atlantic Bight during the autumn of 1987. *Continental Shelf Research*, **13**, 971-997.

Williams, A. J., III, J. S. Tochko, R. L. Koehler, T. F. Gross, W. D. Grant and C. V. R. Dunn (1987) Measurement of turbulence with an acoustic current meter array in the oceanic bottom boundary layer. *J. Atmospheric and Oceanic Technology*, **4, no. 2**, 312-327.

Numerical Simulation of Tidal Currents with TELEMAC for Olympic Games ATLANTA '96

Jean-Marc Janin[1], Michel Benoit[2], Thomas F. Gross[3]

[1] Research Ing., Laboratoire National d'Hydraulique - EDF, 6 Quai Watier, 78 400 Chatou, FRANCE, jean-marc.janin@der.edf.fr

[2] Head of Maritime Group, Laboratoire National d'Hydraulique - EDF, 6 Quai Watier, 78 400 Chatou, FRANCE, michel.benoit@der.edf.fr

[3] Assoc. Professor, Skidaway Institute of Oceanography, 10 Ocean Science Circle, Savannah, GA 31411, USA, tom@skio.peachnet.edu

ABSTRACT :

Several months before the competition, an agreement was contracted between the LNH (Laboratoire National d'Hydraulique) and the FFV (French Yachting Federation) to provide the French yachting team complete information about the tidal currents in the area of the Olympic events. This cooperation led to the organization of two field measurement campaigns and the realization of a numerical model of this complex zone. The area modelled extends 60 km along the shore line and 40 km across. It is actually very complex not because of the topography offshore, which is relatively smooth, but because of the existence of a large macro tidal delta that affects the flow along the coast. This perturbation is particularly significant in Wassaw Sound where the Alpha Circle was positioned (Alpha Circle = Mistral competition). We introduced in the model all the main channels of the sound, which required more than 10,000 nodes to be correctly reproduced. The boundary conditions were provided by C. Le Provost and M. L. Genco from IMG (Mechanical Institut of Grenoble) who built a tidal spectral model of the Northern Atlantic Ocean. Thanks to measurement campaigns, a good validation of the model could be achieved. The tidal signal in this subtropical area is semi-diurnal with a significant diurnal disturbance. We defined and computed 3 typical tidal conditions; a spring semi-diurnal tide, a neap semi-diurnal tide and a mixed diurnal semi-diurnal tide which occurs in between. The choice of these conditions, which were all encountered during the games, was a compromise between a simple use of the data by the athletes and relatively precise information. As a matter of fact, with these charts and an additional current due to wind, best choices could be made to select the routes during the races.

1   INTRODUCTION :

The purpose of the work we are going to describe was to model the area where the sailing competitions for the '96 Olympic Games took place. Then, with the results given with an optimal shape (enough precise but not too complicated to be easily used by the competitors), the French team could select the best sailing routes. As a matter of fact, tidal dominated currents which occur in this part of the world can greatly modify the performance of such sailing boats.

The area of interest is located close to the town of Savannah (Georgia, USA), in the Wassaw Sound and along the Little Tybee Island coast. This area is quite complex with an irregular coast line and a large delta. Consequently the numerical modeling of unsteady tidal flow is not easy.

2   NUMERICAL MODEL :

The domain has been digitized and meshed from sea charts. Then, TELEMAC-2D, the bidimensionnal hydrodynamic software developed at LNH, has been run on a workstation.

TELEMAC-2D software developed at LNH solves the shallow water equations that govern free surface flow, as soon as you assume that velocities are quasi uniform along the vertical. Such an hypothesis is valid for tidal dominated flow like the one encountered in this area. The resolution is achieved via a finite element formulation. Specific details on the software are given in Hervouet and Janin (1994).

2.1 Construction of the model :

The model built covers an area of about 60 km along the coast and 40 km off the shore and includes the main channels of the delta (see figure 1). The following sea charts has been used to generate the model :

- English Admiralty Chart 2801
  (Outer Approaches to Port Royal Sound and Savannah River)
  large scale to get the bathymetry off shore,

- American chart 11512 (Savannah River and Wassaw Sound)
  that covers the central area,

- American chart 11509 (Tybee Island to Doboy Sound)
  that covers the southern area,

- American chart 11513 (Saint Helena Sound to Savannah River)
  that covers the northern area.

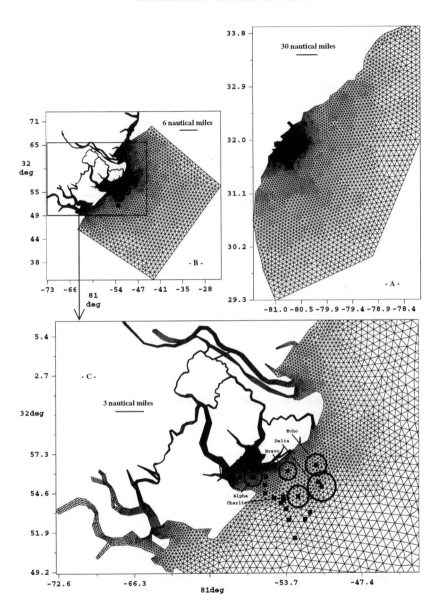

Figure 1 : different meshes used for the simulation, -A- regional mesh, -B- local mesh, -C- zoom of the local mesh in the area of interest

The mesh of unstructured triangles we built contains 16,695 elements and 10,013 nodes. This large number of element is due to the special effort we made to model the delta. The area of interest (Wassaw Sound and Little Tybee Island coast) were also refined and contained about 3,000 nodes. In fact the mesh size was defined as follows :

- in the channels of the delta we have anisotropic elements; the mesh size was set between 50 and 150 m across the rivers in order to guarantee a minimum of 3 elements in that direction; along the flow we have a mesh size between 200 and 500 m chosen in order to have a length/width ratio of about 6;

- in the area of interest (where the competition occurred), we fixed the mesh size at between 100 and 200 m;

- outside of this area we increased the mesh size progressively up to a size of 2 km along the open boundaries

2.2 Boundary conditions :

The boundary conditions we prescribed along the open boundary are tidal waves. We had to be very careful of the quality of this data as these are the main forcing terms in our model (the gravitational attraction of the sun and moon on the water masses inside the domain is totally negligible, considering the size of the domain). The data we used are harmonic parameters of the free surface level given on each node of the open boundary. At each time step we had to prescribe the instantaneous free surface level that we obtained with a recombination of these parameters (our model is actually not a spectral model and solves the primitive shallow water equations).

The data we used has been given by Mr Christian Le Provost and Mrs Marie-Laure Genco from IMG (Institut de Mécanique de Grenoble). They built a fine spectral model of the northern Atlantic Ocean which has calculated the propagation of the 13 main tidal waves. Some intercomparison between large scale models occurred a few years ago and showed that this one gave some of the best results with errors lower than 10 cm on the main tidal wave M2 amplitude (see Le Provost C. and Genco M.L. (1995)).

However, we discovered that the boundary conditions we use were not accurate enough and some spurious reflections appeared in the domain. In fact we had a too large step between the scale of the model used to get the boundary conditions and our model. So, we decided to build an intermediate coarse model that extends from Daytona Beach to Southport (400 km) (see figure 1). We ran this model with boundary conditions given by IMG model as well. Finally, we got a new set of data at the boundary of our local model which behaved much better than the previous one.

2.3 Data for calibration and validation :

Some of the numerical parameters, like bottom friction, have to be tuned to calibrate the model. We performed this task using measurements of both tidal currents and tidal levels at several points inside the domain. Contacts with the NOAA (National Oceanic and Atmospheric Administration) provided us data including tide and current charts for 1995 and 1996. We got especially valuable data around the Savannah National Light Tower located at the entrance of the Savannah River.

In addition two measurement campaigns were conducted in April 1995 and Summer 1995 to complete our data bank, especially with current measurements inside the area of interest. We laid current meters which recorded currents every 5 minutes. Two moorings were selected during the first campaign (Alpha and Bravo circles) and 8 during the second (2 in Alpha, 2 in Bravo, 1 in Charlie, 2 in Delta and 1 in Echo).

When the calibration was completed, we achieved some validation of the results with all the data we got during the 2 campaigns.

The comparison between the model and measured data were in good agreement on the current direction. These results were also consistent with the model of T. Gross (1997) from Skidaway Institute of Oceanography (see fig. 4). On the other hand, the current strength was typically about 1.3 times larger than the modelled currents. The use of several grids, with some interpolations in between that smooth the signal, probably explains this result. However, since this deviation was spatially constant, we could correct for it. Concerning the velocity gradients we got a good behavior of the model, that is to say that we had the right positions of the current maxima and of the recirculation areas, if we refer to T. Gross model.

Globally, results of the model were very satisfactory and appeared to be reliable to be used during the races.

3   MAIN CHARACTERISTICS OF THE FLOW :

3.1 General information :

The currents we observe in this part of the world are dominated by the tides. This means that they are greatly dependent of the kind of tide (semi-diurnal, mixed) and of the strength of the tide (neap, spring).

Most of the other forcing are negligible. The Gulf Stream for instance is located too far from the shore to affect the flow. The fresh water flow that reaches

Wassaw Sound from the 2 channels, Wilmington River and Bull River, creates for sure a constant current, but this one is very low compared to the tidal current.

Only the wind can seriously modify tidal currents. If it blows strongly for several days, it can produce a large deviation of the currents, especially on the surface. We will give some more details about that in paragraph 4.

3.2 Description of the tides :

The tide is a summation of waves with different periods and magnitudes. In the area of interest, 2 classes of waves have a significant magnitude, diurnal waves and semi diurnal waves. The summation of this 2 series of waves can give either a relatively regular semi diurnal, if the diurnal wave is low enough, or a called mixed wave, if not. Such a mixed wave is characterized by 2 consecutive High Water or Low Water levels very different.

Consequently the usual simplified way we have in France to described the tides via the definition of unique coefficient proportional to the tidal amplitude is no longer valid in this part of the world (in France such a coefficient is acceptable because semi-diurnal waves are definitely dominating).

However, as complicated as the tides are along the Georgia coast, we had to find the simplest way to characterize them in order to limit the number of charts the competitors had to deal with. Finally, we chose 3 types of tides which seem to give a satisfying representation of the whole observed tides (see also figure 2) :

- Type I : a weak semi diurnal tide (Low Tide 0.2 m - High Tide 2.0 m)

- Type II : a strong semi diurnal tide (Low Tide -0.2 m - High Tide 2.4 m)

- Type III : a mixed tide (first Low Tide 0.0 m - first High Tide 2.0 m
            second Low Tide 0.0 m - second High Tide 2.3 m)

In chapter 4 we will explain how, from these schematic tidal conditions, we can deduce the current field at anytime.

3.3 Current analysis for each circle :

From an analysis of both the results of the model and the measurements, we can give a general description of the behavior of the flow for each yacht circle. Of course a more complete information will be offered afterwards with current charts calculated every hour, but this description is useful to understand the general behavior of the Wassaw Sound and its vicinity. Also, as we take into account measurements in this general description, we can have an idea of the meteorological effects, that we did not take into account in the model. Later we will discuss more about that.

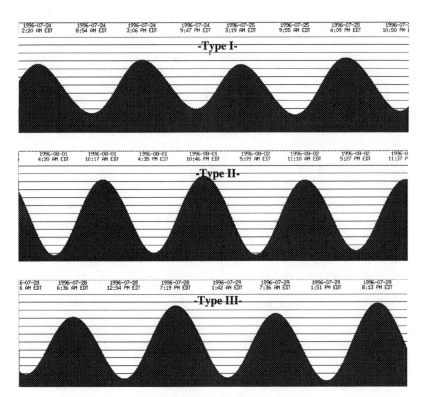

*Figure 2 : different type of tides observed, -Type I- neap tide,
-Type II- spring tide, -Type III- mixed tide*

JOURNEE DU 24 JUILLET 1996

Marée de type I. Pleine Mer à : 02h30 (+2,01 m) Basse Mer à : 08h40 (+0,09 m)
                                    15h10 (+2,16 m)   21h35 (+0,30 m)

8h00	9h00	10h00	11h00	12h00	13h00	14h00
Type I	Type I	Type I	Type I	Type I	Type I	Typε
PM + 5	PM + 6	PM - 5	PM - 4	PM - 3	PM - 2	PM -

14h00	15h00	16h00	17h00	18h00	19h00	20h00
Type I	Type I	Type I	Type I	Type I	Type I	Typε
PM - 1	PM	PM + 1	PM + 2	PM + 3	PM + 4	PM +

Figure 3 : example of daily planning given to the competitors

Alpha Circle (see position on fig.1) :

Currents are mainly dependent on the bathymetry, with fluxes guided by the position of the channels and sandbanks. During the flood, we have a fan-shaped distribution of the currents with 3 possible directions offered by the south, central and north channels. During the ebb, the flow coming from the rivers, due essentially to tidal oscillation and not to fresh water discharge, is distributed mainly on north and south channels. However part of it jumps over sandbanks and reaches the middle of the Sound.

The tides of the 3 defined types are very similar, as only slight differences on velocity intensities are perceptible.

We notice a small delay between the High Water time and the current reversal time. This reversal occurs less that ½ hour later than High Water time given at the Beach Hammoch station (Savannah river entrance). Measurements show a slightly larger delay on the North side.

On the South side, during the flood a current with the following characteristics is encountered : direction WNW - intensity max 0.5-0.8 knot. During the ebb, we have a current : direction ESE - intensity max 0.5-0.6 knot.

On the North side, during the ebb a current with the following characteristics is encountered : direction NW - intensity max 0.5-0.7 knot. During the ebb, we have a current : direction SE - intensity max 0.6-0.7 knot.

Bravo Circle :

This area, like the previous one, is still very close to the shore and the influence of the bathymetry remains important. We observe during the flood that the current is rather facing the shore on the south part of the circle whereas it is parallel to the shore on the north. During the ebb, currents are mainly parallel to the coast all over the circle.

Also, we can see that tides of different kinds still affect the intensity of the flow only.

The reversal is about ½ hour later than High Water. This delay can increase when strong North-Eastern winds are observed.

During the flood the current points at West with an intensity of 0.4-0.5 knot. At the ebb period the flow points at ESE with the same intensity.

During storm conditions (NE wind) observations from NOAA indicate that the reversal occurs more than 1 hour later than High Water. The average flow is about the same as before but lots of fluctuations were measured. In fact, for this circle, close to the shore, we did not get a large modification of the flow due to the storm.

Charlie Circle :

This circle is further from the shore than the 2 previous ones and it is more affected by the general propagation of tidal waves in the north western Atlantic ocean. Consequently, the current is no longer purely rectilinear but rather describes an ellipse during a tidal period.

Larger variations can be observed from one type of tide to the next because diurnal and semi-diurnal tidal waves propagate differently.

The reversal is observed at High Water. During the flood we have the following current : direction NW - intensity max 0.4-0.5 knot, and during the ebb : direction East - intensity max 0.5-0.6 knot.

Delta Circle :

This circle is also far from the shore and like in Charlie Circle, the current traces an ellipse during the tidal period. It goes towards the North West during the flood and turns via North to reach North East and then East during the ebb.

During the flood, the current rotates from 300° to 340° with an intensity between 0.2 and 0.3 knot. During the ebb, the rotation is between 0° and 120° and the intensity 0.3-0.4 knot.

With a north eastern wind storm, flood tidal currents were oriented towards 270° with a strength of 0.4-0.5 knots and ebb tidal currents were at 200° and 0.2-0.3 knots. For this circle the storm definitely affects the flow.

Variations between semi-diurnal and mixed tides were observed in Charlie Circle also occur in Delta Circle.

Echo Circle :

This circle is a bit closer to the coast than the previous ones but not as close as the first two. Consequently we observe a mixed influence of the bathymetry and wave propagation. During the flood, the direction of the current varies between 300° and 320° on the northern and central part of the circle. During the ebb, this direction varies between 75° and 105°.

Variations between the different kinds of tides we simulated are not very important.

3.4 Wind effects :

The wind as we mentioned it in paragraph 1 is the only forcing terms that can largely modify tidal currents, especially on the surface.

As a matter of fact, the wind surface current is definitely not similar to the tidal current (direction and intensity) and the summation of the 2 currents gives a current that can have a large deviation (60°) from the dominant tidal current. This deviation is even larger around the reversal, when the tidal currents are weak.

However, 1 or 2 m under the surface its influence is already greatly decreased. Some measurements has been made during a storm by using both drogues floating on the surface and a current meter 0.8 m under the surface. The modifications of the current compared to the ones with no wind were much larger on the surface than 0.8 m under.

We did not reproduce in our model the wind surface current which is a limitation of our work.

4   USE OF CURRENT CHARTS :

A very practical but important part of the work was to find an easy way to use the results our model produced. We have to keep in mind that the competitors had no time to get instantaneous currents in any complicated way no matter how realistic these were.

In order to get the information quasi immediately, we drew two sets of figures that we will detail blow. The first set contained current charts given every hour for each of the 3 types of tides we had selected. The second set contained daily planning that tells which chart should be used at any time during the day.

4.1 Current charts :

First of all, these charts had to be time referenced. For type I or II, time is referenced from local High Water and charts are given between HW-6h and HW+6h. For type III, time is referenced from the second local High Water and charts are given between HW-17h and HW+6h. As this chart are given every hour, this means that if local High Water happens at 11h00, chart HW-2h should be used between 8h30 and 9h30.

The graphical representation we have adopted to show the currents includes a drawing of vectors and a drawing of colored isosurfaces background. The drawing of vectors gives immediately the direction of the current, it gives also an estimation of its intensity as the size of the vector is proportional to it. However to deduce velocity from the measurement of the size of the vector is not easy and the

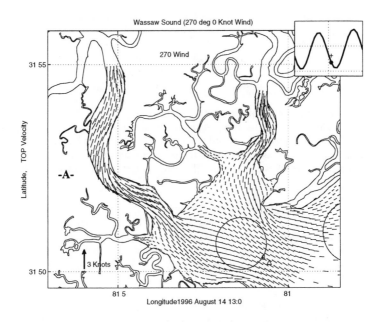

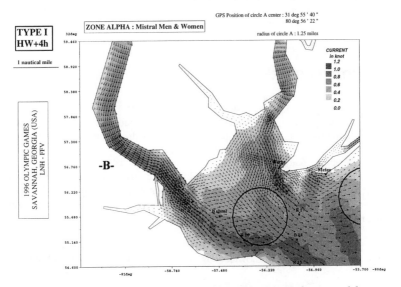

Figure 4 : Current chart during ebb current, -A- Skidaway model
-B- Labratoire National d'Hydraulique model

colored isosurfarces are then much easier to use as they give immediately the range of the velocity (see example figure 4).

In our model tidal flats are included. On these areas, when there is no longer water, no vector is drawn and background is white. So, these areas appear clearly and we could check that their extension is correctly reproduced by the model.

The orientation of the chart is the classical one. The top of it indicates geographic north. The magnetic deviation in this part of the world is low, only 5°, and it can be neglected considering the precision we are looking for. Details that can help the competitors to identify positions on these charts (yacht circles, yellow and NOAA buoys, remarkable landmarks, etc) have been also included.

4.2 Daily planning :

The purpose of this set of figures is to inform very quickly which chart should be used at anytime. We made one figure for each day of the competition (see example fig. 3), from 8h00 AM to 8h00 PM when the races are supposed to take place.

Two types of data must be read in this daily planning. The first is the type of tide encountered on a specific day. The second is the hour period that corresponds to the time of interest. Times of High and Low Water are given in the daily planning as well, with their corresponding water levels.

5   CONCLUSION :

This complete work (modeling, measurements) has been realized in association with French sailing team. We finally provided them the current fields they need to take into account this parameter properly during the competition.

Although, many complex parameters influence the circulation of water masses in this area (mixed tide, delta, tidal flats, etc), we could build an easy tool to get very quickly the desired currents. This point was really appreciated by the team and they have already asked us to perform the same job for the Sydney 2000 Olympics.

In the end, it was a pity than the French team did not succeed during the Atlanta games, but current knowledge is not enough to win such a competition...

ACKNOWLEDGMENT :

This work was partially supported by the French sailing team (FFV). We would like to thank J.M. Gaillard who mainly contributed to this work, M.L. Genco who provided us data from the IMG model and NOAA and SKIO institutes which sent us data.

REFERENCES :

Brown H., Baker D.J. and Wilson W.S., (1994), Tidal Current Tables 1995 - Atlantic Coast of North America, *U.S. Department of commerce - National Oceanic and Atmospheric Administration - National Ocean Service,* Riverdale, MD-USA

Gross T.F. and Werner F.E., (1997), Finite element model of wassaw sound with synthetic marsh flooding boundary conditions, *5th International Conference on Estuarine and Coastal Modeling,* Alexandria, VA-USA

Hervouet J.M. and Janin J.M., (1994), Finite element algorithms for modelling flood propagation, *Work shop on modeling of flood propagation over initially dry areas,* Milano, Italy

Le Provost C. and Genco M.L., (1995), Modèles éléments finis spectraux de la marée sur l'Atlantique nord et l'océan mondial, *Colloque sur la Dynamique Océanique Côtière,* Paris, France (in French)

Schwing F.B., Blanton J.O., Lamhut L., Knight L.H. and Baker C.V., (1984), Ocean circulation and Meteorology off the Georgia Coast - Part 1 and 2, *Georgia Marine Science Center,* Savannah, GA-USA

Surface currents in a 3-D model of Wassaw Sound

Roger Proctor[1] and Rod Carr[2][1]

Abstract

A three-dimensional baroclinic model (Proctor and James, 1996, *J. Mar. Sys.*, *8, 285-295*) is applied to the region surrounding Wassaw Sound, Georgia, USA. The model is used to investigate the role of tides, wind and Savannah River discharge on the circulation. It is shown that the average summer Savannah discharge has a noticeable effect on the surface current ellipses. This effect is small compared to the change in current caused by typical summer winds. However, wind direction can enhance or diminish the impact of the river discharge. Modelled current speed and direction showed consistent agreement with measured currents through the tidal cycle. The model results formed an essential part of the tactical briefing for the British Sailing Team at the 1996 Olympic Games.

Introduction

The use of numerical models to provide tidal current atlases and to aid the interpretation of observations in support of competition yachting is growing. The increasing sophistication of modelling techniques enables detailed hydrodynamic features (e.g. small-scale topographic circulations, resolution of vertical current shear) to be examined and their contributions to spatial and temporal variations in current to be determined, thus providing valuable additional knowledge to the racing strategy. The model study of tidal currents off South Korea, Proctor and Wolf (1990), provided useful information to the British yachting team at the Seoul Olympics in 1988 helping the team to win a Gold Medal in the Star Class.

---

[1][1]Proudman Oceanographic Laboratory, Bidston Observatory, Birkenhead, Merseyside L43 7RA, UK . [2]Royal Yatching Association, RYA House, Romsey Road, Eastleigh, Hampshire SO50 9YA, UK

Tidal currents in Wassaw Sound, Georgia are predominantly semi-diurnal (Werner *et al.* 1993) and, in preparation for the Olympic Games in Atlanta (July/August 1996), a tidal study using a three-dimensional finite element model was carried out (Gross and Werner 1997) and the results made available over the World Wide Web. This clearly showed the variability of tidal currents in the area and the impact of steady winds on the circulation. Prior to this study the region had been subject to little numerical investigation, the model of Werner *et al.* (1993) providing the background tidal dynamics. Gross and Werner (1997) used a barotropic tidal model, making the assumption that baroclinic features, especially the effect of the Savannah River discharge, would not be important to the local circulation. However, pre-Olympic documentation (Windom *et al.*, 1993) clearly shows the Savannah river plume extending into Wassaw Sound. Additionally, at the pre-Olympic competitions in 1994 and 1995, sailors reported that current speed and direction changed markedly within a small area, particularly on race area B (Fig. 1).

The region of Wassaw Sound is characterised by a relatively broad shallow shelf with depths generally less than 10 m, sloping uniformly towards the shelf edge. Within Wassaw Sound, depths are less than 4 m except in the channel of the Washington River where depths exceed 6 m. A large sand bar, parallel to the channel, exists at the mouth of the sound (Fig. 1).

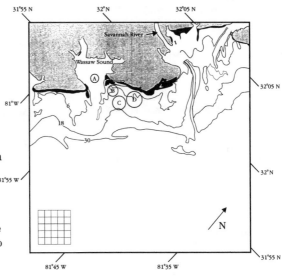

Fig.1 - Regional bathymetry, model domain, race areas, Savannah River (depths in feet)

The shelf to the north is interrupted by the Savannah River navigation channel, which cuts a groove through the gently sloping shelf. The Savanna River is one of the largest rivers discharging into the sea on the US eastern coast.

The British Sailing Team carried out a 14 day study of the currents in the race areas one month before the competition so as to match closely the tidal conditions expected at the games. A grid of 12 fixed buoys were layed and, using 2 fast inflatable boats, current strength and direction were measured at hourly intervals at all buoys. Data were also available from three NOAA buoys, strategically placed by the Olympic organisers.

The work reported here concentrates on the roles that winds and Savannah River discharge have in modifying the tidal motion. The main objective was to anticipate the kind of responses that could be expected if unusual conditions were to occur during the race periods.

The Model

The model, an adaptation of Proctor and James (1996), solves the 3-D incompressible, Boussinesq, hydrostatic equations in spherical polar coordinates with a transformed vertical coordinate, σ. The equations used are

$$\frac{\partial u}{\partial t} = -L(u) + \left(f + \frac{u \tan \varphi}{R} \right) v - \frac{1}{R \cos \varphi} \left(\frac{\partial \Phi}{\partial \chi} - \left[\frac{\partial \zeta}{\partial \chi} + \sigma \frac{\partial D}{\partial \chi} \right] b \right)$$
$$- \frac{1}{\rho_o R \cos \varphi} \frac{\partial P_a}{\partial \chi} + \frac{1}{D^2} \frac{\partial}{\partial \sigma} \left(A \frac{\partial u}{\partial \sigma} \right) \tag{1}$$

$$\frac{\partial v}{\partial t} = -L(v) - \left(f + \frac{u \tan \varphi}{R} \right) u - \frac{1}{R} \left(\frac{\partial \Phi}{\partial \varphi} - \left[\frac{\partial \zeta}{\partial \varphi} + \sigma \frac{\partial D}{\partial \varphi} \right] b \right)$$
$$- \frac{1}{\rho_o R} \frac{\partial P_a}{\partial \varphi} + \frac{1}{D^2} \frac{\partial}{\partial \sigma} \left(A \frac{\partial v}{\partial \sigma} \right) \tag{2}$$

$$\frac{\partial b}{\partial t} = -L(b) + \frac{1}{D^2} \frac{\partial}{\partial \sigma} \left(K \frac{\partial b}{\partial \sigma} \right) \tag{3}$$

where u, v and Ω are the positive velocity components in east (χ), north (φ) and vertical (σ) directions respectively; D is total depth ($= H + \zeta$); ζ is surface elevation above mean sea level; f is the Coriolis parameter; P_a is the atmospheric pressure; $\Phi = $ (total pressure $P - P_a$)$/\rho_o + gz$; b is buoyancy $= g (\rho_o - \rho) / \rho_o$; A is the coefficient of vertical viscosity; and K is the coefficient of vertical diffusivity.

The vertical coordinate $\sigma = (z - \zeta) / (H + \zeta)$, and the nonlinear terms, $L(a)$, are given by

$$L(a) = \frac{u}{R\cos\varphi}\frac{\partial a}{\partial\chi} + \frac{v}{R}\frac{\partial a}{\partial\varphi} + \Omega\frac{\partial a}{\partial\sigma} \qquad (4)$$

with

$$\Omega = -\frac{\sigma}{D}\frac{\partial\zeta}{\partial t} - \frac{1}{DR\cos\varphi}\left[\frac{\partial}{\partial\chi}\left(D\int_0^\sigma ud\sigma\right) + \frac{\partial}{\partial\varphi}\left(D\cos\varphi\int_0^\sigma vd\sigma\right)\right] \qquad (5)$$

also

$$b = \frac{1}{D}\frac{\partial\Phi}{\partial\sigma} \qquad (6)$$

so

$$\Phi = D\int_0^\sigma bd\sigma + g\zeta \qquad (7)$$

Numerical implementation of the model is the same as in Proctor and James (1996) except that here salinity is the prognostic variable, i.e.

$$\frac{\partial S}{\partial t} = -L(S) + \frac{1}{D^2}\frac{\partial}{\partial\sigma}\left(K\frac{\partial S}{\partial\sigma}\right) \qquad (8)$$

with S and T linked to give buoyancy b by an equation of state $b = b(T,S)$ (Gill 1982). In these calculations T is held constant throughout.

One other difference to Proctor and James (1996) is the use on the open sea boundaries of a 'radiation' condition to prescribe the forcing. This takes the form

$$q - (q_T + q_M + q_D) = (c/H)(\zeta - (\zeta_T + \zeta_M + \zeta_D)) \qquad (9)$$

where $c = \sqrt{gH}$.

This allows tidal current and elevation (q_T, ζ_T), meteorologically driven motion (q_M, ζ_M) and density driven motion (q_D, ζ_D) to be input, whilst allowing internally generated gravity waves to propagate out of the model domain. Such a radiation scheme has been used highly successfully in tide and storm surge models (Flather *et al.* 1991). In the present case, the meteorological component is set to zero. On outflow, the density component is specified to be equal to the geostrophic current and elevation given by the density section at the boundary, assuming the current at the sea bed to be zero. On inflow, a zero gradient condition is applied. On closed boundaries there is no normal flow. At the sea surface and sea bed there is zero salt flux.

The Savannah River is represented as a freshwater inflow having a given volume flux at a given salinity extending over a given depth (half the depth at the appropriate grid box). The velocity is calculated from the river width (which may be less than or greater than a gridbox width) to give a momentum flux, which, together with the salinity flux, is input to the advection routines. The corresponding increase in sea level is included in the barotropic routine.

The model has been set up on a 1km by 1km grid of 36 km by 36 km centred on Wassaw Sound. The grid is aligned with the coast at an angle of 045° to North, that is, the direction of alongshore current is oriented positive towards the NE, the onshore / offshore current is positive towards the NW (i.e. onshore). The total depth is represented by 13 σ-levels giving each grid box a thickness, for example, of 1m in 13m of water, 2m in 26m of water. In the race area the water is generally less than 10m deep, so that the surface currents presented here generally correspond to the flows within the top 1m of the water column.

Digital bathymetry for the model domain was kindly provided by Dr. Rick Luettich (University of North Carolina), derived from NOS nautical charts. In some near coastal areas this had to be supplemented by hand using spot depths from high resolution charts (such as 11512 - Savannah River & Wassaw Sound). Tidal forcing around the open boundary of the model was also provided by Dr. Luettich in the form of tidal constants of both surface elevation and depth-averaged current. Seven tidal constants were provided. In most of this study, only one of these, M_2, has been used. This provides the mean tide range such as experienced on 26/27 July 1996 (mid-race period), and allowed comparison with the modelling of Gross and Werner (1997).

Information on the Savannah River outflow was provided by Dr. Jack Blanton (Skidaway Institute of Oceanography). This shows, Fig. 2, that the Savannah discharge is highly variable, with discharges of up to 1400 m³/s occuring (similar to the largest European rivers, the Elbe and the Rhine). However, the July values seem to be consistent with discharges at the lower end of the scale of 200-300 m³/s although occasionally reaching 600 m³/s.

Model Studies

From the beginning of the study it was decided that the results of Gross and Werner (1997), available over the World Wide Web, provided a good description of the mean tidal state, the modelled tidal ellipses agreeing well with our observations. So, to speed up the model development, we chose to concentrate on establishing the mean (M_2) tidal state, tuning the model to

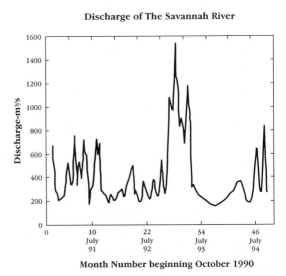

Discharge of The Savannah River

Month Number beginning October 1990

Fig. 2 - Savannah River discharge

provide the best fit with the results of Gross and Werner (1997). These studies with the model have focused on the role that the freshwater inflow and steady state winds might have on the circulation in and around Wassaw Sound.

A) The tidal regime

With temperature and salinity held constant (T=30°C, S=33 psu (Practical Salinity Units)), and zero Savannah inflow, the dominant M_2 tidal motion was established by prescribing the tide around the open boundaries , i.e.

$$q_T = H_{vT} \cos(\sigma_{M2}t\text{-}G_{vT}) \; ; \; \zeta_T = H_T \cos(\sigma_{M2}t\text{-}G_T)$$

where H_{vT}, G_{vT}, H_T, G_T are the amplitude and phase of the component of M_2 current normal to the boundary and M_2 elevation respectively, at grid elements around the open boundary of the model.

The model is initialised to zero motion and then forced by the time-varying open boundary tidal input to generate the tidal motion within the model domain. The model was run for 12 tidal cycles, sufficient to remove the effects of the initial conditions and allow a repeating tidal cycle to be established. The results from the 12th tidal cycle are presented.

Fig. 3 shows a sub-set of the model domain, with approximate race areas (A, B, C and D) marked. Tidal currents at mid ebb (3 lunar hours) are shown. (For computational ease, the model is run in lunar time, so one M_2 tidal cycle takes 12 lunar hours = 12.42 solar hours). Currents in the race area ebb uniformly in an easterly direction with maximum currents (of order 1/2 knot (0.25 ms^{-1})) occuring at mid point of the ebb tide (Fig. 3). There is little variation across the race areas, most variation taking only place in Wassaw Sound at site A, where currents are also smaller. Flood tide currents display a similar pattern, with reversed current direction.

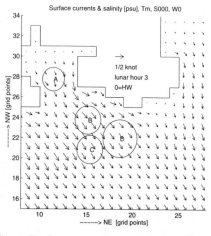

Fig. 3-Surface currents on ebb, M_2 tide

Tidal ellipses at the centre's of the race domains are shown in Fig. 4. (Note: the hour marks shown are lunar hours and that the ellipses are properly oriented with axes directed North and East).

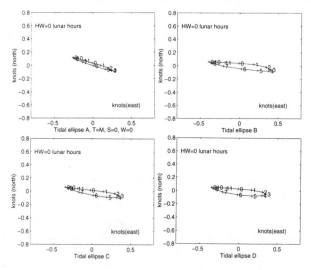

Fig. 4 - Tidal ellipses in race areas, M_2 tide only

Flood currents follow clockwise rotation and appear marginally stronger on the northern side of the race circles (not shown). Race area C does not seem to be affected by the presence of the Wassaw Channel and associated sandbank.

B) Impact of the Savannah River plume on surface currents

For the run including the Savannah discharge, the final values of the 12th tidal cycle were used as initial conditions (to provide a realistic starting point) and the model run for a further 12 tidal cycles with the Savannah inflow of 200m³/s held constant. After this time, the Savannah plume has established a coastal current and the leading front of the plume has passed through the race area. The Savannah discharge, on reaching the sea, was assumed to have a salinity of 20 psu (Blanton, personal communication).

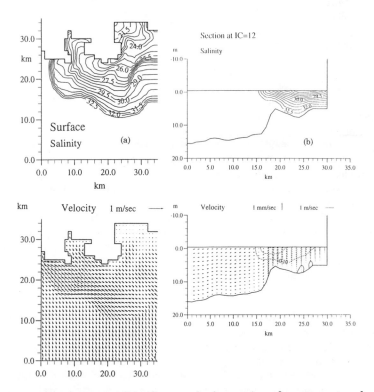

Fig. 5 - Development of the Savannah plume, a) surface current and salinity, b) vertical section of salinity and current across shelf.

The development of the plume is shown in Fig. 5, which gives surface currents and salinities (Fig 5a), and a vertical section across the shelf between race areas A and B at flood tide after 6 tidal cycles (Fig 5b). The sharp front of the plume and the intensification of current near the front can be seen. The vertical section indicates that the plume extends to the sea bed and is confined to the shallowest near shore region.

At the inflow point (column 24, row 35) there is a small increase in the water level (2-3 cm) due to the Savannah River inflow.

Surface currents at mid-ebb are shown in Fig. 6 and current ellipses in Fig. 7. Some variation in surface current is evident. The extent and movement of the plume is also clear. The strongest currents occur at the plume front seawards of the race areas. Tidal straining displaces fresher water offshore with the ebb tide and moves saltier water onshore with the flood. The net effect of the river inflow is the addition of an along shore flow (SW-directed) of 5-10 cm/s (0.1-0.2 knot). The effect though is variable over the tidal cycle with marked cross-race area variations in current. It is also apparent that the Wassaw Channel and sandbank now have an effect on the currents in race areas C and D, particularly close to low water (between lunar hours 4 to 8). Ellipses show more irregular shapes, not returning to the postion at time 0 because of the dynamic imbalance inherent in the unstable river plume. The ellipse at A now rotates anti-clockwise.

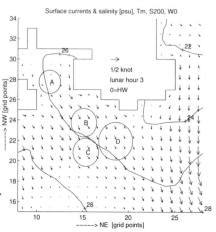

Fig. 6 - Surface currents and salinities on ebb, M_2 tide and Savannah River (200 m³/s)

A simulation was carried out with a discharge value of 600 m³/s to assess the extreme condition. This level of discharge was not expected to occur during the racing periods, the long range weather forecast was for low tropical storm activity over the river catchment, but it was considered better to have an idea of what may happen, even if it didn't! The effect of increasing the outflow to peak rates had little impact on the currents in the race area, because the salinity distribution is similar.

Increasing the discharge increases the area covered by lower salinity water and smoothes and pushes the frontal boundary (between fresh and salt water) further offshore. It seemed reasonable therefore to assume that a variable rate of discharge (within the summer expected range) was unlikely to impact on the currents in the area.

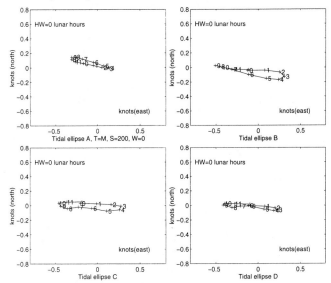

Fig. 7 - Tidal ellipses in race areas, M_2 tide and Savannah River discharge (200 m^3/s).

C) Impact of tidal state - currents at spring and neap tides

Using the mean discharge (200m^3/s) simulations were carried out over separate spring and neap tidal conditions, each simulation run for 12 M_2 tidal cycles. Only two tidal constituents were used to generate the tidal conditions - M_2+S_2 for springs, M_2-S_2 for neaps - accounting for approximately 90% of the variability. Sites A and B have similar ellipses for the three tidal states (neap, mean and spring), with increasing east/west magnitude going from neaps to mean to springs. Sites C and D also display similar ellipse patterns except that spring tides have a mean southerly displacement of ~ 0.1 knot. This arises from the differing salinity pattern at spring tides.

D) The impact of steady winds

Four simulations were also carried out with steady winds added to the tide and mean (200 m^3/s) river discharge. Steady winds of 15

knots from the NE, SE, SW and NW were applied. Current ellipses for
wind directions NE and NW are given in Figs. 8 and 9 respectively.

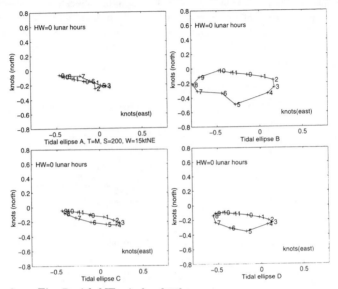

Fig. 8- as Fig. 7 with NE winds of 15kts

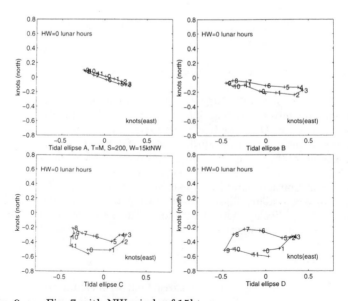

Fig. 9 - as Fig. 7 with NW winds of 15kts

The effect of winds from 2 compass points, SW and SE, is to confine the freshwater from the River to the east of the race area (Fig. 10). This, together with enhanced mixing from the wind, results in the water column being well-mixed and thus similar to the situation shown in Figs. 3 and 4. Winds from the SW are the most frequent in July / August. For the NE wind, most freshwater discharge is confined to the coast but some does pass around to the race area.

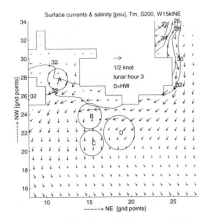

Fig. 10 - as Fig. 6 for NE winds

The most dramatic impact on the river plume is seen with NW winds which cause the plume to stream offshore (Fig. 11), impinging on the race areas C and D. Current response to the winds is generally a displacement in the direction of the wind of about 0.15 knots, approximately 1% of the wind speed, but on occasions (e.g. NW wind at C) up to 0.4 knots. Race area A is least affected under all wind conditions but this must be treated with caution as topographic steering is not included in the windfield and within-estuary conditions could be different to those on the coast.

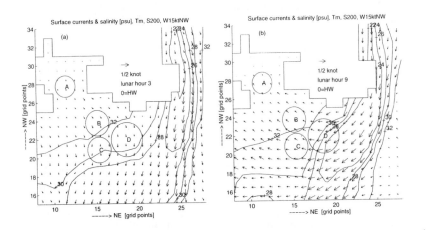

Fig. 11 - as Fig. 6 on (a) ebb and (b) flood for NW winds

A specific point of interest was how an imposed windstress would alter the currents, both initially and during tidal cycles, and how quickly the currents would adjust when the windstress was removed. The above results showed that winds from the NE, SE and SW acted to push the river plume away from the race area, resulting in well-mixed water. Thus it seemed reasonable to consider the current response to wind with no river input in these cases. Fig. 12 shows the surface current response to winds from NE with zero Savannah inflow (S=0). The figure shows the model response to wind over 12 tidal cycles, or 144 lunar hours. This was obtained by subtracting the results of the run without wind from the results of the run with wind. For the first 24 lunar hours the model ran with no wind. After that the steady wind was switched on for 96 hours, then switched off and the model run continued to t=144 hours. The switch on and off points are flagged with 'o'. Also shown are the High Water times at the Savannah River entrance, shown as '*'. This is useful in the explanation of the time series. The figure shows the model response for the 4 race domains. Current directions are consistently in the wind direction (except at A), and are not plotted here.

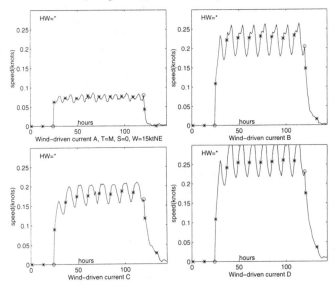

Fig. 12 – Model response to NE winds and zero Savannah discharge

From Fig. 12 it is clear that:

a) current speeds vary with location and wind direction;

b) response to imposed wind stress is rapid, within 1-2 hours whereas the response to removed windstress is slower, 12-15 hours;

c) oscillations of current are present because of the effect of the tidal current, i.e. if the tidal current is opposing the wind direction, the wind-driven current is reduced; if the tidal current is flowing with the wind direction, the wind-driven current is enhanced, hence the semi-diurnal signal.

As expected, oscillations with NE/SW (alongshore) winds are larger than for SE (onshore) winds (not shown).

The situation for NW winds was different, and for that the river flow was retained, so the model was run for NW winds with river input. Fig. 13 shows the surface current response to a NW wind with Savannah inflow (S=200). (Note: Speed scales are different to Fig. 12).

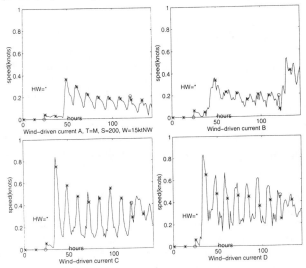

Fig. 13 - as Fig.12 with NW wind and Savannah discharge (200 m³/s)

From Fig. 13, we can see that:

a) currents are larger than in the well-mixed cases because the stratification causes the wind to act on a thinner layer of less dense water;

b) there is a slower response to establishing the wind driven current, especially at A where more than 24 hours is evident;

c) relaxation after the wind is switched off is quite different. This is a result of the stratified water restoring its dynamic balance. This response will be highly dependent on the level of stratification and is

highly time dependent.   Interestingly, at B the current increases
when the wind is switched off. This is because the river plume front is
moved closer inshore and the strong currents associated with the front
pass through this area.

Discussion and Conclusions

The mean (M_2) tidal currents in the race areas are affected by
the presence of the river plume.

The effect is small, with a typical July river discharge of $200m^3/s$
there is a SW-directed alongshore current of 0.1-0.2 knots in addition
to the tidal current, which has values of the order 0.5 knot.

The presence of the river plume introduces greater directionality
into the currents in the race areas and causes more small scale
variability.

The density induced currents may not be significant if winds are
of the order 15 knots, as used in the simulations of Gross and Werner
(1997), where wind-driven currents of 1 knot or more are seen.
However, the scenario of low windspeed is one where the plume
currents may be a significant factor.   Peak discharge has no larger
impact on the race areas.

Neap-spring currents increase linearly in magnitude. Currents at
spring tides are offset to the south by a small amount as a result of the
differing salinity pattern.

With 15 knot steady winds, the wind-driven currents add
approximately 1% of the windspeed to the surface layer velocity, in the
direction the wind is blowing. NW winds produce larger offshore
currents. SE and SW winds act to block the passage of the river water
into the race area, and together with wind-mixing result in a well-mixed
water column. NW winds, on the other hand, have the greatest effect on
the river plume in the race area, strong salinity gradients impacting on
race areas C and D.

The model information was put together with the measured
current data to provide each sailor with an hourly current flow chart to
aid their tactical decisions.   The team were briefed on the basis of
percentage reliability of the current forecast. During the race periods,
the model predictions of current, particularly at times other than the
turn of the tide, were in very good agreement with the measured
currents. At the times of slack water, discrepancies between modelled
and observed currents could be correlated with variable winds and the

number of afternoon thunderstorms which had occured prior to the race, which would have modified the inflow of freshwater to the racing domain. Comparing Fig. 4 (no river flow) and Fig. 7 (mean Savannah discharge) delays of up to 1 hour in the timing of slack water can be seen. Thus, the knowledge that the amount of freshwater inflow to the area was likely to affect slack water timing could be built into the race strategy.

Acknowledgements

Some of this work was carried out during a 10 week visit by RP to the National Institute of Water and Atmospheric Research Ltd. (NIWA), Wellington in June 1996 and the authors gratefully acknowledge their support. We also wish to thank Dr. Rick Luettich (University of North Carolina) and Dr. Jack Blanton (Skidaway Institute of Oceanography) for providing us with important data.

References

Flather, R. A., Proctor, R., Wolf, J. 1991: Oceanographic forecast models. In: *Computer modelling in the environmental sciences.* (Farmer, D. G., Rycroft, M. J. *eds.*) pp. 15-30, Clarendon Press, Oxford.

Gill, A. E. 1982: *Atmosphere-Ocean Dynamics.* International Geophysics Series 30. Academic Press, New York. 662 p.

Gross T.F. and Werner, F.E. 1997: Finite element model of Wassaw Sound with synthetic marsh flooding boundary conditions. This volume.

Proctor, R., Wolf, J. 1990: Modelling the tides for the 1988 Olympic Games. *Ocean Challenge,* 1, 10-14.

Proctor, R., James, I. D. 1996: A fine-resolution three-dimensional model of the southern North Sea. *Journal of Marine Systems 8:* 285-295.

Werner, F. E., Blanton, J. O., Lynch, D. R., Savidge, D. K. 1993: A numerical study of the continental shelf circulation of the U.S. South Atlantic Bight during the autumn of 1987. *Continental Shelf Research, 13, 8/9,* 971-997.

Windom, H., Blanton, J. O., Werner, F. E. 1993: Climatology and Oceanography of the Georgia Coast. Skidaway Institute of Oceanography Report. Unpublished manuscript.

CIRCULATION MODELING AND FORECASTS FOR THE DANISH SAILORS
DURING THE 1996 OLYMPIC YACHTING EVENTS

Jan Borup Jakobsen[1] and Morten Rugbjerg[1]

Introduction

In the spring of 1995 Danish Hydraulic Institute (DHI) was asked, by Dansk Sejl-union (DS – the Danish Sailing Association), to support the Danish yachting team during the Olympic Games held in the summer of 1996 by providing current forecasts for the Sound of Wassaw – the area where the Olympic Yachting Competitions were to take place. The current forecast would, together with weather forecasts made available by the National Weather Service during the competitions, form a solid basis of knowledge about the conditions in which the Olympic sailing events were to be conducted.

The early contact between DHI and DS enabled the sailors to become acquainted with the current forecasts, as the forecasts were not only provided during the Olympic Games, but also (in a simplified form) during the Pre-Olympic Games held in the summer of 1995. Furthermore, measurements made by the yachting team during the Pre-Olympic Games provided DHI with the possibility to validate the current forecast model.

Current Model Setup

DHI's generalized hydrodynamic modelling system, MIKE 21, which is based on the two-dimensional shallow water equations, was used to provide the current forecasts. The solution technique applied in MIKE 21 is the alternating direction implicit (ADI) technique known for its robustness and reliability (Abbott et al. 1981).

To obtain a fine resolution of the flow field in an area of interest, like the area around Wassaw Sound, and at the same time limit the computational efforts

[1] Danish Hydraulic Institute
Agern Allé 5
2970 Hørsholm

required, a nesting facility is available in MIKE 21. Nested areas are dynamically coupled, which means that the governing equations are solved for all areas simultaneously. Thus, not only can information from the coarser grid areas affect the finer grid areas, but also information obtained in the finer grid can propagate back into the coarser grid areas. This nesting facility was applied for the current forecast model as described below.

Another facility available in MIKE 21 and applied in the present forecast model is the possibility of flooding and drying computational cells. This facility is very important in order to satisfactorily describe the flow in shallow water areas, like Wassaw Sound, with significant tidal amplitudes.

In MIKE 21 tidal waves are simulated through the boundary conditions typically prescribed as water level variations at all open boundaries. And as the flow through Wassaw Sound and the surrounding stretches of water is dominated by tide, the location of available tidal stations was the most important factor when determining the location of open model boundaries. Thus an open boundary was placed at Charleston and another at Sapelo Sound as tidal harmonic constituent data was available at these two locations from the United States National Ocean Service, obtained through Admiralty Tide Table (Hydrographer of the Navy, 1996). Four tidal constituents (M_2, S_2, O_1 and K_1) were used to calculate the time varying water levels at these two boundaries (boundaries 1 and 2 in Figure 1), using the method specified in Admiralty Tide Table. The water level boundary conditions at boundary 3 were calculated as an interpolation of the water level at the outermost points of boundaries 1 and 2.

The model covered an area of 92 km by 183 km extending about 80 km away from the coast. Data from sea charts was used to create the model bathymetry. One level of nesting was applied with a grid size of 600 m in the coarse grid and a grid size of 200 m in the fine grid.

The wind applied in the model was constant within the area and varied only in time.

According to Skidaway Institute of Oceanography (1996) the fresh water plume from the Savannah River extends down to Wassaw Sound during certain wind conditions. Furthermore, the plume is influenced by the Coriolis effect, which will deflect the plume to the south of the river mouth. This baroclinic flow component will have a minor effect on the flow pattern in the area off Wassaw Sound. However, as the mixed diurnal/semidiurnal tide is the dominating phenomenon in the area, and the tide is capable of reversing the current direction in the racing area daily (or twice a day), it is assumed that the mixing of the plume and the seawater takes place north of the area of interest and that the density effects can be neglected.

The tidal water level variation in the Savannah River in the city of Savannah is included in the model, thereby taking into account the discharge from the river. The same approach was used for the Colleton River.

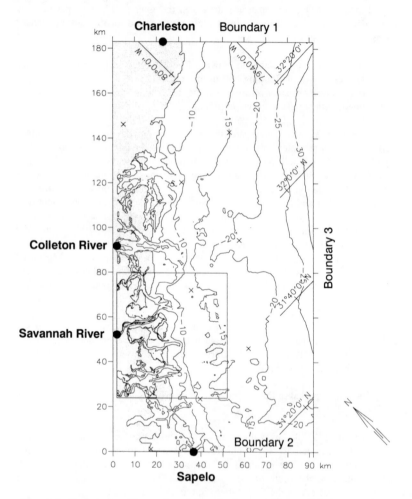

Figure 1: Bathymetry. The enclosed area is the nested area with finer grid spacing. Isolines indicate depth contours in meters.

Initial Model Calibration and Current Forecasts during the Pre-Olympic Games

For the Pre-Olympic Games, the model was run several weeks in advance applying a constant wind of 3 ms^{-1} from 140°, which according to data collected by the coach of the Danish yachting team describes the prevailing wind in the area in July/August. Maps showing hourly flow patterns during the daily competitions were produced on transparencies, which could be put on top of the sea chart covering the area, and were given to the Danish yachting team before they left Denmark. Also tables, one for each race area with hourly current speed and direction, were produced and given to the Danish yachting team.

During the Pre-Olympic Games, a time lag of about 50 minutes between forecasted and measured currents in the racing area was found. This time lag was corrected in the forecasts made a year later for the Olympic Games by gradually changing the prescribed phase for all tidal constituents towards the outermost points for boundaries 1 and 2.

Current forecasts during the Olympic Games

Each morning Danish time (i.e. 02 EDT), before starting the production of the current forecast on DHI, a daily fax from the Danish coach with a weather forecast and comments about the current forecast for the previous day was studied. The weather forecast received by fax was produced by the National Weather Service. As a secondary source, a weather forecast was available from Savannah Airport. Furthermore, on-line measurements of wind and current were available from three buoys placed off Wassaw Sound by NOAA Data Buoy Center (NDBC). All these data were accessible via the Internet and together provided the basis for the wind input to the current forecast.

In order to produce the daily current forecast, the model was run every day for a period of 48 hours. During the first 34 hours the actual measured wind from the buoys and from the airport was used. For the last 14 hours the wind from the weather forecast was used, taking special care in getting the onset of the land breeze properly represented, as it would significantly affect the current conditions in the afternoon when the competitions were taking place. For the whole period hourly wind speeds and directions (constant in space) were specified to the model. The model interpolated the hourly values linearly in time. The last 5 hours of the current forecast covered the actual time of the races.

When the production of the daily current forecast was finished, data were extracted and the hourly current plots and tables produced. An example of the flow field is shown in Figure 2. The current forecast was then faxed to the Danish coach, who received the forecast early in the morning (06 EDT) due to the time difference of 6 hours.

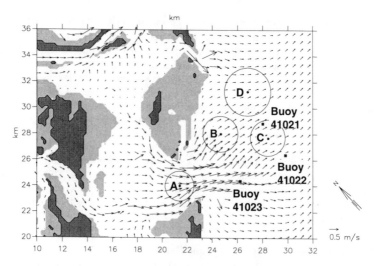

Figure 2: Flow field on 26 July 1996 09 EDT. Circles A,B,C and D indicate the racing areas. Light grey areas represent exposed (dried) mud flats while dark areas represent land areas, which are not subject to flooding and drying.

During the Olympic Games, a daily evaluation of the current forecasts was performed at DHI. Two different sources of information were used for the evaluation:

- A subjective evaluation based on the feedback from the sailors was received, as part of the daily correspondence with the Danish coach.
- An objective comparison of forecasted and measured current speed and direction was possible due to the hourly measurements from the on-line buoys.

Figure 3 shows a comparison between forecasted and measured current speed and direction. Good agreement between forecasted and measured turning of the tide is observed, keeping in mind the sampling rate of one hour. Generally, the forecasted current speeds compared well with the measured ones. However, during certain periods the forecasted current speeds were somewhat smaller than the measured ones as also observed by the sailors.

For comparison a model run was performed for the entire period of the Olympic Games with a 'standard' wind, consisting of a land-sea breeze with speeds ranging from 1 to 4 ms^{-1}. Comparing the current forecast based on the 'standard' wind with the current forecast based on the measured/forecasted wind gives a good indication of the importance of the wind. As seen from Figure 3, the current is, as expected, dominated by the tide, but the difference in wind speed and direction between the measured/forecasted wind and the standard wind can alter the direction of the current up to 20°.

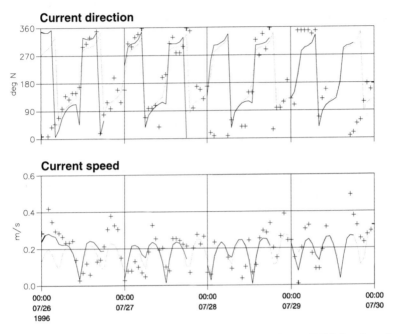

Figure 3: Current direction and speed at location of buoy 41022 for four days. Current forecast with standard wind is shown with a dotted line, current forecast with measured/forecasted wind is shown with a full line (only last 18 hours for each forecast) and measurements of current are shown with crosses.

Given the two facts, that a spatially constant wind was applied and the model was depth-integrated, the discrepancies found between forecasted and measured currents were qualitatively understandable and quantitatively acceptable. The continuing evaluation thus indicated a satisfactory performance of the current model setup and no further adjustments were found necessary.

Conclusions

After the Olympic Games a meeting between the forecast team at DHI and the coach for the Danish yachting team was held in order for the forecasters to get feedback from the sailors regarding the quality and usefulness of the current forecasts. In general the sailors made use of the current forecasts and would not have wanted to be without them.

In order to make this kind of operational setup, it proved to be very useful to have readily and accessible data such as the daily weather forecasts and current measurements from the NDBC on-line buoys off Wassaw Sound. Because of the easy accessibility of these data through the Internet it turned out to be an advantage to do the current forecast at DHI due to the time difference between Denmark and Savannah.

The Danish yachting team won one gold medal during the Olympic Games, and came very close to winning another two medals

References

Abbott, M.B., McCowan A.D., & Warren, I.R. 1981. "Numerical Modelling of Free-Surface Flows that are Two-Dimensional in Plan", Transport Models for Inland and Coastal Waters, Academic Press, New York, pp. 222-283.

Hydrographer of the Navy, 1996. "Admiralty Tide Table", Taunton, U.K.

Skidaway Institute of Oceanography. Savannah, 1996, "Sailing Savannah". Published by Office of the Mayor, City of Savannah, Georgia, USA.

Use of a Hydrodynamic Model for Establishing a Minimum Freshwater Flow to the Lower Hillsborough River Estuary

XinJian Chen[1] and Michael S. Flannery[2]

Abstract

Decreased freshwater discharge from an upstream reservoir has contributed to ecological problems in the Lower Hillsborough River, a narrow tidal river system in Tampa, Florida. On the average, no water is released from the reservoir dam for nearly half the year, resulting in a brackish zone near the dam with low dissolved oxygen concentrations and salinity levels that are harmful to freshwater organisms. In order to keep the river in a biologically healthy condition, it is necessary to establish a minimum freshwater flow released from the dam.

As a step in the process of determining the minimum flow, the effects of the freshwater inflows on salinity distributions in the river are being studied. This study developed a laterally averaged model to simulate hydrodynamics and salinity transport in the river. The model solves the governing equations of hydrodynamics and salt transport in a x-z grid system. It has boundary-fitting ability and can simulate the wetting-drying phenomenon of the river bank. The model was calibrated and verified using measured real-time data collected by the United States Geological Survey (USGS.) The model was used to simulate a series of minimum flow scenarios to evaluate the effects of different freshwater releases on salinity distributions in the river. These scenarios included diverting a portion of the discharge from an artesian spring located 3.5 km below the dam to the base of the dam.

Introduction

The Lower Hillsborough River, located in the City of Tampa, Florida, has a riverine length of about sixteen kilometers (Figure 1). The average depth is around 3 meters (the

1. Surface Water Improvement and Management (SWIM) Section, Southwest Florida Water Management District, 7601 Highway 301 North, Tampa, FL 33637.

2. Environmental Section, Southwest Florida Water Management District, 2379 Broad Street, Brooksville, FL 34609.

deepest area is about 6.2 meters.)
The average width is about 40
meters at mean sea level. The
river starts at the Hillsborough
dam, which releases fresh water
to the river, and ends near Platt
Street to meet Hillsborough Bay.
About 3.5 km downstream from
the dam, water from Sulphur
Springs enters the river laterally
via a short side channel.
Hydrodynamics and salinity
distributions in the river are
controlled by (1) tide at Platt
Street, (2) salinity distribution at
Platt Street, (3) flow over the
dam, (4) flow from Sulphur
Springs, and (5) storm water
runoff below the dam.

        The river is stratified
most of the time due to relatively
weak tides at Platt Street. The   **Figure 1** Lower Hillsborough River
stratification is further enhanced
by the lateral spring water input to the top layer of the river. Near the inflow point of the spring
flow, salinity can increase from almost freshwater at the top to about 15 ppt near the bottom
within a depth of just two meters. Because of the narrowness, a vertical two-dimensional model
should be able to describe the physical characteristics of the river. However, most existing
vertical 2-D hydrodynamic models in the public domain (e.g., CE-QUAL-W2) cannot be
directly used on the Lower Hillsborough River because their turbulence mixing algorithms are
too rough and cannot properly handle the unique features of Sulphur Springs. Hence, this study
decided to independently develop a laterally averaged model for estuaries (LAMFE) which uses
a TKE model (Sheng and Villaret, 1989, Chen, 1994) to calculate vertical eddy
viscosities/diffusivities and can be used to simulate hydrodynamics and salinity transport in the
Lower Hillsborough River. The model code was written and compiled from "scratch", and was
calibrated and verified using field data collected by the USGS along the river. The verified
model was then used to conduct a series of scenario tests for various management options. The
effects of freshwater releases from the dam, spring flow, as well as routing a portion of spring
flow to the base of the dam, on salinity distributions in the river were studied in the scenario
runs.

Field Data

        Real-time stage data with a 15-minute interval collected by the USGS at three stations
(Platt Street, Sligh Avenue, and the 22nd Street) for the period from November 1981 to
September 1982 were available for this study. For the same time period, hourly mid-depth
salinity was also measured by the USGS at Platt Street, Columbus Avenue, Sligh Avenue, and

the 22nd Street. Because Platt Street is near the mouth of the river, measured data at this station were used as boundary conditions in the simulation, while field data at other stations were used to calibrate and verify the model.

The USGS also provided daily spring flow from Sulphur Springs and daily discharge data from the dam. Discharges at the dam were used as the boundary condition for the upstream boundary, while spring flow was input to the model as a point source to the river. Another fresh water input to the river is storm water runoff. In order to take account the effects of runoff during rainfall events on salinity in the river, runoff calculated using rainfall data collected at the Lowry Park station were used in the model simulation.

All these data were graphically presented in Chen (1997). Not all of the data are usable because there exist some bad/missing values in the data set. Since an objective of the study was to evaluate how the release of fresh water from the dam affects salinity distributions in the Lower Hillsborough River, especially for the portion of the river upstream of the spring, periods with large release rates from the Hillsborough dam were not used in the simulation because large release rates keep the most upstream station (22nd Street) fresh all the time. Based upon these considerations, this study chose data collected in November 1981 and a three-month period in 1982 (April through June) for the model calibration and verification. For simplicity, only the three-month data collected in 1982 are presented and discussed (Figures 2 through 4).

Some physical characteristics of the river can easily be seen from the measured discharge, rainfall, salinity and surface elevation data:

1. Tide is the driving force for the river. Surface elevations measured at all three stations show significant tidal variation.

2. From Platt Street to Sligh Avenue, tidal data only show minor damping and time lag, indicating small friction of the river bottom for this river range. From Sligh Avenue to 22nd Street, some small damping and time lag in tidal data can be seen.

3. Salinity in the river also shows strong tidal signals, especially from Sligh Avenue downstream. For the region near the dam, salinity varies with tide more significantly when discharge rates from the dam are small than when discharge rates are large. In the latter case, the discharge from the dam pushes salt water further downstream and keeps water near the dam fresh. If discharge from the dam is increased to about 1000 cfs or greater, fresh water can reach Platt Street.

4. Runoff from rainfall on the drainage basin below the dam can strongly affect salinity distributions in the river. Because the drainage basin for the Lower Hillsborough River is a well-developed urban area (City of Tampa), time of concentration is short and runoff coefficient can be as high as 0.63 (HSW, 1992). As an example, for a storm event with a rainfall intensity of 1 inch per hour and a duration of 1 hour, the direct runoff to the river would be on the order of 1000 cfs for 3 to 4 hours.

5. While discharge from the dam can vary from 0 cfs to thousands of cfs, discharge out of the spring has little variation. Historical data show that discharge from the spring typically varies between roughly 20 cfs and 60 cfs, with an average of 31 cfs.

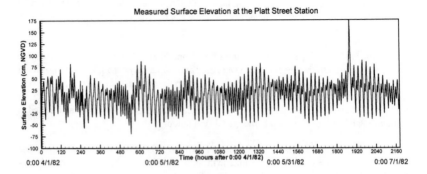

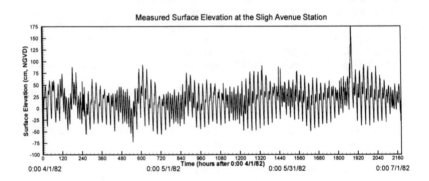

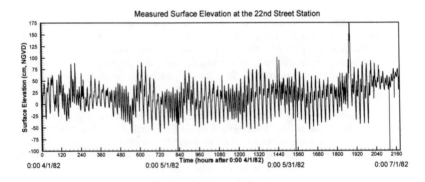

Figure 2 Real-time surface elevation data in the Lower Hillsborough River, April - June 1982.

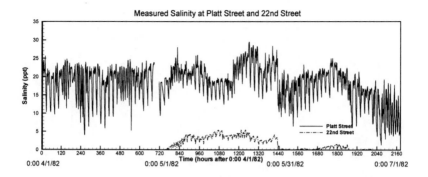

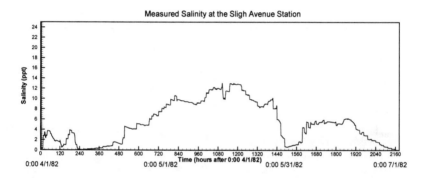

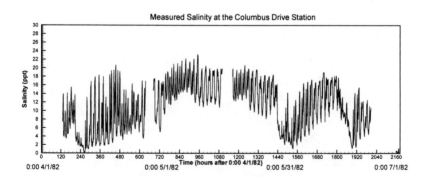

Figure 3 Real-time salinity data in the Lower Hillsborough River, April - June 1982

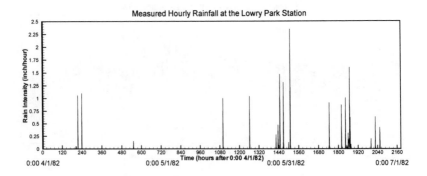

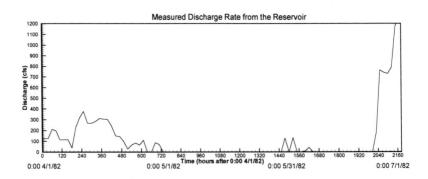

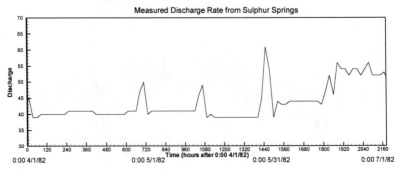

Figure 4 Rainfall, discharge rate from the Hillsborough dam and flow rate from the Sulphur Springs, April - June 1982

Numerical Model

By considering the balance of water, momentum, and salt content in a small cubic with a dimension of $\Delta x \times \Delta y \times \Delta z$, the following governing equations can be derived:

Continuity Equation:

$$\frac{\partial ub}{\partial x} + \frac{\partial wb}{\partial z} = v \tag{1}$$

where u and w are velocities in x- and z-directions, respectively, v is the velocity for lateral input (direct runoff, tributary, etc.), and b is the width of the estuary.

Equation for the free surface is

$$\left(h\frac{\partial b}{\partial h} + b\right)_f \frac{\partial h}{\partial t} = \frac{\partial}{\partial x}\int_{h_o}^{h} ubdz + \int_{h_o}^{h} vbdz + r \tag{2}$$

where t is time, h is the surface elevation, h_o is the bottom elevation, r is the rain intensity in cm/sec, and the subscript 'f' denotes the free surface

Momentum Equation:

$$\frac{\partial u}{\partial t} + u\frac{\partial u}{\partial x} + w\frac{\partial u}{\partial z} = -\frac{\tau_{wx}}{\rho b} - \frac{1}{\rho b}\frac{\partial pb}{\partial x} + \frac{1}{\rho b}\frac{\partial}{\partial x}(\rho bA_h\frac{\partial u}{\partial x}) + \frac{1}{\rho b}\frac{\partial}{\partial z}(\rho bA_v\frac{\partial u}{\partial z}) \tag{3}$$

where ρ is density, A_h and A_v are horizontal and vertical eddy viscosities, respectively, τ_{wx} is the wall shear stress, and p is pressure which depends on elevation z and salinity s:

$$p = g\int_z^h \rho(z,s)dz \tag{4}$$

The wall shear stress τ_{wx} is assumed to follow the quadratic law:

$$\tau_{wx} = \rho C_w u\sqrt{u^2 + w^2} \tag{5}$$

where C_w is the friction coefficient for the wall.

The vertical eddy viscosity A_v is calculated by solving the turbulent kinetic energy equation from the velocity gradient, while the horizontal eddy viscosity A_h is assumed to be time-independent and is limited by cross-section length scale.

Boundary conditions specified in the z-direction are shear stresses. At the free surface, shear stress is specified by wind shear stress. At the bottom, it is assumed that turbulence is fully developed and a log--layer distribution of velocity can be used to calculate the bottom shear stress:

$$\tau_b = \rho [\frac{\kappa}{\ln(z_b/z_o)}]^2 u_b \sqrt{u_b^2 + w_b^2}$$

(6)

where κ is the von Karman constant (0.41), u_b and w_b are horizontal and vertical velocities, respectively, at a level z_b near the bottom, $z_0 = k_s/30$ and k_s is the bottom roughness..

In the x-direction, boundary conditions are specified with either the free surface elevation or velocity. If surface elevation is specified, velocity at the boundary is calculated from Equation (3) with the assumption $\partial u/\partial x = 0$.

Salinity Equation:

$$b\frac{\partial s}{\partial t} + \frac{\partial ubs}{\partial x} + \frac{\partial wbs}{\partial z} = \frac{\partial}{\partial x}(B_h b \frac{\partial s}{\partial x}) + \frac{\partial}{\partial z}(B_v b \frac{\partial s}{\partial z}) + vs_o$$

(7)

where s is salt concentration, B_h is the horizontal diffusivity, B_v is the vertical diffusivity, and s_o represents salt content in tributaries.

For the top layer, the above equation becomes

$$\frac{\partial \eta bs}{\partial t} + \frac{\partial ub\eta s}{\partial x} = (wbs + B_v b\frac{\partial s}{\partial z})_- + \frac{\partial}{\partial x}(B_h \eta b\frac{\partial s}{\partial x}) + \eta vs_o + rb_f s_r$$

(8)

where η is the thickness of the top layer, the subscript '-' denotes the bottom of the top layer, and s_r represents salinity in rainfall (0 for default).

Once salinity is solved, the following state equation is used to calculate the density:

$$\rho = \frac{P+1}{\alpha + 0.698P}$$

(9)

where P and α are functions of temperature (T) and salinity (s):

$$p = 5890 + 38T - 0.375T^2 + 3s$$
$$\alpha = 1779.5 + 11.25T - 0.0745T^2 - (0.38 + 0.01T)s$$

The finite difference method is used to solve the above equations numerically. Rectangular grid with z-level are used to derive the difference equations. Although a z-level model requires a lot of programming efforts due to the complexity in treating the free surface,

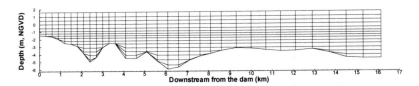

Figure 6. Grid system for the Lower Hillsborough River.

it does have the advantages of reducing numerical diffusion (Sheng et al., 1989) and automatically taking care of the wetting-drying phenomena (Casulli, 1990) of the river bank. Therefore, this study chose to use z-level, instead of σ-level, which has been widely used (e.g., Perrels and Karelse, 1981, Blumberg and Mellor, 1987, Chen and Sheng, 1994). The model first solves the continuity equation to get the new surface elevation, then it updates salinity distribution by solving the salinity equation. Finally, the momentum equation is solved to update the velocity field.

Model Calibration/Verification

To apply the model to the Lower Hillsborough River, a grid system shown in Figure 5 was used to discretize the river. Thirty-two grids ranging from 300 meters to 840 meters in length were used along the river and sixteen vertical grids were used to resolve the depth.

Salinity and surface elevation data measured at the Platt Street station were used as boundary conditions at the downstream end of the river. Because salinity is significantly stratified at Platt Street and was measured only at the mid-depth, measured data can not be directly used in the simulation because it requires real-time vertical salinity profiles to represent the real situation. This study developed a pre-process program to estimate salinity profiles at Platt Street from the mid-depth salinity data based on the assumption that the stratification at Platt Street is a function

Figure 5 Drainage basin of the Lower Hillsborough River and its 17 sub-basins.

of the total freshwater release from the dam and from the spring in the previous day. Using salinity profile data collected during another study (WAR/SDI, 1995), this relationship between stratification at Platt Street and fresh water release was obtained (Chen, 1997.)

Rainfall data at the Lowry Park station was used to estimate runoff below the dam. It is assumed that rains were uniformly distributed over the entire drainage basin. By incorporating a previous study conducted by HSW Engineering, Inc.(1992), the drainage basin was divided into 17 sub-basins to take into account the variations in runoff characteristics among sub-basins (Figure 6). Hourly runoff from each individual sub-basin was calculated from the sub-basin area, hourly rainfall, and its runoff coefficient. The hourly runoff was then distributed during a sixteen-hour period using the unit hydrograph method, which can be estimated from the HSW study(1992). The final freshwater input to the river from an individual sub-basin during the simulation time step was added to its corresponding river segment in the model.

For the upstream boundary, a uniform velocity distribution was assumed and calculated from the discharge data and the instantaneous cross-section area. Based on water chemistry data available for the reservoir, a salinity value of 0.1 ppt was assigned to waters released from the dam.

A uniform discharge distribution was assumed for the spring water inflow. However, the uniform lateral discharge to the river was distributed only at the top portion of the water column (about -2.0 feet, NGVD, and above), because the bottom of the spring is about -2.0 feet, NGVD. Historical data show that the spring has a salinity value varying around 1.5 ppt. Thus, a constant of 1.5 ppt is assigned to the spring flow.

Because there are no synoptic data available for the simulation, an arbitrary initial condition for the salinity/velocity field was assumed. This is acceptable because the problem is mathematically dependent more on boundary conditions than on initial conditions. However, a spin-up run is needed to get a close-to-real initial salinity/velocity conditions. The time period for the spin-up run is about a week. After the spin-up run, effects of the arbitrary initial conditions can be eliminated and the real model run can then be conducted.

The three-month period (April - June in 1982) were divided into four periods, each with 20 to 23 days. The second period (from 0:00, 4/1/1982 to 24:00, 5/10/1982) was used for model calibration, while the rest of the periods were used for model verification. Figures 7 and 8 show the comparisons between simulated and measured surface elevation, while Figures 9 and 10 show the comparisons of simulated and measured mid-depth salinity. Again, for simplicity, only comparisons during the second and fourth periods are presented here.

It can be seen from the comparisons that the model works well. Simulated surface elevations are almost identical with measured data, while simulated salinity agree with field data reasonably well. The calibrated model parameters were used during the verification process and still produced satisfactory agreement between model results and field data. Due to the large amount of bad data in measured salinity at the Sligh Avenue station (Figure 3), no comparison of simulated salinity with data for this station is presented here. Nevertheless, the overall trend of simulated salinity is the same as that of measured data for the Sligh Avenue station (Chen, 1997).

One important finding was that the effects of rainfall on salinity distributions cannot be neglected and runoff below the dam must be considered in the model. From the rainfall data,

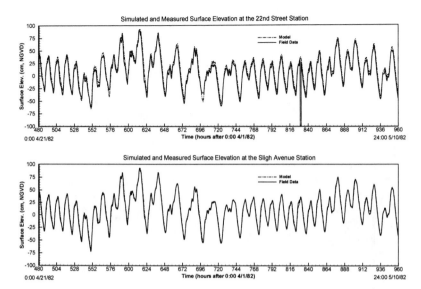

Figure 7 Comparison of simulated and measured surface elevation, April 21 -May 10, 1982.

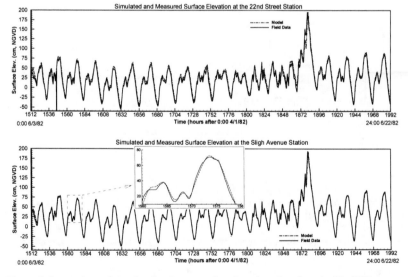

Figure 8 Comparison of simulated and measured surface elevation, June 3 - 22, 1982.

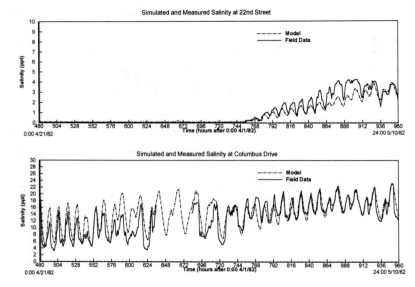

Figure 9 Comparison of simulated and measured salinity, April 21 -May 10, 1982.

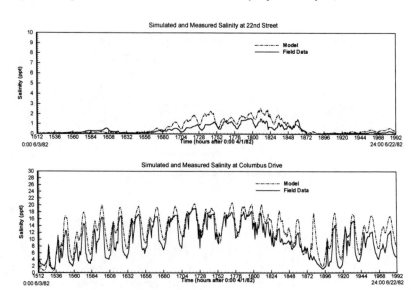

Figure 10 Comparison of simulated and measured salinity, June 3 - 22, 1982

it can be seen that a storm event occurred on June 12th, followed by a few storm events on June 16th and 17th. As a result, data show a significant increase in water level and a steady decrease in salinity at all measurement stations. These changes were due to local runoff as no water was released from the reservoir during this period and spring flow showed little variation. To examine the effect of local runoff on salinity, a model run was conducted that did not include any runoff from the basin below the dam (Figure 11). It can be seen from this figure that without considering the runoff effect, simulated salinity time series at the 22nd Street and Columbus Drive cannot represent the real condition. The slight steady decrease in simulated salinity time series is due to the steady decrease in the salinity boundary condition at Platt Street (measured data). Model experiments show that only when the runoff below the dam was included could a good agreement between data and simulated salinity be reached.

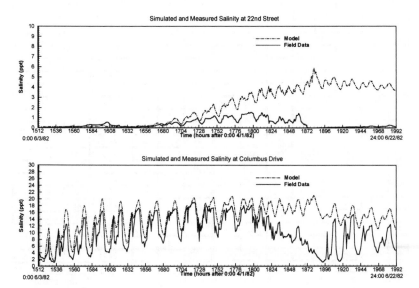

Figure 11 Simulated salinity without considering the effect of stormwater runoff, June 3 - 22, 1982.

Scenario Studies and Conclusions

After the model was verified, it was used to conduct scenario tests to evaluate how different freshwater releases from the reservoir affect the salinity distributions in the river. Based upon possible variations of spring flow, water releases from the dam, as well as the possibility of routing a portion of spring water to the base of the dam, thirty-eight scenario runs were conducted.

All runs used the same model parameters determined in the model calibration process. Again, measured surface elevations and salinity at Platt Street were used as boundary conditions. The time period in which field data were used as boundary conditions was the first

18 days of the third period of the 1982 data (from May 11 through May 28). The reason to use this time period is that very little rainfall was received during this time span and the tide around May 28 was the spring tide period.

In order to isolate the effect of freshwater releases on salinity distributions in the river, runoff caused by rains were turned off in the scenario test runs. The final results for each run include velocity and salinity fields for 48 hours with a 30-minute interval, as well as volumes of water for various salinity ranges. Water volumes of various salinity ranges for each run are presented in Table 1. Figure 12 shows simulated salinity fields at the high and low tides for one of the 38 runs.

River Flow	Sp. Fl. at Dam	Spring Flow	Salinity Ranges					River Flow	Sp. Fl. at Dam	Spring Flow	Salinity Ranges				
			0 - 1	0 - 4	4 - 11	11 - 18	> 18				0 - 1	0 - 4	4 - 11	11 - 18	> 18
(cfs)	(cfs)	(cfs)						(cfs)	(cfs)	(cfs)					
0	0	20	0.0	0.0	298.2	997.8	1518.7	15	20	20	37.5	231.5	486.3	1143.4	960.4
10	0	20	0.0	55.0	405.7	1101.0	1255.6	0	0	31	0.0	4.9	442.3	1098.8	1270.9
20	0	20	28.5	156.9	421.6	1151.3	1089.6	5	0	31	0.0	45.7	461.6	1140.0	1170.8
40	0	20	110.9	299.5	518.0	1104.8	900.6	10	0	31	0.0	130.2	437.4	1153.6	1097.9
80	0	20	283.5	513.9	698.2	887.3	729.5	15	0	31	19.4	171.5	456.5	1161.6	1030.5
100	0	20	342.0	612.0	735.2	805.3	679.0	20	0	31	39.7	209.4	478.3	1147.9	985.4
0	0	40	0.0	21.0	517.5	1151.1	1128.8	30	0	31	84.9	285.6	525.6	1110.5	901.0
10	0	40	3.1	178.5	481.8	1156.9	1003.2	40	0	31	127.3	359.8	570.6	1050.0	844.0
20	0	40	52.0	257.3	526.0	1119.4	919.7	60	0	31	214.4	464.0	665.9	936.4	761.1
40	0	40	159.5	413.9	627.7	986.6	797.5	80	0	31	295.3	557.2	723.1	849.6	700.2
80	0	40	311.2	594.0	733.7	819.4	683.9	100	0	31	361.3	660.9	743.8	771.3	656.6
100	0	40	372.9	697.1	760.2	734.4	641.9	0	10	21	0.0	29.8	422.9	1098.1	1266.3
0	10	10	0.0	12.6	298.7	978.0	1525.7	5	10	21	0.0	101.5	407.6	1144.5	1164.6
5	10	10	0.0	46.5	351.7	1041.7	1376.4	10	10	21	1.6	144.3	422.5	1155.7	1096.8
10	10	10	0.0	89.4	370.7	1105.7	1251.6	15	10	21	28.9	178.8	447.5	1160.8	1033.0
15	10	10	16.3	131.3	387.5	1140.2	1159.5	0	15	16	0.0	47.5	402.9	1095.6	1271.1
0	20	20	0.0	117.7	419.2	1151.8	1129.9	5	15	16	0.0	120.2	398.3	1157.8	1142.0
5	20	20	0.0	157.9	437.5	1164.5	1059.8	10	15	16	3.6	151.6	415.9	1159.4	1092.3
10	20	20	5.6	193.8	463.8	1160.0	1003.0	15	15	16	31.7	183.6	441.1	1162.6	1032.0

Table 1. Water volumes (in 1000 m^3) within various salinity ranges for the 38 combinations of flows (in cfs).

From Table 1, some conclusions can be drawn from these scenario runs:

1. If the release of fresh water from the dam is small (5 cfs or less), no fresh water (salinity less than 1 ppt) will be presented in the river.

2. By increasing the release to 10 cfs, a small freshwater volume can be attained unless the spring flow is very low.

3. Because the spring flow out of Sulphur Springs has an average salinity of 1.5 ppt, fresh water does not exist in the river if no water is released from the dam, even when a portion of spring water is routed to the base of the dam.

4. If there is a water release from the dam, routing a portion of the spring flow to the

base of the dam can affect the fresh water volume in the river. For example, with a 20 cfs release from the reservoir and a 40 cfs from the spring, the freshwater volume in the river increases from 72860 m³ to 99720 m³ if 50 percent of the spring water is routed to the river at the base of the dam.

5. With a release of water from the reservoir at 40 cfs or greater, a freshwater zone (salinity less than 1 ppt from top to bottom) can be maintained in the first 1000 meters downstream of the dam (Chen, 1997).

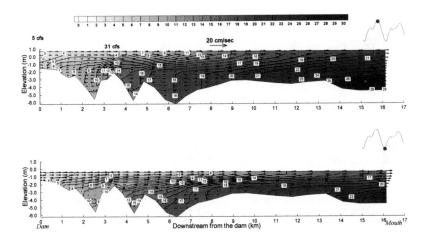

Figure 12 Simulated salinity at high and low tides with a release of 5 cfs from the dam and a flow rate of 31 cfs from Sulphur Springs.

Acknowledgment

We would like to thank Yvonne Stoker and Michael DelCharco of the USGS for providing necessary field data for this study. Support from Granville Kinsman and David Moore of the SWFWMD is also appreciated.

Reference

Blumberg, A. F. and G. L. Mellor, 1987: A description of a three-dimensional coastal ocean circulation model, *Three-Dimensional Coastal Ocean Models*, American Geophysical Union, pp. 1-16.

Casulli, V., 1990: Semi-Implicit Finite Difference Methods for the Two-Dimensional Shallow Water Equations, *Journal of Computational Physics*, Vol. 86, No. 1, pp. 56-74.

Chen, X.-J., 1994: Effects of Hydrodynamics and Sediment Transport Processes on Nutrient Dynamics in Shallow Lakes and Estuaries, Ph.D. Dissertation, University of Florida, Gainesville.

Chen, X.-J. and Y. P. Sheng, 1996: Application of A Coupled 3-D Hydrodynamics- Sediment-Water Quality Model, *Estuarine and Coastal Modeling*, ASCE, pp. 325-339.

Chen, X.-J., 1997: Simulation of Hydrodynamics and Salinity Transport in the Lower Hillsborough River, SWIM Section, Southwest Florida Water Management District.

HSW Engineering, 1992: Tampa Bypass Canal and Hillsborough River Biologic Monitoring Assessment Program; Task I, Tampa.

Perrels P. A. J. and M. Karelse, 1981: A Two-Dimensional, Laterally Averaged Model for Salt Intrusion in Estuaries, *Transport Models for Inland and Coastal Waters*, Academic Press, Inc., pp. 483-535.

Sheng, Y.P., H. K. Lee, and K. H. Wang, 1989: On numerical strategies of estuarine and coastal modeling, *Estuarine and Coastal Modeling*, ASCE, pp. 291-301.

Sheng, Y.P. and C. Villaret, 1989: Modeling the effect of suspended sediment stratification on bottom exchange process, *J. Geophys. Res.* **94:** 14429-14444.

WAR/SDI, 1995:Tampa Bypass Canal and Hillsborough River Hydrological Monitoring Program: Second Interpretive Report. Prepared for the West Coast Regional Water Supply Authority and the City of Tampa, Florida.

Hydrodynamic Flow and Water Quality Simulation of a Narrow River System
Influenced by Wide Tidal Marshes

Gavin Gong, Ken Hickey and Mel Higgins
ENSR
35 Nagog Park, Acton, MA 01720

Abstract

A two-dimensional, laterally averaged hydrodynamic flow and water quality model
was developed for a narrow tidally influenced river system in the southeastern
United States, to assess the impact of an effluent discharge on dissolved oxygen
and other water quality parameters in the river. The presence of wide areas of
tidal marshland adjacent to the river poses a challenge to the modeling effort, since
the shallow marshes have little effect on hydrodynamics but can have a substantial
impact on water quality. The marshes were simulated by imposing a series of
inflow boundaries along the main branch of the river, characterized by low inflow
rates and high BOD concentrations.

Representation of the tidal marshes in this manner was critical to the successful
calibration of the model. Subsequent simulations involving the effluent discharge
and marsh inflows showed that the effluent discharge has a minimal effect on
dissolved oxygen concentrations in the river relative to the tidal marshes, due in
part to the effluent release strategy of discharging during ebb tide only. This result
underscores the importance of accurately simulating the effect of the tidal marsh
areas in this estuarine system, and the ability of this two-dimensional model
application to account for lateral as well as vertical characteristics.

Introduction

A hydrodynamic flow and water quality model has been developed for a tidally
influenced riverine system in the southeastern United States. The objectives were
to evaluate potential impacts of an industrial discharge on dissolved oxygen levels
in the river, and to assess the relative importance of naturally occurring non-point
sources of oxygen-consuming substances to the river. The study was motivated
by summertime dissolved oxygen measurements in the river below the stream
standard of 4.0 mg/L (SCWCS, 1990). The accurate simulation of natural river
processes was central to determining whether the effluent discharge is responsible
for the observed low dissolved oxygen concentrations.

The features of the modeled river system are presented in Figure 1. The model domain extends from the mouth of the river, located at roughly 33.35°N and 79.28°W, to a point eleven miles upstream. It consists of a narrow main branch of the river, four small tributary creeks, and an oxbow near the mouth of the river. The river discharges into a large estuary, which causes the entire model domain to be tidally influenced. Tidal flowrates at the mouth of the river are on the order of 400 m³/s. The effluent is located roughly three miles upstream of the river mouth, and discharges directly into the main branch of the river at an average rate of 1x10⁵ m³/day. One important characteristic of this effluent is that it is released only during the ebb tide of each twelve hour tidal cycle, for a duration of 1.25 hours, which results in an actual effluent flow rate during discharge of 11.2 m³/s.

The physical characteristics of this riverine system influence hydrodynamic flow and water quality in the river. In particular, tidal marshlands exist adjacent to the main river branch, located predominantly between river mile three and river mile nine. The marshes consist of broad areas of shallow, stagnant water, and are most extensive during high tide. They are not expected to affect hydrodynamic flow in the river; however marsh waters do exert a considerable demand on dissolved oxygen, through sedimentary uptake and biological activity. Consequently, water emptying from the marshes into the river during ebb tide may have substantial impacts on water quality in the river.

The tidal marshes introduce a lateral dimension to the otherwise narrow, two-dimensional model domain. However, study constraints prohibited the use of a three-dimensional model, which presented a particular challenge to the simulation of the marshes. A two-dimensional model therefore had to be developed which could effectively simulate both the vertical features of the tidally influenced river and the lateral characteristics of the surrounding marshes.

This paper focuses on the pivotal role played by the wide tidal marshes present in the study area, and the means by which this lateral feature is represented in a two-dimensional, laterally-averaged model. The influence of the marshes was investigated using a fully calibrated and validated hydrodynamic and water quality model of the study area. However, this paper does not attempt to present a comprehensive description of the developed model and study area features. Only the fundamental features of the model setup and calibration are described, with emphasis placed on the representation of the tidal marshes. Simulations using the calibrated model are then described, which compare the relative influence of the marshes and the effluent discharge on water quality in the system.

Model Setup

The two-dimensional, laterally averaged, hydrodynamic and water quality model CE-QUAL-W2 was selected for this application. CE-QUAL-W2 is a free surface, finite difference program developed by Thomas Cole and Edward Buchak at the U.S. Army Corps of Engineers, Waterways Experiment Station, Vicksburg, MS (Cole and Buchak, 1995). The model has been in continuous development since 1975, and has been applied to rivers, lakes, reservoirs and estuaries. The model's

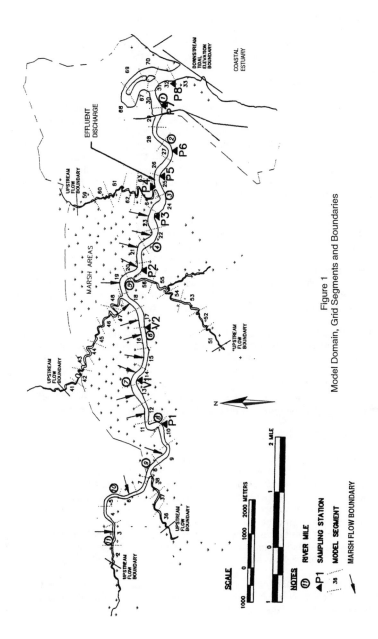

Figure 1
Model Domain, Grid Segments and Boundaries

computational grid consists of 59 longitudinal segments, 32 of which make up the main branch of the river. The remaining segments comprise the four tributary creeks and the oxbow, resulting in a total of six model branches, as indicated in Figure 1. Each model segment is roughly 600 meters in length, which represents a compromise between optimal spatial delineation and computational efficiency. Sixteen model layers are defined, each with a thickness of one meter.

A comprehensive field program was conducted for this system, designed to provide accurate boundary condition values and calibration data (Hickey et al, 1997). The field program included acoustic doppler current profiler (ADCP) measurements, CTD deployments, analytical sampling, and long-term continuous measurements at stations indicated in Figure 1. Data provided by the field study included bathymetry, water velocity, dissolved oxygen, and biochemical oxygen demand (BOD) measurements, which were utilized by both the hydrodynamic and water quality components of the modeling study.

Model boundary conditions are also indicated in Figure 1. Upstream freshwater inflows to applicable branches were specified as flow boundaries. The effluent discharge was specified using a flow boundary entering directly into the main branch at model segment 25. The downstream end of the main branch was specified as a variable-head boundary, to represent the tidally-varying elevations in the estuary. Varying water quality values over a tidal cycle at this boundary were specified using field data at station P8, located near the mouth of the river, to accurately simulate the interaction of the river with the coastal estuary.

All model simulations were performed for roughly 50 days under climactic and tidal conditions representative of the calibration data, which was collected during the summer. The marsh areas and the effluent discharge were active throughout the simulations, and a three day period at the end of the simulation was used for predicted model results. The 50 day duration allowed all model boundary conditions, including the downstream tidal boundary, marshes and effluent, to exert their full effect on the system.

Representation of Tidal Marshes

Representation of the tidal marshes using CE-QUAL-W2 was complicated by the laterally-averaged nature of the model, which prohibits the explicit inclusion of the broad marsh areas in the model domain. One possibility was to assign extremely wide widths to the uppermost model layers for segments where the marshes reside. However, since the model assigns a homogeneous water velocity throughout each model cell, the model would not be able to resolve the hydrodynamic differences between the main branch channel and the stagnant marsh waters within the top model layers. In addition, the resulting sharp gradients between neighboring cell sizes may have introduced numerical instabilities.

Therefore an approach was adopted whereby instead of explicitly modeling the tidal marshes, they would be implicitly modeled by simulating the hydrodynamic and kinetic effects of the marshes on the main branch of the river. The effects of the tidal marshes were represented using a set of 15 flow boundaries entering into various segments along the main branch (see Figure 1). Each flow boundary was assigned a low inflow rate so as not to affect hydrodynamics. However, an elevated BOD concentration was associated with these low inflows, to represent the expected BOD loading to the river from sedimentary and biological activity in the marshes. Thus the kinetic processes occurring within the tidal marshes themselves are not simulated by the model. Instead the marshes are represented as a boundary condition which introduces a direct BOD source loading to the main branch of the river.

Hydrodynamic Calibration

The hydrodynamic features of the river system were modeled first, independent of the water quality features. Calibration of the hydrodynamic model focused on matching predicted tidal flowrates and water surface elevations to measured values at nine stations along the main branch of the river. Final hydrodynamic model calibration parameter and boundary condition values are listed in Tables 1 and 2. The marsh inflows were assigned negligible inflow rates of 0.08 m^3/s, based on the areal fraction of river basin runoff attributed to the marshes. Adjustment of the river bathymetry specification turned out to be the critical calibration parameter for the hydrodynamic model, as predicted tidal flowrates were found to be highly sensitive to relatively small changes in bathymetry.

Figure 2 presents calibrated model flowrate vs. measured flowrate at peak flood and ebb tide. Overall, model predictions compare favorably to measured data at all stations. Peak flood and ebb tide flow rates are high near the mouth of the river

Table 1
Hydrodynamic and Water Quality Model Calibration Parameter Values

Parameter	Value	Units
Horizontal Eddy Viscosity	1	m^2/sec
Model Segment Chezy Coefficient	20.0 - 24.9	$m^{0.5}$/sec
Sediment Oxygen Demand	0.4	gO_2m^2/day
Wind Magnitude	5	meters/sec
Horizontal Eddy Diffusivity	1	m^2/sec
Temperature BOD coefficient	1.017	--
Ammonia Decay Rate	0.25	per day
Nitrate Decay Rate	0.09	per day
River BOD_5 Decay Rate	0.1	per day
River BOD_u/BOD_5	1.57	--
Effluent BOD_5 Decay Rate	0.02	per day
Effluent BOD_u/BOD_5	4.45	--

Table 2
Hydrodynamic and Water Quality Model Boundary Condition Values

Parameter	Marsh Inflows	Upstream Inflows	Tidal Boundary	Effluent Discharge
Tidal Amplitude (m)	--	--	0.594	--
Tidal Period (min)	--	--	743	--
Flow Rate (m³/sec)	0.08	0.45-1.125	--	11.25
Temperature (C)	28	28	28	31
Dissolved Oxygen (mg/L)	2.0	3.0	3.9-4.7	4.0
BOD (mg/L)	4.0-90.0	4.0	1.9	26.3
Salinity (ppt)	0	0	0.04-12.0	0
NH_3 (mg/L)	0	0	0.06	0
NO_3+NO_2 (mg/L)	0	0	0.2	0

(roughly 400 m³/s), and decrease with distance upstream to substantially lower values (less than 10 m³/s) and the far upstream end of the model domain. Flood and ebb tide flow rates are very similar, which reflects the minimal presence freshwater in this tidally dominated system.

Water Quality Calibration

The water quality model was developed while maintaining the calibrated hydrodynamic properties of the system. Water quality model development was aimed at capturing the kinetic processes that affect dissolved oxygen concentrations in this system, e.g., carbonaceous BOD, ammonia oxidation, sediment oxygen demand, and reaeration. The effects of algae were not included, since no diurnal trends in dissolved oxygen concentrations were observed in the measured data.

Calibration focused on dissolved oxygen and BOD concentrations, since all carbonaceous oxygen consuming processes associated with the effluent discharge are expressed in terms of a total BOD concentration in this study. Final water quality model calibration parameter and boundary condition values are listed in Tables 1 and 2. Adjustment of the BOD concentration assigned to each of the fifteen marsh inflow boundaries turned out to be one of the critical calibration parameters for the water quality model, which underscores the pivotal role played by the tidal marshes on water quality in the river. The fifteen marsh inflows were assigned BOD concentrations ranging from 4 mg/L to 90 mg/L, for a total BOD loading of roughly 4450 kg/day from the marshes.

Figure 3 presents water column average concentrations for the calibrated model vs. measured data along the main branch of the river, for dissolved oxygen and BOD. The measured values represent various times throughout the tidal cycle.

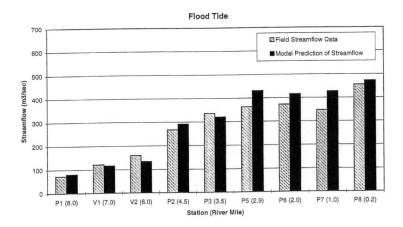

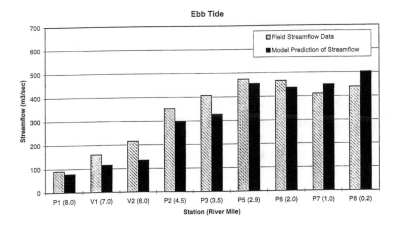

Figure 2
Hydrodynamic Model Calibration
Predicted vs. Measured Tidal Flowrates at Peak Flood and Ebb Tide

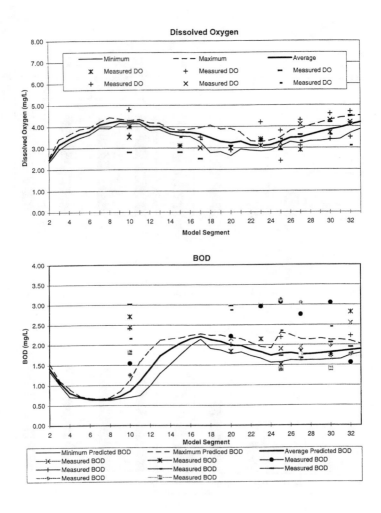

Figure 3
Water Quality Calibration
Predicted vs. Measured Dissolved Oxygen and BOD

The model results represent the minimum, maximum, and average value over a three day period at the end of the 50 day calibration run. Note that both the marshes and the effluent are active throughout the simulation, in order to be comparable to the calibration data. The calibrated model agrees quite well with observed values, which confirms the model's ability to replicate the major processes affecting kinetics and water quality in the system. The observed dissolved oxygen concentrations below stream standards are captured by the model. Both the calibrated model and the measured data show the lowest dissolved oxygen concentrations occurring in the middle reaches of the river, where the marshes reside.

Representation of the tidal marshes has proven to be critical to the successful calibration of the hydrodynamic and water quality model. Simulating the effect of the marshes on the main branch of the river using inflow boundaries was able to incorporate the impact of the kinetic processes occurring within the marshes, while maintaining a negligible impact on model hydrodynamics. The sensitivity of the water quality model calibration to the marsh inflow BOD concentrations confirms the contribution of the marshes to water quality in the river.

Relative Importance of the Tidal Marshes

The calibrated hydrodynamic and water quality model was used to evaluate the relative importance of the tidal marshes to the observed characteristics of the river system. In particular, the contribution of the marshes to the observed dissolved oxygen concentrations below 4.0 mg/L was evaluated. A simulation was performed in which the fifteen marsh inflows were removed from the model. As expected, simulated hydrodynamics were not affected, since the marsh inflows carried negligible inflow rates. In contrast, the water quality simulation exhibited substantial differences, as indicated in Figure 4.

Without the BOD contribution provided by the marshes, predicted BOD concentrations in the main branch of the river decrease dramatically to near zero values throughout the upper half of the river. Consequently, this reduced BOD load results in noticeably higher predicted dissolved oxygen concentrations, well above the stream standard of 4.0 mg/L. The change is most apparent in the middle reaches of the river, where the marsh inflows enter the river. Thus the tidal marshes appear to be an integral component of the water quality in the river. These results indicate that the observed dissolved oxygen concentrations below stream standards in the middle and upper reaches of the river are principally due to natural occurring non-point sources of oxygen demand on the river, i.e., the tidal marshes.

Relative Importance of the Effluent Discharge

The calibrated model was also used to evaluate the relative importance of effluent discharge to the river system, and the extent to which the effluent contributes to the observed low dissolved oxygen concentrations. First, a dye release simulation was performed using the calibrated hydrodynamic model, to track the transport of a

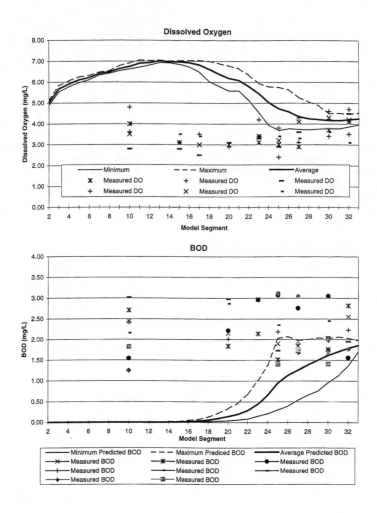

Figure 4
Predicted Dissolved Oxygen and BOD without Marsh Inflows

single effluent plume over time. A conservative tracer concentration of 100 mg/L was assigned to one discharge event. Resulting tracer mass concentrations were used to calculate the percentage of discharged tracer mass leaving the system after each discharge, to determine the extent of net upstream transport of effluent waters, and to assess the long-term buildup of tracer mass throughout the system. Quantification of effluent transport throughout the system can assist in identifying the areas of the river that could potentially be impacted by the effluent.

Figure 5 depicts the tracer mass location and quantity at successive slackwater low and high tides after its initial release. As indicated in Figure 5, the single effluent plume is transported back and forth through the downstream reaches the river. At no point in the simulation does the centroid of the plume travel upstream of river mile four, which indicates that any impact of the effluent discharge is relegated to downstream reaches of the system. Furthermore, total tracer mass decreases rapidly with time as part of the plume exits the system via the mouth of the river during successive ebb tides. Roughly 65 percent of the total tracer mass exits the system during its initial ebb tide release, and virtually all of it has left the river within three days. This simulation demonstrates that a substantial portion of each effluent plume is quickly transported out of the mouth of the river as it is released with ebb tide, and what remains in the system does not reach upstream portions of the river.

A second tracer simulation was also performed in which a tracer concentration of 100 mg/L was assigned to each effluent discharge. This simulation was run for 50 days, and showed that the total tracer mass in the system and tracer concentrations within model segments reached equilibrium values after about 5 days. In addition, tracer mass never traveled upstream of river mile 22. Thus neither long term buildup nor significant upstream transport of constituent mass originating from the discharge effluent does occurs in the system.

The dye release simulations indicate that the effluent exits the river system fairly rapidly, which minimizes the impact of its associated BOD load on water quality within the river. However, this rapid flushing of the effluent plume is exacerbated by the discharge orientation and effluent release strategy, in which the effluent is released into surface waters of the river during ebb tide only. The discharge waters would potentially have a greater impact on water quality in the river if the effluent plume were to remain in the system for a longer period.

An additional simulation was performed in which the effluent was released during flood tide instead of ebb tide. As indicated in Figure 6, this results in substantial increases in BOD and corresponding decreases in dissolved oxygen, compared to an ebb tide release. The change is most noticeable immediately upstream of the discharge location. With a flood tide discharge, virtually all of the effluent plume remains in the river after the initial discharge. Effluent slowly leaves the system over extended periods of time through decay, dispersion, and other gradual transport processes. This effluent release strategy allows the effluent's BOD load to exert its full affect on water quality in the river.

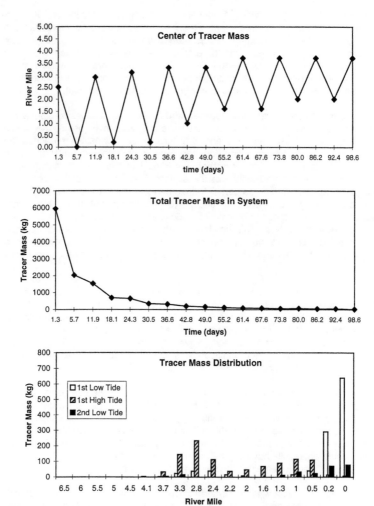

Figure 5
Effluent Discharge Dye Release Simulation Results

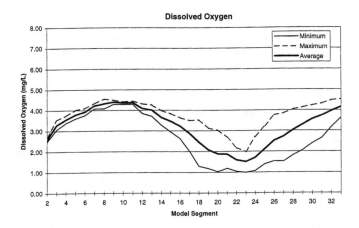

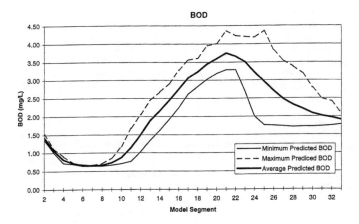

Figure 6
Predicted Dissolved Oxygen and BOD with a Flood Tide Discharge

This simulation indicates that the effluent discharge can potentially be an influential component of water quality throughout the modeled river domain. The importance of the discharge is dependent on how much of the effluent plume remains in the system after its initial release, and where it is carried by hydrodynamic transport. The effluent does not contribute to the observed low dissolved oxygen concentrations, since the current effluent release strategy results in a rapid flushing of the effluent out of the system.

Conclusions

A two-dimensional, laterally averaged hydrodynamic and water quality model was successfully applied to a narrow river system with wide tidal marshes. By representing the kinetics of the marshes on the river as a series of inflow boundaries with low flowrates and high BOD concentrations, the model was able to implicitly account for the lateral features represented by the marshes. Calibration of the model showed that water quality in the river is strongly influenced by the marshes, which underscored the importance of accurately simulating the effect of the marshes on the river.

Subsequent simulations using the calibrated model revealed the marshes to be of principal importance to the observed water quality characteristics of the river. The observed dissolved oxygen values in the river below the stream standard of 4.0 mg/L can be attributed primarily to the BOD load associated with the tidal marshes rather than to the BOD load associated with the effluent discharge. The minimal predicted impact of the effluent is due in part to the ebb tide only release strategy employed by the effluent, which provides rapid and efficient flushing of the effluent plume after its initial release. Predicted effects for a discharge during flood tide showed that the effluent can potentially have a strong influence on water quality throughout the river.

References

Cole, T.M. and E.M. Buchak, 1995. CE-QUAL-W2: A Two-Dimensional, Laterally Averaged, Hydrodynamic and Water Quality Model, Version 2.0. Instruction Report EL-95. US Army Engineer Waterways Experiment Station, Vicksburg, MS. March 1995.

Hickey, Ken, M. Gerath, J. D. Bowen, and G. Gong, 1997. Design of Field Program and Use of Field Measurements to Calibrate a Hydrodynamic and Water Quality Model of an Estuarine River System. Presented at the ASCE 5th International Conference on Estuarine and Coastal Modeling, Alexandria, VA. October, 1997

South Carolina Water Classifications and Standards (Regulations 61-68) and Classified Waters (Regulations 61-68), April 27, 1990.

An Eulerian-Lagrangian Particle Dispersion Modeling for Surface Heat Transport

Seung-Won Suh [1]

Abstract

Buoyant dispersion of surface discharged heat water was simulated using a random walk particle tracking scheme based on an Eulerian-Lagrangian method. Numerical computation was done on a depth averaged two dimensional field adopting a harmonic finite element model as hydrodynamic analysis tool for effective computation and an Eulerian-Lagrangian dispersion model in heat transport. In random walk analogy for heat dispersion, it was assumed that particles have buoyancy and might cause additive horizontal dispersive effect. Most efforts are concentrated onto effective simulation of surface heat water in finite element model using buoyant particles.

The proposed scheme was applied to compute excess temperature distribution for surface discharged heat water of existing coastal power plant. Simulated results were compared with both observed data and other computed model, CORMIX3. The results are in agreement with both of them.

Introduction

The cooling water discharge, an unwanted by-product from power plants, causes harmful effects around power plants, especially the discharged water which is relatively warmer than ambient water and may cause serious effects in farming and fishing areas. Since harmful impacts occasionally arises a few kilometers apart from power plant, understanding of dispersion mechanism of

[1]Department of Ocean System, Kunsan Nat'l University, Chonbuk 573-702, Korea.
e-mail : suh@ks.kunsan.ac.kr

heated water for far field is important. However far field (FF) mechanism is directly dependent on near field (NF) characteristics. So we have to consider both fields simultaneously.

Although a numerical model is believed to be a suitable approach, it may not provide results satisfying both in NF region and in FF region in which the heat dispersion may play a significant role. Thus, numerical results of existing models are concentrated on either NFR or FFR by neglecting the other. But the main interesting domain for most of numerical models based on depth-averaged transport scheme is FFR rather than NFR.

In practice, depth-averaged two-dimensional models with a varying grid system have widely been used to analyze the surface heat dispersion mechanism.

Generally we can classify mass transport models including heat dispersion into two main categories:

- Concentration models: Concentration models can be classified as Eulerian model(EM), Lagrangian model(LM), and Eulerian-Lagrangian model (ELM) (Dimou, 1992). Among these models ELM is believed as a best approach in coastal modeling, however it has some mass unbalance problem. Tang and Adams(1995) dealt with mass conservation in ELM. The main advantage of ELM can be depicted that advection part is only tracked once along characteristic line in periodic flow, which is commonly seen in coastal flow.

- Particle tracking models: Mass is represented by discrete particles and these are tracked on Eulerian grid. At each timestep the displacement of each particle consists of an advective, deterministic component and an independent, random Markovian component (Dimou, 1992). However owing to rapid increase of hardware performance of computers, particle tracking models can replace concentration models. This scheme has advantages of eliminating the numerical dispersion while it has some error on conversion of particles into concentration in FF. This approach is thought more realistic in some cases, e.g., modeling the settling of suspended solids, larvae or fish eggs scattering in ecosystem modeling. Disadvantages of this method are the time consumed in tracking each particle for the whole domain. Until now most approaches for particle tracking are only for passive movement of neutral particles depending on ambient flow.

To represent more detail in NF, some approaches using ELM has been applied by Dimou and Adams(1993). They suggested three-dimensional movement of particles in NF for ocean outfall transports, however they did not account for buoyancy. Zhang (1995) attempted hybrid method of NF into FF model using 3-D model, ECOMsi. In Zhang(1995), NF characteristics of ocean outfalls are predicted by RSB model and it is used as input of FF model, ECOMsi. He concentrated on finding initial trap heights of plume using NF model and taking particle tracking procedure into FF model. Even though he dealt with NF particle tracking for slight buoyant problem he only accounted for passive particles.

Since active particle movement affects ambient flow and it causes modification of hydrodynamic simulation, it is very hard to model the whole mechanism in this manner even though the physical meaning is so clear. In this study a new attempt have been made on particle tracking. It is treated that the particles have buoyancy and act as semi-active particles rather than passive neutral for heat transport simulation of surface discharge. Surface discharged heat water is treated as buoyant particles in front of discharging point and this vertical moving characteristics devotes to horizontal spreading eventually. To accommodate this idea, some assumptions have been made. Since this approach is on testing status we have tested for some limited cases. Also it has restrictions such as the movements of particles do not directly influence ambient flow field, i.e., semi-active movement of particles cannot cause entrainment flow due to initial strong jet momentum in front of discharge point.

Numerical tests were done in a simplified one-dimensional channel under the condition of advection dominated flow for neutral particles as a measure of effectiveness of random walk particle model in a flat bed and sloping bed accounting for pseudo-velocity of particle movement. The results are fairly coincident with analytical solutions around source point where high concentration gradient observed.

Buoyant particle modeling was applied for an imaginary geometry representing sloping bottom in front of surface discharge point. Since we have no analytical solution available, numerical tests were compared with NF model, CORMIX3. In real application, we applied this semi-active random walk particle tracking to the power plant of previous attempt (Suh, 1997). General tendencies of model results are in good agreement compared with field data.

Random walk particle modeling

The general form of heat transport equation is given by the following equation:

$$\frac{\partial T}{\partial t} + \nabla \cdot \mathbf{u}T = \nabla \cdot \mathbf{D} \cdot \nabla T + Q \tag{1}$$

where $T(x,t)$ is the heat concentration, $\mathbf{u}(x,t)$ is the velocity vector, $\mathbf{D}(x,t)$ is the diffusivity tensor and Q represents sources/sinks.

The random walk analogy of particle movement for Eq.(1) has following form of particle deviation;

$$\Delta \mathbf{x} = \mathbf{X}^n - \mathbf{X}^{n-1} = \mathbf{A}(\mathbf{X}^{n-1}, t)\Delta t + \mathbf{B}(\mathbf{X}^{n-1}, t)\sqrt{\Delta t}\mathbf{Z} \tag{2}$$

where $\mathbf{A}$ and $\mathbf{B}$ are given by the expressions:

$$\mathbf{u} = \mathbf{A} - \nabla(\frac{1}{2}\mathbf{BB}^T) \tag{3}$$

$$\mathbf{D} = \frac{1}{2}\mathbf{BB}^T \tag{4}$$

Δt is the timestep, $\mathbf{Z}$ is a vector of random numbers with zero mean and variance one (Dimou, 1992).

Eq.(2) is equivalent to Eq.(1) in the limit of large number of particles N_p and small Δt.

Particle tracking models are also a better choice for concentration models in cases where transport and fate processes are described more physically by attributes of the individual particles rather than their aggregation (Dimou, 1992). Also the reverse hypothesis can be applied for heat transport in NF.

Since we are usually more interested in concentration contours than mere particle locations, final form of model representation should be concentration model in FF. And thus hybrid concept of semi-active buoyant particle tracking in NF and EL concentration in FF should be considered.

In Dimou and Adams (1993) a 3-D hybrid particle tracking ELM is developed for the representation of sources whose spatial extent is small compared to that of the discretization. Sources are simulated by the introduction of particles. Each particle advects and diffuses independently and when the particles have dispersed enough they are mapped onto the numerical grid by identifying the particle density associated with each node. They mainly dealt with the issue of interfacing the particle tracking mode with the concentration mode.

Buoyant particle random walk simulation

Surface discharged heat water is treated as a buoyant mass just in front of discharging point and this physical vertical moving characteristics help horizontal spreading. So this is converted to an additive dispersion in NF until discharged heat water poses buoyancy compared to ambient water. Following steps are assumed in resolving buoyant particles;

First, discharged water through cooling system of power plant has buoyant density differences, $\Delta \rho = \rho_a - \rho_o$, in which ρ_o is cooling water density and ρ_a is ambient water density. This heat water can be treated as a series of particles having mass with buoyancy. Owing to $\Delta \rho$, particles go upward and finally spreads in horizontal plane which cause an additive horizontal dispersion until buoyant particles lose their nature, that is $\Delta \rho$ approaches to zero.

Second, this additive buoyant dispersion, D_b, is a function of space and time. The spatial component, $D_{b_{x_i}}$ can be represented as;

$$D_{b_{x_i}} = C_i e^{-k x_i} \tag{5}$$

$$for \quad x_i \leq x_{i_{NF}}, \quad D_{b_{x_i}} \geq D_{x_i}$$
$$at \quad x_i = 0, \quad D_{b_{x_i}} = D_{max}$$
$$at \quad x_i = x_{i_{NF}}, \quad D_{b_{x_i}} \approx D_{x_i}$$

Which means that this assumption is only valid in NF. Coefficients k and C_i should be determined by field measurements or simulations. Temporal component, D_{bt}, also has similar form just replacing independence variable with time. Strength of D_b is affected by $\Delta \rho$. This empirical relationship should be determined from field measurement.

Thus we can rewrite the Eq.(2) with these new buoyant dispersion $D_{b_{x_i}}$, D_{bt} and get the following semi-active buoyant particle random walk tracking algorithm as following Eqns.(6), (7). Two main components of this scheme are deterministic part representing pseudo-velocity due to spatial variation of depth and dispersion, and random component for particle movement. In which buoyant dispersion included in both components. It increases movements of particles in NF farther while buoyant dispersion reaches to FF dispersion. Hence we get a new Lagrangian random walk model.

$$\Delta x = \left[\frac{D_{bx}}{h} \frac{\partial h}{\partial x} + \frac{\partial D_{bx}}{\partial x} + u \right] \Delta t + \sqrt{2D_{bt}\Delta t} \; Z_{n1} \tag{6}$$

$$\Delta y = \left[\frac{D_{by}}{h} \frac{\partial h}{\partial y} + \frac{\partial D_{by}}{\partial y} + v \right] \Delta t + \sqrt{2D_{bt}\Delta t} \; Z_{n2} \tag{7}$$

Eulerian-Lagrangian simulation

In general, ELM requires hydrodynamic driving force information i.e., we have to take a hydrodynamic model prior to the modeling of EL dispersion. The governing equations for hydrodynamic simulation are the depth averaged continuity and momentum equations. As a numerical scheme Galerkin's finite element method using linear interpolation is applied. A harmonic approach for hydrodynamic equations is to accommodate frequency unknowns. This approach can be interpreted as a realistic treatment in coastal modeling. In these numerical simulation two kinds of linear harmonic models, TEA and FUNDY5 were applied in model tests and actual application.

For EL simulation, we have the following two modes on simulating surface heat water dispersion, that is, the particle tracking mode and the concentration mode.

The particle mode: Heated water mass is represented by particles with buoyancy. These particles are represented according to Eqns.(6), (7). The advection is calculated using a fourth order Runge-Kutta method. Different numerical treatment of the two components does not cause any problem in coastal water problems because the flows are advection dominated. When the particles reach to NF boundary, the particle locations are mapped onto node concentrations. After they are mapped onto node concentrations and the rest of the model computations continue in the concentration mode. In case of a continuous source a certain number of particles is released every timestep. Each unit mass of particles is mapped onto node concentrations if $x_i \geq x_{i_{NF}}$.

The concentration mode: In this research a two-dimensional finite element ELM was adopted. The main advantage of using an ELM is that use of large

Courant numbers, a feature particularly important in surface waters where we are dealing with advection dominated flows. In this case the use of ELM is compatible with the use of a particle tracking model where forward tracking of particles is used.

The governing equation of heat dispersion model is expressed as:

$$\frac{\partial T}{\partial t} + u_i^* \frac{\partial T}{\partial x_i} = K_{ij} \frac{\partial^2 T}{\partial x_i \partial x_j} + \gamma T \tag{8}$$

in which u_i^* represents an apparent velocity

$$u_i^* = u_i - \frac{1}{h} \frac{\partial(hK_{ij})}{\partial x_j} \tag{9}$$

and T is the temperature of surface discharged water, K_{ij} is dispersion coefficients and γ is heat loss coefficient at surface.

Since the governing Eq. (8) includes both hyperbolic and parabolic operators, an operator split EL scheme may be a plausible choice(Baptista et al., 1984). Introducing a temporary variable, T^f, which means the temperature of mid time level between prior computation level $n-1$ and present time level n, yields following separated equations.

$$\frac{T^f - T^{n-1}}{\Delta t} + \left(u_i^* \frac{\partial T}{\partial x_i} \right)^{n-1} = 0 \tag{10}$$

$$\frac{T^n - T^f}{\Delta t} = \left(K_{ij} \frac{\partial^2 T}{\partial x_i \partial x_j} \right)^n + \gamma T^n \tag{11}$$

A discrete form of the pure advective equation (10) can be rewritten as:

$$\frac{DT}{Dt} = \frac{\partial T}{\partial t} + u_i^* \frac{\partial T}{\partial x_i} = 0 \tag{12}$$

This equation implies that the temperature concentration, T, remains constant on the following characteristic lines;

$$\frac{dx_i}{dt} = u_i^*(x, y, t) \tag{13}$$

The heat concentration is traced along the characteristic line and interpolated onto finite elements. It is then used as an input of the dispersion equation.

Model tests for neutral particles

To test the ability of proposed random walk model, numerical tests are done in one dimensional channel where exact solutions are available. In this tests we neglect buoyant characteristic of particles. Solutions in plug flow for continuous source at center point of the geometry as shown in Fig.(1) can be obtained as,

$$C(x,t) = \frac{M}{\sqrt{2\pi}\sigma_x} exp \left\{ \frac{-(x - \bar{x})^2}{2\sigma_x^2} \right\} \tag{14}$$

in which $\sigma_x^2 = \sigma_{ox}^2 + 2Dt$ and $\bar{x} = x_o + ut$, and subscript $_o$ means initial source point. In numerical tests, to account for strong advection we set $u = 10m/s$, $D = 10m^2/s$. The number of particles in this simulation were 1000 and 5000. As shown in Fig.(2) the simulated results are in good agreement with analytical solution. But due to the random characteristic of particles, concentration profiles show somewhat skewed after several hundred meters from source point. However in front of source, concentration profile of particle model coincides with analytical solution, thus it guarantees the proposed scheme can be applied in NF where high concentration gradient are occurred.

Another numerical simulation is done in sloping channel to represent bottom slope of general coastal area. In case of pure advection for an exponential depth variation, $h(x) = h_o exp(ax)$ the mass transport equation (8) has modified form with artificial advection term having pseudo-velocity of $-aD$. Thus its analytical form is represented as like as Eq.(14) merely setting $\bar{x} = x_o - aDt$. Computations were done in $D = 10m^2/s$, $h(x) = 5exp(0.003x)$ for $-5000 \leq x \leq 5000$ in order to account for depth variation from $1.1m$ to $22.4m$ for $10Km$ length of sloping shore bottom. The numerical simulation showed satisfactory results as in previous case. But accuracy also decreased with time being after releasing of particles. However these two tests showed the applicability of this proposed random walk model into real coastal geometry in case of advection dominated flow for a mild slope coast area.

Model tests for buoyant particles

In model test of Lagrangian random walk tracking for 1-D channel we can compare the numerical results with analytical solutions so it is easy to verify the capability of developed model. Since the main purpose of this paper is modeling of buoyant particles, model tests were done in a simplified imaginary rectangular bay consisting of $10Km$ width and $40Km$ length as shown in Fig.(3). We assumed the discharging heat mass is $65m^3/s$ with $7°C$ excess temperature and $\Delta\rho = 1.0Kg/m^3$. Where topography is represented as sloping hill from a mid point of right side wall to account for pseudo-velocity of particles. Even though model tests were done by changing some parameters, D_{max}, Δt and N_p, as an attempt we set $D_{max} = 100\ m^2/s$, $x_{NF} = 300\ m$ and $N_p = 1000$.

Some computed results are shown in Fig.(4) for the zoomed in domain of the model tests area as excess temperatures. They show that overall tendencies are almost same except horizontal spreading. These results are compared with NF model, CORMIX3. It shows that this proposed scheme provides more plume-like spreading in front of discharge point. So even in case of maximum ebb and flood flow state, the offshore dispersion is greater than that of CORMIX3. Also the extent of affecting distance along coastline is shorter than that. However it is not sensitive on the variation of maximum dispersion coefficients and number of particles used in simulation. Thus even though we cannot exactly evaluate the capability of this proposed scheme, it can describe the heat dispersion in sloping topographical coast with satisfaction.

Near Field characteristics and CORMIX3 modeling

In order to model the real heat dispersion for a coastal thermal power plant, Boryong power plant is selected. Which is located at the west coast of Korea, general map is shown in Fig.(5). Thermal discharges can be explained by local densimetric Froude number defined as $F_o' = u_o/\sqrt{g_o' l_0}$, where $l_o = \sqrt{b_o h_o}$ and $g_o' = g\frac{(\rho_a - \rho)}{\rho_a}$. As a results of field experiments, in case of $65m^3/s$ heat discharge with $7°C$ excess temperature , partial surface buoyancy appeared for $F_o' < 2.4$ while $F_o' > 4.7$ showed active vertical mixing. From the analysis of field data increasing the value of buoyant discharge, $B = g_o'Q$ from $1.43 \times 10^{-2}m^4/s^3$ to $2.03 \times 10^{-2}m^4/s^3$ yields deepening of mixing height. Some observed data with maximum mixing depth $h_{max} \approx 0.42l_o F_o'$ and mixing length $x_{max} \approx 15l_o F_o'$ are constructed for further modeling of heat dispersion.

According to the previous study (Suh, 1997), the NF is only limited within $200m$ - $300m$ in this power station. As described the characteristics of NF are extremely different from those of FF, thus a coupling the NF results into

FF was considered. As a coupling method, a simple welding method was considered due to its simplicity. First the CORMIX3 was run for several cases depending on ambient flow conditions. Basically it is a steady model it cannot account for time varying unsteady conditions, so the results of each typical discrete time level are placed as patches on finite element meshes. After putting NF values as patches on elements they are treated as sources and obeying Gaussian puff distribution with time integration.

Real application of proposed scheme

To test the feasibility of developed hybrid model for surface heat transport, this model is applied to Boryong thermal power plant as described. In this simulation, as a trial approach we test 1000, 3000 and 5000 particles with buoyancy. And from the field measurements we set $x_{NF} = 300m$, $\Delta\rho = 1.04 Kg/m^3$. Sensitivity tests of parameters are done. As a reference value, we have $D_{max} = 100 \ m^2/s$, $D_x = 5 \ m^2/s$, and $D_y = 10 \ m^2/s$.

Computed excess temperatures for typical ebb and flood flows are drawn in Fig.(6) with CORMIX3 results. It shows satisfactory tendencies compared with field data. As in the simplified model test, it also shows a little bit blow up spreading to offshore direction, which is not seen in CORMIX3 simulation. However according to the analysis of observed data these results are believed more realistic.

Conclusion

In this study a new method for heat dispersion around coastal power plant was proposed by using buoyant particles tracking method on Eulerian-Lagrangian finite element models. In NF these buoyant particles cause horizontal spreading, which is eventually converted as extra additive dispersion coefficients. Model tests for neutral particles show that numerical results are in good agreement with analytical solutions. For buoyant particles simulation in both ideal geometry and real power plant, computed results represent realistic distribution of excess temperature in comparison with CORMIX3 and field observed data. Thus this proposed scheme can predict both NF and FF heat dispersion with a reasonable accuracy.

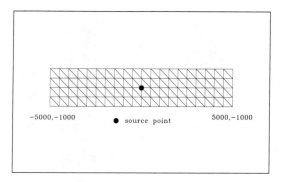

Figure 1: FEM grids for random-walk neutral particle tracking test

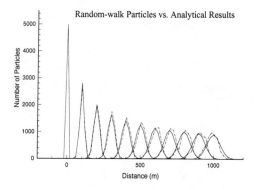

Figure 2: Comparison of random-walk particle distribution with analytical solution for 5000 particles. Solid lines show analytical solution with every 10 seconds, whereas dashdots show corresponding numerical solution. $u = 10\ m/s,\ D_x = 10\ m^2/s,\ h = 10\ m$.

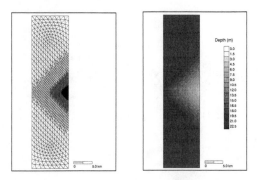

Figure 3: FE meshes and topography for buoyant particle tracking

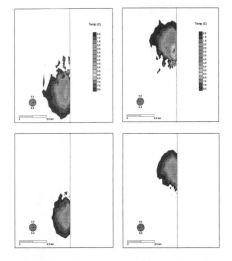

Figure 4: Buoyant random-walk particle tracking results in terms of excess temperature around discharge point at ebb and flood tide. Upper figures represent considering NF buoyant effect in case of $D_{max} = 100 \ m^2/s, D_{x_i} = 10 \ m^2/s$ for $N_p = 500$. Lower ones are for neutral particles.

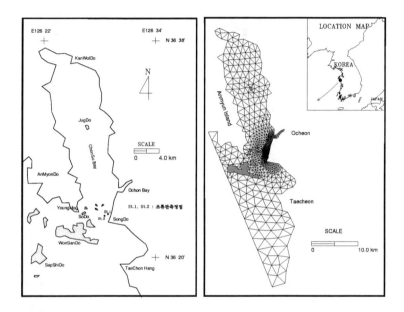

Figure 5: Map showing study area and FE grids for real application of buoyant particle EL modeling

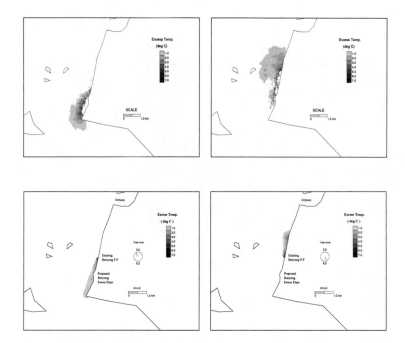

Figure 6: Computed excess isotherms based on buoyant particles at ebb and flood flow comparing with hybrid model results of ELM with CORMIX3.

Acknowledgment

This research is supported, in part, by KOSEF (951-1201-011-2).

References

Baptista, A.M. Adams, E.E. and Stolzenbach, K.D. (1984). Eulerian Lagrangian Analysis of Pollutant Transport in Shallow Water, R296, Ralph M. Parsons lab., MIT.

Dimou, K. (1992). "3-D Hybrid Eulerian-Lagrangian / Particle Tracking Model for Simulating Mass Transport in Coastal Water Bodies", MIT Ph.D. Thesis.

Dimou, K. and Adams, E.E. (1993). "A Random-Walk, Particle Tracking Model for Well-mixed Estuaries and Coastal Waters", **Estuarine, Coastal and Shelf Science, 37**, pp 99-110.

Suh, S.W. (1997). "Two-Dimensional Finite Element Analysis for Surface Discharged Heat Water", *Proceedings of the 10th International Conference for Numerical Methods in Thermal Problems*, Swansea, UK, pp 932-941.

Tang, L. and Adams, E.E. (1995). "Effect of Divergent Flow on Mass Conservation in Eulerian-Lagrangian Schemes", *4th Int'l Conference on Estuarine and Coastal Modeling*, pp106-115.

Zhang, X.Y. (1995). "Ocean Outfall Modeling - Interfacing Near and Far Field Models with Particle Tracking Method", MIT Ph.D. Thesis.

Models for Suspended Sediment Dispersion and Drift

C.G. Hannah[1], Z. Xu[1], Y. Shen[1] and J.W. Loder[1]

Abstract

Extensions of a model for the dispersion and drift of suspended material in the benthic boundary layer are described. These include continuous (as well as bulk) release of material, horizontal variability in the ocean environment, and an improved representation of the near-bottom region. Exploratory applications to the fate of drilling wastes on Georges Bank are presented, using current profiles from moored measurements and a 3-d numerical circulation model. The sensitivity of dispersion and drift to the material's settling velocity, to spatial and temporal variability in the environment, and to the duration of the release is discussed.

Introduction

A set of models for the dispersion and drift of suspended sediment in the benthic boundary layer on the continental shelf is being developed. The motivation is the estimation of the fate of wastes discharged from hydrocarbon drilling, and their potential impacts on benthic organisms such as the sea scallop on Georges Bank (see Gordon et al. 1993 and Muschenheim et al. 1995 for background). The primary interest is the near-bottom concentration of dense material in the near field around a release site.

The formulation and initial testing of the basic model, referred to as bblt for benthic boundary layer transport model, is described by Hannah et al. (1995, 1996). Key features of the model include the important physical process of vertical shear dispersion (horizontal dispersion due to vertical mixing and vertical shear in horizontal currents), and the flexibility to be forced by time-varying current profiles from either observations or a 3-d numerical model. Herein we report on several recent model extensions. The basic model, which assumes no horizontal variations in the ocean environment, has been extended to allow continuous as well as bulk releases of sediment, and is now referred to as 'local bblt'.

---

[1]Fisheries and Oceans Canada, Bedford Institute of Oceanography, Dartmouth, N.S., B2Y 4A2, Canada

A new version, referred to as 'spatially-variable `bblt`', which allows full spatial variability in water depth, currents and bottom stress, is under development and will be briefly described. The representation of the near-bottom sublayer has been improved, and diagnostics have been identified for the interpretation of model results. We will use these enhancements to investigate the sensitivity to temporal and spatial variability of the currents, to the duration of release, and to settling velocity of the suspended sediment.

The model extensions and sensitivities will be illustrated using exploratory applications to the Northeast Peak of Georges Bank. Hypothetical cases of either a bulk or continuous release of material will be considered. Forcing by moored current measurements or a 3-d circulation model will be used. The applications are for summer when the tidal and mean flows dominate the currents on Georges Bank.

Methods

Basic Model

The basic features of `bblt` are as follows. The sediment load is partitioned into discrete pseudo-particles or 'packets' each with mass m and settling velocity w_s. These packets are advected horizontally and mixed vertically. The overall (horizontally-averaged) vertical distribution of the sediment is governed by a specified quasi-equilibrium concentration profile which is used to derive a probability density function for the vertical position of the packets. Vertical mixing is represented by random exchange (shuffling) of the packets, which is strongly influenced by a specified vertical mixing time scale, t_m. With the value of $t_m = 3$ hr used here, shear dispersion associated with the vertical shear in the semi-diurnal tidal current is near its maximum value, while dispersion associated with the vertical shear in the mean current is near the low end of its range. See Hannah et al. (1995, 1996) for further details.

The concentration profile used herein is a modified Rouse profile

$$c(z) = c_a(a/z)^p \quad ; \quad p = w_s/(\kappa u_*) \tag{1}$$

where c_a is the concentration at the reference height a, z the vertical coordinate ($z \geq a$ and positive upwards), κ the von Karman constant and u_* the friction velocity (e.g. Rouse 1937, Dyer 1986). The reference height a is taken as the lower boundary of the suspended sediment profile, with a value here of 1 cm. Bedload transport is presently neglected.

The time-varying velocity is specified at a number of levels in the vertical, either from current meter observations or a numerical model. Linear interpolation is used to estimate the velocity between these levels. Below the bottom level, a logarithmic boundary sublayer is used with the current speed given by:

$$u(\xi) = u_b + \frac{u_*}{\kappa} \ln \xi/\xi_b \quad ; \quad a \leq \xi \leq \xi_b \tag{2}$$

where ξ_b is the height above the actual seafloor of the lowest model level or the bottom current meter, and $u_b = u(\xi_b)$. The current direction in this sublayer is taken as vertically-uniform.

For the simulations discussed here, the friction velocity is calculated using a quadratic bottom stress law $u_*^2 = C_d u_b^2$ where C_d is the drag coefficient. If the lowest current data level is 1 m above the sea floor then $C_d = 0.005$ (Lynch and Naimie, 1993); otherwise C_d is adjusted to yield an equivalent stress (based on the assumed logarithmic boundary layer). The near-bottom logarithmic sublayer and the quadratic stress law are extensions beyond Hannah et al. (1995) whose significance will be discussed elsewhere.

Local bblt

The basic development of bblt, and its extensions and evaluations in areas such as the near-bottom sublayer, temporal variability and vertical mixing parameterization, have been carried out using the local version which neglects horizontal variations in water depth, currents and bottom stress (Hannah et al. 1995, 1996). This simplification results in a major reduction in the computational demand of bblt, while retaining the effects of vertical and temporal variations in the current which are generally the primary factors in short-term dispersion and transport at local positions. Another major advantage of local bblt is that it can be readily forced with velocity-profile time series from either a circulation model or moored measurements, without requiring the difficult horizontal interpolation and extrapolation of observed currents for the typical situation of sparse horizontal sampling. The primary evaluations in this report use local bblt.

One key extension of local bblt is the capability to consider a continuous release or input of sediment, as well as the bulk or instantaneous release considered in Hannah et al. (1995, 1996). This extension involves allowing the number of packets N to increase as a function of time. Otherwise, the implementation is consistent with the bulk release version.

Spatially-Variable bblt

Another key extension of bblt is the ongoing development of a spatially-variable version which allows horizontal variations in the water depth, currents, and bottom stress. This version requires a (time-varying) 3-d velocity field, generally derived from a numerical circulation model. The primary changes are that: 1) the velocity field must be evaluated at each packet's location; and 2) the calculation of a packet's vertical position requires that the vertical probability density function be evaluated based on the bottom stress at the packet's horizontal location. These changes lead to a significant increase in computational cost.

Discharge Cases and Forcings

For illustrative purposes the exploratory model applications presented here will all consider the discharge of 53 tonnes of material into the benthic boundary layer. The discharge will be prescribed as either a bulk release at the start time or a uniform continuous (hourly) release over 5 days. Model results will

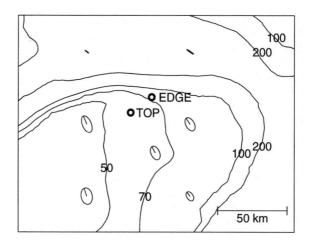

Figure 1: The Northeast Peak of Georges Bank showing the Edge and Top sites in relation to bathymetry (in m). Also shown are the M2 tidal ellipses (scaled to the actual tidal excursion).

be presented for settling velocities of 0.2 and 1.0 cm/s which approximate the range expected for flocculated drilling wastes, based on recent laboratory studies (Muschenheim and Milligan 1996).

Results with observational current forcing will be presented for a frontal-zone site (42° 05'N, 66° 48'W) near the northern edge of Georges Bank (Figure 1), where there are summer current meter observations at 4 heights above the bottom (3 m, 10 m, 32 m, 56 m) in 67 m of water (Loder and Pettipas, 1991). The value of the drag coefficient adjusted for the 3-m height is $C_d = 0.0035$. This 'Edge' site is near the NEP2 site used in Hannah et al. (1995, 1996).

Results with forcing by currents from a 3-d circulation model will be presented for the Edge site and for a 'Top' site (water depth 60 m) on the central Bank (Figure 1), using both the local and spatially-variable versions of bblt. The 3-d model currents are the seasonal-mean, M2 and M4 flow components for September-October from the finite-element model solutions of Naimie (1995). The model is prognostic with an advanced turbulence closure (Naimie 1996; Lynch et al. 1996), and is forced by M2 tides and observed seasonal-mean wind stress and ocean density fields.

For the Edge site, three different forcing families are used: 1) the circulation model currents; 2) current time series reconstructed from harmonic analysis of the observations (using tidal constituents M2, N2, S2, M4, K1, O1 and the mean Z0); and 3) the observed half-hourly current time series. For the local bblt applications with observationally-based forcings, 5-day periods were selected around the times of maximum and minimum currents (in July 1988), which we call spring and neap tides respectively.

Statistical Measures and Interpretative Diagnostics

Bulk measures of the horizontal dispersion and transport of the sediment can be obtained from the statistical moments of the vertically-integrated horizontal distribution. The mean drift velocity of the horizontal center of mass is denoted $(\bar{U}, \bar{V})$. An effective horizontal diffusivity can be computed as

$$K' = \frac{1}{2} \frac{\partial \bar{\sigma}}{\partial t} \tag{3}$$

(e.g. Csanady 1973) where σ is the horizontal variance of the packet positions and the overbar denotes a time averaging which smoothes out short-term fluctuations such as the tides. For two horizontal dimensions we compute the diffusivities along the axes of maximum (K'_{max}) and minimum (K'_{min}) variance.

The instantaneous value of the vertical center of mass, Z, provides a measure of the extent of the (horizontally-averaged) vertical distribution.

Analytical solutions for idealized advection-diffusion problems point to key scaling parameters which can be used to help interpret the bblt results. For a bulk release, the 2-d diffusion equation (no advection), indicates that the concentration, C, at the source should scale as $C \propto 1/D$, where $D = \sqrt{K'_{max}K'_{min}}$ (Fischer et al. 1979). With advection, the concentration following the (horizontal) center of mass can be expected to follow the same scaling.

For a continuous release, consider the steady-state solution to the 2-d advection-diffusion equation, in the limit that the diffusion in the direction of the advection can be neglected. The concentration levels scale as $C \propto 1/\sqrt{D_y u}$ (Fischer et al. 1979), where u is the (constant) mean drift velocity in the x direction and D_y is the (constant) diffusivity in the y direction. Herein we approximate $D_y \sim \sqrt{K'_{max}K'_{min}}$ and use the the mean drift speed for u, where the values are taken from the corresponding bulk release. At time t a steady state should be established at a distance $x = ut$ downstream from the source.

These scalings were derived for the depth-integrated concentration and ignore changes in the vertical distribution of the sediment. Thus, they can only be approximate indices of near-bottom concentration. However for most of the cases examined here, the sediment settles to the near-bottom region once or twice every M2 tidal cycle, so the scalings are also representative of maximum concentration averaged over a specified near-bottom interval. Here we take this interval as 1 to 20 cm above the bottom (where the lower limit is the reference height a) and refer the average concentration over this interval as the near-bottom concentration.

Temporal Variability

In this section we examine how the dispersion and transport vary with settling velocity and temporal variability in the velocity forcing and sediment discharge, using local bblt.

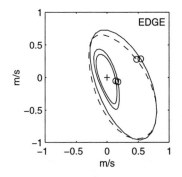 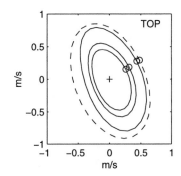

Figure 2: Velocity profiles at the Edge and Top sites comprising the mean, M2 and M4 components. The track of the head of the velocity vector during an M2 tidal cycle is shown for 4 heights above the bottom: z = (1, 3, 37) m and surface (dashed) at the Edge site, and z = (1, 3, 30) m and surface (dashed) at the Top site. The circles indicate a common time.

Table 1: Summary statistics for bulk releases at the Edge Site using local `bblt` and different current forcing. The range of Z over the simulation and the speed and direction of $(\bar{U}, \bar{V})$ are given.

forcing	Z	K'_{max}	K'_{min}	speed	direction	$1/D$
	(m)	(m²/s)	(m²/s)	(cm/s)	(deg. T)	(s/m²)
$w_s = 0.2$ cm/s						
circulation model						
M2+Z0+M4	22/33	106	60	17	116	0.012
observations - harmonic reconstruction						
spring tide	11/26	160	92	17	115	0.008
neap tide	3/26	110	50	16	122	0.014
$w_s = 1.0$ cm/s						
circulation model						
M2+Z0+M4	0.02/13	92	13	7.1	147	0.029
observations : harmonic reconstruction						
spring tide	0.02/18	198	28	10.4	146	0.013
neap tide	0.01/9.6	87	7.2	5.6	152	0.040

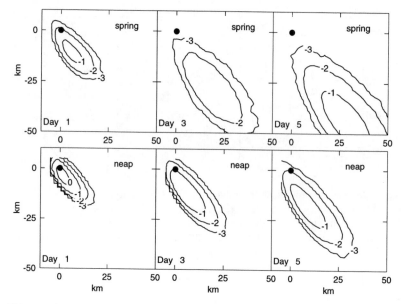

Figure 3: Snapshots of near-bottom concentration for a bulk release at the Edge site with $w_s = 1$ cm/s, and observational current forcing (harmonic reconstruction) for spring and neap tides. The release site is at the origin, and the isolines are $\log_{10}$ of the concentration in g m^{-3}. Aliasing of the high-frequency fluctuations in Figure 4 is apparent in the spring tide snapshots.

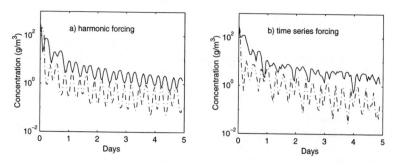

Figure 4: Near-bottom concentration following the horizontal center of mass for bulk releases at the Edge site with $w_s = 1$ cm/s, and observational current forcing. The solid lines are neap tides and the dash-dot lines are spring tides.

The vertical structure of the mean, and M2 and M4 tidal currents is illustrated in Figure 2 using the current ellipses from representative vertical levels in the 3-d model field (Naimie 1995). The clockwise-rotating M2 component dominates at both sites with peak speeds near 0.9 m/s and larger amplitudes at the Top site, particularly in the lower water column. The Edge site also has a strongly-sheared southeastward mean current with peak speed near 0.3 m/s, and a small near-bottom M4 component.

Bulk Releases at Edge Site

The long-term evolution of the spatial distribution of sediments for bulk releases is illustrated in Figure 3 using observational forcing for spring and neap tides, $w_s = 1$ cm/s, and near-bottom concentrations. After one day the basic shape approximating the tidal ellipse (Figure 2a) is established, and then the sediment drifts and spreads out (Hannah et al., 1995, 1996). The spreading is faster, and the concentrations lower during spring tides. The patterns are similar for the depth-averaged concentrations, but the values are lower.

For the near-bottom concentrations there are large variations over the tidal period superimposed on the longer-term dilution (Figure 4). These variations arise from expansion and collapse of the vertical distributions due to time variability in the bottom stress. There is a rapid decrease in concentration over the first day and then a slow decrease with higher-frequency fluctuations. In general, the rate of the long-term dilution and the amplitude and structure of the high-frequency fluctuations depend on the details of the observational forcing. For the present site and season where the tidal and mean currents dominate, the long-term dilution rate is similar for the reconstructed and original observational time series (Figure 4). The primary sensitivities in these cases are the increased overall dilution rate and the smaller minimum concentrations (because of the greater vertical extent of the sediment) during spring tides.

Summary statistics for representative bulk release cases at the Edge site are given in Table 1 for settling velocities $w_s = 0.2$ and 1 cm/s. The range of the vertical center of mass reflects the vertical distribution, while the diffusivities and mean drift indicate the dispersion and transport. The depth-averaged concentration following the horizontal center of mass is well predicted by the $1/D$ scaling parameter for these cases, so its value is included as a concentration index (it is also a rough index of the maximum near-bottom concentration but the latter has additional scatter associated with differences in vertical structure). The results can be summarized as follows:

1. The dispersion rates are higher and the long-term concentrations are substantially lower for the lower settling velocity, due to greater shear dispersion in the upper water column at this site (associated with the sheared mean flow).

2. The mean drift is also higher (and shifted in direction) for the lower settling velocity due to the near-surface intensification (and directional shift) of the

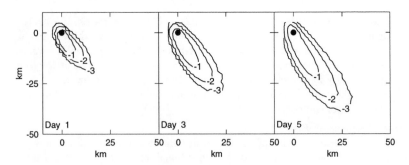

Figure 5: Snapshots of near-bottom concentration for a continuous release at the Edge site with neap-tide forcing (harmonic reconstruction) and $w_s = 1$ cm/s. The isolines are $\log_{10}$ of the concentration in g m^{-3}.

mean current (Figure 2).

3. The model results are bracketted by the spring and neap tide results, pointing to approximate agreement of the model and observed tidal and mean currents.

4. The long-term concentrations increase by a factor of 2-3 between spring and neap tides, associated with current speed decreases of about 40%. For the lighter sediment the change is primarily associated with the reduced current shears (and associated shear dispersion) during neap tides, while for the heavier sediment, the larger change in dilution rate is associated with both the reduced shears and reduced vertical extent of the sediment.

Continuous Releases at Edge Site

The evolution of the spatial distribution for a continuous release is illustrated in Figure 5 for the same neaps forcing case as in Figure 3. The material spreads out over the tidal ellipse with similarities to the bulk release case (Figure 3), but the concentrations are generally lower due to the protracted release and the area of peak concentration remains connected to the discharge position during the 5-day release period. Based on the mean drift for the corresponding bulk release case (Table 1), the 2-d advection-diffusion solution noted above suggests that a steady state should be established at 24 km from the source after 5 days (at 5.6 cm s^{-1}). One can see that the leading edge of the 10^{-1} isoline which is 20-25 km from the source has only a limited advance between days 3 and 5, supporting the applicability of the continuous-release scaling parameter described above.

Summary statistics for continuous release cases with $w_s = 1$ cm/s and the observational current forcings are shown in Table 2. The scaling parameter computed using diffusivities and drift velocities from the corresponding bulk release cases (so that these quantities are reflective of the water movement) is

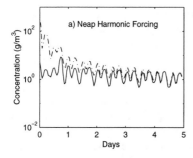
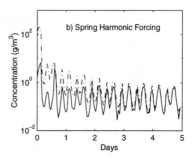

Figure 6: Comparison of the near-bottom concentration following the horizontal center of mass for bulk (dash-dot lines) and continuous (solid lines) releases at the Edge site with $w_s = 1$ cm/s (all from harmonic reconstruction).

Table 2: Summary statistics for continuous releases cases at the Edge site ($w_s = 1$ cm/s). C_f is the maximum near-bottom concentration over the last day of the simulation. The values for D and u are taken from the corresponding bulk release cases reported in Table 1. Other columns are as in Table 1.

	K'_{max} (m²/s)	K'_{min} (m²/s)	speed (cm/s)	direction (deg T.)	C_f (g/m³)	$1/\sqrt{Du}$
observations - harmonic reconstruction						
spring tide	307	15.6	5.0	145	0.64	0.35
neap tide	102	5.0	2.9	153	2.3	0.84
observations - half-hourly time series						
spring tide	314	18	5.8	139	0.71	0.37
neap tide	90	2.9	2.7	151	2.5	1.0

included, and shows rough agreement with the day 5 concentrations. There is again a factor of 3 increase in the long-term concentrations between the spring and neap cases, and little difference between the results for the reconstructed and original current time series. Figure 6 shows that, for a particular forcing, the concentration (following the center of mass) in the continuous release case is initially much lower than in the bulk release case, since there is much less sediment in the water. However by the end of the continuous release period, the difference has largely disappeared.

Comparison of the summary statistics for the continuous release cases (Table 2) with those for the corresponding bulk release cases in Table 1 points to substantial differences in the effective dispersion and drift rates *during* the continuous release period. The effective dispersion rate along the major axis (K'_{max}) is substantially increased by the protracted discharge through the release of sediment into different water parcels because of the mean and tidal drifts. On the other hand, the effective drift is substantially reduced due to the delay in release of much of the sediment. The magnitude and longer-term significance of these differences should increase with the mean current strength.

Spatial Variability

Local bblt: Top vs Edge Sites

The pronounced horizontal gradients in the mean and tidal currents on Georges Bank (e.g. Lynch and Naimie 1993; Naimie 1996) provide the potential for both strong horizontal variations in dispersion rates and significant influences of horizontal structure on the magnitude of the rates at particular positions. Although the current strength differences between the Edge and Top sites (Figure 2) are small compared to bank-scale variations, these sites illustrate the potential significance of horizontal structure.

We start with applications of local bblt to bulk discharges at these sites, using the two different settling velocities and currents from the 3-d model. The summary statistics in Table 3 indicate that there are significant differences between the sites, with the magnitude and even the sign of the difference depending on the settling velocity. For the lower settling velocity where the sediment is distributed over the water column, dispersion and drift are greater at the Edge site because of the much stronger (sheared) mean current and the greater vertical extent of the tidal current shear. In contrast, for the higher settling velocity where the sediment is much more confined to the lower water column, drift is greater at the Edge site due to the stronger mean current but the overall dispersion rate is slightly higher at the Top site (see the scaling parameter in Table 3) because of the larger near-bottom tidal current shear (Figure 2). These examples point to the importance of horizontal variations in the current's vertical structure.

Spatially-Variable bblt

The various local bblt applications indicate that, within a few days, there can be significant spreading and/or drift of the sediment over distances of 20-40 km (Figures 3 and 5). On the northern edge of Georges Bank there are significant spatial changes in both the mean and tidal currents over such distances, as illustrated in Figure 7a,b. To investigate the quantitative importance of these changes to dispersion and drift, we consider bulk-release applications of spatially-variable bblt for the same two sites and settling velocities, using the spatially-variable currents, bottom stress and depths from the 3-d circulation model.

The summary statistics (Table 3) indicate that spatial variability can have significant influences on both drift and dispersion, and these influences depend on site (local current structure) and settling velocity. For the Edge release site, the inclusion of spatial variability results in an increase in the overall dispersion rate (as measured by $1/D$) and a decrease in the drift rate for both settling velocities. The increased dispersion is qualitatively consistent with the expected effect of the horizontal spreading being enhanced by horizontal current structure in the frontal zone through an additional shear dispersion mechanism involving

Table 3: Summary statistics for bulk releases forced with the mean + M2 + M4 currents from the circulation model. Otherwise as for Table 1.

w_s (cm/s)	Z (m)	K'_{max} (m²/s)	K'_{min} (m²/s)	speed (cm/s)	direction (deg T.)	$1/D$ (s/m²)
Edge site: local bblt						
0.2	22/33	106	60	17	116	0.012
1.0	0.02/13	92	13	7.2	147	0.029
: spatial bblt						
0.2	28/45	165	55	13	123	0.010
1.0	0.08/15	70	39	5.0	155	0.019
Top site: local bblt						
0.2	24/28	30	12	3.0	78	0.053
1.0	0.6/15	81	23	3.2	80	0.023
: spatial bblt						
0.2	25/31	61	36	4.9	110	0.021
1.0	1.4/15	62	21	2.8	115	0.028

horizontal mixing across horizontal shear (in both the tidal and mean flows). The decrease in drift rates for the Edge site is consistent with the smaller mean velocities in the downstream direction to the southeast of the release site, where the frontal-zone jet weakens (Figure 7a). There is also a change in the shape of the spreading sediment patch (e.g. Figure 7d-f vs. Figure 7c), which is qualitatively consistent with the increased dispersion and the downstream weakening and turning of the mean current.

For the Top release site the inclusion of the spatial variability results in increases in drift and dispersion for the smaller settling velocity, and decreases in drift and dispersion for the larger settling velocity. These changes illustrate the potential for opposing influences from the inclusion of spatial variability, and hence for opposite results at different sites or for different vertical sediment distributions. While an enhancement of local dispersion can generally be expected with the inclusion of spatial variability in flows without strong convergence zones, there may be additional tendencies for the dispersion rate of a particular sediment patch to either increase or decrease depending on whether it drifts into areas with greater or lower dispersion than at the release site. Applications of local bblt at various sites on the Northeast Peak (not reported here) indicate that the dispersion and drift rates decrease to the south of the Top site, which can explain why the net drift and dispersion rates from spatially-variable bblt can decrease in some cases, such as for the higher settling velocity in Table 3.

Further applications and evaluations with spatially-variable bblt are clearly required to determine the magnitude and spatial extent, and understand the mechanisms of horizontal structure influences on the dispersion and drift patterns and rates. The present exploratory applications point to important influences

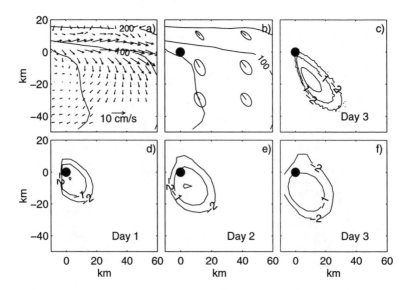

Figure 7: Comparison of local and spatially-variable bblt results for a bulk release at the Edge site with $w_s = 1$ cm/s: a,b) near-bottom mean currents and depth-average tidal ellipses from the 3-d circulation model; c) snapshot of near-bottom concentration at day 3 from local bblt; and d,e,f) snapshots of near-bottom concentration from spatially-variable bblt. The isolines in c-f) are $\log_{10}$ of the concentration in g m^{-3}.

and the potential value of the model extension in this area.

Discussion

The bblt models are computational tools for evaluating the significance of vertical sediment distribution, and spatial and temporal ocean variability to the horizontal dispersion and transport of suspended sediment, and for estimating the longer-term fate of dense discharges in different settings. The bblt results presented here and previously (Hannah et al., 1995, 1996), and our general understanding of shelf current structure and associated dispersion (e.g. Fischer et al. 1979; Zimmerman 1976), indicate that dispersion and transport involve a complex interplay of the sediment properties affecting its vertical distribution, and spatial and temporal structure in the ocean currents and vertical mixing (parameterized here through the bottom stress and the mixing time scale).

The results from local bblt are most relevant to the first few tidal periods when the horizontal sediment extent (for an initial point distribution) approximates the tidal ellipse and horizontal current changes are generally limited. Vertical shear is the dominant spatial current structure during this early phase and, in combination with vertical mixing of the sediment (by turbulence or sinking), results in rapid horizontal spreading through vertical shear dispersion (e.g. Fischer et al. 1979). The local bblt applications presented here illustrate how the spreading rate can depend on the vertical sediment distribution (settling velocity) in relation to vertical shear in both the mean and tidal currents, how the spreading rate can vary with spring-neap tidal changes through both shear strength and effects on vertical mixing, and how vertical distribution affects the drift rate. The bulk and continuous release comparisons point to important effects on short-term concentrations, but to more limited effects after the cessation of a protracted release over several tidal periods.

The results from spatially-variable bblt indicate that, in complex and energetic areas such as Georges Bank, horizontal dispersion and drift rates can be either increased or decreased depending on the current structure and vertical distribution of the sediment. Horizontal shears in the currents provide a general tendency for increased dispersion, even during the first several tidal periods, because the initial vertical shear dispersion can be sufficient to spread the sediment across these shears so that an additional (horizontal) shear dispersion process involving horizontal mixing and horizontal shear becomes significant (Zimmerman 1976). On the other hand, the sediment can drift and disperse into regions with either higher or lower drift and dispersion rates, so that the magnitude and even the sense of the net effect can vary from site to site.

These results illustrate the hierarchy of spatial current structures involved in benthic boundary layer dispersion on the continental shelf. Additional complexity can be expected with additional temporal current structure such as associated with variable winds. Ongoing sensitivity evaluations, model improvements and observational validations are required for realistic and reliable applications to dispersion issues in a spatially- and temporally-variable environment such as the continental shelf.

Acknowledgements

We are grateful to Chris Naimie for providing the 3-d circulation model solutions; Don Gordon and Kee Muschenheim for ongoing input; Kee Muschenheim and Brian Petrie for internal reviews; and the (Canadian) Federal Panel for Energy, Research and Development (PERD) for funding.

References

Csanady, G. T. 1973. *Turbulent Diffusion in the Environment*. D. Reidel Publishing Company.

Dyer, K. R. 1986. *Coastal and Estuarine Sediment Dynamics*. Wiley and Sons.

Fischer, H. B., E. J. List, R. C. Y. Koh, J. Imberger and N. H. Brooks. 1979. *Mixing in Inland and Coastal Waters*. Academic Press.

Gordon, D. C., P. J. Cranford, D. K. Muschenheim, J. W. Loder, P. D. Keizer and K. Kranck. 1993. Predicting the environmental impacts of drillings wastes on Georges Bank scallop populations. In P. M. Ryan, editor, *Managing the Environmental Impact of Offshore Oil Production*, pages 139–147. Canadian Society of Environmental Biologists.

Hannah, C. G., J. W. Loder, and Y. Shen. 1996. Shear Dispersion in the Benthic Boundary Layer. In: Estuarine and Coastal Modelling, Proceedings of the 4th International Conference, edited by M. Spaulding and R. Cheng, pp. 454-465. ASCE, New York.

Hannah, C. G., Y. Shen, J. W. Loder and D. K. Muschenheim. 1995. bblt: Formulation and Exploratory Applications of a Benthic Boundary Layer Transport Model. Can. Tech. Rep. Hydrogr. Ocean Sci. 166: vi + 52 pp.

Loder, J. W. and R. G. Pettipas. 1991. Moored Current and Hydrographic Measurements from the Georges Bank Frontal Study, 1988-89. Can. Tech. Rep. Hydrogr. Ocean Sci. 94: iv + 139 pp.

Lynch, D.R., J.T.C. Ip, C.E. Naimie, and F.E. Werner. 1996. Comprehensive coastal circulation model with application to the Gulf of Maine, *Continental Shelf Research*, **16**, 875-906.

Lynch, D.R., and C.E. Naimie. 1993. The M2 tide and its residual on the outer banks of the Gulf of Maine. *J. Phys. Ocean.*, **23**, 2222-2253.

Muschenheim, D. K. and T. G. Milligan. 1996. Flocculation and accumulation of fine drilling waste particles on the Scotian Shelf. *Mar. Pollut. Bull.* **32**, 740-745.

Muschenheim, D. K., T. G. Milligan and D. C. Gordon. 1995. New technology and suggested methodologies for monitoring particulate wastes discharged from offshore oil and gas drilling platforms and their effects on benthic boundary layer environment. Can. Tech. Rep. Fish. Aquat. Sci. 2049: x + 55 p.

Naimie, C. E. 1995. Georges Bank Bimonthly Residual Circulation – Prognostic Numerical Model Results. Report # NML 95-3, Numerical Methods Laboratory, Dartmouth College, Hanover NH.

Naimie, C. E. 1996. Georges Bank residual circulation during weak and strong stratification periods - Prognostic numerical model results. *J. Geophys. Res.* **101**, 6469–6468.

Rouse, H. 1937. Modern concepts of the mechanics of turbulence. *Trans. Am. Soc. Civ. Eng.*, **102**, 463–543.

Zimmerman, J.T.F. 1976. The tidal whirlpool: a review of horizontal dispersion by tidal and residual currents. *Neth. J. Sea Res.* **20**, 133-154.

Wave and Sedimentation Modeling for
"Bahia San Juan de Dios", Mexico

Claudio Fassardi[1] and J. Ian Collins[2], Member ASCE

Abstract

A description is given of the approach and the different numerical models that were used to investigate the coastal processes that affect the "Bahia San Juan de Dios" on the Pacific coast of Mexico. These processes affect, in particular, the entrance to a small craft harbor which is subjected to sedimentation and wave agitation. The lack of wave information required the derivation of an average local wave climate from offshore storms. The wave input was used to model the bay's shoreline evolution and compute the sediment transport rates at the harbor entrance.

Introduction

During the summer and autumn months the entrance to a harbor, located in the "Bahia San Juan de Dios" on the Pacific coast of Mexico, suffers from the occasional appearance of 2-3 meter waves on otherwise apparently calm days. In combination with the shoaling of the entrance these waves break in the entrance and present a hazard to vessels attempting to enter or leave the harbor.

The precise reasons for the appearance of large waves at the harbor entrance were not known. It was hypothesized that they were swells from distant storms, focused at the harbor entrance by the presence of offshore islands and associated shoals. Even under such conditions, these waves would be a minor hazard to navigation unless local breaking was induced by shoals in the harbor entrance. During the first few years of existence of the harbor it was noted that shoaling developed in the entrance as a result of longshore sediment transport. The origins and magnitude of this transport had to be

1 Senior Engineer, *Scientific Marine Services, Inc.,* 101 State Place - Suite N, Escondido, CA 92029
2 Vice President, *Scientific Marine Services, Inc.,* 101 State Place - Suite N, Escondido, CA 92029

determined in order to evaluate the merits of alternative mitigation measures to decrease the sedimentation and therefore the severity of the wave agitation at the entrance.

No direct measurement of the local wave climate was available. Numerical modeling was undertaken to determine the conditions that could produce the observed sedimentation and local wave shoaling. It was concluded that the arriving swell during the summer and autumn months was dominated by waves generated by tropical storms that run parallel to the Pacific coast of Southern Mexico. A statistical analysis revealed that the annual storm activity in an average year could be closely represented by a combination of hurricane and tropical storms traveling approximately 200 nautical miles offshore in a shore-parallel direction.

Typical parameters for these storms were used in a numerical model to generate time varying pressure and wind fields. These fields were then used to generate time series of the offshore wave conditions using a time dependent, discrete, spectral model based on the energy balance equation. The typical offshore waves generated in the storms were modified to determine their intensity in the vicinity of the beach and harbor entrance by means of a linear refraction-diffraction model (a parabolic mild slope approximation).

Historical nautical charts dating back to 1945 and more recent aerial photographs were used to calibrate and verify a shoreline response model. Finally, use was made of this model to determine the sediment transport rates that affect the harbor entrance.

Site Observations

A chart of "Bahia San Juan de Dios", located on the Pacific coast of Mexico, is shown in Figure 1.

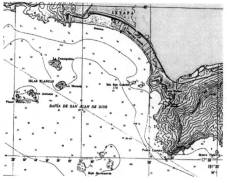

Figure 1. Bahia San Juan de Dios, Mexico.

The bay, located approximately 200 km north of Acapulco at 17° 39' N and 101° 37' W, features a rocky coast on the east side and a sandy 2,400 m long "pocket" beach that ends at the harbor entrance on the west side. The "Islas Blancas" are situated less than a nautical mile offshore from the harbor entrance.

During a three day visit to the site in the month of October, small amplitude swell was arriving at the bay. It was observed that the swell direction changed with time, arriving initially from approximately 190 degrees, slowly turning and arriving from 220 degrees at the end of third day. This change in wave direction was consistent with what would be produced by storms that run parallel to the Pacific coast of Mexico. The typical sequence of arrival of waves generated by these storms to the bay would start with intense swell from 160 to 190 degrees. Then, a period of somewhat less intense waves would follow as the swell is blocked by the presence of the offshore islands. As the storm moves towards the northwest waves would increase in intensity and arrive from 220 to 230 degrees.

The consequence of this wave sequence is that initially the sediment is transported along the beach towards the west by the first set of waves that arrive from the east side of the "Islas Blancas". As the wave direction shifts towards the west, the littoral drift decreases due to the blocking action on the waves by the islands and the sediment settles in the vicinity of the harbor entrance. As the waves arrive from the west side of the islands, the sediment is moved back to the beach with a fraction of it being transported into the harbor entrance. A sequence of storms would eventually build a sand bar at the entrance which, in combination with the storm swell, would produce the high waves and breakers that are occasionally observed at the harbor entrance.

Based on this hypothesis, the first objective would be to prevent the arrival of the sediment at the harbor entrance and secondly, to reduce the wave agitation. Wave measurements and observations were non-existent for this area and therefore a wave hindcast from storms was performed to derive offshore wave conditions. The wave climate in the vicinity of the beach and harbor entrance for the sediment transport study was derived from the offshore wave conditions by performing a refraction-diffraction analysis.

Offshore Wave Analysis

The "Global Tropical/Extratropical Cyclone Climatic Atlas" (1994) was used as the source of storms that affect the area. Storm data between 1980 to 1992 was used since it was noted to be the most complete for modeling purposes. Between those years an average of 10 storms per year have passed by the site. Of those 10 storms, approximately 4 become "Tropical Storms" (wind speed > 34 knots) and 2 reach Hurricane status (wind speed > 64 knots) while the other 4 are classified as "Tropical

Depressions". Earlier storm records show a much smaller number of storms each year owing to lack of observation techniques.

Figure 2 shows the tracks of some storms that passed by the "Bahia San Juan de Dios" during the period 1980-92. Note that the general trend of the tracks is that they run parallel to the coast and within a band that extends between 100 to 400 nautical miles from the coast.

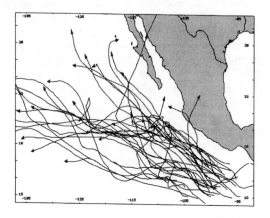

Figure 2. Storm tracks during the period 1980-92.

Figure 3 shows a statistical analysis of the storms that cross a line perpendicular to the coast. Data is presented in terms of number of storms per year per nautical mile versus offshore distance to the storm's track.

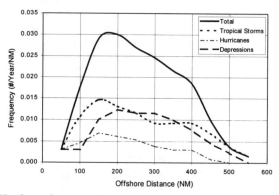

Figure 3. Number of storms that cross a line perpendicular to the coast versus offshore distance to the storm's track.

Note in Figure 2 that a storm can occasionally move towards the coast. Historically, these storms caused significant damage due to high waves and intense rain. For the purposes of this study these rare events are not of major importance. The objective of the study was to determine the average wave conditions that move sediments along the coast towards the harbor and produce agitation at the entrance.

A detailed analysis of the courses and intensities of the storms in the period 1980-92 was performed. Table 1 summarizes this analysis and shows that the number and kind of storms varies significantly each year.

Year	Tropical Depressions	Tropical Storms	Hurricanes
1980	1	4	0
1981	2	1	1
1982	4	3	2
1983	0	6	5
1984	3	7	3
1985	6	3	1
1986	5	3	3
1987	2	6	1
1988	2	5	0
1989	7	4	1
1990	8	6	0
1991	2	5	3
1992	1	4	5
Average	3	4	2

Table 1. Summary of storm activity offshore "Bahia San Juan de Dios" between 1980-92.

Within the time and budget constraints of this study it was not practical to determine the offshore annual wave conditions from historical data for each individual storm. In addition, many of the storms were in different stages of development which makes the determination of these conditions more difficult. Therefore, a simplified approach was used where the annual offshore wave conditions were computed from a set of storms which could be considered as representative of a typical year. A typical year was then defined as composed of 2 hurricanes and 4 tropical storms traveling parallel to the coast (approximately SE-NW) at 200 nautical miles offshore. Table 2 shows the storm parameters used for the hindcast of offshore wave conditions.

Paremeter	Tropical Storm	Hurricane
Forward Speed (km/h)	13.89	13.89
Radius to Maximum Wind (km)	22.5	22.5
Maximum Wind Speed (knots)	60	100
Central Pressure (mb)	990	960
Neutral Pressure (mb)	1012	1012

Table 2 - Parameters for tropical storm and hurricane.

Time varying pressure and wind fields were generated for a hurricane and a tropical storm according to the parameters described in Table 2. The wind fields were used to produce time series of the corresponding offshore wave conditions by means of a time dependent, discrete, spectral model based on the energy balance equation (DHI MIKE 21-OSW, version 2.6, 1996). In the description of the spectral energy, an upper bound is applied representing the saturated sea state. The limiting spectral shape is a Pierson-Moskowitz spectrum, where the scaling factor depends on the water depth and total energy.

Figures 4 and 5 show the time series of significant wave height and direction for the tropical storm and hurricane described in Table 2, respectively, offshore "Bahia San Juan de Dios".

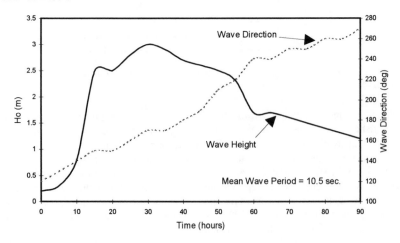

Figure 4. Significant wave height and wave direction time series for the tropical storm described in Table 2, traveling parallel to the coast and 200 nautical miles offshore.

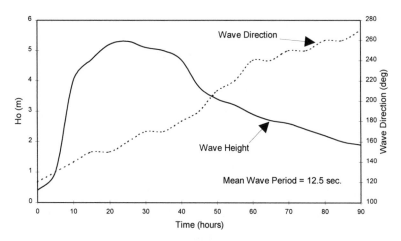

Figure 5. Significant wave height and wave direction time series for the hurricane described in Table 2, traveling parallel to the coast and 200 nautical miles offshore.

Local Wave Conditions

The bathymetry in "Bahia San Juan de Dios", the islands, coastline and shoals indicate that significant wave transformation may occur by means of refraction, diffraction and shoaling as the storm swell travels from deep to shallow water. In order to determine what the wave climate would be in the vicinity of the beach and harbor entrance a linear refraction-diffraction model based on the parabolic approximation to the elliptic mild slope equation was used (DHI MIKE 21-PMS, version 2.6, 1996). The model takes into account the effects of refraction and shoaling due to varying depth, diffraction along the perpendicular to the predominant wave direction, energy dissipation due to bottom friction and wind. It also takes into account the effect of frequency and directional spreading using linear superposition.

The offshore significant wave height and wave direction time series shown in Figures 4 and 5 were used as input to the model defined above to determine the wave conditions at: a) on the access channel to the harbor at 10 m water depth (referred to as "Entrance" and b) at the 13 m water depth contour on a line perpendicular to the coast and 400 m towards the east along the coast from the harbor entrance (referred to as "Beach"). The local significant wave height and wave direction time series generated by the hurricane and tropical storm defined in the previous section at the "Entrance" area are shown in Figure 6.

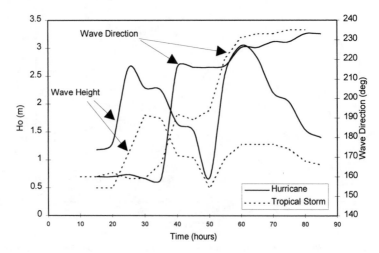

Figure 6. Significant wave height and wave direction time series generated by hurricane and tropical storm at the "Entrance" area.

The local significant wave height and wave direction time series generated by the hurricane and tropical storm defined in the previous section at the "Beach" area are shown in Figure 7.

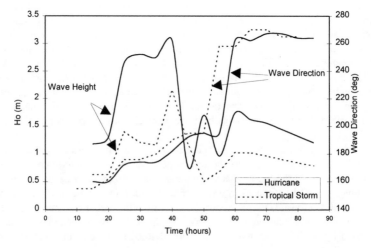

Figure 7. Significant wave height and wave direction time series generated by hurricane and tropical storm at the "Beach" area.

Figures 6 and 7 show that with these "average" storms it takes about 20 hours for the first set of swell waves to arrive. 20 hours after the arrival of the first set of waves, the wave direction turns towards the west and the blockage effect of the islands becomes evident as the wave heights drop sharply and the wave direction shows a sudden step for both "Entrance" and "Beach" locations.

Figures 8, 9 and 10 show a partial view of the model grid and the wave patterns for hurricane swell arriving at the coast when the offshore wave directions are 170 degrees, 210 degrees and 240 degrees, respectively.

Figures 8 through 10 illustrate the effect that the storm waves could have on sediment transport. As the swell arrives from 160-180 degrees during the first hours, the littoral drift would move the sediment towards the harbor entrance. As shown also in Figures 6 and 7, this is followed by a relatively calm period of approximately 15 hours where the wave heights are lower and the seas are confused as the swell is blocked by the offshore islands. Finally, the swell arrives from the west side of the islands with a direction of 220-235 degrees thereby producing a littoral drift that would move the sediment to the east. The harbor entrance may act as a sediment trap catching a fraction of the sand that is moved back to the beach.

The time series shown in Figures 6 and 7 were used to create an annual wave climate which was composed of 2 hurricanes and 4 tropical storms. This annual wave climate was used as input for a coastline evolution and sediment transport model. Historical nautical charts dating back to 1945 and more recent aerial photographs were used to calibrate and verify a shoreline response model. Finally, use was made of this model to determine the sediment transport rates that affect the harbor entrance.

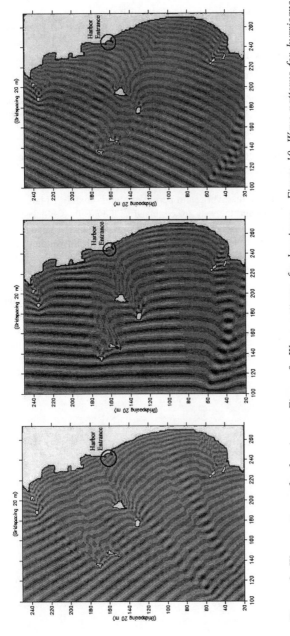

Figure 8. Wave patterns for hurricane swell, offshore mean wave direction = 170 degrees.

Figure 9. Wave patterns for hurricane waves, offshore mean wave direction = 210 degrees.

Figure 10. Wave patterns for hurricane waves, offshore mean wave direction = 240 degrees.

Shoreline Response and Sediment Transport

The littoral processes modeling system used in this study was GENESIS developed by US Coastal Engineering Research Center (CERC, 1989). GENESIS simulates shoreline changes by spatial and temporal differences in longshore sand transport. The main utility of the modeling system lies in simulating the response of the shoreline to structures sited nearshore.

The equation governing shoreline change is formulated by conservation of sediment volume. It is assumed that the beach profile translates seaward or shoreward along a section of coast without changing shape when a net volume of sediment enters or leaves the section during a time interval.

The nautical chart published by the US Navy Hydrographic Office Chart, number 21318 of 1947 (now obsolete) was examined and it was found that the shoreline compared well with an aerial photograph taken in 1945. This shoreline was compared to the one shown in a more recent photograph taken in 1993 and it was noted that the shoreline adjacent to the harbor entrance (east jetty) had moved seaward approximately 100 m. This variation of shoreline position can only be taken as a guideline since the water levels at the time the photographs were taken and the wave conditions during previous days were not known. However, it provides useful information to calibrate the shoreline response and sediment transport model.

GENESIS was applied to model the shoreline changes between 1945 and 1995. During the period between 1945 and 1975 the beach was only bound by the rocky headlands to the west and east. During 1975 the small entrance was build by the west headland featuring two 200 m long jetties. In combination with the channel that was dredged, this construction improved the flushing of a lagoon behind the beach. In 1991-92 the harbor was constructed and the jetties were extended approximately 60 m seaward. This extension appears to be insufficient to prevent sedimentation in the entrance as evidenced by the present sedimentation and agitation.

Three time periods were modeled with GENESIS including approximately 2,200 m of beach to the east of the harbor entrance:

- from 1945 to 1974: shoreline evolution with no structures
- from 1974 to 1992: shoreline evolution with 200 m jetties perpendicular to the shore
- from 1992 to 1995: shoreline evolution with 260 m jetties perpendicular to the shore

The annual wave climate produced by 1 hurricane and 4 tropical storms as defined in Figures 6 and 7 was used in the simulations, the time step was 2 hours and the grid resolution was 25 m (measured along the original beach). The K_1 and K_2 constants used in the CERC formulation that produced the best fit between the predicted and observed shorelines were 0.3 and 0.6 respectively. Figure 11 shows the shoreline evolution simulation results for the periods 1945-74, 1974-92 and 1992-95.

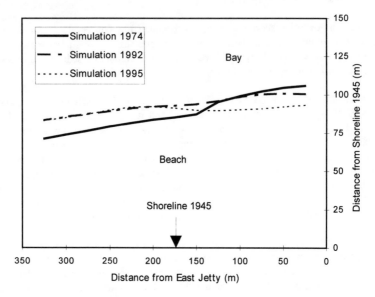

Figure 11. Shoreline evolution simulation results in the vicinity of the east jetty.

Figure 11 shows the large seaward growth of the beach during the period 1945-74 which is consistent with the magnitude of the growth observed in the aerial photographs. There is a relatively small growth in the period 1974-92 in the presence of the new jetties and a small retreat that occurred during 1992-95 after the harbor entrance was constructed.

Figure 12 shows the position of the shoreline in 1945 (baseline), the shoreline position based on beach surveys performed in 1993 and 1995, and the shoreline position as predicted by GENESIS for 1995.

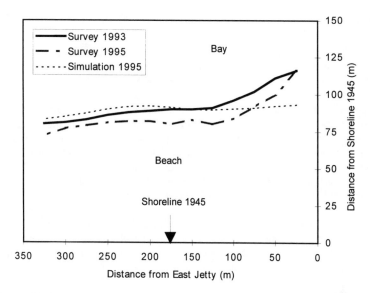

Figure 12. Comparison of shoreline positions from beach surveys and simulation.

Figure 12 shows that the correlation between the simulated and observed shoreline positions is good except for an area close to the jetty where the model fails to reproduce waves-sediment-structure interactions such as cross-shore transport. An additional refinement, using the actual storms since 1992 rather than average ones, may produce closer agreement.

The general agreement shown in Figure 11 (a 50 year simulation) justifies the calibration coefficients (K_1 and K_2) used in the shoreline evolution model as well as the approximation made for the wave climate selected. These were used to estimate long term annual sediment transport rates for the actual beach configuration and the harbor entrance at the design depth. It was determined that the sediment transport rate to the west is 12,000 m³/year while the transport rate to the east is 2,000 m³/year. Since the coast to the west of the harbor entrance is rocky and does not show evidence of sand accretion it can be concluded that the harbor entrance works as a sediment trap. The net 10,000 m³/year sedimentation of the harbor entrance is consistent with the volume of sand accreted there, derived from annual bathymetric surveys.

Conclusions

This study provides estimated storm wave statistics on the Pacific coast of Southern Mexico derived from historical storm data and a time dependent, discrete, spectral model based on the energy balance equation (DHI MIKE21-OSW).

The presence of offshore islands and shoals, the varying direction of arrival of storm waves and the focusing and unfocusing of waves at the harbor entrance made it difficult to compute wave conditions and the wave climate in its vicinity by means of typical handbook techniques. A linear refraction-diffraction model based on the parabolic approximation to the elliptic mild slope equation was used (DHI MIKE21-PMS) to transform the offshore wave climate into the wave climate in the vicinity of the harbor entrance for use in sediment transport modeling.

The computed wave climate, in conjunction with the coastline evolution model GENESIS, produced results that compared well with observations and surveys. In addition, it provided confidence in predicting the coastline evolution which would be modified by the presence of planned mitigation alternatives.

References

Danish Hydraulic Institute, MIKE 21 - OSW, Offshore Spectral Wind-Wave Model, version 2.6, 1996.

Danish Hydraulic Institute, MIKE 21 - PMS, Parabolic Mild Slope Wave Model, version 2.6, 1996.

Department of the Army, Waterways Experiment Station, Corps of Engineers, (1989) GENESIS: Generalized Model for Simulating Shoreline Change.

U.S Navy - Department of Commerce, (1994): "Global Tropical/Extratropical Cyclone Climatic Atlas," National Climatic Data Center.

Design of Field Program and Use of Field Measurements to Calibrate a
Hydrodynamic and Water Quality Model of an Estuarine River System

Ken Hickey, Mark Gerath, and Gavin Gong
ENSR
35 Nagog Park, Acton, MA 01720

James D. Bowen
Department of Engineering Technology
University of North Carolina at Charlotte

Abstract

A study of impacts on summer-time dissolved oxygen levels of an industrial discharge loading BOD to an estuarine river was performed. Hydrodynamics were of particular interest because of the presence of strong, tidally reversing currents and a discharge release scheme dictating a high volume release for a short period of time during ebb tide. Water quality challenges included quantification of relative impacts of the discharger and extensive upriver marshes on ambient dissolved oxygen levels in the study area.

A field program was designed and implemented to obtain measurements in support of specific regulatory requirements for this complex estuarine river. Field program components included long-term, continuous measurement of water velocity, tide, temperature, salinity, and dissolved oxygen at locations throughout the study area. In addition, three synoptic surveys featured acoustic doppler current profiling (ADCP), CTD deployments, and analytical sampling throughout the study area. Surveys were scheduled to capture a range of conditions from worst-case to average in terms of summer-time dissolved oxygen levels. The goal of the field program was first to definitively characterize the system hydrodynamics so that the tidal excursion was well known. In this way, different sources of BOD loadings could be resolved and the water quality model properly constrained.

The two-dimensional, laterally averaged finite-difference model CE-QUAL-W2 (Cole, T.M. and Buchak, E.M., 1995; see Gong et al. 1997) was selected to perform hydrodynamic and water quality simulations. Calibration of the hydrodynamic model included comparison of model predictions to synoptic streamflow measurements collected at flood and ebb tide using the ADCP and long-term continuous measurements of water velocity collected at the river mouth. Calibration of the water quality model involved adjustment of loadings and reaction rates to capture observations of long-term continuous and synoptic surveys of salinity and dissolved oxygen and analytical laboratory results from samples collected throughout the study area.

The field program was successful in supporting development of a hydrodynamic and water quality model of the estuarine river. The model was effectively applied to evaluate the relative influence of the discharged effluent and other sources on dissolved oxygen concentrations throughout the study area during summer-time conditions. In particular, detailed knowledge of the hydrodynamic regime allowed successful resolution of BOD loadings from the industrial discharge and upstream non-point sources.

Introduction

A study of impacts of an industrial discharge loading BOD to an estuarine river on summer-time dissolved oxygen levels was performed. The estuarine river (Figure 1) has strong tidally reversing currents and drains approximately 260 square kilometers of coastal lowlands in the southeastern United States. The mouth of the river enters an estuarine bay at a point 19 km (12 miles) from the Atlantic Ocean. The study area extended from the mouth to rivermile 11 and included four small tributaries. Distances upstream of the mouth of the river are expressed as "rivermiles", consistent with the USGS protocol, and are denoted in encircled numbers in Figure 1. River widths vary from 100 to 200 meters and depths vary from 9 meters, where dredged near the mouth, to 5 meters in some upriver reaches. Extensive marshes border the river from miles 3 through 9.

An industrial discharger, located at mile 2.9, releases approximately 100,000 cubic meters per day of effluent with a BOD concentration of approximately 26 mg/l. The discharger employs a hold and release scheme whereby the entire daily discharge is released for a period of 75 minutes during each of two daily ebb tides. Thus, the total daily discharge is released over a period of 2.5 hours at a volumetric flowrate of 11.1 m^3/sec.

The study was mandated by a state regulatory agency and was motivated by frequent measurements of summer-time dissolved oxygen concentrations below the water quality standard of 4.0 mg/l. The agency was concerned, based on the perception from historical dye studies that the effluent was subject to net upstream transport, that low summer-time dissolved oxygen concentrations were largely due to the BOD load associated with the effluent. In response to these concerns and perceptions, a strategy was developed that required a hydrodynamic characterization rigorous enough to determine conclusively the transport of the effluent released at ebb tide as well as the tidal excursions along the river length.

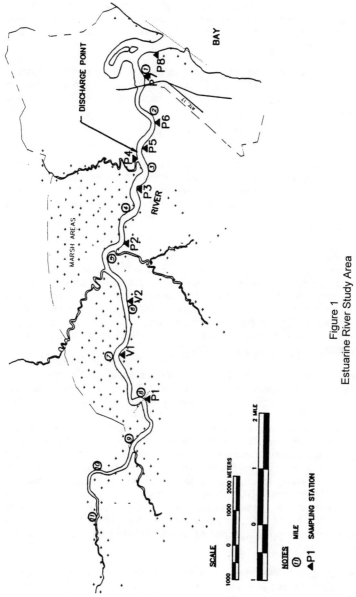

Figure 1
Estuarine River Study Area

Water quality-related tasks, such as quantification of the impact of the industrial discharge effluent and extensive upriver marshes on low summer-time dissolved oxygen levels, could only be performed after a well-constrained hydrodynamic model was established.

A conceptual approach, including field measurement and numerical modeling components, was developed to meet study objectives. This paper focuses on quantification of two key dynamic parameters of this study, tidal flowrates and dissolved oxygen.

Field Program Design

A strategy employing both intensive synoptic surveying and long-term continuous monitoring was applied to characterize hydrodynamics and water quality dynamics of the estuarine river. Synoptic surveys were one week in duration, while long-term monitoring deployments spanned two month periods. Three synoptic surveys and two long-term monitoring deployments were performed during summer and fall of 1995 and 1996.

Hydrodynamics
A boat-based ADCP was utilized to characterize the tidal flowrates throughout the study area. An ADCP transmits sonar signals and processes acoustic returns to calculate water velocity based on Doppler frequency shifts of small particles throughout the water column. A boat-based ADCP is used to obtain velocity measurements throughout the water column and, while underway, across the waterbody. Volumetric flowrate measurements (e.g., tidal flowrates) were obtained by transecting the river using a boat-based ADCP.

In order to capture strong, tidally-reversing flows, synoptic ADCP surveys were performed during maximum flood and ebb tide. Tidal flowrate measurements were obtained by transecting the river with the ADCP at 10 sampling locations throughout the study area (Figure 1). A synoptic ADCP survey was typically performed in 70 minutes, thus approximating a "snapshot" of tidal flowrates throughout the study area. This method was applied during each of three summer-time surveys and effectively quantified maximum tidal flowrates. Long-term water velocity measurements, collected at the river mouth, were used to quantify the tidal water velocity pattern, to characterize long-term variations, and to provide a context for the synoptic ADCP surveys. Current meter measurements were supplemented with tide gages deployed over two month periods at 3 locations along the river. In this way, the volumetric flow rate, both at the times of the synoptic surveys and over longer periods, was known along the river's axis with a high degree of certainty.

A dye study was performed to provide a direct measurement of effluent transport. Rhodomine WT dye was added to the effluent plume as it entered the river at ebb tide. The dye was tracked using Turner fluorometers until dye was no longer measurable in the study area. Agency perception that net upstream transport of the discharge effluent was occurring was based on previous dye studies. Review of the previous dye studies revealed that both dye studies were performed by releasing dye at flood tide rather than at ebb tide concurrent with the effluent release. This critical distinction was captured in the modeling effort and highlighted in discussions with the agency.

Water Quality
Accurate spatial and temporal characterization of dissolved oxygen dynamics was required to determine the relative impacts of point and non-points sources on summer-time low DO concentrations and to evaluate the effects of photosynthesis on DO concentrations. To capture spatial variations, synoptic CTD surveys were performed in which vertical profiles of dissolved oxygen were collected during all four tide stages (flood, high water slack, ebb, and low water slack). During these surveys, dissolved oxygen profiles were collected at 10 sampling stations. To capture temporal variations, continuous recording dissolved oxygen meter arrays were deployed at five locations throughout the study area, concurrent with the synoptic surveys and for a period of one week. This method was applied during each of three synoptic surveys. Laboratory evaluation of water quality (including confirmation of field DO measurements, nutrients, BOD_5, BOD_{30}, and BOD_u) was performed at high water slack and low water slack at 8 locations. Finally, BOD, DO, and temperature impacts were evaluated by synoptic surveys of the tributaries in an attempt to quantify marsh conditions.

To evaluate the likelihood of naturally occurring low summer-time dissolved oxygen concentrations in the estuarine river, a statistical analysis of DO measurements collected in the river and in neighboring estuarine rivers was performed. Recent, DO measurement data was obtained from the U.S. EPA STORET data base system. This analysis revealed that the river was not unusual among neighboring estuarine rivers in having common DO excursions below 4.0 mg/l during summer and fall conditions. This analysis was helpful in persuading agency officials to consider the possibility that low summer-time DO concentrations in the river were naturally occurring.

Summary of Field Measurements

Hydrodynamics
Long-term water velocity measurements indicated that currents were strong and tidally reversing. Figure 2 presents one week of water velocity measurements collected at the river mouth during August 1996. The magnitude of velocity is indicated by the length of the "stick" and the direction is indicated by the angle, where the north is in the "+y" direction. The water velocity sensor was located at mid-depth near mid-channel in a reach of the river that is dredged to a uniform depth of 9 meters. Water velocity dynamics were observed to be remarkably consistent from one tide cycle to the next throughout this week and throughout the

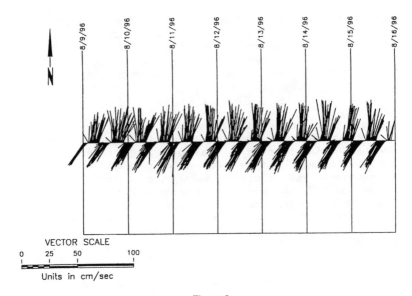

Figure 2
Long-term Continuous Water Velocity Measurements at Station P8
August 9 - 16, 1996

four month study period. Tidal forcings were observed to dominate, while freshwater inflows and wind had little influence on water velocity dynamics, during the study period. Small freshwater inflows would be expected based on the small basin area and are supported by a total freshwater inflow estimate of less than 2 m^3/sec from a previous study and personal correspondence with USGS hydrologists. Also, significant wind influences would not be expected, excepting for major storm events, in the estuarine river due to its physical configuration (i.e., very small fetch, generally sheltered).

Tidal flowrate measurements obtained using the boat-based ADCP during field surveys of August 1996 are summarized is Figure 3. The figure contains tidal flowrate magnitudes only. The sampling location is identified on the horizontal axis and tidal flowrate is on the vertical axis. Sampling rounds are identified in the legend. Tidal flowrates measured at ebb and flood tide were opposite in direction during all sampling. Ebb and flood tidal flowrates were found to be relatively uniform from the mouth to approximately mile 4.5 (near station P2). At Station P7, survey vessel passage and concurrent flowrate measurement was impeded by a large docking facility. This likely accounts for lower flowrate measurements collected at Station P7. Tidal flowrates declined significantly from miles 4.5 to 8.0.

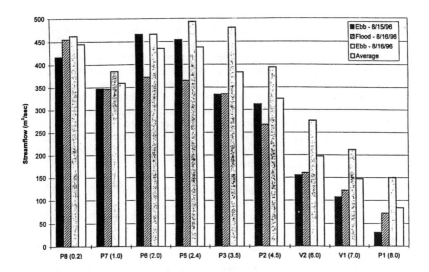

Figure 3
ADCP Tidal Flowrate Measurements
Collected During the August 1996 Synoptic Survey

The overall decline in tidal flowrates was due to both decreased water velocities and reduced cross-sectional area in the upper reaches. A decline in water velocities to 20 cm/sec from 65 cm/sec was typically observed when moving from the lower 4 miles of the river to the upriver reaches. In addition, no significant lateral or vertical currents were observed in the study area. This pattern of discharge and water velocity was important in selecting an appropriate model, conceptualizing DO impacts, and performing model calibration.

Tidal flowrate measurements, collected over a four month period, confirmed that strong, tidally reversing currents were present throughout the study area. Tidal flowrates measured during ebb and flood tides were similar in magnitude and opposite in direction, indicating that freshwater inflow was not significant. Based on the sharp decline in tidal velocities above mile 4.5, tidal excursions will be far longer in the lower reaches than the upper reaches of the river. Due to closer proximity to the mouth and greater tidal excursions, waters of the lower reaches are expected to experience far more favorable flushing than in the upper reaches of the river.

A dye study was performed in September 1996. Dye was released with the discharge effluent at ebb tide and transport of the dye was tracked as it moved from the discharge location (mile 2.9) to the river mouth. The center of dye mass reached the mouth of the river in 2 hours and was no longer detectable in the river soon thereafter. This result is consistent with tidal excursion estimates made based on typical ADCP water velocity measurements (e.g., 65 cm/sec) collected in the lower reaches of the river. The dye study provides quantification of favorable flushing of the waters of the lower reaches of the river. Previous dye studies, with dye released with the flood tide, showed that flushing in the upper reaches of the river is poor. Dye that reached river mile 4.5 and above, stayed in the river for long periods of time (i.e., several weeks). This result is also consistent with tidal excursion estimates based on typical ADCP water velocity measurements (e.g., 20 cm/sec) collected in the upper reaches of the river.

The hydrodynamic surveys concluded that a consistent pattern of strong, reversing currents was present in the river. Further, tidal excursions were found to be far longer and flushing far more favorable in the lower reaches than the upper reaches of the river.

Water Quality
Figure 4 contains depth-averaged dissolved oxygen measurements collected during each tide stage. Little variation in DO concentrations was evident with depth. The sampling location is identified on the horizontal axis and DO concentration is on the vertical axis. DO concentrations ranged from 2.4 mg/l to 4.8 mg/l. The highest DO concentration measurements were collected near the river mouth, especially at high tide when estuarine bay water of higher DO enters

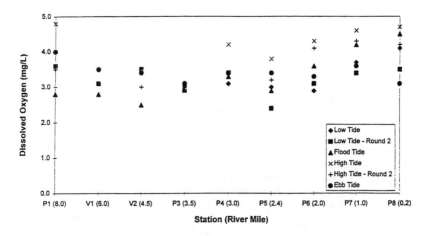

Figure 4
In-Situ Dissolved Oxygen Measurements
Collected During the August 1996 Synoptic Survey

the river system. The lowest DO measurements were collected between miles 2.9 and 6.0, adjacent to extensive tidal marshes. Typical dissolved oxygen measurements in the upper reaches of the river were approximately 1 mg/l lower than typical dissolved oxygen measurement towards the mouth.

Figure 5 contains typical continuously recorded dissolved oxygen measurements collected during a four day period at sampling location P5 (mile 2.9) and at sampling location P1 (mile 8.0). DO concentration measurements are presented on the y-axis (left side) and time is presented on the x-axis. Tide height measurements, collected concurrently at stations P5 and P1, are displayed below DO measurements and with relative elevations provided on the y-axis (right side). DO measurements ranged from 3.0 mg/l to 6.2 mg/l and were strongly influenced by tide. At location P5, DO measurements were collected by two sensors, located at 25% and 75% of the water column, and were nearly identical, indicating a well mixed water column. This observation is consistent with synoptic in-situ profile measurements. At location P1, measurements were collected at mid-depth.

In reviewing these continuous DO figures, the reader should disregard excursions in DO levels recorded at high water and low water slack. These slack water drops were recorded because of the presence of stagnant water in the immediate vicinity of the sensor membrane and are not representative of dissolved oxygen concentrations in the river during slack waters. This is especially prevalent at Station P1 where water is slack longer.

DO concentrations were observed to be strongly influenced by tide. Based on the hydrodynamic characterization, water masses are known to be transported up and down the river with the tide cycle. Thus, the correlation between DO and tide measurements indicates that water masses upstream and downstream a specific sensor have significantly different DO. DO measurements at locations P1 and P5 were both strongly influenced by tide, but were 180° out-of-phase from one another (Figure 5). At location P5, DO concentration measurements are in-phase with the tide height; that is, when tide height was high, DO concentrations were high and when tide height was low, DO concentrations were low. Thus, the continuous DO concentration record at location P5 indicates that waters upstream of location P5 have DO concentrations that are significantly lower (i.e. nearly 2.0 mg/l lower) than waters downstream of location P5. Conversely, at location P1, DO concentrations are 180° out-of-phase with tide height indicating that waters upstream of location P1 have DO concentrations significantly higher than water downstream. Thus, the DO measurements presented in Figure 5 indicate that the water mass between these locations (i.e., P1 and P5) has significantly lower DO than water masses upstream and downstream of these locations. This difference in behavior was helpful in model conceptualization and calibration.

Continuous and synoptic measurements concluded that the waters in the mid-reaches (i.e. miles 2.9 to 8.0) of the river have significantly lower DO

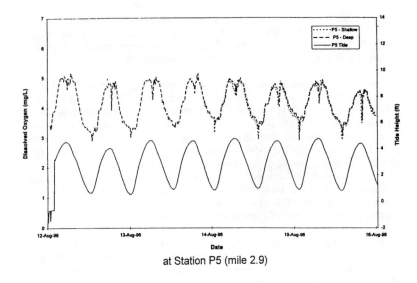

at Station P5 (mile 2.9)

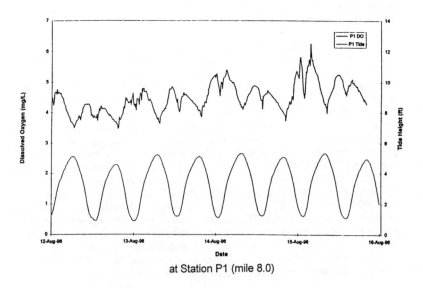

at Station P1 (mile 8.0)

Figure 5
Continuous Dissolved Oxygen Measurements
at Stations P5 and P1 on August 12 - 16, 1996

concentrations than waters near the mouth or upstream of mile 8.0. If discharge effluent containing high BOD loadings were transported to and resided in this section for extended periods of time, then the discharge effluent could be responsible for the DO depletion observed in this section of the river. Based on the hydrodynamic characterization and subsequent numerical modeling, however, the discharge effluent is not predicted to be transported upriver in significant quantities or for significant periods of time. Extensive marshes reside along the river between miles 3 and 8 and are known to be sources of oxygen demanding substances. It appears likely that BOD loadings from these marshes and long water residence times are primarily responsible for the depletion of DO concentrations observed in waters of this section of the river.

Use of Field Measurements to Calibrate a Numerical Model

A two-dimensional, laterally averaged, finite-difference hydrodynamic and water quality model, CE-QUAL-W2, was applied to the estuarine river system. A laterally averaged model is appropriate for this application because no significant lateral circulation was observed during ADCP surveys. A detailed description of model development and application is provided in a companion paper (Gong, et al., 1997). Some results of the hydrodynamic and water quality calibration are presented and discussed below.

Figure 6 presents simulated vs. measured tide heights for a 12 hour period at the river mouth (location P8), and at rivermile 8.0 (location P1). The simulated tide height time series is similar to measured values. The comparison provided in Figure 6 contains a representative comparison of simulated tide height times series to the tide height measurement record. Both simulated tidal amplitudes and wavelenghts were well-matched to the tide height measurement record. At the river mouth (location P8), the simulated and measured tide waves were nearly in-phase. At rivermile 8 (location P1), however, the simulated tide wave lagged the measured tide wave by nearly one hour. Delay in propagation of the tide wave upstream is likely primarily due to the serpentine configuration of the river. The numerical model, CE-QUAL-W2, is not capable of accounting for bends in the simulated water body. Thus, the inability of the model to reproduce the observed tide time lag is likely due to the model's oversimplification of the physical configuration of the water body.

Figure 7 presents predicted vs. measured tidal flowrates at peak flood and ebb tide. Overall, hydrodynamic model predictions compare favorably to field measurements at all locations. Validation of the hydrodynamic model to field measurements from a second synoptic survey also resulted in a favorable match. The resulting hydrodynamic model was successful because it was based on a strong set of field measurements. Hydrodynamic parameter values (e.g., tidal flowrates) applied to the simulations were specified with a relatively high degree of certainty. Thus, hydrodynamic simulations likely provide accurate representations

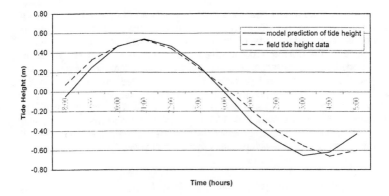

at Station P8 (river mouth)

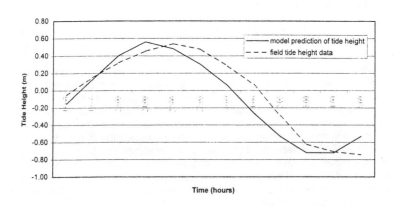

at Station P1 (mile 8.0)

Figure 6
Hydrodynamic Model Calibration
Predicted vs. Measured Tide Heights for a Typical 12-hour Period

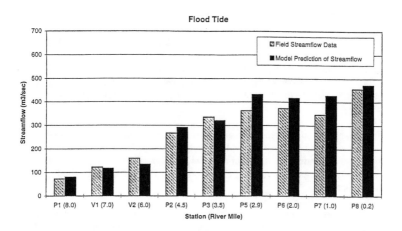

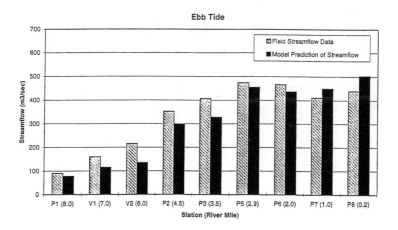

Figure 7
Hydrodynamic Model Calibration
Predicted vs. Measured Tidal Flowrates at Peak Flood and Ebb Tide

of the hydrodynamic regime of the estuarine river.

The hydrodynamic model predicted tidal flowrates and water velocities that result in long tidal excursions and favorable flushing in the downstream reaches and shorter tidal excursions and poor flushing in upstream reaches. Further, model simulations of dye releases resulted in agreement with both recent and historic dye studies and provided detailed specification of tidal excursions and residence times throughout the study area.

Figure 8 presents a comparison of predicted vs. measured water column average DO concentrations along the river. Model predictions of DO concentrations are presented as lines indicating maximum, minimum, and mean DO values over the tidal cycle. Field measurements of DO are presented as points and identified by sampling event.

Predicted and measured DO concentrations throughout the river are similar. DO concentrations are lowest in the mid-reaches of the river and higher at the mouth and in upper reaches. Predicted dissolved oxygen concentrations over the course of the tidal cycle generally fall within the range of measured values throughout the river. Calibration of DO and BOD required that large, distributed BOD loading terms representing the extensive marshes be added to the model. Initial calibration attempts made without marsh BOD loading terms resulted in very poor agreement with field measurements especially in reaches adjacent to marshes. The necessity of marsh BOD loading terms to calibrate the water quality model is a consequence to the hydrodynamic prediction that the discharge effluent is not transported upriver and results in a well-constrained quantification of marsh BOD loading.

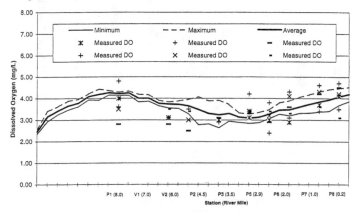

Figure 8
Water Quality Calibration
Predicted vs. Measured Dissolved Oxygen Concentrations

A strategy employing hydrodynamic characterization as a means of differentiating BOD sources impacting DO in an estuarine river was successfully applied. Through an thorough hydrodynamic field program and numerical modeling, the hydrodynamic regime was accurately characterized and it was possible to quantify DO impacts from the effluent discharge and the extensive marshes distinctly. The importance of this approach is increased by the degree of uncertainty and the paucity of estimates of BOD loadings from tidal marshes in the southeastern United States.

References

Cole, T.M. and E.M. Buchak, 1995. CE-QUAL-W2: A Two-Dimensional, Laterally Averaged, Hydrodynamic and Water Quality Model, Version 2.0. Instruction Report EL-95. US Army Engineer Waterways Experiment Station, Vicksburg, MS. March 1995.

Gong, G., K. Hickey, and M. Higgins, 1997. Hydrodynamic Flow and Water Quality Simulation of a Narrow River System Influenced by Wide Tidal Marshes. Presented at the ASCE 5th International Conference on Estuarine and Coastal Modeling, Alexandria, VA. October, 1997.

Data Assimilation Experiments
with the Navy Layered Ocean Model.

Igor Shulman [1], Ole Martin Smedstad [2]

ABSTRACT

Development of an oceanic prediction system is under way at the Naval Research Laboratory (Carnes et. al., 1996). The data assimilation component of the system includes: optimum interpolation of altimeter sea surface heights; statistical inference of the subsurface fields and nudging of this information into the model. The system provides a way to project the surface information into subsurface fields of the Navy Layered Ocean Model (NLOM). In this paper, we investigate the effect of using energy integrals derived from the model to constrain subsurface fields estimated using the statistical inference technique. In this case, the projection of surface information into subsurface fields is divided into two steps: the first guess of the nudging terms is obtained by the statistical inference technique; then, a simple variational problem is used to correct these terms in accordance with the current model dynamics. This approach combines the statistical and dynamical information and provide a continuous feedback from the model to the inference of subsurface fields. Results from data assimilation experiments with a Sea of Japan version of the NLOM are presented.

1. INTRODUCTION

Assimilation of altimeter derived data is becoming a key component of prediction systems. Many different approaches have been developed for assimilation of altimeter derived data into oceanic models. One of the important components of these data assimilation methods is the projection of surface information into subsurface fields. It is clear that there is no unique solution when reconstructing three dimensional subsurface fields from two dimensional surface information. For this reason, many researchers use various additional physical knowledge or physical transformation in order to constrain the projection and eliminate the null space, e.g. Hurlburt 1986; De Mey and Robinson 1987; Hurlburt et al. 1990; Mellor and Ezer 1991; Ezer and Mellor 1994; Haines 1991; Ghil and Malanotte-Rizzoli 1991; Reinecher and Adamec 1995; Oschlies and Willebrand 1996. Reviews of these methods can be found in Ghil and Malanotte-Rizzoli (1991), Reinecher and Adamec (1995) and De Mey (1997). We brifiely describe some of the methods related to our research.

---

[1]Institute of Marine Sciences, Univ. of Southern Mississippi, Stennis Space Center, MS 39529.

[2]Planning Systems, Inc., Stennis Space Center, Mississippi, 39529

Hurlburt (1986) developed a dynamical approach to infer the subsurface fields from satellite altimeter data. The dynamics of the numerical ocean model were used to transfer the surface information to the deep ocean. As noted in Hurlburt (1986) and Hurlburt et al. (1990), the primary limitation of this dynamical technique is the convergent time scale and the dependence on the update interval (when there is a large disagreement between the observations and the model, it can delay downward transfer of surface information and, therefore, introduce phase errors).

Statistical methods can be used to overcome these problems (Hurlburt et al. (1990), Mellor and Ezer (1991) and Ezer and Mellor (1994)). In (Hurlburt et al. (1990)), long-time model runs were used to derive the statistical relationships between surface and subsurface pressures. A statistical approach was also used in Mellor and Ezer (1991) and Ezer and Mellor (1994) to assimilate the sea surface height and sea surface temperature into the Princeton Ocean Model. The preprocessed correlations between surface and subsurface fields, model error estimates, and an optimal interpolation approach were the basis for the projection of the surface information into the deep ocean. Note that in Hurlburt (1990), EOFs (empirical orthogonal functions) are used. In the Mellor and Ezer approach, time-averaged data at one point on the sea surface were used to calculate the correlation functions. In Hurlburt (1990), the subsurface fields were inferred instantly and were not dependent on the updating interval. However, in this case, we need data which is compatible with the model "to avoid serious dynamical imbalances in the model that can cause incorrect eddy propagation speeds and errors in evolution of current systems" (Hurlburt, 1990). Moreover, statistical inference is complicated by the weak correlation between the surface and subsurface fields.

During recent years, a variational approach to data assimilation have been intensively developed (see recent review by De Mey, 1997). Although the variational approach is certainly a valid one, and should be further developed, at the present stage it requires a tremendous amount of computer time and memory, significant additions and changes to the dynamical model code (e. g. the development and the integration of adjoint code), and some a priori hypothesis about the statistical properties of data errors. "The nonlinear turbulent character and the large number of degrees of freedom of an eddy field make the descent-space topology non-quadratic" (De Mey, 1997). It is possible that multiple minima exist in variational methods. A practical solution might be to use the statistical inference to obtain the first estimate of the subsurface fields and then, a simplified variational method can be used to correct the statistically inferred fields according to the model dynamics (see also, De Mey, 1997).

Development of an oceanic prediction system is under way at the Naval Research Laboratory (Carnes et. al., 1996, Smedstad et. al., 1997). The data assimilation component for the system includes: optimum interpolation of altimeter sea surface heights; statistical inference of the subsurface fields (Hurlburt et al., 1990) and nudging of this information into the model (Smedstad and Fox, 1994; Smedstad et. al., 1997). The system provides the way to project the surface information into subsurface fields of the Navy Layered Ocean Model (NLOM). As noted in Smedstad and Fox (1994) the inclusion of the statistical inference technique of Hurlburt et. al. (1990) dramatically improved the data assimilation system, but nudging of statistically inferred subsurface fields might produce dynamical imbalances. In such cases, a very gentle nudging should be used, to avoid spurious gravity waves. Nudging of statistically inferred subsurface fields into the model equations may impose an incorrect distribution of the energetics of the dynamical modes of the model.

In this paper, we investigate the effect of using energy integrals derived from the model to constrain subsurface fields estimated using the statistical inference technique. In this case, the projection of surface information into subsurface fields is divided into two steps: the first guess of the nudging terms is obtained by the statistical inference technique; then, a simple variational problem is used to correct these terms in accordance with the current model dynamics. This approach will combine the statistical and dynamical information and provide a continuous feedback from the model to the inference of subsurface fields.

In section 2, the NLOM model and the assimilation scheme are briefly described. Section 3 provides the description of the approach. Results of data assimilation experiments with the model of Sea of Japan are described in section 4.

2. NLOM AND THE DATA ASSIMILATION SYSTEM

The Navy Layered Ocean Model (NLOM) is used in the assimilation experiments. This model is based on the primitive equation model of Hurlburt and Thompson (1980) but with expanded capability (Wallcraft, 1991).

The model has a free surface corresponding to the sea surface height (SSH), a variable observed by satellite altimetry. A realistic bottom topography is included in the lowest layer (finite depth version). The model variables are the layer thickness, the density and the transports in each of the active layers.

The satellite observations are assimilated track by track into the model. The first step is an Optimal Interpolation (OI) of the difference between the model and the observations around the satellite tracks. The numerical experiments discussed here use a homogeneous error covariance function with a spatial decorrelation scale of 115 km in the east-west direction and 110 km in the north-south direction and a time decorrelation scale of two days. Different length and time scales were tried, but the results were not very sensitive to (reasonable) alternative choices for them. The second step is the update of the pressure fields in the lower layers by using the statistical inference technique of Hurlburt et. al., (1990). There is only a weak correlation between the upper and abyssal layer pressure fields at each point of the model, a much higher one between the surface layer pressure field and the intermediate layer pressure fields. Hurlburt et. al., (1990) showed that the abyssal layer pressure fields could be better inferred by relating them to empirical orthogonal functions derived from an array of points in the upper layer. This approach has been extended to all of the subsurface layers. The change in pressures determined from the statistical inference is inverted into the corresponding changes in layer thicknesses.

The velocity fields in all layers are updated using a geostrophic correction calculated from the pressure changes. The third step is to nudge the obtained corrections to the layer thicknesses and velocities into the governing equations (Smedstad and Fox, 1994). This is done over a time period of one day. The model is not very sensitive to this period.

3. APPROACH

We introduce the following quantities:
Kinetic energy density e_k:

$$e_k = \frac{1}{2}\rho_k \mathbf{v}_k \mathbf{v}_k \tag{1}$$

where $\mathbf{v}_k$ is the velocity in layer k, ρ_k is the density in layer k.
Kinetic energy per area unit E_k:

$$E_k = \frac{1}{2}\rho_k \mathbf{v}_k \mathbf{v}_k H_k \tag{2}$$

where H_k is the layer thickness.
Available potential energy PE:

$$PE = \frac{1}{2}g((\rho_1 - \rho_a)\eta^2 + \sum_{k=1}^{N-1}(\rho_{k+1} - \rho_k)h_k{}^2) \tag{3}$$

where η is sea surface elevation, ρ_a is density of air, h_k is the anomaly in kth layer thickness.
Potential vorticity PV_k:

$$PV_k = (\zeta_k + f)/H_k \tag{4}$$

where ζ_k is a relative vorticity, f is the Coriolis parameter.

Suppose $\delta\eta$ is the difference between the observed and model sea surface elevation, which is obtained by an Optimal Interpolation of the difference between the model and the observations around the satellite tracks. The corresponding correction to the first layer pressure has the form: $\delta p_1 = g\delta\eta$, where g is the gravitational constant. According to the data assimilation scheme described in Section 2, the statistical inference technique is used to estimate from δp_1 the corresponding anomalies of subsurface pressure (δp_k, k=2,M, where M is a number of layers in the NLOM model).

The corresponding anomalies in layer thicknesses δh_k^s can be obtained from these subsurface pressure anomalies. Also, the velocity corrections $\delta\mathbf{v}_k^g$ are calculated from the geostrophic relations. The corrections δh_k^s and $\delta\mathbf{v}_k^g$ are nudged respectively into the layer thickness and momentum equations of the NLOM. Nudging is performed over the period between two consecutive times of updating of the $\delta\eta$ field (in the experiments described in section 4 one day is used). We propose to correct the δh_k^s and $\delta\mathbf{v}_k^g$ before nudging them into the governing equations. Several variational problems will be considered where these corrections are constrained by conserving one or more of the quantities (1)-(4).

Optimization problem A (minimization of the norm of the nudging terms):

Estimated from the statistical inference technique, the corrections to the layer pressure δp_k are linearly related to the corrections in layer thicknesses δh_k^s:

$$\mathbf{A} \cdot \delta\mathbf{h}^s = \delta\mathbf{p} \tag{5}$$

where $\delta\mathbf{h}^s$ is a vector of corrections to the layer thicknesses, $\delta\mathbf{p}$ is the vector of corrections to the layer pressures, and $\mathbf{A}$ is a linear matrix. Our intent is to choose a solution to linear system (5) which has the minimal norm:

$$\min_{\delta h_k^s} J = |\delta\mathbf{h}^s| \qquad (6)$$

where J represents the norm of the vector $\delta\mathbf{h}^s$. In this case, the gentle nudging will be introduced, and the changes in quantities (3) and (4) (potential energy of the model and potential vorticity) will be minimal. The Singular Value Decomposition (SVD) method can be used to solve the problem (5)-(6). In this case, an approach should be developed in order to choose a small number μ as a threshold for the small singular values (w_j) of the matrix $\mathbf{A}$.

$$if \qquad w_j \le \mu \sum_{k=1}^{M} w_k \qquad then \qquad w_j = 0 \qquad (7)$$

Suppose $\delta\mathbf{h}_\mu^s$ is the solution of (5)-(6) obtained by SVD with the threshold μ. In this case, the product $\mu\frac{\partial \mathbf{h}_\mu^s}{\partial\mu}$ can be interpreted as the first member of the Taylor series of the difference between the "exact" solution and the solution $\delta\mathbf{h}_\mu^s$. The product represent the error in estimation of $\delta\mathbf{h}^s$ caused by neglecting the small singular values. Therefore, one would like a value μ which provides:

$$\min_{\mu}[|\mu\frac{\partial \mathbf{h}_\mu^s}{\partial\mu}|/|\mathbf{h}_\mu^s|] \qquad (8)$$

The results of using criteria (8) in the inversion of (5) are demonstrated in section 4.

Optimization problem B (conservation of the kinetic energy).

In the data assimilation scheme considered, the most reliable information is the surface information, $\delta\eta$, the difference between the satellite and the model SSH. Therefore, we might want to project this information instantaneously into the deep ocean in such a manner as to conserve the kinetic energy of the dynamical model and let the model dynamically adjust between the two consecutive times of updating of the $\delta\eta$ field. In this case, we have the following optimization problem:

$$\min_{\delta\mathbf{v}_k} J = \frac{1}{2}\int_{L_k} (\delta\mathbf{v}_k - \delta\mathbf{v}_k^g)^2 dx \qquad (9)$$

$$\int_{L_k} \rho_k \mathbf{v}_k \delta\mathbf{v}_k H_k dx = 0 \qquad (10)$$

The solution $\delta\mathbf{v}_k$ of (9)-(10) will conserve the total kinetic energy of the kth layer of the model, and at the same time, will be the "closest" correction to the geostrophically estimated $\delta\mathbf{v}_k^g$. The solution of problem (9)-(10) can be obtained by using the Lagrangian method (Fletcher, 1987), and has the following form:

$$\delta\mathbf{v}_k = \delta\mathbf{v}_k^g + \lambda_k^e \rho_k \mathbf{v}_k H_k \tag{11}$$

where the Lagrangian multiplier λ_k^e can be determined by substituting (11) into constraint (10).

Optimization problem C (conservation of the potential energy)

The representation of potential energy in (3) has two terms. The first term is a change in potential energy caused by a change in the sea surface elevation. The second term accounts for contributions of the layer thicknesses to the potential energy. Again, we can transfer the sea surface information $\delta\eta$ into subsurface fields in such way as to suppress the contribution of the second term of (3) (contribution of changes in the layer thicknesses to the total balance of potential energy). In this case, we can consider the following problem:

$$\min_{\delta h_k} J = \frac{1}{2} \int_{L_k} \left(\delta h_k - \delta h_k{}^s\right)^2 dx \tag{12}$$

$$\int_{L_k} (\rho_{k+1} - \rho_k) \delta h_k h_k dx = 0 \tag{13}$$

The solution δh_k of (12)-(13) will conserve the total (integrated over the layer) contribution of the second term of (3) in the change of potential energy, and at the same time, will be the "closest" correction to the statistically estimated anomalies δh_k^s. The solution of the problem (12)-(13) can be obtained by using the Lagrangian method (Fletcher, 1987), and has the following form:

$$\delta h_k = \delta h_k{}^s + \lambda_k^p (\rho_{k+1} - \rho_k) h_k \tag{14}$$

where the Lagrangian multiplier λ_k^p can be determined by substituting (14) into constraint (13).

Problem C can be solved together with problem B. In this case, we will project the surface information to the subsurface fields by conserving the total kinetic energy for each model layer , as well as the second term in (3) (total change in potential energy caused by changes in the layer thicknesses)

We would like to note that an idea of using energetics to assimilate altimeter data into an eddy-resolving primitive equation models has also been proposed in (DYNAMO, 1997). Two approaches were introduced:1) the minimization of the model energy; 2) the minimization of the change in the model energetics resulting from the assimilation process.

Optimization problem D (conservation of the potential vorticity):

In Cooper and Haines (1996) and Haines (1991), the potential vorticity conservation method is used to transmit sea surface elevation measurements into the deep layers. According to (4), we have

$$\delta PV_k = \frac{\delta\zeta_k}{H_k + \delta h_k} - \frac{(f + \zeta_k)\delta h_k}{H_k(H_k + \delta h_k)} \tag{15}$$

where δPV_k and $\delta \zeta_k$ are changes in the potential vorticity and relative vorticity caused by changes in the layer thicknesses (δh_k) and velocities $(\mathbf{v}_k)$. The relative change in the potential vorticity after ignoring the second order terms will have the following form:

$$\frac{\delta PV_k}{PV_k} = \frac{\delta \zeta_k}{f + \zeta_k} - \frac{\delta h_k}{H_k} \tag{16}$$

Now, we can consider the following optimization problem:

$$\min_{\delta h_k} J = \frac{1}{2} \int_{L_k} (\delta h_k - \delta h_k{}^s)^2 dx \tag{17}$$

$$\int_{L_k} \frac{\delta h_k}{H_k} dx = 0 \tag{18}$$

The solution δh_k of (17)-(18) will provide the conservation of the second term in the equation for the relative change in the potential vorticity (integrated over the layer) and will be "closest" to the statistically inferred field.

The solution of the problem (17)-(18) can be obtained by using the Lagrangian method (Fletcher, 1987), and has the following form:

$$\delta h_k = \delta h_k{}^s + \lambda_k^v \frac{1}{H_k} \tag{19}$$

where the Lagrangian multiplier λ_k^v can be determined by substituting (19) into constraint (18).

4. NUMERICAL RESULTS

The experiments described in this paper use a $1/8°$ 4-layer finite depth Sea of Japan version of NLOM. The model domain extends from $127.5°$E to $142°$E and from $35.5°$N to $49°$N. Figure 1 shows the model domain and the bottom topography. At the solid boundaries a no-slip boundary condition is used. The model boundary follows the 200 m isobath with a few exceptions like the shallow straits connecting the Sea of Japan and the Pacific Ocean. The model has three open boundaries, one inflow port at the Tsushima Strait, and two outflow ports, one at the Tsugaru Strait and one at the Soya Strait. The inflow through Tsushima Strait has an annual mean transport of 2.0 Sv with a seasonal variation of 1.3 Sv peak-to-peak. Two thirds of the inflow exits through the Tsugaru Strait, while the remainder exits through Soya Strait. At all the open boundaries 2/3 of the flow is distributed through layer 1 and the remaining 1/3 through layer 2. The model was spun up with climatological windstresses, until equilibrium. The model was then forced with ECMWF daily winds for the period 1981 through 1994.

This coarse resolution version of the model does not give the most realistic picture of the Sea of Japan. Hogan and Hurlburt (1997) have shown that a much higher resolution is necessary to be able to model the complex eddy field and the deep circulation of the Sea of Japan. Their best results were with a $1/32°$ version of the model. We decided to use the coarse resolution model in order to be able to perform several long data assimilation experiments.

BOTTOM TOPOGRAPHY

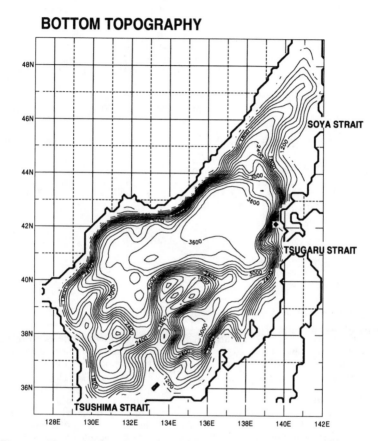

Figure 1: The geometry and bottom topography of the Sea of Japan ocean model.The grid resolution is 1/8 degree in latitude and 45/256 degree in longitude. The contour interval is 150 m.

The experiments performed are identical twin experiments, which means that the "observations" are themselves model solutions. The year 1982 was chosen to represent the real ocean in order to provide us with the "observed" SSH and pressure distributions in the model layers. SSH along Topex/Poseidon tracks were extracted from the model solution and assimilated into the model. The initial conditions for the assimilation experiments were taken to be the model field from the 5th day of 1991 from the ECMWF integration. Fig. 2 a and b show the initial and the final SSH field for the "observed" ocean. The comparisons between the different data assimilation schemes were made after one year of assimilation at the 3rd day of 1992.

The following relative model skill in prediction of SSH is used to compare the results:

$$\zeta_1 = \mathsf{max}\left(SSH^{model} - SSH^{obs}\right) \qquad \zeta_2 = \mathsf{min}\left(SSH^{model} - SSH^{obs}\right) \quad (20)$$

where SSH^{model} and SSH^{obs} are model and observed SSH.

The performance of the assimilation algorithm is also monitored by calculating the root-mean-square (RMS) error for the pressure fields. The RMS error is given by

$$RMSP_k = \left(\frac{\sum\limits_{i=1}^{N_x} \sum\limits_{j=1}^{N_y} (p_k^{model} - p_k^{obs})^2}{N_x N_y} \right)^{\frac{1}{2}} \tag{21}$$

where $N_x N_y$ are the number of ocean points in the entire model domain and $RMSP_k$ is the RMS error in the kth layer.

First, we tested the application of criteria (8) for the inversion of the pressure corrections into the corresponding changes in layer thicknesses (optimization problem A). At the beginning of the simulations, we considered a set of small μ values, and for each value, SVD was applied. The $\mu^o = 1.5 \cdot 10^{-4}$ was chosen to satisfy the criteria (8). For the initial model state (Fig. 2 a), the estimates (20) are $\zeta_1 = -10.1$ cm, and $\zeta_2 = +15.2$ cm. The use of the data assimilation scheme with the μ^o produced a significant improvement in prediction of SSH: $\zeta_1 = -5.9$ cm, $\zeta_2 = +5.9$ cm. Runs with the other values of μ gave worse results. For example, the application of data assimilation scheme with the $\mu = 1.0 \cdot 10^{-8}$ produced the following values of the estimate (20): $\zeta_1 = -15.1$ cm, and $\zeta_2 = +6.1$ cm.. Also, the values of $RMSP_k$ in (21) were significantly higher than the corresponding values for $\mu = \mu^o$: two times higher for the first layer, two and a half times higher for the second and third layers and one and a half times higher for the fourth layer.

In the next set of experiments, the applications of optimization problems B, C and D were tested. In all these experiments, the value $\mu = \mu^o$ was used. Therefore, the main goal of these experiments was to determine if the application of optimization problems B, C or D improves the performance of data assimilation scheme. Five different experiments were performed. In Case 0 the model was integrated without assimilating any observations, Case 1 used the original assimilation scheme with $\mu^o = 1.5 \cdot 10^{-4}$ determined from the

INITIAL SEA SURFACE HEIGHT

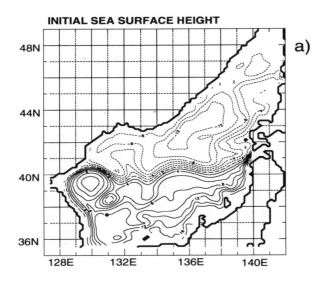

a)

"OBSERVED" SEA SURFACE HEIGHT

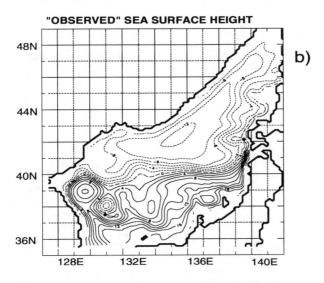

b)

Figure 2: These two fields represent the model runs used as a) the initial sea surface height and b) "observed" sea surface height in the identical twin experiments. The contour interval is 2 cm.

Name	Assimilation Method	ζ_1 (cm)	ζ_2 (cm)	$\delta RMSP_1$	$\delta RMSP_2$	$\delta RMSP_3$	$\delta RMSP_4$
Case 0	No assimilation	-11.0	13.1	200	260	136	5.6
Case 1	Optimization A	-5.9	5.9	—	—	—	—
Case 2	Optimization B	-3.9	4.3	-21.3	-11.2	-3.	-5.3
Case 3	Optimization B and C	-3.7	3.6	-20.	-9.0	-0.7	32.8
Case 4	Optimization D	-4.7	4.0	-8.2	-1.	-1.8	0.0

Table 1: Results from the experiments

optimization problem A; Case 2 used the optimization problem B (conservation of the total kinetic energy of the model); Case 3 used the optimizations problem B together with problem C; and Case 4 used optimization problem D. In addition to estimate (20), the following relative model skill in prediction of layer pressure fields was used:

$$\delta RMSP_k = (RMSP_k - RMSP_k^o) * 100 / RMSP_k^o \qquad (22)$$

where $RMSP_k$ is the RMS difference between observed and model pressure fields for the kth layer, $RMSP_k^o$ is the RMS difference between the "observed" and the model pressure fields for the kth layer in the Case 1 run. Therefore, $\delta RMSP_k$ represents the percentage of RMS error increase (positive) or decrease (negative) relative to the Case 1 run. All estimates are calculated after one year of assimilation. The results from the experiments are listed in Table 1.

In comparison to the Case 1 run, there is an improvement in predictions for the two top layers, but a smaller improvement for the two lowest layers in Case 2. At the same time, results (Table 1) show that in all cases of assimilation of SSH there is a weak influence on bottom layer dynamics. This is probably due to the use of the statistical inference technique to estimate the changes in bottom layer pressure. Finally, Fig. 3a shows the SSH field at the 3rd day of 1992 when the model was run without assimilating any observations (Case 0), while Fig. 3b shows the same field after assimilating the identical twin topex data (Case 3). As can be seen, the assimilated field is close to the "observed" field shown in Fig. 2b.

5. CONCLUSIONS and DISCUSSION

Previous results of applications of the Data Assimilation and Rapid Transition system of the Naval Research Laboratory showed significant improvement predictions in the Pacific Ocean in comparison to the model runs without assimilation (Carnes et. al., 1996, Smedstad et. al., 1997). At the same time, as noted in (Carnes et. al., 1996), further improvements in assimilation parameters and techniques are needed in order to reach the required prediction skill of the system. This is especially important for inferring subsurface fields from surface information and tuning nudging parameters. In the framework of the NRL data assimilation system, we propose to use variational problems in order to constrain the statistically inferred subsurface fields according to the current model dynamics. This approach combines the statistical and dynamical information and provide a continuous feedback from the model to

NO ASSIMILATION

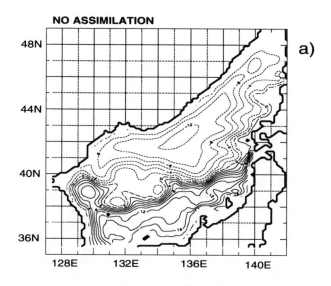

a)

TOPEX ASSIMILATION

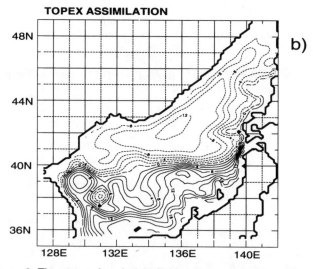

b)

Figure 3: The sea surface height field at the end of a) the model run without assimilation and b) assimilation of the identical twin topex data. These figures should be compared to the "observations" in Figure 2b. The contour interval is 2 cm.

the inference of subsurface fields. Four variational problems are introduced: (A) minimization of the norm of the nudging terms; (B) conservation of the kinetic energy of the model; (C) conservation of the potential energy and (D) conservation of the potential vorticity. This approach has been tested in identical twin experiments with the $1/8°$ 4-layer finite depth Sea of Japan version of the NLOM. The use of the optimization problem A with the optimal value of μ produced a significant improvement in prediction of SSH and the layer pressures (Table 1). Combinations of problem A with the problems (B), (C), or (D) provided the improvement in the predictions for the top layers of the model (see Table 1), but a smaller improvement for the bottom layer. Results show that in all cases of assimilation of SSH there is a weak influence on bottom layer dynamics. This is probably due to the use of the statistical inference technique to estimate the changes in bottom layer pressure; also, the $1/8°$ model does not have enough resolution for realistic predictions of the circulation in the Sea of Japan (Hogan and Hurlburt, 1997). Overall, the obtained results are very promising and experiments with the real data (Topex/Poseidon) will be conducted.

6. ACKNOWLEDGMENTS

This work is a contribution to the 6.2 Basin Scale Prediction System project, Data Assimilation and Rapid Transition (DART) and Large-Scale Modeling tasks sponsored by the Office of Naval Research (program element number 62435N) as part of the Naval Ocean Modeling and Prediction program. Shulman was supported by the Naval Ocean Modeling and Prediction program (contract N00014-97-1-0171). We thank Drs. Hurlburt, Wallcraft, and Mr. Rhodes for assistance and help. This work was supported in part by grants of Cray C90 time from the DoD Major Shared Resource Centers at U. S. Naval Oceanographic Office, Stennis Space Center, Mississippi.

7. REFERENCES

Carnes M.R., D.N. Fox, R.C. Rhodes and O.M. Smedstad, 1996: Data assimilation in a North Pacific Ocean monitoring and prediction system. In : Modern approaches to data assimilation in ocean modeling. P. Malanotte-Rizzoli, ed., Elsevier Science Publishers 67, 319 -345.

Cooper, A. and K. Haines, 1996: Data assimilation with water property conservation. J. Geophys. Res., 101, 1059-1077.

De Mey, P. and A. R. Robinson 1987: Assimilation of altimeter eddy fields in a limited area quasi-geostrophic model. J. Phys. Oceanogra., 17, 2280 -2293.

De Mey P., 1997: Data Assimilation at the Ocean mesoscale: a review. In: Data Assimilation in Meteorology and Oceanography: Theory and Practice., 305 -317.

DYNAMO, 1997: Dynamics of North Atlantic Models: Simulation and assimilation with high resolution models, 1997 Scientific report Nr. 294, 334 pp. EC MAST DYNAMO, contract no. MAS2-CT93-0060.

European Centre for Medium-range Weather Forecasts (ECMWF), 1994: The description of the ECMWF/WCRP level III-A global atmospheric data archive, ECMWF, Reading/Berks, UK, 72pp.

Ezer, Tal and G. L. Mellor, 1994: Continuous assimilation of Geosat altimeter data into a three-dimensional primitive equation Gulf Stream model. J. Phys. Oceanogra., 24, 832 - 847.

Fletcher, R., 1987: *Practical Methods Of Optimization*. John Wiley and Sons, New York, 436 pp.

Ghil, M. and P. Malanotte-Rizzoli, 1991: Data assimilation in meteorology and oceanography. Vol. 33, *Advances in Geophysics*, Academic Press, 141-266.

Haines, K., 1991: A direct method of assimilating sea surface height data into ocean models with adjustments to the deep circulation. *J. Phys. Oceanogr.*, **21**, 843-868.

Hogan, P. J. and H. E. Hurlburt, 1997: Sea of Japan circulation dynamics via the NRL layered ocean model. *Proceedings of the CREAMS International Symposium*, 28-30 January, Fukuoka, Japan, 109-112.

Hurlburt, H. E. and J. D. Thompson, 1980: A numerical study of Loop Current intrusions and eddy shedding. *J. Phys. Oceanogr.*, **10**, 1611-1651.

Hurlburt, H. E. , D. N. Fox, and E. J. Metzger, 1990: Statistical inference of weakly correlated subthermocline fields from satellite altimeter data. J. Geophys. Res., 95, 11375 -11409.

Hurlburt, H. E., 1986: Dynamic transfer of simulated altimeter data into subsurface information by a numerical ocean model. *J. Geophys. Res.*, **91**, 2372-2400.

Mellor, G. L. and T. Ezer, 1991: A Gulf Stream model and an altimetry assimilation scheme, *J. Geophys. Res.*, 96, 8779-8795.

Oschlies, A. and J. Willebrand, 1996: Assimilation of Geosat altimeter data into an eddy-resolving primitive equation model of the North Atlantic Ocean, *J. Geophys. Res.*, 101, 14175 - 14190.

Reinecher, M. M. and D. Adamec, 1995: Assimilation of altimeter data into a quasi-geostrophic ocean model using optimal interpolation and eofs. *J. Mar. Sys.*, 6, 125-143.

Smedstad, O. M. and D. N. Fox, 1994: Assimilation of altimeter data in a 2-layer primitive equation model of the Gulf Stream. *J. Phys. Oceanogr*, **24**, 305-325.

Smedstad, O. M., D. N. Fox, H. E. Hurlburt, G. A. Jacobs, E. J. Metzger and J. L. Mitchell, 1997: Altimeter data assimilation into a $1/8°$ eddy resolving model of the Pacific Ocean. *Journal of the Meteorological Society of Japan*, **75**(1B), 429-444.

Wallcraft, A. J., 1991: The Navy Layered Ocean Model users guide, NOARL Report **35**, Stennis Space center, MS, 21pp.

Evaluation of subtidal water level in NOAA's Coastal Ocean Forecast System for the U.S. East Coast

John R. Schultz[*]
and
Frank Aikman III[**]

ABSTRACT

A three-dimensional, baroclinic, primitive-equation ocean model is used to make 24-h forecasts of coastal water level along the East Coast of the United States. The model has 6 to 10 km horizontal resolution in the near-coastal ocean and is driven by forecast momentum, heat, and moisture fluxes derived from NOAA's 80 km resolution atmospheric Eta model. The model results are compared to observations of water level from NOAA's coastal water level gauges. For subtidal water level in the synoptic-band (2-10 days), from Portland, ME to St. Augustine, FL, the coherence-squared between model and data is above 0.75. On average, the subtidal root-mean-square difference between forecasts and data is ~.11 m and the forecasts have an approximate 66% success rate at predicting high and low water events greater than one standard deviation.

[*] Now at Neptune Sciences, Inc., 11341 Sunset Hills Road, Reston, VA 20190-5205.
JRS conducted this work at NOS/CSDL while a University Corporation for Atmospheric Research (UCAR) Postdoctoral Fellow, supported through the NOAA Coastal Ocean Program Office. [**] NOAA, National Ocean Service, Coast Survey Development Laboratory, 1315 East-West Highway, Silver Spring, MD 20910

1. Introduction

Along the East Coast of the United States, much of the interest in sea level studies has been focused on estuarine and coastal environments. Studies have examined estuarine buoyant discharge off the Delaware coast because these flows have important effects on the transport of terrestrial material into the coastal ocean (Garvine, 1996). Along the North Carolina coast the influence of large along-isobath density gradients has been examined as a mechanism for driving the offshore detachment of low-salinity shelf-water into the deep ocean (Gawarkiewicz, et al. 1996). It has long been known that the Gulf Stream encroaches onto the shelf-break in the South Atlantic Bight (Bane and Brooks, 1979), influences the circulation in the Middle Atlantic Bight (Csanady and Hamilton, 1988), and Churchill and Cornillon (1991) have measured expelled Gulf Stream water over the continental margin well north of Cape Hatteras. Atmospheric cyclones are known to be the dominant meteorological forcing during winter time (Mooers et al., 1976), and the synoptic-band response of the subtidal sea level at the coast has been examined by Noble and Butman (1979), Wang (1979), and Lee et al. (1984). The experimental forecast system described here may be useful to address these issues because the inherent errors in the system may be corrected to varying degrees by refinement in the lateral open boundary conditions, by the assimilation of satellite altimeter data, as well as the assimilation of temperature data from remote and in situ sources, and by the addition of a nowcast cycle.

In this paper we investigate the ability of a three-dimensional, baroclinic, primitive-equation ocean model to account for coastal water level variability at subtidal frequencies. The development of the Coastal Ocean Forecast System (COFS) is the result of the cooperative efforts by NOAA's National Centers for Environmental Prediction (NCEP), National Ocean Service (NOS), and Geophysical Fluid Dynamics Laboratory, and Princeton University with support provided by NOAA's Coastal Ocean Program. The model is driven by 24-h forecast momentum, heat, and moisture fluxes derived from NOAA's regional, atmospheric Eta model. In previous studies by Ezer and Mellor (1994b, 1994c, 1997), the basic experimental methodology has been prepared to forecast the Gulf Stream region, and to also include coastal regions with complicated topography in the calculation. Here we ask what is the skill of the 24-hour forecast system with respect to subtidal coastal sea level. The question is important because models with simpler vertically averaged dynamics (Greatbatch et al., 1996) may be able to reasonably simulate coastal sea level, but these models cannot produce the nonlinear dynamics associated with the three-dimensional density field.

The work in this paper extends that of Aikman et al. (1996) and Ezer and Mellor (1994c), by concentrating on subtidal time scales, and by evaluating the sea level produced by an experimental coastal ocean forecast system. The intent of this paper is to quantify the skill of the basic COFS modeling system (i.e. provide a benchmark) with which to compare future versions. We evaluate the set-up and set-down of wind-driven coastal water levels in a forecast model without tides and without a nowcast (or hindcast) to establish initial conditions before the forecast is run. COFS presently runs a continuous series of daily 24-h forecasts, without the ocean model ever being reinitialized. In Section 2 we describe the experimental/operational methodology of the COFS. This is followed by a description of the observations and methodologies used for model evaluation. The model subtidal water level evaluation results are presented in Section 4, and Section 5 presents a summary of our results and some discussion.

2. System Description

The Coastal Ocean Forecast System has been producing experimental daily 24-h simulations of water levels and 3-dimensional temperature, salinity, and currents for the U.S. East Coast on an operational basis at NCEP since August 1993 (Aikman et al., 1996; Kelly et al., 1997). The system consists of the Princeton Ocean Model (Blumberg and Mellor, 1987) forced at the surface by forecast momentum, heat, and moisture fluxes derived from the 80 km resolution Eta atmospheric model (Black, 1994). The ocean model has been used in studies of bays and estuaries, and for basin-scale and high-latitude phenomenon. The numerical formulation of the present work is very similar to that used in recent studies of the Gulf Stream system (Ezer, 1994; Ezer and Mellor, 1994b, 1994c).

The vertical resolution is represented by 19 sigma levels. The horizontal grid is a coast following, curvilinear orthogonal system. The bottom topography is shown in Fig. 1 along with the mass transports assigned along the open boundary. The bottom topography is based on DBDB5 except in coastal regions where the NOS 15" topography is used because of better resolution and accuracy of the bottom topography and coastal geometry. The grid resolution is approximately 6-10 km in coastal regions and 10-20 km for the deep ocean. The shallowest depth of the bottom topography in coastal regions is 10 m. Extending the bottom topography in coastal regions to depths as shallow as 5 m did not seem to improve the forecasts of coastal water level.

The fixed mass transport along the southern open boundary in the Straits of Florida is determined by STACS data (Leaman et al., 1987), and east of the Bahamas the transport is based on Lee et al., 1997. Mass transport along the eastern boundary is estimated from climatology and diagnostic calculations (Ezer and Mellor, 1994c). Open boundary temperature and salinity is derived from Levitus' (1996) monthly and annual climatology, respectively. The surface atmospheric pressure, evaporation minus precipitation, 10 m winds, and heat used to drive the model are derived from forecasts made by the 80 km resolution atmospheric Eta model. The fields are bilinearly interpolated to the COFS grid. Climatological monthly river runoff is added as a freshwater flux for the principal rivers. The atmospheric pressure loading due to the spatial distribution of the surface pressure is added to the forecast sea level because it improves water level at open ocean stations like Bermuda. Further information about the COFS experimental design and system descriptions may be found in Aikman *et al.* (1996), Lobocki (1996), and Ezer and Mellor (1994a).

3. Observations and Methods

Appropriate sea level data were obtained from NOS water level gauges which recorded two years of uninterrupted observations. Atmospheric pressure and wind data were obtained from National Data Buoy Center (NDBC) buoys. The two-year evaluation period is from October 1, 1993 through September 31, 1995, and the locations of the stations are shown in Fig. 1. The hourly observed water level data were demeaned, and a 30-h low-pass filter was applied to remove the tides. Similarly, the hourly forecasted water level from the model were demeaned and subject to a 30-h low-pass filter, although the COFS ocean model reported on here does not include tides.

The evaluation methods used here are considered standard and their usefulness has been described by Anthes (1983). Techniques such as root-mean-square (rms) differences between the forecasted values and the observations are a measure of the accuracy of the forecasts since the water levels from the NOS gauges undergo extensive analysis and verification and are a standard measurement of coastal water level. Correlation coefficients determine the degree of linear relationship between observations and forecasted values, and help to determine if the forecast could be used as an acceptable substitute for the observations, if the correlation coefficients are particularly high.

The mariner navigating inland and near-coastal waters is interested in techniques that reliably predict both high and low water events. The degree of reliability of COFS may be assessed by defining a "success rate" and a "false alarm rate". The success rate determines the number of the times that an event was observed and it was also forecasted. The false alarm rate is a statistic that determines the number of times that an event was forecasted, but was not observed. The success rate and false alarm rate are defined by a specific threshold for a high or low water event. For this study, the high or low water thresholds are defined as ± 1 σ, where σ is the standard deviation of the observations at the station. For example, high or low water events at the ± 1 σ threshold occurred about 12% of the time during the two-year evaluation period. A success rate counts the number of times that the forecast is correct when the observations indicate a high or low water event has occurred. A false alarm rate measures the number of times that the model has predicted a high or low water event, but the observations did not confirm the prediction. Implicit in this calculation is a window of time in which the analysis is performed. To compute the success rate for a 1 σ high water event, the window of time was defined as follows. When a 1 σ high water event occurred, the forecast was searched 3 hours before the event and 3 hours after. Then, if the forecast predicted a value above the threshold within the ± 3 hr window, a success was recorded, and the search loop was exited for that event.

Power spectra, coherence-squared, and transfer amplitude plots are employed (Bendat and Piersol, 1971) and serve as an aid in interpreting the realism of the forecasts, because these techniques illustrate the frequency content of the forecast compared to the observations. These analysis tools are used here to relate the forecasts to the observations and to examine the alongshore relationships of the water level signals in both the observations and the predictions.

4. Subtidal Water Level Evaluation

Sample subtidal sea level time series are shown in Fig. 2. The observations suggest higher amplitude at the northern stations in winter compared to stations south of Cape Hatteras, and greater variability in the winter compared to summer (not shown). The model exhibits similar behavior. In general, the subtidal sea level forecasts are in closer agreement with the observed subtidal sea level north of Cape Hatteras, where the direct wind forcing is greater than in the South Atlantic Bight. A feature of the sea level data worth noting is that some events occur at all stations or seem to propagate between stations. For example, on January 4, 1994, an event is found at all gauges. A similar event is found on February 4, 1995. The model, shown by the thin dashed line, reproduces these events with similar amplitudes and phases.

Fig. 3 shows plots of the rms differences, correlation coefficients, and ratio of model standard deviation to the observed standard deviation at all the coastal water level gauges (see also Table 1, columns 1-3). The meridional average of the rms difference is 0.11 m, the average correlation coefficient is 0.75, and the average ratio of the modeled standard deviation to the observed is 0.98. The model forecasts tend to have smaller rms differences north of Cape Hatteras. Also, the forecasts tends to be most highly correlated to the observations in the Middle Atlantic Bight. The model variability is greater than the observations north of Atlantic City, and is less variable than the observations south of there. Taken together, these statistics suggest that the forecasts are better to the north of Cape Hatteras, where the wind forcing is stronger.

Table 1 (columns 5-8) illustrates the ability of the model to capture specific high and low water events, which are both important to the mariner navigating coastal waters. The results suggest that the forecast has reasonable success rates and perhaps tolerable false alarm rates for high and low water events measured at the ± 1 σ threshold. The statistics suggest that a ± 6 h window (not shown here) does not greatly improve the results relative to a ± 3 h window. The average success rate for a ± 3 hr window is 63% for 1 σ high water event, 68 % for a low water event, and both high and low water events average a 38% false alarm rate. The statistics also show that the forecast is more useful at some locations as opposed to others. For example, the forecast is highly accurate near Lewes, DE with a 70% success rate at the ± 3 h window, while the results at Newport, RI show that the forecast is unreliable there. Sandy Hook, NJ is a region where the forecast is accurate for low water events, but needs improvement near Chesapeake Bay Bridge Tunnel (CBBT), VA, and Duck, NC. In each location, the quality of the forecast comparison to the observations depends on the exposure of the water level gauge, the local geometry and bathymetry, and on the distance from the gauge to the model grid point (which can vary from ~10-30 km.) However, distance from the gauge to the model grid point is not a strong factor in the reported errors. The overriding factor is the degree to which the gauge is sheltered from the open ocean.

Power spectra reveal the distribution of variance as a function of frequency at three representative observational sites (Fig. 4). The interpretation, as deduced from the standard deviations, that the model under predicts the water level in the South Atlantic Bight, and over predicts in the Middle Atlantic Bight and Gulf of Maine, applies to the lowest frequency behavior. The model captures the "redness" of the spectrum (i.e., most of the large-scale energy in the ocean occurs at low-frequencies) and many of the general details.

Fig. 5 shows the coherence and phase between the model and observations for the same three stations. This illustrates the ability of the model to capture events in the observations. In the synoptic band, the coherence is high, tending to be above 0.75 in all cases, and in the 3 - 5 day period band the coherence-squared approaches 0.9 at Portland. Small phase differences exists in the synoptic-band, and approach zero at Lewes in the 3 to 5 day period band. In general, the coherence-squared tends to be greater in the wind-driven band than suggested by the correlation coefficient.

The passage of a 990 mb cyclone through the coastal waters of the Middle Atlantic Bight and Gulf of Maine illustrates the influence of the storm on subtidal water level variability (Fig. 6). The storm traveled northward at a speed of ~16 m/s, nearly the average ~15 m/s speed reported by Mooers et al. (1976). Between Lewes and Sandy Hook there is a simultaneous piling up of water against the coast. The sea level signal propagates northward from this region with the storm, and the amplitude diminishes. Between Lewes and Duck the sea level propagates southward, consistent with low-frequency shelf wave propagation. The speed is ~1200 km/d, within a factor of two of the speed determined by Wang (1979), which was ~600 km/d.

The model coherence along the shelf suggests that the model describes the northward and southward propagating sea level events in the synoptic band (Figs. 7a and 7b). A positive phase lag (converted from degrees to time difference in hours) suggests a northward propagating event in the synoptic band because the southern-most station in the pair leads the northern-most station. Likewise, a negative phase lag suggests that the southern-most station lags the northern-most station, consistent with the suggestion of southward phase propagation. A comparison between six sets of adjacent stations from Portland, ME to St. Augustine, FL shows similar coherence-squared, phase, and transfer function for both the model and the observations in the synoptic wind-driven band. Between Portland and Sandy Hook (Fig. 7a) both the model and observations suggest predominantly northward propagating events. Between Sandy Hook and Lewes (Fig. 7a) both the model and observations suggest a simultaneous response. Between Lewes and Duck (Fig. 7b) both the model and observations suggest northward propagating sea level around a ten day period, and southward propagating sea level at a five day period. Between Duck and Fort Pulaski (Fig. 7b) there is no significant coherence in either model or observations in the synoptic band. Between Fort Pulaski and St. Augustine (Fig. 7b) both the model and observations suggest southward propagation around a ten day period, and northward propagation at periods of five days and faster.

A number of sensitivity tests have been conducted to examine the impacts on the model forecasts of grid resolution, bathymetry, variable versus fixed Florida Straits transport, and other forecast wind fields (NCEP global aviation forecast model versus Eta). None of these factors appears to improve the model performance. However, a 12 month examination of the subtidal water levels in a version of COFS with the addition of tides (Chen and Mellor, 1997) suggests that the non-linear interaction of the tides and the wind-driven component results in an approximate 10% improvement in the subtidal water level statistics (Aikman et al., 1998). In addition, a nowcast/forecast version of COFS has been tested that uses a 24-hour nowcast that is forced at the surface by Eta analyzed wind fields to establish the ocean model initial state before the 24-h forecast is run. The results of a six month test suggest an approximate 20% improvement in the subtidal water level forecasts when a nowcast precedes the forecast (Aikman et al., 1998). It appears that the implementation of tides in the ocean model at NCEP (now operational), and the future implementation of a nowcast/forecast cycle should result in significant improvement to the model subtidal water level forecast skill.

5. Summary and Discussion

We have described results from a three-dimensional, free-surface, primitive-equation, sigma-coordinate model, driven by 24-h forecast momentum, heat, and moisture fluxes derived from NCEP's atmospheric Eta model. The model shows skill in predicting the subtidal water level variability in the synoptic band along the entire East Coast of the United States. The coherence squared is above 0.75 between the model and data in the synoptic band and the subtidal rms difference between forecasts and data is ~.11 m. The average success rate is 66% for subtidal water level, with a false alarm rate of 38%.

This study provides a benchmark with which to evaluate the improved accuracy of future versions of the COFS. These include a presently operational version that incorporates tides and an experimental (parallel operational) version that assimilates SST. Eventually, COFS will implement an operational version that includes both of the important wind-driven and tidally-driven coastal water effects as well as the assimilation of SST and altimeter-derived sea surface height data in a nowcast/forecast cycle.

Acknowledgments: We are grateful to many people whose efforts have contributed to the results reported on here. They include George Mellor and Tal Ezer (Princeton); D.B. Rao, Larry Breaker, John Kelly, and Lech Lobocki (NCEP); and Kate Bosley, Eugene Wei, Charles Sun, Richard Schmalz, and Phil Richardson (NOS).

References

Aikman, F. III, E.J Wei and J.R. Schultz, 1998. Water Level Evaluation for the Coastal Ocean Forecast System. Preprint: *AMS Second Conference on Coastal Atmospheric and Oceanic Prediction and Processes*, 11-16 January 1998, Phoenix, AZ.

Aikman, F., III, G.L. Mellor, T.Ezer, D. Sheinin, P. Chen, L. Breaker, and D.B. Rao, 1996. Towards an operational nowcast/forecast system for the U.S. East Coast, In: *Modern Approaches to Data Assimilation in Ocean Modeling*, P. Malanotte-Rizzoli, Ed. Elsevier Oceanography Series, 61, 347-376.

Anthes, R.A., 1983. Regional models of the atmosphere in middle latitudes. *Monthly Weather Review*, 111, 1306-1335.

Bane, J.M., and D.A. Brooks, 1979. Gulf Stream meanders along the continental margin from the Florida Straits to Cape Hatteras. *Geophysical Review Letters*, 6, 280-282.

Bendat, J.S. and A.G. Piersol, 1971. *Random Data: Analysis and Measurement Procedures*. Wiley-Interscience, New York, 407p.

Black, T.L., 1994. The new NMC mesoscale Eta model: Description and forecast examples. *Weather and Forecasting*, 9, 265-278.

Blumberg, A.F. and G.L. Mellor, 1987. A description of a three-dimensional coastal ocean circulation model. in *Three-Dimensional Coastal Ocean Models*, 4, edited by N. Heaps, American Geophysical Union, 1-16.

Chen, P. and G.L. Mellor, 1997. Determination of tidal boundary forcing using tide station data. Chapter in Coastal Ocean Prediction. Christopher N.K. Mooers, editor. CRC Press, Inc., 23p, in press.

Churchill, J.H., and P.C. Cornillon, 1991. Gulf Stream water on the shelf and upper slope north of Cape Hatteras. *Continental Shelf Research*, 14, 409-431.

Csanady, G. T. and P. Hamilton, 1988. Circulation of slope water. *Continental Shelf Research*, 8 (5-7), 565-624.

Ezer, T. 1994. On the interaction between the Gulf Stream and the New England seamount chain. *Journal of Physical Oceanography*, 24, 191-204.

Ezer, T. and G. L. Mellor (Eds.), 1994a. A Coastal Forecast System: A Collection of Enabling Information. COFS Technical Note 94-1. Princeton University, Princeton, NJ 08544-0710.

Ezer, T. and G.L. Mellor, 1994b. Continuous assimilation of Geosat altimeter data into a three-dimensional primitive equation Gulf Stream model. *Journal of Physical Oceanography*, 24, 832-847.

Ezer, T. and G.L. Mellor, 1994c. Diagnostic and prognostic calculations of the North Atlantic circulation and sea level using a sigma coordinate ocean model. *Journal of Geophysical Research*, 99(C7), 14159-14171.

Ezer, T. and G.L. Mellor, 1997. Data assimilation experiments in the Gulf Stream: How useful are satellite-derived surface data for nowcasting the subsurface fields? *Journal of Atmospheric and Oceanic Technology*, in press.

Garvine, R.W., 1996. Buoyant discharge on the inner continental shelf: A frontal model. *Journal of Marine Research*, 54, 1-33.

Gawarkiewicz, G., T.G. Ferdelman, T.M. Church, and G.W. Luther III, 1996. Shelfbreak frontal structure on the continental shelf north of Cape Hatteras. *Continental*

Shelf Research, 16, 1751-1773.

Greatbatch, R.J., Y. Lu, and B. DeYoung, 1996. Application of a barotropic model to North Atlantic synoptic sea level variability. *Journal of Marine Research*, 54, No. 3, 451-469.

Kelly, J.G.W., F. Aikman III, L.C. Breaker and G.L. Mellor, 1997. Coastal Ocean Forecasts. *Sea Technology*, 38 (5), 10-17.

Leaman, K. D., R. L. Molinari and P. S. Vertes, 1987. Structure and variability of the Florida Current at 27°N: April 1982-July 1984. *Journal of Physical Oceanography*, 17, 565-583.

Lee, T.N., W. Johns, R. Zantropp and E. Fillenbaum, 1997. Moored observations of western boundary current variability and thermohaline circulation at 26.5° N in the subtropical North Atlantic, *Journal of Physical Oceanography*, 26(6), 962-983.

Lee, T.N., W.J. Hou, V. Kourafalou, and J.D. Wang, 1984. Circulation on the continental shelf of the southeastern United States. Part I: Subtidal response to wind and Gulf Stream forcing during winter. *Journal of Physical Oceanography*, 14, 1001-1012.

Levitus, S., 1996. Climatological atlas of the world of the ocean. NOAA, U.S. Department of Commerce, Washington, D.C.

Lobocki, L., 1996. Coastal Ocean Forecast System description and users guide. COFS Technical Note 96-2. OMB Contribution No. 127. National Centers for Environmental Prediction, Camp Springs, MD., 69p.

Mooers, C.N.K., J. Fernandez-Partagas, and J.F. Price, 1976. Meteorological forcing fields of the New York Bight (first year). University of Miami (RSMAS) Tech. Rep. TR76-8, 151p.

Naval Oceanographic Office, W. Hemi. Bathymetry, DBDB5 BATHY, Center for Air Sea Technology, Stennis Space Center, MS.

Nobel, M. A. and B. Butman, 1979. Low-frequency wind-induced sea level oscillations along the East Coast of North America. *Journal of Geophysical Research*, 84, 3227-3236.

Wang, D.P., 1979. Low-frequency sea level variability on the Middle Atlantic Bight. *Journal of Marine Research*, 37, 683-697.

TABLE 1.

Root-mean-square (rms) error, correlation coefficients, the ratio of the model-to-observed standard deviations, and the observed standard deviation (columns 1 to 4, respectively) for two years (October 1993 to September 1995) of 24-h forecasted and observed *subtidal* water levels, for the same two years, columns 5 to 8, respectively, show the success and false alarm rates for high (and low) water events greater (or less) than one standard deviation (1 σ).

Station	RMS error (m)	Corr. Coeff.	Model/ Observ. Stn. Dev.	Stn. Dev. of obs. (m)	1 σ high water success rate	1 σ high water false alarm rate	1 σ low water success rate	1 σ low water false alarm rate
Port.	10	67	1.07	.12	.60	.58	.71	.49
Newp.	10	.72	1.08	.12	.56	.58	.73	.38
Sn. Hk.	11	81	1.06	.17	.67	.41	.85	.32
Atl. C.	11	80	0.98	.18	.62	.34	.76	.32
Lewis	11	80	0.97	.18	.70	.33	81	.32
CBBT	13	73	0.96	.17	.64	.31	.56	.49
Duck	11	73	0.86	.17	.59	.26	.46	.42
Ft. Pu.	12	79	0.89	.19	.65	.26	.56	.31

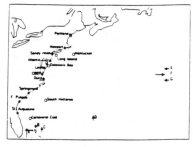

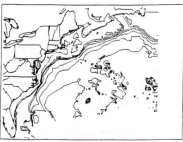

Fig. 1. Top diagram. Geographical locations of NOS water level gauges represented by solid circles and NDBC moored buoy locations are shown as open squares. Along the southern and eastern open boundaries the imposed mass transport is shown as arrows labeled A through G. The units of the mass transport are Sverdrups (1 Sv = 10^6 m³ s⁻¹). A mass transport of 30 Sv enters the domain in the Florida Current (A), 38 Sv exits the domain in the deep western boundary undercurrent (B), 5 Sv enters the domain in the Antilles Current (C), and a broadly distributed weak inflow of 33 Sv enters the domain in the subtropical gyre. Along the eastern boundary 30 Sv enters the domain north of the Gulf Stream (E), 90 Sv exits in the Gulf Stream (F), and 30 Sv enters in the subtropical gyre (G). Bottom diagram. Bathymetry. The shallowest depth plotted is the 50 m isobath. Isobaths deeper than 200 m begin with the 1000 m depth and extend to 5000 m in increments of 1000 m.

SUBTIDAL WATER LEVEL FLUCTUATIONS

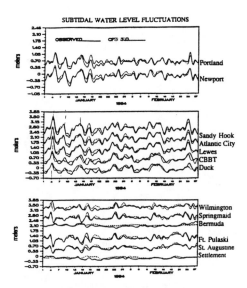

Fig. 2. A portion of the demeaned and 30-h low-pass filtered water level records (solid line - observed; dashed line - forecasted) during the winter (January and February) of 1994. Stations are positioned from north to south. See Fig. 1 for locations.

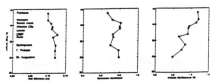

Fig. 3. Subtidal water level statistics. Shown in the first panel is a plot of the rms difference between observations and the forecast. The correlation coefficient between observations and the forecast is shown in the second panel. The third panel shows the ratio of the standard deviation of the forecasts to the observations (see Table 1, columns 1 - 3).

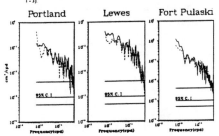

Fig. 4. The power spectrum of subtidal water level is shown at three representative stations. The forecast is represented as a dashed line and the observations are shown as a solid line. The 95% confidence intervals are indicated by the three horizontal lines.

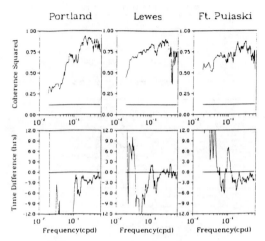

Fig. 5 The coherence-squared and phase difference in hours between the data and the forecast at the same three representative stations. The 95% test of significance is represented by the horizontal line in the coherence-squared plot. Negative phase implies the forecast lags the data.

ATMOSPHERIC PRESSURE
AND WATER LEVEL

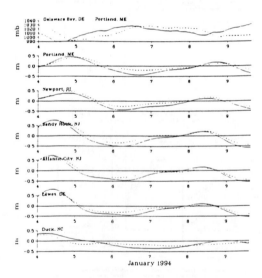

Fig. 6 Top frame: The hourly time series of atmospheric pressure from the Delaware Bay buoy is shown in the solid line. The dashed line represents the atmospheric pressure from the Portland buoy. Lower frames: Subtidal water level. Forecast water level is shown by the dashed line, data is shown by the solid line.

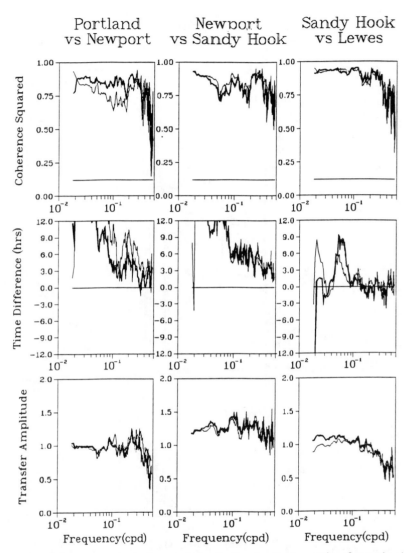

Fig. 7a. Coherence-squared, phase, and transfer function for three pairs of water level stations along the U.S. Northeast Coast. The data is represented by the thick solid line, and the forecast is represented by the thin line. The 95% confidence level is the horizontal solid line in the coherence-squared plot. Negative phase lag means the second series in the title lags the first series.

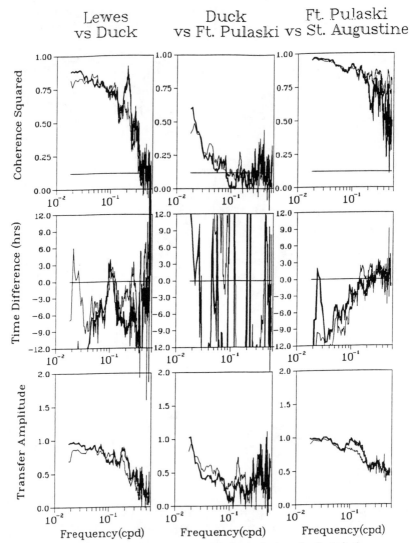

Fig 7b. Coherence-squared, phase, and transfer function for three pairs of water level stations along the U.S. Southeast Coast, as in Fig. 7a.

Sediment Transport Dynamics in a Dredged Channel

Sean O'Neil[1] and David P. Podber[2]

Abstract

The flow field and resulting resuspension, transport and deposition of sediment in a dredged, Great Lakes tributary are examined for the purpose of investigating the effects of flow reversal and sediment deposition in dredged tributaries. Hydrodynamics were simulated by dynamically coupling a laterally averaged, vertical-plane model, which includes a turbulence closure sub-model, and a one-dimensional river model. The transport of suspended sediment was modified by the inclusion of source/sink terms for resuspension, deposition and settling. The model study has been performed using conditions which are typical of the fall season for the region, and in this case a single sediment grain size was used. The dredged portion of the model domain contains two sills that form as the result of the interaction of storm-induced long-waves at the lake end of the tributary and increased sediment load due to runoff upstream. The model results show flow reversals and significant lake water intrusion upstream even for moderate long-wave amplitudes.

Introduction and Objectives

Harbor dredging is the necessary result of long term, persistent deposition of watershed-derived sediments. Periodic redredging is required to ameliorate the occurrence of sills and bars which form as a result of the interaction of wave climate, channel geometry, tributary flow and littoral drift. As opposed to the persistent and predictable tidal forcing on coastal harbors, the harbors on the Great Lakes are moderated by random, long-wave effects derived from storms. The storm surges and resulting seiches, coupled with a high sediment influx from watershed runoff, often conspire to yield a two-sill bottom configuration which motivates annual maintenance dredging.

[1]Graduate Research Associate, [2]Visiting Assistant Professor. Department of Civil and Environmental Engineering and Geodetic Science, The Ohio State University, 470 Hitchcock Hall, 2070 Neil Avenue, Columbus, OH 43210-1275. email: sean@niagara.eng.ohio-state.edu.

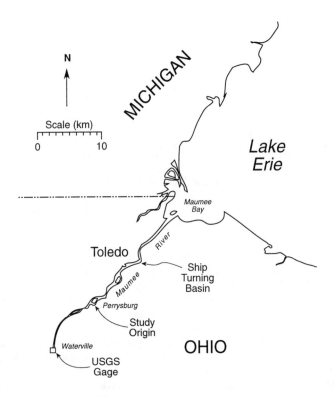

Figure 1: The Maumee River region.

This study is designed to be exploratory and uses a combination of realistic and idealized boundary condition data, therefore no comparisons were made with field-collected data. The model runs are designed to delineate the effect of dredging in tributaries and the resulting physical processes responsible for the formation of sills and/or bars in an estuarine setting. The models developed for these simulations will also be used to enhance the tributaries' portion of Great Lakes Forecasting System (e.g., Bedford and Schwab, 1994) which currently produces nowcasts four times daily and twice-daily 24-hour forecasts of the physical state of Lakes Erie and Ontario. This contribution details some of the suspended sediment dynamics resulting from conditions marked by flow reversals and stratification.

Study Site

Toledo Harbor, on the Maumee River, is the third busiest port in the Great Lakes shipping arena. The site of the model investigation, shown in Figure 1, extends along the river, through the Maumee Bay, along a dredged navigation channel. The Maumee River delivers the single largest tributary-derived sediment load to Lake Erie, contributing 44% of the total annual load (Kemp et al., 1976). This high load is a result of nearly 85% of the watershed land being used for agriculture. The extreme shallowness of Maumee Bay, with an average depth 1.5 m, and the Western Basin of Lake Erie, average depth 7.6 m, necessitates the maintainance of the 152 m wide, 8.5 m deep navigation channel (the U.S. Army Corps of Engineers currently specifies the channel to be 500 ft wide and 32 ft deep). The dredged portion of the channel extends from a ship turning basin 9 km upstream in the river, to approximately 6 km into the bay. Figure 2 shows a profile of the modeled portion of the river and navigation channel including the double-silled bathymetric configuration near the mouth of the river, and the form which the bathymetry-following, sigma-coordinate system. Note that the vertical and horizontal grid spacing lengths have been exaggerated for clarity.

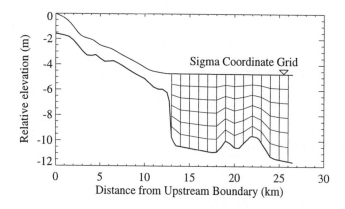

Figure 2: Schematic of model domain profile.

Physical Setting

The system comprised of the Maumee River, Bay and Western Basin of Lake Erie displays all of the physical behavior which might be found in a typical marine estuarine system, excluding tidal regularity. Storm events in the Lake Erie region occur with a frequency of 5-7 days during the spring and fall, with typical durations of 1-2 days. The corresponding increase of river flow rate and stage attains values that are significant fractions of the ambient levels.

A typical storm track during these seasons will follow the major axis, west to east of Lake Erie, producing a significant storm surge at the eastern end of the lake. The storm surge will decay into a lakewide seiche with frequently observed 14.4, 9.1, 5.9 and 4.2 hour longitudinal modes. The seiche typically produces water elevation

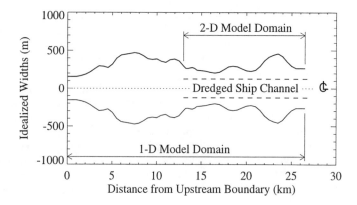

Figure 3: Schematic of model domain.

changes of more than 1 m, which is a significant fraction of the average water depth
at Toledo Harbor. After the storm event the decaying seiche may take 3-4 days to
completely disappear. The narrow and deep dredged channel (see Figure 3) results in
a "pipelining" of the excess flow due to runoff into the lake. The oscillations combined with the seasonal variations in water density gradients, here due to temperature
differences between river and lake water, result in a system which behaves like a typical estuary during storms and like a river during calm conditions. Flow reversals are
frequently noted as is stratification and internal waves.

Numerical Models

The numerical methods and schemes employed for this set of model runs have all
been well documented and used by several researchers under a variety of conditions.
Therefore only a brief description of the models and an outline of the dynamic coupling performed will be described. A complete listing of the equations, assumptions
and model validation results can be found in the references given below. A schematic
of the model coupling is shown in Figure 4.

The one-dimensional hydrodynamics are calculated using an abbreviated version
of the model originally developed by Bedford et al. (1983), subsequently extended by
the Environmental Laboratory at the U.S. Army Corps of Engineers, Waterways Experiment Station, called CE-QUAL-RIV1H. The core of the model employed here
solves the hyperbolic de St. Venant equations using a four-point finite difference
scheme and the resulting system of equations is solved iteratively using Newton's
method. A compact two-point, fourth-order accurate, advection scheme is employed
to ensure that sharp gradients are properly resolved. The model is employed from

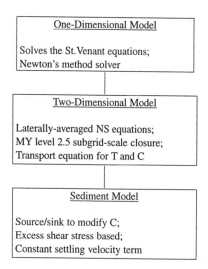

Figure 4: Model coupling schematic.

a point downstream of the Waterville, Ohio gauging station, past the last rapids, to a point 6 km into Maumee Bay. The model boundary conditions are the volumetric flow rate at the upstream end and the water elevation at the downstream end.

In a mode splitting sense, the computed discharge and stage at each node are passed from the one-dimensional model subroutines to the two-dimensional model. The laterally-averaged Navier-Stokes (NS) equations, employing the hydrostatic and Boussinesq approximations, and transformed to a bathymetry-following (sigma) coordinate system, are used to solve for the velocities and pressure of the internal mode. The advection-diffusion equation for temperature, which uses a third-order upwind advection scheme, is also modeled and the vertical turbulent eddy coefficients are computed from a laterally-averaged version of the Mellor-Yamada (MT) level 2.5 closure submodel (e.g., Blumberg and Mellor, 1987). The two-dimensional model development and testing has been discussed in Podber and Bedford (1994).

The model is employed in the dredged portion of the channel (see Figures 2,3) and requires upstream and downstream boundary conditions which are here assumed to be a logarithmic velocity profiles for flows into the domain and a zero-flux (radiation) condition at the downstream end if the net flow is into the lake. At the upstream boundary the logarithmic velocity profile is constructed to preserve the flow rate passed from the one-dimensional model, and the temperature is allowed to remain constant through the depth. Downstream the temperature has a zero-flux condition for outflow and an assumed profile for flow reversals. At the surface no flux of momentum or heat is assumed; while at the bottom there is no flux of heat and a shear stress is imposed proportional to the square of the horizontal velocity just above the bottom.

The sediment transport component of the model solves the advection-diffusion equation for suspended sediment concentration using the same third-order upwind advection scheme as mentioned above, and source/sink terms for the constant settling, erosion, and deposition of sediment. The erosion and deposition terms are parameterized using the model of Sheng and Lick (1979), where deposition is proportional to the sediment concentration and erosion is proportional to an excess shear stress as compared to a shear stress for sediment resuspension. Boundary conditions are specified to be no sediment flux through the water surface, a well mixed upstream condition giving a constant input sediment concentration, and a zero-gradient outflow condition or an assumed lake-like concentration profile imposed downstream. The empirical parameters in the erosion and deposition terms are based on samples collected from the Western Basin of Lake Erie and reported in Sheng and Lick (1979). Though these parameters are likely dependent on grain size, their determination was based on mixtures of sediment though they are used here for the monoculture of grain size used in this model.

Conditions and Assumptions

The specific conditions and assumptions which are applied for a fall season case study are outlined. The boundary conditions at the two ends of the model would normally come from data, or possibly, the output from other models; however, in this case the conditions were obtained from Shindel et al. (1993). For the fall conditions of interest, a steady river inflow was applied upstream of 40 m³/s, that agrees with the values obtained from flow hydrographs in Pinsak and Meyer (1976) during September. To approximate the effect of the lake seiche after a storm, a downstream sinusoidal water elevation with a moderate amplitude of 0.61 m and a period of 14 hours, corresponding to the primary mode, was applied for four complete cycles. Given the short term nature of the simulation, the bottom was assumed to be fixed so that the sills were not moving or changing shape. This is justified by the fact that (with the exception of the spring snow melt discharge) the bottom probably does not evolve much under the influence of a single storm event, but does over the course of a season or longer.

The river temperature was assumed to be a constant upstream value of 17°C and the lake was assumed to have a temperature of 24°C (Shindel et al., 1993). Since the flow in the river prior to the dredged ship turning basin has been assumed to be well mixed, the vertical line of nodes at the upstream end of the two-dimensional model was prescribed with a constant sediment concentration of 40 mg/L. At the downstream end of the model a logarithmic profile was constructed with a maximum concentration near the bottom of 10 mg/L for flow reversals and a zero-gradient condition for outflows. The form of the profile was found not to greatly affect results in trials with different profile shapes. Model runs were performed for a single sediment size with a Stokes' settling velocity determined to be 9.0×10^{-3} m/s for sandy-silt with a median diameter of 0.1 mm.

Results

The model runs examine the result of seiche-induced flow reversals as opposed to regular down-channel flow during inter-event (storm) periods for the fall season case. Figure 5 depicts the flow-field over the dredged protion of the channel, and only every third column of nodes is shown for clarity. The value of the vertical velocities throughout the domain have been multiplied by a factor of fifty during plotting to clearly portray the flow reversal and gyres which the model produces. The velocity field has been overlayed on the suspended sediment concentration contours. Again, for clarity, only five contour grades are used. The results are presented for the fourth cycle at every 1/8 of a seiche cycle as indicated by the small sinusoid in the upper right of the plots. The depth sense of the sinusoid is correct in that generally an west-to-east passing storm creates a storm surge at the eastern end of Lake Erie, drawing down the water at the mouth of the Maumee River. Only one cycle is presented since the conditions in the model domain do not change visually over the last two cycles.

The velocity fields in Figure 5 show that imposing a moderate seiche amplitudes can cause flow reversals and significant upstream intrusion of lake water. This might be expected when the upstream flow rate is relatively small as it is here at 40 m^3/s. However, in the previous studies of O'Neil et al. (1996) and Podber and Bedford (1994) it was shown that flow reversals can also occur for the much higher flow rates and increased sediment loads of 140 m^3/sec and 1000 mg/L, respectively, corresponding to a spring storm. In the previous studies it was seen that these increased input levels prevented the lake water intrusion from getting further upstream than the peak of the downstream sill.

The figures also depict a persistent vertical gyre which remains located above the downstream sill, moving from directly above the peak to just above the downstream extent of the slope of the sill. The location of the gyre would seem to indicate that the sill disturbs the flow in such a way that the relatively smaller velocities in the gyre could allow for enhanced deposition on the sill itself. This may be an important agent in sill formation and migration.

The temperature difference between the relatively colder river water and warmer lake water shows the expected stratification. The denser, sediment-laden river water enters the dredged portion or the flow system and dives below the lake water. Due to the relatively small stream flow entering the dredged portion of the channel from upstream, the less dense lake water intrudes nearly the whole extent of the dredged channel after three seiche cycles. Also of note is the variation in the profile of the stratification, where the sediment density contours show the same pattern along the channel length as the bathymetry.

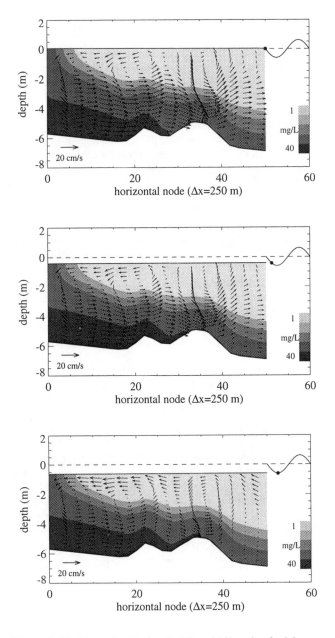

Figure 5: Model run for silt size; 0, 1/8, and 1/4 cycle of seiche.

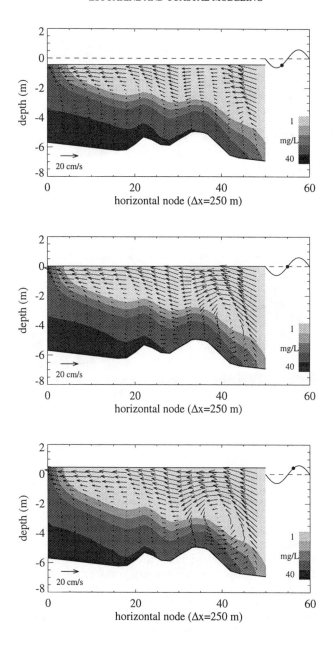

Figure 5: (continued) Model run for silt size; 3/8, 1/2, and 5/8 cycle of seiche.

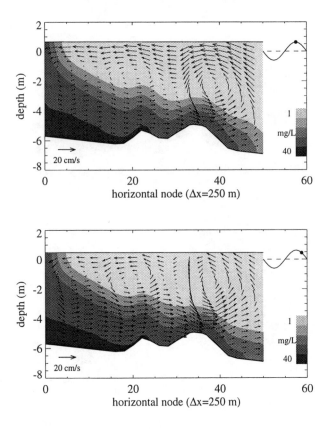

Figure 5: (continued) Model run for silt size; 3/4 and 7/8 cycle of seiche.

Conclusions

An idealized set of model tests has been devised and tested for application in a dredged tributary in the Great Lakes. The riverine system, when examined using the idealized storm conditions coupled with the seasonal temperatures and sediment loads, exhibits several of the features found in typical marine estuaries. Published data for the temperature, suspended sediment concentration, upstream volumetric flow rate, and downstream water level variation, extracted for the season of interest were used as input boundary conditions. The resulting dynamics reveal flow reversals near the mouth of the system, a prominant gyre forming above a sill, suspended sediment and temperature stratification. The model also exhibits the formation of an internal wave, a density wedge with dense, sediment-laden and cooler river water flowing along the bottom, and the less dense lake water intruding 4/5 of the distance upstream in the dredged channel model domain.

Acknowledgements

This research was supported in part by the Coastal Ocean Program of NOAA. The Ohio Supercomputer Center provided computing resources under OSC Grant No. PAS891-1. The support is very much appreciated. This is a contribution from the Great Lakes Forecasting System.

References

Bedford, K. W. and Schwab, D. J. (1994). The Great Lakes Forecasting System: An Overview, in G. V. Cotroneo and R. R. Rumer (eds), *Hydraulic Engineering '94, Proceedings of the Conference*, ASCE, Buffalo, NY, pp. 197–201. See URL http://superior.eng.ohio-state.edu/.

Bedford, K. W., Sykes, R. M. and Libicki, C. (1983). Dynamic advective water quality model for rivers, *J. Environ. Eng.,* ASCE, **109**(3): 535–554.

Blumberg, A. F. and Mellor, G. L. (1987). A description of a three-dimensional coastal ocean circulation model, in N. S. Heaps (ed.), *Three-dimensional Coastal Ocean Models*, Coastal and Estuarine Sciences 4, American Geophysical Union, pp. 1–16.

Kemp, A. L. W., Thomas, R. L., Dell, C. I. and Jaquet, J.-M. (1976). Cultural impact on the geochemistry of sediments in Lake Erie, *J. Fish. Res. Board Can.,* **33**: 440–462.

O'Neil, S., Bedford, K. W. and Podber, D. P. (1996). Storm-derived bar/sill dynamics in a dredged channel, *Proc. Int. Conf. Coastal Eng.*, Vol. 25, ASCE, pp. 4289–4299.

Pinsak, A. P. and Meyer, T. L. (1976). Environmental baseline for Maumee Bay, *Maumee River Basin, Level B Study, MRB Series No. 9*, Great Lakes Environmental Research Laboratory.

Podber, D. P. and Bedford, K. W. (1994). Tributary loading with a terrain following coordinate system, in M. L. Spaulding, K. W. Bedford, A. F. Blumberg, R. T. Cheng and J. C. Swanson (eds), *Estuarine and Coastal Modeling III, Proceedings of the 3rd International Conference*, American Society of Civil Engineers, Oak Brook, IL, pp. 475–488.

Sheng, Y. P. and Lick, W. (1979). The transport and resuspension of sediments in a shallow lake, *J. Geophys. Res.,* **84**(C4): 1809–1926.

Shindel, H., Klingler, J., Mangus, J. and Trimble, L. (1993). Water Resources Data – Ohio, 1992. Volume 2, St. Lawrence River Basin, *Technical report*, U.S. Geological Survey, Water Resources Division.

Hydrodynamical Modelling of Nontidal Coastal Areas Assessing the Morphological Behaviour

H. Weilbeer and W. Zielke[1]

Abstract

Morphological processes in nontidal coastal areas are mainly wave–dominated. In order to consider the hydro– and morphodynamical conditions, a coupling of a shallow water wave model with a two–dimensional vertical integrated flow model has been carried out. The results obtained from these hydrodynamical models will be included in a morphological model, which determines the sediment transport rates due to currents and waves, and thus the morphological response of the simulated region.

Stationary simulations for waves and currents are carried out and combined for a large range of typical meteorological situations in order to assess how climate changes affect the hydro– and morphodynamical conditions in this coastal area. These cases were defined due to a statistical analysis of the output fields of a coupled ocean – atmosphere global circulation model. The wave parameters of this surface wave model are used as input to the coastal area models.

Introduction

The prediction of the morphological behaviour of a coastal zone, for present or future environmental conditions, is a task for a coastal area morphodynamic modelling system. A system of this kind consists of coupled wave, flow and sediment transport components, which are able to describe the dynamical behaviour of the simulated area due to the feedback of morphological changes to the hydrodynamical conditions (de Vriend et al., 1993b, Nicholson et al. 1997).

In the area presented in this paper, a quantification of sediment transport and of the morphological behaviour under changed environmental conditions is planned. It is impor-

[1] Institut für Strömungsmechanik, Universität Hannover, Appelstr. 9A, 30167 Hannover, FRG

tant to note that this area is located at a coast nearly without any tidal movements. Thus the definition of boundary conditions for the hydrodynamical models is very difficult due to the absence of periodical water level oscillations.

The investigated domain covers an area of about 90 km by 60 km, and is located at the German coast of the Baltic Sea (Figure 4). Attention is concentrated on the morphological behaviour of the outer coast formed by the peninsula Fischland, Darß and Zingst. Observations have shown that the Western coast of Fischland and the Northern coast of Zingst are erosion zones. At Darßer Ort, which is found between these erosion zones, accretion is observed.

Waves and wave–driven currents are highly important for the morphodynamics of this region, so emphasis was laid on the simulation of the hydrodynamical conditions in the coastal area, especially in the nearshore zone. For that purpose, a coupling of the free surface model *TELEMAC–2D* (Hervouet, 1993) and the wave model *HISWA* (Holthuijsen and Booij, 1989) due to wave–induced forces has been carried out.

The idea is to discern the effect of hydrodynamical forces on various coast sections for specific wind situations by wave and flow modelling on a large scale. The results calculated by the wave and the flow model are suitable for sediment transport calculations with regard to erosion as well as accretion zones. By combining calculated typical situations it is intended to develop scenarios which are to show the sensitivity of any particular coast section to changing conditions.

In the following text, hydrodynamical simulations will be described and preliminary scenario results will be presented. The method will also be used for the sediment transport calculations still to be made.

Hydrodynamical Simulations

Flow and wave conditions in the area under research must be thoroughly understood, since the results will be decisive for the qualitative and quantitative representation of the sediment transport results, which in turn are required for the morphological findings. In order to obtain experience in the use of coupled models and in the spatial resolution required in the models, initial studies in a small subdomain have been carried out.

Along a stretch of the coast before Zingst, crosshore profile measurements were available. These data were sufficient as a basis for the generation of computational grids used in the hydrodynamic models. Wave calculations were carried out for various directions and intensities, and then analyzed with regard to the resulting wave forces and the flow conditions they induced.

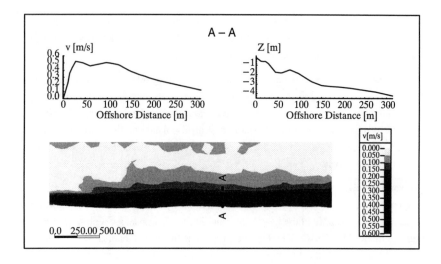

Figure 1: Wave–driven longshore currents

A significant influence of grid–scale on the hydrodynamic results was evident in these coupled wave and flow computations. The flow model as well as the wave model requires a spatial resolution of at least 10 m in order to reproduce satisfactory results. This resolution is vital since the waves, particularly in nearshore regions, undergo a number of transformations and have a high spatial variability. Furthermore, it was evident that more reliable results were obtained, if the wave–driving forces were calculated using the dissipation formulation (Dingemans et al., 1987).

Figure 1 gives, as an example, a result obtained from these small area investigations. The isoarea plot shows the values of the wave–driven longshore currents as well as a cross–section in the approximate centre of this area indicating the topographic profile and the longshore currents. Measurements were not available, but an estimation of the longshore current, obtained by analytical approximations, yields similar values and distributions.

These findings were then implemented in the model of the outer coast. For the wave model, high resolution is necessary in the nearshore region and demands a so–called nesting of grids (Figure 2). Consequently, an elaborate nested grid system for the wave model has been developed and also the FE–mesh of the flow model has been highly resolved along the shoreline in order to meet all requirements (Figure 3).

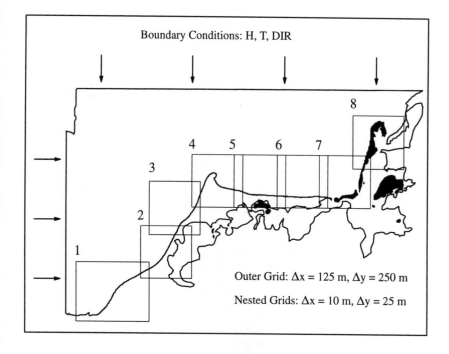

Figure 2: Nested grid system for the wave model

First the wave model is run for the entire area with a spatial resolution of approximately $\Delta x/\Delta y = 125/250$. The boundary conditions for the eight nested grids defined in advance are retrieved from the model. Next the nested grids are calculated with a spatial resolution of approximately $\Delta x/\Delta y = 10/25$. The wave parameters (wave height, period, wave length, dissipation etc.) required for the successive models are interpolated onto the nodes of the FE–mesh.

Defining the boundary conditions for the wave and the flow model presented a considerable problem for the region under consideration. The flow model requires boundary conditions at two open boundaries. It has been mentioned above that there is no periodicity as in tidal areas. Consequently, the boundary conditions have to be interpolated either by water levels measured in the area, or from a larger scale hydrodynamic model of the Baltic Sea.

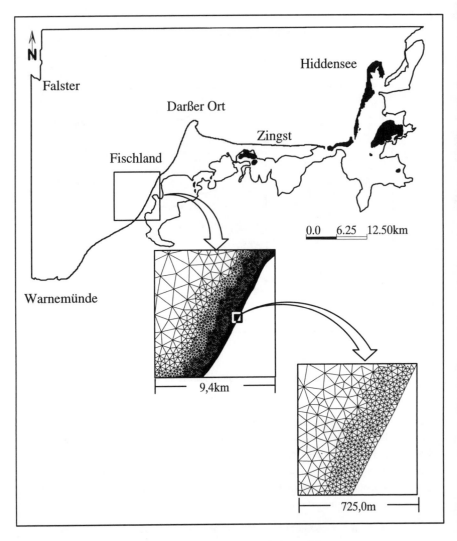

Figure 3: FE–mesh of the flow model

However in this particular application the free surface elevation at the open boundaries in the flow model were kept constant, since noticeable water level elevations will only occur during an extreme storm event. It is true that such events will affect the bottom morphology of the area, but they cannot be completely represented by applying a method like the one used here. Instead, attention was paid to the definition of boundary conditions, i.e. wave heights H, wave periods T and incident wave direction DIR along the open boundary of the wave model.

Figure 4 presents the procedure for the compled model application. The climate input comes from global climate computations carried out at the German Climate Computation Centre (DKRZ). Wind fields calculated for a span of 13 years (1981 – 1994) were extracted and taken as input data for a sea motion model of the Baltic Sea (*HYPAS*, Günther et al., 1979, Kolax, 1996). The wave data output from this model are used as basis for defining typical situations.

First, the events were classified into 8 wind velocity classes and 12 wind velocity directions. For the required values H/T/DIR the above hindcast results, considering regional wave measurements, were used as boundary conditions for the wave model (*HISWA*). In addition, a constant wind field was assumed corresponding to the direction and the velocity classes.

As a consequence of the geometry, the waves can be included only in half the number of direction classes, reaching from the Southwest to the Northeast section. Since, however, the predominant wind direction in the region is West, the main directions are accounted for. The other flow fields are generated exclusively by wind–induced forces.

Unfortunately, the computation times necessary for the high–resolution models limit their applicability. Continuous simulations cannot be carried out with these coupled models. A recoupling of calculated water levels and current velocities – and so a real interaction between currents and waves – is equally difficult to realize. Instead, boundary conditions and driving forces were kept constant for each situation, until stationary flow conditions had been established.

Figures 5 and 6 show the current pattern near Darßer Ort, which is established when the wind and wave direction is from West and Northeast at a wind velocity of 12.3 m/s and a wave height of approximately 1.5 m. The strong longshore currents in front of Fischland as well as the protecting effect of Darßer Ort are clearly visible.

Measured time series of wind data or data obtained from climate models can be used to form statistical wind distributions in accordance with the above classification. The number of occurences (described by a Weibull distribution) leads to calculations of factors for the weighting of typical meteorological situations. As long as no changes of bathymetry are taken into account, the individual results can be combined at random, thus helping to find the resulting flow.

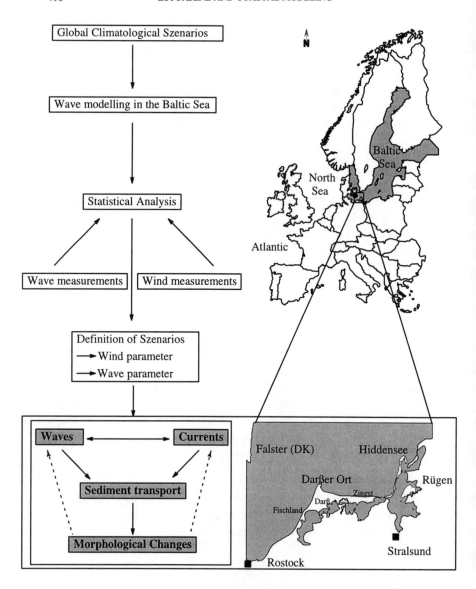

Figure 4: Modelling concept to assess the morphological behaviour

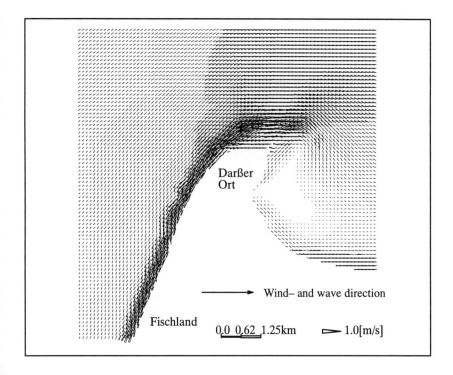

Figure 5: Current pattern near Darßer Ort

This procedure was carried out with a wind data series gathered over 20 years. Figure 7 shows the weighting factors for each wind situation. The upper graph represents the average distribution of the time range (1970 – 1990), the middle graph gives the distribution for one year (1990) with extremely much Western wind, and the lowest one covers a contrasting year (1976) with extremely little Western wind.

The flows resulting from these distributions are shown in Figure 8. The flow direction (represented by the arrows) is always East, as can be observed in nature. The varying flow velocities are clearly recognizable. It is particularly satisfying that the highest velocities, and consequently the probably highest hydrodynamical forces are found in those regions which are known to be erosion zones. This interpretation regarding the hydrodynamical forces is a rather positive evidence. Further analyses of the hydrodynamic results, e.g. regarding bottom shear stresses, will follow.

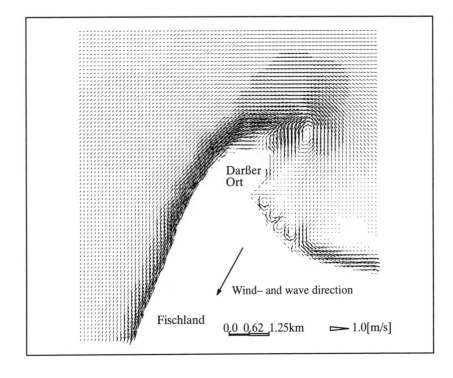

Figure 6: Current pattern near Darßer Ort

Sediment transport:

This method is to be applied for the calculation of sediment transport as well. It is practicable, in this case, to use a so–called ISE–Model (*Initial–Sedimentation–Erosion*) (de Vriend et al., 1993b). Starting from the wave and flow conditions already found, the sediment transport can be calculated in a separate run completely decoupled from the hydrodynamics. As a first step, the formulations of e.g. Bijker or Van Rijn on potential sediment transport due to currents and waves are used for the calculations (Van Rijn, 1989). Then, by weighting the individual events in the way already described above, the possible resulting sediment transport is calculated. The varying wind distributions are found from measured wind data series, or else extracted from model calculations for different climate scenarios (e.g. doubling or tripling the carbone dioxide (CO_2) content in the atmosphere)

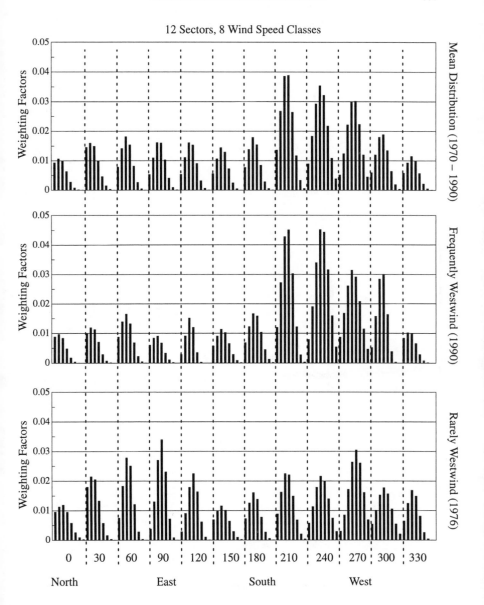

Figure 7: Weighting factors for different scenarios

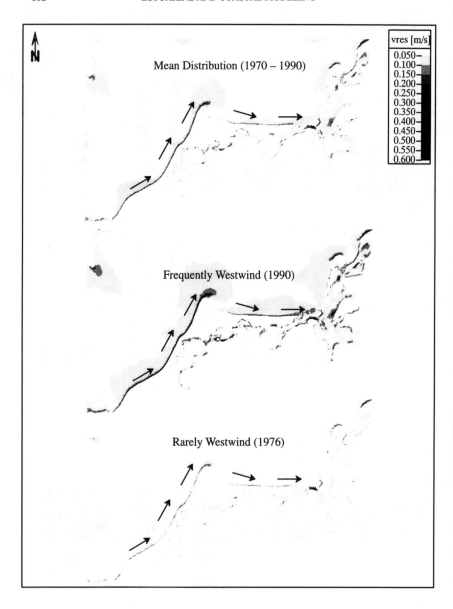

Figure 8: Resulting current pattern for different scenarios

It would, generally speaking, be possible to carry out morphodynamic calculations, but in reality they would not be feasible due to the size of the area. Application of models of this kind is limited regarding time scale and spatial scale. Succesfull calculations have been carried out in small coastal areas only, in order to find out, for example, about the changes in the bathymetry, due to constructions, such as breakwaters or groynes (Nicholson et al., 1997). The boundary conditions again are to be considered, since the effect of wave chronology, e.g. the succession of events may be decisive for a morphological development (Southgate, 1995). Therefore, the described way will be continued to show the sensitivity of this coastal area to changing environmental conditions.

Conclusions

A method for carrying out two–dimensional hydrodynamical numerical simulations in a large non–tidal area has been presented, with the intention to give forecasts on possible morphological trends due to changed environmental conditions. By coupling a wave model with a flow model with (wave–) boundary conditions coming from a larger numerical model, the hydrodynamical situation for single events can be calculated. By combining such situations, scenarios can be developed which can be used for investigating the behaviour of this particular section of the coast.

This procedure was first tested with the aid of hydrodynamic results. The resulting flow direction and the zones of higher hydrodynamic activity coincided with the trends observed in the nature. So, it may be assumed that the conception meets the expectations and that future sediment transport studies will prove this.

Acknowledgements

This work has been undertaken as part of the project "Klimawirkung und Boddenland–schaft" which is part of the project "Klimaänderung und Küste" funded by the BMBF.

References

M.W. Dingemans, A.C. Radder, H.J. de Vriend: Computation of the driving forces of wave–induced currents, *Coastal Engineering 11*, 1987.

DKRZ (Deutsches Klimarechenzentrum Hamburg): The ECHAM3 atmospheric general circulation model. In: Report Nr. 6 des DKRZ Hamburg, 1993.

H. Günther, W. Rosenthal, T.J. Weare, B.A. Worthington, K. Hasselmann, J.A. Ewing: A hybrid parametrical wave prediction model. *Journal of Physical Oceanography*, Vol. 84, p. 5727 – 5738.

J.–M. Hervouet: TELEMAC–2D Version 2.0, Principle Note, Rapport EDF HE–43/92/13, 1993.

L. Holthuijsen, N. Booij : A Prediction Model for Stationary, Short–Crested Waves in Shallow Water with Ambient Currents, *Coastal Engineering 13*, 1989.

M. Kolax: The Climate Impact on the Surface Wave Energy Distribution of the Baltic Sea, AG Klimaforschung des Meteorologisches Institut der Humboldt–Universität Berlin, 1996.

J. Nicholson, I. Broker, J.A. Roelvink, D. Price, J.M. Tanguy, L. Moreno: Intercomparison of coastal area morphodynamic models, *Coastal Engineering 31*, 1997.

P. Péchon, F. Rivero, H. Johnson, T. Chesher, B. O'Connor, J.–M. Tanguy, T. Karambas, M. Mory, L. Hamm: Intercomparison of wave–driven current models, *Coastal Engineering 31*, 1997.

H. N. Southgate: The effects of wave chronology on medium and long term coastal morphology, *Coastal Engineering 26*, 1995.

L. C. van Rijn: Sediment transport by currents and waves, Report H 461, Delft Hydraulics, 1989

H. J. de Vriend: Mathematical Modelling and large–scale coastal behavior–Physical processes and Predictive Models, *Journal of Hydraulic Research*, Vol. 29, No.6, 1991.

H. J. de Vriend, M. Capobianco, T. Chesher, H.E. de Swart, B. Latteux, M.J.F. Stive: Approaches to long–term modelling of coastal morphology: a review, *Coastal Engineering 21*, 1993.

H. J. de Vriend, J. Zyserman, J. Nicholson, J.A. Roelvink, P, Péchon, H.N. Southgate: Medium–term 2DH coastal area modelling, *Coastal Engineering 21*, 1993.

NUMERICAL SIMULATION OF HYDRODYNAMICS FOR PROPOSED INLET, EAST MATAGORDA BAY, TEXAS

Adele Militello[1], Nicholas C. Kraus[2], M. ASCE

ABSTRACT: A new inlet to East Matagorda Bay, Texas, called the Southwest Corner Cut (SW Cut), has been proposed as an environmental enhancement. East Matagorda Bay is a large, shallow estuary acted upon by strong seasonal winds and by weather fronts in the fall and winter. A two-dimensional depth-averaged hydrodynamic model was applied to investigate the consequence of the presence of the SW Cut on the existing inlet, Mitchell's Cut, and on navigation safety in the Gulf Intracoastal Waterway (GIWW). This paper discusses the dynamic wind-generated motions measured in East Matagorda Bay as part of this study, calibration of the model over 30 days covering passage of several winter fronts, and predicted engineering consequences of opening the SW Cut. With the SW Cut in place, hydrodynamic conditions at Mitchell's Cut are predicted not to change, so that its natural stability will not be compromised. Also, a significant benefit will result through a predicted 25% decrease of peak current in a dangerous section of the GIWW due to flow captured by the SW Cut.

INTRODUCTION

East Matagorda Bay, Texas, is a pristine, elliptic-shaped estuary about 6 km wide on average and extending about 37 km from Caney Creek on the east to the Colorado River Navigation Channel on the west (Fig. 1). The long axis is oriented approximately east - west, but tilted 27 deg counterclockwise, an orientation that results in frequent wind dominance of the hydrodynamics. East Matagorda Bay presently has one direct opening to the Gulf of Mexico through an unmaintained artificial inlet, called Mitchell's Cut, dredged open in May, 1989 as a flood-relief channel for the low-lying upland towns. East Matagorda Bay is also indirectly

---

1) Research Oceanographer, U.S. Army Engineer Waterways Experiment Station, Coastal and Hydraulics Laboratory, 3909 Halls Ferry Road, Vicksburg, MS 39180-6199.
2) Research Physical Scientist, U.S. Army Engineer Waterways Experiment Station, Coastal and Hydraulics Laboratory.

connected to the Gulf of Mexico through long and circuitous navigation channels (Colorado River Navigation Channel and Gulf Intracoastal Waterway (GIWW)) that follow the perimeter of the bay on its west and north sides.

The County of Matagorda has proposed installation of a water-exchange cut or inlet that would connect the southwestern end of the bay to the Colorado River Navigation Channel, near to the Gulf of Mexico. This study investigated whether the new cut would cause harmful flows elsewhere in the bay-channel system, such as might promote closure of Mitchell's Cut, or would exacerbate already dangerous flows along the GIWW near the Colorado River Navigation Channel.

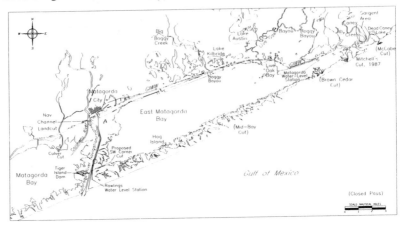

Figure 1. Site map of study area.

STUDY SITE

East Matagorda Bay became isolated from the western and larger part of Matagorda Bay and from Pass Cavallo, located in the southwest corner of Matagorda Bay, by a prograding delta that crossed the bay from the mainland and joined to Matagorda Peninsula (Bouma and Bryant 1969, Morton et al. 1976). The delta formed rapidly starting in 1929, when a log raft and massive sediments that had been entrapped on the Colorado River were freed by local interests concerned with flooding of low-lying inland areas. The delta reached Matagorda Peninsula in 1935 and formed a new bay called East Matagorda Bay.

In 1992, the Colorado River was re-routed to discharge into Matagorda Bay rather than into the Gulf of Mexico, and East Matagorda Bay was connected to the old river channel that reached the Gulf. The strong cross current at the intersection of the river and the GIWW often disrupts barge traffic along the GIWW. A pair of locks on the GIWW at the intersection mitigates the cross current. These locks, which are normally closed except to allow passage of vessels along the GIWW, were assumed to be closed for the modeling study described below.

Price (1952) observed that Texas bays and lagoons typically possess major freshwater inflows at their northern ends and an inlet on their southern ends. East Matagorda Bay presently does not have such an opening in its southwest corner because of the formation of the Colorado River delta. The proposed SW Cut is perceived as such a channel that would restore a natural circulation pattern. It was permitted by the U.S. Army Corps of Engineers to be 3.2 km long, 30 m wide at the bottom, and 1.5 m deep with respect to mean water level.

The GIWW is presently routed along the northern perimeter of East Matagorda Bay, sheltered from wind waves on the bay by islands composed of dredged material. Typical bay water depth ranges between 0.6 and 1.2 m, whereas the GIWW is maintained to a depth of 3.6 m, with advance and over-dredging potentially adding another 0.6 m of depth, and the waterway has a design bottom width of 38.5 m and top width of 91.4 m. Water flow in the GIWW is thus efficient as compared to that in the shallow bay and must be accounted for in the hydrodynamic analysis.

The opening of SW Cut calls into consideration issues such as inlet stability of both the SW Cut and Mitchell's Cut, and alterations to the current in the navigation channel and GIWW. In the present study, field measurements were made and numerical modeling performed to determine whether the SW Cut would remain open and to focus on critical issues related to possible alterations in the current in East Matagorda Bay and GIWW that would make the cut unacceptable (Kraus and Militello 1996).

FIELD MEASUREMENTS AND ANALYSIS

Because East Matagorda Bay is not navigable (except for the GIWW) and published bathymetric data old, a hydrographic survey by dual-frequency echosounder and hand-held staff (in very shallow areas) was conducted. The survey was controlled horizontally by differential GPS and covered approximately 220 km of transect lines. Depths were transformed to a common datum by reference to a local tide gauge to serve as bathymetry for the hydrodynamic model.

Water level and current were measured at two platforms, one located at the eastern end of the bay (EMAT) across from Brown Cedar Cut, and one at the southwestern end of the bay (SWEMAT) near the proposed site for the SW Cut. An anemometer located at the EMAT gauge provided wind measurements.

Wind

Annual wind measured at the EMAT platform is predominantly from the southeast and east-southeast (120 to 150 deg). Strong winds associated with winter frontal activity and storms originate from north-northwest to east-northeast (330 to 30 deg). During most of the year, the wind has an easterly component that drives water from the eastern side to the western side of East Matagorda Bay.

The wind speed and direction for the hydrodynamic simulation period are shown in Fig. 2. The direction north is either 0 deg or 360 deg, and wind from the east has direction 90 deg. Weather fronts that periodically (about 5.2-day interval by spectral analysis) move across the bay are apparent as abrupt shifts from 0 to 360 deg, interspersed with periods of east-southeast to southeast wind (150 deg). Wind speed shows sharp peaks associated with the passing northern fronts.

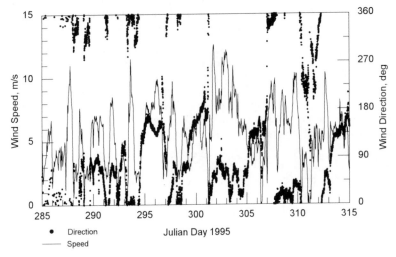

Figure 2. Wind speed and direction during the modeling period.

Water Level

The demeaned water levels at EMAT and SWEMAT for the modeling period are shown in Fig 3. The small daily fluctuations in the water level correspond to the astronomical tidal forcing from the Gulf of Mexico, which has a range of 1.4 m at Galveston. In contrast, the tidal range for EMAT is only 10 cm. A striking behavior in the two water level records appears as numerous simultaneous pairs of inverted spikes for which the water level at EMAT is set down and the water level at SWEMAT is set up. For example, on JD301, the difference in water level between the eastern and western ends of East Matagorda Bay was almost 0.6 m. As seen from Fig. 2, this substantial tilt in water level was produced by an impulsive northeast front with wind speed of about 12 m/s. The frontal wind impulse had been preceded by several days of moderate wind from the southeast, after which the wind turned sharply and blew from the northeast. A 0.6-m tilt in the water level over some 20 km between measurement stations is remarkable considering that this bay, surrounded by marshes and wetlands, is only 1.3 m deep in its deeper regions.

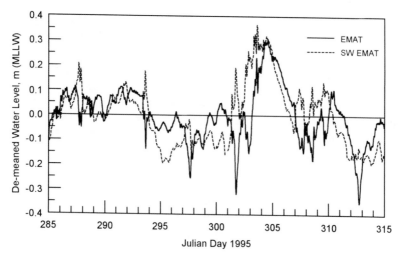

Figure 3. De-meaned water levels at EMAT and SWEMAT during modeling period.

Current

Because of the approximate east-to-west orientation of East Matagorda Bay and the frequent occurrence of wind with an easterly component, there is substantial movement of water along the major (E-W) axis of the bay. The E-W component of the current at EMAT has a typical maximum in the range of 15 cm/s, whereas the current at SWEMAT has a range less than 5 cm/s. In contrast, in Mitchell's Cut, Caney Creek, and the navigation channel, which are narrow channels oriented approximately N-S, the N-S component of the current is strongest.

NUMERICAL SIMULATION OF HYDRODYNAMICS

Water level and circulation in East Matagorda Bay were simulated with a depth-averaged, two-dimensional finite-difference model applied over a variably spaced, rectilinear grid. The persistent and typically strong wind blowing over the shallow water indicates a well-mixed system and appropriateness of a depth-averaged model. The applied model, M2D (Militello 1998), is a finite-difference approximation of the mass continuity and momentum equations given by

$$\frac{\partial \eta}{\partial t} = -\frac{\partial(hu)}{\partial x} - \frac{\partial(hv)}{\partial y} \tag{1}$$

$$\frac{\partial u}{\partial t} = -g\frac{\partial \eta}{\partial x} - u\frac{\partial u}{\partial x} - v\frac{\partial u}{\partial y} + fv - C_b\frac{u|U|}{(h+\eta)} + C_d\frac{\rho_a}{\rho_w}\frac{W^2\cos(\theta)}{(h+\eta)} \tag{2}$$

$$\frac{\partial v}{\partial t} = -g\frac{\partial \eta}{\partial y} - u\frac{\partial v}{\partial x} - v\frac{\partial v}{\partial y} - fu - C_b\frac{v|U|}{(h+\eta)} + C_d\frac{\rho_a}{\rho_w}\frac{W^2\sin(\theta)}{(h+\eta)} \tag{3}$$

where h = still-water level referenced to a specified datum, η = deviation in water level from h, t = time, u = current speed parallel to the x axis, v = current speed parallel to the y axis, U is the total current velocity, g = the acceleration due to gravity, f = Coriolis parameter, C_b = empirical bottom friction coefficient, C_d = wind stress (drag) coefficient, ρ_a and ρ_w = density of air and of water, respectively, W = wind speed, and θ = wind direction. The friction coefficient is calculated by

$$C_b = \frac{g}{C^2} \tag{4}$$

in which C_b = Chezy coefficient given by $C = (R^{1/6})/n$, R = hydraulic radius, and n = Manning coefficient. The hydraulic radius is given by

$$R = \frac{A}{P} \tag{5}$$

where A is the cross-sectional area of flow, and P is the wetted perimeter. The wind stress coefficient is given by (Hsu 1988)

$$C_d = \left(\frac{0.4}{14.56 - 2\ln W_{10}}\right)^2 \tag{6}$$

where W_{10} is the wind speed at an elevation of 10 m.

The explicit finite-difference scheme is central in space and calculates variables on a staggered grid. The momentum equations apply values of the velocity and water-surface elevation from the previous time step to calculate new velocity values. The approximation for the continuity equation incorporates updated values of velocity from momentum equation calculations and applies those values to the calculation of the water-surface elevation. The advective terms are spatially and temporally averaged.

Four grids were generated (existing condition; SW Cut installed; existing condition with Mitchell's Cut closed; SW Cut installed with Mitchell's Cut closed). Each grid consisted of approximately 8,000 active computational cells with minimum cell-size dimension of 33 m and maximum dimension of 505 m. Figure 4 shows the model domain and bay bathymetry with the SW Cut included. Ocean boundaries were forced with water level measured at a NOS-operated tide gauge at Galveston. Open, non-forcing boundary conditions were established at grid cells that reside on channels extending beyond the grid domain (Caney Creek and the GIWW). At these open boundaries, the water-level gradient was specified to be spatially constant normal to the boundary.

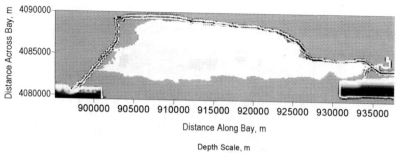

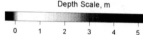

Figure 4. East Matagorda Bay bathymetry and model domain.

Wind data taken at the EMAT platform were applied over the model domain (spatially-constant wind forcing) for simulations including the local wind. Set up and set down caused by far-field wind forcing along the coast was implemented through the ocean water-level forcing, taken from measurements. Thus, the ocean boundary condition propagated both tidal and far-field forcings into the model domain.

Calibration

The bottom friction coefficient was the only parameter adjusted for the calibration, which took typical values based on the bottom and side bank conditions, with the Manning coefficient ranging between 0.022 and 0.028 $s/m^{1/3}$. Larger values of the Manning coefficient, up to 0.1 $s/m^{1/3}$, were specified in the vicinity of the mouth of the navigation channel to represent transition losses at the entrance.

Figs. 5 and 6 show calculated water levels obtained with the calibrated model and the measured water level fluctuations at EMAT and SWEMAT, respectively. The calculated water-level fluctuations follow closely those of the measurements, and both the short-period (tidally induced) and long-period (wind-induced) motions are reproduced by the model. The model captures the influence of the wind on the water-level fluctuation and the current (next paragraph), which indicates that the wind stress formulation employed (Eq. 6) works well in extremely shallow water. The root-mean-square (rms) differences between calculated and measured water levels at EMAT and SWEMAT were 4.8 and 4.2 cm, respectively.

Comparisons of the calculated E-W current speed and the measurements are shown in Fig. 7 and Fig. 8 for EMAT and SWEMAT, respectively. The measured currents were low-pass filtered with a cutoff frequency of 3 cycles/day to remove the wind waves and ambient noise. The simulated currents generally track the measured currents and indicate that the model calibrated well. Deviations between the measured and calculated currents are typically 2 to 3 cm/s at EMAT and 1 to 2 cm/s for SWEMAT, with the measured current at SWEMAT being very weak (-2 to 5 cm/s). The rms differences between calculation and measurements were 1.4 and 4.9 cm/s, respectively, at EMAT and SWEMAT.

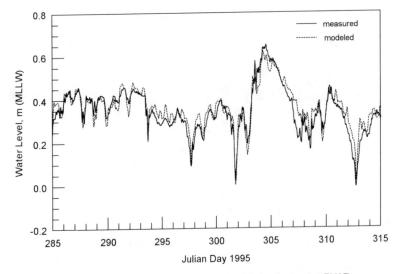

Figure 5. Comparison of measured and simulated water level at EMAT.

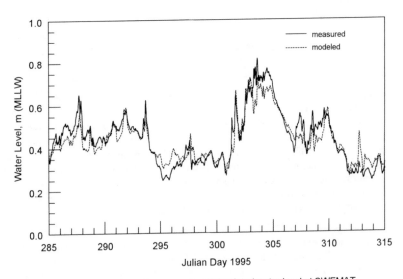

Figure 6. Comparison of measured and simulated water level at SWEMAT.

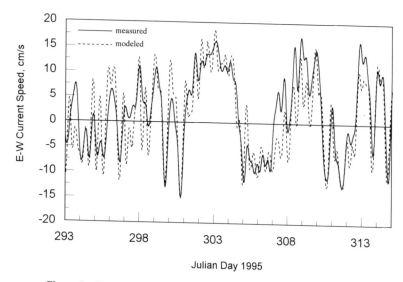

Figure 7. Comparison of measured and simulated current at EMAT.

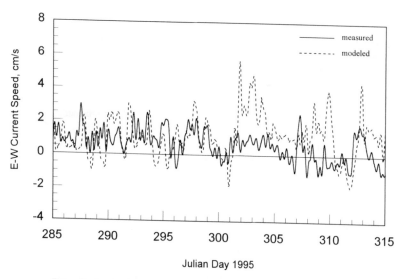

Figure 8. Comparison of measured and simulated current at SWEMAT.

Simulation Results

Predictions of water-level fluctuations with and without the SW Cut were examined for several locations in channeled portions of the study site. The navigation channel reach located seaward of the SW Cut was calculated to have a reduced tidal range with the cut installed. The reduction in tidal range diminished in the Gulfward direction as the influence of the Gulf forcing increased. During flood tide, the SW Cut will allow water to flow into the bay, reducing the hydraulic head from the point where the SW Cut meets the navigation channel to the mouth of the navigation channel. During ebb tide, the SW Cut will allow water to flow from the bay to the navigation channel and increase the hydraulic head from the point of connection to the Gulf. The flow of water out of the bay through the SW Cut will be increased by winds with an easterly component and will keep the hydraulic head at the Cut higher than it would be without the wind.

The SW Cut is predicted to take a portion of the flow that normally passes through the navigation channel, resulting in reduced water-level fluctuations in the channel upstream of the Cut. Reduction in water-level range occurs in the navigation channel and in the western reach of the GIWW along the northern perimeter of the bay. The tide range gradually approaches that of the existing condition toward the eastern end of the GIWW and is only slightly altered (decreased) in the vicinity of Caney Creek.

An important benefit of the Cut will be a reduction in peak flow speed at the intersection of the GIWW and the navigation channel of as much as 25%, which would improve navigability in the GIWW. The reduction in current speed upstream of the cut owes to the SW Cut carrying a portion of the flow that would otherwise travel through the navigation channel. Mitchell's Cut is predicted to experience no perceptible change in current speed with the SW Cut installed, and the change in current at the GIWW in the vicinity of Caney Creek also was negligible.

With the SW Cut installed, the discharge in the navigation channel at Rawlings (north of the Cut, see Fig. 1) was reduced by approximately 20 to 25%. On average, the discharge diverted through the SW Cut was 30% of the total flow rate through the navigation channel south of the Cut for the month-long simulation period. The reduction in flow occurs for both flood and ebb tide.

Flow in SW Cut and Mitchell's Cut

Entrance and exit losses at the SW Cut were accounted for by assigning values of Manning's n of 0.08 and 0.06 s/m$^{1/3}$ at two cells on the ends of the confined portion of the SW Cut. The value of 0.08 s/m$^{1/3}$ was applied at the outermost cells, relative to the confined region of the SW Cut, and the value of 0.06 s/m$^{1/3}$ was applied to the adjacent inner cells. All remaining cells in the confined portion of the SW Cut were assigned values of $n = 0.025$ s/m$^{1/3}$.

Magnitude and direction of the flow through the SW Cut have implications for scour or deposition in the Cut and for exchange of water between East Matagorda

Bay and the navigation channel. Fig. 9 plots the calculated current velocity in the middle of the SW Cut. Positive values denote flooding into the bay from the navigation channel, and it is seen that there is a bias for flow to ebb out of the bay. The peak flow speed typically reaches 40 cm/s on flood or ebb, with the stronger ebb currents peaking past 80 cm/s as a result of wind-driven set up. A net outward flow is expected because of the frequent winds with an easterly component and resultant wind setup on the western end of East Matagorda Bay, with a water-elevation gradient induced between the bay and navigation channel. The mean discharge through the cut will be directed out of the bay for most of the year, and the net (outward) discharge of the flow for the modeling period varied between about 9 and 11.5 m³/s, depending on the value of the friction coefficient assigned to the Cut. The outflow through the cut is balanced by increased flow into the system through the GIWW, Caney Creek, and Mitchell's Cut.

The value of a full two-dimensional model in calculation of the hydrodynamics of this multiple-inlet system is demonstrated with simulations performed with and without local wind forcing. Far-field wind forcing was maintained in the calculations through set up and set down contained in the water-level forcing at the ocean boundaries. Figure 9 also compares calculations with and without local wind forcing for the SW Cut, where only a portion of the model period is shown for readability. Although qualitative trends in variation of the current are reproduced, the ebb flow is notably reduced without local wind forcing. The general (downward) shift toward ebb, present when local wind forcing acts, is absent in the current calculated without local wind. This key feature, ebb dominance, would therefore be lost without incorporation of local wind forcing. With local wind, the mean discharge through the SW Cut was 11.4 m³/s (ebbing), whereas without local wind the mean discharge was 0.1 m³/s (ebbing).

The wind event that occurred from JD301 through JD303 had strong winds peaking at greater than 12 m/s from the northeast. The current through the SW Cut shown in Fig. 9 contains non-tidal variance in the signal. This variance corresponds to changes in the coastal water level caused by the wind event. Set up or set down on the coast will change the gradient in head between the bay and the ocean. Consequently, the signature of this non-tidal variance is seen in the SW Cut current.

Current velocities in excess of 60 cm/s will readily erode sand and clay (typical sediments in East Matagorda Bay), and stable tidal inlet entrances are known to be those that possess maximum flows on the order of 1 m/s for an open coast (e.g., Jarrett (1976); also, Bruun (1990) for a review) and 30 cm/s for wave-sheltered coasts and limited longshore transport (Riedel and Gourlay 1980). In the present situation, currents of 1 m/s magnitude were measured in this study both at the mouth of Mitchell's Cut and at the mouth of the navigation channel, and flow speeds often exceeding 60 cm/s were also calculated by the hydrodynamic model for the SW Cut for its design depth of 1.5 m. In further simulations, the current speed was found to exceed 1 m/s if the channel scoured to a nominal depth of 3.7 m (which is the approximate depth at its connection with the navigation channel).

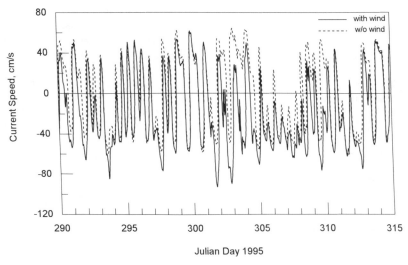

Figure 9. Current speed in the center of the SW Cut, with and without local wind.

As opposed to a classical tidal inlet, the SW Cut will act more as a tidal river mouth, with the river flow replaced by a quasi-steady, wind-driven out-flow having tidal fluctuations superimposed on it. No flood tidal delta in the bay will form because of the ebb bias in the current. On a continuing basis, fine sand, silt, and clay will be transported out of the SW corner and into the navigation channel. Suspended fine-grained sediment might increase water turbidity, and the additional material would eventually be deposited in shoals off the mouth of the navigation channel. The volume of additional material is expected to be very small compared to the several hundreds of thousands of cubic yards a year dredged at the mouth (Heilman and Edge 1996) and not alter navigability of the channel.

Change in current velocity and discharge at Mitchell's Cut with installation of the SW Cut is of concern for stability of Mitchell's Cut. However, Fig. 10 shows that there will be no significant change in discharge through Mitchell's Cut with installation of the SW Cut. The net change in discharge at Mitchell's Cut with the SW Cut installed will slightly increase, and typically occur when the tide is flooding. Over the modeling period, the average change in discharge at the mouth of Mitchell's Cut was approximately 3 m³/s (net flooding), which is less than 2% of the typical daily peak discharge of about 200 m³/s. The increased flow into Mitchell's Cut is a response to water flowing out of East Matagorda Bay through the SW Cut.

Mitchell's Cut has remained open since it was created in May, 1989, and it evidently owes its longevity to the adequate flow from the GIWW and Caney Creek. It is suspected, but was not verified in this study, that infrequent heavy precipitation and subsequent strong discharges from Caney Creek may be a significant factor in maintaining the stability of Mitchell's Cut. Therefore, a drought would promote closure of the Mitchell's Cut. Mitchell's Cut is an ephemeral inlet; because it is not stabilized by structures, it has a finite life. If Mitchell's Cut were to close after opening of the SW Cut, the results of the present study indicate that the closure process will be unrelated to the hydraulics in the SW corner of East Matagorda Bay. However, the water level in the bay would be lowered if Mitchell's Cut closed, which could have a deleterious environmental impact.

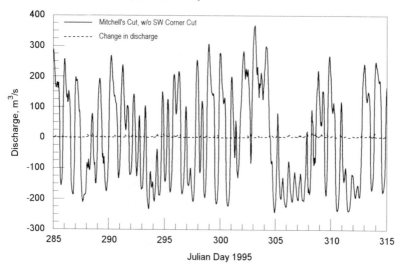

Figure 10. Discharge at Mitchell's Cut for the existing condition and the change in discharge with the SW Cut emplaced.

CONCLUDING DISCUSSION

This study confirms the general observation of Price (1952) that the southwest corner of a bay or estuary is a hydraulically efficient location for an inlet on the Texas coast. Therefore, a strong wind-dominated ebb flow is predicted that will keep the SW Cut open. Main conclusions of this study are as follows.

1. Based on the hydrodynamic analysis, the SW Cut, if opened, will remain open unless artificially closed. The flow in the cut will be ebb dominated because of a bias introduced by the wind, and the flow speed will regularly reach 0.6 to 1 m/s if the design dimensions of the Cut are maintained. Because of the expected strong outflows, scour-prevention measures should be taken in building the Cut.

2. With the SW Cut open, the peak flow speed and discharge will decrease at the intersection of the GIWW and the navigation channel land cut. A maximum decrease of 25% in peak flow speed is expected and will improve navigability in a section of the GIWW presently experiencing safety problems. If the Cut is opened, there will be a slight increase in both ebb and flood peak flow speed at the mouth of the navigation channel and will act in favor of maintaining the navigation channel entrance.

3. The stability of Mitchell's Cut, an important flood-relief channel for communities bordering the eastern end of the bay, will not be altered with opening of the SW Cut.

ACKNOWLEDGEMENTS

The original study was sponsored by the Texas Department of Transportation and the Texas Parks and Wildlife Department while the authors were staff members at Texas A&M University-Corpus Christi. This paper was prepared as part of activities of the Coastal Inlet Modeling System work unit of the Coastal Inlets Research Program conducted for Headquarters, U.S. Army Corps of Engineers (USACE), at the Waterways Experiment Station. Permission was granted by Headquarters, USACE, to publish this information.

REFERENCES

Bouma, A. H., and Bryant, W. R. 1969. "Rapid delta growth in Matagorda Bay, Texas." Lagunas Costeras, Un Sinposio, 171-189.

Bruun, P. (1990). *Port Engineering, Volume 2: Harbor Transportation, Fishing Ports, Sediment Transport, Geomorphology, Inlets, and Dredging* (Fourth Edition). Gulf Publishing Co., Houston, TX.

Heilman, D. J., and Edge, B. L. 1996. "Interaction of the Colorado River Project, Texas, with longshore sediment transport." *Proceedings 25th Coastal Engineering Conference*, ASCE, NY, 3309-3322.

Hsu, S. A. 1988. *Coastal Meteorology*. Academic Press, San Diego, CA.

Jarrett, J. T. 1976. "Tidal prism-inlet area relationships." GITI Report No. 3, U.S. Army Corps of Engineers, Coastal Engineering Research Center, Vicksburg, MS.

Kraus, N. C. and Militello, A. 1996. Hydraulic Feasibility of Proposed Southwest Corner Cut, East Matagorda Bay, Texas. Technical Report TAMU-CC-CBI-96-03, Conrad Blucher Institute for Surveying and Science, Texas A&M University-Corpus Christi, Corpus Christi, TX.

Militello, A. 1998. "Hydrodynamics of shallow, wind-dominated lagoons." Doctoral Dissertation in preparation, Division of Marine and Environmental Systems, Florida Institute of Technology, Melbourne, FL.

Morton, R. A., Pieper, M. J., and McGowan, J. H. 1976. "Shoreline changes on Matagorda Peninsula (Brown Cedar Cut to Pass Cavallo), an analysis of historical changes of the Texas Gulf shoreline. Geological Circular 76-6, Bureau of Economic Geology, The University of Texas at Austin, Austin, TX.

Price, W. A. 1952. "Reduction of maintenance by proper orientation of ship channels through tidal inlets." *Proceedings 2nd Conference on Coastal Engineering*, ASCE, NY, 243-255.

Riedel, H. P, and Gourlay, M. R. 1980. "Inlets discharging into sheltered waters." *Proceedings 17th Coastal Engineering Conference*, ASCE, NY, 2550-2564.

Stage-frequency estimates for the Louisiana Coast

by

S. Rao Vemulakonda[1]
Norman W.Scheffner[1]
David J. Mark[1]
and
Mitchell E. Brown[2]

Abstract: This paper reports on the development of
stage-frequency relationships (under storm conditions)
for Lake Pontchartrain and the coast of Louisiana. Both
tropical and extratropical storms were considered. The
hydrodynamic model ADCIRC-2DDI was used for the study.
It was calibrated for Hurricane Betsy and verified for
Hurricane Andrew. There was excellent agreement with
observed surges at several gages. The model was next
used to compute surges for a set of 19 tropical and 11
extratropical storms. Results were used in a statistical
procedure, called the Empirical Simulation Technique, to
compute stage-frequency relationships at several project
locations.

Introduction

The coast of Louisiana (Figure 1) is characterized
by generally low land elevations in relation to the Gulf
of Mexico, extensive marshes and interconnected bayou
systems. Historically, this area has been prone to
extensive flooding and property damage due to the action
of storms, both tropical (hurricanes) and extratropical.
Over the years, a series of levees and control structures
have been constructed to reduce the impact of storms.
Recently the US Army Engineer District, New Orleans

---

[1]US Army Engineer Waterways Experiment Station, Coastal
and Hydraulics Laboratory, 3909 Halls Ferry Rd, Vicksburg,
Mississippi 39180-6199.

[2]Mevatec Corp., Vicksburg, Mississippi 39180.

(CEMVN) embarked on two related studies: the Lake
Pontchartrain Study to investigate storm impacts along
the Lake Pontchartrain and Lake Borgne areas as well as
the open coast of Louisiana, and the Morganza Study on
the present and future uses of the area from Morganza to
the Gulf of Mexico. Both studies require estimates of
stage-frequency relationships at different locations for
storms. Time series of storm water levels also are
needed at several locations to drive a detailed unsteady
flow hydraulic model known as the UNET (Barkau, 1993).
For brevity, this paper describes only storm surge
modeling performed in support of the Lake Pontchartrain
Study. Similar simulations were also performed for the
Morganza Study.

To assist the CEMVN, the Coastal and Hydraulics
Laboratory of the US Army Engineer Waterways Experiment
Station (WES) applied the 2D finite element, long wave
hydrodynamic model ADCIRC-2DDI, (hereafter called ADCIRC
for convenience) (Westerink, et al. 1992) to the study
area. The model was modified by Luettich and Westerink
(1995) to account for wetting and drying of land cells
due to flooding. A previously used global grid of the
entire Gulf of Mexico was modified to provide fine
resolution in areas of interest to the study. ADCIRC was
calibrated and verified with field data for surges
obtained at several locations during Hurricanes Betsy
(1965) and Andrew (1992). There was excellent agreement.
Subsequently, the model was applied to several historic
storms and computed maximum surges as well as time series
of surges were saved at gage locations of interest.

A stage-frequency analysis was performed for the
gage locations using the Empirical Simulation Technique
(EST) (Scheffner, Borgman, and Mark 1996). In this
statistical approach, the storm characteristics are used
as input and maximum elevations as the response. This
technique was applied previously to Long Island
(Scheffner and Butler 1995) and the coast of Delaware
(Scheffner, Borgman, and Mark 1996).

ADCIRC Numerical Model

ADCIRC is a 2D depth-averaged model which uses the
Generalized Wave-Continuity Equation (GWCE) approach to
solve the equations of momentum and continuity. It
employs numerical discretizations using the finite
element method in space and finite difference method in
time. It allows for extreme grid flexibility which
permits simultaneous regional/local modeling and is

highly accurate and efficient. Additional details are
provided by Westerink, et al. 1992.

Details of Wet/Dry Algorithm

The following description is reproduced from
Luettich and Westerink (1995). The scheme operates on a
fixed grid, with whole individual elements being either
wet or dry. Conceptually, a dry element has barriers
along all of its sides. As it is flooded, the barriers
are removed. Nodes may be classified as 'dry',
'interface' or 'wet'. Interface nodes are connected to
both dry and wet elements. They are similar to the
standard land/water boundary nodes. A 'no slip' boundary
condition is implemented at these nodes. A node dries
when the depth H falls below some user specified minimum
depth HMIN (e.g., 0.3 ft). The only exception to drying
is that wet nodes must remain wet for a minimum number of
time steps (e.g., 10 to 20) to avoid numerical noise. A
similar constraint is imposed for wetting of dry nodes.
An interface node wets if the following conditions are
met : (i) the water level gradient, and (ii) the vector
sum of the water level gradient and the wind stress both
favor water movement towards all the dry nodes connected
to the interface node, and (iii) enough water accumulates
in the wet elements along the flood/dry interface that a
favorable water level gradient exists. When the
interface node is wet, simultaneously all the dry nodes
connected to it are made interface nodes. In
implementing the scheme in ADCIRC, the dependent variable
in the GWCE is changed from the water level at the new
time level to the change in water level from the previous
time step. For shallow water, the option of a depth-
dependent drag coefficient, which works similar to
Manning's formulation, is provided. Testing with
idealized cases has shown that the wet/dry algorithm
works quite well, the numerical noise generated by
wetting and drying of elements is generally damped out by
friction and the solution is smooth and stable.

The Numerical Grid and Forcing

To develop the study grid, we started with an
existing grid of the Gulf of Mexico, which was previously
verified for tides and storm surge with the ADCIRC model
(Scheffner, et al. 1994), and added finer resolution in
the study area. The result is the grid shown in Figure
2. It has 25,732 nodes and 50,215 elements. The
advantage to this telescoping grid approach is that
boundary conditions are clearly posed and away from the
project area of interest. In this case, there are two

boundaries, one across the Strait of Florida and the other across the Yucatan Channel. A blow-up of the refined grid in the project area is shown in Figure 3. The area of grid refinement extends from the Gulf of Mexico shoreline south of New Orleans to Interstate Highway 10 approximately 60 miles north. It includes Lake Pontchartrain and Lake Borgne as well as Atchafalaya Bay to the west. Minimum element dimensions are on the order of 200 ft and the smallest elements are located near Chef Menteur Pass which connects Lake Borgne with Lake Pontchartrain. The complex bathymetry (Mean Sea Level Datum) for the study area, shown in Figure 4, clearly indicates the levees, low-lying land areas, etc.

Tides

Tidal forcing was used, in addition to wind forcing, only for calibration and verification of ADCIRC to historic hurricanes. Five tidal constituents (M_2,S_2,K_1,O_1 and P_1) were modeled. For each constituent, amplitude and phase (nodal factor and equilibrium argument) were furnished at the two open boundaries, using the results of a larger hydrodynamic model (Westerink, Luettich, and Scheffner 1993) that included in its domain the Western North Atlantic Ocean, the Gulf of Mexico and the Caribbean Sea. In addition, ADCIRC used tidal potential forcing at all grid nodes.

Winds

For tropical storms, the National Hurricane Center's HURricane DATa (HURDAT) database (Jarvinen, Neumann, and Davis 1988) was used to determine the track and characteristic parameters (eye location, maximum wind speed, forward speed, minimum pressure, etc.) of a historic hurricane as a function of time. This information was used as input to the Planetary Boundary Layer (PBL) model (Cardone, Greenwood, and Greenwood 1992) which in turn computed and interpolated wind speed components and atmospheric pressure to ADCIRC grid nodes, on an hourly basis. The information was archived to data files and used to force ADCIRC.

For extratropical storms, wind information (velocity components) was obtained from the US Navy's Fleet Numerical Meteorological and Oceanographic Center's database every 6-hr at 2.5 deg lat-long spacing and used as input to ADCIRC, which interpolated the winds to grid nodes as needed in time.

Empirical Simulation Technique

The EST is a statistical procedure for modeling
frequency-of-occurrence relationships for non-
deterministic, multi-parameter systems. In the present
case, it is used to develop the frequency relationship
for the storm response as given by the maximum surge as a
function of storm characteristic parameters. Because of
their natural differences, tropical and extratropical
storms are considered separately. For tropical storms,
for a given station, the characteristic storm parameters
are taken to be the distance from the eye to the station,
the maximum winds, the central pressure deficit, forward
speed, and the radius to maximum winds, at the time the
eye of the storm was closest to the station. From
historic storms which impacted the study area, a limited
'training set' of storms is selected and simulated with
ADCIRC, and the computed maximum surges are archived for
stations where frequencies are to be computed. Tides are
modeled separately from storms and added subsequently in
the procedure. Because each surge has an equal
probability of occurring at high tide, MSL during peak
flood, low tide, and MSL during peak ebb, each of the
computed surge elevations is linearly combined with the
tidal elevations for the four phases to generate a larger
probability set of events. The EST approach is used to
generate multiple (e.g., 100) statistical simulations of
a N-year (e.g., N=200) sequence of storms at each of the
selected stations. Because the procedure is repeated
multiple times, it is possible to determine an average
stage-frequency relationship as well as a measure of
uncertainty such as standard deviation, for each station.
This coupled with the fact that the EST preserves the
statistical relationships inherent in the historic data,
without relying on assumed parametric relationships or
assumed parameter independence, makes the method superior
to traditional techniques such as the joint-probability
method. Additional details are provided by Scheffner,
Borgman, and Mark (1996).

ADCIRC Calibration and Verification

The ADCIRC model was calibrated with water levels
for Hurricane Betsy and verified with water levels for
Hurricane Andrew because extensive information was
available for these severe storms. No separate
calibration was performed for extratropical storms.
Typically, for tropical storms, simulations were started
approximately two days before landfall of the storm and
lasted for a total of four days. For extratropical
storms, the simulations were performed for 8 days because

of the longer duration of the storms. Generally, a time
step of 5 seconds was used. In the wet/dry routine, HMIN
was set to 0.3 ft, and wet and dry nodes were kept in
their respective states for a minimum of 12 time steps.
At startup, a ramp function was used to apply the forcing
gradually and smoothly to its full value. Because the
primary interest of the study was in maximum surges at
different gage locations, time series information on
surges was saved at intervals of 10 min to resolve the
maxima. Also, surges over the entire grid were saved at
the same interval and the solution was examined using
visualization software. Maximum computed surges (maxima
for the duration of the storm) were determined at all the
grid nodes and contours of maximum surge were plotted and
examined. Past experience indicates some uncertainty in
the values for radius to maximum winds computed by the
PBL model. Therefore, the radius was used as a tuning
parameter, within reasonable limits, to tune ADCIRC
results during calibration. Where the winds passed over
land, the PBL computed winds over water were reduced
locally to a fraction (40 percent) of the computed values
before using them in ADCIRC so as to realistically
represent the land drag effect. This fraction was
determined by trial-and-error during calibration.

Table 1 shows a comparison of computed and observed
peak surges at select stations for calibration and
verification. The locations of the stations are shown in
Figure 4 with closed square symbols and station numbers
corresponding to Table 1. These stations were selected
for presentation because they are distributed around the
project region. The agreement is excellent. One of the
difficulties encountered during the study was in
determining the correct vertical datums for various
prototype gages. The uncertainty in datum levels may be
on the order of 1 ft for some gages. The primary
interest of the study was in the magnitudes and not
timing of peak surges, so no special attempt was made to
match the times. However, a comparison of computed time
series of water surface elevations with observations for
Hurricane Betsy at Biloxi (Sta. 2), and Rigolets (Sta. 7)
is shown in Figure 5 to demonstrate reasonable agreement
in the timing. Because of the excellent match of peak
elevations at widely distributed stations shown in Table
1, and the overall matching of computed maximum surge
patterns with observations, the calibration and
verification were considered successful.

Results and Discussion

Following calibration and verification, 19 (17
historic and 2 synthetic) hurricanes (Table 2) and 11
extratropical storms (Table 3) were simulated using
ADCIRC and the results were archived for use in the EST.
The two synthetic storms were obtained by moving the
track of the 1915 hurricane of record west- and eastward.
Wherever possible, model results were checked against
available prototype data on surges.   There was good
agreement. Deviations are attributable to uncertainties
in gage datums and differences in bathymetry used in
model and that existing at the time of a particular
storm. Next, tide only was simulated over the region for
a period of 15 days and results were saved at the same
locations as before for use in the EST.

One interesting and challenging aspect of the study
was the extensive application of the wetting and drying
algorithm over very large areas of inundation.   Overall,
the wet/dry technique worked well. The results provided
sufficiently accurate estimates for computing stage-
frequency relations for areas that are normally dry and
impacted only by storm surge.

For the EST, the storms shown in Tables 2 and 3
formed the two training sets. At each gage location of
interest, 100 statistical simulations of a 200-year
sequence of storms were performed, and the results were
used to develop separate frequency-of-occurrence
relationships for tropical and extratropical storms.
Finally, the two relations were combined into one
relationship applicable to both storms. As an example,
the results obtained for the Rigolets station (Sta. 7)
are shown in Figure 6.

Summary and Conclusions

For the Lake Pontchartrain Study, the 2D
hydrodynamic model ADCIRC together with the statistical
procedure EST was used to compute stage-frequency
relationships at several locations. An existing global
grid of the Gulf of Mexico was refined to provide
sufficient resolution in the area of interest, with the
smallest grid cell dimensions on the order of 200 ft. To
our knowledge, this is the first time that storm modeling
at this resolution on such a large region has been
attempted in this particular geographic area. The model
ADCIRC was calibrated with observed surges for Hurricane
Betsy and verified to observed surges for Hurricane

Andrew. There was excellent agreement at gages distributed widely over the study area. The model was used to compute surges for a training set of 19 tropical and 11 extratropical storms. Tides only were simulated separately over the region. These results were used in the EST for 100 simulations, each of 200 years, to obtain stage-frequency relationships. ADCIRC results were archived and furnished to supply boundary conditions to the UNET model.

Even though the procedure and the tools described in this study were used for previous WES studies, the Lake Pontchartrain Study posed special challenges in terms of the complexity of the region (lakes, levees, waterways, extensive marshes, etc.) and the bathymetry and topography as characterized by extensive shallow and low-lying areas. A special wet/dry routine was developed and implemented in ADCIRC for this study. Overall the routine worked well, even though there is scope for improvement. Considering the excellent agreement with observations obtained, the model ADCIRC performed very well for this complex region. On the whole, the modeling study met all of its objectives and is considered successful.

Acknowledgments

The simulations reported in this paper were performed for the Lake Pontchartrain Study funded by the US Army Engineer District, New Orleans. Permission to publish this paper was granted by the Chief of Engineers.

References

Barkau, R.L. 1993. "UNET One-dimensional Unsteady Flow Through a Full Network of Open Channels," US Army Corps of Engineers, Hydrological Engineering Center, Davis, CA.

Cardone, V.J., Greenwood, C.V., and Greenwood, J.A. 1992."Unified Program for the Specification of Hurricane Boundary Layer Winds over Surfaces of Specified Roughness," Contract Report CERC-92-1, September 1992, US Army Engineer Waterways Experiment Station, Vicksburg, MS.

Jarvinen, B.R., Neumann, C.J., and Davis, M.A.S. 1988. "A Tropical Cyclone Data Tape for the North Atlantic Basin, 1886-1993: Contents, Limitations, and Uses," NOAA Technical Memorandum NWS NHC 22.

Luettich, R.A., Westerink, J.J., and Scheffner, N.W.

1992. "ADCIRC: An Advanced Three-Dimensional Circulation Model for Shelves, Coasts, and Estuaries, Report 1 : Theory and Methodology of ADCIRC-2DDI and ADCIRC-3DL," Technical Report DRP-92-6, November 1992, US Army Engineer Waterways Experiment Station, Vicksburg, MS.

Luettich, R.A., and Westerink, J.J. 1995. "Implementation and Testing of Elemental Flooding and Drying in the ADCIRC Hydrodynamic Model," Final Contractor's Report under Contract No. DACW39-94-M-5869, July 1995, prepared for the US Army Engineer Waterways Experiment Station, Vicksburg, MS.

Scheffner, N.W., et al. 1994. "ADCIRC: An Advanced Three-Dimensional Circulation Model for Shelves, Coasts, and Estuaries, Report 5 : A Tropical Storm Database for the East and Gulf of Mexico Coasts of the United States," Technical Report DRP-92-6, August 1994, US Army Engineer Waterways Experiment Station, Vicksburg, MS.

Scheffner, N.W., and Butler, H.L. 1995. "Storm Stage-Frequency Computations for Long Island, NY: A Comparison of Methodologies," Proceedings of the 4th Estuarine and Coastal Modeling Conference, ASCE, San Diego, CA, pp.80-91.

Scheffner, N.W., Borgman, L.E., and Mark, D.J. 1996. "Empirical Simulation Technique Applications to Tropical Storm Surge Frequency Analysis for the Coast of Delaware," ASCE Journal of Waterways, Ports, Coastal and Ocean Engineering, March 1996.

US Army Engineer District, New Orleans 1980. "Grand Isle and Vicinity, Louisiana, Phase II General Design Memorandum."

Westerink, J.J., et al. 1992. "Tide and Storm Surge Predictions Using a Finite Element Model," Journal of Hydraulic Engineering, ASCE, 118, pp. 1372-1390.

Westerink, J.J., Luettich, R.A., and Scheffner, N. 1993. "ADCIRC: An Advanced Three-Dimensional Circulation Model for Shelves, Coasts, and Estuaries, Report 3 : Development of a Tidal Constituent Database for the Western North Atlantic and Gulf of Mexico," Technical Report DRP-92-6, June 1993, US Army Engineer Waterways Experiment Station, Vicksburg, MS.

TABLE 1. Comparison of Maximum Surges (ft) at Select
Stations for Hurricanes Betsy and Andrew

		Betsy		Andrew	
No.	Name	Obs.	Comp.	Obs.	Comp.
1	Pascagoula	5.4	5.0		
2	Biloxi	7.6	7.7		
7	Rigolets at US90	7.0	7.5	3.7	3.8
12	Causeway/North	5.1	6.0		
13	Causeway/Midlake	3.5	4.6	4.4	4.2
18	Blind River	4.2	5.1		
19	Ruddock	10.2	9.2		
29	Westend	5.6	5.1	4.1	3.3
32	Seabrook	4.4	3.5		
42	Chef Menteur@US90	9.1	8.3	4.2	4.5
46	Shell Beach	9.3	7.4	5.3	5.2
49	Breton Sound			5.3	5.4
52	Phoenix	8.3	8.7		
57	Catfish Lake			5.9	5.6
58	Port Eads	3.0	3.4		
70	Grand Isle	4.5	4.7	3.9	4.2
72	Leeville	5.5	6.2	5.8	5.1
79	B. Petite Caillou			8.3E[*]	7.1
97	Eugene Island			3.8	3.4

Obs. = Observed,   Comp. = Computed,   [*]E = estimated

TABLE 2. Tropical Storms

Event/HURDAT #	Name	Start Date
1 - 127	Not Named	13 Aug 1901
2 - 187	Not Named	18 Sep 1909
3 - 214	Not Named	27 Sep 1915
4 - 232	Not Named	5 Aug 1918
5 - 241	Not Named	20 Sep 1920
6 - 324	Not Named	31 Jul 1933
7 - 397	Not Named	3 Aug 1940
8 - 471	Not Named	1 Sep 1948
9 - 562	Flossy	22 Sep 1956
10 - 565	Audrey	25 Jun 1957
11 - 634	Hilda	2 Oct 1964
12 - 639	Betsy*	8 Sep 1965
13 - 703	Edith	14 Sep 1971
14 - 731	Carmen	6 Sep 1974
15 - 775	Bob	9 Jul 1979
16 - 838	Juan	26 Oct 1985
17 - 899	Andrew**	23 Aug 1992

* Calibration   **Verification

TABLE 3. Extratropical Storms

Event	Start Date	Event	Start Date
1	3 Dec 1977	7	9 Nov 1980
2	11 Jan 1978	8	23 Oct 1985
3	4 Feb 1978	9	11 Sep 1988
4	9 Mar 1978	10	15 Dec 1991
5	26 Dec 1978	11	11 Mar 1993
6	3 Feb 1980		

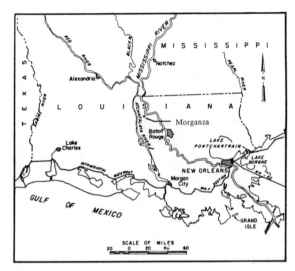

Figure 1. Study area (after US Army Engineer District, New Orleans, 1980)

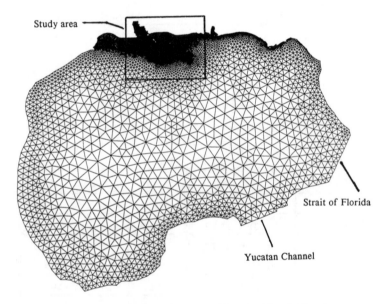

Figure 2. ADCIRC grid for the Gulf of Mexico

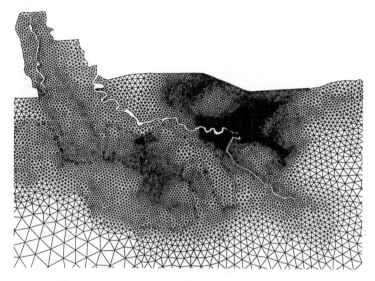

Figure 3. Blow-up of grid in study area

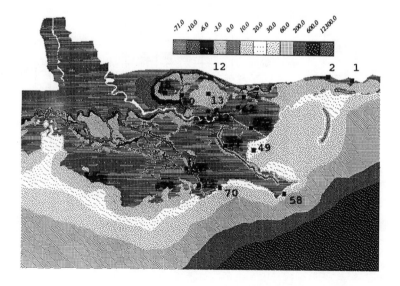

Figure 4. Local bathymetry in ft, and station locations

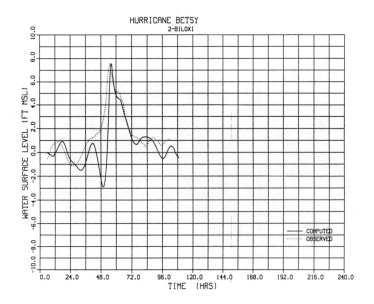

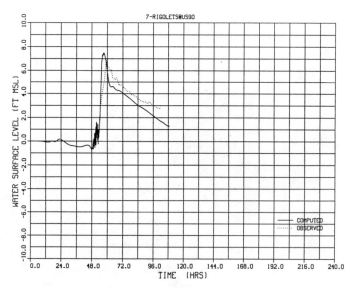

Figure 5. Comparison of water surface elevations

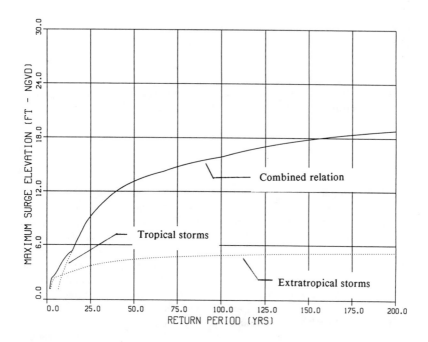

Figure 6. Stage-frequency curves for Rigolets at US 90

An Analysis of Grid Convergence for the Dynamic Grid Adaptation Technique Applied to the Propagation of Internal Waves

David P. Podber, [1], Keith W. Bedford, M. ASCE[2],

Abstract

The dynamic adaptive grid (DGA) technique was developed and implemented. The technique clusters grid nodes in regions of rapid variation in the temperature field thereby capturing motions of the thermocline. The model was run in adaptive and non-adaptive mode to reveal enhancements due solely to increased localized resolution. A series of numerical experiments were devised to incite the formation and propagtion of an internal wave on an idealized thermocline.The results are compared qualitatively and the predicted internal wave speed is compared to the analytic solution for the idealized thermocline. The DGA technique showed marked enhancement both qualitatively and quantitatively especially with a small number of grid nodes in the vertical. The uniform grid placement method showed better quantitative results when the number of grids was large enough to fully resolve the vertical structure.

Introduction and Objective

Numerical models applied to stratified flow have three major difficulties to overcome. The primary problem is that of resolution. When strong stratification occurs in deep water an inordinately large number of vertical grid nodes must be used to resolve the gradients, or grid nodes must be appropriately clustered which requires a priori knowledge of the location of the stratification. The second difficulty is that regions of sharply varying bathymetry can cause problems in the computation of the pressure gradients that drive the flow field in models using terrain following coordinates (Haney, 1991; Messinger, 1982; Messinger and Janjic, 1985; Podber and Bedford, 1993; Podber, 1991). The third major

---

[1] Visiting Assistant Professor, Department of Civil and Environmental Engineering and Geodetic Science, The Ohio State University, Columbus, Ohio, 43210;

[2] Professor, Department of Civil and Environmental Engineering and Geodetic Science, The Ohio State University, Columbus, Ohio, 43210;

difficulty for numerical approaches has been inconsistencies created by using the mode splitting technique. The vertically averaged velocity (U) that is computed using free mode routines tends to diverge from the vertical average of the three dimensional velocity field ($\int udz$) computed by the internal mode routines. This should not be surprising since the velocities are computed with different governing equations and different numerical techniques. An attempt to remedy this inconsistency was suggested in Mellor (1990), but requires shifting the velocity structure by subtracting out the difference between U and $\int udz$ from the velocity field (u). This changes the velocity structure, in particular it will move the velocity structure relative to the temperature structure which will affect the development of the temperature field in a spurious manner.

In order to improve the computation of the velocity fields and scalar transport in strongly stratified flow, a modeling procedure that solved these problems was devised. The improvements can be summarized as,

    1) increasing resolution with grid adaptation,

    2) computing the pressure in the physical coordinate thereby minimizing pressure instabilities, and

    3) avoid the use of mode splitting.

The technique used to place the grids, the governing equations and the numerical procedure are all described in the following sections. The model is tested by simulating the long term response of a stratified fluid to a wind shear event. The model results are compared to each other and measured against an analytic internal wave speed. A grid convergence test is performed for runs that use uniform grid placement, and runs that use a clustering approach.

Grid Node Placement

The procedure used to compute the grid node locations is called the Dynamic Grid Adaption (DGA) and is modeled after the Continuous Dynamic Grid Adaptation method as described by Dietachmayer and Droegemeier (1992). The DGA method uses a variational grid placement technique as described in Thompson et al. (1985), which requires the forumulation of a user defined weight function. This function can be tailored to fit the application of interest. In this case the weight function is given as,

$$W(x, z, t) = W_1 \left| \frac{\partial T}{\partial z} \right| + W_2 \left| \frac{\partial^2 T}{\partial z^2} \right| . \tag{1}$$

Where T is temperature, and W_1 and W_2 are constants that can be determined by the user. The following functional is then minimized,

$$\text{minimize} \int \left(1 + \lambda W^2 \right) \left(\frac{\partial z}{\partial \zeta} \right)^2 d\zeta , \tag{2}$$

where λ is a parameter that controls the degree of clustering. The solution of the minimization problem will be the desired grid placement, the details of the procedure can be found in Podber (1997) or in Thompson et al. (1985).

An example application of the procedure is shown in Figure 1, which shows a typical Great Lakes temperature profile (Saylor et al. (1981)). The thermocline is overlayed with a uniform grid in the left plot, and a grid that was placed using the adaptive grid technique in the right plot. Along the right vertical axis of each plot, is listed the distance between the grid nodes. The adaptive procedure pushes the nodes into the region of the thermocline at while stretching the grids out near the bottom.

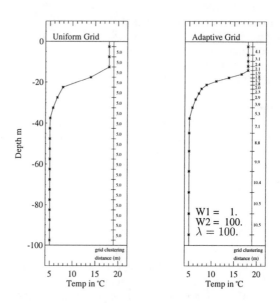

Figure 1: Grid Placement on a Thermocline

The Governing Equations in Rectangular Coordinates

The equations for conservation of mass, linear momentum, and heat that are used in this model are respectively,

$$\frac{\partial u}{\partial x} + \frac{\partial w}{\partial z} = 0.0 \tag{3}$$

$$\frac{\partial u}{\partial t} + u\frac{\partial u}{\partial x} + w\frac{\partial u}{\partial z} = -\frac{1}{\rho_0}\frac{\partial P}{\partial x} + \frac{\partial}{\partial x}\left(A_H\frac{\partial u}{\partial x}\right) + \frac{\partial}{\partial z}\left(A_V\frac{\partial u}{\partial z}\right) \tag{4}$$

$$\frac{\partial T}{\partial t} + u\frac{\partial T}{\partial x} + w\frac{\partial T}{\partial z} = \frac{\partial}{\partial x}\left(K_H\frac{\partial T}{\partial x}\right) + \frac{\partial}{\partial z}\left(K_V\frac{\partial T}{\partial z}\right) \tag{5}$$

The equation for conservation of linear momentum includes the Boussinesq approximation, and Reynolds averaging. The equation is for an idealized vertical plane, that is without lateral averaging.

The variables A_H and A_V are the horizontal and vertical eddy viscosities respectively. ρ_0 is the background density of water, and the pressure P is assumed to be the hydrostatic pressure. K_H and K_V are the horizontal and vertical eddy diffusivities.

The boundary conditions are no horizontal velocity at the left and right hand side, no heat flux through the surface, bottom, or sides, no surface stress, and the bottom stress given as,

$$\tau_b = c_d \, u_{bottom} \, |u_{bottom}| \tag{6}$$

where $c_d = 0.0025$.

The Governing Equations in Generalized Coordinates

Since the grid node placement varies throughout the computation, a set of fully general coordinates must be chosen, namely $x \rightarrow \xi$ and $z \rightarrow \zeta$. The full details of the mapping can be found in Podber (1997).

Partial derivates of grid related variables will be expressed using the subscript notation. For example, the partial derivative of ξ with respect to x, will be written as ξ_x. While partial derivatives of the variables representing flow or fluid properties will be expressed as fractions of the type $\frac{\partial u}{\partial \xi}$. This convention will help to visually separate the differentiation that occurs due to grid variables and differentiation that is to be performed on flow variables. By distinguishing the partial derivatives used in grid transformation from those required to satisfy the governing equations the resulting equations retain the "look and feel" of the original ones.

The transformed conservation of mass equation is written as,

$$\xi_x \frac{\partial u}{\partial \xi} + \zeta_x \frac{\partial u}{\partial \zeta} + \xi_z \frac{\partial w}{\partial \xi} + \zeta_z \frac{\partial w}{\partial \zeta} = 0.0 \tag{7}$$

The transformed conservation of linear momentum equation is written as,

$$
\begin{aligned}
\frac{\partial u}{\partial \tau} + u_e \frac{\partial u}{\partial \xi} + w_e \frac{\partial u}{\partial \zeta} = {} & -\frac{g}{\rho_o} \frac{\partial}{\partial x} \left(\int_z^0 \sigma_t(x, z^*) \, dz^* \right) - g \frac{\partial \eta}{\partial x} \\
& + \xi_x \frac{\partial \left(A_H \left(\xi_x \frac{\partial u}{\partial \xi} + \zeta_x \frac{\partial u}{\partial \zeta} \right) \right)}{\partial \xi} + \zeta_x \frac{\partial \left(A_H \left(\xi_x \frac{\partial u}{\partial \xi} + \zeta_x \frac{\partial u}{\partial \zeta} \right) \right)}{\partial \zeta} \\
& + \xi_z \frac{\partial \left(A_V \left(\xi_z \frac{\partial u}{\partial \xi} + \zeta_z \frac{\partial u}{\partial \zeta} \right) \right)}{\partial \xi} + \zeta_z \frac{\partial \left(A_V \left(\xi_z \frac{\partial u}{\partial \xi} + \zeta_z \frac{\partial u}{\partial \zeta} \right) \right)}{\partial \zeta}
\end{aligned}
$$

where,

$$u_e = [(u - x_\tau)\xi_x + (w - z_\tau)\xi_z] \quad \text{and} \quad w_e = [(u - x_\tau)\zeta_x + (w - z_\tau)\zeta_z] \tag{8}$$

Table 1: Governing Equations and Solution Techniques

Equation	Numerical Technique
Grid Node Placement	Gauss-Seidel
Momentum Equation 8	Runge-Kutta 4th order time: 4th order centered space
Scalar Equation 9	Runge-Kutta 4th order time: 4th order centered space
Continuity Equation 7	2nd order
Free Surface	Implicit

and represent the velocity in the transformed, or computational, space.

The transformed equation for temperature is written as,

$$
\frac{\partial T}{\partial \tau} + u_e \frac{\partial T}{\partial \xi} + w_e \frac{\partial T}{\partial \zeta} = \xi_x \frac{\partial \left(K_H \left(\xi_x \frac{\partial T}{\partial \xi} + \zeta_x \frac{\partial T}{\partial \zeta} \right) \right)}{\partial \xi} + \zeta_x \frac{\partial \left(K_H \left(\xi_x \frac{\partial T}{\partial \xi} + \zeta_x \frac{\partial T}{\partial \zeta} \right) \right)}{\partial \zeta}
$$

$$
+ \ \xi_z \frac{\partial \left(K_V \left(\xi_z \frac{\partial T}{\partial \xi} + \zeta_z \frac{\partial T}{\partial \zeta} \right) \right)}{\partial \xi} + \zeta_z \frac{\partial \left(K_V \left(\xi_z \frac{\partial T}{\partial \xi} + \zeta_z \frac{\partial T}{\partial \zeta} \right) \right)}{\partial \zeta} \tag{9}
$$

where u_e and w_e are given above.

Since the pressure gradient is computed in the physical domain which is an unusual treatment, the process will be described in more detail. When transformed, the pressure gradient can be written as,

$$
\frac{\partial P}{\partial x} = \longrightarrow \xi_x \frac{\partial P}{\partial \xi} + \zeta_x \frac{\partial P}{\partial \zeta} \ . \tag{10}
$$

Using the transformed pressure term can lead to errors in the numerical calculation when the terrain of the basin being modeled is steep. This effect has been well documented (Messinger (1982), Haney (1991)). To avoid this problem the pressure gradient is calculated in the non-transformed space as was suggested by Mahrer (1984). This requires interpolation of the pressure field, which increases the computational effort, however it leads to a model that can be implemented in steep basins without conditioning of the bathymetery. For this model the pressure gradient is computed as follows,

$$
-\frac{1}{\rho_o} \frac{\partial P(x,z)}{\partial x} = - \underbrace{\frac{g}{\rho_o} \frac{\partial}{\partial x} \left(\int_z^0 \sigma_t(x,z^*)\, dz^* \right)}_{\text{computed in physical domain}} \ -g \frac{\partial \eta}{\partial x}. \tag{11}
$$

Model Implementation

The computational techniques used to solve the equations that make up this model are listed in Table 1. The grid used in the computational domain is an Arakawa C type grid (Arakawa and Lamb (1977)) The turbulent mixing in the horizontal is set to a constant 20 m^2/s for the eddy viscosity and to a constant

2 m^2/s for the eddy diffusivity. These values were observed in Lake Ontario during the IFYGL study (Aubert and Richards (1981)). The vertical mixing of momentum and heat is set by following the Richardson number based procedure of Pacanowski and Philander (1981), with the parameters adjusted so that the vertical eddy diffusivity and viscosity would remain within the range observed for the Great Lakes.

The numerical experiments performed are listed in Table 2. The DGA results were generated using adaptation while the Σ results set the grid stretching parameter to zero so that a uniform grid placement was performed. All of the tests used the same geometry of a flat bottom, 100 meters deep with a 1 kilometer horizontal spacing. Test 0 has a uniform temperature of 4 °C and the grids were placed uniformly in the vertical. Test 1 and Test 2 have the DGA method turned on, while Test 3 and Test 4 are for a uniform grid placement, which is tantamount to a Σ coordinate for a flat bottom. Both the DGA and the Σ tests were run for weak, intermediate, and strong wind cases. In the strong wind case the wind stress was set to $\tau_s = 0.30$ Pa which corresponds roughly to a 14 m/s wind. The intermediate wind case had the wind stress set to $\tau_s = 0.03$Pa which corresponds to approximately a 5 m/s wind, and in the weak wind case $\tau_s = 0.01$ Pa corresponding to approximately a 3 m/s wind. The weak wind case excites the internal wave without causing an upwelling of the cold hypolimnion water which in turn weakens the density gradient and therefore will reduce the speed of the internal Kelvin wave. The intermediate wind case causes an internal wave that steepens to from an internal bore. The strong wind event was included because it more accurately depicts the type of events seen in the Great Lakes, where upwelling fronts appear after storms pass (see Beletsky et al. (1997)).

A uniform depth of 100 meters was used for both cases so that bathymetric effects would not appear and differences in the results could be attributed solely to different gridding. The model thermocline that was imposed consists of a top mixed layer that is 5 meters deep and 20°C and bottom hypolimnion that begins at 15 meters deep and has a uniform temperature of 5°C. The thermocline is between 5 and 15 meters and is a constant gradient between 20°C and 5°C (see Figure 2)

The wind stress was ramped up using a $\sin^2$ function over a 1.2 day period. The function was multiplied by a scaling so that the time integrated wind stress applied in the DGA and Σ models could compare with the results of Beletsky et al. (1997). A graph of the functional form of the windstress is shown in Figure 3. The wind stress is found by multiplying the value of the form by the appropriate τ_s. The heat flux is set to zero for these experiments.

Since this modeling effort is geared toward creating a result that can be used for computations of real lakes and estuaries, none of the simplifications that help to generate exact solutions were made. This means that verification of the model results depends on a blend of qualitative assestment and gross feature representations. An example of the latter would be the computation of the in-

Table 2: Experiments and Internal Wave Speeds

	Wind Stress	Thermocline	Grid_n n = vertical grid nodes	Internal Wave Speed (m/s)
Test 0	$\tau_s = 0.03$ Pa	No	Uniform	
Test 1	$\tau_s = 0.01$ Pa	Yes	DGA_20	
Test 2	$\tau_s = 0.03$ Pa	Yes	DGA_20	0.321
Test 3	$\tau_s = 0.30$ Pa	Yes	DGA_20	
Test 4	$\tau_s = 0.01$ Pa	Yes	Σ_20	
Test 5	$\tau_s = 0.03$ Pa	Yes	Σ_20	0.279
Test 6	$\tau_s = 0.30$ Pa	Yes	Σ_20	
Grid Convergence Tests			
Test 7	$\tau_s = 0.03$ Pa	Yes	Σ_40	0.324
Test 8	″	″	Σ_80	0.336
Test 9	″	″	Σ_160	0.346
Test 10	″	″	DGA_40	0.333
Test 11	″	″	DGA_80	0.333
Test 12	″	″	DGA_160	0.330

ternal wave speed, which was performed for this model. Computing the internal wave speed accurately is good evidence that model is behaving credibly, but clearly it is not the whole story. However, the internal wave speed is the only objective measure used to analyse the performance of the model. Considering that the internal wave speed computation is closer to an analytically computed wave speed allows us to interpret differences between the Σ and DGA as error.

Free Surface Analytic Test : Test 0

Test 0 is a flow case with no temperature structure. A wind stress of 0.03 Pa was applied at the surface. The free surface was tipped in response to the wind and a flow similar to a driven cavity flow was formed. The free surface response at steady state can be computed analytically following Dean and Dalrymple (1984). The comparison between an analytic solution at steady state and the computed free surface after 18 hours of simulation showed a 4% difference at 0 km and 100 km between the analytical and numerical solutions. This difference may be error, but there are some assumptions used in the analytic solution that are absent from the numerical solution. The extent of accuracy between the analytic and numerical solution, tends to indicate that the surface stress boundary condition is working correctly.

Internal Bore Case: Test 2 compared with Test 5

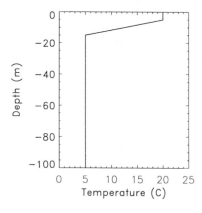

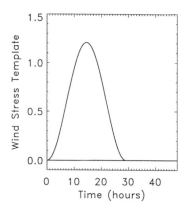

Figure 2: Initial Temperature Profile   Figure 3: Wind Stress Form

Figure 5 show the response of the DGA model to wind stress at the time of peak forcing. Figure 6 show the model response after the wind has died off and the internal bore has set up. The temperature contours are in degrees C and the velocity scale is listed at the bottom of the plots. The vertical scale is the depth in meters and the horizontal scale is the length in kilometers. The clustering of grids near the surface is clearly noticeable in the plots. Not only does the clustering resolve the temperature well, but it also resolves the velocity boundary layer at the surface. Figure 5 shows that the return flow is more tightly confined to the epilimnion in the DGA case. This physical phenomenon also speaks to the necessity to model the temperature structure correctly. An improperly placed thermocline will cause the return flow to be centered at the wrong depth, which would be a miscalculation of an important physical phenomenon. After 90 hours one can still notice that there is quite a bit more momentum remaining in the epilimnion in the DGA case than in the Σ case.

One way to compare the differences between the temperature field that is calculated with the DGA method and the field computed with the Σ grid placement method is to set up a difference, or anomaly plot. The difference plots shown in Figures 9 and 10, and show the evolution of the difference between the two methods at 18 hours and 90 hours. Please note that the only thing that was changed between the two runs was the gridding method.

After 18 hours the differences are uniform horizontally, with a slight indication of dip or increased error near the right boundary. As time progresses, a 2 °C difference is noted as an internal extremum. This difference gets sharper and is transported, first from right to left, then at 108 hours it hits the left hand wall

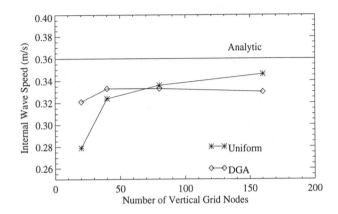

Figure 4: DGA and Σ internal wave speeds

and bounces back toward the right. This difference represents the difference in the internal wave speed. By examining the images in the vector plots of the Σ and DGA results the phase shift is evident. The horizontal span of the region of error is also growing as the simulation progresses. this represents the increasing distance between the wave fronts as time passes. It is important to note that while both of the solutions have similar quantitative results the phase error is large and growing larger.

Internal Wave Speed and Grid Convergence

In all runs the DGA and the Σ results are qualitatively similar. The wind stress sets up a free surface slope and a velocity boundary layer, which in turn depresses the thermocline at the downwind side and elevates it on the upwind side. Once the wind dies off the internal wave begins to propagate. The speed of the internal wave is easily measured and can be used as a benchmark for the model's performance.

The internal wave speed for the DGA and Σ models was computed by placing numerical current meters at various locations in the flow field and then tracking the local temperature maxima and minima as they progressed through the domain. Table 2 shows computed and analytic results for the internal wave speed for the initial temperature structure given. The number listed in the analytic column, 0.36 m/s, is computed from a formulation given by Csanady (1982) for a two layer body of water with a linear gradient between two layers of uniform density.

The DGA result is much better than the Σ result, producing an internal

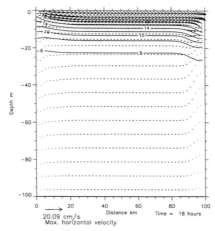

Figure 5: DGA method for the internal bore case Test 2 at 18 hours.

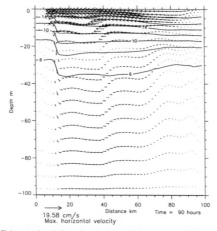

Figure 6: DGA method for the internal bore case Test 2 at 90 hours.

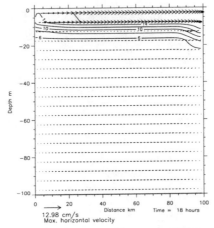

Figure 7: Σ grid for the internal bore case Test 5 at 18 hours.

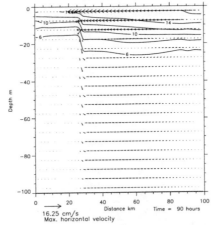

Figure 8: Σ grid for the internal bore case Test 5 at 90 hours.

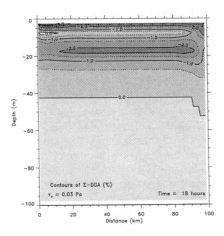

Figure 9: Σ-DGA difference plot, (Test 5 - Test 2) at 18 hours.

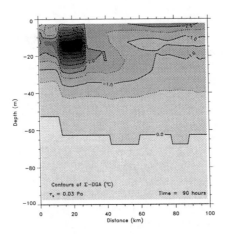

Figure 10: Σ-DGA difference plot, (Test 5 - Test 2) at 90 hours.

wave speed approximately 90% of the analytic solution. The improvement of the internal wave speed can be directly attributed to the enhanced vertical resolution, since the only parameters changed between tests 2 and 5 is the vertical resolution. It could be that the enhancement is due to a better computation of the velocity field as well as better representation of the temperature structure.

To test for grid convergence the Σ and the DGA model were run with 20, 40, 80, and 160 grid nodes in the vertical. With only 20 nodes, the DGA was able to compute the internal wave speed to 89% of the analytic, while the uniform grid computed approach computed 78% of the analytic wave speed. The wave speed computed for each vertical resolution is plotted in Figure 4. The figure shows that for the lower resolutions of 20 and 40 grid nodes in the vertical, the DGA method performed comparably to the Σ method with twice as many grids. However, for the higher number of grid nodes the Σ method showed better results and better convergence characteristics. This is due to the fact that the DGA technique, as implemented here produces a non-orthogonal grid system, and such systems are prone to error when they stray far from orthogonality (Thompson et al. (1985)). The Σ approach does well because the vertical structure is resolved with 160 nodes in the vertical.

For most geophysical applications, the horizontal scales are such that, using 160 nodes in the vertical is not compuationally possible because of the large horizontal scales. Many applications report using even less than 20 nodes in the vertical, however if the vertical structure can be resolved with a uniform or other type of orthogonal grid system, clearly that would be the better choice.

Computational Efficiency

The computational advantages of using an adaptive grid come in the form of increased accuracy with fewer grid nodes. These efficiencies are described in general by Dietachmayer and Droegemeier (1992). For the model presented here, we can see from Figure 4 that the DGA using 20 nodes in the vertical, compares with the Σ using 40 nodes. This does not mean that the DGA is twice as efficient, because of the extra cpu time required to compute and manage the grid information. The ratio of fluid cpu time to grid cpu time will depend on the type of machine used for the computation. When highly vectorized on the Cray-YMP, a single grid adjustment took 3.77 times the cpu of a single fluid step. On a scalar machine, the ratio is about 1. Using the Cray and adjusting the grid every 10 minutes with a 24 second time step, the grid procedure was seen to increase the cpu cost by 15%. This would imply an 85% speed up using the DGA. In future work, if the grids were to be computed in parallel, then the gridding procedures would have only a nominal cost.

Conclusions

Better numerical results can be obtained by placing higher resolution in areas where the variables of interest are changing rapidly. These tests were applied diagnostically to an idealized thermal structure for the purpose of establishing the potential benefits of this new modeling procedure. That the DGA technique outperforms a uniform grid is a necessary condition for further research into this type of approach. The results clearly indicate that grid clustering is a valuable numerical approach. Further tests of the efficacy of the DGA approach for modeling thermoclines would certainly include a test based on thermocline formation due to meteorological forcing for a region with field data available. Other tests of interest could include sharp tidal fronts, plume outfalls, or air-sea interactions where gradients are sharp in the horizontal as well as at the interface.

References

Arakawa, A. and Lamb, V. (1977). *Methods in Computational Physics*, Academic Press N.Y., chapter Computational Design of the UCLA General Circulation Model, pp. 174–267. ed. J. Chang.

Aubert, E. and Richards, T. (eds) (1981). *IFYGL - The international field year for the Great Lakes*, National Oceanic and Atmospheric Administration, Great Lakes Environmental Research Laboratory, Ann Arbor, Michigan.

Beletsky, D., O'Connor, W., Schwab, D. and Dietrich, D. (1997). Numerical simulation of internal kelvin waves and coastal upwelling fronts, *Journal of Physical Oceanography* 27(7): 1197–1215.

Csanady, G. (1982). *Circulation in the Coastal Ocean*, D. Reidel Publishing Company, Dordrecht, Holland.

Dean, R. and Dalrymple, R. (1984). *Water Wave Mechanics for Engineers and Scientists*, Vol. 2 of *Advanced Series on Ocean Engineering*, World Scientific, pg. 353.

Dietachmayer, G. and Droegemeier, K. (1992). Application of continuous dynamic grid adaption techniques to meteorological modeling. Part I: Basic formulation and accuracy, *Monthly Weather Review* 120: 1675–1706.

Haney, R. (1991). On the pressure gradient force over steep topography in sigma coordinate ocean models, *Journal of Physical Oceanography* 21: 610–619.

Mahrer, Y. (1984). An improved numerical approximation of the horizontal gradients in a terrain-following coordinate system, *Monthly Weather Review* 112: 918–922.

Mellor, G. L. (1990). *User's Guide for A three-dimensional, primitive equation, numerical ocean model,* The Atmospheric and Oceanic Sciences Program, Princeton University, Princeton NJ.

Messinger, F. (1982). On the convergence and error problems of the calculation of the pressure gradient force in sigma coordinate models, *Geophys. Astrophys. Fluid Dynamics* **19**: 105–117.

Messinger, F. and Janjic, Z. (1985). Problems and numerical methods of the incorporation of mountains in atmospheric models, *Lectures in Applied Mathematics* **20**: 82–120.

Pacanowski, R. and Philander, S. (1981). Parameterization of vertical mixing in numerical models of tropical oceans, *Journal of Physical Oceanography* **11**: 1443–1451.

Podber, D. (1991). *Modelling Great Lakes tributary flow and transport,* Master's thesis, The Ohio State University, Columbus, Ohio.

Podber, D. (1997). *Modeling Strongly stratified flow with the Dynamic Grid Adaptation (DGA) technique,* PhD thesis, The Ohio State University, Columbus, Ohio.

Podber, D. and Bedford, K. (1993). Tributary loading with a terrain following coordinate system, In *Estuarine and Coastal Modeling III*, pp. 475–488. eds. Spaulding, M. L., K.W. Bedford, A. Blumberg, R. Cheng, C. Swanson.

Saylor, J., Bennett, J., Liu, P., Pickett, R., Boyce, F., Murthy, C. and Simons, T. (1981). *IFYGL - The International Field Year for the Great Lakes,* The National Oceanic and Atmospheric Administration, Great Lakes Environmental Research Laboratory, Ann Arbor, Michigan, chapter Water Movements, pp. 247–324.

Thompson, J., Warsi, Z. and Mastin, C. (1985). *Numerical Grid Generation: Foundations and Applications,* North-Holland pg. 483.

Tidal Circulation and Larval Transport Through a Barrier Island Inlet

Richard A. Luettich, Jr.[1], James L. Hench[1], Crystal D. Williams[1], Brian O. Blanton[2], and Francisco E. Werner[2]

Abstract

As a part of the NOAA South Atlantic Bight Recruitment Experiment (SABRE), a detailed modeling study is in progress to define likely larval transport pathways in the vicinity of Beaufort Inlet, North Carolina. Modeled tidal circulation near the inlet has been verified using data from a NOAA tide study conducted in the mid 1970s. Particle tracking results indicate that larval ingress is most effective close to shore and that a substantial residual eddy exists to the west of the inlet and just offshore of the western ebb tidal delta. This eddy may play a significant role in trapping and preventing larvae from entering the inlet. Results also indicate a substantial difference in the way that transport occurs on the east and west sides of the inlet due to the inlet geometry and the configuration of primary channels that lie immediately inland. The latter results are independent of any assumed simple day/night behavior and are remarkably consistent with larval and drifter data from a multidisciplinary field study in Beaufort Inlet during March of 1996.

Introduction

A number of species of marine fish have early life cycles that consist of open ocean spawning followed by primarily hydrodynamic transport across the continental shelf and into estuarine nursery areas. In North Carolina coastal waters, these

[1] University of North Carolina at Chapel Hill, Institute of Marine Sciences, 3431 Arendell St., Morehead City, NC, 28557
[2] University of North Carolina at Chapel Hill, Department of Marine Sciences, CB 3300, 12-5 Venable Hall, Chapel Hill, NC, 27599-3300

species constitute nearly 90 percent of the annual commercial catch (Miller, 1984). The NOAA South Atlantic Bight Recruitment Experiment (SABRE) was initiated to develop process level understanding of the linkages between population variability in these species of fish and variability in the physical and biological environment in which they exist.

A critical part of the recruitment journey takes place near the estuary mouth where exchange occurs between continental shelf and estuarine waters. Along the southeastern Atlantic and Gulf of Mexico coasts, many estuaries are fronted by barrier islands and exchange occurs via relatively small tidal inlets. An example is Beaufort Inlet, North Carolina, which lies along the U.S. Atlantic coastline roughly midway between Cape Hatteras and Cape Fear (Figure 1a). This inlet connects the Atlantic Ocean to a complicated system of sounds and estuaries, including Bogue Sound to the west, the Newport River Estuary to the north and Back Sound to the east. Water flowing through Beaufort Inlet communicates with estuarine nursery areas via the Morehead Channel to the northwest, the Radio Island Channel to the north and the Shackleford Channel to the east (Figure 1b). The Morehead Channel and Radio Island Channel are periodically dredged to target depths of 13.7 m and 4.2 m, respectively; the Shackleford Channel is not dredged. A well-developed ebb tidal delta exists to the west of the inlet and a smaller one exists to the east (Figure 1d).

As a part of SABRE, a physical and biological field sampling program was conducted immediately inside Beaufort Inlet in March, 1996, to determine larval transport over several tidal cycles. Detailed results from this field program will be presented in a special issue of Fisheries Oceanography (Blanton et al., 1998, Churchill et al., 1998, Forward et al., 1998). The modeling described in the present paper represents an initial attempt to understand and replicate some of the observations obtained in this field program. Specifically, we are interested in a plausible explanation for why observed larval densities were an order of magnitude greater in Shackleford Channel than in the much larger Morehead Channel that runs directly through the center of the inlet.

Tidal Circulation Model

Hydrographic data (Klavans, 1983 and Blanton et al., in review) indicate minimal vertical salinity stratification close to Beaufort Inlet except perhaps during early stages of the flood tide. The flow is primarily tidal and is dominated by the M2 astronomical component.

We have modeled the primary tidal circulation using the fully-nonlinear, vertically-integrated, two-dimensional, finite element hydrodynamic model ADCIRC, (Luettich et al., 1992; Westerink et al., 1994). Simulations were

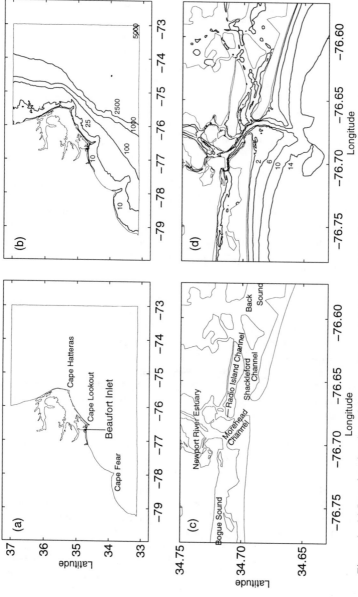

Figure 1. (a) Location drawing of the complete model domain. (b) Bathymetric contours (meters) for the complete model domain (c). Location drawing in the immediate vicinity of Beaufort Inlet, North Carolina. (d) Bathymetric contours (meters) in the immediate vicinity of Beaufort Inlet.

performed on a domain that stretches from approximately the North Carolina - Virginia border to the South Carolina - Georgia border and includes water from the deep ocean (~5000 m depth) to tidal flats that wetted and dried during the calculation (Figure 1). Element sizes varied from over 10 km in the deep ocean to less than 25 m near shore (Figure 2). A standard quadratic friction coefficient of 0.0025 and a constant lateral viscosity of 2 m^2 s^{-1} were used in the simulations. The model was forced along the open boundary by the M2 astronomical tide extracted from a tidal database generated from a much larger ADCIRC simulation of the western North Atlantic, Gulf of Mexico and Caribbean Sea. Model results were harmonically analyzed to extract the M2 component, the Eulerian residual field, and the M4, M6 and M8 overtides.

An initial verification of the model has been made against observed water surface elevation and velocity data collected in the late 1970s by the National Ocean Survey (NOS), (Klavans, 1983). Locations of the data stations are shown in Figure 3. Comparisons between NOS and model results are presented in Tables 1 and 2. Elevation errors suggest the model is roughly 10 percent high and a few degrees late in phase. Tidal ellipses indicate that the velocity direction and phase are in agreement with observations although the magnitude of the ellipse major semi-axis in the model is about 25 percent low near the inlet mouth. At all observational stations the flow is nearly rectilinear making the minor semi-axis and rotation sense extremely sensitive to any error in the field data or model results. Since the bathymetry and geometry of North Carolina coastal waters can be highly variable in time, considerable uncertainty exists in the accuracy of the model bathymetry in many parts of the sounds and estuaries, particularly in relation to the conditions that existed at the time of the NOS tidal survey. We have made no effort to improve the model vs data comparison by calibrating the model's bathymetric depths, friction coefficient or the open boundary conditions. Rather, we consider the model to be reasonably well verified in its present state.

The overall characteristics of the M2 tide are shown in Figures 4 and 5. Strong bottom friction due to shallow water depths and high current velocities rapidly attenuates and retards the M2 wave as it propagates inside the inlet (Figure 4). Maximum velocities in the inlet reach approximately 1 m s^{-1}. Flow through the inlet on flood tide is reasonably well organized along streamlines with some indication of separation occurring along the headland to the west of the inlet. On ebb tide there is evidence of much stronger flow across streamlines particularly in areas where excess westward momentum from the Shackleford Channel tends to carry water part way across the inlet mouth. Another characteristic of the velocity field is the phase difference between the flow immediately inside the inlet in the Shackleford Channel and in the Morehead Channel. Specifically, the velocity field turns approximately 45 minutes earlier in the Morehead Channel than in the Shackleford

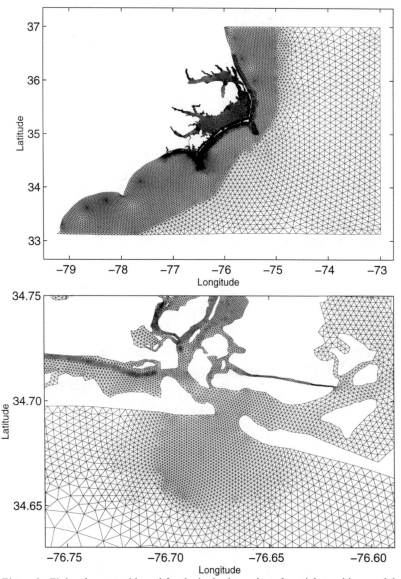

Figure 2. Finite element grid used for the hydrodynamic and particle tracking model simulations. The grid contains 32,218 nodes and 58,641 elements ranging in size from greater than 10 km in the deep ocean to less than 25 meters near shore.

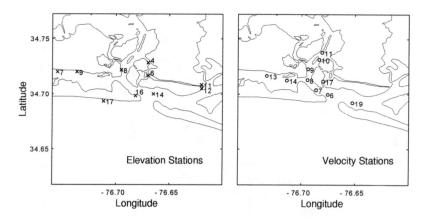

Figure 3. Location of water level stations and current meter moorings from the
NOS tidal study, (Klavans, 1983).

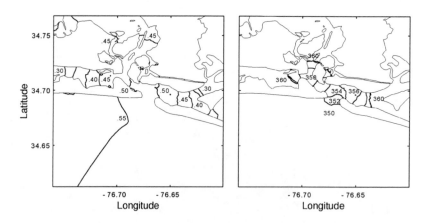

Figure 4. Co-amplitude and co-phase lines for the M2 tidal constituent.

Table 1. Comparison between NOS tidal elevation data and ADCIRC model results for the M2 astronomical tide.

	Elevation Amplitude (m)		Phase (deg)	
Station	NOS	Model	NOS	Model
4	0.415	0.471	14	2.8
6	0.414	0.467	11	0.7
7	0.281	0.273	25	19.9
8	0.437	0.495	6	358.4
9	0.330	0.361	16	9.0
11	0.294	0.238	25	22.5
12	0.273	0.252	31	19.2
14	0.452	0.508	358	353
16	0.450	0.499	358	355.4
17	0.536	0.551	356	348.2
RMS Difference		0.044		7.9

Table 2. Comparison between NOS tidal ellipse data and ADCIRC model results for the M2 astronomical tide.

	Direction (deg)		Phase (deg)		Major Semi Axis (m/s)		Minor Semi Axis (m/s)		Rotation	
Station	NOS	Model	NOS	Model	NOS	Model	NOS	Model	NOS	Model
6 top	319	314	317	318	1.020	0.732	0.012	0.005	CCW	CW
6 bottom	326		317		1.020		0.018		CW	
7 top	320	311	323	311	0.768	0.469	0.013	0.016	CCW	CCW
7 bottom	307		322		0.925		0.018		CCW	
8 top	326	317	325	321	0.614	0.495	0.026	0.033	CCW	CCW
8 bottom	326		325		0.585		0.018		CCW	
9	27	3.4	309	304	0.489	0.549	0.049	0.003	CW	CCW
10 top	40	35	312	305	0.696	0.610	0.006	0.003	CCW	CW
10 bottom	45		312		0.681		0.005		CW	
11	42	15	290	308	0.547	0.374	0.010	0.001	CW	CCW
13	292	283	353	348	0.756	0.770	0.002	0.005	CW	CW
14	87	91	131	162	0.705	0.874	0.009	0.043	CW	CW
17	21	18	317	312	0.601	0.599	0.039	0.002	CCW	CW
19	310	308	155	155	0.683	0.521	0.010	0.004	CCW	CW
RMS Difference		13		13		0.18		0.022		

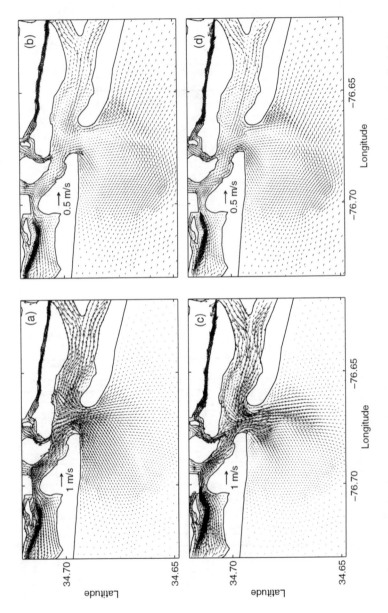

Figure 5. Vertically-integrated current field for the M2 tidal constituent at time of (a) maximum flood, (b) slack before ebb, (c) maximum ebb, and (d) slack before flood.

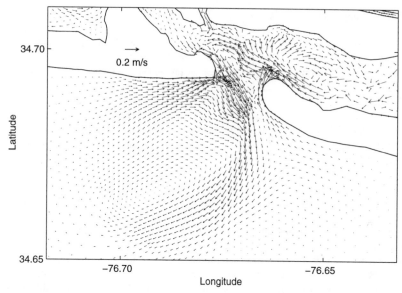

Figure 6. Steady, vertically-integrated Eulerian residual currents from the hydro-dynamic model run.

Channel (Figure 5). This difference in phase in the model is in close agreement with current meter data collected during the Beaufort Inlet field program in March, 1996 (Blanton et al., 1998).

The steady Eulerian residual velocity field from the model is shown in Figure 6. Typical of a classical inlet system, it indicates net flow toward the inlet near shore and net outflow in the center of the inlet. This flow pattern forms part of a strong (~10 cm s^{-1}) gyre over the ebb tidal delta to the west of the inlet while no similar eddy exists to the east of the inlet. Within the inlet throat, residual velocities are inward along the east side of the inlet and outward along the west side and are consistent with the observational data from Logan (1995) and Churchill et al. (1998). There is also a significant residual velocity (~15 cm/s) directed from the Shackleford Channel into the mouth of the Morehead Channel.

The water entering Beaufort Inlet on a flood tide is split so that approximately 55 percent is transported down the Morehead Channel, 41 percent down the Shackleford Channel and 4 percent down the Radio Island Channel.

Larval Simulations

To examine possible larval transport pathways in the vicinity of Beaufort Inlet, we utilized the hydrodynamic flow fields (steady, M2, M4 constituents) described above as input to a particle tracking model (Foreman et al., 1992; Baptista et al., 1984). The particle tracking model was initiated with a rectangular array of 200 particles in three different starting positions (west, center and east) outside of the inlet (Figure 7). For each starting position, 10 model runs were made varying the release time in 10 evenly-spaced intervals over the M2 tidal cycle. All runs lasted for 10 M2 tidal cycles which corresponds to a typical period of time a larvae of this life stage would remain in the water column (R. Forward, personal communication).

In a first set of runs particles were assumed to exhibit no behavior and therefore to float passively with the currents. The paths traversed by the particles released at slack before ebb for the three initial positions are shown in Figure 7. It is clear that particles starting on the west side of the inlet are transported directly into the Morehead Channel and up into the Newport River Estuary and Bogue Sound. Particles that start in the center of the inlet appear to be transported into the Morehead, Radio Island and Shackleford Channels. Particles that start out to the east of the inlet are primarily transported into the Shackleford Channel, although a significant number of them also progress up the Radio Island and Morehead Channels. The pathlines also show the tendency for particles to become trapped in the residual eddy located on the outer part of the western ebb tidal delta (Figure 7). Finally, animations of the particle trajectories show that net transport toward the inlet is most efficient close to shore in areas where the Eulerian residual circulation is toward the inlet.

To better quantify these results we placed "numerical nets" across the Morehead, Radio Island and Shackleford channels at the approximate positions of plankton sampling stations during the March 1996 field program. We then counted the number of particles that crossed each net during the course of each model run. To present the results we have separated the data by starting location but ensembled it for the 10 different release times (Figure 8).

For the starting location to the west of the inlet, approximately 45 percent of the particles never make it across any net but rather are trapped offshore in the residual eddy. Those particles that do cross one of the net locations are carried almost entirely down the Morehead Channel and are virtually excluded from both of the other channels. For the starting location across the center of the inlet, about the same number of particles never make it across any net, however the particles that do cross one of the nets are nearly equally divided between the Shackleford and Morehead Channels. A smaller number are transported down the Radio Island Channel due to the much lower flow volume that enters this channel compared to the

Figure 7. Starting locations and particle transport paths for the three different starting locations considered with the particle transport model.

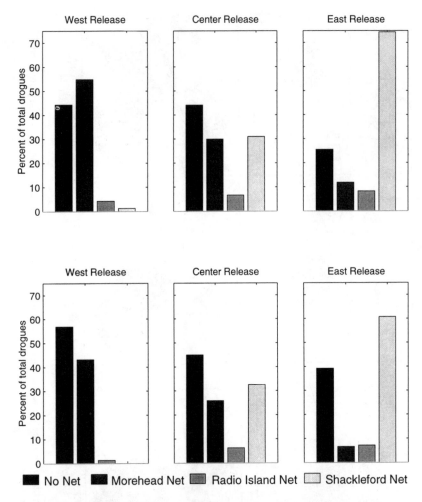

Figure 8. Summary results for net crossings from the particle tracking model in
which particles are assumed to have no behavior (passive transport) or a simple
day/night behavior. In each graph bars from left to right represent particles crossing
no nets, particles crossing the Morehead Channel net, particles crossing the Radio
Island Channel net and particles crossing the Shackleford Channel net, respectively.

other two. For the starting location to the east of the inlet, only about 25 percent of the particles are lost before crossing any of the nets. The remaining 75 percent of the particles almost all cross the Shackleford net. However, approximately 10 percent also cross the Morehead net. These latter results are remarkably consistent with the order of magnitude larger larval catches observed in the Shackleford Channel versus the Morehead Channel and are suggestive of a larval source entering the inlet from the east. It is also interesting to note that nearly as many particles cross the Radio Island net as cross the Morehead net despite the order of magnitude lower volume flux into the Radio Island Channel.

A second set of particle tracking runs was performed in which the particles were specified to have a simple day/night behavior. In this case, particles were assumed to move passively with the current during the night and remain stationary during the day (simulating the ability of many larvae to stay on the bottom to avoid visual predation during the day). Daytime and night time were each assumed to last 12 hours which is representative of the division of light and dark hours at this latitude during the month of March. For these runs the same particle starting locations and the same set of 10 evenly spaced release times during the tidal cycle were used as in the first set of runs. However, in addition the time of sunrise was varied over 12 even increments through the tidal cycle. We summarize the results by ensembling the net crossings for all 120 runs (10 release times x 12 sunrise times) for each starting location (Figure 8).

In general the addition of this day/night behavior has relatively little effect on the particle simulation results. In all cases the number of particles that do not cross a net is increased, presumably because the particles are now only moving with the flow for 5 out of the 10 tidal cycles and therefore have less of a chance to make it into the inlet. However, the spatial distribution of the transport pathways is essentially the same as for the no behavior case. The reason for this is that the length of daylight corresponds fairly closely to the length of the tidal cycle. Therefore, after a particle freezes for 12 hours, it will begin moving with the flow at approximately the same phase of the tide as when it stopped 12 hours earlier. Since the particle tracking simulations are relatively short (approximately 5 days), the total shift in the relative phases of the tidal cycle and the day/night cycle are only 2 hours from the corresponding median value for the run with no day/night behavior.

Conclusions

The two-dimensional, vertically-integrated version of the ADCIRC finite element model is able to provide a highly realistic representation of the dominant M2 tidal circulation within Beaufort Inlet in an essentially uncalibrated model application. A "standard" quadratic friction coefficient (0.0025), the lowest possible lateral viscosity for model stability ($2\ \mathrm{m}^2\ \mathrm{s}^{-1}$), bathymetric depths taken from NOS survey data and nautical charts and boundary condition forcing from a larger regional model all were used without modification. Closer agreement with observational data could undoubtedly be achieved with a systematic effort to calibrate these inputs if so desired.

Results from a particle tracking model that utilizes the vertically-integrated hydrodynamic flow fields suggest very distinct transport pathways into and through the sound/estuary system depending on where particles start outside the inlet. Tracks for particles that start to the east of the inlet are highly consistent with larval catches from a field experiment during March 1996 and therefore suggest a possible larval source to the east of the inlet. Although unverified by observation data, the model suggest that many larvae may be trapped in a large eddy that exists over the ebb tidal delta to the west of the inlet. Also, the model suggests that the near shore region outside the inlet is a particularly important area for larval transport. We are continuing to understand the processes effecting larval ingress by introducing additional tidal constituents and characteristic wind forcing for the region and by coupling this near inlet model with a model of larger scale transport along and across the shelf.

References

Baptista, A.M., E.E. Adams and K.D. Stolzenbach, 1984. Eulerian-Lagrangian analysis of pollutant transport in shallow water, MIT R.M. Parsons Lab T.R. No. 296, Massachusetts Institute of Technology, Cambridge, Ma.

Blanton, J.O., J. Amft, R.A. Luettich, and J.L. Hench, 1998. Tidal and sub-tidal fluctuations in temperature, salinity and pressure for the winter 1996 larval ingress experiment - Beaufort Inlet, N.C., Fisheries Oceanography, in review.

Churchill, J.H., R.B. Forward, R.A. Luettich, J.L. Hench, W.F. Hettler, L.B. Crowder and J.O. Blanton, 1998. Circulation and larval fish transport within a tidally dominated estuary, Fisheries Oceanography, in review.

Foreman, M.G.G., A.M. Baptista and R.A. Walters, 1992. Tidal model studies of particle trajectories around a shallow coastal bank, Atmosphere-Ocean, 30(1):43-69.

Forward, R.B., K.A. Reinsel, D. Peters, R.A. Tankersley, J.H. Churchill, L. Crowder, W.F. Hettler, S.M. Warlen, and M. Green, 1998. Transport of fish larvae through a tidal inlet, Fisheries Oceanography, in reveiw

Klavans, A.S., 1983. Tidal hydrodynamics and sediment transport in Beaufort Inlet, North Carolina. NOAA Technical Report NOS 100, April 1983, 119 pp.

Logan, D.G., 1995. Oceanographic processes affecting larval transport in Beaufort Inlet, N.C., MS Thesis, North Carolina State University, Raleigh, N.C., 129 pp.

Luettich, R.A. Jr., J.J. Westerink and N.W. Scheffner, 1992. ADCIRC: An Advanced Three-Dimensional Circulation Model for Shelves, Coasts and Estuaries, Report 1: Theory and Methodology of ADCIRC-2DDI and ADCIRC-3DL, DRP Technical Report DRP-92-6, Department of the Army, US Army Corps of Engineers, Waterways Experiment Station, Vicksburg, Ms., November 1992, 137pp.

Miller, J.M., J.P. Reed, L.J. Pietrafesa, 1984. Patterns, mechanisms and approaches to the study of migrations of estuarine-dependent fish larvae and juveniles. In Mechanisms of Migration in Fishes, J.D. McCleave, G.P. Arnold, J.J. Dodson and W.H. Neill [eds.], Plenum Publishing Corp, NY, pp. 209-225.

Westerink, J.J., R.A. Luettich, Jr. and N.W. Scheffner, 1994. ADCIRC: An Advanced Three-Dimensional Circulation Model for Shelves, Coasts and Estuaries, Report 2: Users Manual for ADCIRC-2DDI, DRP Technical Report DRP-92-6, Department of the Army, US Army Corps of Engineers, Waterways Experiment Station, Vicksburg, Ms., January 1994, 156pp.

Subject Index

Page number refers to the first page of paper